33 UAB1150-1

VOLUME I

LAUNCHING THE EUROPEAN ORGANIZATION
FOR NUCLEAR RESEARCH

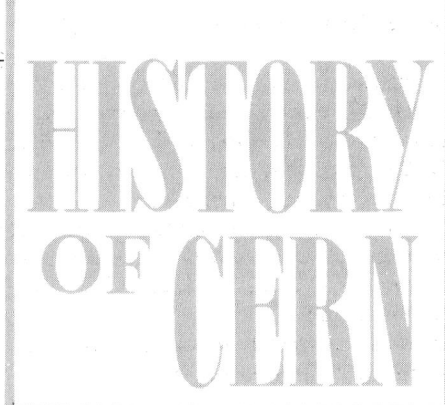

ARMIN HERMANN
JOHN KRIGE
ULRIKE MERSITS
DOMINIQUE PESTRE

and with a contribution by
LANFRANCO BELLONI

Study Team for CERN History

NORTH-HOLLAND

1987   Amsterdam · Oxford · New York · Tokyo

Elsevier Science Publishers, B.V., 1987

All rights reserved. No part of this publication may be reproduced, stored in a retrieval system, or transmitted, in any form or by any means, electronic, mechanical, photocopying, recording or otherwise, without the prior permission of the publisher, Elsevier Science Publishers B.V. (North-Holland Physics Publishing Division), P.O. Box 103, 1000 AC Amsterdam, The Netherlands.

Special regulations for readers in the USA: This publication has been registered with the Copyright Clearance Center Inc. (CCC), Salem, Massachusetts. Information can be obtained from the CCC about conditions under which photocopies of parts of this publication may be made in the USA.

All other copyright questions, including photocopying outside of the USA, should be referred to the publisher.

ISBN 0 444 87037 7

Published by:

North-Holland Physics Publishing

a division of

Elsevier Science Publishers B.V.
P.O. Box 103
1000 AC Amsterdam
The Netherlands

Sole distributors for the USA and Canada:

Elsevier Science Publishing Company, Inc.

52 Vanderbilt Avenue
New York, N.Y. 10017
USA

Library of Congress Cataloging-in-Publication Data

History of CERN.

Bibliography: p.
Including indexes
Contents: v. 1. Launching the European Organization
for Nuclear Research.
  1. European Organization for Nuclear Research--
History. 2. Nuclear physics--Research--Europe--Societies,
etc.--History. 3. Particles (Nuclear physics)--Research
--Europe--Societies, etc. I. Hermann, Armin, 1933–
II. Belloni, Lanfranco. III. European Organization for
Nuclear Research.
QC793.H57 1986   539.7'07204   86-24390
ISBN 0-444-87037-7 (v. 1 : U.S.)

Printed in The Netherlands

Preface

As early as 1958 the Director-General of CERN, Cornelis Bakker, thought 'that it might be a good idea to set down in a proper way the history of CERN'. He raised the matter on 2 December 1958 in the Committee of Council which agreed to set up an 'editing committee'. This would consist of 'distinguished persons who had been instrumental in founding CERN, such as Professors Auger and Amaldi, Sir Ben Lockspeiser, [and] Mr. Picot'.

The fruit of this first initiative was a fourteen-page report by Lew Kowarski, himself one of CERN's pioneers. Entitled 'An account of the origin and beginnings of CERN' and published by the organization in April 1961, it was based primarily on the personal recollection of events which, Kowarski felt, were 'still fresh' in his mind. The committee as well as the Director-General accepted the report as 'satisfactory'.

In subsequent years those who built CERN had several more opportunities to give their versions of its history. To take a few examples, Pierre Auger spoke on this topic in 1970 at a colloquium celebrating the 20th anniversary of the Centre Européen de la Culture, in 1972 both Edoardo Amaldi and Lew Kowarski dealt with it in their lectures at the International School of Physics in Varenna, while three years later Denis de Rougemont published a pamphlet on the organization's prehistory.

The first 'non-pioneer' to work on the emergence of CERN was Robert Jungk. Jungk came across some twelve boxes at UNESCO House in Paris containing papers and correspondence concerning the prehistory and early history of the laboratory. And although he based his book *The Big Machine* mainly on interviews, he realized that the written sources would be extremely valuable to future historians. On his instigation, copies were deposited in the CERN archives.

At the Hundredth Meeting of the Committee of Council held on 18 May 1972, itself an historical occasion, John Adams, Director-General of CERN Laboratory II, presented what later came to be called the CERN History Project. What we have in mind, he said, is 'to commission the writing of an official history of the Organization from its origins to the present day'. CERN estimated that 'some two or three man years of work' would be needed. One delegate thought that 'it might be possible to find a student who would do the work free of charge', another that 'CERN was still at a stage where it should make history rather than write it'. Nevertheless the proposal was approved and some six months later Margaret

Gowing—at that time lecturer in contemporary history at the University of Kent, and well known for her book on *Britain and Atomic Energy*—was asked to write the official history of CERN. She was joined by Lew Kowarski, who had just retired from CERN, and assisted by CERN staff Monique Senft and Simon Newman.

This group collected a considerable amount of source material including 55 interviews with personalities who had played an outstanding part in the establishment and development of the organization. However, despite the work that they did, the envisaged history of the organization was never written. All material, including part of a narrative history, was handed over to the CERN archives.

When this project was abandoned in 1977, some members of the Committee of Council felt that another attempt should be made to investigate the history of the laboratory. Such a study should be carried out by a team of European historians and financed outside the CERN budget. It was decided to set up an Advisory Committee for CERN History and to launch a preliminary study to determine the objectives of a detailed history.

This 'feasibility study' was carried out by Armin Hermann, professor for History of Science and Technology at the University of Stuttgart, and was financed by the Volkswagen Foundation. In his report he estimated that sixteen man-years would be required to investigate the pre-history and creation of CERN and the first ten years of operation of the laboratory. The history would comprise about 1200 pages in two volumes. On the recommendation of the Advisory Committee, the Committee of Council accepted this proposal.

After sufficient funds had been made available from institutions in several member countries, the newly established Study Team for CERN History started work on 1 October 1982. A year later it comprised Armin Hermann, John Krige, Ulrike Mersits, and Dominique Pestre; Lanfranco Belloni and Laura Weiss were part-time collaborators. For its part CERN provided offices and some infrastructural support.

* * *

We present here the first of the Team's two volumes on the history of CERN from its inception in the late 40s up to the mid-60s. The nature of the object of study, as well as that of the documentation, has led us to divide these 17–18 years into roughly two successive periods. One deals with the birth and official establishment of the organization—and thus covers the years 1949–1954—while the other studies the life of the European laboratory during the first twelve years of its existence.

The most notable feature of the foundation period was that the project remained in the hands of a rather small group of people—say 30 to 40—for about five years. These people, who were dispersed through Europe, had rather different backgrounds. They included diplomats, high-ranking science administrators, and some two dozen nuclear physicists and cosmic-ray specialists. Whatever their differences, they had a strong common aim, that of setting up a new type of scientific organization, an international laboratory equipped with the most sophisticated apparatus Europe could build. Granted their centrality, we have devoted a good deal of attention to describing the group and its activities, the reasons which motivated them, and the forces which may have shaped or influenced their behaviour. The study is thus mainly chronological, rather exhaustive, and centred around the decision-making process.

Probing further, three remarks are apposite:
1) Decisions of a purely scientific nature, decisions in which scientific experts played the determinant role, were taken during the five years of gestation. Since scientific stakes are not reducible to socio-political considerations, their importance had to be assessed independently. A presentation of the state of nuclear and of cosmic-ray physics in the early fifties was thus needed.
2) However, during this period the political dimension was at least as important as the scientific one. This was a consequence of the nature of the project itself: being international, it implied signature of a treaty and called for the presence of representatives of ministries of foreign affairs; being linked with nuclear matters, it demanded decisions at the highest level of the state apparatus; being ambitious and expensive, it required, in each country, consultation between groups whose interests often diverged. Hence the need to enlarge the field of our enquiry.
3) Finally, two broad types of behaviour appeared clearly between 1949 and 1954. One, active and innovating, characterized the nucleus of 30 to 40 people who elaborated the project and kept it away from governments as long as possible. The other, reactive but not passive, typified the behaviour in political circles and scientific communities in each country asked to give their opinion on the project. By and large these last groups played a limited role in shaping the organization.

The last remark explains why the events described in this volume are presented from two complementary points of view. On the one hand, the account is written from the vantage point of the central actors, discussing the group's internal tensions and disputes, the way it built its proposals, and the way it negotiated with the world at large. On the other, the events are presented from the point of view of those who were not part of the nucleus but who had to assess, to comment, to recommend, or to decide. The interest of this second approach is of course that it enables us to understand how the decision-making process operated at the beginning of the 50s in the various European countries.

This volume is structured in the light of these considerations and comprises four main parts and a conclusion.

The first deals with those aspects of the technical and scientific background which informed the scientific aims of CERN—briefly, to build a giant accelerator enabling Europe to compete with the Americans. Put differently, part one describes the emergence of the new field of high-energy physics.

Parts two and three are mainly written from the point of view of the central actors, the former presenting what was called the prehistory of CERN, part three what was known as the provisional CERN. The date dividing them is 15 February 1952, the date on which an intergovernmental agreement was signed establishing a provisional Council of Representatives of European States.

To elaborate a little, part two describes the first ideas and suggestions for collaborative ventures in 1949 and 1950, their convergence on a project for a big synchrotron, the way the negotiations were handled to draw in European governments, and the limits of the Agreement finally reached in February 1952. Part three deals with the setting up of the main scientific, administrative, and political structures of CERN, with the design of both accelerators, and with the definition of the main rules prescribing the relationship between the organization and the member states. In this part events *contained* within the life of the provisional CERN—like the drafting of the Convention or the choice of a site—are dealt with in some depth. By contrast, what is only *initiated* during this period, and hence what will blossom only afterwards—like the plans for the accelerators or the contract policy with industry—is merely presented in broad outline. Such issues are the subject of specialized chapters in volume II.

Part four is written primarily to understand the specificity of national situations, and the features of the national decision-making processes in particular. It shows the reactions of scientific communities—which were far from homogeneous—to the envisaged European laboratory project, the attitudes of political circles—which were not yet familiar with long-term research and development projects—and the ways information was transmitted and collective decisions were taken. In principle, every national case would have been worth studying. Switzerland for example, because she joined, for the first time, an organization comprising primarily western European countries, thus deviating from her traditional attitude of neutrality. Yugoslavia, to see how in the early fifties her approaches to CERN were part and parcel of a general reorientation of her foreign policy. The Scandinavian countries and the Netherlands, to investigate at greater length the nuances of opinion about the desirability of the project.

Unfortunately, we present only four detailed case studies—on France, Germany, Italy, and the United Kingdom. This does not mean that we have ignored the other countries: in fact we deal with them as and when necessary in parts I, II, and III. However, we had to make choices, and we settled on these four countries because they were of crucial importance for the project, either in its earliest stages, or by virtue of their scientific and financial contributions, because rather different situations prevailed in each, as the reader will discover, and because their histories were already familiar to four of the authors.

In the conclusion we attempt to pull together the various threads which have emerged in the book. Here we try to spell out a global interpretation of the birth of CERN, describing the process whereby the organization was launched, analyzing the factors which made that possible, and posing some 'grand historical questions'.

One final remark about language. We have written this volume in English so as to secure it the widest possible readership. This has placed a heavy burden on all the Team members, only one of whom is a native English speaker. We have also decided to give the quotations in English, making the translations ourselves. As for the text itself, some of us have chosen to write directly in English, although it is our second language. Others have preferred to have their work translated. As a result the level of the language is somewhat uneven between the chapters.

* * *

It goes without saying that a collaborative venture of this nature would not have been possible without the help, encouragement, and advice of individuals and institutions throughout Europe, and indeed beyond.

To begin with we would like to thank Dr. Günther Lehr (Bundesministerium für Forschung und Technologie, Bonn) who took the initiative in the Committee of Council, and the CERN History Advisory Committee—Paul Levaux (chairman), Alfred Günther (secretary), John Adams, Edoardo Amaldi, Axel Horstmann, David Jones (now replaced by Geoff Oldham), Michel Paty, Olaf Pedersen, and Roman Sexl (now replaced by Herbert Pietschmann)—which helped solve many financial, administrative, and scholarly problems.

A full list of those who sponsored the venture is given on page xi; we would like to thank especially the Stiftung Volkswagenwerk in Hanover, the Joint Committee of the Science and Engineering Research Council and the Social Science Research Council in the United Kingdom, the Centre National de la Recherche Scientifique in France, the Fonds zur Förderung der Wissenschaftlichen Forschung in Austria, and the Netherlands Organization for the Advancement of Pure Research.

We have been fortunate in having the opportunity to discuss our work with several CERN pioneers, who have given us valuable new information and have made useful criticisms of preliminary versions of our texts. In this connection we would particularly like to thank Pierre Auger, Edoardo Amaldi, Jan Hendrik Bannier, Alexander King, Jean Mussard, François de Rose, and Denis de Rougemont.

Many members of the CERN staff supported this work in a variety of ways. Without the help of Roswitha Rahmy of the CERN archives, and Yves Felt, her assistant, we would never have found our way through the thousands of documents at our disposal in boxes which we have been free to consult with a minimum of formality.

Above all we have appreciated the untiring efforts of Alfred Günther, who devoted much of his time, before and after his retirement, to the project.

Other archivists and librarians all over the world have also assisted us. Some deserve special mention: Marion Bird at the Science and Engineering Research Council, Swindon; Klaus Henning of the archive of the Max-Planck-Gesellschaft; Cécile Pozzo di Borgo at the Ministry of Foreign Affairs in Paris; Ken Ryan at the Brookhaven National Laboratory; Robin Rider of the Bancroft Library, Berkeley, and Joan Warnow at the American Institute of Physics, New York.

Finally, we wish to thank Rita French, the Team's Secretary, for her dedication and patience, David Stungo and his group at CERN, particularly Lise Karen-Alun, for their assistance with the production, and last but not least, Arie P. de Ruiter and Jane Kuurman of North-Holland Physics Publishing for their amiable co-operation.

Study Team for CERN History
Armin Hermann
John Krige
Ulrike Mersits
Dominique Pestre

This project has been supported financially by the following institutions:

Austrian Foundation for the Advancement of Scientific Research
Centre National de la Recherche Scientifique, France
Consiglio Nazionale delle Ricerche, Italy
Danish Natural Science Research Council
Federal Ministry for Science and Research, Austria
Istituto Nazionale di Fisica Nucleare, Italy
Joint Committee of the Science and Engineering Research Council and the Social
  Science Research Council, United Kingdom
LK-NES Foundation, Denmark (private industry)
Netherlands Organization for the Advancement of Pure Research (Z.W.O.)
Norwegian Research Council
République et Canton de Genève, Département de l'Instruction Publique
University of Milan, Italy
Volkswagen Foundation, Federal Republic of Germany

Contents

Preface v
Contents xiii
Remarks on the notes and bibliographies xviii
List of archival sources consulted xix
List of abbreviations xxi

PART I. THE POST-WAR EMERGENCE OF HIGH-ENERGY PHYSICS

1 From cosmic-ray and nuclear physics to high-energy physics 3
Ulrike Mersits

1.1 The scientific situation in 'elementary-particle physics' around 1945/46 4
1.2 Institutional changes in nuclear physics due to the war 11
1.3 The post-war accelerator programmes 13
1.4 Experimental particle physics: developments from 1946 to 1953 24
1.5 The theoreticians: from the first Shelter Island conference to the Rochester conferences 32
1.6 Concluding remarks 41

PART II. THE PREHISTORY OF CERN, 1949–FEBRUARY 1952

2 The first suggestions, 1949–June 1950 63
Dominique Pestre

2.1 The years following the war 64
2.2 Two proposals for European collaboration in November 1949 69
2.3 The European Cultural Conference, Lausanne, 8–12 December 1949 72
2.4 Attempts to implement the suggestions 75
2.5 The fifth General Conference of UNESCO, Florence, June 1950 82
2.6 The situation in June 1950 88

3 The fusion of the initiatives, June–December 1950   97
Dominique Pestre

3.1 The promotion of the Florence project, June–October 1950   98
3.2 The meeting between Auger and Dautry and the decision to call a gathering of scientists in Geneva in December   101
3.3 The Geneva meeting of 12 December at the Centre Européen de la Culture   109
3.4 Immediate reactions to the 12 December meeting   112
3.5 Conclusion: Ideas and motivations behind Amaldi's, Auger's, and Dautry's proposals   116

4 The period of informed optimism, December 1950–August 1951   123
Dominique Pestre

4.1 The implementation of the resolution of 12 December: French, Italian, and Belgian roles in the organization of the project, December 1950–May 1951   124
4.2 Meeting of consultants at UNESCO in Paris, May 1951   130
4.3 Discussions arising out of the consultants' report, May–July 1951   134
4.4 The intergovernmental meeting is called   140
4.5 A marriage of convenience   141

5 The period of conflict, August–December 1951   147
Dominique Pestre

5.1 The gradual emergence of an alternative programme, August–October 1951   148
5.2 Second meeting of consultants, Paris, 26 and 27 October 1951   157
5.3 Tension peaks, late November–early December 1951   163
5.4 The roots of the division   169

6 The establishment of a Council of Representatives of European States, December 1951–February 1952   179
Dominique Pestre

6.1 The intergovernmental conference in Paris, 17–20 December 1951   180
6.2 Negotiations between December and February   188
6.3 The second session of the conference, Geneva, 12–15 February 1952   194
6.4 Concluding remarks on the respective roles of scientists and politicians in the process of establishment of the Council   198

PART III. THE PROVISIONAL CERN, FEBRUARY 1952–OCTOBER 1954

7 Survey of developments 209
John Krige

7.1 Establishing the study groups and the secretariat 211
7.2 Fixing the energies of the accelerators 211
7.3 The discovery of the strong-focusing principle 213
7.4 The choice of a site for the new laboratory 213
7.5 Britain takes the plunge 215
7.6 Consolidating the scientific work 216
7.7 The Convention and its signature 219
7.8 The new mood in CERN 222
7.9 Putting down roots in Geneva 223
7.10 The nomination of the first Director-General 225
7.11 The ratification of the Convention 226
7.12 The philosophy of the organization 228
Appendix 7.1 Scale of percentage contributions to the permanent organization applicable during the period to 31 December 1956 231
Appendix 7.2 Rates of exchange used for accounting purposes (a) as from 1 December 1952 and (b) as from 1 January 1954 231
Appendix 7.3 Dates of deposit of instruments of ratification of the Convention at UNESCO House in Paris 232
Appendix 7.4 Official delegates to the nine Council sessions 232

8 Case studies of some important decisions 237
John Krige

8.1 The choice of a site 238
8.2 The Convention. Two key clauses 246
8.3 Financing the interim organization 252
8.4 The nomination of the first Director-General and its aftermath 261
8.5 The strong-focusing principle: the decision and its early consequences 273
8.6 Planning the future laboratory 282

Bibliography for Parts II and III 293
John Krige and Dominique Pestre

Collection of photographs and documents of historic interest

PART IV. NATIONAL DECISIONS TO JOIN CERN

9 French attitudes to the European laboratory, 1949–1954 303
Dominique Pestre

9.1 The French political and diplomatic context, and nuclear policy 305
9.2 The French scientific context, 1945–1955 309
9.3 French initiatives regarding the UNESCO project, the decision-making process, 1949–May 1952 317
9.4 Towards the signing of the Convention, February 1952–July 1953 328
9.5 The debate surrounding French ratification, July 1953–September 1954 334
9.6 Concluding remarks 342

10 The Italian scenario 353
Lanfranco Belloni

10.1 CNR and the 'Years of Reconstruction' 354
10.2 Rome physicists' involvement in the European project 359
10.3 CNR and Italian support of the European laboratory project 369

11 Germany's part in the setting-up of CERN 383
Armin Hermann

11.1 Up to the UNESCO conference in Paris 384
11.2 Heisenberg's appointment as German delegate 393
11.3 Heisenberg's rôle at the UNESCO conference 399
11.4 The ratification 405
11.5 German positions 413

12 Britain and the European laboratory project: 1951–mid-1952 431
John Krige

12.1 The period of detachment 433
12.2 The UNESCO May report and the mounting opposition to it on the continent and in Britain 438
12.3 Forging the alliance: the offer of the Liverpool cyclotron and the Paris conference 445
12.4 Confronting the new question: Should Britain join the Council of Representatives? 454
12.5 The Cabinet Steering Committee reconsiders the case 462

13 Britain and the European laboratory project: mid-1952–December 1953　475
John Krige

13.1　The emergence of the Harwell group　477
13.2　The polarization of the scientific community　482
13.3　Progress at governmental level　487
13.4　Refocusing the issue: the 'discovery' of the alternating-gradient principle　490
13.5　Britain joins CERN　495
Appendix 13.1　A short survey of the committees involved in the CERN decision　503
Appendix 13.2　Decision-making at the science–government interface: some general comments　506

PART V.　CONCLUDING REMARKS

14 The how and the why of the birth of CERN　523
John Krige and Dominique Pestre

14.1　A brief narrative account　524
14.2　A classical interpretation of CERN's origins　525
14.3　The first group of actors: the physicists　529
14.4　The second group of actors: high-level science administrators and some diplomats　530
14.5　The activities of the 'CERN lobby'　532
14.6　The reactions of the member states　535
14.7　The motivations in governmental circles　536
14.8　CERN, an American puppet?　537
14.9　CERN, an organization of military importance?　539

Appendix 1.　Who's who in the foundation of CERN　545
Armin Hermann

Appendix 2.　Chronology of events　567
John Krige and Dominique Pestre

Name index　587
Thematic subject index　594

Remarks on the notes and bibliographies

The text of each chapter is immediately followed by its own notes.

There are six major bibliographies—at the end of part one, at the end of part three (covering both parts two and three), and at the end of each chapter in part four. The bibliographies are always situated after the notes and items are listed alphabetically.

The following reference system has been used:

- Official UNESCO documents are referred to by their official labels followed by the date of the document. They are easily identified for they always have the form UNESCO/... . All are available in UNESCO's archives.

- The same applies to the minutes of the CERN provisional Council and its committees, and to the papers prepared for them. They are referred to as CERN/..., and are in CERN's archives. Further information is only given when we have reason to believe that it is needed to retrieve the document easily. Note that the date given is that of the document, not of the meeting.

- The location of other documents is specified in brackets using the following format : (archive-file name). Thus (CERN-DG20328) means box 20328 in the series DG (Director-General) in CERN's archives; (AEF–80) means box 80 in the series Affaires Atomiques of the French Ministry of Foreign Affairs, Paris. A list of archival sources consulted follows these remarks.

- The notes refer to items in the bibliographies using the following format : Author or subject (year). Thus Amaldi (1972) or Scheinman (1965). The full reference is given in the associated bibliography.

List of archival sources consulted

In compiling the list we have included the abbreviations used in the notes and references. In general, the abbreviation identifies a specific collection of documents in an archive, which is italicized in the text below.

AA	–	Auswärtiges Amt, Politisches Archiv, Bonn.
AEF	–	Ministère des Affaires Etrangères, Paris: *Affaires Atomiques* file.
AEI	–	Ministero per gli Affari Esteri, Archivio Storico-Diplomatico, Rome.
AF	–	Archives de France, Paris: AF–F17bis: *Ministère de l'Education Nationale* file AF–307AP: *Archives privées, Raoul Dautry* file AF–CNRS: *'Versement effectué par le CNRS le 22/8/80, cote 80/284'*.
AIP	–	American Institute of Physics, New York City.
AST	–	Archivio di Stato di Torino: *Gustavo Colonnetti*'s papers.
BNL	–	Brookhaven National Laboratory, Long Island (NY).
CC–CHAD	–	Churchill College, Cambridge (UK): *Sir James Chadwick*'s papers.
CE	–	Council of Europe, Historical Archives, Strasbourg.
CEC	–	Centre Européen de la Culture, Geneva. Denis de Rougemont permitted us to consult the archives of the Centre, and assisted us. Important for the period up to mid-1951.
CNR	–	Consiglio Nazionale delle Ricerche, Ufficio di Presidenza, Rome.
DFG	–	Deutsche Forschungsgemeinschaft, Bonn-Bad Godesberg.
DM	–	Deutsches Museum, Sondersammlungen der Bibliothek, Munich: mainly *Max von Laue*'s papers.
LBL	–	Lawrence Berkeley Laboratory, Berkeley, California.
MPG	–	Bibliothek und Archiv zur Geschichte der Max-Planck-Gesellschaft, Berlin: mainly *Walther Bothe*'s papers.
MPI/K	–	Max-Planck-Institut für Kernphysik, Heidelberg: mainly *Wolfgang Gentner*'s papers.
MPI/P	–	Max-Planck-Institut für Physik und Astrophysik, Munich: mainly *Werner Heisenberg*'s papers.
MR	–	Archivio Mario A. Rollier, Milan.
NBI	–	Niels Bohr Institute, Copenhagen.

PRO	–	Public Record Office, Kew, London: PRO–AB: *UK Atomic Energy Authority* file PRO–CAB: *Cabinet Office* file PRO–DSIR: *Department of Scientific and Industrial Research* file PRO–ED: *Department of Education and Science* file PRO–FO: *Foreign Office* file
SERC	–	Science and Engineering Research Council, Hayes, Middlesex: *NP* file.
SU	–	Stanford University Archives, Stanford, California: series SC 303, *Felix Bloch's* papers.
UN	–	United Nations Organization, Geneva: collection of official documents available at the library.
UNESCO	–	UNESCO, Paris: *European Organization for Nuclear Research (CERN), Official Correspondence* file. The folders are in chronological order. No more precision than the date has accordingly been given in the notes. Essential for the period up to mid-52.
CERN	–	CERN, Geneva. Amongst dozens of series, we would like to mention: • The file labelled *CHIP/...*, collected in the 70s for an earlier CERN History Project; • the folders called *Mussard* file, comprising documents deposited by Jean Mussard, Auger's right-hand man in UNESCO. Essential; • The *Kowarski interview* file; • Five series of personal documents collected by our team now in CERN's archives. They were kindly lent to us by Pierre Auger, Edoardo Amaldi, J. Hendrik Bannier (on the Netherlands), Paul Leveaux (on Belgium), and André Mercier (on Switzerland). We have referred to them as *CHS, Auger (or Amaldi...) file, CERN*. • Finally, a collection of interviews made in the mid-70s by Margaret Gowing, Lew Kowarski, and their collaborators.

List of abbreviations

Organizations and Committees

BUC	'British UNESCO Committee'. More precisely, British Committee for Co-operation with UNESCO in the Natural Sciences
CEA	Commissariat à l'Energie Atomique (F)
CEC	Centre Européen de la Culture
CERN	European Council for Nuclear Research (until 6 October 1954) European Organization for Nuclear Research (after 6 October 1954)
CISE	Centro Informazioni Studi ed Esperienze (I)
CNR	Consiglio Nazionale delle Ricerche (I)
CNRN	Comitato Nazionale per le Ricerche Nucleari (I)
CNRS	Centre National de la Recherche Scientifique (F)
DFG	Deutsche Forschungsgemeinschaft (FRG)
DSIR	Department of Scientific and Industrial Research (UK)
ECOSOC	United Nations Economic and Social Council
ECSC	European Coal and Steel Community
EDC	European Defence Community
EURATOM	European Atomic Energy Community
FOM	Institute for Fundamental Research into Matter (NL)
ICSU	International Council of Scientific Unions
IIPN	Institut Interuniversitaire de Physique Nucléaire (B)
IISN	Institut Interuniversitaire des Sciences Nucléaires (B)
INFN	Istituto Nazionale di Fisica Nucleare (I)
IOC	Cabinet Steering Committee on International Organizations (UK)
IUPAP	International Union of Pure and Applied Physics
MPG	Max-Planck-Gesellschaft (FRG)
NATO	North Atlantic Treaty Organization
NPC	Nuclear Physics Committee (UK)
OEEC	Organization for European Economic Cooperation
OLD	Overseas Liaison Division of the DSIR
OSR	Overseas Scientific Relations Committee (UK)
UNESCO	United Nations Educational, Scientific and Cultural Organization
UNO	United Nations Organization
ZWO	Organization for the Advancement of Pure Research (NL)

Countries

A	Austria
B	Belgium
CH	Switzerland
DK	Denmark
F	France
FRG	Federal Republic of Germany
G	Greece
GDR	German Democratic Republic
I	Italy
L	Luxembourg
N	Norway
NL	The Netherlands
S	Sweden
UK	United Kingdom
USA	United States of America
USSR	Union of Soviet Socialist Republics
Y	Yugoslavia

Currencies

BF	Belgian francs
FF	French francs
SF	Swiss francs
MSF	*Million* Swiss francs (and other currencies)

PART I

The post-war emergence
of high-energy physics

CHAPTER 1

From cosmic-ray and nuclear physics to high-energy physics[1]

Ulrike MERSITS

Contents

1.1 The scientific situation in 'elementary-particle physics' around 1945/46   4
    1.1.1 Cosmic-ray physics   5
    1.1.2 Nuclear physics   8
1.2 Institutional changes in nuclear physics due to the war   11
1.3 The post-war accelerator programmes   13
    1.3.1 The principle of phase stability   14
    1.3.2 The United States   15
    1.3.3 Great Britain—the leading country in Europe   19
    1.3.4 Continental western Europe   21
    1.3.5 The AG focusing principle   23
1.4 Experimental particle physics: developments from 1946 to 1953   24
    1.4.1 The weak nature of the mesotron and the detection of the pi-meson (1946/47)   25
    1.4.2 The artificial production of charged and uncharged pi-mesons (1948/49)   26
    1.4.3 The complexity of the mass spectrum   28
1.5 The theoreticians: from the first Shelter Island conference to the Rochester conferences   32
1.6 Concluding remarks   41
    Notes   44
    Bibliography   52

By the autumn of 1952 the protagonists of the European laboratory had agreed that it should be a centre for research in high-energy physics equipped with two accelerators: a giant proton synchrotron of 25 GeV, and a synchro-cyclotron of 600 MeV. In this, the first chapter, we want to focus on the scientific roots of this choice, we want to understand why it was decided to build these facilities rather than others. Our analysis is restricted in two ways: we concentrate on the *immediate* post-war period, up to the early fifties, and we devote particular attention to events relevant to the launching of CERN. It is therefore *not* our intention to give a complete description of the development of high-energy physics in the post-war years.

The first two sections below deal with the situation at the end of the Second World War. Here we survey the state of play in the two scientific fields devoted to investigating the basic constitution of matter, i.e. cosmic-ray physics and nuclear physics, and describe the institutional changes in the latter which had occurred due to the war.

The three sections that follow cover the period after 1946. They discuss the more-or-less independent factors which collectively shaped elementary-particle physics in the fifties, namely, the development of accelerators, the new data on elementary particles produced by cosmic-ray and nuclear physicists, and the innovations made in theory in response to the experimental results.

With this information at our disposal we present in the conclusion (section 1.6) a more comprehensive picture of the whole situation, showing how the various factors we have identified overlapped and interacted with one another. By doing so, we are able to see roughly when, how, and why high-energy physics using accelerators became an independent discipline—a discipline at once prestigious in the eyes of governments and fascinating to a new generation of physicists.

A quick word of warning is needed before we begin. The meaning of terms like nuclear physics and high-energy physics is not static, and has changed considerably over the past 40 years. To avoid misunderstanding we periodically provide definitions of such terms which give the meanings they had to physicists at the time.[2]

1.1 The scientific situation in 'elementary-particle physics' around 1945/46

In the immediate post-war period concepts like elementary-particle physics or high-energy physics were not used as such. However, to study the origins of this discipline we have to take into account two different research activities, namely cosmic-ray physics and nuclear physics. This distinction only really applied to experimentalists, as theoreticians were taking basic experimental data from both sources.

The first big international conference in Europe in the field of particle physics took place about a year after the end of the Second World War. It was held from 22 to 27 July 1946 at the Cavendish Laboratory in Cambridge and was entitled *Fundamental Particles and Low Temperature.*[3] Its task was both to provide a meeting place for the international community of physicists and to help orientate them in the rapidly growing field, especially after the long break caused by the war.

The war had rendered scientific communication in Europe rather difficult and the conference offered a welcome opportunity to establish new contacts or to re-establish old ones. In all some three hundred physicists attended it, one third of them from foreign countries. Many European states (but not Germany for well-known reasons), as well as other countries like the USA, the USSR, China, and India were represented.

The conference was divided into two sections which dealt with completely different subjects; only the first is of interest to us here. As the preface to the proceedings explained: 'the subject of Fundamental Particles [had become] so wide that the arrangement of the programme was not easy. The time available for the whole Conference could well have been filled by the discussion of any one of a number of topics which in fact were compressed into a single session or less. But, at the first post-war gathering, one was reluctant to narrow down the field too much.'

This first section of the conference was thus meant to give as broad an overview as possible on the subject of fundamental particles, and comprised five sessions, namely 'General introduction and survey', 'Mesons and cosmic rays', 'Experimental techniques', 'Nuclear forces and relativistic particles', and 'Theory of Heisenberg's S-matrix'. We now wish to study the different approaches to these topics by both the cosmic-ray physicists and the nuclear physicists. In this way it will be possible for us to see the 'state of the art' in these fields.

1.1.1 COSMIC-RAY PHYSICS[4]

Before going on to speak of 'cosmic-ray physics', we should clarify what this term meant in the mid-forties. Basically one could divide the work of these researchers into two broad categories according to the type of question they were trying to answer:

1. What are the constituents of cosmic radiation?
2. What is the origin of cosmic radiation and what are its effects on Earth?

The investigations dealing with the first question are of primary interest to us. We shall not deal with the second, though we should mention that it became the main

Notes: pp. 44 ff.

task of cosmic-ray physicists when the 'particle aspect' of their work was taken over by physicists using accelerators.

Let us begin by looking at the field from a scientific angle with a summary of the particles known by 1945. Altogether there were eight, namely the electron, the positron, the proton, the neutron, the photon, the neutrino, and the positive and negative mesotron (also referred to as the cosmic-ray meson and named mu-meson by Powell in 1947). It should be mentioned here that, due to the similarity in mass, the mesotron was (wrongly) thought at the time to be Yukawa's predicted field quantum mediating the strong interaction (later identified as the pi-meson).[5]

At the Cambridge conference twelve talks were given in the session devoted to mesons and cosmic radiation and physicists like P.M.S. Blackett, G. Bernardini, B. Ferretti, L. Janossy, L. Leprince-Ringuet, and J.G. Wilson took the opportunity to survey some of the work they had done during the war years and to present their current investigations. Among the main topics were the production of mesons (i.e. to look for the primary particles in cosmic radiation) and their decay. The latter also embraced the question of why the positively-charged particles decayed into a positron and one neutrino (as was thought at the time), while the negative ones were easily absorbed by atomic nuclei. Cosmic-ray showers were also discussed and, last but not least, a quarter of all talks was devoted to the problem of meson-mass measurements.

Both the purely scientific content of the talks and the comments show the difficulties physicists faced in carrying out experiments, and bring out clearly the advantages accelerators were to offer. We want to discuss three of them in some detail.

1. Inaccuracy in measurement

This problem existed from the very beginning of cosmic-ray work and meant that the results obtained were perhaps of a more qualitative than quantitative nature. As J.G. Wilson said, 'identification of all these [detected particles] depends upon the limitations of our technique, which have up to now allowed only individual measurements correct to about 10%'.[6] This impeded particle identification and made it difficult to draw precise conclusions. Accordingly the work provided little guidance for theory, and physicists spent a great deal of time developing better detectors. These, and the experience gained in using them, were later of decisive importance to physicists who worked with accelerators.

2. Low reproducibility

Owing to the low particle-flux density of cosmic radiation, some events occurred too arbitrarily and infrequently to provide good statistics. Thus we find Leprince-Ringuet closing his talk with the statement that 'this photograph is in favour of [a cosmic-ray particle of 990 m_e mass], but is not sufficient to confirm, the existence of a heavy meson',[7] largely because it was only a single photograph. Two years later in September 1948 Heitler again referred to this problem in a talk at a conference on *Cosmic Radiation* in Bristol, saying: 'Each of these items of evidence [for heavy mesons] is rather convincing, [...], but as long as there is only one example for each event we cannot be quite sure about the existence of these particles'.[8] A further direct consequence of this was that research providing reliable results proved somewhat time-consuming.

3. Lack of theoretical background

The third difficulty which occurred and which was clearly recognized was the lack of theoretical background for the experimental work. Since Yukawa's prediction in 1935 that a field quantum mediated the strong interaction, no other predictive theory had been put forward. As already mentioned, this theory had been erroneously 'satisfied' in 1937 by the discovery of the mesotron in cosmic radiation.[9] Yukawa put it thus in his Nobel speech in 1949: 'At that time [1937], we came *naturally* [our emphasis] to the conclusion that the meson which constituted the main part of the hard component of cosmic rays at sea level was to be identified with the mesons which were responsible for the nuclear force'.[10] Expressed differently, many of the physicists did not 'expect' a variety of further elementary particles, largely because there was no 'need' for them, for they were not necessary to explain anything. This attitude also partially accounts for the fact that the erroneous identification of the mesotron and Yukawa's meson persisted for so many years (1937–1947), although even at the time some experimental results seemed not to fit very well into the picture of strong interactions.[11]

The question of whether or not to believe in the existence of many more particles of different masses emerged clearly at the Cambridge conference in 1946. J.G. Wilson started his talk by saying that 'a crucial topic concerning fundamental particles today is that of *existence* [our emphasis], particularly with reference to the meson group. Are mesons of a unique mass? Are there several types of mesons differing in mass? Or, is a continuous distribution of masses possible? These are questions for which the theoretician is not able to offer much assistance, and so they must primarily be the concern of the experimental physicist'.[12]

What consequences did these three problems have for the work of the cosmic-ray physicists? Both inaccuracy in measurement and low reproducibility meant that their results provided a rather poor basis for theoretical work. However, theoretical guidance was necessary for evaluating data and partly also for conceiving new experiments. The ensuing weak interplay between experiment and theory would start changing only when accelerators became the main research tool in particle physics.

To conclude, we will summarize the main contributions of cosmic-ray physics to particle physics. Without doubt the central task occupying physicists was to establish the law describing the nuclear force. Accordingly, two main areas of investigation were fundamental. Firstly, one had to understand how the cosmic-ray mesons were produced, to measure their main properties like mass, charge, spin, and lifetime, and to study their interaction with matter. Here, it was assumed that these mesons were responsible for mediating the nuclear force, so that such studies could throw light on the nature of strong interactions. Secondly, nucleon–nucleon interactions were extensively investigated. Though considerable information was collected on each of these aspects, physicists were not able to establish definite interconnections between them. Their work at that time could thus be compared to the attempts to complete a puzzle without knowing what the final picture would look like, although some of its main pieces were already found.

Despite all the difficulties mentioned above, we should not overlook the considerable number of results presented at the Cambridge conference. It is true that physicists were distressed because much of their data was rather inaccurate and perhaps not sufficiently situated within a theoretical framework. However, this was the only way to get information on the fundamental building blocks of matter and the field had a very stimulating influence on many of them. It was therefore of crucial importance.

1.1.2 NUCLEAR PHYSICS[13]

To grasp the notion of 'nuclear physics' as used in 1946, we wish to contrast it with its present-day meaning. The major difference is possibly the fact that at the time nuclear physics had a much broader meaning. Unlike today, the term was not so closely associated with the idea of a specific energy region. Rather it was the type of phenomena investigated and the kind of questions asked which determined whether the work was regarded as nuclear physics. In fact, the whole complex of research relating to questions on the basic structure of matter and the laws behind it, i.e. nuclear forces, mesons, field theory, etc., belonged at that time to nuclear physics.

Eight talks on *experimental nuclear physics* were presented at the Cambridge conference, all but one at the session entitled 'Experimental techniques'. As the scientific details of these talks were of secondary importance to the development of elementary-particle physics, we prefer to restrict ourselves to giving a general impression of the situation in this field. From the United States Herbert Anderson, Enrico Fermi, Leona Marshall, and Walter Zinn described part of the work on neutron physics they had performed under contract with the Manhattan Project at the University of Chicago. Similar papers by British and Italian participants (E. Broda, N. Feather, D.H. Wilkinson, D.L. Livesey, C.F. Powell, G.P.S. Occhialini, E. Amaldi and his group at Rome) dealt with nuclear fission and neutron physics.

To summarize the situation as revealed by these papers, we should like to stress three characteristic points. To begin with, we find a somewhat selective presentation of the war work since many results were still classified at the time. In addition, little basic research was described. No mention was made of investigations pursued before the war using existing accelerators, such as nucleon–nucleon scattering or measurements of the main properties of the deuteron. This is understandable since fundamental research in this field ceased to a great extent during the war, although a large number of questions still remained unanswered. Finally, all talks were on fission or neutron physics demonstrating that these fields were flourishing. Indeed, owing to the wartime atomic-energy project in the United States and Britain, physics related to fission, and thus neutron physics, work on uranium isotopes, etc. had been the major topics of research for many of the physicists. About a hundred new unstable isotopes had been created using accelerators (which at that time achieved a maximum energy of roughly 20 MeV), and a number of new elements had also been found, of which the most well known was plutonium-239.

Turning to the contribution of *theoretical nuclear physics* to the 1946 Cambridge conference, to judge by the number of talks given, this was the dominant part. Nearly three complete sessions out of five were devoted to it, and Niels Bohr, Max Born, and many other leading European theoreticians were among the speakers.

Before discussing the papers presented, we want to make a general remark. Already after the war we can identify two distinct directions of research in theoretical nuclear physics. One aimed at an understanding of 'the structural and phenomenological complexities of the nucleus', the other sought 'a more and more refined ultimate analysis of matter'.[14] This of course did not mean that there was a strict division within the community of nuclear physicists, and some in fact worked for quite a while in both fields. However, joining the first group and staying in it meant remaining a nuclear physicist (in today's sense), while those in the second ultimately became elementary-particle physicists.

Let us now turn to the scientific content of the talks.

The first session was mainly devoted to quantum field theory, in particular to the problem of formulating quantum electrodynamics in a consistent manner. The opening address was given by Niels Bohr with a talk on 'Problems of elementary-particle physics'.[15] He dealt rather generally with a number of pre-war concerns: the self-energy problem of point-like charges, electron-pair production, and the treatment of field quanta such as mesons, though he offered no solutions. The other speakers in this first session were Wolfgang Pauli on 'Difficulties of field theories and the field quantization', P.A.M. Dirac on 'The difficulties of quantum electro-dynamics', and Max Born on 'Relativistic quantum mechanics and the principle of reciprocity'. As the preface to the proceedings of the conference emphasized the operative word was 'difficulty'. It added that 'the passage of the years [had] not altered the basic fact that the subject under discussion transcend[ed] the limits of well-established theory and that no entirely satisfactory new point of departure [was] known'. Thus theory was presented as having reached something of an impasse.

The fourth session on 'Nuclear forces and relativistic particles', which was almost entirely theoretical, was opened by Leon Rosenfeld. To introduce his talk he gave the following general description of the situation in theoretical physics with respect to the problem of nuclear forces:

Notes: pp. 45 ff.

> A survey of the problem of nuclear forces must at the moment remain inconclusive. On the theoretical side, progress is hampered by the present imperfection of the quantum theory of fields of forces; besides the fundamental problems of self-energies and proper magnetic moments which affect any treatment of the elementary particles themselves, peculiar difficulties are encountered, as we shall see, even in the derivation of the law of interaction between such particles. On the empirical side, the situation is perhaps more hopeful, inasmuch as we can expect information of decisive value from experiments extending only slightly further than those at present available; but the existing evidence is not yet sufficient to allow us to make a choice between a number of theoretical possibilities.[16]

The subsequent talks covered topics like strong-coupling meson theories, the collisions of neutrons and protons with deuterons, possible new equations for fundamental particles, and the possible existence of mass spectra of fundamental particles. They were given by G. Wentzel, H.S.W. Massey, R.A. Buckingham, A. Proca, and C. Møller.

The last session was entirely devoted to the *S*-matrix theory proposed by Werner Heisenberg in 1943 as an answer to the seemingly unsolvable divergence difficulties in relativistic quantum theory of fields. His aim and hope was the construction of a theory containing only *observables,* i.e. scattering cross-sections, energy levels.[17] W. Heitler, C. Møller, and E.C.G. Stueckelberg presented talks on this subject.

The topic of weak interactions was not dealt with at all at this conference. In 1934 Enrico Fermi[18] had formulated the first theory of nuclear beta decay. This was the only example of a weak decay until the detection of the mu-meson in 1946. As its application to physical reality was quite satisfactory and no really new experimental results had been found, it was not further discussed on this occasion.

Summarizing, we can say that the general aim was to find a consistent formalism to describe the 'behaviour' of elementary particles, and so to develop quantum field theory to the stage at which it would have overcome its current difficulties and could be used to understand cosmic-ray phenomena, to treat field quanta such as mesons, and to describe how particles were transformed into each other. But there was no real prospect of rapid solutions. On the other hand, on reading these papers something else emerges. We frequently find the more or less explicitly stated hope that the needed help for the theoretical problems would come from the experimentalists. Max Born, for example, expressed this very clearly in the introduction to his talk stressing that he 'expect[ed] the real solution of [their] difficulties from the discovery of new facts in the domain of high energy which [would] be experimentally explored in the near future'.[19] Similarly Leon Rosenfeld remarked that theoreticians 'would require evidence from a domain of energies still much higher than those we have been considering, but not outside the range of the modern accelerators'.[20] In other words what was needed were well-defined experiments with sufficiently accurate results to serve as a basis on which the theoreticians could proceed.

1.2 Institutional changes in nuclear physics due to the war[21]

On looking at the manner in which high-energy physics has been done since the fifties, two main characteristics stand out. Firstly, investigations in this field are generally undertaken in big laboratories in which large, very expensive, and technically highly-sophisticated equipment is concentrated. Secondly, increasingly large collaborations of physicists and technicians are involved. In this section we will briefly sketch the origins of this approach, thus the start of 'big science', whose roots lie in the Second World War.

As is well known, the detection of nuclear fission by O. Hahn and F. Strassmann in 1938/39 together with the outbreak of the war brought about an abrupt change in the whole character of nuclear physics. Investigations hitherto regarded as basic research were now considered primarily from the point of view of practical applicability.[22] Two extensive war projects were launched in the allied countries, namely, the development of radar and research in connection with the making of an atomic bomb. A large number of American physicists and emigrants from Europe were involved in one or the other of these. Big collaborations were set up with the best available staff, and previously unthinkable amounts of money were placed at the physicists' disposal.

We suggested above that this new style of working had an enormous influence in the post-war period. We shall therefore investigate both its positive features, underlined by the 'success' achieved during the war, and its negative ones, which caused considerable anxieties within the physics community.

To start on the positive side, a large number of very good physicists working under virtually no financial constraints, equipped with a variety of tools, all concentrated in one place and dedicated to a specific goal, proved to be a very good basis for highly efficient work. In addition close collaboration was established between various disciplines like theoretical physics, experimental physics, and electrical engineering. The cross-fertilization between these different fields led to the relatively rapid development and improvement of the research tools required.

The best known example of a big war-time laboratory is of course Los Alamos (New Mexico), a special weapons laboratory established by General Leslie R. Groves and under the scientific leadership of J. Robert Oppenheimer.[23] It was set up in March 1943, and by spring 1945 its staff numbered 2000. This included many of the best physicists in the world, both theoreticians and experimentalists, who gathered there to work on the development of the bomb. The average age of the staff was very low so that a whole new generation of physicists and technicians was produced—and returned to 'normal' life with very high expectations for their future work, expectations which were different in three main respects from those they had had before the war.

Their *financial expectations* had escalated. The pre-war threshold for the cost of an experiment had now been exceeded by several orders of magnitude. Many

Notes: p. 46

physicists wanted to continue to work on this scale, despite its expense, as they had seen the many exciting possibilities which adequate funding opened up.

Secondly, their *position vis-à-vis the government* had changed. Scientists had proved to themselves and to the government that such large-scale projects could work, and work efficiently. Even more important was that in the war nuclear physics had demonstrated its practical applicability for peaceful and particularly also for military purposes. Strong support could thus be expected from official quarters.

Finally, the *relationship between industry and physics* had been reinforced. For industry, on the one hand, the interest in co-operation lay in the possible practical application of the results obtained by pure research. For their part, physicists became increasingly aware that a fruitful exchange of ideas with industry could be rather useful. In particular, it could lead to some of the important items of equipment they needed being available off the shelf.

Having briefly sketched the more positive effects of the war on the style of research and on the attitudes of the scientists, we now intend to contrast them with the doubts and anxieties expressed at the time. A central question for many physicists was of course that of their moral responsibility once they had seen the consequences of their work. We do not wish to discuss this here preferring to concentrate on the way scientists thought about their future in nuclear physics.

Throughout the war, physicists had been doing mission-oriented research and now felt that a great deal of fundamental physics remained to be done. As the Director of the American Institute of Physics expressed it, 'it [was] serious enough that the best efforts of physicists [had] been for several years diverted from fundamental research. The advance of basic physics [had] probably been retarded by the war, whatever [might] be said about the stimulation of its application to industrial purposes'.[24]

The question of how to re-start doing 'real' physics again was therefore a very pertinent one. This is one reason why in May 1944 a conference on the issue was held in Philadelphia.[25] It was attended by some sixty American physicists, who discussed topics such as physics education, relations between industry and physics, and problems of the future funding of science. It is interesting to see that many considered these changes occurring in the style of research to be rather dangerous for the future of nuclear science. Their comments can be grouped around three topics.

1. *Difficulties of running large-scale institutions under 'normal' conditions*

Some physicists foresaw difficulties in drawing up a programme for a very big institution, in organizing it, and in finding unanimity among physicists to work along the lines agreed. During the war there was pressure from without, i.e. from the government, coupled with the inner desire to help win the war. The physicists were struggling together to reach their aim. In peace-time these pressures would no longer be present, and this could lead to disorientation within a big organization.

2. *Financial dependence of science*

Scientists planning big research projects for the post-war period needed financial support from industry or the state. This fact raised questions like 'Who is going to own the results of research?' and 'Will there be enough freedom to do investigations in whatever field one chooses?'.[26] L.C. Dunn of Columbia University expressed this even more explicitly in his talk entitled 'To what extent is government financial aid for training and research desirable?'. He concluded by asking whether this was really 'the chief problem [...] or whether it [was] not the fundamental question of the divorce of support and control, that is, the maintenance of freedom for research workers and the prevention of local interference and favoritism which everyone fears as the government enters new fields of science'.[27]

3. *Centralization of research*

This topic covers two slightly different issues. There was the complaint that there would be no place for individualism in the new way of doing research. For some people work in smaller groups seemed more fruitful. Complementing this, the necessity for decentralization was stressed because only in this way would people 'keep up their sense of responsibility'.[28]

We do not want to go into further details on these aspects. Our purpose was simply to give an impression of the situation in nuclear physics at the end of the war. In parts of the following section we will see how this new way of working became a reality, especially in the United States.

1.3 The post-war accelerator programmes[29]

To describe the main developments in accelerator construction between 1945 and 1952, we obviously have to start with the invention of the principle of phase stability in 1945. This revolutionary idea combined with the fact that during the war nuclear physics had become a research topic of high prestige, triggered to varying degrees the construction of new accelerators. We shall study three cases in detail: the United States, Great Britain, and continental western Europe.

In the case of America, the leading country, we will focus on the accelerator programmes at Brookhaven and Berkeley, two big research establishments which played an important role as models for European physicists. Concerning the British accelerators we will show that in terms of machine energy they did not rank particularly high among accelerators in the world at the time of their completion. On the other hand when originally planned some of them were the largest, or embodied technically novel features. Great Britain was thus the only European country whose evolution in this respect can be compared to that of the USA. Lastly, in continental Europe, two features will emerge as characteristic: that only a few countries pursued an extensive accelerator programme and that the machines built

Notes: p. 46

were all in an energy region limited to doing pure nuclear physics and not particle physics.

Our period terminates with the development of another new concept in accelerator design—the alternating-gradient focusing principle. It provided the basis for accelerators of even higher energies which could be built within reasonable financial limits. This idea, formulated by a group at Brookhaven in the summer of 1952, is of particular interest as it was the principle upon which the CERN proton synchrotron was built.

One last remark. When reading the following we should keep two things in mind. Firstly, from what was said in section 1.1.1, it is clear that any decision on the energy of an accelerator was based on rather vague scientific grounds.[30] Secondly, it would be anachronistic to think that the wish to construct bigger and bigger accelerators grew logically out of the problems cosmic-ray physicists had. The launching of accelerator programmes is better seen as an autonomous development. Of course for the majority of physicists the final aim of these machines was to study cosmic-ray particles under laboratory conditions, but we should remember that the task of building such accelerators was in itself very challenging from a technical point of view.

1.3.1 THE PRINCIPLE OF PHASE STABILITY

In 1945 four types of accelerators were in use: Van de Graaff, Cockcroft–Walton, cyclotron, and betatron. The cyclotrons, which were able to produce the highest energies, soon reached their energy limit. This limitation, as Bethe and Rose had already pointed out in 1937, was due to the relativistic mass increase at very high particle velocities, which resulted in a change in the frequency of the revolving particle. This frequency fell out of phase with the frequency of the accelerating field, so that the particles were decelerated. In practice this limit lay at roughly 25 MeV for protons.[31]

A solution to this problem was proposed in 1945 by E.W. McMillan and V. Veksler independently. It came to be known as the principle of phase stability.[32]

The main idea behind it was to compensate for the relativistic mass increase either by changing the accelerating high-frequency voltage or the magnetic field strength during the acceleration of the particles. Hereby it was possible to accelerate particles into the GeV region without encountering major difficulties. Consequently, one could not only operate cyclotrons at higher energies, thus converting them into synchro-cyclotrons, but also build a completely new type of accelerator, the *synchrotron*.

The principle of this new machine was to keep the particles on a path of constant radius by varying both the magnetic field strength and the frequency of the accelerating voltage with increasing particle energy.[33] One therefore needed only a ring of magnets, and not huge disc-like magnets as in the cyclotron. This had both technological and financial advantages, the former because it was easier to achieve a homogeneous magnetic field and a good vacuum in the small region of the ring, the latter because less material was needed for construction. However the most striking advantage was that this kind of accelerator could be used for accelerating both protons and electrons.

This idea, as we will see below, opened up completely new possibilities and made it feasible to build enormous machines like the Bevatron at Berkeley and the Cosmotron at Brookhaven.

1.3.2 THE UNITED STATES

As mentioned above, here we shall consider mainly the machines built at Brookhaven and at Berkeley.

The idea of Brookhaven arose late in 1945 in discussions between Isidor I. Rabi and Norman F. Ramsey.[34] The plan was to have a big collaborative research centre, and nineteen of the major East-Coast research institutes and two industrial laboratories were invited to attend a meeting in January 1946. About two months later the decision was taken: nine of the universities joined together to establish a research laboratory for nuclear science.[35] The aim was not only to do physics, but also chemistry, biology, medicine, and technical research, using both atomic piles and accelerators.

Two piles were envisaged. A conventional pile, constructed as a first step, would allow physicists to start experiments fairly soon, to gain experience and to pursue the necessary development work for the second reactor. That was to be a rather advanced type providing a high neutron flux. The equipment would have various applications, e.g. in medicine for neutron therapy, for neutron physics 'to get a better insight into the nature of the atomic nuclei', and for the production of isotopes, which would be used in chemistry, physics, biology, and medicine. Remarkably, it was also decided that the piles would not be isolated from residential areas, as elsewhere, the idea being to find a solution to this delicate problem which could serve as a model for the power industry in the construction of atomic power stations.[36]

The accelerator programme proposed in the Initial Programme Report of December 1946 was the following: a Van de Graaff of 3.5 MeV, a 60-inch cyclotron, a synchro-cyclotron with 0.6–1 GeV energy, a 1–2 GeV electron synchrotron, and a 'Super High Energy Accelerator'.

The plans for Brookhaven were thus somewhat ambitious, the hope being to cover a wide energy range with accelerators. But in fact only three of them were finally built: the Van de Graaff, the cyclotron and the big accelerator (now known as the Cosmotron).

What were the reasons given in 1946 for the construction of the cyclotron and the big accelerator?[37] The former was a conventional machine to complement the reactor in the production of isotopes, and to study the interaction of fundamental

Notes: pp. 46 ff.

particles and the mechanism of nuclear reactions. By contrast, the idea of having a 'Super High Energy Accelerator' in the region of 10 GeV was initially rather vague. No indication was given of what particles were to be accelerated and it was feared that technical problems would arise at such high energies.

Clarification of these ideas took roughly a year. Meanwhile, as we shall see, W.M. Brobeck at Berkeley had also planned a 10 GeV accelerator. Rabi brought this proposal to Brookhaven, trying to convince the scientists of the importance of such a project. Apparently some of them favoured the newly-started 600 MeV synchro-cyclotron project (abandoned soon after), while others thought that 1 GeV or a little more would be sufficient to do good meson physics. In the event it was decided to have a proton synchrotron either in the region between 2 and 3 GeV or between 8 and 10 GeV, and design studies were launched for both.[38] With a machine of this type a great deal of meson physics could be done, such as multiple meson production experiments, which would provide a greater understanding of the nuclear forces. More significantly, above a threshold energy of 5.6 GeV, nucleon–antinucleon pair production could probably be studied. This would be a step into a completely new field of particle physics.[39]

As it was intended to do particle physics during the years of design and construction of the accelerators, a cosmic-ray programme was launched. In fact in 1947/48 the whole of the particle-physics division consisted mainly of scientists and technicians who had been working on cosmic rays at Aberdeen Proving Grounds. They had transferred a lot of their material like cloud-chambers, detectors, etc. to Brookhaven specifically to study primary cosmic radiation by balloon flights up to about 100,000 feet. The importance of this group naturally diminished when the machines started to work.[40]

We now turn to developments at Berkeley during the same period.[41]

Throughout the war years, Berkeley's Radiation Laboratory had been incorporated into the Manhattan Project, mainly for work on isotope separation, i.e. the separation of U-235 from U-238. This work was done with the 184-inch magnet, which had been installed with a view to building a giant new cyclotron.[42] With this machine, the so-called 'calutron', and with several others modelled on it, the majority of the U-235 required for the bomb was produced. As a consequence of this war effort, the number of scientists grew enormously, the financial situation improved and equipment was developed to a high standard.

Not surprisingly, when the war ended the plans to continue basic nuclear research were rather ambitious. E.O. Lawrence's first aim was to convert the calutron into a *synchro-cyclotron,* based on the newly developed principle of phase stability. This took only about a year and on 1 November 1946 the 184-inch synchro-cyclotron produced its first beam. From then on 195 MeV deuterons and 390 MeV alpha particles were available. (In the course of 1949 the machine was modified to produce 350 MeV protons.) With this machine it thus became possible to produce mesons, until then detected only in cosmic radiation.[43]

In addition to this machine, the laboratory planned to enter completely new territory by constructing a synchrotron. McMillan's idea was to have an *electron synchrotron* of several hundred MeV, which was to be the model for an even bigger machine of that type to be built later. This machine was completed by the beginning of 1949 and achieved an energy of 335 MeV.

The third item in Berkeley's accelerator programme was Luis Alvarez's *proton linear accelerator*. It was the first machine of this type and offered a rather promising alternative to the synchrotrons. Alvarez had worked on the development of radar at MIT's Radiation Laboratory[44] and had realized that linear accelerators could be powered by searchlight-directed radar sets which, he estimated, could produce output energies of roughly 1 GeV. The necessary radar sets he hoped to get from army equipment. The final version of his accelerator was 40 feet long, and produced 32 MeV protons. It was completed in 1948 and produced a high-intensity (0.4 μA), well-collimated beam.[45]

Finally, the highlight of all the facilities was to be a *10 GeV proton synchrotron*, planned in 1946 by W.M. Brobeck.[46] This would provide Berkeley with an outstanding tool, and without any doubt make it the leading institution in the world. At that time (1946), the only accelerator designed to reach energies in the GeV region was the planned 1.3 GeV proton synchrotron at Birmingham. Of course the question of funding such a huge accelerator had to be faced immediately.

On 1 January 1947 the Atomic Energy Commission (AEC) took over the USA's nuclear-energy programme. When the proposal to build a 10 GeV accelerator was submitted to it, it soon became clear that the costs would be too high. Thus the energy was first lowered to 5 GeV, and then immediately raised to 6 GeV, so being at least above the threshold of nucleon-pair production (5.6 GeV). On 8 March 1948 the final decision on the energy of the accelerators was taken: a 6 GeV proton synchrotron (called the Bevatron) would be constructed at Berkeley and a 3 GeV machine of the same type (called the Cosmotron) at Brookhaven.[47]

In all these decisions on accelerators three interesting details should be noted. On the purely scientific level, it was at this time difficult to decide exactly what sort of physics one would be able to perform with what type of accelerator. We should keep in mind that it was only in March 1948, i.e. just when the decision on the accelerators was taken, that the first mesons were artificially produced in the 184-inch synchro-cyclotron at Berkeley. Moreover, very little was known about heavy mesons. All the same, for the big accelerators two energy regions seemed to be particularly interesting: one at roughly 3 GeV, well above the threshold of meson production, thus making possible the investigation of these processes at various energies, the other between 6 and 10 GeV, for the study of antiproton production. In addition it was often argued that anyway it was 'inevitable that new phenomena [would] be discovered when energy ranges [were] extended'.[48]

A second noteworthy feature of the decisions to build these giant accelerators is that on the technological level, results from war research in areas like radar were

Notes: pp. 47 ff.

being incorporated into these programmes, and were in fact absolutely crucial contributions.

Finally, the complexity of obtaining funds for such large-scale projects is clearly demonstrated. Neither purely scientific considerations nor ambitious plans were decisive. A long process of consideration and reconsideration of the project was necessary until a consensus on the financial and scientific levels could be reached, not forgetting that the state's interests were also involved.

The Cosmotron was completed in July 1952 and was thus the first accelerator producing particles with an energy of more than 1 GeV and the first operating proton synchrotron. It opened up the field of heavy mesons to the particle physicists, whose only source of information until then had been cosmic radiation. The Bevatron followed early in 1954 and was the biggest proton accelerator in the world for the next few years. The next generation of such machines would already benefit from another new principle, the alternating-gradient focusing principle (see section 1.3.5).[49]

Table 1.1
Synchrotrons and synchro-cyclotrons in the United States[50]

Laboratory	Energy [MeV]	Date of completion
Synchro-cyclotron		
University of Chicago	450	1951
Carnegie Inst. of Technology	440	1952
Columbia University	385	1950
Berkeley	350	1946
University of Rochester	250	1948
Harvard University	125	1949
Electron synchrotron		
California Inst. of Technology	500	1952
M.I.T.	340	1950
Berkeley	335	1949
Cornell University	300	1951
University of Michigan	300	1953
Purdue University	300	1954
Proton synchrotron		
Brookhaven Cosmotron	3000	1952
Berkeley Bevatron	6400	1954

Finally, as we have restricted the discussion to Berkeley and Brookhaven, we present a list (table 1.1) of all synchro-cyclotrons and synchrotrons with energies of at least 100 MeV completed or under construction in the USA by 1952. This gives us an idea of the breadth of the accelerator programme and the intensive efforts made in this direction.

1.3.3 GREAT BRITAIN—THE LEADING COUNTRY IN EUROPE

To understand the difference between the situation in Great Britain and that on the continent in the post-war period, we must go back to the war years. Many British physicists had been very active both in microwave techniques and, together with the Americans, in developing the bomb. Besides that they did nuclear-physics research in the Anglo–Canadian Project in Montreal. Thus when the war ended Britain was in a situation similar to that of the United States, i.e. nuclear physics was a research subject of outstanding prestige and importance, especially with respect to military applications.

In the light of this, it is not surprising that as early as 29 October 1945 the Prime Minister announced that the government was 'to set up a research and experimental establishment covering all aspects of the use of atomic energy'. The responsibility for this project lay with the Ministry of Supply and Sir John Cockcroft was appointed as Director in January 1946. Harwell was chosen as the site for the Atomic Energy Research Establishment (AERE) and in April 1946 the building started.[51]

What was the research programme for this centre? The basic idea of Harwell was to do applied nuclear physics, and so piles had priority. They would fulfil a double task, namely, they would enable physicists to do basic research and, at the same time, to produce fissile material. While not wanting to go into detail on this part of the programme, we nevertheless wish to mention the reactors built. The first was a graphite low-energy experimental pile, to be called Gleep, which went critical in 1947. The second, called Bepo, was constructed with a view to reaching higher neutron-flux densities and was thus more suitable for the production of isotopes. These isotopes found a ready market all over the world, and were mainly used for medical and biological purposes.[52]

We turn now to our central point of interest, the accelerator programme. A variety of different types of accelerators was planned at Harwell, amongst them a Van de Graaff, a 110-inch synchro-cyclotron, an electron synchrotron, and a linear accelerator.[53]

- The *Van de Graaff* was standard equipment for a nuclear-physics laboratory. It was largely built in the Engineering Unit at the Telecommunication Research Establishment in Malvern, before being erected at Harwell. It reached an energy of 5 MeV.
- The *110-inch synchro-cyclotron* was the first machine in Great Britain to be built incorporating the principle of phase stability. (The initial plan had been to build a 72-inch standard cyclotron.) The machine went into operation in December 1949 and was able to produce protons with a kinetic energy of 175 MeV. For roughly

two years it was the largest operating accelerator in Europe. Its disadvantage was that the machine's energy was just at the threshold of meson production and so the research programme had to concentrate on pure nuclear physics.[54]

- The second cyclic accelerator built was an *electron synchrotron*. The work on this machine was influenced by the view that it should serve as a prototype for the planned 300 MeV electron synchrotron at Glasgow. As a first step towards this goal an existing betatron was modified. In August 1946 the conversion was completed and the world's first electron synchrotron produced 8 MeV electrons. The second step was to construct a 30 MeV machine of the same type (with an 8-inch diameter), which went into operation in October 1947.
- The fourth machine in the Harwell programme was a *linear accelerator* for electrons, whose main feature was that it worked on the travelling wave principle, another world first. The idea was to make particles 'ride' on the crests of very short electromagnetic waves produced by radar equipment—without which the whole scheme would have collapsed. The first device embodying this principle was 2 m long and produced a high-intensity electron beam of 4 MeV energy.

To summarize, we note that the Harwell programme involved the construction of a wide variety of accelerators, and that Britain had both the technicians with adequate skills and the industrial resources to pursue such a programme (which set her apart from many other European countries). Last but not least, from the 'choice'[55] of the machines we can see that the orientation of research was more towards nuclear physics, covering a wide range of aspects, and not towards meson physics.

Harwell's accelerators were only a part of the British nuclear-physics programme. Early in 1946 it was also decided to provide the universities with adequate nuclear-physics equipment. The Nuclear Physics Committee of the Ministry of Supply sent circulars to thirty universities and other institutions asking for an outline of their nuclear-physics programme. The only five universities to reply with requests for large-scale equipment were Birmingham, Cambridge, Glasgow, Liverpool, and Oxford. All of them received grants from the Department of Scientific and Industrial Research for the construction and maintenance of the following accelerators:[56]

- Birmingham: 1.3 GeV proton synchrotron
- Cambridge: 300–400 MeV electron linear accelerator[57] (later abandoned)
- Glasgow: 300 MeV electron synchrotron
- Liverpool: 400 MeV proton synchro-cyclotron
- Oxford: 140 MeV electron synchrotron

Let us now have a closer look at these five accelerators. As early as 1943 M.L. Oliphant had suggested that a big proton accelerator be built at Birmingham. His idea at the time was to use only a ring-shaped magnet for proton acceleration, but we do not know exactly how he intended to achieve this.[58] When the war ended in 1945 he actually designed a powerful *proton synchrotron,* incorporating the principle of phase stability. Oliphant was then in advance of Berkeley and Brookhaven but, because of inadequate resources at Birmingham and owing to problems of housing the machine, the accelerator reached an energy of 1 GeV only in July 1953 (and then with a very low intensity).[59] By this time the 3 GeV Brookhaven machine had been completed.

The two *electron synchrotrons* in Glasgow and Oxford were both built with the advice of the Malvern–Harwell group. The final version of the smaller machine at Oxford only got to 125 MeV, and was not generally regarded as a success. The Glasgow machine, on the other hand, was finished in 1954 and reached an energy of 350 MeV, 50 MeV above the design energy. This machine was referred to as 'most carefully designed and engineered' and as incorporating 'all the best features of the early machines' of its kind.[60]

The Liverpool machine was a *156-inch synchro-cyclotron* corresponding to a proton energy of 400 MeV. It had originally been conceived as a 60-inch cyclotron, but when news of the principle of phase stability arrived from America its energy was increased in steps. It went into operation in 1954 and was Europe's biggest synchro-cyclotron until 1957 when the CERN synchro-cyclotron (600 MeV) was completed.

To summarize, three of her four university accelerators under construction provided Great Britain with the opportunity to do meson physics, although only when machines of much higher energies were already available in the USA. Nevertheless, they provided a good basis for launching a research programme in particle physics.

1.3.4 CONTINENTAL WESTERN EUROPE[61]

We have seen the clear leadership of the United States in the building of accelerators and also the efforts undertaken by Great Britain to pursue a rather ambitious accelerator programme. Activities of this kind were more sporadic in continental western Europe. In what follows we discuss the accelerator programme in various countries on the continent around 1951 because at this date major discussions took place on the future European laboratory.[62]

We start with *Sweden,* since this was the country where there was most activity in building accelerators. By 1951 Sweden had two cyclotrons producing 7 and 25 MeV deuterons, both at the Nobel Institute for Physics. The latter machine, finished in 1951, was one of the biggest conventional cyclotrons ever built. At the Royal Institute of Technology an electron synchrotron of 35 MeV and a betatron of 5.3 MeV were in operation. In addition, there were four electrostatic generators operating at other institutes. However, the most outstanding of all these tools was the 200 MeV synchro-cyclotron at the Gustaf Werner Institute for Nuclear Chemistry in Uppsala. When completed at the end of 1951 it was Europe's biggest synchro-cyclotron (until the 400 MeV synchro-cyclotron at

Liverpool went into operation in 1954). Although its energy theoretically lay just above the threshold for meson production, the actual meson output was much too low to do meson physics.

The next two countries we shall consider, namely the Netherlands and Switzerland, have one characteristic in common. Their accelerator programme was 'backed' by large industrial firms like Philips in the case of the Netherlands and Brown Boveri and Oerlikon in the case of Switzerland.

In the *Netherlands* the principal laboratory for atomic-energy development work was the Institute for Nuclear Research in Amsterdam established in 1946. Before and during the war C.J. Bakker and F.A. Heyn had already designed a cyclotron while at Philips in Eindhoven. This was changed in 1946 to a synchro-cyclotron of 71-inch diameter able to produce 28 MeV deuterons, which was erected at the University of Amsterdam and came into operation in 1949. Indeed Philips continued to build several accelerators of this type, as well as cascade generators, which were mainly sold abroad. In addition to this synchro-cyclotron, two further electrostatic generators (0.3 and 1 MeV) were installed in the Netherlands.[63]

Seven accelerators were in use in *Switzerland* by 1951, namely a cyclotron of 7.5 MeV protons and two Van de Graaffs of 0.85 and 1.5 MeV all at Zurich, two betatrons of 32 MeV, one at Zurich (since 1948) and one at Baden, a 1 MeV electromagnetic accelerator in Basel (finished in 1949) and a 1 MeV tensator at Zurich. In addition there were plans for synchrotrons of 100 MeV and 400 MeV at Brown Boveri in Baden.[64]

Norway and France are the next two countries we would like to group together since both put more emphasis on their reactor programmes.

In 1951 *Norway* had four electrostatic accelerators reaching maximum energies of 1.5 MeV and one betatron producing 47 MeV electrons (at Bergen). Apart from these, further electrostatic generators with energies between 1 and 4 MeV were the only machines planned.[65]

By the end of the war, *France* had four electrostatic generators, the largest having an energy of 3 MeV, and one cyclotron of 32 inch producing 6.8 MeV protons. The French Commissariat à l'Energie Atomique (CEA) was founded late in 1945, its goal being the use of atomic energy for scientific, industrial, and defence purposes. Only two accelerators were planned in 1948—a standard cyclotron of 65 inch (producing 26 MeV deuterons) and a 3.5–4 MeV Van de Graaff. The building of these machines was primarily an engineering task and they were not in fact part of an experimental nuclear-physics programme using accelerators.[66]

As for *Belgium,* in 1951 she had only one operating accelerator, namely a Cockcroft–Walton accelerator at Brussels (built by Philips). Apart from that three others were under construction: a 25–30 MeV proton linear accelerator, a 12 MeV cyclotron (deuterons), started in 1949 and finished in 1952, and a 1.2 MeV electromagnetic generator (built by Philips; finished in 1952). No further accelerators were planned for the time being.[67]

The last two countries to be mentioned are Italy and Germany. By the end of the war *Italy* had only one accelerator, namely a 1 MeV Cockcroft–Walton and no further projects in this direction were undertaken until the mid-fifties. By 1951 *Germany* was in possession of two cyclotrons, one at Bonn producing 2 MeV protons and one in Heidelberg producing 9 MeV deuterons. Further she had two betatrons of 6 and 15 MeV and some five electrostatic accelerators all with energies around 1 MeV. However, due to the restriction of German nuclear work by the allied countries, which prohibited construction of reactors and high-energy accelerators bigger than 100 MeV until the mid-50s, no really big projects could be implemented.

Summarizing, it may be said that the energy of continental research accelerators was appropriate only for nuclear-physics experiments. None of them, not even the large synchro-cyclotron in Uppsala, could go significantly above the threshold for meson production. In short, in 1951, no single continental country was equipped to do particle physics.

One last remark. The situation in the field of accelerators described above changed drastically in the middle of the fifties. All over Europe machines in the GeV range were envisaged or even in some cases under construction: four electron synchrotons, two in Germany at Bonn (1/2 GeV) and at Hamburg (DESY, 6 GeV), one in Italy at Frascati (1 GeV) and one in Sweden at Lund (1.2 GeV), an electron linear accelerator in France at Orsay (1.3 GeV) and two proton synchrotrons one in France at Saclay (SATURNE, 3 GeV) and one in Britain (NIMROD, 7 GeV).[68]

1.3.5 THE AG FOCUSING PRINCIPLE

By 1952 the major part of the post-war American accelerator programme had been realized. The synchro-cyclotrons and electron synchrotrons built at various universities were finished, the big 3 GeV Cosmotron in Brookhaven went into operation, and the 6 GeV Bevatron in Berkeley was well on the way. However, although the techniques used for all these machines could, in principle, be applied in even bigger machines, there were limits imposed by the fact that the costs rose exponentially with the energy.

A way out of the situation was proposed in a paper by E.D. Courant, M.S. Livingston, and H.S. Snyder of Brookhaven entitled 'The Strong-focusing Synchrotron—A New High Energy Accelerator' which was sent to the *Physical Review* in August 1952.[69] Before going into details about this principle, it should be mentioned that the idea had been proposed as early as March 1950 by a Greek engineer, Nicolas Christofilos, who tried to patent it in the United States.[70] Unfortunately, the value of this work was not initially realized.

We now wish to describe briefly the main features of the strong-focusing or alternating-gradient (AG) focusing principle. If one wanted to construct even

Notes: p. 49

bigger synchrotrons within financially reasonable boundaries either the radius of the machine or the cross-section of the magnets (which depend on the size of the vacuum chamber needed) had to be decreased. Owing to deviations from the equilibrium orbit caused by angular and energy spread in the injected beam, scattering by the gas, magnetic inhomogeneities, and frequency errors, betatron and synchrotron oscillations usually occur in such accelerators. These can only be stabilized by a radially decreasing magnetic field, and one with a field gradient $n = -(r/B)(dB/dr)$ which lies between 0 and 1. The idea of strong focusing is to build the magnet ring out of sections with a very high field index (field gradient n) of at least 100. The arrangement of these sectors is such that in the first the magnet field drops with increasing radius (positive n), so that the beam is vertically focused while being at the same time radially defocused. The following sector has a magnet field rising with increasing radius (negative n) which has vertical defocusing and radial focusing as a consequence. Keeping up this sequence over the whole ring one achieves overall focusing of the beam. The decrease of the oscillation amplitude achieved in this way enables one also to decrease the size of the vacuum chamber used. This has primarily two consequences: a good vacuum is easier to obtain and the size of the magnets can be reduced. As calculated in 1952, assuming ideal conditions, for a 30 GeV accelerator, the aperture would be 1 × 2 inches instead of roughly 18 × 6 inches (as indicated in the first design of the CERN proton synchrotron). Financially this meant that, at a given cost, a machine with a ten times higher energy could be built.[71]

What we have described here is the basic idea underlying the new focusing principle. At this stage we do not want to explore the doubts, the problems, etc. which occurred as soon as attempts were made to put those ideas into practice. Some of these are discussed briefly in part III of this volume.

1.4 Experimental particle physics: developments from 1946 to 1953 [72]

In our study of the development of particle physics from 1946 to 1953, we were particularly struck by the fact that the number of new particles, and the information on them, grew very rapidly, and formed the basis on which particle physics was to become an increasingly important part of physics.

The evolution of this domain of physics in the eight years following the war can be divided into three essential stages. The first was the proof of the weak nature of the mesotron and the detection of the pi-meson in cosmic radiation in 1947. The second was the artificial production of these particles in 1948, and the last was the realization that the mass spectrum of elementary particles was more complex than had previously been thought. To illustrate the progress made in this field the *International Congress on Cosmic Rays* held in Bagnères-de-Bigorre (France) in July

1953 is typical.[73] It will help us to see the effects of these different developments and the consequences of them for cosmic-ray physicists in particular.

1.4.1 THE WEAK NATURE OF THE MESOTRON AND THE DETECTION OF THE PI-MESON (1946/47)

In section 1.1.1 we briefly referred to the fact that in 1937 when a particle 200 times heavier than the electron was detected in cosmic radiation it was identified with Yukawa's field quantum mediating the strong interaction. Although there was some disagreement between theory and experiments, this identification was maintained. Before this 'meson-puzzle' could be solved, two steps were necessary. Firstly, some features of the known cosmic-ray meson (the mesotron) had to be studied more closely. Secondly, crucial improvements to the detection devices were needed.

The first step was taken by M. Conversi, E. Pancini, and O. Piccioni late in 1946. They carried out an experiment on the absorption of magnetically-separated mesons of different charge in both iron and carbon plates. It demonstrated that in iron all positive mesons disintegrated, whereas the negative ones were absorbed. However, using carbon as absorber a great deal of negative decay electrons coming from negative mesons were observed.[74] Thus, it was concluded that the mesotron interacted much too weakly with the nuclei to play the role of the meson predicted by Yukawa.

Much of the credit for achieving success in the second stage goes to C.F. Powell. The British were leading in the development of nuclear emulsions. During the war Powell and his group had used the photographic method of studying nuclear processes extensively. Just after the war he was prominent in a small panel set up by the Ministry of Supply which commissioned the photographic firm Ilford to produce emulsions for nuclear research. Consequently, as early as mid-1946, emulsions in four grain sizes and three sensitivities were available. With these it was even possible to detect particles which produced very little ionization.[75]

It was these newly-developed emulsions which enabled Powell's group to detect the first two examples of the pi-meson in cosmic radiation early in 1947.[76] They saw one meson stop and another meson leave the same spot with an energy of a few MeV. This was interpreted as the decay of the Yukawa meson, which they called the pi-meson, into the mesotron (as it had hitherto been called), which they now termed the mu-meson. A few months later 40 of these events had been detected. Some major points could therefore be regarded as settled: there were two mesons pi and mu; the former decayed quickly into the latter, which interacted only weakly, and a unique neutral particle; finally, pi-mesons, when stopped, would produce 'stars', mu-mesons not.

Despite the discovery of the pi-meson and the knowledge of its mass and decay scheme, many problems still remained. Of course, the particle transmitting the nuclear force had been found, but immediately the question arose as to 'what to do' with the mu-meson. Firstly, this particle did not fit into any known theory and, secondly, the decay of the pi-meson into the mu-meson did not correspond to the simple beta-decay scheme proposed by Yukawa.

1.4.2 THE ARTIFICIAL PRODUCTION OF CHARGED AND UNCHARGED PI-MESONS (1948/49)

This crucial step in the development of elementary-particle physics took place roughly a year after the detection of pi-mesons in cosmic radiation. Using the newly-built 184-inch synchro-cyclotron at Berkeley, it had become possible to accelerate alpha particles up to energies of around 350 MeV. Bombarding a carbon target with these particles, C.M.G. Lattes and E. Gardner succeeded in producing and detecting charged pi-mesons, the first artificially produced mesons.[77] It should be mentioned, parenthetically, that this was made possible by the recent development of nuclear emulsions.[78]

On 9 March 1948 the university and the AEC officially announced the success at Berkeley. This event represented an important scientific success, as from now on one could start detailed investigations of these mesons. The resulting enthusiasm can be gauged from the reactions at various conferences where the new results were presented, e.g. at the second Shelter Island conference held from 30 March to 2 April 1948 and at the first post-war Solvay congress held from 27 September to 2 October 1948.[79] At both it was R. Serber, the leader of the theoretical group for the design and the planning of experiments at the 184-inch synchro-cyclotron, who presented the results. On both occasions the talk was received with great interest. Many detailed questions were asked on the lifetime of the meson, on the meson cross-section of the nucleus, on the statistics in π^+ decay, etc. It should also be observed at this point that the problem of reproducibility, which was perhaps the most important one in cosmic-ray work, had vanished. Only thirty seconds of bombardment with the cyclotron produced 100 times as many mesons as Lattes had been able to photograph in 47 days of cosmic-ray observation.[80]

Furthermore this Berkeley success was used to justify the huge amounts of money already spent on the accelerator programme, and was a good argument for further funding. E.O. Lawrence, for example, immediately insisted that 'to exploit the knowledge which the meson may provide it will be necessary to construct super-giant cyclotrons'. *The Times* reporting this event, hinted that the study of mesons might 'lead in the direction of a vastly better source of atomic energy than the fission of uranium'. This again shows the high expectations generally connected with any sort of research even loosely related to applied nuclear research in the post-war years.[81]

What exactly did this event mean for the cosmic-ray physicists? Their feelings were in fact rather ambivalent. In one respect, others had trespassed on their domain of physics and had replaced their rather unreliable source of particles with machines offering advantages such as reproducibility, high particle-flux densities, and the possibility of well-planned experiments. Powell expressed this feeling very clearly at a conference on *Cosmic Radiation* held at Bristol in September 1948. He mentioned that 'before the observations [on pi-mesons] were completed, however, news came of the artificial production of mesons at Berkeley. It seemed clear that the possibility of making measurements of much higher precision would thus become available, and further experiments in this field were temporarily abandoned at Bristol.' The time had come to rethink the work to be done.[82]

On the other hand, he tried at the same time to emphasize the role which cosmic-ray physics would still play: 'At the present time, the importance of the study of cosmic radiation appears to be associated with the fact that the individual particles of which it is composed are often found to be many thousands of times more energetic than any we can generate artificially in the laboratory. In spite of the relatively small number of particles in the incoming stream, we are thus enabled to investigate new types of processes which cannot be observed in any other way'.[83]

Anyway, it had become clear that cosmic-ray physicists had to leave part of their investigations, namely low-energy meson physics, to the physicists working with accelerators.

A further important related event took place early in 1950. The electron synchrotron had been completed at Berkeley and with this accelerator it became possible to produce the first neutral pi-mesons. *Bremsstrahlung* photons resulting from the 335 MeV electron beam were directed onto a carbon target giving the reaction:

$$\gamma + p \rightarrow p + \pi^0$$
$$\longmapsto 2\gamma.$$
$$\gamma + n \rightarrow n + \pi^0$$

The π^0 was then identified by J. Steinberger, W. Panofsky, and J. Steller[84] via its two decay photons. This was the first particle produced by an accelerator which had not previously been established in cosmic-ray experiments. The hypothesis that such a particle existed had already been proposed in 1938 on the basis of considerations on the charge independence of nuclear forces and by 1940 a decay of it into photons had been suggested.[85]

Although the neutral pi-meson was also detected in cosmic radiation soon afterwards, from 1950 onwards all precise information on the properties of pi-mesons and mu-mesons, like mass, lifetime, and spin was obtained with accelerators (see table 1.1 for the six American accelerators able to produce mesons by 1951 and table 1.2 for the status of knowledge at the time). In fact 1950 'was the first year in which the accelerator physicists outdistanced the cosmic-ray physicists in so far as their contributions to our understanding of the properties of mesons, the

Notes: pp. 49 ff.

character of their production, and the nature of their interaction with nucleons [were] concerned. The fact that much greater intensities of artificially produced mesons were available was beginning to make itself felt in a significant and impressive way.'[86]

Table 1.2
Main properties of the pi- and mu-meson (1951) [87]

Particle	Mass [m_e]	Spin	Lifetime [s]	Decay scheme
π^+	276	0	2.6×10^{-8}	$\pi^+ \to \mu^+ + \nu$
π^0	265	0	$< 5 \times 10^{-14}$	$\pi^0 \to 2\gamma$
π^-	276	0	2.6×10^{-8}	$\pi^- \to \mu^- + \nu$
μ^+	210	1/2	2.22×10^{-6}	$\mu^+ \to e^+ + 2\nu$
μ^-	210	1/2	2.22×10^{-6}	$\mu^- \to e^- + 2\nu$

1.4.3 THE COMPLEXITY OF THE MASS SPECTRUM

The first hint that there could be other particles heavier than the mesotron came as early as 1944. L. Leprince-Ringuet and M. Lhéritier reported the detection of a new particle in cosmic radiation with a mass about 990 times the mass of an electron, and consequently much heavier than the mesotron. However, as already pointed out in the first section, since there was only one photograph of this event and the measurement was rather inaccurate, the result was not particularly convincing.[88]

In fact it took another three years before other new heavy unstable particles were detected in cosmic radiation (in 1947). The hunt for new particles was now on and would reach its climax in the fifties.

1.4.3.1 The 'V-particles'.[89] In 1947 G.D. Rochester and C.C. Butler working at Manchester University published a paper on a new kind of event they had found in their Wilson chamber in the course of investigation of cosmic-ray induced penetrating showers. A neutral particle entered the chamber and decayed into two charged particles, an event having a characteristic V-shaped track. In addition incident charged particles left a V-shaped track when decaying. The measured mass of the neutral particle was between 800 and 1000 m_e while that of the charged particle was estimated as close to 1000 m_e. These new events were followed with great interest by the physics community, and promised to reveal something new and hitherto completely unknown.[90] In fact these were the first examples of the (present) particles K^0 and K^+, so the first members of the family of 'strange particles' (as they were later called). However, although the Wilson chamber continued in operation at Manchester, no further events of this form were detected in the following three years.[91]

Meanwhile, in the United States the search for these particles had started and by 1950 Seriff et al. had found a further 34 V-shaped tracks,[92] nearly all of them at their high-altitude laboratory at White Mountain. This kind of particle could now be reliably regarded as established. Being asked to name the new particles, P.M.S. Blackett in consultation with C.D. Anderson baptized both the charged and the neutral particles 'V-particles'.

In the same year, the Manchester Group transported their cloud chamber to the top of the Pic du Midi de Bigorre and by March 1951 they had detected 36 V^0 and 7 V^+. Starting more systematic investigations on these V-particles, Armenteros and his group[93] came to the conclusion that there were in fact two different neutral V-particles each with its own mass and decay scheme. The V_1^0, a kind of superneutron, with a mass of roughly 2250 m_e, decayed as follows

$$V_1^0 \rightarrow p + \pi^-.$$

The V_2^0, with a mass of 950 m_e, decayed thus:

$$V_2^0 \rightarrow \pi^- + \pi^+.$$

However, it was still possible (although the mass values differed so much) to assume that there was only one V^0 undergoing two different three-body decays, namely:

$$V^0 \rightarrow p + \pi^- + \text{neutral particle}$$

$$V^0 \rightarrow n + \pi^+ + \pi^-.$$

This example illustrates clearly the kind of difficulties cosmic-ray physicists were constantly meeting. It was hardly possible to obtain exact results on momenta and angles of decay products, and any conclusions were accordingly rather tentative.

By 1952 the two-body decay of the V_2^0 could be regarded as the most probable one and so these two particles (V_1^0 and V_2^0) were definitely established. No other new neutral unstable particles were found in the period described by us.

While information about the neutral V-particles increased rapidly,[94] little was known about the mass, disintegration scheme, and lifetime of the charged ones. The reason for this was that the ratio of detected charged to neutral V-particles was roughly 1:6. Moreover only positively charged V-particles had been found by 1952. Their mass was determined as roughly 1100 m_e, but with error-margins of 20–30%. This indicated at least that these particles were certainly not the charged counterparts of the V_1^0. On the other hand, some charged V-events seemed to contain pi-mesons or mu-mesons in their decay products and would therefore be similar to the decay products of the K-particles (χ and $\varkappa$), to which we will come later.[95] This was all the information which could be gained on positively charged V-particles, whose status was still in doubt at the time.

The first example of a negatively charged V-particle decaying into a neutral V-particle and a meson was detected by the Manchester Group 1952:[96]

$$V^- \rightarrow V^0_{1 \text{ or } 2} + \text{light meson}.$$

As there was no evidence as yet for a charged V-particle of mass greater than the proton it was assumed that the neutral V-particle in the decay would be a V^0_2. But in fact this was not the case, it was a V^0_1, and this event represents the first example of the 'charged hyperon' Ξ^-, as we now call it.

1.4.3.2 The 'heavy mesons'. In January 1949, the Bristol group published a photograph of the following event: a heavy particle (lighter than a proton) of mass roughly 1000 m_e came to rest and emitted three particles, a negative pi-meson and two other mesons. This photograph was only possible because of the progress made in emulsion techniques. (In the meantime electron sensitive emulsions had been developed, with which any particle of charge one could be recorded, even if it was moving with the velocity of light.) This new particle was called the τ-*meson*, being the first example of what we now call a $K_{\pi 3}$ decay.[97] When a further event of that kind was found in 1950 this particle's existence was regarded as established.

The number of charged particles soon increased further. In 1951 O'Ceallaigh identified the first $\varkappa$-*meson*, as he called it, in the photographic emulsion while studying the decay electrons from mu-mesons. In fact he measured two different mass values, 1320 and 1125 m_e. But as they were so close he supposed that there was only one new particle decaying into a mu-meson and unobserved neutral particles.[98] However, investigating the $\varkappa$-decay closer, it was discovered that there were in fact two different decay modes and thus it was supposed that they came from two different particles with mass values close to each other. The particle having a three-body decay involving a mu-meson kept the name $\varkappa$-meson (nowadays called $K_{\pi 3}$), the other, having a pi-meson among the decay products, was named χ-*meson*, nowadays called $K_{\pi 2}$.[99]

An additional heavy charged particle was found in 1951 and 1952 by a group at M.I.T. using multiplate cloud chambers. It was named the *S-particle* and had a mass of roughly 1400 m_e.[100]

Let us stop here and have a look at what has happened. From 1947 onwards the contribution of cosmic-ray physicists to finding new particles had been enormous (see table 1.3). A rather complex subnuclear world, full of unexpected features had been revealed. It had become clear that particle physics had not stopped with the discovery of the pi-meson; this had only been a start. What we have presented here is only a small part of the work done, only the outstanding and well-known results reported. The cosmic-ray physicists had, within five years, more than doubled the number of particles, although we should state here that at the time different decay modes of the same particle were often taken to be different particles.

Table 1.3
List of V-particles and heavy mesons known in 1952 [101]

	Mass [m_e]	Mode of decay	Present name	Date of detection
V_1^0	2250 (2174.5)	$V_1^0 \to p + \pi^-$	Λ^0	1951
V_2^0	800 (970)	$V_2^0 \to \pi^+ + \pi^-$	K_S^0	1947
V^+	1000 (962)	$V^+ \to ? + ?$	K^+	1947
V^-	? (2585.7)	$V^- \to V_{1\ or\ 2}^0 + $ meson (π,μ)	Ξ^-	1952
τ	969	$\tau \to \pi^+ + \pi^+ + \pi^-$	$K_{\pi3}$	1949
$\varkappa$	1300 (962)	$\varkappa \to \mu^+ + ? + ?$	$K_{\mu3}$	1951
χ	1100	$\chi \to \pi + ?$	$K_{\pi2}$	1952
S	1400	S $\to$ meson $(\mu, \pi) + ? + (?)$	$K_{\mu2}$	1952

(The mass values in brackets are the present values.)

1.4.3.3 The Bagnères-de-Bigorre conference (1953). From 6 to 12 July 1953 the *International Conference on Cosmic Rays* took place at Bagnères-de-Bigorre, a small village near the Pic du Midi (France). The conference lasted six days and its aim was to draw as coherent a picture as possible of all these new particles.

The reason for our choosing this conference in particular is that it was the only rather big international cosmic-ray conference (about 180 participants) which fell at the end of the period discussed by us. Therefore it enables us to gain an impression of the situation in particle physics in the early fifties seen from the point of view of cosmic-ray physics. More than 70 talks were given and a considerable amount of detailed information on all these new particles found in cosmic radiation was presented.

Without entering into details, we would like to mention some important results achieved: the question of neutral cosmic-ray particles was practically settled in various discussions; the V^- decay (Ξ^-) was firmly established by presenting further events; and the first examples for a so-called superproton (later Σ^+) were shown. However, it is much more interesting for our purposes to see that here the first efforts were made to build up a general vocabulary. The talk by Bruno Rossi on the last day was devoted to summarizing the experimental results presented during the preceding days and he took this opportunity to recall the 'nomenclature more frequently used during the conference'. The particles were classified according to their mass in L- (light), K- (mass between pi-meson and proton), and H- (mass between neutron and deuteron; hyperons) particles. The events were divided into two categories, namely V- and S-events. Finally the use of small Greek letters for mesons and capital ones for hyperons was also suggested.[102] This nomenclature and division remained a general rule for quite some time. What is noteworthy is not simply the way the division was made nor that it was kept for quite a while, but the fact that it was made at all. This was a sign that the field of elementary particles had grown in its scientific content to the extent that some order needed to be imposed on it, and that the community of scientists working in the field in many different places had become so large that if international collaboration and communication in particle physics was to be possible, a common vocabulary had to be found.

However, on this same occasion, when cosmic-ray physicists demonstrated their dominant contribution to elementary-particle physics, M. Schein also reported on the plans for 'the artificial production of V^0 by 227 MeV π-mesons generated in the Chicago Cyclotron'.[103] In 1951 and 1952 the Chicago and the Pittsburgh synchro-cyclotrons had been completed and, accelerating protons to energies of respectively 450 MeV and 440 MeV, they were able to produce pi-mesons of higher energy than before. The Brookhaven Cosmotron of 3 GeV energy was completed in 1952 and soon promised to deliver interesting results at a very high energy.

In fact, in the mid-fifties cosmic-ray physics had to some extent reached its zenith in regard to its contribution to particle physics, as many physicists realized. The accelerators were slowly but surely taking over the leading role in this field. As Leprince-Ringuet put it in his closing speech at the conference:[104]

> Perhaps we can say that the future of cosmic rays, in the field of nuclear physics, is tied to that of the accelerators, and more particularly to the rate at which strong focusing is developed. On the other hand this is probably a somewhat limited perspective. We do have exclusive access to phenomena—admittedly rather rare—occurring at far greater energies and, to the extent that our own techniques will be developed, we will be somewhat protected against the appropriation of the field by accelerators of increasing energy.

1.5 The theoreticians: from the first Shelter Island conference to the Rochester conferences

The period 1946–1952 was very eventful for theoretical particle physics. Even if many questions remained unsolved and even if many more arose from new experimental results, it was the time of the first successes and the time when the self-confidence of the theoreticians increased. In what follows, we will not cover topics which belong to nuclear physics, although a lot of important work was done around 1950, for example on models for the nucleus.[105]

To understand these developments, we will study the three Shelter Island conferences[106] and the first three Rochester conferences in detail. While we are aware that there were other very important conferences in this field, we want to restrict ourselves to these because the former were meetings oriented strictly towards theoretical work and because we have here a *series* of conferences. Thus it is possible for us to get a better grasp of the evolution of theoretical physics during these years.

The Rochester conferences thereafter revealed the changes theoretical physics was undergoing in the early fifties, particularly the establishment of a very close collaboration between theoreticians and experimentalists. Of course we will also include major theoretical achievements even if they were not discussed at these conferences.

The initial idea of holding the Shelter Island conferences—as they were later called—came from Duncan MacInnes. He was a physical chemist at the Rockefeller Institute and his idea was that the National Academy of Sciences should sponsor several small conferences on various subjects. He therefore contacted K.K. Darrow, Secretary of the American Physical Society. They envisaged something like the Solvay Congress, in other words to gather a rather select group of phycisists and to publish the discussions later. This was the proposal Darrow presented to Wolfgang Pauli, who was at that time at the Institute for Advanced Study in Princeton. Pauli, however, was more in favour of something on a larger scale, rather like the big conference held in Cambridge in 1946, as he was convinced that this would be necessary to re-establish communal values in physics after the war. However, it was finally decided to hold a small informal conference without proceedings. The aim was to convene a group of 25-30 physicists, most of them of the younger generation. The topic was 'The Quantum Mechanics of the Electron' and the date was fixed for 2-4 June 1947 on Shelter Island (near Long Island, New York).

Reading the list of participants presented in table 1.4 three things are worth noting: it was mainly an American conference, no less than four future Nobel prizewinners were present, namely Bethe, Feynman, Lamb, and Schwinger, and only a few experimentalists were invited.

Table 1.4
List of participants[107]

Chairman:	K.K. Darrow	
Discussion Leaders:	H.A. Kramers	
	J.R. Oppenheimer	
	V.F. Weisskopf	
Participants:	H.A. Bethe	A. Pais
	D. Bohm	L. Pauling
	G. Breit	I.I. Rabi
	E. Fermi	B. Rossi
	H. Feschbach	J. Schwinger
	R.P. Feynman	R. Serber
	W.E. Lamb	E. Teller
	D.A. MacInnes	G.E. Uhlenbeck
	R.E. Marshak	J.A. Wheeler
	J. von Neumann	J.H. Van Vleck
	A. Nordsieck	

Notes: p. 51

Three delegates had been asked to prepare general papers for discussion: Kramers, Oppenheimer, and Weisskopf.[108] As no proceedings were published these are our only sources of information on the topics which were dealt with at this conference. Let us therefore discuss them briefly.

Weisskopf proposed to divide the discussion into three main parts:

A. The difficulties of quantum electrodynamics (QED);
B. The difficulties of nuclear and meson phenomena;
C. The planning of experiments with high-energy particles.

In the section on QED, the problems of infinite self-energy, vacuum polarization, subtraction mechanism for infinities, high-energy limits to QED, and the infra-red catastrophe were to be discussed. Section B was to comprise an overview of the actual attempts to formulate meson theories and their representation of nuclear forces, and also a discussion of the experiment of Conversi, Pancini, and Piccioni. It is interesting to see that, regarding part C, Weisskopf mentioned that it 'could become the most useful part of the conference. A number of very powerful accelerators in the energy region of 200–300 MeV [were] near completion, and it [was] time to inaugurate a systematic program of research, with experiments which [could] be reliably interpreted'. He thus stressed the importance of a close co-operation between theoreticians and experimentalists to prepare in time the future programme to be pursued with accelerators.

In his outline of topics for discussion, Oppenheimer concentrated more on the theoretical interpretation of cosmic-ray results. His aim was to find out if it would be possible to reconcile 'on the basis of usual quantum mechanical formalism the high rate of production of mesons in the upper atmosphere with the small interactions which these mesons subsequently manifest in traversing matter', a contradiction which had appeared from the Conversi et al. experiment. He went on to say that 'to date no completely satisfactory understanding of this discrepancy exists, nor is it clear to what extent it indicates a breakdown in the customary formalism of quantum mechanics'. He thus suggested that topics like the theory of multiple meson production should be studied in greater detail.

We can therefore see that both Oppenheimer and Weisskopf showed considerable interest in current cosmic-ray work. In particular, the experiment showing that the mesotron was definitely not a strong interacting particle attracted much attention in the community of theoreticians. Analyzing this experiment, Fermi, Teller, and Weisskopf had pointed out early in 1947 that the time of capture from the lowest orbit of carbon was in disagreement with theoretical predictions by at least ten orders of magnitude—a discrepancy which was completely unacceptable.[109]

Kramers, on the other hand, devoted his entire general outline to QED and gave a review of the problems which had occurred from its inception in 1927. He stressed two main problems:

– the treatment of the behaviour of an electron with experimental mass in its interaction with the electromagnetic field, and

- unsatisfactory features in the relativity treatment of free particles.

Generally speaking, as far as experimental results were concerned, the work of Piccioni and his group on the decay of the mesotron, and that of Lamb and Retherford on the fine structure of the 2s–2p levels of hydrogen played a key role at the conference. The latter experiment had been reported at the end of April 1947, thus only very shortly before the meeting (and so was not included in any of the discussion proposals). It was the first indication that, contrary to Dirac's theory, the $2s_{1/2}$ state was higher than the $2p_{1/2}$.[110]

Two well-known papers resulted from these three days of discussion in June 1947. The first was the work by Bethe and Marshak on the 'two-meson hypothesis'.[111] Their basic idea was that two kinds of mesons with different masses existed. The heavy type was supposed to be produced with large cross-section in the upper atmosphere and to be responsible for the nuclear forces. The light mesons were regarded as a decay product of the heavy ones, and it was assumed that, as they were observed at sea level, they interacted weakly with matter. It should be mentioned here that the news of the detection of the pi-meson announced by Powell in May 1947 was not known in America by the time of the Shelter Island conference.

The second important paper was also by Hans Bethe.[112] He succeeded in showing that in the hole-theoretic formulation of quantum electrodynamics a non-relativistic calculation of energy-level *differences* led to a finite result, and that this was in agreement with the results of Lamb and Retherford. This proved clearly that the Lamb-shift was a purely quantum-electrodynamical effect.

Lamb and Retherford's experiment together with Bethe's success were a major stimulus for theoreticians' new interest in quantum-electrodynamical problems. The aim was to construct a theory for calculating all possible processes involving electrons, positrons, and photons, with two main features: it had to be a covariant relativistic theory and it had to be renormalizable, so enabling one to treat the infinities occurring due to radiative corrections at very high and very low energies.

The first step in this direction was a paper by Julian Schwinger submitted to the *Physical Review* as early as December 1947,[113] in which he explicitly spelt out the meaning of mass and charge renormalization for the first time:

> The electromagnetic self-energy of a free electron can be ascribed to an electromagnetic mass, which must be added to the mechanical mass [bare mass] of the electron. Indeed, the only meaningful statements of the theory involve this combination of masses, which is the experimental mass of a free electron. [...] The polarization of the vacuum produces a logarithmically divergent term proportional to the interaction energy of the electron in an external field. However [...] such a term is equivalent to altering the value of the electron charge by a constant factor, only the final value being properly identified with the experimental charge.

Nevertheless Schwinger's formulation was not covariant (relativistically invariant) and therefore not satisfactory.

It should be mentioned here that the Japanese physicist Sin-itiro Tomonaga had already published a covariant renormalized version of QED in 1943 in a Japanese journal.[114] It was translated into English late in 1946 and became widely known only after the second Shelter Island conference.

This was the state of the art when J. Robert Oppenheimer started to organize the second Shelter Island conference. Financial support again came from the National Academy of Sciences. Although the first conference had been such a success, it was decided to keep the number of participants small. The meeting was held at Pocono Manor from 30 March to 2 April 1948 and 27 physicists participated. All but 4 of these had been at Shelter Island, and there were six new guests, namely P.A.M. Dirac, Aage and Niels Bohr, Walter Heitler, Gregor Wentzel, and Eugene Wigner.

The discussion centered around three topics: pi-mesons, the precise measurements of atomic levels, and new developments in QED. As far as the first topic was concerned two important things had happened since the last conference. The news of Powell's detection of the pi-meson had reached the United States, and heavy mesons had been produced for the first time (in March 1948) by the 184-inch synchro-cyclotron at Berkeley—'the most exciting event of the year' as Feynman put it. This meant that these particles 'could be studied under controlled conditions and in large numbers'.[115]

Although enthusiastic about the progress in experimental physics, Feynman was rather pessimistic about the chances of resolving some of the theoretical problems with the description of these new facts. He summarized these feelings in a review article on the conference, saying that 'the theoretical physicists admitted that they were unable to bring appreciable order into the picture, and certainly not to predict what kind of particles would be discovered next, or any new properties for particles already discovered. The future of these problems lies almost completely in the hands of the experimentalists'.[116]

Indeed only in QED did theoretical physicists achieve considerable success. Both Schwinger and Feynman presented their calculational schemes which were relativistically invariant. Schwinger's presentation, which was a very complete and massive mathematical one, was an immediate success; Feynman's, which was a more original, much more intuitive presentation, was not so readily accepted. He was trying to reconstruct quantum mechanics and electrodynamics from first principles so as to arrive at simple rules for the direct calculation of physically observable quantities. He had also started to introduce visualization (later called Feynman diagrams) into the until then purely abstract quantum field-theoretical calculations.

As with the first Shelter Island conference, the results were only seen afterwards. After Pocono Feynman wrote his two crucial papers on 'The Theory of Positrons' and 'Space-Time Approach to Quantum Electrodynamics'[117] while Schwinger started to write a cycle of three papers on QED entitled 'I. A Covariant Formulation', 'II. Vacuum Polarization and Self-Energy', and 'III. The Electromagnetic Properties of the Electron—Radiative Corrections to Scattering'.[118]

To summarize this conference we would like to repeat what Richard Feynman wrote shortly afterwards:

> The conference showed that just as we were apparently closing one door, that of the physics of electrons and photons, another was being opened wide by the experimenters, that of high-energy physics. The remarkable richness of new particles and phenomena presents a challenge and a promise that the problems of physics will not all be solved for a very long time to come.[119]

Let us turn away briefly from QED research and look at some events taking place in the field of weak interactions. As already mentioned in section 1.1.2, the theory of beta decay, formulated in 1934 was with slight modifications satisfactorily applied to experimental results. But the experiment by Conversi, Pancini, and Piccioni in 1946, together with the detection of the pi-meson a year later, had clearly shown the existence of another weakly interacting particle, the mu-meson.

By mid-1948 several physicists, among them B. Pontecorvo, O. Klein, and G. Puppi,[120] realized that *all* weak processes investigated thus far seemed to be due to the same interaction, in which the same coupling constant g always appeared. This proved to be analogous to the appearance of the coupling constant e^2 in *all* processes due to electromagnetic interaction. This was most clearly stated by Puppi: 'Comparing, however, the coupling constant [...] for the β decay of the μ meson with the Fermi constant for the capture process [$\mu^- + p \to n + \nu$] and with that for ordinary disintegration [$\pi \to \mu + \nu$] they all come out to be approximately equal to 1×10^{-33} and this may permit us to conclude that the Fermi constants between Dirac particles [spin 1/2] are all equal.' His was the first attempt to formulate the *universal Fermi interaction* visualizable by the so-called 'Puppi triangle' (fig. 1.1).

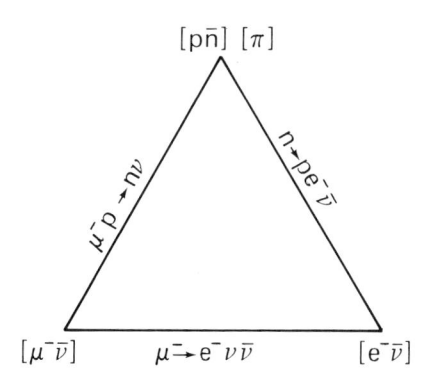

Fig. 1.1. The Puppi triangle

A year later the third and last conference of this series took place. It was held from 11 to 14 April 1949 at Oldstone-on-the-Hudson. Robert Oppenheimer was again chairman. 24 participants were present, mainly those who had attended in previous years. Hideki Yukawa was invited as a guest.

The topic of this conference was 'Fundamental Physics', and subjects like mesons, field theory, the relations of the elementary particles to nuclear forces, and cosmic rays were discussed. Both Richard Feynman and Freeman Dyson,[121] the latter attending this series of conferences for the first time, presented their calculational schemes in quantum electrodynamics. For Feynman this was an opportunity to present his ideas in a more elaborate and convincing manner than at Pocono. The importance of Dyson's talk was twofold. Firstly, he tried to present 'a unified development of the subject of quantum electrodynamics, [...], embodying the main features of the Tomonaga–Schwinger and of the Feynman radiation theory'. But he also carried his work further by discussing higher-order radiative corrections and vacuum polarization phenomena.

It was now clear that the covariant renormalized version of QED would be satisfactory to explain all electromagnetic processes then known, and Feynman, Schwinger, and Tomonaga were awarded the Nobel prize for their work in 1965.[122]

The day after the conference Oppenheimer summarized the achievements of the past years and the future trends as thus:

> The two years since the first conference have marked some changes in the state of fundamental physics, in large part a consequence of our meetings. The problems of electrodynamics which appeared so insoluble at our first meeting, and which began to yield during the following year, have now reached a certain solution; and it is possible, though in these matters prediction is hazardous, that the subject will remain closed for some time, pending the accumulation of new physical experience from domains at present only barely accessible. The study of mesons and of nuclear structures has also made great strides; but in this domain we have learned more and more convincingly that we are still far from a description which is either logical, consistent or in accord with experience.[123]

The three Shelter Island conferences were the precursors of the big Rochester conferences. They had been small, intimate, elitist, and mainly theoretical. The latter conferences were different in many respects, partly because a whole new era in physics had really started. Apart from the size and the choice of much broader topics, the major difference lay in the fact that they were attended by equal numbers of theoreticians and experimentalists. Robert Marshak, the 'father' of these conferences, described the situation as follows:

> The performance of several important experiments on these accelerators, including the artificial production of pions for the first time in 1948, and the anticipation of many more to come had persuaded me that a new series of conferences should be inaugurated in which experimentalists would be given 'equality' with the theorists.[124]

The first Rochester conference took place on 16 December 1950. It lasted only one day, and was attended by some 50 participants, all from America. A group of Rochester industries provided the financial support. As no proceedings of this first, very short, and rather informal meeting were published, we do not know much about the contents. All we know is that there were three sessions on the topics nucleon elastic scattering, meson production by nucleons, and meson production by photons. The centre of interest had thus apparently shifted towards the understanding of strong interactions and pion physics.[125]

The second Rochester conference was already attended by some 90 physicists. It was held from 11 to 12 January 1952,[126] and so fell during a very turbulent period for particle physics. On the one hand a tremendous amount of data on pi-mesons had been collected by physicists working with the numerous accelerators already completed. On the other hand cosmic-ray physics had yielded an ever-increasing number of new particles. Thus the talks given were grouped around three topics:
- interaction of π-mesons with complex nuclei
- interaction of π-mesons with hydrogen and deuterium
- megalomorphs (V,v,K,τ)[127]

The first two topics occupied half a day each, the third the whole of the second day.

Of special interest among the talks on the first day was the report by Herbert Anderson on the 'Absorption of π^+ and π^- mesons in hydrogen and deuterium' and the 'Discussion of these results' by Enrico Fermi. This experiment had recently been carried out with the Chicago machine[128] and had shown that the cross sections $\sigma(\pi^+ + p \rightarrow \pi^+ + p), \sigma(\pi^- + p \rightarrow \pi^0 + n)$ and $\sigma(\pi^- + p \rightarrow \pi^- + p)$ had the relative ratio of 9:2:1. Fermi explained this exciting result as follows:

> If one assumes charge independence, i.e. that the Isospin is a good quantum number, the two possible isospins, namely $I=3/2$ and $I=1/2$ scatter independently. If moreover, one assumes that the isospin $I=1/2$ does not scatter at all one gets the ratio 9:2:1. [...] This conclusion is independent of angular momentum, spin correlation, or anything else. One can therefore interpret the experimental results by postulating the existence of a broad resonance level $I=3/2$ in the band of energy 100–200 MeV, with the consequence that practically all the scattering comes through $I=3/2$ in this energy region.[129]

Theoretical work on this topic was presented by Keith A. Brueckner who interpreted the result in terms of the *first nucleon resonant state,* with isospin $I=3/2$ and angular momentum $J=3/2$, known as the (3,3) pion–nucleon resonance.[130] (In fact in the mid-forties Pauli had already elaborated similar ideas on behalf of cosmic-ray observations.) This marked the beginning of the establishment of numerous hadronic resonances and it showed the theoretical importance of the concept of isospin invariance.

The second day was used to discuss the new heavy unstable particles detected in cosmic radiation. The main questions concerned the production of particles, explanations of their lifetimes, and especially the kind of interactions involved in these processes. However, the most interesting talk (for our purposes) was given by Abraham Pais on 'An ordering principle for megalomorphian zoology'.[131]

'The following is an attempt,' he began, 'to decouple the production of these heavy particles from their decay, talking for the moment only of the observed species'. He tried to find an ordering by assigning to the protons, neutrons, V's, π's and v's [132] a 'mass number' as indicated in the table 1.5.

Table 1.5
'Elementary-particle' families

N_0 (nucleons) $(p = N_0^+, n = N_0^0)$	π_0 (pions) $(\pi^0 = \pi_0^0, \pi^+ = \pi_0^\pm)$	e
N_1 (heavy V particles) $(V_1^0 = N_1^0; V^\pm = N_1^\pm)$	π_1 (light v particles) $(v_2^0 = \pi_1^0, v^\pm = \pi_1^\pm)$	μ
$(N_2?)$ ('resonance' in π-nucleon scattering)	?	?

In column one of this table we find the 'nucleon-family N': the nucleons, the V_1^0 and $V^\pm$ particles (at the time often referred to as the superneutron and the superproton) and possibly the pion–nucleon resonance. The 'pion-family π' is in the second column, consisting of pions, v_2^0 and $v^\pm$ particles (kaons as they are called nowadays). Finally we have the 'lepton-family' containing e and μ.

Pais then introduced an interaction matrix of the form $(N_i N_j \pi_k)$ were i, j, and k were either all equal to zero or to one. Further he proposed a selection rule which, though it did not follow from the available theories, did not contradict the experiments, namely the interaction '$(N_i N_j \pi_k)$ can only be strong if $i+j+k$ is even'.

In fact, this particular idea did not lead very far. However, it was very important as it showed some rather interesting general features in the development of theoretical high-energy physics. The number of particles had grown to such an extent that not all of them could possibly be equally fundamental. Pais suggested therefore to speak 'about *families of elementary particles* [our emphasis] rather than elementary particles themselves,' and to divide physical processes into classes obeying certain selection rules. In the mid-fifties these ideas would become a central point in theoretical particle physics.

To summarize, the situation at this second Rochester conference can be characterized as follows. Strong links between the experimentalists and the theorists were beginning to emerge, a fact which was certainly connected with the rapid developments in particle physics performed with accelerators. Each experimental presentation was followed by a theoretical discussion, asking more detailed questions and trying to give possible approaches to a particular problem. As Robert Marshak described it:

> experimentalists were unfolding a strange new world on the subnuclear level, suggestive theoretical phrases were coined ('live parent', 'heavy brother') and the science of high-energy physics was entering a period of tremendous vitality.[133]

With the third Rochester conference which took place eleven months later, from 18 to 20 December 1952 the developmental process in this series of conferences can be regarded as nearly complete.[134] While the second Rochester conference was still called a 'Conference on Meson Physics', the third was entitled 'High Energy Nuclear Physics'. This was not all that had changed. The number of participants had grown to 140 and for the first time foreign guests were also present—from Great Britain, France, Italy, Australia, Holland, Japan, etc. The conference had become truly international. On the financial side the National Science Foundation was co-sponsor, which meant that support came from a government agency.

Let us have a brief look at the scientific contents of the conference. In these three days seven sessions were held trying to cover the whole of high-energy physics at the time. They were on charge independence and the saturation of nuclear forces, pion production and pion–nucleon scattering, V-particles, superheavy mesons, theoretical approaches to the pion problem, pion–nucleon phase shifts, and megalomorphs. Quite a large number of the presentations were based on experiments performed with accelerators in America since the war. Of course information on further new heavy particles still came from cosmic-ray research, although it was diminishing steadily.

To conclude we should like to make a few remarks about the change in the relationship between theoretical and experimental physics in the decade after the war. In general the development of theory is guided by its own internal logic, aiming at perfection, simplicity, and beauty in its description of physical reality. In this sense it is a self-contained discipline. However it is influenced to a high degree by the stage of development in experimental physics. In the forties the opportunities for a fruitful interplay between theory and experiment were somewhat limited, all the more so since the quality of the data was poor, and there were few results to go on. The three Shelter Island conferences illustrate both the way in which theory develops (QED for example), and the fact that little weight was put on extensive collaboration with experimentalists. In the fifties, however, the attitude of theoreticians seems to have been quite different. The fact that experimental particle physics had changed its style through the building of accelerators had considerable influence on the way theorists perceived their work. The enormous output of precise data, the opportunity to do well-conceived experiments to test theory, and the often implicit demand for theoretical guidance by experimentalists led to a new way of working—and to the setting up of a formal collaboration between theorists and experimentalists such as that at the Rochester conferences.

1.6 Concluding remarks

So far we have restricted ourselves to giving detailed, separate reviews of accelerator programmes, and theoretical and experimental particle physics. We now wish to recapitulate and partly interconnect these items of information to form a more coherent picture. In this way we will be able to see how high-energy physics slowly developed into an autonomous field of investigation and to understand the attraction this field had for scientists.

In the first section of this chapter we studied the state of the art in 1946. We looked at the different parts of physics which dealt in one way or another with the basic concepts of the constitution of matter, considering the particles it comprises and the forces between them. Two related attitudes towards these fundamental particles emerged as characteristic. On the one hand, it was relatively easy to be satisfied with the particles already discovered and not to believe in additional ones. Indeed one could even ask why further particles should exist. They were not 'needed', there was no part of a theory which made sense of them. Related to this there was a tendency not to ask fundamental existential questions, but 'simply' to collect information, out of which, one hoped, a kind of order or hierarchy would ultimately emerge.

Of course, the approaches used to improve the understanding of elementary particles were quite different for cosmic-ray physics, experimental nuclear physics, and theoretical physics. By 1946 the experimental data on new particles came almost entirely from cosmic radiation. This kind of research was 'little science' in the narrowest sense. The daily work was done in small groups, the instrumentation was relatively simple and, as a consequence, not expensive. But, as we have shown in detail, this approach had primarily three weak points: the measurements were inaccurate, the theoretical background poor, and the results were difficult to reproduce—this last point being perhaps the most basic handicap.

For its part, nuclear physics had advanced to the forefront of research through its crucial role in the Second World War. Not only had the image and the prestige attached to it changed, but also the size and cost of projects and the complexity of the instrumentation at its disposal. Things that were impossible before the war were now feasible. On the other hand as never before physicists had to consider public opinion and the civil and military affairs of state.

On the technological side, the invention of the principle of phase stability signified a tremendous advance in the construction of accelerators. It allowed the building of machines which would be big enough to produce artificially those particles previously found only in cosmic radiation.

As for the theoreticians, they could not claim any considerable advance from the immense effort and time expended during the war, and so immediately after they found themselves in a rather weak position. They were confronted with many open questions and again and again the experimental results raised new ones. Those who were often expected to be the backbone of the research field had to admit that they were unable to offer solutions, to give help, or to survey the field in a comprehensive way.

By 1948 the situation had changed drastically. In America nuclear physics was in a prosperous state. Brookhaven had been founded. The Atomic Energy Commission had decided to sponsor two enormous proton synchrotrons, one of 6 GeV at Berkeley and one of 3 GeV at Brookhaven. In addition, a variety of machines in the several hundred MeV range was being built. In Europe the activities in these energy regions were restricted to Great Britain. The atomic-energy programme in the rest of Europe covered only the low-energy region and was not comparable in size to the British or American projects.

In the field of cosmic-ray physics considerable development had also taken place. Conversi, Pancini, and Piccioni had shown the weak nature of the mesotron, Powell had found the pi-meson in cosmic radiation, and the first two examples of V-particles had been detected by Rochester and Butler. These events were significant, and proved to be essential for the evolution of the basic concepts in particle physics. The discovery of the pi-meson had the indirect consequence that the mu-meson lost its *'raison d'être'*. The discovery of V-particles was a clear sign that even more of these new particles could be expected at higher energies.

At this time too the theoreticians were well on the way to improving their situation. The first two Shelter Island conferences had already hinted that the efforts to formulate quantum electrodynamics in a renormalized relativistically invariant way would lead to success. This success would strengthen the confidence of the theoretical physics community in its ability to tackle the numerous other questions in meson physics.

As far as nuclear physics was concerned March 1948 represented a key date. The artificial production of mesons had its *première*. Here for the first time cosmic-ray physics and nuclear physics 'met' on the scientific level. It rapidly became clear that particle physics had now entered a completely new era and that basic research could be approached from an entirely different angle. With accelerators one could now perform more controlled, more accurate, and more searching experiments under laboratory conditions.

However, this was only a first step. It was still not possible to define clearly, for example, the term 'particle physicist'. Was he a cosmic-ray physicist who dealt with questions relating to the constitution of cosmic radiation, or was he a nuclear physicist working with a machine of sufficient energy to produce cosmic-ray particles? In other words, for the time being, the field of fundamental particles was still tackled by two groups in parallel. Each of them offered particular advantages, the nuclear physicists the reliability of their machines, the cosmic-ray physicists the high energies of their particles.

Finally, let us move on to the early fifties. In this period the tasks of nuclear physics and cosmic-ray physics with respect to their contribution to 'particle physics' became more clearly defined. In general terms, one could say that it was at this time that work in the entire field began to be reconsidered. Terms like 'nuclear physics' were beginning to be too broad, too unspecific, and thus new terms like 'low-energy nuclear physics' or 'classical nuclear physics' and 'high-energy nuclear physics'

Notes: p. 52

began to be used in books, in articles, and at conferences. The field was about to be reorganized. In the United States (by 1952), nine accelerators already had energies high enough to produce pi-mesons and an ever increasing part of very essential pi- and mu-meson physics was carried out there. The Cosmotron and the Bevatron were to be finished in 1952 and 1954 respectively, and were destined to take over another part of the cosmic-ray work.

Meanwhile, the cosmic-ray physicists had found a large number of new particles in the 1 GeV region and were busy trying to determine their lifetimes, decay modes, and other properties. The information had grown to such an extent that it became necessary to draw up the first common nomenclature, since this field of physics needed its own particular vocabulary. Nevertheless, at the same time cosmic-ray physicists were obliged to accept that this part of their work would soon also be done by accelerators.

On the theoretical side, the work on formulating quantum electrodynamics had borne fruit. The first ideas of ordering principles, selection rules, etc. were developed and a major effort was made to impose a structure on the chaos of information. It was hoped that by understanding these elementary particles and their interactions one could perhaps also find answers to open questions at the atomic level.

As we have seen, elementary-particle physics thus evolved out of two different disiplines, that of nuclear physics and that of cosmic-ray physics, incorporating the basic research tools (i.e. accelerators) and the research style of nuclear physics on the one hand, and the research topics and the techniques (i.e. detection devices) of cosmic-ray physics on the other. By the early fifties it had taken shape as an autonomous field of physics and had become an attractive and prestigious activity largely because of the fundamental nature of its research objectives.

Notes

1. In 1980 and 1982 conferences were held dealing with the history of 'elementary-particle physics' from the thirties up to the forties and fifties respectively. They are a rich source of general information. See Brown & Hoddeson (1983) and Colloquium-Paris (1982). See also Kayas (1982), 141–161 and Schweber (1985).
2. Used in the contemporary sense, *nuclear physics* means the study of the atomic nuclei, of the reactions occurring when they are bombarded with particles like protons, neutrons or photons, of radioactive decays, etc. The decisive energies are therefore those given by the binding energy of the protons and neutrons in the nucleus, i.e. several MeV (up to roughly 200–300 MeV). *Elementary-particle physics* means the study of the particles building up the nucleus and those found elsewhere in nature and their interactions with each other. The spatial dimensions of the objects under study are smaller than in the case of nuclear physics so that the energies needed for studying them are much higher than those necessary there. They start at some 300 MeV (threshold for meson production) and have in fact no upper limit. Two further remarks

a – In fact no clear dividing line between high-energy physics and nuclear physics exists. There is *intermediate-energy nuclear physics* between the two, i.e. working with energies high enough to create pi-mesons, but the investigations are largely undertaken from a nuclear physics point of view.

b – As mentioned above, as the dividing line between high-energy physics and nuclear physics one takes the energy beyond which meson physics is feasible. The experimental threshold energies for the production of pi-mesons lie at roughly 160 MeV when bombarding a carbon target with protons, and at roughly 260 MeV when using alpha particles as projectiles. These given energies are, as stated, threshold energies and therefore only a negligible number of pi-mesons are produced. To do meson physics effectively the energy should be roughly above 250 MeV for protons and 350 MeV for alpha particles (when using a carbon target). Therefore 300 MeV represent something of an average value. See Thorndike (1952) and note 77 (third paragraph).

3. Conference-Cambridge (1946).
4. The following books were consulted for some general information on the status of cosmic-ray physics in the forties: Janossy (1948), Montgomery (1949), Heisenberg (1946). For a more historical approach see Sekido & Elliot (1982), Amaldi (1977, 1979), Cassidy (1981), Stuewer (1979). See also the Nobel lectures of Blackett (1948) and Powell (1950b).
5. Yukawa (1935).
6. J.G. Wilson, 'Problems concerning the measurement of mass of mesons', in Conference-Cambridge (1946), 73–74.
7. L. Leprince-Ringuet, 'Mass measurements of mesons by the method of elastic collision', in Conference-Cambridge (1946), 46.
8. W. Heitler, 'Cosmic-ray mesons and meson theory', in Symposium-Bristol (1948), 120.
9. Early in 1937 both Anderson & Neddermeyer (1937) and Street & Stevenson (1937a,b) reported on the experimental evidence for the existence of particles less massive than protons, but more penetrating than electrons, obeying the Bethe–Heitler theory. The first real measurement of the mass of this particle is published in Anderson & Neddermeyer (1938). A few months later in June 1937 these observations were connected by several physicists with Yukawa's theoretical prediction of a particle of roughly 200 m_e mass. We want to quote here some examples:
Stueckelberg (1937) mentioned at the beginning of his paper that in fact Yukawa had predicted a particle similar to the one found in cosmic radiation. He concluded: 'it seems highly probable that Street and Stevenson, and Neddermeyer and Anderson have actually discovered a *new elementary particle,* which has been predicted by theory.' Also Yukawa (1937) himself drew the conclusion that the particle found in cosmic radiation would be the one predicted by him.
Oppenheimer & Serber's (1937) conclusion is slightly different. They expressed the hope that one could bring the mass of this new particle 'into connection with the length which plays so fundamental a part in the structure of nuclei: the "size" of the proton and neutron: the range of nuclear forces.' But since the authors doubted that the theory of Yukawa was the final answer to the question of strong interactions, they were reluctant to connect this cosmic-ray particle definitely to Yukawa's predicted particle. Nevertheless, they also expected the cosmic-ray meson to be the one mediating the nuclear forces, and that it would therefore be strongly interacting.
10. Yukawa (1949), 130.
11. The lifetime of the mesotron was one of the features which did not fit into the theoretical predictions. The decay time of the mesotron had first been measured in 1940 and the value (confirmed throughout the following years) was 2×10^{-6} sec. Yukawa's theoretical prediction had been 2×10^{-7} sec. Thus there was a disagreement by an order of magnitude. Moreover the scattering cross-section of cosmic-ray mesons by nuclei was much smaller than theoretically predicted.
12. J.G. Wilson, 'Problems concerning the measurement of mass of mesons', in Conference-Cambridge (1946), 73.
13. For a general historical overview of the field of nuclear physics see Weiner (1972).

14. *Ibid.*, 79.
15. The title of Niels Bohr's talk was one of the very few occasions (at this time) when the term 'elementary-particle physics' was used.
16. L. Rosenfeld, 'The two-nucleon problem', in Conference-Cambridge (1946), 133.
17. Heisenberg (1943).
18. Fermi (1934).
19. M. Born, 'Relativistic quantum mechanics and the principle of reciprocity', in Conference-Cambridge (1946), 14.
20. L. Rosenfeld, 'The two-nucleon problem', in Conference-Cambridge (1946), 142.
21. We wish to mention here a few of the many books and articles dealing with the role of physics during the war and the influence this had on the way science was practised thereafter: Kevles (1978), Snow (1982), De Solla Price (1963), and E.L. Goldwasser, 'How Little Science Became Big Science in the U.S.A.', in Colloquium-Paris (1982), 345–354.
22. In January 1939 the famous paper on the production of an isotope of barium by neutron bombardment of uranium, by Otto Hahn & Fritz Strassmann (1939), was published. These results were interpreted by Lise Meitner & Otto Frisch (1939) as the fission of uranium atoms upon impact by neutrons. By the end of 1939, nearly one hundred articles were published on the question of fission. The December 1939 issue of *The Reviews of Modern Physics* contained, for example, an article reviewing the work done so far.

    A crucial step in the investigation into the possibility of using this knowledge to construct a bomb was the work of Peierls and Frisch, which clarified two major points. First, they pointed out that it is U-235 which undergoes fission and secondly that it should be feasible to separate the necessary amount of U-235 from U-238. It was estimated that several kilos of U-235 were required to build a bomb, i.e. to achieve the critical mass (Kevles (1978), 325 and BIS (1945), 482). See also the work of Flügge (1939) presenting a schematic theory of a chain reaction. For a rather extensive description of the American War programme see Hewlett & Anderson (1962).
23. Some books and articles dealing with the history of the Manhattan Project and in particular of Los Alamos: Baxter (1968), Frisch (1979), Feynman (1985), Groueff (1967), Hawkins (1983), Truslow & Smith (1983), S.R. Weart, 'The Road to Los Alamos', in Colloquium-Paris (1982), 301–313.
24. AIP-Report of the Director (1944), 217.
25. Conference-Philadelphia (1944).
26. *Ibid.*, 318.
27. *Ibid.*, 319.
28. *Ibid.*, 322.
29. Some books and articles used as general background literature for this chapter: Livingston & Blewett (1962), Livingston (1952a,b, 1969), Thomas, Kraushaar & Halpern (1952), Wilson (1958, 1981).
30. To be able to calculate theoretically the threshold for the production of a particle one has to be clear about at least two factors—the reaction in which the particle is produced and the mass values of the particles involved. Since for cosmic-ray particles often neither the mass values nor the exact reaction in which such particles were produced were known exactly, it proved very complicated to lay down the design energies for accelerators.

    A short example to illustate:

    When the threshold energy for mesotron production was calculated in 1946 two assumptions were made:

    1 – the mesotron is the Yukawa meson;
    2 – mesons are produced in pairs.

    Both assumptions were wrong. On the one hand the mass value used should have been bigger, while on the other hand pi-mesons were not produced in pairs. Fortunately, therefore, these two mistaken assumptions 'cancelled' one another out.
31. Bethe & Rose (1937). The article begins: 'It is the purpose of this note to show that a very serious difficulty will arise when the attempt is made to accelerate ions in the cyclotron to higher energies than obtained thus

far. This difficulty is due to the relativistic change of mass which has the effect of destroying either resonance or focusing.' In the article, the inevitable problems of using cyclotrons for acceleration of relativistic particles were clearly demonstrated. Briefly, they were: the mass m of a particle, which is moving with a velocity v, increases by the factor $1/\sqrt{1 - v^2/c^2}$ (c = velocity of light). As the frequency of a particle revolving in a magnetic field is

$$\omega = QB/m$$

(Q = charge of the particle, B = magnetic field strength, ω = frequency) it decreases proportionally as the mass increases. A numerical example: A proton of 25 MeV energy has a velocity 0.23 times that of light which causes a mass increase of roughly 3%. This value represents roughly the upper limit of what could be tolerated.

32. McMillan (1945), received by the journal on 5 September 1945; Veksler (1945), received by the journal on 1 March 1945.

33. As electrons have a very small rest mass they already reach 99% of the velocity of light at 3 MeV. Since the injection energy into a synchrotron is generally higher than this, the additional relativistic mass increase which occurs with further acceleration is negligible and so the accelerating frequency need not be varied in an electron synchrotron.

34. Ramsey (1966).

35. The nine universities which joined together to establish Brookhaven were: Columbia, Cornell, Harvard, Johns Hopkins, Massachusetts Institute of Technology, University of Pennsylvania, Princeton, Rochester, and Yale.

36. See *Initial Program Report of Associated Universities Inc. Brookhaven National Laboratory,* December 1946, pp. 6-9 (BNL; copy CERN-CHS/ARCH/10064).

37. *Ibid.,* Appendix B, Part II.

38. In fact, two preliminary designs were prepared in parallel, one for a 2.5 GeV and one for a 10 GeV machine. Both of them were submitted to the AEC. See *Scientific Progress Report January-July 1948,* BNL, p.15 (BNL; copy CERN-CHS/ARCH/10064).

39. See *Scientific Progress Report, July-December 1947,* BNL, pp. 12-19 (BNL; copy CERN-CHS/ARCH/10064).

40. See *Scientific Progress Report July 1, 1947,* BNL (BNL; copy CERN-CHS/ARCH/10064).

41. The information on the history of the Radiation Laboratory is taken from Heilbron, Seidel & Wheaton (1981) and Seidel (1983), unless otherwise stated. For a short technical description of the accelerators at Berkeley see Chew & Moyer (1950).

42. In 1940 Lawrence asked the Rockefeller foundation to sponsor the construction of a 184-inch cyclotron for the purpose of studying mesons (the mesotron). He suggested that the energy aimed for should be 150 MeV. Apparently he seemed not to be worried by, or to be aware of, the basic problems which would arise when attempts were made to produce such highly-energetic particles in a conventional cyclotron (see note 31). Indeed the huge magnet was manufactured, but was used as an isotope separator throughout the war.

43. See Lawrence (1948).

44. For a short history of MIT's Radiation Laboratory see DuBridge (1946).

45. A short comparison of proton-beam intensities (for the machines built around 1950):

linear accelerator	$0.4\ \mu A$
synchro-cyclotron	$1\ \mu A$ (only for the internal beam)
proton synchrotron	$10^{-4}\ \mu A$

Therefore, a proton linear accelerator could be a real alternative to the proton synchrotron from the point of view of beam intensity. Synchro-cyclotrons could not compete with these two machines since a) they were limited in energy to roughly 1 GeV owing to financial considerations and b) the beam intensity went down considerably when the attempt was made to extract the beam from the machine.

The success of the linear accelerator together with the increasing efforts to investigate new atomic weapons (triggered by the explosion of the first Russian atomic bomb in the summer of 1949) lead Lawrence to propose even larger devices (Mark I and II) to produce explosive material like uranium-233 or plutonium. Mark I was constructed and operated from 1952 for roughly a year. The machine was 60 feet long, producing 50 μA of 30 MeV deuterons. Mark II which was planned to be a quarter of a mile in length was never built (see 'A Neutron Foundry', in Heilbron, Seidel & Wheaton (1981), 66–71).

46. Brobeck (1948). This is the first published design for the 10 GeV machine.
47. *op. cit.*, note 38, 16. The AEC approved the two machines on 14 April 1948 [see Seidel (1983), 397].
48. L.J. Haworth, *Super-High Energy Accelerators,* 16 September 1948, p.2 (BNL; copy CERN-CHS/ARCH/10063).
49. For a description of the Cosmotron see Cosmotron (1953), Livingston et al. (1950), and Blewett (1956), 52–58.

    Information on the Bevatron is to be found in Smith (1951) and Blewett (1956), 59–63.
50. The information for this table is taken from Accelerators (1951a,b), Livingston (1952a), Thomas, Kraushaar & Halpern (1952), Livingston & Blewett (1962), and Thorndike (1952).

    The date of completion given in column three does not always correspond to the year when the design energy was reached, but generally to the year when the first beam was produced. A short remark concerning the CalTech e-synchrotron: This machine was first constructed from the magnet iron and vacuum chamber of the 1/4-scale model of the Berkeley Bevatron. In 1954 the construction of a bigger machine reaching 1.2 GeV was started (Barton (1961), 55).
51. See Harwell (1952), 10.
52. Skinner (1948); Harwell (1952), 12–14 and 34–42.
53. The information on the Harwell machines is taken from: Harwell (1952), 43–55 and Gowing (1974), 203–259.
54. In Harwell (1952), 49–50 we find the following remark about the 110-inch synchro-cyclotron: 'At the full energy of 180 MeV it should be possible to produce mesons [...]. However, the yield of mesons at that energy is known to be small and it is not intended to spend a large proportion of operating time on meson experiments.'

    For a short review of the Harwell programme in the early fifties see Pickavance & Cassels (1952).
55. In fact Pickavance and Adams would have liked to increase the Harwell synchro-cyclotron to 150 inch diameter (corresponding to a proton energy of about 300 MeV). Their machine would have then been well above the threshold for meson production. Gowing (1974), 258.
56. The information on the five machines built at universities is taken from Gowing (1974), 224–226. For the energy values, see M.S. Livingston, *Electro nuclear machines in England,* 1947 (Berkeley: Bancroft Library, Lawrence papers, box 1, folder 8) and Accelerators (1948).
57. Letter Wallace to Verry, 13/2/53 (PRO-DSIR17/562). According to this letter construction of the linac started in 1952. It was finally abandoned in 1954.
58. Oliphant's plan to build a ring-shaped proton accelerator is frequently referred to in the literature on accelerators, but details are never given. See, for example, Livingston (1969), 50–51.
59. For further and more technical information on this machine see Birmingham (1953), Hibbard (1950), and Blewett (1956), 48–51.
60. Livingston & Blewett (1962), 400–401.
61. A general review of the world situation by 1952 in atomic energy is to be found in Atomic Energy (1952). In this article the programmes of various countries are presented and the basic information for this chapter has been derived from it. Additional information is indicated at the appropriate places. The exact energies for the accelerators are taken from Accelerators (1948, 1951a, b). Our account is concerned only with the more powerful devices, of interest for physics research.
62. As we will not deal with the reactor programmes pursued in Europe, we wish briefly to mention the main activities (Weinberg (1954) and Atomic Energy (1952)):

    Sweden's first reactor went into operation by the end of 1953. It was of the research type, fuelled by uranium and air-cooled using heavy water as moderator.

Norway completed its first pile, a heavy water reactor built at Kjeller, in 1950. Because of the lack of uranium, this project was converted into a Dutch–Norwegian venture, with the Netherlands providing the uranium (see Randers (1953)).

In France in mid-December 1948 the 5 kW reactor ZOE at Chatillon went critical for the first time. The second reactor P-2 (1500 kW) followed in October 1952.

63. Casimir (1983), 190.
64. M.E. Regenstreif, *Rapport sur mon voyage en Suisse, 28-30 juin 1951*, (UNESCO/NS/MEMO/7984).
65. McReynolds (1955).
66. For a short overview of the French activities in nuclear physics after the war see section 9.2.
67. M.E. Regenstreif, *Rapport sur mon voyage en Belgique, 4-7 juillet 1951* (UNESCO/NS/MEMO/240168).
68. This information is taken from Barton (1961).
69. Courant, Livingston & Snyder (1952).
70. Christofilos (1950).
71. The value for the aperture is taken from Courant, Livingston & Snyder (1952) and from the Minutes of the third Council Meeting, 4-7/10/52, CERN/GEN/4, the financial estimate from *Report to the Secretary, covering visit to USA by Goward, Wideröe and Dahl in August 1952*, CERN-PS/S4, 25/8/52 (CERN-DG20551).
72. Some books and review articles written on meson physics around 1950: Thorndike (1952), Marshak (1952a,b), Powell (1950a), Snyder (1949), Camerini & Muirhead (1950), Gray (1948).
73. Congrès-Bagnères-de-Bigorre (1953). Two conferences of the same kind had been held in 1947 and 1950. The proceedings of these conferences are a very valuable source of information, as they give an impression of the development in cosmic-ray physics at this period. Symposium-Krakow (1947) and Conference-Bombay (1950).
74. Conversi, Pancini & Piccioni (1947). For a historical presentation see: O. Piccioni, 'The observation of the leptonic nature of the "mesotron" by Conversi, Pancini, and Piccioni', in Brown & Hoddeson (1983), 222-241. The theory forming the background for this experiment had been put forward by Tomonaga & Araki (1940).
75. See Powell et al. (1946) and Rotblat (1950).
76. Lattes et al. (1947), Lattes, Occhialini & Powell (1947).
77. Gardner & Lattes (1948). In this paper, meson tracks found in photographic emulsions exposed near carbon and other targets bombarded by 380 MeV alpha rays were reported. About 2/3 of the tracks ended in stars (thus were negative π-mesons). A meson intensity roughly 10^8 times that available in cosmic rays was obtained. The mass of the meson was measured to be $313 \pm 16\ m_e$.

Three further remarks. In the same year a paper by Occhialini & Powell (1948) reported the measurements of small-angle scattering of positive and negative pi-mesons from cosmic radiation in photographic emulsions, and showed that they have a mass identical with those produced by Lattes and Gardner. This publication shows that it was not obvious to physicists that artificially produced mesons were identical to those produced in cosmic radiation

Our second remark concerns the strong energy dependence of pi-meson production. This is illustrated nicely in the following paragraph to be found in Heilbron, Seidel & Wheaton (1981), 57: 'They found no pions when they sent 165 MeV protons against their carbon target; the yield at 200 MeV was but one percent, and that at 300 MeV less than half of the yield at 345 MeV. For alpha particles, the number of pions observed fell by two thirds as the energy of projection declined from 380 to 342 MeV, and to fewer than seven events (less than one percent of the maximum) at 260 MeV.'

Lastly, we want to list here some reactions in which pi-mesons are produced:

$$p + p \rightarrow p + n + \pi^+$$
$$p + p \rightarrow p + p + \pi^0$$
$$p + n \rightarrow p + p + \pi^-$$

50   *From cosmic-ray and nuclear physics to high-energy physics*

78. The physicists at Berkeley had already tried to detect pi-mesons produced in their synchro-cyclotron with the help of photographic plates before the arrival of Lattes. They failed because they were not able to develop the plates in the way needed.
79. For details on the Shelter Island conferences see section 1.5.

    Some brief information on the Solvay congresses: they were the most significant and important meetings of the European theoretical physics community during the pre-war period. The 8th Solvay Congress planned for 1939 was not held because of the political situation in Europe, and so the one from 22 to 29 October 1933 in Brussels under the title 'Structures et Propriétés des Noyaux Atomiques' turned out to be the last before the war. After the war the Solvay congresses lost their special position among the conferences in theoretical physics. Many of the leading European theoreticians had left Europe in the thirties and early forties. Accordingly, other conferences like those of Shelter Island began to play a rather central role.

    R. Serber, 'Artificial Mesons', in Conseil-Solvay (1948), 89–110. From the Shelter Island conference we have no record of Serber's talk as no proceedings were published.
80. Gray (1948), 35.
81. Heilbron, Seidel & Wheaton (1981), 57.
82. C.F. Powell, 'Properties of the π- and μ-mesons of Cosmic Radiation', in Symposium-Bristol (1948), 87.
83. C.F. Powell, 'Introduction', in Symposium-Bristol (1948).
84. Steinberger, Panofsky & Steller (1950). Already in September 1949 Bjorklund et al. (1950) at Berkeley had observed high-energetic photons coming from a target bombarded by protons from the 184-inch synchro-cyclotron (proton energies higher than 175 MeV). They suspected that the decaying particle producing these photons would be a neutral pi-meson, but they could not make the necessary confirming measurements because of the concrete shielding of the synchro-cyclotron.
85. Fröhlich, Heitler & Kemmer (1938) and Sakata & Tanikawa (1940).
86. Marshak (1952b), 1.

    A short example of the advantages of using accelerators: With the Berkeley synchro-cyclotron it became possible to measure the pi/mu mass ratio with sufficient precision to identify the neutral recoil particle with a neutrino. Hitherto, the ratio had been measured too high and it had therefore been supposed that the neutral particle had a considerable mass.
87. The data for this table are taken from Thorndike (1952). In 1952 one thought of having only one kind of neutrino for both electrons and mu-mesons.
88. Leprince-Ringuet & Lhéritier (1944).
89. For an overview on the V-particles see Rochester (1951) and Rochester & Butler (1953). For a more historical presentation see Rochester (1984).
90. Rochester & Butler (1947). Before the detection of these V-particles by the British physicists, there had already been some sporadic hints about their existence, but their importance had not been realized. To illustrate the difficulty of this kind of work, we should mention that the two photographs by Rochester and Butler of the V-particles were taken from a collection of 5,000.
91. In 1948 the findings of Rochester and Butler were discussed at length at two international cosmic-ray conferences, namely at the *Symposium on cosmic rays* held in honour of Millikan's 80th Birthday at the California Institute of Technology in June and at the *Cosmic-ray Symposium* in Bristol in September. This led to an increasing interest in this subject both in Europe and the United States.

    See Symposium-Pasadena (1948) and for a summary of this conference B.R. (1948); Symposium-Bristol (1948).
92. Seriff et al. (1950); 34 V-shaped tracks had been found in 11,000 photographs taken. 30 were neutral V-particles and only 4 were charged ones.
93. Armenteros et al. (1951a,b).
94. A review of the information on the neutral V-particles by 1951 can be found in Butler (1952).
95. Rochester & Butler (1953); see also the remarks by O'Ceallaigh, note 98.

96. Armenteros et al. (1952): 21 charged V-decays had been observed, but among those there was only one single photograph of the type described. For a while this particle was referred to in the literature as the 'cascade particle' since it decayed into a proton via a V_1^0.

97. For the detection of the tau-meson see Brown et al. (1949); for further information see Fowler et al. (1951).

98. O'Ceallaigh (1951). The author remarked in this paper that 'the $\varkappa$-particles are about 1.3 times more massive than the τ-particles, but the inaccuracies in the experimental value are such that the possibility cannot be excluded that they represent alternative modes of decay of particles of the same type. Further, they may also be of the same type as the unstable charged particles of Rochester and Butler [V_2^0] which are observed to decay in flight.'

99. These experiments, which led to the discovery that the $\varkappa$-meson was in fact two particles (the $\varkappa$- and the χ-meson), were carried out throughout 1952. 60 photographs were evaluated for this purpose. The results are presented in M. Menon & C. O'Ceallaigh, 'Observations on the mass and energy of secondary particles produced in the decay of heavy mesons', in Congrès-Bagnères-de-Bigorre (1953), 118–124.

100. Bridge & Annis (1951); Annis et al. (1952).

101. The information for this table is taken from R.E. Marshak, 'Particle Physics in rapid transition: 1947–1952', Brown & Hoddeson (1983), 397, Rochester & Butler (1953), 403 and from the original papers cited above.

102. 'Texte de la conference du Prof. B. Rossi à la séance de clôture' in Congrès-Bagnères-de-Bigorre (1953), 259–269. This nomenclature was largely disseminated by physicists from all countries through various publications after the conference.

103. M. Schein, 'On the artificial production of V^0-particles by 227 MeV pi-mesons generated in the Chicago Cyclotron', in Congrès-Bagnères-de-Bigorre (1953), 166–168.

104. 'Discours de clôture par L. Leprince-Ringuet' in Congrès-Bagnères-de-Bigorre (1953), 288.

105. A very good insight into the status of theoretical nuclear physics in the low-energy region around 1950 can be gained from Blatt & Weisskopf (1952). For a short review on the status of meson theory see Bethe (1954).

106. The information on the three Shelter Island conferences is taken from: Schweber (1984, 1986), and R.E. Marshak, 'Particle Physics in rapid transition: 1947–1952', and J. Schwinger, 'Renormalization theory of quantum electrodynamics: an individual view', both in Brown & Hoddeson (1983), 376–401 and 329–353. In June 1983 many participants of the first Shelter Island conference met again. A short summary can be found in GBL (1983).

107. The list of participants is taken from Schweber (1984), 141.

108. These papers for discussion are entirely reprinted in Schweber (1984), 151–157.

109. Fermi, Teller & Weisskopf (1947). Weisskopf (1947) presented at Shelter Island a solution to the difficulty of reconciling the high rate of production of mesons with their subsequent weak interaction with matter. He postulated that the primary cosmic-ray proton would convert a normal nucleon (which is in an 'air' nucleus) into an 'excited' nucleon, which would be capable of emitting mesons—a theory which was not regarded as particularly convincing.

110. Lamb & Retherford (1947). This experiment, which represented the most precise test of QED theory, had only become possible through the great war-time advances in microwave techniques at about the three centimetre wavelength.

111. Bethe & Marshak (1947).
Before the experiment of Conversi et al., two Japanese theoreticians Sakata and Inoue (1946) published a paper in which they clearly doubted the fact that the Yukawa meson was to be identified with the mesotron. They suggested that the mesotron would be a decay product of the Yukawa meson. (This paper remained more or less unknown in America and Europe.)

112. Bethe (1947).

113. Schwinger (1948a).

114. For the Japanese contribution to QED see: Tomonaga (1946, 1948). For an overview see: J. Schwinger, 'Two shakers of physics: memorial lecture for Sin-itiro Tomonaga', in Brown & Hoddeson (1983), 354–375.
115. Feynman (1948a), 8.
116. *Ibid.*, 9.
117. Feynman (1949a,b). Before these two papers Feynman (1948b) had presented his space-time approach to non-relativistic quantum mechanics.
118. Schwinger (1948b, 1949a, c).
119. Feynman (1948a), 10.
120. Pontecorvo (1947), Klein (1948), Puppi (1948).
121. For Dyson's contribution to QED see Dyson (1949a,b,c).
122. Feynman (1966); Dyson (1966).
123. Letter Oppenheimer to Richards, 15/4/49, reprinted in Schweber (1984), 150.
124. Marshak (1970), 93.
125. Marshak (1985), 1.
126. Conference-Rochester 2 (1952).
127. Megalomorph: An expression for the new elementary particles used by Fermi, referring to the variety of different forms in which these particles occur.
128. The results of this experiment on the pion cross sections were used as an argument to fix the energy for the CERN synchro-cyclotron at 600 MeV. At this energy it would also be possible to measure these very fundamental cross sections at higher energies. See Mersits (1984).
129. Conference-Rochester 2 (1952), 26.
130. K.A. Brueckner, 'Theoretical Interpretation of meson scattering and photomesic production', Conference-Rochester 2 (1952), 26–39.
131. Conference-Rochester 2 (1952), 87–93.
132. The particles to which the symbol v was attributed are the charged and neutral V-particles with smaller mass (see table 1.3). They had generally been referred to as V^+ and V_2^0. The big Vs stand for the heavy charged V-particle found by the Manchester Group in 1952 and the V_1^0.
133. Marshak (1970), 93.
134. Conference-Rochester 3 (1953). From the title of the Rochester Conferences and from the way the time was divided between theoreticians and experimentalists (both cosmic-ray physicists and those working with machines), one can follow the changes in this field of physics over the years. The second Rochester Conference was called 'Conference on Meson Physics' and the cosmic-ray physicists were well represented on the experimental side. The following conference was called 'High Energy Nuclear Physics' and this designation was kept until the 8th Rochester Conference held in 1958 at CERN under the title 'high-energy physics'. It was not only the title that changed. At the fourth conference in 1954, the part of cosmic-ray physics had already shrunk to one out of five sessions and at the sixth conference, R.B. Leighton remarked that 'next year those people still studying strange particles using cosmic rays had better hold a rump session of the Rochester Conference somewhere else.' (Brown & Hoddeson (1983), 399).

Bibliography for Chapter 1

Accelerators (1948)   H.H. Goldsmith, *List of High Energy Installations,* BNL-L-101, July 1948.
Accelerators (1951a) B.E. Cushman, *Bibliography of Particle Accelerators, July 1948 to December 1950,* University of California, UCRL-1238, March 1951.

Accelerators (1951b)	S. Shewchuck, *Bibliography of Particle Accelerators, January to December 1951*, University of California, UCRL-1951, September 1951.
Accelerators (1954)	F.E. Frost and J.M. Putman, *Particle Accelerators*, University of California, UCRL-2672, November 16, 1954.
AIP-Report of the Director (1944)	'American Institute of Physics—Report of the Director for 1944', *Review of Scientific Instruments*, **16** (1945), 217-222.
AIP-Report of the Director (1945)	'American Institute of Physics—Report of the Director for 1945', *Review of Scientific Instruments*, **17** (1946), 235.
Amaldi (1977)	E. Amaldi, 'Personal Notes on Neutron Work in Rome in the 30's and Post-war European Collaboration in High-energy Physics', in C. Weiner (ed.), *History of Twentieth Century Physics*, Proceedings of the International School of Physics 'Enrico Fermi', Course LVII, Varenna 1972 (New York: Academic Press, 1977).
Amaldi (1979)	E. Amaldi, 'The Years of Reconstruction', in C. Schaerf (ed.), *Perspectives of Fundamental Physics* (London: Harwood Academic Publishers, 1979).
Amaldi (1984)	E. Amaldi, 'From the Discovery of the Neutron to the Discovery of Nuclear Fission', *Physics Reports*, **111** (1984), 1-332.
Anderson & Neddermeyer (1937)	C.D. Anderson and S.H. Neddermeyer, 'Note on the Nature of Cosmic-ray Particles', *Physical Review*, **51** (1937), 884-886.
Anderson & Neddermeyer (1938)	C.D. Anderson and S.H. Neddermeyer, 'Cosmic-ray Particles of Intermediate Mass', *Physical Review*, **54** (1938), 88-89.
Annis et al. (1952)	M. Annis, H.S. Bridge, H. Courant, S. Olbert, and B. Rossi, 'S-Particles', *Nuovo Cimento*, **9** (1952), 624-627.
Armenteros et al. (1951a)	R. Armenteros, K.H. Barker, C.C. Butler, A. Cachon, and A.H. Chapman, 'Decay of V-particles', *Nature*, **167** (1951), 501-503.
Armenteros et al. (1951b)	R. Armenteros, K.H. Barker, C.C. Butler, and A. Cachon, 'The Properties of Neutral V-Particles', *Philosophical Magazine*, **42** (1951), 1113-1135.
Armenteros et al. (1952)	R. Armenteros, K.H. Barker, C.C. Butler, A. Cachon, and C.M. York, 'The Properties of Charged V-particles', *Philosophical Magazine*, **43** (1952), 597-612.
Atomic Energy (1952)	'World Progress in Atomic Energy', *Nucleonics*, **10** (1952), 7-35.
Barton (1961)	M.Q. Barton, *Catalogue of High Energy Accelerators*, 6 September 1961, BNL 683 (T-230).
Baxter (1968)	P. Baxter, *Scientists Against Time* (Cambridge: The MIT Press, 1968).
Bethe (1947)	H.A. Bethe, 'The Electromagnetic Shift of Energy Levels', *Physical Review*, **72** (1947), 339-341.
Bethe (1954)	H.A. Bethe, 'Mesons and Nuclear Forces', *Physics Today*, **7** (1954), 5-11.
Bethe & Marshak (1947)	H.A. Bethe and R.E. Marshak, 'On the Two-Meson Hypothesis', *Physical Review*, **72** (1947), 506-509.
Bethe & Rose (1937)	H. Bethe and M.E. Rose, 'The Maximum Energy Obtainable from the Cyclotron', *Physical Review*, **52** (1937), 1254-1255.
Birmingham (1953)	'Proton Synchroton of the University of Birmingham', *Nature*, **172** (1953), 704-705.
BIS (1945)	British Information Services, 'Statements Relating to the Atomic Bomb', *Reviews of Modern Physics*, **17** (1945), 472-490.
Bjorklund et al. (1950)	R. Bjorklund, W.E. Crandall, B.J. Moyer, and H.F. York, 'High Energy Photons from Proton-Nucleon Collision', *Physical Review*, **77** (1950), 213-218.
Blackett (1948)	P.M.S. Blackett, 'Cloud Chamber Researches in Nuclear Physics and

	Cosmic Radiation', Nobel Lecture, 13 December 1948 in *Nobel Lectures Physics 1942-1962* (Amsterdam: Elsevier Publishing Company, 1964).
Blatt & Weisskopf (1952)	J.M. Blatt and V.F. Weisskopf, *Theoretical Nuclear Physics* (New York: Wiley, 1952).
Blewett (1956)	J.P. Blewett, 'The Proton Synchrotron', *Reports on Progress in Physics,* **19** (1956), 37-79.
Bridge & Annis (1951)	H.S. Bridge and M. Annis, 'A Cloud-Chamber Study of the New Unstable Particles', *Physical Review,* **82** (1951), 445-446.
B.R. (1948)	B.R., 'Cosmic Rays', *Physics Today,* Sept. (1948), 27-28.
Brobeck (1948)	W.M. Brobeck, 'Design Study for a Ten-BeV Magnetic Accelerator', *Review of Scientific Instruments,* **19** (1948), 545-551.
Brown et al. (1949)	R. Brown, U. Camerini, P.H. Fowler, H. Muirhead, C.F. Powell, and D.M. Ritson, *Nature,* **163** (1949), 82.
Brown & Hoddeson (1983)	L.M. Brown and L. Hoddeson (eds.), *Birth of Particle Physics* (Cambridge: Cambridge University Press, 1983). These are the proceedings of the International Symposium on the History of Particle Physics, Fermilab, May 1980.
Butler (1952)	C.C. Butler, 'Unstable Heavy Cosmic-ray Particles', in *Progress in Cosmic-ray Physics* (Amsterdam: North-Holland Publishing Company, 1952), 67-123.
Camerini & Muirhead (1950)	U. Camerini and H. Muirhead, 'Mesons', *Physics Today,* May (1950), 16-21.
Casimir (1983)	H.G.B. Casimir, *Haphazard Reality* (New York: Harper & Row, 1983).
Cassidy (1981)	D.C. Cassidy, 'Cosmic-ray Showers, High-energy Physics, and Quantum Field Theories: Programmatic Interactions in the 1930s', *Historical Studies in the Physical Sciences,* **12** (1981), 1-39.
Chew & Moyer (1950)	G.F. Chew and B.J. Moyer 'High Energy Accelerators at the University of California Radiation Laboratory', *American Journal of Physics,* **18** (1950), 125-131.
Christofilos (1950)	N. Christofilos, 'Focussing System for Ions and Electrons', U.S. Patent No.2,736,799, reprinted in *Livingston (1966).*
Colloquium-Paris (1982)	*International Colloquium on the History of Particle Physics: Some Discoveries, Concepts, Institutions from the Thirties to Fifties,* 21-23/7/1982, Paris. (Paris: Edition de Physique, supplement *Journal de Physique,* Tome 43, décembre 1982).
Conference-Bombay (1950)	*International Conference on Elementary Particles* held at the Tata Institute of Fundamental Research, Bombay, on 14-22 December 1950, under the patronage of IUPAP with the financial assistance of UNESCO.
Conference-Cambridge (1946)	*Report on an International Conference on Fundamental Particles and Low Temperature,* Cavendish Laboratory, Cambridge, 22-27 July 1946, (London: Taylor and Francis, Ltd, 1947).
Conference-Philadelphia (1944)	The proceedings of the Conference held in Philadelphia are published as 'Report of National Research Council—Conference of Physicists', *Review of Scientific Instruments,* **15** (1944), 183-328.
Conference-Rochester 2 (1952)	A.M.L. Messiah and H.P. Noyes (eds.), *Proceedings of Second Rochester Conference, January 1952.*
Conference-Rochester 3 (1952)	H.P. Noyes, M. Camac, and W.D. Walker (eds.), *Proceedings of the Third Annual Rochester Conference, December 18-20, 1952* (New York: Interscience, 1953).

Congrès-Bagnères-de-Bigorre (1953)	*Congrès International sur le Rayonnement cosmique,* organisé par l'université de Toulouse sous le patronage de l'UIPPA avec l'appui de l'UNESCO, Bagnères-de-Bigorre, 6–12 juillet 1953.
Conseil-Solvay (1948)	*Rapport du 8 Conseil Solvay de Physique, Particules Elementaires, 27 septembre au 2 octobre 1948* (Brussels: Secretaires du Conseil, 1950).
Conversi, Pancini & Piccioni (1947)	M. Conversi, E. Pancini, and O. Piccioni, 'On the Disintegration of Negative Mesons', *Physical Review,* **71** (1947), 209–210.
Cosmotron (1953)	Collection of articles by various authors on the Cosmotron, *Review of Scientific Instruments,* **24** (1953), 723–870.
Courant, Livingston & Snyder (1952)	E.D. Courant, M.S. Livingston, and H.S. Snyder, 'The Strong-Focussing Synchrotron—A new High Energy Accelerator', *Physical Review,* **88** (1952), 1190–1196.
De Solla Price (1963)	D.J. De Solla Price, *Little Science, Big Science* (New York: Columbia University Press, 1963).
DuBridge (1946)	L.A. DuBridge, 'History and Activities of the Radiation Laboratory of the Massachusetts Institute of Technology', *Review of Scientific Instruments,* **17** (1946), 1–5.
Dyson (1949a)	F.J. Dyson, 'The Radiation Theory of Feynman'; talk given at the *International Congress on Nuclear Physics and Quantum Electrodynamics* held in Basel from 5–9 September 1949, published by Helvetica Physica Acta (Basel: Birkhäuser, 1950).
Dyson (1949b)	F.J. Dyson, 'The Radiation Theories of Tomonaga, Schwinger and Feynman', *Physical Review,* **75** (1949), 486–502.
Dyson (1949c)	F.J. Dyson, 'The S-Matrix in Quantum Electrodynamics', *Physical Review,* **75** (1949), 1736–1755.
Dyson (1966)	F.J. Dyson, 'Tomonaga, Schwinger, and Feynman Awarded Nobel Prize for Physics', *Science,* **150** (1966), 588–589.
Fermi (1934)	E. Fermi, 'Versuch einer Theorie der β-Strahlen', *Zeitschrift für Physik,* **88** (1934), 161–177.
Fermi, Teller & Weisskopf (1947)	E. Fermi, E. Teller, and V. Weisskopf, 'The Decay of Negative Mesotrons in Matter', *Physical Review,* **71** (1947), 314–315.
Feynman (1948a)	R.P. Feynman, 'Pocono Conference', *Physics Today,* **1** (1948), 8–10.
Feynman (1948b)	R.P. Feynman, 'Space-Time Approach to Non-Relativistic Quantum Mechanics', *Reviews of Modern Physics,* **20** (1948), 367–387.
Feynman (1949a)	R.P. Feynman, 'The Theory of Positrons', *Physical Review,* **76** (1949), 749–759.
Feynman (1949b)	R.P. Feynman, 'Space-Time Approach to Quantum Electrodynamics', *Physical Review,* **76** (1949), 769–789.
Feynman (1966)	R.P. Feynman, 'The Development of the Space-Time View of Quantum Electrodynamics', *Science,* **153** (1966), 699–708. (Reprint of Nobel Lecture).
Feynman (1985)	R.P. Feynman, *Surely You're Joking, Mr. Feynman!* (New York: W.W. Norton & Company, Inc., 1985).
Flügge (1939)	S. Flügge, 'Kann der Energieinhalt der Atomkerne technisch nutzbar gemacht werden?', *Naturwissenschaften,* **27** (1939), 402–410.
Fowler et al. (1951)	P.H. Fowler, M.G.K. Menon, C.F. Powell, and O. Rochat, 'Masses and Modes of Decay of Heavy Mesons, II. τ-Particles', *Philosophical Magazine,* **42** (1951), 1040–1050.
Frisch (1979)	O. Frisch, *What Little I Remember* (Cambridge: Cambridge University Press, 1979).

Fröhlich, Heitler & Kemmer (1938)	H. Fröhlich, W. Heitler, and N. Kemmer, 'On the Nuclear Forces and the Magnetic Moments of the Neutron and the Proton', *Proceedings of the Royal Society,* **A166** (1938), 154–177.
Gardner & Lattes (1948)	E. Gardner and C.M.G. Lattes, 'Production of Mesons by the 184-inch Berkeley Cyclotron', *Science,* **107** (1948), 270–271.
GBL (1983)	GBL, 'Seeds of History sown on Shelter Island', *Physics Today,* Sept. (1983), 67–69.
Gowing (1974)	M. Gowing, *Independence and Deterrence, Britain and Atomic Energy, 1945–1952,* Vol. 2 (London: The MacMillan Press Ltd, 1974).
Gray (1948)	G.W. Gray, 'The Ultimate Particles,' *Scientific American,* June (1948), 27–38.
Groueff (1967)	S. Groueff, *Manhattan Project: The Untold Story of the Making of the Atomic Bomb* (Boston: Little, Brown and Co., 1967).
Hahn & Strassmann (1939)	O. Hahn and F. Strassmann, 'Über den Nachweis und das Verhalten der bei Bestrahlung des Urans mittels Neutronen entstehenden Erdalkalimetallen', *Naturwissenschaften,* **27** (1939), 11–15.
Harwell (1952)	*Harwell—The British Atomic Energy Research Establishment 1946–1951* (London: Her Majesty's Stationery Office, 1952).
Hawkins (1983)	D. Hawkins, *Project Y: The Los Alamos Story, Part I: Towards Trinity* (Los Angeles: Tomash, 1983).
Heilbron, Seidel & Wheaton (1981)	J.L. Heilbron, R.W. Seidel, and B.R. Wheaton, 'Lawrence and his Laboratory: Nuclear Science in Berkeley', *LBL News Magazine,* **6,** (Fall 1981), special issue.
Heisenberg (1943)	W. Heisenberg, 'Die beobachtbaren Grössen in der Theorie der Elementarteilchen I, II', *Zeitschrift für Physik,* **120** (1943), 513–539 and 673–702.
Heisenberg (1946)	*Cosmic Radiation—Fifteen Lectures edited by W. Heisenberg* (New York: Dover Publications, 1946). This series of symposia on cosmic-ray problems was held in 1941 and 1942 at the Kaiser Wilhelm Institut for Physics and were published in 1943 in German by Springer Verlag.
Hewlett & Anderson (1962)	R.G. Hewlett and O.E. Anderson, *The New World 1939/1946. A History of the United States Atomic Energy Commission, Vol. 1* (State Park, Pen.: Pennsylvania State University Press, 1962).
Hibbard (1950)	L.U. Hibbard, 'The Birmingham Proton Synchrotron', *Nucleonics,* **7** (1950), 30–43.
Janossy (1948)	L. Janossy, *Cosmic Rays* (Oxford: Clarendon Press, 1948).
Kabir (1963)	P.K. Kabir (ed.), *The Development of Weak Interaction Theory* (New York: Gordon and Breach, 1963).
Kayas (1982)	G.J. Kayas, *Les Particules Elementaires de Thales à Gell-Mann, These.*
Kevles (1978)	D.J. Kevles, *The Physicists: The History of a Scientific Community in Modern America* (New York: A.A. Knopf, Inc., 1978).
Klein (1948)	O. Klein, 'Mesons and Nucleons', *Nature,* **161** (1948), 897–899.
Lamb & Retherford (1947)	W.E. Lamb and R.C. Retherford, 'Fine Structure of the Hydrogen Atom by a Microwave Method', *Physical Review,* **72** (1947), 241–243.
Lattes et al. (1947)	C.M.G. Lattes, H. Muirhead, G.P.S. Occhialini, and C.F. Powell, 'Processes Involving Charged Mesons', *Nature,* **159** (1947), 694–697.
Lattes, Occhialini & Powell (1947)	C.M.G. Lattes, G.P.S. Occhialini, and C.F. Powell, 'Observations on the Tracks of Slow Mesons in Photographic Emulsions', *Nature,* **160** (1947), 453–456 and 486–492.

Lawrence (1948)	E.O. Lawrence, 'High-energy physics', *American Scientist,* Winter Issue (1948), 41–49.
Leprince-Ringuet & Lhéritier (1944)	L. Leprince-Ringuet and M. Lhéritier, 'Existence probable d'une particule de $990m_e$ dans le rayonnement cosmique', *Comptes rendus de l'Academie des Sciences,* **219** (1944), 618–620.
Livingston et al. (1950)	M.S. Livingston, J.P. Blewett, G.K. Green, and L.J. Haworth, 'Design Study for a Three-BeV Proton Accelerator', *Review of Scientific Instruments,* **21** (1950), 7–22.
Livingston (1952a)	M.S. Livingston, 'Synchrocyclotrons', *Annual Review of Nuclear Science,* **1** (1952), 163–168.
Livingston (1952b)	M.S. Livingston, 'Proton Synchrotron', *Annual Review of Nuclear Science,* **1** (1952), 169–174.
Livingston (1966)	M.S. Livingston, *The Development of High-Energy Accelerators,* Classics of Science, Volume III (New York: Dover Publications, Inc., 1966).
Livingston (1969)	M.S. Livingston, *Particle Accelerators: A Brief History* (Cambridge: Harvard University Press, 1969).
Livingston & Blewett (1962)	M.S. Livingston and J.P. Blewett, *Particle Accelerators,* in International Series in Pure and Applied Physics (New York: McGraw-Hill Book Company, Inc., 1962).
Marshak (1952a)	R.E. Marshak, *Meson Physics* (New York: McGraw-Hill Book Company, Inc., 1952).
Marshak (1952b)	R.E. Marshak, 'Meson Physics', *Annual Review of Nuclear Science,* **1** (1952), 1–42.
Marshak (1970)	R.E. Marshak, 'The Rochester Conferences: The Rise of International Cooperation in High-energy Physics', *Bulletin of the Atomic Scientists,* **26** (1970), 92–98.
Marshak (1985)	R.E. Marshak, 'Scientific Impact of the First Decade of the Rochester Conferences (1950–1960)', VPI–HEP-85/8. Talk given at *International Symposium on Particle Physics in the 1950's: Pions to Quarks,* Fermilab, 1–4 May 1985 (to appear in proceedings).
McMillan (1945)	E.M. McMillan, 'The Synchrotron—A Proposed High Energy Particle Accelerator', *Physical Review,* **68** (1945), 143–144.
McReynolds (1955)	A.W. McReynolds, 'Nuclear Research in Norway', *Physics Today,* **8** (1955), 13–16.
Meitner & Frisch (1939)	L. Meitner and O.R. Frisch, 'Disintegration of Uranium by Neutrons: A New Type of Nuclear Reaction', *Nature,* **27** (1939), 239–240.
Mersits (1984)	U. Mersits, *Construction of the CERN Synchro-cyclotron* (Geneva: CERN, CHS-13, 1984).
Montgomery (1949)	D.J.X. Montgomery, *Cosmic-ray Physics* (Princeton: Princeton University Press, 1949).
Occhialini & Powell (1948)	G.P.S. Occhialini and C.F. Powell, 'The Artificial Production of Mesons', *Nature,* **161** (1948), 551.
O'Ceallaigh (1951)	C. O'Ceallaigh, 'Masses and Modes of Decay of heavy Mesons, I. $\varkappa$-Particles', *Philosophical Magazine,* **42** (1951), 1032–1039.
Oppenheimer & Serber (1937)	J.R. Oppenheimer and R. Serber, 'Note on the Nature of Cosmic-ray Particles', *Physical Review,* **51** (1937), 1113.
Pickavance & Cassels (1952)	T.G. Pickavance and J.M. Cassels, 'High-Energy Nuclear Physics at Harwell', *Nature,* **165** (1952), 520–523.
Pontecorvo (1947)	B. Pontecorvo, 'Nuclear Capture of Mesons and the Meson Decay', *Physical Review,* **72** (1947), 246–247.

Powell et al. (1946)   C.F. Powell, G.P.S. Occhialini, D.L. Livesey, and L.V. Chilton, 'A New Photographic Emulsion for the Detection of Fast Charged Particles', *Journal of Scientific Instruments,* **23** (1946), 102–106.

Powell (1950a)   C.F. Powell, 'Mesons', *Reports on Progress in Physics,* **13** (1950), 350–424.

Powell (1950b)   C.F. Powell, 'The Cosmic Radiation', Nobel Lecture, 11 December 1950, in *Nobel Lectures Physics 1942–1962* (Amsterdam: Elsevier Publishing Company, 1964).

Puppi (1948)   G. Puppi, 'Sui mesoni dei raggi cosmici', *Nuovo Cimento,* **5** (1948), 587–588. An English translation is published in *Kabir (1963)*.

Ramsey (1966)   N.F. Ramsey, *Early History of Associated Universities and Brookhaven National Laboratory,* Brookhaven Lecture Series (Brookhaven: BNL 992 T-421, 1966).

Randers (1953)   G. Randers, 'The Dutch-Norwegian Atomic Energy Project', *Bulletin of the Atomic Scientists,* **9** (1953), 369–371.

Rochester (1951)   G.D. Rochester, 'Heavy Unstable Particles', talk given at the *Cosmic-ray Colloquium,* Dublin, September 1951.

Rochester (1984)   G.D. Rochester, 'The Discovery of the V-particles'. Talk given at the Conference *Fifty Years of Weak Interactions: From Fermi Theory to the W,* Racine, Wis., USA, May 1984 (to appear in proceedings).

Rochester & Butler (1947)   G.D. Rochester and C.C. Butler, 'Evidence for the Existence of New Unstable Particles', *Nature,* **160** (1947), 855–857.

Rochester & Butler (1953)   G.D. Rochester and C.C. Butler, 'The New Unstable Cosmic-ray Particles', *Reports on Progress in Physics,* **16** (1953), 364–407.

Rotblat (1950)   J. Rotblat, 'Photographic Emulsions', *Progress in Nuclear Physics,* **1** (1950), 37–72.

Sakata & Inoue (1946)   S. Sakata and T. Inoue, 'On the Correlation between Mesons and Yukawa Particles', *Progress of Theoretical Physics,* **1** (1946), 143–150.

Sakata & Tanikawa (1940)   S. Sakata and Y. Tanikawa, 'The Spontaneous Disintegration of the Neutral Mesotron (Neutretto)', *Physical Review,* **57** (1940), 133.

Schweber (1984)   S.S. Schweber, 'Some Chapters for a History of Quantum Field Theory', in B.S. Dewitt and R. Stora (eds.), *Relativity, Groups and Topology II,* Les Houches, Session XL, 27 June–4 August 1983 (Amsterdam: North-Holland Publishing Company, 1984).

Schweber (1985)   S.S. Schweber, 'Some Reflections on the History of Particle Physics in the 1950's'. Talk given at the *International Symposium on Particle Physics in the 1950's: Pions to Quarks,* Fermilab, 1–4 May 1985 (to appear in proceedings).

Schweber (1986)   S.S. Schweber, 'Shelter Island, Pocono and Oldstone: The Emergence of American Quantum Electrodynamics after World War II', *Osiris,* vol. 2 (1986).

Schwinger (1948a)   J. Schwinger, 'On Quantum-Electrodynamics and the Magnetic Moment of the Electron', *Physical Review,* **73** (1948), 416–417.

Schwinger (1948b)   J. Schwinger, 'Quantum Electrodynamics. I. A Covariant Formulation', *Physical Review,* **74** (1948), 1439–1461.

Schwinger (1949a)   J. Schwinger, 'Quantum Electrodynamics. II. Vacuum Polarization and Self-Energy', *Physical Review,* **75** (1949), 651–679.

Schwinger (1949b)   J. Schwinger, 'On Radiative Corrections of Electron Scattering', *Physical Review,* **75** (1949), 898–899.

Bibliography for Chapter 1

Schwinger (1949c)	J. Schwinger, 'Quantum Electrodynamics. III. The Electromagnetic Properties of the Electron—Radiative Corrections to Scattering', *Physical Review*, **76** (1949), 790–817.
Schwinger (1958)	J. Schwinger (ed.), *Selected Papers on Quantum Electrodynamics* (New York: Dover, 1958).
Seidel (1983)	R.W. Seidel, 'Accelerating Science: The Postwar Transformation of the Lawrence Radiation Laboratory', *Historical Studies in Physical Sciences*, **13** (1983), 375–400.
Sekido & Elliot (1982)	Y. Sekido and H. Elliot, *Early History of Cosmic-ray Studies* (Dordrecht: D. Reidel Publishing Company, 1982).
Seriff et al. (1950)	A.J. Seriff, R.B. Leighton, C. Hsiao, E.W. Cowan, and C.D. Anderson, 'Cloud-Chamber Observations of the New Unstable Cosmic-ray Particles', *Physical Review*, **78** (1950), 290–291.
Skinner (1948)	H.W.B. Skinner, 'The Work of the Harwell Establishment', *Bulletin of the Atomic Scientists*, **4** (1948) 107–109.
Smith (1951)	L. Smith, 'The Bevatron', *Scientific American*, Feb. (1951), 20–25.
Snow (1982)	C.P. Snow, *The Physicists—A Generation that Changed the World* (London: Papermac, 1982).
Snyder (1949)	C.W. Snyder, 'Current Ideas about Mesons', *Nucleonics*, **4** (1949), 42–52.
Steinberger, Panofsky & Steller (1950)	J. Steinberger, W. Panofsky, and J. Steller, 'Evidence for the Production of Neutral Mesons by Photons', *Physical Review*, **78** (1950), 802–805.
Street & Stevenson (1937a)	J.C. Street and E.C. Stevenson, *Bulletin of the American Physical Society*, **12** (1937), 13.
Street & Stevenson (1937b)	J.C. Street and E.C. Stevenson, 'Penetrating Corpuscular Component of the Cosmic Radiation', *Physical Review*, **51** (1937), 1005 (abstract).
Stueckelberg (1937)	E.C.G. Stueckelberg, 'On the Existence of Heavy Electrons', *Physical Review*, **52** (1937), 41–42.
Stuewer (1979)	R.H. Stuewer, *Nuclear Physics in Retrospect: Proceedings of a Symposium on the 1930s* (Minneapolis: University of Minnesota Press, 1979).
Symposium-Bristol (1948)	'Symposium on Cosmic Radiation', Bristol, September 1948; Proceedings published as F.C. Frank and D.R. Rexworthy (eds.), *Cosmic Radiation* (London: Butterworths Scientific Publications, 1949).
Symposium-Krakow (1947)	*Symposium on Cosmic Rays*, Krakow, October 1947, International Union of Pure and Applied Physics, Document R. C. 48-1.
Symposium-Pasadena (1948)	'Symposium on Cosmic Rays', Pasadena, 21–23 June 1948, *Reviews of Modern Physics*, **21** (1949), 1–18.
Thomas, Kraushaar & Halpern (1952)	J.E. Thomas, W.L. Kraushaar, and I. Halpern, 'Synchrotrons', *Annual Review of Nuclear Science*, **1** (1952), 175–198.
Thorndike (1952)	A.M. Thorndike, *Mesons—A Summary of Experimental Results* (New York: McGraw-Hill Book Company, Inc., 1952).
Tomonaga (1946)	S. Tomonaga, 'On a Relativistically Invariant Formulation of Quantum Theory of Wave Fields', *Progress of Theoretical Physics*, **1** (1946), 27–42. The original version of this paper was published in Japanese in 1943.
Tomonaga (1948)	S. Tomonaga, 'On Infinite Field Reactions in Quantum Field Theory', *Physical Review*, **74** (1948), 224–225.
Tomonaga & Araki (1940)	S. Tomonaga and G. Araki, 'Effect of the Nuclear Coulomb Field of the Capture of Slow Mesons', *Physical Review*, **58** (1940), 90–91.
Truslow & Smith (1983)	E.C. Truslow and R.C. Smith, *Project Y: The Los Alamos Story, Part II: Beyond Trinity* (Los Angeles: Tomash, 1983).

Veksler (1945)	V. Veksler, 'A new Method of Acceleration of Relativistic Particles', *Journal of Physics, USSR,* **9** (1945), 153–158.
Weinberg (1954)	A.M. Weinberg, 'A Nuclear Journey through Europe', *Bulletin of the Atomic Scientists,* **10** (1954), 215–217.
Weiner (1972)	C. Weiner, *Exploring the History of Nuclear Physics* (New York: American Institute of Physics, 1972). This book contains the proceedings of two conferences held by the American Academy of Arts and Sciences on this subject, namely one in May 1967 on 'Emergence and Growth of Nuclear Physics as a Research Field' and one in May 1969 on 'The Role of Theory in the Development of Nuclear Physics'.
Weisskopf (1947)	V.F. Weisskopf, 'On the Production Process of Mesons', *Physical Review,* **72** (1947), 510.
Wilson (1958)	R.R. Wilson, 'Particle Accelerators', *Scientific American,* **198–199** (1958), 65–76.
Wilson (1981)	R.R. Wilson, 'U.S. Particle Accelerators: An Historical Perspective', *AIP Conference Proceedings,* **92** (1982), 298–327.
Yukawa (1935)	H. Yukawa, 'On the Interaction of Elementary Particles I', *Proceedings of the Physical-Mathematical Society of Japan,* **17** (1935), 48–57.
Yukawa (1937)	H. Yukawa, 'On a Possible Interpretation of the Penetrating Component of the Cosmic Rays', *Proceedings of the Physical-Mathematical Society of Japan,* **19** (1937), 712–713.
Yukawa (1949)	H. Yukawa, 'Meson Theory in its Development', Nobel Lecture, 2 December 1949, in *Nobel Lectures Physics 1942–1962* (Amsterdam: Elsevier Publishing Company, 1964).

PART II

The prehistory of CERN
1949–February 1952

CHAPTER 2

The first suggestions[1]
1949–June 1950

Dominique PESTRE

Contents

2.1 The years following the war   64
    2.1.1 The UN Economic and Social Council   64
    2.1.2 The UN Atomic Energy Commission   66
    2.1.3 The European Movement   68
2.2 Two proposals for European collaboration in November 1949   69
2.3 The European Cultural Conference, Lausanne, 8–12 December 1949   72
2.4 Attempts to implement the suggestions   75
    2.4.1 The European Movement initiative   76
    2.4.2 The French attempts   79
2.5 The fifth General Conference of UNESCO, Florence, June 1950   82
    2.5.1 Rabi's resolution   82
    2.5.2 First reactions to Rabi's resolution   84
    2.5.3 A standard misconception about Rabi's resolution   87
2.6 The situation in June 1950   88
    Notes   90

The official date of the foundation of CERN is easily had*. The Convention setting up the organization was signed on 1 July 1953 and was ratified by France and Germany on 29 September 1954. By then nine states had done so, and CERN came officially into being. However, the interest of this date is rather limited. It ended a process initiated several years before and understanding such a process is more illuminating than having a date of birth.

In this chapter we will look back to the years 1946–1950. It is not that we think that CERN's history begins at this time and we certainly do not have the search for 'ancestors' in mind. But a whole series of events which took place in the years following the war were not without significance for the emergence of CERN at the beginning of the 50s. And if we want to understand this process it is helpful to know the atmosphere which prevailed in scientific circles in the late 40s, the attitudes common among scientific administrators and politicians (particularly as far as international scientific collaboration was concerned), and the networks of relations which linked these various groups of people to one another.

This is why, by way of introduction, we will first describe the circles where scientific co-operation was discussed in the immediate post-war years.

2.1 The years following the war

2.1.1 THE UN ECONOMIC AND SOCIAL COUNCIL

The first ideas for international laboratories were put forward as early as 1946 inside the United Nations Organization.[2] In October that year the French delegation to the UN Economic and Social Council submitted a draft resolution inviting the 'Secretary General to consult UNESCO and the other specialized Agencies concerned and to submit to the Economic and Social Council if possible during the next session, a general report on the problem of establishing UNITED NATIONS RESEARCH LABORATORIES'.[3] Accepted after a long discussion, this resolution led to a 'Report of the Secretary General on establishing United Nations Research Laboratories' on 23 January 1948. Before writing this report a wide range of people and organizations had been contacted. The number and variety of their responses were indicative of the prevailing interest which scientists in many fields had in

---

* I would like to thank John Krige for his untiring help in improving the English of my texts. But for the many days he spent on them, the quality of the language would have been awful.

international initiatives, even if some of them were against the idea of international laboratories as such.

In the following years lists of priorities were established and discussed at length, the only clear result being the proposal to set up an International Computation Centre. We do not need to describe this rather complicated matter, even less to give the reasons for the Centre's failure. But some remarks are called for.

The first point to stress is that, immediately after the war, a group of people inside the UN agencies was very keen on international scientific co-operation. What is more they wanted to transcend the limits of the by-then conventional forms of co-operation. As the French resolution said, 'it would seem that in the general interest of humanity, something more than mere coordination is needed and it may be asked whether certain fields of scientific research ought not to be taken over by the central organs of the UN or by specialized agencies'. In other words, there was a desire to build *new kinds* of international scientific structures. Among those who had this aim were Henri Laugier, French biologist and assistant Secretary-General in the UN Economic and Social Council, Julian Huxley, first Director-General of UNESCO (1946–1948), Joseph Needham, first Director of the Exact and Natural Sciences Department of UNESCO (1946–1948), and Pierre Auger, his successor.[4]

What motivated these people? What state of mind was reflected in these projects? The *Statement of Reasons* given by the UN draft resolution E/147 submitted on 2 October 1946 throws some light on this.

Firstly, the setting up of UN research laboratories would help to alleviate the suffering of mankind. As the resolution said, 'nobody questions today the decisive importance of scientific research in the development of human knowledge, social progress and the improvement of human conditions generally', and 'many branches of scientific research connected with the promotion of human knowledge, particularly in the field of public health, would yield considerably more effective results if they were conducted on an international plane'. The French specified that, to be as efficient as possible, the collaboration must be *rationally organized*. Elaborating, they gave the examples of astronomy and astrophysics: 'it is quite evident that considerable advantage would be gained by replacing haphazard collaboration by systematic international action in order to locate well-equipped observatories at those sites of the globe selected as most suitable for the development of research.' In other words, there were fields of science 'in which research work can only be organized efficiently and in a rational and disinterested way on an international basis'.[5]

These two reasons — using science to improve the living conditions of mankind and acting as rationally as possible by planned international co-operation — strike a familiar chord. They are related to the rationalistic and positivistic spirit inherited from the Enlightenment and the XIXth century, a spirit which was still very much alive among scientists, perhaps particularly among the French.[6]

However a more specific motive also played a role: the resolution aimed to

Notes: pp. 90 ff.

promote the growth of an international spirit. Behind that we must read the rejection of war, the determination to prevent any new conflict. Of course these feelings were also part of the rationalistic vision of science. As the resolution stated: 'if all scientists and research workers of the United Nations are given an opportunity of establishing close relationships with one another on the working level, and if every scientific victory is really made a joint triumph, scientific research will have acquired *its true meaning*'. On the other hand the immediacy of the horrors of war revived such hopes and gave them more weight.[7]

A last point is to be stressed about the UN initiatives. Little importance was attached to physics—and to nuclear physics in particular—in the various lists of priorities established by the UN or UNESCO experts. For example, let us listen to Pierre Auger—until recently himself active in cosmic-ray physics—during the 5th meeting of the Committee of Scientific Experts convened in August 1949 by the UN: 'If physics, which was scarcely mentioned in the report, had not been dealt with at greater length, it was because that science had reasonably good resources at its disposal and had received considerable assistance from Governments and Foundations, so that its immediate needs were less than those of other sciences.' Consequently, only cosmic-ray studies were suggested as a field in which co-operation would be of great value, though the minutes concluded by saying that 'Professor Auger did not think it [was] necessary to give [this] proposal high priority, any more than in other fields of physics'.[8]

The idea of close *intergovernmental* collaboration in the scientific field, the idea on which CERN is based, was then partly rooted in lengthy discussions initiated immediately after the war. To survive it required the determination of several people, who as early as 1946 set out to spread the word and to bring such a scheme to fruition. However CERN is devoted to nuclear physics, and we have seen that this field was not often mentioned by the United Nations experts. Which leads us to turn towards other circles where the problem of international collaboration in this specific domain of science was raised.

2.1.2 THE UN ATOMIC ENERGY COMMISSION

The UN Atomic Energy Commission was one such circle. Established in June 1946 with the aim of bringing atomic weapons under some form of international control, this Commission was run by diplomats even though their scientific advisers were present at all meetings. As Kowarski commented, 'it was a pleasure to watch the diplomats grapple with the difference between a cyclotron and a plutonium atom; we had to compensate by learning how to tell a subcommittee from a working party, and how—in the heat of a discussion—to address people by their titles rather than names. Each side began to understand the other's problems and techniques; a mutual respect grew in place of the traditional mistrust between eggheaded pedants and pettifogging hairsplitters'.[9]

Initially, the scientists tended to keep contacts with their own diplomats (at the national level) distinct from their discussions with scientific colleagues (from different countries). But after a while bridges between scientists and diplomats of different nations were built. It is difficult to know how widespread such linkages were, but there was at least one important one. François de Rose, French diplomat and representative on the United Nations Atomic Energy Commission, rapidly became a great friend of J. Robert Oppenheimer, who used to invite him to spend weekends in Princeton. As a result de Rose thought of organizing meetings between his French scientific advisers and the eminent physicist. During meals at de Rose's, Perrin, Kowarski, and Auger had the opportunity to discuss with the Americans the particular situation of science in Europe. It became clear to them, according to de Rose and Kowarski, that the various countries of the old continent did not individually have the ability to build the big apparatus required by modern science, and that it would be more worthwhile for them to do so together on the European level.[10]

More precisely, these men were thinking of building nuclear scientific equipment. According to Kowarski, that meant that 'while never forgetting meson physics as a scientifically more worthwhile field, they [the European physicists] usually sought to combine the accelerator and the reactor aspects. In both these fields the need for European collaboration was obvious; their combination in a single project appeared natural, even though the early promoters were aware of the slight, but possibly decisive difference in the public attitude towards the two questions'.[11]

This remark probably gives a not too-inaccurate impression of the state of mind of physicists in the late 40s regarding the question of what should be done to promote nuclear physics in Europe. In fact, we must not forget that at the time the meaning of 'nuclear physics' was different from what it is now. In particular, the spheres of high- and low-energy physics were not clearly separated. Institutionally speaking for example, the laboratories created immediately after the war and dealing with nuclear research often had both reactor and accelerator departments. This was the case for the Brookhaven National Laboratory (BNL) on the east coast of the United States, for the Atomic Energy Research Establishment (AERE) at Harwell, and for the French Commissariat à l'Energie Atomique (CEA) near Paris. Similarly, if we look at the proceedings of international conferences held in the late 40s, we notice that generally particle physics, field theory, quantum electrodynamics, or nuclear physics (as we now call them) were discussed in the same frameworks. Only cosmic-ray conferences dealt primarily with the high-energy field, but here very little was said about accelerator techniques. In fact the division between 'big science' and science without large and expensive equipment was at least as important as the division between low-energy physics and what we now call particle physics. Of course distinctions between the fields gradually emerged during the late 40s/early 50s. But it would be a mistake, though a rather common one, to project backwards what was only true from the mid–50s onwards.[12]

Notes: p. 91

The problem of European collaboration in the nuclear field (in the sense just explained) was then raised around the UN Atomic Energy Commission. We have mainly mentioned French people, but it is likely that other physicists—and perhaps diplomats—took part in these informal discussions. As Lew Kowarski has written, 'from Rabi through Bernardini [then at Columbia University] and Amaldi, similar ideas began to be discussed in Italy, through Kramers [himself one of the leading UN experts], in Holland and Denmark'. Unfortunately it is impossible to be more explicit because of a lack of documentation.[13]

To conclude with this circle of people, it is of interest to notice that a network of nuclear physicists and of diplomats was built up, that both groups learned from each other how to handle complex situations which combined scientific knowledge, technical skills, large amounts of money, and the necessary approval of governments, and that they were already thinking of European co-operation in nuclear matters. But nothing was actually done, everything remained informal, and there was no progress beyond the level of talks and hopes.

2.1.3 THE EUROPEAN MOVEMENT

Our quick survey of the circles where international co-operation was discussed would be incomplete without our saying a few words about the people working within the *European Movement*. As is well known, numerous attempts were made immediately after the war to foster various forms of European unity. Let us recall only the most important ones: the birth of the Organization for European Economic Co-operation (OEEC) in the first months of l948; the Congress of Europe held in The Hague in May of the same year, which was attended by 800 personalities; the creation of the Council of Europe one year later, thanks to the activities of the European Movement which was established in the months following the Hague Congress; and, from 1950 onwards, the first examples of European community organisms (European Coal and Steel Community, European Defence Community).[14]

Among these initiatives, those taken by the European Movement are of particular interest to us because some important scientific administrators were involved. For the moment we would like to mention only three of them: Gustavo Colonnetti, former member of the Italian Constituent Assembly (1946–1948) and president of the Italian Consiglio Nazionale delle Ricerche since 1944; Raoul Dautry, former minister and Administrator-General of the French Commissariat à l'Energie Atomique from its inception; and Jean Willems, Director of the Belgian Fonds National de la Recherche Scientifique since its foundation in l928 and first President of the Belgian Institut Interuniversitaire des Sciences Nucléaires which was set up in 1947. By virtue of their office these men were directly interested in the management of science; by virtue of their ideological and political convictions they were interested in European unity. This accounts for their trying to do something for European science, trying to imagine new ways of collaborating to satisfy the financial demands of 'big science'. Their importance lay in their power to obtain money for expensive projects and to influence and convince ministers, if not a whole government.[15]

Their involvement in projects for European laboratories was helped by a fourth man, Denis de Rougemont, a Swiss writer. He was at that time one of the most active people in European circles. In 1947 he had met Einstein in Princeton and had discussed with him the interest of 'linking the ideas of European unity with the control of nuclear energy'. According to de Rougemont, on his return to the continent he went and saw Dautry to resume the discussion. The brain drain and the needs of Europe were their main preoccupations. These discussions bore fruit in December 1949: a commission of the European Cultural Conference held in Lausanne from the 8th to the 12th of that month proposed the creation of a European institute for nuclear science.[16]

2.2 Two proposals for European collaboration in November 1949

This introduction brings us to the end of 1949 when, within the space of a few months, there was a surge of activity aimed at achieving European collaboration in nuclear physics. There is no single reason for this but the preparations for the European Cultural Conference we have just mentioned played a role.[17]

The most active person during the preparatory stages of its scientific component was Raoul Dautry. Administrator-General of the French CEA, as we have said, he also was President of the French Executive Committee of the European Movement. To prepare the Cultural Conference Dautry wrote to several of his close colleagues including some of the CEA research directors (Guéron and Kowarski) and Pierre Auger at UNESCO. We do not have copies of the letters he sent but the replies help us to get an idea of what Dautry had in mind. He hoped that the Lausanne conference could take the lead and propose the creation of a 'European centre for atomic research'.[18] Indeed for Dautry a form of collaboration in nuclear matters was required which covered not only fundamental and applied research, but even industrial scale production. Responding to this Guéron spelt out some of the production problems faced by the European states. In the present circumstances, he said, 'the factories manufacturing the components for nuclear equipment cannot be used to full capacity by one country alone [...] A single group of installations could thus meet the need of several countries' of Europe. As examples he gave the production of Norwegian heavy water and 'the fabrication of pure calcium [in France, which] will have to stop if no outlet is found'.[19] For his part, Kowarski replied by sending a longer and more comprehensive report entitled 'Note on European Atomic Co-operation', a report which reflected the informal conversations which had been held within the French delegation to the UN Atomic Energy Commission.[20]

Finally, and independently of these new steps in France and in European circles, two other initiatives were underway at the end of 1949. We have little evidence for the one, but according to Amaldi 'the idea of the European laboratory was further discussed among Italian physicists' at this time. The other is somewhat better documented. On 11 November 1949, on the occasion of the inauguration of a Dutch cyclotron in Amsterdam, representatives from Belgium, Denmark, the Netherlands, Sweden, and Switzerland got together to discuss ways of implementing some form of co-operation in the field of nuclear physics.[21]

We will start our study of this set of proposals and resolutions by looking at Kowarski's report and the minutes of the Amsterdam meeting. The main reason for this choice is that the proposals made therein are quite similar and that they fit into the same framework of thought, both having been formulated by nuclear physicists.

According to Kowarski's report the situation prevailing since 1945 was characterized by the 'absence of international agreement' between the USA and the USSR on the one hand, and by the special position occupied by the USA and the UK within the western world on the other. The 'Anglo-Saxon' front was based on the fact that these two nations were ahead of all the others in the nuclear field and on their oft-repeated determination to limit the dissemination of their knowledge, excluding even their European allies. For Kowarski the British were not as opposed as was the American State Department to helping atomic research in western Europe, but '*in practice*' they could only 'follow the American example' and respect American policy. This was why they never responded favourably to requests for aid from France, Norway, Belgium, and the Netherlands during the latter part of the 1940s. From this Kowarski concluded that in the nuclear field, 'continental western Europe could only count on its own resources'. Kowarski did, however, stress the other reason why the British would not be interested in co-operating in a European project: they were much too far ahead in nuclear matters for any collaboration with continental Europe to be worth their while.[22]

Having thus identified the geographical zone, Kowarski presented the basic principle underlying his idea: to set up 'a rational division of labour' in Europe utilizing existing specializations in different countries. As the aim of the co-operative scheme was to give Europe the various research tools required by modern science, he suggested that the contribution of each member state be clearly spelt out within the context of an overall plan. In this way each country could do its bit towards the realization of the common project. For example, the Swiss contribution could be the construction of a 'really big synchro-cyclotron' available to scientists of all the other collaborating countries, while small French nuclear reactors would no longer remain specifically French, nor would the Copenhagen Institute for Theoretical Physics be restricted to the Danes. National substrata would of course continue to exist but the more 'specialized' level would represent each country's contribution to a 'supra-national' venture. In the long term Kowarski also envisaged the possibility of a third level consisting of 'Super-Saclay or super-Brookhaven' type laboratories but this idea was only sketched in outline.

The scientists who met in Amsterdam on 11 November 1949 were motivated by the same kind of considerations as Kowarski, i.e. that of co-ordinating programmes and building large-scale research tools: 'The undersigned, invited by F.O.M. [the Dutch Institute for Fundamental Research into Matter] as unofficial representatives of the countries Belgium, Denmark, the Netherlands, Sweden and Switzerland, judging that in the field of nuclear physics a joint effort may lead to the realization of scientific programs and the creation of research tools that are outside the scope of each of the countries separately, move that a joint committee be formed containing two representatives from each of the countries concerned in order to investigate the possibilities of such a co-operation'. They also suggested appointing an 'International Committee' whose main tasks would be to 'discuss the possibilities of building in common a small reactor (a zero-order pile) or a similar one in each country' and to 'study the possibilities of jointly building large accelerators'.[23]

In his 'Note' Kowarski tried to envisage the inevitable obstacles to such projects, of which he stressed two. The first was inherent in the very nature of the work itself: in all countries nuclear physics was subject to very strict control by governments. As he said, 'all national resources in the atomic field are inextricably connected with defence secrets and with a mentality which is in no way open-minded to large projects and supra-national realizations.' The second obstacle was linked to the international political context and to America's attitude in particular: as several countries still hoped for support from the USA, they hesitated to get involved in any multilateral collaboration of which the Americans might not approve, particularly one with countries like France or Italy where the Communists were powerful. Thus, unwilling to risk losing the possibility of American support (described by Kowarski as a trap, '*un leurre*'), no joint action was taken and European states continued to dream of bilateral collaboration with the UK or the USA. Kowarski did however believe that 'there [had] been a considerable change since the announcement made by Truman on 23 September' [1949], in which the American President revealed that the Russians had exploded an atomic bomb that summer. Kowarski felt that 'the pretence that there was an important "secret" in pre-1948 atomic science and technology' could no longer be sustained and that the attitudes of the Americans and the British were bound to be more relaxed as a result. This in turn might open the door to a new era of co-operation in Europe. For him the Amsterdam meeting was a positive step in this direction.

The importance of Truman's announcement cannot be overestimated: it was a point to which Kowarski returned again and again.[24] Indeed, its psychological impact helps us to understand why rather vague proposals emanated from Paris and Amsterdam. In fact, not enough time had elapsed since September for any proper talks to have taken place: we are dealing here with immediate reactions. This also explains why nothing specific was to come of these proposals. And although discussions continued during the next few months, no concrete action was actually taken until May 1950. As for those European states who gathered in Amsterdam, although some delegates were appointed as called for by the resolution (for example by Sweden), the collaboration envisaged by them never materialized.[25]

2.3 The European Cultural Conference, Lausanne, 8–12 December 1949[26]

In the months following the Congress of Europe in The Hague in May 1948 six out of the seven ideological movements which had gathered there founded the European Movement. This body organized four specialized conferences to study more precise topics — in Brussels and Westminster on economical and political problems, in Lausanne on cultural matters, and in Rome on the social situation in Europe. For our purposes only the Lausanne conference (8–12 December 1949) is of interest, since the scientific situation in Europe was on its agenda.[27]

The size of the conference and the social importance of some of the people who attended it deserve some introductory words. Up to now we have met small groups of people, mainly scientists. Here we are dealing with a large conference, properly organized, prepared well in advance, and attended by 170 people from 22 countries (including Germany). Very influential people participated — Paul-Henry Spaak and Duncan Sandys, for example — but also former ministers, senators, members of parliament, rectors of universities, deans of faculties, ... During the five days of the meeting three commissions were set up, one of them called the Commission of Institutions. This in turn had a sub-committee of about 15 people whose role was to discuss scientific matters and to present a draft resolution.

We have no list of the people who formed this sub-committee, but using a general list of participants and an article published in 1975 by the Centre Européen de la Culture, it is possible to gain a rather good idea of its members. Raoul Dautry, Gustavo Colonnetti, and Jean Willems were there. We have already mentioned the influence which these important science administrators had in their own countries. There were also scientists present: Max von Laue, physicist and Nobel prizewinner; Paul Montel, professor of classical mechanics (*mécanique rationnelle*) and mathematics, former dean of the faculty of sciences in Paris; Cyril Darlington, Fellow of the Royal Society, and Director of the John Innes Horticultural Institution; Charles Manneback, theoretical physicist, professor at Louvain University; and André George, French physicist and philosopher, student and friend of Louis de Broglie. Though these were all obviously important and respected people not one was a high ranking physicist in nuclear science, and not one was at the head of a decisive field of research in physics. Max von Laue and Paul Montel were rather old, André George was more of a philosopher than an active physicist. Indeed of the physicists only Charles Manneback was still active in research.[28]

Denis de Rougemont presented a general report to the conference. It identified the two main questions which were to form the basis of the discussion: 'Firstly "the material and moral conditions of cultural life in Europe" and secondly, a consideration of the "institutions and reforms" which [were] desirable with a view to the development of a European outlook'. Studying the obstacles and limitations to the free circulation of people and ideas in Europe, the report spent some time discussing the particular situation in science. Here 'discoveries of immense importance [were] kept as secrets [...]. The State [exercised] a tight control over researches in the field of nuclear physics, and [regarded] them with a suspicion which [suggested] the policeman. The interests of the scientist [were] entirely subordinated to political and military considerations'. For this reason the report suggested 'the drawing-up and adoption of a limited number of practical resolutions, all aiming simply at the suppression, pure and simple, of all obstacles to freedom of movement'. Typically, it proposed the preparation of 'plans for European collaboration (and not merely controlled and reticent relationships between national bodies)', 'the immediate creation of a European fund for scientific research, which would be under the direct control of the competent organs of the Council of Europe' and the creation of a 'European Centre for atomic research connected with the fund'.[29]

On the same day Raoul Dautry read a message from Louis de Broglie, who was unable to attend the meeting. It was couched in rather general terms drawing attention to the value of combining forces to develop science in Europe. However de Broglie did suggest the possibility of establishing 'a laboratory or institution where it would be possible to do scientific work, but somehow beyond the framework of the different participating states. Being the product of a collaboration between a large number of European countries, this body could be endowed with more resources than national laboratories and could, consequently, undertake tasks which, by virtue of their size and cost, were beyond their scope. With the help of scientists from different nations, it would serve to co-ordinate research and its results, to compare methods, and to adopt and implement work programmes'. As a model, de Broglie gave the International Bureau of Weights and Measures set up in Sèvres, near Paris. He noted that the international situation did not yet allow research centres to be established on a world scale, but that it should be possible to do so at the European level. He concluded with the hope that this kind of co-operation—easier in the scientific field than in others because of the 'true nature' of science—would point the way towards similar activities in other spheres of cultural life.[30]

Raoul Dautry, who probably had asked Louis de Broglie personally to write this message, made two further concrete proposals after he had read it out. He suggested that the conference study the ways of strengthening collaboration in two fields: in

Notes: pp. 92 ff.

astronomy and astrophysics, by building powerful telescope(s) and all the necessary auxiliary material, and in the atomic energy field, by setting up a centre with all the requisite modern apparatus. Dautry concluded: 'what each European nation is unable to do alone, a united Europe can do and, I have no doubt, would do brilliantly'.[31]

Of course, the words 'atomic energy' having been uttered, everyone was on his guard. Let us listen to Denis de Rougemont recalling these days 25 years later:

> On the eve of the first session — and this will give you some feeling for the atmosphere at the time — I had had a very lively discussion with the leaders of the European Movement who attended the conference. One of them, radically opposed to any public discussion of nuclear matters, asked me if I wanted to reveal all our atomic secrets to the Russians. Indeed to speak of atomic research at that time was immediately to evoke, if not the possibility of blowing up the whole world, then at least preparations for a third world war, major acts of espionage, state secrets ... . In fact journalists who heard de Broglie's message and Dautry's speech pestered the members of the scientific commission to such an extent that, on the second day, I was obliged to lock our fifteen scientists in a hall of the Federal Court where the conference was being held, and to make them telephone the secretariat if they wanted to leave it [...].[32]

Despite these inconveniences, on concluding its work, the scientific sub-committee submitted a resolution which was unanimously accepted. Two levels of collaboration were recommended. On the one hand, the directors of national research organizations 'should meet periodically in a spirit of disinterested collaboration'; on the other 'specialized European Institutes in close touch with the corresponding national organisations and with those of UNESCO' should be created. To illustrate 'the Commission suggest[ed] that the foundation of an Institute of Nuclear Physics in its application to daily life should be taken into consideration immediately'.[33]

In conclusion, we would like briefly to compare what Kowarski and the physicists gathered in Amsterdam proposed with the resolution passed in Lausanne. A difference is worth stressing. In November we had nuclear physicists who wanted to improve their equipment in fundamental research. They were aware that some forms of collaboration were required if they were to reach international standards in the major fields of big science. No single country would have the money, the knowledge, the manpower to compete with the giants such as America and even the United Kingdom. Hence the search for allies. For various reasons the small countries of Europe were thinking of working together, while Kowarski planned to share the load between continental European nations. In both cases the aim of the collaboration was given beforehand: providing the *tools* for *fundamental* research in *nuclear* and *corpuscular* physics. However, the limits of the geographical framework were ill-defined: it might be Europe, but not necessarily the whole of Europe.

In Lausanne the problem was posed in a slightly different way. What was central was the will to build Europe, and science was primarily seen as one *means*, among others, to this end. Perhaps it was a particularly good one because of its 'true nature', as de Broglie had said, but for the majority of the people present scientific development was not an end in itself. Thus one person might be content with the mere co-ordination of national policies, while another might think of European centres in physics, or in biology, or in other fields of science.

All the same only nuclear science was mentioned in the final resolution. Of course this was not a matter of chance: it reflected the mood of the time. For most delegates the words 'nuclear' or 'atomic' were flags around which to rally forces, they were part of the mythology which surrounded 'the Atom', they alluded to the power of science and of scientists over nature—they meant the New Science. A collaboration in the field would also be the best symbol of the success of the European ideal: as was said during the conference, nuclear physics was strictly controlled by the national states and a joint effort in this field would be a wonderful victory for the 'Europeans'.

To avoid being one-sided we must add that for people like Colonnetti, Dautry, or Willems other considerations played a role. They knew how difficult it would be to build the big apparatus required by the new science. They were aware of its financial, scientific, technical, and industrial demands, and from this point of view the advantages of collaboration were clear to them. As science administrators, they had no illusions about what 'big science' meant; for them the word 'nuclear' probably had less mythical connotations.

It is also noteworthy that these men were oriented as much towards the industrial and economic needs of Europe as towards those of fundamental research. The latter were not absent from their minds—Dautry, for example, thought of astrophysics—but applied research remained of paramount importance. The resolution proposed an institute devoted to applications to everyday life, Dautry was always speaking of nuclear *energy*, and he concluded his speech by saying 'one day, perhaps not 20 years hence, the material life of Europe will no longer be based on millions of tons of coal but on a few tons of uranium. By that time the physiognomy of the world economy will have changed, and if European industries find themselves condemned to using only today's sources of energy they will have no choice but to close down'.[34] Similarly, nobody mentioned the high-energy field or particle physics in Lausanne. By contrast, in Amsterdam, as in Kowarski's report, fundamental research in both low- and high-energy physics was the primary concern.

2.4 Attempts to implement the suggestions

In the first half of 1950 both the French CEA scientists and the people in the European Movement tried to follow-up what they had discussed at the end of 1949.

Notes: p. 93

It seems that they acted more or less independently of each other even if there were some clear connections between the two groups.

2.4.1 THE EUROPEAN MOVEMENT INITIATIVE

In the months following the Lausanne conference Denis de Rougemont, who was directing preparations for setting up a Centre Européen de la Culture, contacted the main scientific administrators who had been in Lausanne in December 1949. On 10 March 1950 he wrote to Dautry asking for a short meeting with him: 'I believe,' he said, 'that there is no more time to waste if we want to put into practice what was decided during the Lausanne conference, particularly in the field of scientific research'. Four days later they both agreed to write first to Colonnetti, George, and Willems. Thus commenced an important exchange of letters between these five people.[35]

To study this exchange we will not follow a chronological order since doing so would lead to numerous repetitions; rather we prefer to present what was discussed in a thematic way.

All the letters dealt with a central point: how to organize the next meeting of 'the people responsible for scientific research in their own countries and convinced of the necessity of the project proposed in Lausanne'. According to de Rougemont this group would have to 'set up a concrete project which could be submitted either to the Council of Europe or to the various governments'. Three problems were then regularly discussed: Who was to be invited? What concrete proposal should be made? When should this meeting be?[36]

Who was to be invited? Some phrases used by Dautry and de Rougemont reveal what kind of people they had in mind. In March Dautry spoke of 'the group of people responsible for scientific research' and in April de Rougemont suggested finding 'some scientists and research administrators', some 'people actually responsible and capable, in the future, of making the project a success in their domain'.[37] Asked to give some names in Great Britain, the Netherlands, Switzerland, and the Scandinavian countries, Jean Willems and Gustavo Colonnetti answered along the same lines. On 25 April Willems suggested that H.A. Kramers, one of the leading European physicists and then President of the International Union of Pure and Applied Physics (IUPAP), and J.H. Bannier, Director of the Dutch Organization for the Advancement of Pure Research (ZWO) be invited from the Netherlands; for Great Britain he proposed Sir Ben Lockspeiser, Secretary of the Department of Scientific and Industrial Research (DSIR), Alexander King, former Director of the Scientific Secretariat of the Lord President's office and, failing them, Sir John Cockcroft, Director of the Atomic Energy Research Establishment, Harwell; for Switzerland, Paul Scherrer, the most influential Swiss physicist. One week later, Colonnetti sent his names: Perez, Director of the French Centre National de la Recherche Scientifique (CNRS), 'Sir Simon, Head of the Service for Scientific

Research for the UK', and M. Kreuit, [Kruyt in fact] 'President of pure scientific research in the Netherlands'. If we put all these together, we get a list of important managers of science in Belgium (Willems), France (Dautry, Perez), Great Britain (Lockspeiser, King, Simon), Italy (Colonnetti), and the Netherlands (Bannier). We also have the names of some scientists who had a leading role both at home and in the international physics community (Kramers, Scherrer), or who were at the head of a big science centre (Cockcroft). Clearly high-ranking scientific administrators were expected to play the leading role in this European study group.[38]

What concrete proposal should be submitted to the various governments or to the Council of Europe? Apparently this question was never actually resolved in spite of de Rougemont's efforts to get clear answers. However it seems that there were two different approaches, two different kinds of suggestion.

On the one hand, Dautry thought that the main aim of the meeting should be to study the possibilities of setting up European scientific *institutes.* Our documents suggest that he never developed his idea. Thus, on 18 April 1950, de Rougemont asked him to make some suggestions for the agenda of the meeting. 'It will be easier', he said, 'to invite people if we have something precise to put on the table'. Apparently Dautry did not propose anything more detailed, and a letter written five months later by de Rougemont to Scherrer was still rather vague, and without an agenda. Furthermore, as late as October 1950 Dautry wrote to Auger in the following terms:

> Dear friend,
> At the meeting of its Centre Culturel held last week in Geneva, the European Movement gave top priority to the question of European scientific centres (nuclear energy, astrophysics, fluid mechanics, etc....). Another meeting will be held on 12 December.
> In this connection I would like to know what UNESCO wants to do so as not to thwart its activities. If you are free for a few minutes one day, could you pass by rue de Varenne [CEA offices] so that we can have a chat about this?
> Cordially,
> R. Dautry

Obviously then, Dautry was repeating what he had said in Lausanne ten months before; he had not refined his ideas, particularly about how to implement the resolution.

Additional evidence is provided by a report written by M.P. Levy, representative of the Secretary-General of the Council of Europe at the meeting held to inaugurate the Centre Européen de la Culture on 7 and 8 October 1950. He wrote that he had 'asked Dautry during the meeting just what problems his group was dealing with, but the answer was awfully vague. De Rougemont thanked me in private for having put this question, and assured me that he too had only managed to get rather nebulous information from this working group'.[39]

Notes: p. 93

Dautry was not the only person consulted on the matter. Colonnetti and George gave their own opinions at the beginning of May and their projects were rather different. For Colonnetti, 'the fundamental problem [was] that of the relations to be established between the research centres of different countries with the aim of exchanging researchers and of coordinating the efforts. My ideal would be the foundation of an *International Office for Research* in the framework of the Centre Européen de la Culture'. He made no mention on this occasion of European *laboratories*. For his part André George proposed to proceed with caution, not to be too ambitious. The idea of European laboratories seemed to him very good in principle, but only applicable in the long term. In the particular case of a centre devoted to nuclear physics, the obstacles seemed to him even more formidable. Accordingly he suggested that one began with a short-term project, a project 'easily realisable' — for example, building a documentation centre, or an office for the exchange of ideas. As de Rougemont said when sending the short text written by George to Colonnetti 'it expresses very well the opinion of the "moderates" about this problem of co-ordination of research. It seems to me that his suggestions are in accordance with those proposed by you for an International Office for Research'.[40]

We can therefore conclude that:

- this group was not at all homogeneous as far as the concrete proposals to be put forward were concerned. Some people were thinking merely in terms of co-ordinating existing national institutions, at least initially. Others, on the contrary, would have liked to discuss the establishment of new structures, of European laboratories. We must also remember that this difference was already present in Lausanne;

- nevertheless, nothing was clear for either group. No real progress had been made since December 1949 and everybody was more or less repeating what had been said several months before. The positive counterpart to this was that these people remained open to all suggestions;

- only one thing is sure: Denis de Rougemont was eager to set up the scientific committee of the CEC. He worked ceaselessly to that end. The others, while agreeing to do what they could, were probably too busy and overwhelmed with other tasks to really tackle the problem.

When should the meeting be? Not surprisingly de Rougemont found it difficult to convene the meeting. The first one, proposed by Colonnetti for May, was postponed to the end of June: George and Dautry were not free and more time was needed to contact the people recommended by Willems and Colonnetti. For the same reasons, the June meeting was in turn postponed to October, and apparently it did not take place even then. Overall we have the impression that only a short discussion occurred (during the inauguration of the CEC on 7 and 8 October 1950), when the reports of the various working groups of the Centre were presented. In the proceedings, it is simply said: 'M. Dautry (group of scientific researches) recommends to go ahead quickly. He has made contact with UNESCO and will inform the Centre how rapidly matters are proceeding'.[41]

2.4.2 THE FRENCH ATTEMPTS

Two reports and a few letters written between April and July 1950 lead us to believe that several people in France continued discussing the possibility of setting up a European facility devoted to atomic research. We have already looked at the November 1949 report by Kowarski. It was obviously a step towards a project to be worked out in greater detail.

A short letter by Kowarski to Auger dated 20 June 1950 gives us the names of some of the people involved:

> My dear Pierre [Auger],
> Francis [Perrin] ought to have spoken to you about the attached paper which I wrote for de Rose and which was subsequently adopted by us four as an unofficial expression of the views of the 'scientific departments of the Commissariat [the CEA]'. Dautry is informed and de Rose has had a few copies circulated among his entourage.
> Best wishes
> Lew [Kowarski]

At least four people discussed the matter then: three scientists (Kowarksi, Perrin, Auger) and a diplomat whom we have already met, François de Rose. Apparently Dautry was in a slightly different position: we do not know if he was invited to discuss the text and, if he was, to what extent he was involved in drafting it. The expression 'scientific departments of the Commissariat' seems to indicate that he was only informed without taking an active part in the writing.[42]

The two reports we have just mentioned bear the same title: 'Note on the creation of a co-operative organism for atomic research in occidental Europe', they are both signed by Kowarski, and they are dated April and April/May 1950.[43] The first text, which was apparently already the result of an exchange of ideas, looks like a draft version of the second which is built on exactly the same model. The latter is only a little more detailed and a long *discussion des motifs* (statement of reasons), written in a classical diplomatic style and aimed at convincing laymen of the necessity and the importance of the project, is added at the beginning.

The preamble to the second text (the 'statement of reasons') raised again the considerations exposed in November as far as the USA and the UK were concerned: the USA was now leading research in fundamental physics—and particularly in the domain of big science; Great Britain, 'America's partner during the war' was successfully maintaining an independent effort which allowed her to keep abreast of the progress in atomic science, though because of 'the international situation' she had to stand apart from, and could not serve as leader for, continental Europe. The latter had therefore to make its own way if it wanted to catch up with the most advanced countries and was not to lose its best scientists.

Notes: p. 93

Even if some successes (the French pile, the Dutch cyclotron) had shown that building a fairly modern apparatus was still within the capabilities of a single European country, the construction of a big centre equipped with several such machines transcended them. Thus the need to collaborate. There were two ways of doing this: either the work could be shared between various countries, as was suggested by Kowarski in the November 1949 report, or intergovernmental centre(s) could be established. Now, in both 1950 reports, only the second was said to be promising: the division of labour envisaged in the former approach would be difficult to realize, the result might be too great a specialization of the countries (and of the scientists themselves), no enthusiasm would be engendered for the scheme, and anyway some facilities would still be too great a burden for one country alone to bear.

A last point was stressed in the introduction: the aim of the laboratory must be 'essentially scientific', if need be dealing with 'one or two problems like the production of power from an atomic pile, though all studies for immediate applications should be left to other organizations'. This statement is important because while the projects previously mentioned shared the principle of having no connection at all with military activities, they did not face the problem of possible industrial applications. And of course, in the framework of international collaboration in nuclear science (including reactors), this question was of paramount importance. Here, for the first time, the problem was raised and it was suggested that one avoid everything dealing too closely with immediate applications.

After this preamble specially written for some French officials and people outside the CEA, the laboratory to be set up was described.

The apparatus should be an accelerator in the 500 MeV–1 GeV range and a research reactor of 10 to 20,000 kW. A radiochemical treatment unit should be included. Lew Kowarski commented that with this laboratory, 'the west side of the European continent would still be far from being able to compete with the USA', but 'the danger of a complete lack would be avoided'.

A provisional estimate of costs was also made. Five to ten billion French francs per year would be required for 4 or 5 years. Kowarski suggested that France should pay three tenths, i.e. around 3 billion a year. He added: 'Let us note that England spends around 30 billions per year for atomic research and production and the US 300'.

Finally an administrative structure was proposed. A 'senate' comprising delegates nominated by governments would have the final decision-making power; an 'executive structure' would deal with the daily management, and a 'council of scientific experts' would propose the scientific projects.

Having read these documents two remarks can be made. The first is that the French, and mainly the nuclear physicists, were thinking seriously of what should be done. These reports indicate that regular discussions had taken place in some circles since November 1949 and that a project had been defined. The second point to stress is that these men were still thinking in terms of both low- and high-energy facilities. For them high-energy physics was not thought of as a possible autonomous candidate for the laboratory. In other words, the apparatus required by meson physics was still perceived as part of—if not an addition to—more general equipment centred around reactors.

The French were eager to have their proposal discussed in Europe.[44] The question was first raised at the Ministry of Foreign Affairs. François de Rose met Alexandre Parodi, Secretary-General of the Ministry who had been his head in the United Nations Organisation in New York from 1946 to 1948. According to de Rose, Parodi was in favour of having France contact other European countries to study together the possibility of establishing such a nuclear centre. From Kowarski's reports we learn that three scenarios were considered: 'either to use the OEEC research commission, or send a letter on Quai d'Orsay notepaper to European scientists, or approach the various national atomic committees'.

After further discussion, and doubtless because none of these suggestions seemed to be really satisfactory, de Rose and the physicists of the CEA thought it would be better first to test the ground by meeting some well-known European scientists privately. Their main reason was that the opinion of such people would be decisive, in particular in getting the agreement of the various governments. As de Rose put it, 'after canvassing the possibilities, it was decided that Kramers was the most important man to contact first; so Francis Perrin and myself went to see Kramers in Holland [...] We had a long discussion with him. We expressed our view, we said that the French government would be ready to take an initiative but we wanted to know what his reaction would be. I cannot remember whether he immediately voiced reservations or whether he decided that he would have to think it over. Anyway, I know that his final view was negative'. Disappointed by this answer, the two French officials more or less gave up the idea, feeling that there was nothing more that could be done. 'It was a great pity', de Rose concluded, 'but that was that'.[45]

In fact, another initiative took place more or less independently. The April/May report was probably sent to several people, notably to several officials in the French government. A copy of it arrived on the Minister of Finance's desk, who devoted part of his speech at the 102nd session of the OEEC Council on 7 July 1950 to scientific research. Mentioning the increasing role of science in modern production, and drawing attention to its escalating cost for every nation, he suggested that one should consider the advantages that scientific co-operation would have. He proposed two ways of doing this: either by creating a European Centre, or by signing an agreement 'in terms such that each of them [the participating countries] agree to devote resources to a specific domain, and all agree to exchange the results of their work'. Without mentioning the nuclear field, he took over the proposals made by Kowarski.[46]

Notes: pp. 93 ff.

Apparently this speech had no concrete results either, so that all the French initiatives taken at this time were stillborn. This leads us to ask a last question: what were the reasons for the failure? The answers are of several kinds. On the one hand and fundamentally, the project proposed by the French included co-operation in the field of nuclear reactors. We have already noted the difficulties raised by the links between this field of research and military and industrial activities. What is more the controls of the USA over this kind of scientific and technological knowledge were still in force and the hopes for bilateral collaboration with America or Great Britain were still attractive to some European countries like the Netherlands. It is also likely that financial arguments played a role. According to François de Rose, Kramers' answer was inspired partly by his feeling that the European laboratory could have the money only at the expense of existing national laboratories. This would lead to an intolerable situation because these were already starved of resources.[47] It should be added, however, that the presence of influential Communists in the French CEA as well as the fear of having a centre dominated by the French (in the absence of the British) also may have put a man like Kramers off the idea. After all, the links (and particularly the scientific links) of the Netherlands and the Scandinavian countries were stronger with Great Britain than with France.

We see then that the prospect for those favouring widespread European co-operation seemed rather gloomy. Despite several attempts there was little reason to be optimistic. Truman's announcement had apparently changed nothing—at least as far as western Europe was concerned. June 1950, however, was to prove a turning point.

2.5 The fifth General Conference of UNESCO, Florence, June 1950

A new protagonist appeared in the prehistory of CERN in June. At the time the fifth General Conference of UNESCO was being held in Florence (Italy). Nothing dealing with European collaboration in nuclear physics had been put on the agenda (even though, by this time, responsibility for studying the establishment of international scientific laboratories had passed from the United Nations to UNESCO). Nevertheless a resolution of that sort was voted unanimously on 7 June 1950 at the suggestion of Nobel prizewinner Isidor I. Rabi, who was a member of the American delegation to a UNESCO conference for the first time.[48]

2.5.1 RABI'S RESOLUTION

Rabi has briefly described what led him to propose a resolution on 7 June 1950. As a member of his national delegation, he was surprised that its programme did not recommend any really constructive action. Deeming the conference to be an appropriate occasion, he suggested a resolution to the head of his delegation. Its nature was so controversial that it was decided to phone the USA immediately. Washington quickly gave the green light. As Rabi did not know how one put an item on the agenda of the conference, he took advice from his fellow scientists, particularly Auger, Thomson, and von Muralt from Switzerland. He also spent a long time with Amaldi discussing several points with him. Eventually the resolution was written without the help of the other members of the rather hostile American delegation, and officially discussed on 7 June in the presence of (among others) Sir John Maud, chairman of the Executive Board, Huxley (UK), Kruyt (NL), von Muralt (CH), Severi and Colonnetti (I), and Auger (UNESCO). Unanimously agreed on, the text was registered as resolution 2.21.[49]

The text which Rabi submitted appears quite neutral, and it is difficult to detect something original in it. It asks the Director-General 'to assist and encourage the formation and organization of regional research centres and laboratories in order to increase and make more fruitful the international collaboration of scientists in the search for new knowledge in fields where the effort of any one country in the region is insufficient for the task'.[50]

In presenting the resolution Professor Rabi was more explicit about some points. According to the official debate recorded by UNESCO, he first underlined the fact that his initiative 'was primarily intended to help countries which had previously made great contributions to science' and that the 'creation of a centre in Europe where many experts in the various branches of science were available would be a small beginning which might give the impetus to the creation of similar centres in other parts of the world'. Then, making it clear that he was 'acutely aware that in certain fields of research the United States and perhaps the United Kingdom held almost a monopoly' he expressed the idea that 'the initial venture might take place in the field of physics'. In other words Rabi suggested that this first laboratory could be the fruit of a predominantly inter-European collaboration and could focus primarily on research in physics—a field which demanded a lot of money and in which the continent had been forced to give up the leadership to more powerful allies.[51]

Shortly thereafter, Rabi held a press conference. UNESCO published a kind of summary of it on 9 June, but unfortunately it is obscure on the point of interest to us. Concerning the scientific aim of the laboratory it reads:

> Professor Rabi said he hoped the centre in Western Europe would be able to afford a large Cyclotron of the type which exists in the United States in such places as the Massachusetts Institute of Technology, the University of Chicago, Columbia University in New York, and Berkeley, California. He thought it should also have a synchrotron: [sic] material for expensive biology such as exists at Woods Hole, Massachusetts, and elaborate computing laboratories such as the one at Harvard University.[52]

We can now draw a general picture of Rabi's proposal. Scientifically speaking, he apparently thought of a multipurpose laboratory, like Brookhaven National Laboratory, working at the forefront of research in a number of fields. Three were mentioned as prime candidates: high-energy physics, biology, and computing. Two remarks are apposite here. Firstly, Rabi seems never to have mentioned nuclear reactors, which is hardly surprising. He was a prominent member of the American scientific establishment and he knew very well that reactor physics was politically sensitive.[53] Secondly, when speaking of high-energy physics, he proposed to build medium-sized accelerators (cyclotrons according to the press-release, but more probably synchro-cyclotrons), and not primarily a big proton synchrotron comparable in energy to the Berkeley Bevatron (6 GeV) or the Brookhaven Cosmotron (3 GeV) then under construction.

As far as the geographical framework was concerned, Rabi explicitly thought of *western* Europe. On the eve of the Korean war this restriction by a representative of the United States also seems quite natural. And all the more so since Rabi had just participated in the International Science Policy Survey Group set up by the US Department of State which had published a report called *Science and Foreign Relations* in May that year. Written in the context of the cold war, it stressed that it was urgent for the United States to help western Europe rebuild her fundamental research.[54] In this sense, Rabi was only partly right when he said, in 1973, that he had had the idea for his resolution on the boat sailing to Europe. Of course, he took the opportunity of his presence in Florence to draw up and present a resolution, but he did not suddenly discover the value for Europe of co-operation in scientific matters. He was not a newcomer in the field of international science policy and the fact that he expressed in his press conference 'the opinion that some money could come from the United States' and that the centres he proposed were 'one of the best ways of saving western civilization' would confirm — if it were needed — that Rabi's initiative is to be situated in the broader context of American foreign policy.[55]

A last point on the geo-political context. Apparently, the United Kingdom was not to be part of the scheme, as shown by Rabi's remark on the special position of the UK *vis-à-vis* continental Europe, and his explicit reference to 'France, Italy, Holland, Belgium, Switzerland, and West Germany' in his press conference. Anyway, the lead which the British had in nuclear physics, as well as their traditional reluctance for any involvement on the continent, made it most unlikely that they would join in.

2.5.2 FIRST REACTIONS TO RABI'S RESOLUTION

In the weeks and months after Florence some of the participants in the conference referred to Rabi's resolution and to the discussions it provoked. We propose first to present their observations, and then to comment on them.[56]

1 – On 15 July in a memorandum sent to Ivan Matteo Lombardo, Italian Minister of Foreign Trade, Mario Rollier wrote:

During the recent UNESCO conference held at Florence from the 22nd of May to the 18th of June a proposal was approved which demanded of the governments participating in this international organization, the finances necessary for the installation *in continental Europe of a centre for nuclear energy studies* which would have the capability of similar institutions now existing in America and Great Britain, institutions which no country in continental Europe could possibly finance alone. The proposition was accepted in principle and it remains to decide whether the centre will be located in Holland or in Switzerland (our emphasis).

2 – On 21 July, having to sum up Rabi's proposition, Auger spoke in a meeting of 'the creation in Europe of a regional laboratory dealing with *corpuscular physics*' (our emphasis). According to the proceedings (which he did not write himself), he said that the contribution from some European countries to this laboratory could consist 'in production knowledge; for others in costly material such as providing uranium or building an isotope separator; and finally for others (e.g. Denmark) in providing specialized scientific personnel'.

3 – On 23 October Sir George Thomson, presenting his report to the British Committee for Co-operation with UNESCO in the Natural Sciences said: 'the conference had given favourable consideration to a proposal from the US, intended primarily for Western Europe, that a regional laboratory, *possibly for nuclear physics,* should be set up by groups of states to enable scientists of those countries to enjoy the use of expensive apparatus which each would be unable to purchase independently' (our emphasis).

In addition to these three 'reports' on Rabi's proposal, the implications of the resolution were discused on (at least) two further occasions—the meeting of IUPAP's Executive Committee held in Cambridge, Mass. on 7 and 8 September, when it was referred to by Edoardo Amaldi, and the Harwell International Nuclear Physics Conference held in the Clarendon Laboratory in Oxford (UK) from 7 to 13 September, where Auger was invited to explain what he intended to do as Director of the Exact and Natural Sciences Department at UNESCO about the Florence vote. In both cases, it has been impossible to find any proceedings of the debates, though some documents do hint at them and at the resolution passed in Florence[57]:

4 – During the IUPAP Executive Committee Amaldi proposed that a discussion be held about what could be done to implement the proposition of 7 June for 'the building of a *European Laboratory in Nuclear Physics*', to use the terms of the letter he sent to Auger on 3 October. In this letter Amaldi commented: 'After some discussion, the Executive Committee decided to have two reports prepared on this issue: one to be written by Rabi *who should make his thoughts as clear as possible,* the other one by me who has been given the responsibility of getting in touch with European physicists so as to get, if possible, an agreement on several fundamental points' (our emphasis).[58]

5 – During a meeting in Geneva on 12 December 1950, Auger mentioned his visit in Oxford. In the official proceedings one reads: 'M. Auger [pointed] out that at a recent meeting of physicists in Oxford, two tendencies arose:

- not to limit too severely the field of research at the beginning,
- as proposed by Niels Bohr, to begin by building a big apparatus to accelerate particles (a thousand million volts) and to gather people around it'.[59]

6 – Reporting to the Consiglio Nazionale delle Ricerche (Italy) on 18 December 1950 about the same meeting Bruno Ferretti wrote: 'Regarding the project [advocated in Florence] which received the support of important American scientific personalities, different tendencies emerged during international congresses. The first, supported by the French, would like the laboratory to contain, not only big equipment for nuclear physics, accelerators, etc, but also atomic piles. The second, supported by Bohr, envisaged the construction of big accelerating machines but without the installation of atomic piles, the construction of which implies a complex set of economico-political questions which escape the grasp and the competence of scientists'.[60]

One thing is striking when we read these texts: that what people retained of Rabi's resolution and of the debates it aroused was rather different, not to say contradictory. How could this have happened? And is it indicative of a particular problem? Three comments help answer these questions.

To begin with, Rabi's resolution was vague, as we have said. It is certain that this often happened at UNESCO where the drafting of a text which was to be put to the vote required a great deal of caution. This can account for the variety of reactions, for the fact that the participants had no reason to go home with the same idea of the laborator(ies) Rabi had suggested be built.

In addition, all our information indicates that there were a number of quite separate conversations in Florence. Rabi spent two days with Amaldi, but he met Thomson and Auger on other occasions. The discussions were therefore not necessarily centred on the same aspects of the project. After all, the participants had other things on their minds, not only concerning the resolution itself (were eastern European countries to be excluded? how could the text submitted to the vote be integrated into the general programme of UNESCO? and so on) but also concerning the other subjects discussed during the 26 days of the meeting.[61]

Finally, many people interested in the idea of a European laboratory were not in Florence—people like Dautry, de Rose, and Kowarski, just to mention a few Frenchmen. And though they welcomed the resolution, this did not prevent them continuing to argue the case for *their conception* of the project. This happened, for example, in the case of Kowarski who went on explaining his ideas at a meeting in July with Belgian physicists and, according to Ferretti, also at the Oxford Conference in September.[62]

In fact then, after Florence, if some boundaries had been laid down, much had still to be settled. Rabi had made proposals but nobody was—nor felt—bound by them. He had perhaps initiated a new process in inscribing his resolution into the official programme of an intergovernmental organization, but *as far as the scientific aim(s) of the laborator(ies) were concerned,* he had merely added an opinion to an already long European list—and no final decision had yet been taken.

2.5.3 A STANDARD MISCONCEPTION ABOUT RABI'S RESOLUTION

We have stressed the reactions to Rabi's text for a particular reason. A 'standard version' of the June events exists, a kind of ritualized account of origins which contradicts what we have established and which has played an important role in the mythology surrounding the birth of the organization.

The first trace of this 'standard version'—and probably a key moment in its genesis—is found as early as December 1950. In the meeting held in Geneva on the 12th, Auger declared that 'During this conference in Florence, Professor Rabi had specified *orally* that he envisaged [...] the creation of a nuclear physics laboratory *for studying particles of high energy*' (our emphasis).[63] In the following years, this version of the June events was regularly repeated. In 1961 (to give just one of numerous examples) Kowarski wrote in an article entitled 'An account of the origin and beginnings of CERN', '[...] meson (or high-energy) physics was the most obvious and attractive candidate [for the new laboratory], clearly recognized as such by June 1950, and discussed as such between Amaldi and Rabi on the eve of the latter's official statement at the General Conference of UNESCO held in Florence'. Six years later he made the same point: 'In the peculiarly abstract language of international proceedings, any mention of what field of research, or even what region, were expected to be given prior attention, was again scrupulously excluded; but neither Amaldi (with whom Rabi had a long discussion on the eve of his declaration) nor Auger—at that time Director of Natural Science at UNESCO—needed to be given any further hints. A major doubt had been removed, and an official framework was found for the first concrete steps to be taken; *accelerator-based fundamental physics was the obvious first target*' (our emphasis).[64]

This 'standard version' of the June events immediately provokes a number of related questions. Whence the unanimity in attributing to Rabi the proposition to build a laboratory dedicated *only* to high-energy physics? Why a consensus that he was the 'true father' of CERN?[65] And why did it emerge so quickly—only six months after the Florence conference and at a time when most physicists and all governments still had to be convinced, and all the money still had to be found?

In reply we would like to suggest two lines of thought. Firstly, it is well known that the human memory tends in general to reorganize past events, to make them more coherent, more rational, more simple. In other words, it tends to reread the past in the light of the present. Now, as we shall see, the project to have a laboratory centred around a giant accelerator was clearly formulated as early as December 1950. Hence the possibility that Auger connected this decision with Rabi's proposal toward the end of the year.

Notes: pp. 94 ff.

In this case, however, this kind of explanation has a particular weakness: why did this 'reinterpretation' take place so *quickly*? Our second argument removes this difficulty. Before presenting it we need to make a detour. Reading the proceedings of the Geneva meeting quoted above, we were struck by the way in which Bohr's name was used to sustain the idea of a laboratory centred around a big accelerator. We know (from what occurred in 1951 and 1952) that Bohr was *not* enthusiastic about the construction of such enormous equipment in Europe. So, even if he had explained to someone in September 1950 that reactors should be avoided, even if he had said, or agreed, that the building of a big accelerator would be a possibility, he was probably not an ardent champion of the idea. In fact, if his name was mentioned by Auger and Ferretti, it is because he was Bohr, a prestigious figure in scientific circles. In other words, the use of his name served as a kind of rhetorical appeal to authority: if Bohr was in favour, the project should be implemented.

Our idea about Rabi is that the same kind of thing happened with him. Because he was a leading scientist of course, but also—and this was more important in this context—because he was an *American,* because his proposal could be seen as saying that the United States, or at least some of her senior and politically influential scientists, were no longer against European co-operation in nuclear matters. Let us remember Kowarski's fears at the end of 1949 about a possible American veto, or the distrust of the physicists of the small countries of Europe. *After Florence there was less reason to feel this any longer.* And according to all the protagonists, the Rabi proposal was immediately understood in this way. Hence the importance of attributing the paternity of CERN to Isidor I. Rabi.

To end this point, we would like to add that if America's blessing—through Rabi—was essential for the Europeans, other considerations motivated more by self-interest also played a role in the ready acceptance of Rabi's resolution and Rabi's paternity. Many people knew that the execution of an ambitious project would require technical knowledge and skills exceeding those then available on the continent. Accordingly they hoped that the proposal meant that help could be expected from the United States. And it is true that, at least as far as money was concerned, Rabi told some people during the conference 'that special US funds might be available for this purpose' and that he indeed tried to raise these. Later he and his American colleagues would also support the Europeans scientifically and technically. One can see then all the implicit connotations of resolution 2.21 and one can grasp why it was so frequently and so quickly referred to as being the true beginning of CERN's history.[66]

2.6 The situation in June 1950

It is now possible to draw some conclusions about the initiatives which began in November 1949.

The first point to remember is that a model seems to underlie all proposals, namely to build a type of 'super-Brookhaven' for Europe, to quote Kowarski, namely a centre for *nuclear science* built *in common* around *heavy equipment.*

The aspects 'in common' and 'around heavy equipment' are easy to understand. The interest of such a centre lay as much in the sharing of the necessary financial means as in the pooling of scientific and technical knowledge and skills. This implied the construction of *expensive* and *sophisticated* machines which no single European state could afford. The fear then often expressed was that, if such equipment was not quickly set up in Europe, the brain-drain—already particularly severe in some countries—would get worse and spread throughout the continent.

As for the apparatus, many favoured building a reactor. To use two anachronistic abbreviations, we would say that people thought more often in terms of a mix of EURATOM and CERN than in terms of CERN itself. This conception is not surprising when we bear in mind the state of 'nuclear' science, the existing institutions in this domain, and the situation of Europe at the time. But thinking along these lines led to a trap: to speak of reactors was to allude directly to military potentiality, to speak of reactors was to evoke important economic interests, interests which would undermine the possibility of a collaborative pure research facility.

The second point we would like to stress is that these several independent initiatives had more or less the same objective. While European nations had made several attempts at *bilateral* agreements with the USA or the UK in the previous years (1946-1949), we find four or five proposals for *multilateral* collaboration within Europe in only *eight months* around the end of 1949 / beginning of 1950. These initiatives were relatively autonomous of one another, and it is pointless to look for THE father of THE project. Better to visualize this period as comprising a *number of parallel events* only loosely associated with each other, than as *a chain of successive events,* each consciously connected to its predecessor. This is also why the reasons underlying the various proposals differed, scientific considerations being more important in some cases, political or economic ones in others. However a conjunctural element probably catalyzed the sudden flurry of proposals—the explosion of the Soviet bomb. It shook the western world and showed the poverty of the experts' predictions. Suddenly some 'secrets' did not seem as important, and it was recognized that the role of Europe alongside the United States in the East-West conflict had to be increased. In particular it became essential that she be associated with the best developments in science.

Our third and last point. We have said that Rabi's resolution was important because it was submitted by the American delegation. Behind that, one could—and in fact did—read the Americans' blessing and possible help in the future. However, there is another reason why this resolution was so crucial: during Rabi's speech, and during the debate which followed, Pierre Auger was explicitly mentioned as being the one to implement it. And as von Muralt said during the debate at Florence, the real interest of resolution 2.21 was that it 'was coupled with a man's name'. Without that, he added, 'probably nothing would be done'.[67] Doubtless he was right. After all, many resolutions were voted in the United Nations or in UNESCO without any tangible consequence. Here, unusually, 'marching orders' were given to someone who welcomed them and who would try to avoid the project being bogged down in UNESCO's bureaucracy. If then, as subsequent events revealed, the Florence resolution was important, it was because a group of people felt themselves responsible for it and because one of them was *officially* in charge of its implementation, had the administrative means to do so (as Director of the Exact and Natural Sciences Department of UNESCO), had the opportunity to act directly at the *governmental level*. In short, the scheme now had its 'product champion' — something which had never occurred before.

Notes

1. The archival material used in parts II and III was collected by a rather systematic exploration of *a priori* relevant collections (see the list given at the outset of this volume). We nevertheless did this only for France, Germany, Italy, United Kingdom and international organizations. On Sweden, see Nyberg & Zetterberg (1977).

   In this chapter, we particularly used:
   - a collection of documents issued by the United Nations Economic and Social Council (referred to below as UN ECOSOC). A copy is available at the United Nations library, Geneva.
   - a collection of documents issued by the United Nations Educational, Scientific and Cultural Organization (referred to below as UNESCO), Paris.
   - the archives of the Centre Européen de la Culture (referred to below as CEC) in Geneva. Denis de Rougemont allowed us to consult them. We would like to thank him here.
   - the personal archives of Raoul Dautry (referred to below as AF-AP307) in the Archives de France, Paris.
   - the CERN archives in Geneva, and particularly the papers of Lew Kowarski kept in the Kowarski interview files.
   - different collections of personal papers collected by our team in 1983, notably personal documents of Edoardo Amaldi and Pierre Auger. They are now in the CERN archives and are referred to as CERN History Study (CHS), Amaldi (or Auger) file, CERN.

   Several interviews were made and copies of the transcripts are available in the CERN archives:
   1. Edoardo Amaldi by M. Gowing and L. Kowarski, 25 July 1973.
   2. Pierre Auger by A. Günther and A. Hermann, 10 December 1982.
   3. Isidor I. Rabi by L. Kowarski, 6 November 1973.
   4. Isidor I. Rabi by E. Amaldi, 2 March 1983.
   5. François de Rose by M. Gowing and L. Kowarski, 4 January 1974.
   6. Denis de Rougemont by M. Gowing, L. Kowarski, and M. Senft, 16 January 1975.

   The author also had several conversations with Auger in 1983, and one with de Rose (October 1983). A. Hermann, J. Krige, and D. Pestre met de Rougemont in 1983.

   Some articles have already been written about the early CERN. The more interesting ones are: Amaldi (1977), CEC (1975), Kowarski (1955), Kowarski (1961), Kowarski (1977a). Jungk (1968) is a more journalistic story, though still worthwhile. Herzog (1977) is a thesis; its first part is an account of the

beginnings of CERN. For more detail, see the bibliography for parts II and III (at the end of part III) and chapter 14, note 3.
2. The main documents used here are:
   - ECOSOC, Draft Resolution submitted to the Economic and Social Council by the French Delegation on the Establishment of United Nations Research Laboratories, doc E/147, 2 October 1946.
   - ECOSOC, Report of the Secretary-General on Establishing United Nations Research Laboratories, doc. E/620, 23 January 1948, 356 p.
   - ECOSOC, Committee of Scientific Experts on International Research Laboratories, Summary table of proposals regarding the Establishment of United Nations Research Laboratories (Prepared by the United Nations Secretariat), doc E/CONF.9/PC/ 3, 4 August 1949.
   - ECOSOC, Meetings of the Committee of Scientific Experts convened by the UN Secretary-General held at UNESCO House, Paris, l6–24 August 1949, doc E/CONF.9/PC/SRl-SR11.

   For more general remarks on UN Agencies, see Noyes (1946), Noyes (1947), Noyes (1950), Ging-Hsi (1950), Malina (1950).
3. ECOSOC, E/147, 4.
4. For details see for example Auger (1973), Huxley (1970), Laves & Thomson (1957), 294-296, Needham (1976). For a more detailed list of people, see for example ECOSOC,E/CONF.9/PC/SRl-SR11.
5. ECOSOC, E/147, 1 and 2.
6. This rationalistic spirit (in the sense of the spirit which prevailed in the French *Cahiers Rationalistes*) is well illustrated by Auger (1956). See also my study of the scientism and rationalism of the French scientists in Pestre (1984a), in particular chapter 6, 'Les convictions culturelles et épistémologiques des physiciens français, étude d'un socle archéologique', 171-207.
7. ECOSOC, E/147, part IV 'Creation of an International Spirit', 3. In this connection, we could also remind the reader of the importance of the scientists in the after-war movement for a 'World Government'. For an ardent plea, see for example Urey (1949). For a devastating criticism, see Niebuhr (1949). See also Gilpin (1962).
8. ECOSOC, E/CONF.9/PC/SR5, 6.
9. Kowarski, Draft version of an article: 'The making of CERN: memories and conclusions' dated March 1967, unpublished, 17pp., here 2-3 (CERN-Kowarski interview file). For more general information on the UN AEC, see the *Bulletin of the Atomic Scientists,* from volume 1 (1946) to volume 5 (1949); in particular **2**(3-4), 1946, 3-12; **3**(1), 1947, 10-15 and **4**(7), 1948, 205-210. For a selection of articles from the *Bulletin of the Atomic Scientists,* see Grodzins & Rabinowitch (1963).
10. Kowarski (1955), 355; de Rose, interview, 12-13. In 1958 de Rose had already mentioned what he had done in New York in a *Note pour le Ministre,* 25/6/58 (AEF-75). Having recalled the ideas of Oppenheimer about the situation inherited from the war, he added: 'During my stay in the United States (1946 -1950) he [Oppenheimer] often spoke to me about it, and I took the initiative to organize several meetings between him and men like Pierre Auger and Francis Perrin.'
11. Kowarski (1955), 355.
12. For the institutions, see for example Ramsey (1966), Harwell (1952), CEA (1958), Heilbron et al. (1981). About the scientific conferences, see Amaldi (1977), 326-35, interview of Hans Bethe by Ch. Weiner and J. Mehra, 27 October 1966; of W.K.H. Panofsky by Ch. Weiner, 6 March 1974; of R.E. Peierls by Ch. Weiner, 11-13 August 1969; of Victor Weisskopf by Ch. Weiner and G. Lubkin, 22 September and 5 December 1966 (AIP-interviews). Finally, see chapter 1 of this volume.
13. Kowarski, *op. cit.* note 9, 3; Amaldi (1977), 336.
14. See for example Teitgen (1963), Lecerf (1965), Grosser (1972). For a *short* presentation, see the 25 introductory pages of Polach (1964). For a more precise chronology, see appendix 2 of this volume.
15. Colonnetti (no date), Dautry (1952), Willems (no date).
16. CEC (1975), 7-8; interview of Denis de Rougemont, 1-3. The quotation is from de Rougemont's speech at the 25th anniversary of the CEC (CEC (1975), 7).

17. The European Cultural Conference is studied in detail in section 2.3 below. For the following two paragraphs, see letter Auger to Dautry, 21/10/49; Kowarski, *Note sur la coopération atomique européenne,* November 1949; Note Guéron to Dautry, 6/12/49 (AF–AP307/212). These three documents were replies to a letter from Dautry.
18. Letter Auger to Dautry, 21/10/49 (AF–AP307/212).
19. Note Guéron to Dautry, 6/12/49 (AF–AP307/212).
20. A shorter version is in Dautry's papers (AF–AP307/212). A longer one (8 pages instead of 4) is in Kowarski's papers (CERN-Kowarski interview file). Both have the same title and the same date.
21. About the Amsterdam initiative, see *Résumé du compte-rendu de la réunion tenue à Amsterdam le 11 novembre 1949 par la FOM* (CEC), and Nyberg & Zetterberg (1977). About Italy, see Amaldi, *Notes for CERN History,* received by CERN archives on 29/1/81 (CHS, Amaldi file, CERN).
22. Quotations from Kowarski, *Note sur la coopération atomique européenne,* November 1949 (CERN, Kowarski interview file). About the relations between continental Europe and Great Britain or the USA, see Gowing (1974), particularly vol.I, 329–348. She describes the attempts at bilateral co-operation proposed by the French, Norwegians, Dutch, and Belgians to the Americans and the British between 1946 and 1950, and explains why they failed.
23. Quotations from *Résumé du compte-rendu..., op. cit.* note 21. In his own text, Kowarski suggested that the laboratory 'genre super-Brookhaven ou super-Saclay', have a 10,000 kW pile.
24. About the shock created by Truman's announcement, see the whole issue of the *Bulletin of Atomic Scientists* of October 1949 (**5**(10), 1949). About the hopes raised by the Truman speech in Europe, let us merely quote from the report sent by the correspondent of *Le Monde* to Paris on 23 September (the day of the Truman declaration). Under the title: 'First reactions in Washington, Release of "atomic" secrets to Canada and Great Britain', he wrote: 'A first consequence seems certain: the disappearance of all parliamentary opposition to the release of atomic secrets to Great Britain and Canada. The democratic leader of the senate, Mr. Lucas, declared that from now on it would be "a purely academic question" and that co-operation would be established in this domain.' Accordingly there was some hope that 'secrets' would be passed to the countries of western Europe.
25. Nyberg & Zetterberg (1977), 11.
26. The main documents on this conference are: *General Report* submitted by the CEC, 19 pages; *Présentation du Rapport Général* by Denis de Rougemont, 8/12/49, 8 pages; *Message de Monsieur de Broglie,* read by Dautry, 9/12/49, 3 pages; *Discours de Monsieur Dautry,* 10(?)/12/49, 3 pages; *Liste des membres de la Conférence,* 9/12/49, 11 pages; *Résolutions proposées par la Commission des Institutions,* 11/12/49, 8 pages. These documents are in the CEC archives. To them must be added: *Mouvement Européen, Conférence Européenne de la Culture, Lausanne, 8–12 décembre 1949, Résolutions et déclaration finale,* edited by Bureau d'Etudes pour un Centre Européen de la Culture, 1950(?)(CEC); CEC (1975); interview of Denis de Rougemont.
27. See *Relazione del Prof. Mario Rollier al Congresso Culturale per l'Unità Europea, La Commissione di Co-operazione scientifica del Centro Europeo della Cultura et il Laboratorio Europeo di Fisica Nucleare,* Milano, 18–19 marzo 1951, 5 pages (MR; copy in CHS, Rollier file, CERN).
28. A list of the people who attended the sub-committee for natural sciences is given in CEC (1975), 12–13, but apparently some mistakes were made on that occasion. Mario Rollier and 'le physicien Ferretti, représentant du Professeur Amaldi' are said to have been there. But neither Rollier nor Bruno Ferretti are on the lists we have and they never claimed to have been in Lausanne. A *Giovanni* Ferretti is on the list but anyway he was not sent by Amaldi, who was not involved in the Lausanne conference. We have the impression that this meeting was being confused with that called by the CEC to deal with the same topics in Geneva in December *1950.* In 1950 both Rollier and Bruno Ferretti took part, the latter representing Amaldi.
29. *General Report..., op. cit.* note 26, quotations from pp.1, 8–9, 11, 12, 18; last quotation from CEC (1975), 12.

30. *Message de Monsieur de Broglie*, 1–2.
31. *Discours de Monsieur Dautry*, 1–2.
32. CEC (1975), 13.
33. *Mouvement Européen, Conférence...*, op. cit. note 26, 5–6.
34. *Discours de Monsieur Dautry*, 2.
35. Letter de Rougemont to Dautry, 10/3/50; personal diary of de Rougemont, March 1950 (CEC).
36. The quotations are from letter de Rougemont to Colonnetti, 3/4/50 (CEC).
37. Personal diary of de Rougemont, 14/3/50, in which Dautry's words are reported; letter de Rougemont to Willems, 18/4/50; letter de Rougemont to Dautry, 18/4/50 (CEC).
38. Letter Willems to de Rougemont, 25/4/50; letter Colonnetti to de Rougemont, 2/5/50 (CEC).
39. Letter de Rougemont to Dautry, 18/4/50; letter de Rougemont to Scherrer, 22/9/50 (CEC). Letter Dautry to Auger, 14/10/50 (CHS, Auger file, CERN). *Rapport au Secrétaire général sur la mission à Genève du 6 au 8 octobre 1950,* by M.P. Levy, 9/10/50, 8 pages (CE-19229, vol. I).
40. See letter George to de Rougemont, 23/4/50; letter Colonnetti to de Rougemont, 2/5/50; *Note sur les objectifs du groupe de travail Recherches Scientifiques,* by George, 5/5/50; letter de Rougemont to Colonnetti, 16/5/50 (CEC). It is interesting to notice that Colonnetti used the 5th General Conference of UNESCO in June to make a similar proposal. According to the proceedings, he presented a resolution (5C/35) 'to establish international relations between research centres in different parts of Europe which had hitherto managed to make bilateral arrangements for the exchange of persons and documentation.' The proposal was withdrawn without a vote being taken. See *Records of the General Conference of UNESCO, fifth session, Florence 1950; Summary Records of the Meeting of the Programme and Budget Commission,* 16th Meeting, 7/6/1950, p.m., section 2.2, 365.
41. According to several letters (de Rougemont to Scherrer, 22/9/50; to Dautry, George, Willems, 23/9/50...), it was planned to gather the working group on 'Scientific Researches' in the afternoon of 8 October. We have no evidence that this group met, either from the CEC archives, or in the report written by M.P. Levy for the Council of Europe. We only know that the problem was raised during the plenary session of the 'Conseil Supérieur du Centre Européen de la Culture' which met on 7 October. The quotation is from the proceedings of this session (CEC).
42. The letter from Kowarski to Auger accompanied a report written by Kowarski dated April/May 1950. The expression 'adopté par nous quatre' is not clear. We have the impression that it implies *four scientists in the CEA* including Francis Perrin (High Commissioner) and Lew Kowarski. In this case the others would be Goldschmidt (chemistry) and Guéron (physical chemistry) who were the two other Directors of departments with Kowarski (piles and accelerators). If this assumption is right six people were involved and de Rose played an important role in asking that this document be written.
43. In CERN, Kowarski interview file. All the quotations in the following pages are from the April/May report.
44. It would be interesting to ask why the French (but also the Italians and the Belgians through the CEC) were so eager to collaborate in nuclear science. Two hypotheses can be proposed. On the one hand, it is noteworthy that throughout the French elites at that time there was the feeling we have already met and which resulted from the war: the feeling that the future of the world required mutual understanding, that the future of Europe required unity. However, in the particular field of scientific co-operation, another kind of reason must be stressed: Belgian, French, and Italian science was rather behind that of most of the northern European countries (and of Great Britain of course). The war did enormous damage, laboratories were still partly destroyed or badly equipped, the budgets for science were meagre, and industry found it difficult to help build modern apparatus. This was particularly so as regards technology linked with accelerators. See for the first hypothesis, Grosser (1972), for the second, Châtelet (1960), Colloque Caen (1957), Amaldi (1979). About the accelerator programmes, see Rapport CEA (1951), 24–26; letter from Amaldi to Mussard, 30/6/51; memo 240168 from Regenstreif to Auger, *Rapport sur mon voyage en Belgique (4–7 juillet 1951)* (UNESCO). For more documented studies see sections 1.3 and 9.2.

45. Quotation from the April Kowarski report, 3; interview with F. de Rose, 13-14.
46. *Extrait du discours prononcé par M. Petsche, ministre des finances et de l'économie nationale au Conseil de l'OECE le 7 juillet,* copy sent by de Rose to Auger in a letter dated 8/7/50 (Mussard file, CERN).
47. Interview with F. de Rose, 14.
48. *Records of the General Conference of UNESCO, fifth session, Florence 1950; Resolutions,* section B, resolution 2.21, 38; *Summary Records of the Meeting of the Programme and Budget Commission,* 16th Meeting, 7/6/1950, p.m., section 2.2, 364-65; *American scientist expresses views on UNESCO-sponsored science research centre,* UNESCO Press Release 311, 9/6/50 (AF-AP307/212).
49. This is written from the interviews with Amaldi (3-7), Auger (11-15), Rabi (interview by Kowarski, 1-6), Rabi (interview by Amaldi, 1-3), and the letter Rabi sent to Amaldi after this interview, 30/5/83 (CERN interview files).
50. *Records of..., op. cit.* note 48, 38.
51. *Summary Records..., op. cit.* note 48, 364-65.
52. *American scientist..., op. cit.* note 48.
53. See for example Bernstein (1975).
54. International Science Policy Survey Group, Report, *Science and Foreign Relations,* [US] Department of State, 3860, May 1950. See also Berkner report (1951), Drechsler (1951). For more general feelings on science and politics at the time, see Conant (1950). For one historical survey see Ronayne (1984), chapter 1.
55. Quotations from *American scientist..., op. cit.* note 48. For a more detailed analysis of this aspect, see section 14.8.
56. The following documents refer to this meeting: Mario Rollier, *Promemoria sull'urgenza di predisporre una legislazione italiana sull'energia nucleare,* sent to On. Ivan Matteo Lombardo, Ministro del Commercio Estero, Roma, 15/7/50 (MR, copy in CHS, Rollier file, CERN); *Procès-Verbal de la séance du 21 juillet 1950 au CEA à Paris,* point I. Examen d'une collaboration sur l'initiative de l'UNESCO; point II, *Point de vue officieux français sur la création d'un organisme coopératif européen* (CHS, Auger file, CERN); Sir George Thomson, *Report on the 5th General Conference of UNESCO,* 23/10/50, in Minutes of the 7th meeting of the British Committee for Co-operation with UNESCO in the Natural Sciences, doc. NS (50)21, 3/1/51 (PRO-ED157/301).
57. In spite of several attempts we could not find the proceedings of these two meetings. For the Harwell conference, we have a document called 'International Nuclear Physics Conference, September 1950, Harwell, Atomic Energy committee, Proc. of Harwell Nucl. Phys. Conference, edited by E.W. Titterton, reference AERE-/M/68' but the talk of Pierre Auger is not included. Nor is it referred to in an article subsequently written by Titterton giving a short account of the conference. See E.W. Titterton, 'The Harwell Nuclear Physics Conference, 1950', *Nature,* **4226**, 28/10/50, 709-711.
58. Letter Amaldi to Auger, 3/10/50, reproduced in Amaldi (1977), 349-351.
59. Centre Européen de la Culture, *Compte rendu analytique de la réunion du 12 décembre 1950,* 18/12/50, signed J. P. de Dadelsen (CEC); quotation p.3.
60. Consiglio Nazionale delle Ricerche, seduta del 18/12/50, 154-159 (CNR); quotation p.155.
61. The most surprising account is that of Mario Rollier who only mentioned 'nuclear energy studies'. Why should that be? We can put forward two suggestions. On the one hand, we do not know exactly to what extent Rollier was involved in the private conversations held in Florence. Neither Amaldi, nor Auger, nor Rabi, mentioned having had a talk with him and he did not speak during the official debate, if the proceedings are to be believed. So his report might have been his own interpretation of a text in the writing of which he played no role. On the other hand, Mario Rollier was at that time dedicated to the development of nuclear energy in Italy. This is why he wrote the letter to his minister Matteo Lombardo. And he may have slightly distorted what happened in Florence so as to urge his government to set up an official body responsible for domestic nuclear energy in Italy.
62. What Ferretti called 'The French'.

63. *Compte-rendu analytique..., op. cit.* note 59, 3.
64. Kowarski (1961), 3; Kowarski, *op. cit.* note 9, 4.
65. When the first agreement at governmental level was signed on 15 February 1952, the agreement which established the provisional CERN, the European delegates sent Rabi the following telegram (15/12/52, UNESCO):

    'WE HAVE JUST SIGNED THE AGREEMENT WHICH CONSTITUTES THE OFFICIAL BIRTH OF THE PROJECT YOU FATHERED AT FLORENCE. MOTHER AND CHILD ARE DOING WELL, AND THE DOCTORS SEND YOU THEIR GREETINGS'.

    A full list of the signatories is given in note 41, chapter 14.
66. About money see Thomson's report referred to in note 56 and the UNESCO Press Release referred to in note 48. In the latter one reads:

    'He [Rabi] expressed the opinion that some of the money could come from the United States, and the point was made by the United Kingdom delegate to the Committee, in the course of the discussion of the proposal, that steps might be taken to get financial help from the United States Economic Co-operation Administration.' Recently, Rabi added that he went to Paris some time after June 1950 to try to get money through the Marshall Plan (interview, p.8). See also Krige (1983).
67. *Summary Records..., op. cit.* note 48, 366.

CHAPTER 3

The fusion of the initiatives[1]
June–December 1950

Dominique PESTRE

Contents

3.1 The promotion of the Florence project, June–October 1950   98
3.2 The meeting between Auger and Dautry and the decision to call a gathering of scientists in Geneva in December   101
    3.2.1 Why a meeting with the CEC?   102
    3.2.2 The list of participants   104
    3.2.3 The agenda for the meeting   107
3.3 The Geneva meeting of 12 December at the Centre Européen de la Culture   109
    3.3.1 The participants   109
    3.3.2 The meeting   109
    3.3.3 The text of the resolution   112
3.4 Immediate reactions to the 12 December meeting   112
    3.4.1 Reactions of the participants and of their countries   112
    3.4.2 The British reaction   114
    3.4.3 The press   114
3.5 Conclusion: Ideas and motivations behind Amaldi's, Auger's, and Dautry's proposals   116
    Notes   118

In the previous chapter we saw how different steps towards the creation of European centres for nuclear physics were taken between the end of 1949 and the middle of 1950. These moves reflected both a number of cultural or scientific interests, sometimes shared sometimes not, and their encounter with realities of a geographico-political (Europe, France, Italy...) or economic (atomic energy) nature. The proposal made by Isidor I. Rabi in Florence in June 1950, however, deserved special mention. It was generally recognized at the time as being a key development because it was proposed by an American, because it was inscribed in the working programme of an official international organization, because it was taken charge of by Auger and because it sparked off a new chain of scientific thought, with several participants abandoning the idea of a nuclear reactor.[2]

Before the Florence meeting the evolution of the several schemes had been held back for lack of specific scientific aims, and by the politico-nuclear context bequeathed by the war. Moreover, the groups behind the different projects worked quite separately. By contrast the next six months were to see the beginning of a different phase marked by: 1) the meeting of the different networks of people; 2) the first precise definition of a scientific project—a laboratory for fundamental research built around a giant accelerator; 3) the first investment decisions, the first collection of money from the interested countries; and finally, 4) the first characterization of the procedure to be followed for establishing official contact with the various states, and for involving them in the project.

3.1 The promotion of the Florence project, June–October 1950[3]

Two men in particular took on the responsibility of following up the resolution passed on 7 June 1950, namely Amaldi and Auger. The first, Edoardo Amaldi, left Europe for New York on 30 June. He arrived in America on 7 July, where he met friends and colleagues including Bernardini, Rossi, Pontecorvo, Fano, Goudsmit, Weisskopf, Placzek... At the end of August he went to Long Island to visit Brookhaven where he saw the progress on installing the Cosmotron. His diary records how impressed he was by the laboratory: 'Passo la giornata a visitare Brookhaven. Colossale!' The following week, while working with the IUPAP Executive Committee, he suggested that the organization consider what it could do to help put the Rabi resolution into practice. After discussion it was decided to

> commission two reports on the subject: one by Rabi, who should describe his idea in as much detail as possible, and the other by me [Amaldi], who was instructed to contact European physicists, in order to reach agreement on several basic issues, if possible.[4]

Back in Italy, Amaldi saw Ferretti who told him about Auger's speech at Oxford some weeks previously and the discussion which had followed. Sympathizing with Auger's point of view, Amaldi wrote to him on 3 October asking Auger for further details in order, he said, 'that the actions of UNESCO and IUPAP should complement each other effectively'. His own idea was to write to various European physicists to ask their opinions on the location, management, and funding of a possible European laboratory. However he added that he would like to know Auger's and Rabi's opinions before writing so as to be able to put forward a specific programme from the beginning. Otherwise, he said, 'the sceptics will kill the thing before it's born'.[5]

A few weeks later, not having received a reply from Auger, Amaldi wrote to his friend Cacciapuoti (then at UNESCO in Paris) asking him to see Auger, and to let him know what transpired during their conversation. Cacciapuoti's reply arrived at the beginning of November: the project was taking shape, Auger was working on it, but progress was slow. Amaldi's response to this showed signs of irritation. Of course he understood the problems in Paris, but 'Auger could still have replied to my letter, especially as it was not just a private letter. Its main aim was to try and start some form of collaboration'. We shall have to explain why Auger did not reply personally to Amaldi's letter, but it is too early to do so here. Let us simply take note of the Italian physicist's firm determination to see the Florence resolution put into practice.[6]

Let us now turn to Auger. It is difficult to trace all his movements, but as a director at UNESCO there is no doubt that he travelled widely and took advantage of this to talk about the Florence resolution. We know of several occasions on which he did so. Firstly, the visit of Belgian nuclear physicists to Paris in July provided an opportunity to mention Rabi's project. Auger noted subsequently that the four Belgian delegates were in favour of a European regional laboratory 'but being impatient to collaborate with the French, [they proposed] a bilateral convention initially, formulated in such a way that it could [be merged] into the European laboratory as soon as it [was] founded'. At the beginning of September Auger was in England. According to the handwritten note we have just quoted he met Cockcroft and Tizard, of whom he wrote that both agreed on the idea of a regional European laboratory. Cockcroft 'intimated that British scientists [would] be very much in favour but that the Government [would] have reservations about the contributions in money and in personnel which could be envisaged'. For his part Sir Henry Tizard said that he intended 'to speak about it to the committee of the Royal Society of which he [was] president'. Auger also gave a talk on the subject at the international nuclear physics conference held in Oxford from 7 to 13 September, though as Kowarski commented some years later 'after Auger's address, the expressed comments tended to explain why the proposed institute could not possibly work, rather than how it could'.[7]

This did not discourage Auger unduly. He met other people, perhaps Swedish or

Notes: pp. 118 ff.

Yugoslav representatives at UNESCO.[8] He undoubtedly had discussions with the French. In his handwritten note he said: 'CNRS: director Dupouy has been informed and is in favour'; 'Dautry, very much in favour'; 'the CEA: Perrin, Kow[arski], Gué[ron], Gold[schmidt], very much in favour whilst making no secret of the political and industrial problems'. On 18 October he met Willems. He noted: the Belgians 'are beginning [a] machine, but will join Dutch centre; will definitely agree to join European lab. if it is [for] the construction of an apparatus (synchrotron)'. Lastly, towards the end of October he met a British scientist who suggested that Great Britain might put one of the machines she was building at the disposal of the European laboratory. Evidence of this conversation—apart from Pierre Auger's own recollections—is to be found in the minutes of the meeting held in Geneva in December 1950, which record that 'M. Auger mentioned that Britain might be prepared to hand over either the Liverpool or the Birmingham cyclotron to the Laboratory [...]'.[9]

As we see then, Auger tried to find out what other people thought of the situation. However it was impossible to remain on such an informal level. Towards the beginning of October he therefore jotted down a few lines taking stock of what he had achieved so far, and setting out his thoughts about the future. He envisaged three steps. Firstly, to continue consultations, both unofficially with scientists and, more officially, by sending to different European countries 'a letter signed D.G. [Director-General of UNESCO] requesting interested governments to nominate a body or person with whom official preliminary discussions [could] be held'; thereafter, to draw up a preparatory project to be discussed by 'delegates from interested countries'; finally, once this project was sufficiently detailed, to submit the proposals officially to the different governments concerned.[10]

By way of conclusion two points deserve particular mention. For some four months Auger—the so-called 'travelling salesman' for the Florence project—and Amaldi played the leading roles. They publicized its existence and tried to involve European nuclear physicists in it. Above all they tried to promote the *idea of a laboratory financed jointly by European states.* In doing this they often mentioned Rabi's name, which is easily comprehensible in the light of the previous chapter. Of course Auger always showed a great deal of interest in the laboratory's equipment, but his main problem was finding *the political and financial means* of putting resolution 2.21 into practice.

The second point to stress is that the task of 'travelling salesman' does not seem to have been an easy one. While Auger often noted that a particular person was in favour—doubtless recording immediately any remark of this kind—everyone seems to remember strong resistance. We have already quoted Kowarski's remark about the attitudes at the meeting in Oxford. Auger expressed similar sentiments in a recent interview. When asked whether a lot of people were sympathetic to the scheme in these early days he replied:

Some people, not lots [...] In fact, in general, it was difficult to get the scientific community of different countries [...] to accept the idea and to consider that it was viable, possible and in fact good.

As for Amaldi, in 1974 he said the following about that period: 'I would say that most of the people were not prepared; I remember they were not against it but it was new to them[...]'. And his regret at this rather cold attitude is evident in his description of the IUPAP meeting in September 1950: 'The chairman and the president of IUPAP at that time was Kramers but Kramers was not well. I am sure that if Kramers was there, probably he would have taken more positive action [than the chairman, Darwin]'. While this may not be true, there is no doubt that Amaldi was very disappointed.[11]

3.2 The meeting between Auger and Dautry and the decision to call a gathering of scientists in Geneva in December[12]

In the course of his consultations Auger met Dautry in Paris between 14 and 19 October 1950. This meeting is of particular historical importance, since it was the first time the paths of two of our relatively independent groups crossed. One of these was the Centre Européen de la Culture which included de Rougemont and three important scientific administrators, Willems (Belgium), Dautry (France), and Colonnetti (Italy); the other was the group of nuclear physicists who supported the Florence project, namely Amaldi, Auger, and those they had managed to convince of its merits.

The official inauguration of the CEC took place on 7 and 8 October 1950. Raoul Dautry had stated on this occasion that he was in contact with UNESCO and that he would keep the Centre informed about the progress made by his study group on scientific problems. Auger returned from the United States at the beginning of October, and Dautry wrote to him on the 14th:

I would like to know before then [the next meeting of his study group on 12 December] what UNESCO wants to do and to act in accordance with that. If you have a few minutes to spare one day could you pass by rue de Varenne so that we can discuss the matter.[13]

The two men duly met and considered various possibilities for European laboratories and agreed to use the opportunity provided by the forthcoming meeting at the CEC on 12 December to have a wide exchange of views. At this point it was arranged for the two groups to meet: Auger would invite several scientists and Dautry would try and bring together those members of the European Movement who were interested in scientific collaboration. Dautry even said, in a letter to de Rougement and Silva of 19 October, that 'it [was] absolutely essential that no one [missed] the meeting on the 12 December because it [could] produce important and immediate results'. It seems then that both Auger and Dautry came away from their discussion favourably impressed.[14]

During the following weeks arrangements were made for the meeting. On 7 November Auger submitted a list of seven people to be invited on 12 December. On the 9th Dautry confirmed his letter of the previous month, and the CEC began sending out letters of invitation on 13 November. A month later the meeting opened at 9.45 a.m. in Geneva.

Before describing the meeting itself, we should like to clarify three issues of particular interest. Firstly, why did Auger, a director at UNESCO entrusted the previous June with the specific task of implementing resolution 2.21, consider it necessary to take part in a meeting formally arranged by the Centre Européen de la Culture? Secondly, we want to establish, by studying the list of participants, what kind of meeting Auger, Dautry, and the CEC had in mind. Finally we shall explore how, and by whom, the agenda of the meeting was drawn up.

3.2.1 WHY A MEETING WITH THE CEC?

In order to answer this question we must first look back at the way events unfolded. By mid-October Auger was nearing the end of the 'consultative' phase. He had met leading scientists from several countries and had collected together opinions and proposals. It was now time for him to move to a more formal stage, to arrive at a synthesis of these opinions, to transform them into a basic programme. There were three kinds of possibilities open to him. Either he could draft a text himself, remaining within the ambit of UNESCO and using the organization to contact interested countries or scientists designated by them; or he could work with IUPAP, through Amaldi, and prepare a joint document; or he could accept Dautry's offer and co-operate with the CEC.

Pierre Auger 'chose' the third option. We can imagine several possible reasons why he did so. For one thing there were special personal relationships involved. As early as 1945 Auger and Dautry had collaborated in the establishment of the French Commissariat à l'Energie Atomique (CEA) which they managed together with the Joliots and Perrin. They remained in contact afterwards, notably after November 1949 in connection with Dautry's and Kowarski's proposals. Auger thus knew Dautry very well and it would seem quite natural for him to have preferred to work with him in Paris.[15] All the same we suspect that personal considerations were not the only reason. In particular, one must bear in mind the *nature* of the CEC, of IUPAP, and of UNESCO. The Centre Européen de la Culture was essentially an organization whose aim was to further co-operation in *western Europe,* it had a *flexible administrative structure,* and it was able to act as a *fund-raising intermediary.* UNESCO and IUPAP, on the contrary, had many disadvantages—at least while the project was still at the embryonic stage, where a single unwise move could cause it to miscarry. These 'objective' reasons require further explanation.

Consider first the fact that the CEC was by definition western European. In the autumn of 1950 Auger was acting principally in his capacity as Director of the UNESCO Department of Exact and Natural Sciences, mandated by resolution 2.21 which referred only to the possibility of assisting in the creation of regional laboratories. Undoubtedly Rabi, by way of example, had verbally suggested 'beginning' with Europe, certainly Auger had proposed that a western European physics laboratory be included in UNESCO's programme at a meeting of the organization's Executive Committee in November 1950, but at some stage it was necessary to draw up a specific list of the countries which might take part, a *list* which would have to be compatible with the political options available in 1950—if the project was to succeed. This was where the CEC, rather than IUPAP or UNESCO, could be of assistance. The latter organizations were *required* to give equal treatment to all their member states, and there was no obvious reason for excluding Czechoslovakia, Hungary, or Poland, for example, or for including the Federal Republic of Germany, which at that time was not a member of either IUPAP or UNESCO.[16]

This freedom of choice in the political complexion of the states one wished to include was reinforced by a second trait peculiar to the CEC: it was composed of only a small group of people and did not have a restrictive bureaucratic structure. There was no need to obtain repeated authorization from finicky legal directors, and therefore no need to fear further delays. In fact everything was decided the day Auger met Dautry, and Auger was free to invite whomever he thought fit, without having to negotiate from the outset with people who were not necessarily convinced of the importance or the urgency of the scheme. Obviously he was aware that an official intergovernmental conference would have to be called, but Auger realized that the time would only be ripe for this when the project had assumed a definite shape and had been approved by the scientific community itself. He maintained this more flexible and informal attitude throughout 1951.[17]

The third reason which Auger had for preferring the CEC was probably the most important. He knew money to be the sinews of war, but UNESCO had given him very little (2,000 or 5,000 dollars according to our sources), and had made it clear that no more would be provided. This was insufficient to allow several people to work full-time on the project, to enable them to travel to the United States, to Great Britain, to Sweden to find out what type of machine to construct, or to pay the travel expenses of European scientists who would have to be consulted. Auger therefore needed money. The CEC was not rich, far from it, but—and Dautry probably confirmed this—it would contact very high-level scientific administrators and money would be found. Indeed in November the French Prime Minister's office was contacted, partly thanks to Dautry who was able to tell de Rougemont in December that France could provide 50 million francs, while Colonnetti wrote in similar vein to de Gasperi on 18 December and got an advance of 2 million lire from the Italian CNR the same day.[18]

Notes: p. 119

Dautry, for his part, was equally willing to co-operate with Auger. He did not want to present a project which would compete with UNESCO's, he hoped to implement the resolution passed in Lausanne in November 1949—and for him the Rabi resolution followed from the same principle—and he had some ideas of his own that he wanted to pursue within the framework of his association with Auger. Hoping to develop closer co-operation with Italy in the nuclear field as a whole, he asked the CEC to send invitations for the Geneva meeting on 12 December to Senator Casati and to Mario Rollier, a fervent advocate of an atomic energy commission in Italy and personal advisor on nuclear matters to the minister Ivan Matteo Lombardo. To sum up the situation, the meeting between Auger and Dautry had every chance of being mutually beneficial.[19]

We can now suggest an answer to the question we raised earlier: why did Auger not reply to Amaldi's offer of collaboration made on behalf of IUPAP? What we have just said helps us to throw some light on the matter. Auger did not choose to work with IUPAP because it offered no advantages over UNESCO: it was not European, it had no money, and it would not enable Auger to meet a group of men he did not already know. Indeed it would cause further complications: there would have to be a meeting of members appointed by the Executive Committee or by the member states, the discussion would have to begin all over again, those who were against the project would have to be won over, and the end result would probably be nothing more than a vague compromise document which would not allow one to make any progress. As for involving Amaldi *personally* in the scheme, the Italian physicist's name headed the list of people to be invited on 12 December which Auger sent to Dautry.

3.2.2 THE LIST OF PARTICIPANTS

In addition to Auger and Dautry, a total of seventeen people were invited to the 12 December meeting: five by the CEC itself, two by Dautry, seven by Auger, and three who were suggested as replacements.

The CEC invited the members of the scientific commission who had met at the Lausanne conference, namely Gustavo Colonnetti, Jean Willems, André George, Charles Manneback (at Willems' specific request), and Max von Laue. The only member of the 1949 commission not to be invited was Cyril Darlington, the English biologist. Dautry invited Casati and Rollier, as we already know.[20]

This brings us to Auger's list. It comprised Edoardo Amaldi (Italy), Marc de Hemptinne (Belgium), Hendrik Kramers (Netherlands), Paul Scherrer (Switzerland), Gunnar Randers (Norway), and Karl Manne Siegbahn or Ivar Waller (Sweden). To see what significance, if any, to attach to Auger's list, it is interesting to compare it with the one which Amaldi sent to him on 3 October 1950. The latter was built in the following way:

Britain	J.D. Cockcroft
France	?
Switzerland	P. Scherrer
Germany	W. Heisenberg
Belgium	M. Cosyns
Holland	C.J. Bakker
Denmark	N. Bohr
Sweden	K. Siegbahn
Norway	?
Spain	?
Austria	?

The first thing to notice is the way in which Amaldi has drawn up his list, not using names as a starting point but instead the western European members of IUPAP with the addition of Germany and Austria. Let us do the same to compare the two lists.[21]

Opposite the word France Auger wrote the name of Francis Perrin.[22] This choice is easy to understand, since he was probably the most 'senior' French nuclear physicist, at least since Joliot's dismissal in April 1950. Indeed he was appointed Haut-Commissaire of the CEA the following year. However, Perrin was not on the list which Auger sent to Dautry for the meeting on 12 December because he was to be in India at the time. Dautry would represent France in Geneva.

For Switzerland and Sweden both Auger and Amaldi suggested Scherrer and Siegbahn, two of the most influential physicists in their respective countries, particularly in nuclear physics.[23] The names they put forward for Belgium and the Netherlands were different, however. In the case of Belgium, Amaldi suggested Cosyns, a specialist in cosmic rays, whereas Auger suggested de Hemptinne, director of the nuclear physics centre at the important University of Louvain. For the Netherlands Auger preferred Kramers, president of IUPAP, a close colleague of Bohr and his likely successor, to Amaldi's suggestion of Bakker, an accelerator expert. Lastly Auger proposed Randers for Norway, who was incontestably the leading scientific administrator in the country in terms of nuclear physics.

This gives us our first insight into Auger's view of the meeting. For him its main purpose was not to bring together cosmic ray or accelerator experts, but a less specialized gathering of high-level people, influential if possible in the field of nuclear physics, who were able, if not to commit their countries, then at least to speak with sufficient authority to give weight to the resolution which would be adopted. Colonnetti and Amaldi for Italy (Dautry had probably at least mentioned the names of Colonnetti and Willems to Auger), Willems and de Hemptinne for Belgium, Kramers for the Netherlands, Siegbahn (or Waller) for Sweden, Randers for Norway, Scherrer for Switzerland, Dautry for France, and Auger for UNESCO, make up a list of the foremost scientists and administrators of the time.[24]

Notes: p. 120

There remain three countries for which Auger put forward no name, and which would therefore probably not be represented: West Germany, Denmark, and the United Kingdom. These were certainly not second rate countries, particularly in nuclear physics. Why then were they 'excluded'? When we asked Auger recently why he did not have the CEC invite Bohr or Heisenberg, whose names had been suggested by Amaldi, he simply said that these scientists were far too important for such a meeting. He went on to say that it was only a short meeting to draw up a preliminary programme and that he did not wish to trouble Bohr, for example, with it. Of course his opinion would be vital, but only at a later stage. This seems quite plausible, all the more so as Auger did not know these people or any of their countrymen very well. However, in the case of Germany, we would like to put forward a more specific reason. Not that we think that Auger harboured anti-German or 'French-revanchist' feelings, certainly not. All of his previous behaviour, both before and after the war, bears adequate testimony to this. Nevertheless there was still an objective German 'problem' in 1950. The Federal Republic of Germany had been officially created the year before, but many issues remained to be solved: she was still not a member of UNESCO and the possibility of her involvement in a European laboratory for nuclear physics was bound to be a sensitive issue. Two brief examples of this will suffice here. In May 1951 Ernest Lawrence, director of the Radiation Laboratory in Berkeley, wrote to Niels Bohr:

> [...] Another institution is interested in inviting Heisenberg and I have been asked as to the advisability of so doing. There have been some indications that because of his conduct in the war he no longer is held in high esteem. I should certainly not favor inviting him over unless you still regard him as a friend and scientific colleague who is welcome in your institute. I would greatly value your confidential views and would appreciate a cable at our expense.

So too the French Communist newspaper *L'Humanité*, reporting on a press conference held by Francis Perrin in Bombay on 22 December 1950, during which he mentioned the possible involvement of Germany in the European laboratory in accordance with the resolution passed on 12 December in Geneva, ran the headline: 'Nazi scientists to participate in atomic research in Paris; the reason for Joliot-Curie's dismissal becomes clear'.[25]

There was, then, a German 'problem'. Which suggests to us that Auger, good tactician that he was, preferred to prevent this sensitive issue from interfering with his plans *at this stage*. The most important thing was to define a project, for which it would be difficult enough to gain approval without providing additional reasons for its opponents to nip it in the bud. Germany would be invited to participate in due course (in fact this was done in 1951 following her admission to UNESCO), but for the moment diplomacy dictated that the question of German participation be set aside: the political future of the project required it.

Finally we come to Great Britain. We know that Auger met several British scientists on various occasions and yet none was invited. It was not because Auger did not know the British nuclear physicists: Blackett was one of his closest friends, and he knew Cockcroft well too. So why were no invitations issued?

The answer would seem to be that he knew—or had good reason to believe—that the British would not be interested in participating fully in such a project. The fact that Cockcroft had expressed 'an opinion firmly in favour of the idea of a European laboratory' did not mean that he considered the United Kingdom a potential participant. It meant primarily that it would be excellent for the Europeans... on the continent. Similarly, while the offer of the use of the Liverpool cyclotron showed Britain's good will, it would be foolish to infer from this that the United Kingdom intended to participate to the same extent as other countries. Typically, as Chadwick put it in March 1951: 'The opinions of [British nuclear physicists] about the European Centre are somewhat diverse but the balance is definitely against our taking any active part in it [...] At the same time I do not see why we should not help when we can and especially in the preparatory stages [...]'.[26]

If we assume, therefore, that Auger knew of or guessed at these attitudes, the fact that the British were not invited indicates that he did not want the doubts about their participation to be confirmed too openly at the meeting. His policy can therefore be summed up as follows. On the one hand he wanted a small commission, preferably composed of high-ranking science administrators and nuclear physicists. On the other hand he wanted as few problems as possible to arise—in other words, he wanted a 'good' commission. Not that he thought he would avoid all problems—the cases of Germany and Great Britain would be dealt with in due course. But there were priorities and the problems had to be dealt with one at a time.[27]

3.2.3 THE AGENDA FOR THE MEETING

The agenda for the meeting on 12 December seems to have been discussed by Auger and Dautry on the one hand, and by Amaldi and Ferretti on the other. According to the report which Ferretti presented to the Consiglio Nazionale delle Ricerche on 18 December 1950 'Professor Amaldi in agreement with Professor Ferretti [had] outlined a project [for the 12 December meeting] broadly in line with Bohr's views', a project which was discussed during the meeting and which envisaged a laboratory built around a large particle accelerator. It was also Ferretti who presented the first estimate of the cost of the laboratory during the meeting, based on the cost of the Brookhaven Cosmotron. Although we have not been able to find Ferretti's and Amaldi's preliminary project there undoubtedly was one, and it proves that the Italians clearly had in mind a laboratory based initially on an accelerator and *without a nuclear reactor*.[28]

For his part Auger presented a report which came to the same conclusion, recognizing 'the need to construct a laboratory for the study of high-energy elementary particles in Western Europe'. The fact that Auger was thinking along

Notes: p. 120

these lines from the beginning of the meeting is confirmed by the way he rephrased resolution 2.21 when drawing up the preliminary draft for UNESCO's 1952 programme. This is dated 29 October 1950 and reads: 'the establishment of a high-energy physics laboratory for Western Europe, a study of which will have been made in 1951, will be continued'.[29]

The Centre Européen de la Culture as such seems to have been only marginally involved in the *scientific* definition of the project. On 14 October, in accordance with the Lausanne resolution of December 1949, Dautry was still talking about 'nuclear power, astrophysics, fluid mechanics, etc. ...'; on the 19th he mentioned having discussed with Auger 'the establishment of several scientific laboratories already thought of [those just quoted] and others we [CEC] have not yet considered'. On 30 October he suggested to Auger that the meeting on 12 December should discuss the laboratories put forward by the experts of the United Nations Economic and Social Council; lastly in mid-November the CEC's letter of invitation referred to the idea proposed by the scientific commission in Lausanne, namely a 'project for a European pool for atomic research'. Apparently then there were still many options under consideration by Dautry and the CEC, even though only a 'European nuclear physics laboratory based on a large accelerator of elementary particles' was retained on 12 December 1950.[30]

To sum up one can say that it was Auger who, paradoxically, turned out to be the principal organizer of the CEC meeting. The programme he wished to have adopted had been clear in his mind for some two months, and was supported by Ferretti who, in agreement with Amaldi, had arrived at the same results. Moreover, Auger expected the meeting to have concrete results both as regards the confirmation of the geographical scope of the project (specification of which states to invite...) and as regards its funding. This might prompt one to ask whether Auger did not act in a rather autocratic manner, trying to remain in control of the entire operation? The question is justified for one detects a clear wish on Auger's part that the project should not fall into other hands. There are several reasons for this, including of course Auger's personal ambitions, but one concern seems to us to have been paramount: to ensure the success of the project, which was far from guaranteed at this stage. Paraphrasing Amaldi, the sceptics could easily kill the thing before it was born. To avoid that, the project had to be taken charge of by someone who believed in it, and who was determined to succeed. This was the role that Auger took upon himself.[31]

Related to this one further comment could be made about Auger's attitude toward the end of 1950. He had made up his mind to push the scheme to equip Europe with a giant accelerator. He was less sure, however, about the best institutional framework to be used to further that end. He was aware of the limitations of UNESCO—hence his willingness to 'co-operate' with Dautry and the CEC. That body was of course not problem-free either, and when he went to Geneva, Auger was probably open to all possibilities. The one he would finally choose would be that which provided him with the resources and the authority to advance the project now adopted as his own.

3.3 The Geneva meeting of 12 December at the Centre Européen de la Culture[32]

3.3.1 THE PARTICIPANTS

The first point that strikes one when studying this meeting is how different its composition was from that originally envisaged—only eight of the nineteen people invited were present. Could this have had any consequences for the Auger or Amaldi-Ferretti projects? Let us first quickly discuss this point.

To begin with, we note that none of the people who were close to the CEC and who had been invited by it attended the gathering. Dautry and Colonnetti in particular were ill and sent their apologies a few days before the meeting.[33] Of the two people proposed by Dautry, only Rollier was present and five of the scientists suggested by Auger were absent (Amaldi, de Hemptinne, Scherrer, and the two Swedes Siegbahn and Waller). However most of the five made sure that they were represented. Amaldi sent Ferretti, Willems and de Hemptinne asked the CEC to invite Verhaeghe and Capron while Scherrer sent his colleague Preiswerk.[34]

The eight people who met in Geneva were therefore Auger, Capron, Ferretti, Kramers, Preiswerk, Randers, Rollier, and Verhaeghe. Two characteristics of this group are important as far as the likely reception of Auger's project was concerned. Firstly, it was very young. Excluding Kramers, who was 56, the average age of the others was only 42! By way of comparison, that of some of the people absent (Colonnetti, Dautry, Scherrer, Siegbahn, von Laue, Waller, and Willems) was 62. Auger was therefore faced with rather young but influential active scientists who were likely to be open to the idea of being once more in a position to compete on the international stage. Secondly, all were specialists in nuclear physics, most (with the notable exception of Kramers) were interested in accelerators and involved in big science. Thus a project with the aim of building a very large accelerator, and that outside the confines of a single country, could not but attract them and arouse their enthusiasm. This is in fact precisely what happened. But before discussing this more carefully, we must return to the meeting itself.

3.3.2 THE MEETING[35]

The meeting opened at 9.45 am with a report on the Centre Européen de la Culture presented by Denis de Rougemont, followed by a brief account of the Lausanne and Florence conferences. Auger then described the purpose of the

Notes: pp. 120 ff.

meeting. Referring to the UNESCO resolution he first drew attention to the long-term funding of the laboratory: this would not be by UNESCO but by the participating states. Next, recalling the consultations held since June, and particularly those which took place at the Oxford conference in September, he proposed that the laboratory should be devoted to the study of elementary particles. Finally, having outlined the principles on which collaboration between the CEC and UNESCO could be based, Auger opened the discussion, insisting that it should close with a resolution specifying the countries to be contacted, the envisaged construction programme, and the administrative structure needed in the short and long term.

The first issue debated was that of which states to invite. After an exchange of views, a list of twelve countries was agreed upon who were UNESCO members and politically aligned with western Europe, thus including Greece. Turkey was not considered, and neither were Spain, Portugal, and Finland, which were not members of UNESCO. However Yugoslavia was added at Auger's request, as was the Federal Republic of Germany, though the latter too was not a member of UNESCO. It was understood that the countries were not to be contacted officially at this stage, but only after discussions within the European scientific community to be held during the following months.

The scientific programme was then considered. Presented as being under the patronage of Bohr, the idea of having a laboratory for the study of elementary particles using a large accelerator was kept.[36]

At the end of the morning the problem of finance was tackled. It soon became clear that United Nations regulations would be a good model on which to base a scale of national contributions. Someone was accordingly sent to the information centre of the European office of the United Nations in Geneva to fetch the relevant statistics.

The afternoon session again considered the question of finance. Taking into account the reported British 'wish' to participate only for form's sake[37] and acknowledging France's strong desire to have something done, the following scale of contributions was agreed: France 30%, Federal Republic of Germany and Italy 12.5% each, Sweden 10%, Switzerland 7.5%, Belgium and the Netherlands 6.5% each, and the remaining countries less than 5% (Great Britain, Denmark, Austria, Greece, Luxemburg, Norway, Yugoslavia). The total cost of the project was then looked into. Ferretti presented a preliminary estimate and pointed out that the cost of Brookhaven (Cosmotron and auxiliary apparatus) came to 10 million dollars per annum for five years. Allowing for the difference between American and European prices a provisional figure of 5 million dollars per annum for five years was adopted, and it was decided to make this figure public. Thereupon the representatives of France and Switzerland let it be known that the amounts agreed on corresponded to what their governments would be prepared to pay. Finally the meeting turned its attention to the site of the laboratory and the need for a training programme for theorists. Regarding the former, several ideas were put forward: Geneva, Copenhagen, and the Basle/Mulhouse area were mentioned. Some participants even remember maps being brought out. In the event it was agreed that the choice of a site was not the most pressing problem (even if it was the most thorny), and that it should be made at a later stage by the participating governments. Instead the participants preferred to draw up a list of criteria for shaping the decision.[38]

At the end of the meeting a short-term schedule was established. A study office of three or four people was proposed, to include Bruno Ferretti and a Frenchman. Its tasks were to draw up construction plans for the laboratory, to prepare its work programme, and to send a fact-finding mission to Brookhaven. As soon as the project was sufficiently precise it was to be submitted to the Director-General of UNESCO, who would call an intergovernmental conference with the aim of adopting the project and of making the necessary financial arrangements. It was suggested that this meeting might take place in April 1951. The 'construction of the large machine' would begin during the following months, or in 1952 at the latest. The laboratory could begin operation in 1955.

A resolution was then drawn up and copies were sent officially to the Committee of Ministers of the Council of Europe, to the Cultural Affairs Committee of the European Consultative Assembly, and to UNESCO.

As we have implied, the programme under discussion aroused the enthusiasm of the participants. As Pierre Auger wrote twenty years later, 'I still have a vivid memory of the meetings of December 1950 [...]. The exchange of views about our chances of success and the equipment programme of the proposed laboratory were at once enthusiastic and realistic. There was a sense of wonder at being able to make such ambitious plans with a real chance of success which was something completely new'. This impression is confirmed both by the correspondence between the participants during the following weeks (and that of their close friends such as Colonnetti, Dautry, Willems, and Amaldi), and by a reading of the minutes of the meeting written in the days following it. Kramers was the only exception. He feared that the European centre might remove the best scientists from national laboratories and that their funds would be cut back in order to finance the new scheme. Nor was he convinced of its usefulness (or of its urgency), particularly for young physicists. In his view priority should be given to a European teaching centre. However, such doubts—which were to be given forcible expression during 1951—were only voiced by a minority. It was the enthusiasts and the convinced who had the upper hand. Their reply to Kramers was that the future lay in a new type of scientific work, that if Europe did not face up to this she would see her best talents emigrate to the United States, that she would 'be reduced to studying secondary problems in experimental physics neglected by American research', and that in the long term even theoreticians would be unable to work in Europe, lacking the stimulation from unexpected experimental results.[39]

Notes: p. 121

3.3.3 THE TEXT OF THE RESOLUTION

We would like to end our examination of the documents relating to this meeting with a few words on the resolution, particularly because it is the official version of the proceedings, and the only text really written in common.[40]

The commission began by associating its work with that of the European Cultural Conference held in Lausanne in December 1949, and with those of UNESCO and of the United Nations Economic and Social Council (ECOSOC). It then *recommended* that:

- 'a laboratory' be established based on the construction of a large machine for accelerating elementary particles 'whose power' should 'be greater than that envisaged for machines presently under construction', meaning those at Brookhaven (3 GeV) and at Berkeley (6 GeV). The fact that it was not intended merely to copy these machines but to build something bigger is a measure of the ambitiousness of the project;

- 'a European fund be set up for the construction and operation of this laboratory', with contributions from the founder states. The figure of 5 million dollars per annum was quoted, but the scale of contributions was not specified;

- a site be chosen which satisfied five criteria: proximity to an important research centre, availability of specialized labour and energy sources, ease of access, central location, and the possibility of extra-territorial status;

- a summer school for theoretical physics be founded, in accordance with a plan proposed by Cécile Morette (the future Mme De Witt);[41]

- 'a study office be created immediately in Paris, in liaison with UNESCO'.

3.4 Immediate reactions to the 12 December meeting

Before summing up the situation in December 1950, it is useful to examine the different reactions expressed at the end of December and at the beginning of 1951. Three examples of these will enable us to cover a wide spectrum.

3.4.1 REACTIONS OF THE PARTICIPANTS AND OF THEIR COUNTRIES

We have said that the resolution was accepted with some enthusiasm. The Italians were the first to turn this sentiment to some account. The national council of the Consiglio Nazionale delle Ricerche met one week after the 12 December meeting. The president immediately asked that Italian participation in the study office proposed in Geneva be discussed. A letter was read out from Colonnetti dated 13 December in which he recommended 'immediate participation by means of a contribution, however small, the more so as the inclusion of our physicists has already been agreed to and one of them, Ferretti, will hold an important post [in the office]'. Then came his telegram to the CNR of the 15th: 'Professor Ferretti will submit to you my proposal to allocate one million French francs for the purpose of our participation in the study office [...]. Request the approval of the council and the Giunta'. After this Ferretti presented his report, and the proposal to contribute a million francs was approved on condition that the office was in fact created, and that it was known which positions would be effectively reserved for Italian scientists in the organization.

Without wishing to describe Italian actions in further detail, it is worth noting that Colonnetti contacted the government on 18 December, that he wrote to Auger and to Dautry at the same time, that the three men dined together in Paris in January, and that the promised money arrived at UNESCO in February.[42]

A similar process was under way in Belgium. On 10 February 1951 Verhaeghe, president of the scientific commission of the Belgian Institut Interuniversitaire de Physique Nucléaire officially informed Auger that the Institute had approved the Geneva project in principle, that he would recommend Belgian participation to the relevant authorities, and that Paul Capron was prepared to travel to Paris. On 23 February the Institute's Management Committee met and decided to allocate fifty thousand Belgian francs to the study office. At the same time Verhaeghe, who was due to go to the United States from 25 April to 25 May, was also invited to Paris for discussions with the office.[43]

In France Auger contacted François de Rose at the Ministry of Foreign Affairs. On 19 December he explained the financial needs of the study office to him and suggested that one of the states, 'in this case France', should advance some funds. Referring to previous discussions, Auger seemed to think that the money could be made available quickly. In fact it took somewhat longer, and it was only in May that he was officially notified of an offer of 2 million francs—and not 50 or 15 million as had been stated in Geneva. It should be noted though that various scientists, as well as Raoul Dautry, were keeping a close watch on developments.[44]

Lastly Waller from Sweden, Randers from Norway, and Preiswerk from Switzerland kept in touch with Auger. On 13 February Waller said how much he was interested in the laboratory, and that he intended to pass on what he knew to the Swedish Atomic Committee which was to meet the following week; on 13 March he noted that many young physicists were particularly interested and that the Swedes were waiting for Auger's opinion before asking their government for financial assistance. As for Switzerland and Norway, we know from Waterfield, the British Scientific Attaché in Paris, that Auger had told him in January that the study office would 'probably consist of Ferretti, a Norwegian and a young Frenchman' and that he 'hoped to obtain contributions from Switzerland, Norway, and Belgium soon'.[45]

Notes: pp. 121 ff.

3.4.2 THE BRITISH REACTION[46]

The British reaction presents a striking contrast: the project was heavily criticized. Scientifically speaking it was deemed by Skinner to be 'one of the high-flown and crazy ideas which emanate from UNESCO', while Blackett's first reaction was: 'This is quite crazy! If France can afford all that money, why don't they finance their present research and build up their Physics again into a decent state'. He also doubted that the problems of building a large accelerator were essentially financial, that it would be possible to do better than the Americans in less than three years, and that it was urgent to begin training theoreticians. His two concluding questions to Cockcroft revealed his astonishment: 'Who is behind the scheme? Is it seriously intended?'. British scientists also wondered whether it was wise to build a laboratory from scratch, and whether it was right to invest so much money in an accelerator in the face of other obvious needs.[47]

Apart from this the British raised another kind of problem. At a meeting of a committee of the Department of Scientific and Industrial Research held in January 1951 the authority of the Centre Européen de la Culture for sponsoring the resolution passed on 12 December was questioned, as well as its right to issue mandates 'or for M. Auger to undertake them'. The answer was provided by R. Crivon of the Secretariat-General of the Council of Europe: 'The European Centre is a private organization and had any UK representative gone to this meeting it would, of course, have been in a personal capacity'. For his part A.S. Halford of the Council secretariat added 'that while there is a Recommendation from the Consultative Assembly that the official patronage of the Council of Europe be bestowed on this body [the CEC], the Recommendation has not yet been considered by the Committee of Ministers which alone can say yea or nay'.[48]

3.4.3 THE PRESS[49]

The two levels of scepticism which we have just described were to be found in the press, though aggravated here by a fundamental confusion between high-energy research, research with reactors, and applied atomic energy research—with all the political problems which that implied. The result was a considerable amount of indecision and doubt.

Questions about the nature of the CEC arose on 26 December. One week before, *Le Monde* had concluded its account of the work done in Geneva by saying: 'the committee has also decided to establish an international study office [...] in Paris immediately [...]'. Reacting to this report on the 26th the UNESCO Secretariat asked for it to be made clear that 'the committee could not take such a *decision,* and that it had only *recommended* that the Centre Européen de la Culture, which had convened it, should contact UNESCO about the immediate establishment of an international study office'. However, most of the doubts and ambiguities concerned the scientific character of the project. The best documented article was one in *L'Observateur,* and we shall base our discussion on it.[50]

According to Etienne Gilbert, the author of the article, *two* projects were involved in (or were lurking behind) the meeting in Geneva. Some of his informers said that a 'European nuclear physics laboratory' was all that was contemplated, and in particular the construction of large particle accelerators. Others thought that some combination or 'pooling' of atomic energy studies or establishments was at stake. The former project, he noted, '[did] not present the same serious drawbacks as the other' even if 'it [needed] looking at more closely': Great Britain was certainly not interested ('I am not giving away any secrets' wrote Jérome Cardan in the *Tribune des Nations* of 29 December, 'when I say that [...] when sounded out on this the British refused point-blank'), and the financial burden would fall essentially on the shoulders of France. Moreover, the participation of Germany seemed to be desired, which represented a serious political problem *(L'Humanité, La Tribune des Nations).*

Nevertheless the most insistent questions concerned the future 'atomic energy pool'. Here the usual criticisms were raised: problems about secrecy and American control, industrial results, military implications, consequences for the French CEA ... . Concluding, Gilbert mentioned the numerous reservations he had heard, those expressed by the British of course, but also those of 'a number of continental physicists' who remained coldly cautious—a caution illustrated by the conference at Harwell, where Auger '[did] not receive much encouragement', and by the Joliots' distrust.

We may well ask who these journalists' 'informers' were, and why some of them talked of a possible European pool for atomic energy in the same breath as the Geneva resolution. About the sources we can think of certain scientists at the CEA who were in favour of (or opposed to?) the project submitted by Kowarski and approved by the 'scientific departments of the Commissariat' in April/May 1950; or of some of Dautry's associates, who were most certainly still in favour of the project proposed at Lausanne and who, in so far as they knew exactly what Auger wanted to have accepted at Geneva, might have considered this as a step towards the more ambitious project they had set their hearts on. As for the sources' reasons, they were not necessarily machiavellian. It is just that their ideal, their long-term project, was different from that of Auger and Amaldi. For the latter the Geneva meeting was not a step towards something else, it was an end in itself. For others whose involvement in the Geneva meeting could be traced back to other concerns, it was a different matter. It is not surprising then that there were different 'readings' of the 12 December resolution.

Indeed two official texts published on the 16 December reveal the actual state of uncertainty which existed in some people's minds. These were the two CEC press releases intended to give an account of the meeting held four days earlier. Let us quote the beginning of both to illustrate this: 'For the study and the *peaceful*

Notes: p. 122

development of atomic energy, Western Europe can organize a *'pool'* which, by 1954, will be equipped with a cosmotron as big as or bigger than the largest and *unique* machine of this kind now being built in the United States *(1 billion volts)'* (our emphasis).[51] There then followed a more than doubtful description of a cosmotron (indeed both wrong and mixed up), a few words expressing the usual faith in industrial applications and a comment on the uncertain future of sources of industrial energy in Europe. The main reasons for such a mediocre text lay partly in the prevailing confusion in non-scientific circles surrounding the magic words 'atomic' and 'nuclear' but partly too in the dominant state of mind at the CEC: for the last year they had been thinking of nuclear physics 'in its application to daily life', as the Lausanne Conference had put it, they had been thinking in terms of a European pool for research into 'atomic energy'. It is therefore understandable that the press did not know what to believe and tried to get more information.

3.5 Conclusion: Ideas and motivations behind Amaldi's, Auger's, and Dautry's proposals

Now that we have presented all the elements, we can reconsider the overall view of the situation in December 1950, immediately after the Geneva meeting.

We concluded our previous chapter by stressing the particular importance of the June meeting at UNESCO. Resolution 2.21, passed at this meeting, had the advantage of being proposed by an American, which implied that the United States was not (or was no longer) against European collaboration in certain areas of nuclear research. The June resolution was also explicitly linked with Auger's name, which meant that it stood a chance of not getting lost in the labyrinth of committees convened by UNESCO. Finally, since it had been passed by the General Assembly of this organization, it became an integral part of its programme for 1951. The project thus gained official status, and a possible channel for stimulating action at government level at some future date.

The importance of the meeting in Geneva on 12 December lay in the fact that it provided an opportunity for Auger to broaden the institutional basis of his activities, to consult with top level people likely to agree with his scheme, and to draw up an 'official' statement of what was to be done next. More particularly it provided him with a text stipulating that the laboratory should be dedicated to the study of high-energy particles—for which a large accelerator would be built. It accepted that the centre had to be funded by European states—of which a provisional list was drawn up—and not by UNESCO or other international bodies. It also recognized that a timetable had to be drawn up and that a study office associated with UNESCO and the CEC had to be created immediately. A step forward had then been taken.

This summary of the new situation demands further precision: we think it right to stress the importance of general political considerations in the 12 December decision.

Conclusion

The main concern of those who met in Geneva was that *something had to be done quickly* for Europe if the United States were not to be allowed to dominate world science in the future. European states needed to be shaken up and made to understand that they were entering a new technico-scientific era; that science needed money, a lot of money, that it was a long-term investment—but a profitable one; that it involved a kind of wager but a wager they had to take. For most of the signatories of the resolution, as long as the area invested in was an interesting one—and obviously, for Amaldi and Auger, two cosmic-ray physicists, the domain of high-energy was a perfect choice—the scientific detail was secondary. There would be time enough to discuss more precisely the kind of accelerator to build, the costs and the location; or to decide to establish other laboratories if the need arose. For the time being, this was not the most urgent thing to do.

The corollary of this remark is that the decision was not primarily informed by a comparative estimate of European needs and that the data 'justifying' the 12 December resolution remained vague. As a result certain people—among them Dautry and some CEC leaders—went on thinking that the 'choice' of high-energy physics was probably too narrow and that Europe needed something more important. Others—like Kramers or numerous British physicists—felt, on the contrary, that the project was too ambitious, too divorced from reality, not adequately grounded from a scientific point of view.

It seems therefore that Amaldi, Auger, and their friends would have to go on defending and adapting their 'choice'. After all, only one or two dozen people felt committed by the December 1950 resolution, and its force of conviction remained rather uncertain. Contrary to the current teleological ways of telling CERN's story, it must be concluded that 'Geneva 1950' was only a prudent step devised to allow things to move quietly forward. In this sense, it did not necessarily prefigure what CERN was to be.

A last question remains which deserves some consideration. It was raised by two scholars fifteen years ago and could be formulated like this: was France not playing a somewhat disproportionate role in this affair? Was there not a primarily French project behind it all? Or, to be more incisive, was it not an operation mounted by France in her own interest?[52] Indeed everyone knew of the weakness of French high-energy research. There were of course a few groups working on cosmic-ray physics (Leprince-Ringuet's and Auger's) but the lacunae in accelerator technology were important. Only the CEA was building a machine at the time, and it was merely a low-energy cyclotron. What is more, France was prepared to take on 30% of the costs, Dautry had promised 50 million francs in the short term—which was really too large a sum, as Ferretti noted in his report—and had organized the Geneva meeting with Auger, another Frenchman. Can we then believe that 'l'intérêt supérieur de la France' was absent from their minds, and Dautry's in particular, whose rank and function were practically those of a minister?

Notes: p. 122

If this description is true—and it is apparently supported by strong arguments—then a few points still need to be clarified. Firstly, it must be stressed that it applies rather to certain individuals than to France 'in general', and to Dautry rather than to Auger. Not that Auger never thought of the spin-off from the project for French physics: like everyone else he knew that helping to put European science back on its feet would also contribute to the rebirth of various national science efforts. But there is no reason to suggest that Auger's actions—or those of Amaldi *vis-à-vis* Italy—were primarily aimed at helping first and foremost 'France'—or 'Italy'—and particularly in their competition with Great Britain.[53] As far as we can judge, Auger's main motive was to give *physicists*—be they French or Danish—an accelerator comparable in size to the American ones.

On the other hand it seems more certain that Dautry had more complex motivations and that he more consciously articulated the European programme with the French CEA programme—even if it is doubtful that any manipulation was involved.[54] In his letter of 18 December 1950 to de Gasperi, head of the Italian government, Colonnetti also stressed that the government had to anticipate spending money to develop a national substratum in high-energy physics if Italy was to make full use of the European laboratory. The reply to our original question would therefore be that France indeed wanted to shape the organization to ensure that it fulfilled her needs, as did Italy for example, and as Britain would do in the future. After all, this seems to be rather common science policy practice[55]. But it would be misleading to see in Auger's or Dautry's behaviour a systematic campaign mounted primarily or exclusively in France's own interest. If further proof were needed, the lack of co-ordination between Auger, Dautry, and the 'scientific departments of the Commissariat' would provide it.

Notes

1. The documents consulted for this chapter are to be found mainly in the following archives:
    - UNESCO archives, Paris, *CERN, Official Correspondence* file (referred to as UNESCO)
    - Centre Européen de la Culture, Geneva (referred to as CEC)
    - Raoul Dautry personal archives, Archives de France, Paris (referred to as AF-AP307).
    - personal documents collected by our team in 1983, including documents of Edoardo Amaldi and Pierre Auger, which are now in the CERN archives. These are referred to as: CERN History Study (CHS), Amaldi (or Auger) file, CERN.
2. See section 2.5.
3. The key documents used for writing this first part were: Amaldi's Diary (CHS, Amaldi file, CERN); letter Amaldi to Auger, 3/10/50, reproduced in Amaldi (1977), 349–51; notes made by Auger, probably at the beginning of October 1950 and entitled: *Conversations about the Physics Laboratory,* 3 pages; note made by P. Auger on 18 October 1950 and entitled: *Willems, Belgium, National Found.,* 1 page (both in CHS, Auger file, CERN).

4. Letter Amaldi to Auger, 3/10/50, *op. cit.* note 3. The IUPAP meeting took place at Cambridge (Mass.) on 7 and 8 September. With regard to this meeting and to the meeting held in Oxford (UK) from 7 to 13 September during which Auger spoke, see section 2.5.
5. All this information is taken from the letter quoted in note 3.
6. Letters Amaldi to Cacciapuoti, 27/10/50, 16/11/50, 27/11/50 (CHS, Amaldi file, CERN); reply from Cacciapuoti between 27/10/50 and 16/11/50 (we do not have this letter); quotation from the letter dated 16/11/50.
7. Extracts from Auger's notes quoted in note 3; Kowarski (1977a). Let us remember that Sir John Cockcroft was Director of Harwell, while Henry Tizard was at that time chairman of the Defence Policy Research Committee (part of the Ministry of Defence) and of the Advisory Council on Scientific Policy.
8. In *Compte-rendu analytique de la réunion du 12 décembre 1950,* written on 18/12/51 by J.P. de Dadelsen (CEC), we can read: 'the Committee, having heard Mr Auger state that Yugoslavia, a UNESCO member, would certainly welcome the opportunity of letting its researchers take part in the work being considered, decided...'. Thus it seems likely that Auger had met some Yugoslav representatives. As regards Sweden, our information is taken from Nyberg & Zetterberg (1977). According to these authors, Professor Gustafson had recently reported having met Amaldi, Auger, Kowarski, and Jacobsen at a meeting in Paris in autumn 1950. However, the description given seems to correspond to a meeting held in autumn *1951*. Moreover, to our knowledge, there was no meeting of these five people in Paris in 1950. It therefore seems unlikely that this information is correct.
9. Notes by Auger quoted note 3; the last quotation is taken from the report of de Dadelsen quoted note 8, p.6. The name of Auger's 'source' is not mentioned in the documents of the time.
10. Notes by Auger quoted note 3.
11. Interview with Auger, Paris, 10/12/82, p.15; interview with Amaldi, 22/1/74, pp. 3-4 (CERN).
12. The letters between Auger, Dautry, and the Centre Européen de la Culture, which we mention in this part, can be found in the CEC archives, unless otherwise stated. We refer the reader to sections 2.3 and 2.4 for details on the role of the CEC in 1949 and in early 1950. Here let us simply recall the Lausanne conference (December 1949) and the attempts to get a study group together under the chairmanship of Raoul Dautry to put its resolutions into effect.
13. Letter Dautry to Auger, 14/10/50 (CHS, Auger file, CERN).
14. The account of the meeting is taken from letter Dautry to de Rougemont and Silva, 19/10/50 (CEC).
15. A more detailed analysis of the French situation is given in chapter 9. Let us simply note here that Auger and Dautry discussed this kind of plan as early as November 1949, particularly with regard to preparations for the Lausanne conference. See, for example, letter Auger to Dautry, 21/11/49 (AF-AP307/212).
16. As regards the resolution proposed by Rabi see section 2.5. Regarding the meeting of the Executive Committee of UNESCO, see *Preliminary draft programme for 1952 put forward by the Director General and submitted to the Executive Committee,* 29/10/50, item 5 of the agenda, draft resolution 2.2. This preliminary draft was discussed at the 24th session of the Executive Committee held from 2 to 10 November (UNESCO, 24EX/SR1-11, in particular pp. 23-26). It became draft 25EX/2 dated 5/1/51, and was adopted by the Executive Committee at the 25th session held from 15 to 27 January, and submitted to the General Assembly of June 1951 as the programme for 1952.
17. De Dadelsen explained Auger's presence in Geneva as follows *(op. cit.* note 8): 'Nevertheless, if UNESCO had wanted to implement this [Rabi] resolution alone, it would have had difficulty in: a) defining the region, i.e. establishing a list of participating nations; b) making some of its non-European members understand that they would benefit indirectly from the laboratory's achievements; c) drawing up a budget'. a) and b) refer to the reason we have just suggested, c) refers to the third reason. About the 'restrictive bureaucratic structure' of UNESCO, see section 6.4.
18. According to Auger in February 1951, 5,000 dollars were voted at Florence in June 1950 (letter Auger to de Rougemont, 16/2/51, CEC). This is, moreover, the sum asked for by Rabi when he proposed his resolution on 7 June. However, according to Waterfield, British Scientific Attaché in Paris, UNESCO had

offered 2,000 dollars. See letter McMillan to Awbery, 30/1/51 (PRO-DSIR17/559). For Dautry's contacts with the French government, see for example the letter from the *President du Conseil* (Prime Minister) to the Minister of Foreign Affairs, 20/11/50 (AF-AP307/211); for Italy, see notes 28 and 42 below.

19. Regarding the belief amongst members of the CEC that the Florence resolution followed on from the Lausanne resolution of December 1949, see for example letter de Rougemont to George, 23/9/50 (CEC): 'As you know, UNESCO voted in Florence for a proposed European pool for nuclear research, on an American motion, supported by all the European governments. It is identical to our Lausanne plan.' Regarding collaboration with Italy, see letter Dautry to Malvezzi, 19/10/50 (AF-AP307/211): 'Mr Colonnetti came to see me today [...]. I believe that we are about to start a fruitful collaboration between France and Italy and I am delighted.' See also letters Rollier to Dautry, 7/11/50 and Rollier to Lombardo, 16/11/50 (MR, copy in CHS, Rollier file, CERN).
20. Correspondence in CEC archives.
21. Czechoslovakia, Hungary, Poland, and Finland were also members of IUPAP.
22. The original of Amaldi's letter to Auger, dated 3/10/50, is in the UNESCO archives.
23. For biographical information, see appendix 1 of this volume.
24. With regard to the 'authority' which these men enjoyed, see for example what McMillan, an official in the DSIR, wrote to Crivon at the Council of Europe on 5/1/51 (CE-O695): 'In view of the distinguished group of eminent European scientists present at the meeting, and the fact that at an earlier discussion at UNESCO the U.K. raised no objection, I am somewhat diffident to express my own personal views'.
25. Telegram Lawrence to Bohr, 31/5/51 (Berkeley: Bancroft Library, Lawrence papers, box 3, folder 3); *L'Humanité* of 22/12/50, front page. See also section 9.5.
26. Chadwick's opinion is taken from a letter he sent to Cockcroft on 13/3/51 (PRO-AB6/912). Just as we said that Auger had 'nothing' against the Germans, so we can say that he had 'nothing' against the British. Early in 1951 he got in touch with Cockcroft again, concluding: 'I am very glad to see that a real interest is shown by the countries constituting the "region" and if some interest is shown concretely by the UK it will be even more gratifying' (letter Auger to Cockcroft, 2/3/51, PRO-AB6/912).
27. One last remark is needed before closing this study. We reached these conclusions by comparing the various lists drawn up by different people. They should then be read more as plausible suggestions than as certainties. Indeed it is possible that Amaldi's and Auger's lists might have been somewhat different if they had been drawn up six months earlier or later. For example, Amaldi might have proposed Kramers; or, if Auger had known the Danish physicists better, he might have proposed the name of one of Bohr's collaborators. This analysis is therefore limited in its scope, especially since our knowledge of the personal relations between Amaldi and Auger and the European physicists is imperfect.
28. Amaldi probably came to write this report because he was given the task of doing so at the meeting of the Executive Committee of IUPAP in September 1950 and because, knowing that he would be absent on the 12 December and having taken the project to heart, he wanted to make sure that his ideas and the ideas of his Italian colleagues would be discussed. Ferretti's report was submitted to the Council of the CNR on 18 December (CNR-Minutes of Consiglio di Presidenza, 8/12/50).
29. Auger's report is summarized in Rollier, *Rapporto sulla reunione del Groupe d'Etudes et de Recherches Scientifiques del Centre Européen de la Culture, tenutosi a Ginevra il 12 dicembre 1950, 20/12/50* (MR, copy CHS, Rollier file, CERN). The first quotation is taken from this report. References to the UNESCO preliminary draft programme are given in note 16.
30. Letters Dautry to Auger, 14/10/50, 30/10/50 (CHS, Auger file, CERN); letter Dautry to de Rougemont and Silva, 19/10/50 (CEC). The letters of invitation sent by the the CEC (for example, the letter sent to Amaldi, 25/11/50) are in the CEC archives. The final quotation is taken from the *Résolution* passed on 12 December (CEC).
31. See for example what Mussard wrote to Amaldi on 26/6/51: 'so far Prof. Auger has only consulted people with whom he had had personal contacts about the European Nuclear Physics Laboratory. You realize that we must behave with some caution, since the project is at a preliminary stage' (UNESCO).

Notes

32. There are a great number of documents about this meeting. There is a report by Rollier (see note 29), a report by Ferretti and summarized in the Italian CNR's minutes (given note 28), a *Note sur la résolution du 12 décembre* drafted by de Dadelsen on 13/12/50, 8 pages, a *Compte-rendu analytique* drafted by de Dadelsen on 18/12/50, 9 pages, two press communiqués drafted by the CEC and dated 16/12/50, and finally a *Résolution* passed by the group assembled in Geneva and therefore dated 12/12/50 (CEC archives). These documents were all drafted either on the day of the meeting itself or during the following week. It should be noted that the analytical report of 18 December seems to be a revised version of the note of the 13th, which cut out some mistakes and added some extra details of the sort 'Mr Auger stated...', 'Mr Ferretti pointed out...', 'Mr Kramers expressed concern about...', etc. Some time later Rollier submitted a report in which he dwelt at length on the Geneva conference: *The European Atomic Energy Pool,* report to the Congress held in Genova from 12 to 14 September 1952, under the auspices of the local Chamber of Commerce (MR). Finally, see CEC (1975), 18–41.
33. According to the correspondence in the CEC archives, Willems had to leave for the United States, Max von Laue had to be in Stockholm for the Nobel prize 50th anniversary celebrations, André George had already promised to give a series of lectures at this time, and Charles Manneback had only just returned from the United States and was unwilling to leave Brussels again.
34. So, only Siegbahn, who was ill, and Waller who was in Stockholm for the Nobel prize celebrations, did not send representatives, though from the beginning of 1951 Waller showed an interest in the progress of the project.
35. The only 'chronological' report is the one by Rollier quoted note 29. We are following his account here, with the addition of details taken from other reports.
36. Regarding Bohr's 'patronage', see section 2.5; regarding Yugoslavia, see section 14.8 and appendix 2 of this volume.
37. This was reported by Rollier in his report (*op. cit.* note 29). It was apparently Auger who said that Britain would co-operate purely 'as a matter of form'.
38. From this point forward the problem of the site of the laboratory seems to have become a contentious issue. There are signs of this in a letter from de Rougemont to Auger dated 24/2/51: '[...] When we were drafting the press communiqué on our meeting at the Centre on 12 December we agreed not to mention the sites chosen, so as to avoid premature anxieties or jealousy' (CEC).
39. The Auger quotation is taken from the speech he made in 1975 for the twenty-fifth anniversary of the Centre Européen de la Culture (CEC (1975), 32); Kramers' fears and the answers which were given are in de Dadelsen, *Compte rendu analytique,* 18/12/50, pp. 6–7 (CEC).
40. According to letter de Rougemont to Auger, 24/2/51 (see note 38), the press communiqué was also drafted jointly. We shall discuss this point later.
41. We shall not dwell on this subject because the initial plan for a theoretical physics summer school was independent of both the CEC and the UNESCO initiatives, and was subsequently (from early 1951) taken over by the French CNRS. Accordingly it had nothing more to do with the European laboratory. The reason why Cécile Morette came to Geneva was that she was looking for a source of money for her scheme—later known as Ecole des Houches.
42. Minutes of the Council of the CNR described in note 28; letter Colonnetti to de Gasperi, 18/12/50 (AST-copy in CHS, Colonnetti file, CERN); letter Dautry to Auger, 9/1/51 (CHS, Auger file, CERN).
43. Exchange of letters between Verhaeghe and Willems on the one hand, and Auger on the other (UNESCO). Two million lire were 14/15,000 SF; 50,000 BF were roughly 4;500 SF.
44. For detail and references, see section 9.3. It is interesting to note that, before the Geneva meeting, Dautry informed de Rougemont that France would be ready to offer money immediately. According to letter McMillan to Awbery, 30/1/51 (PRO-DSIR17/559), 10 to 15 million francs were mentioned. According to Ferretti's report (Minutes of Consiglio di Presidenza del CNR, 18/12/50) 50 million were advanced. This led the Italians and the Belgians to offer a contribution for the study office as well. The irony is that the

French money arrived much later (in September 1951) and that it was reduced to 2 million French francs (around 25,000 SF).

45. Letters Waller to Auger, 19/2/51, 13/3/51 (UNESCO); letter from Waterfield reproduced in letter McMillan to Awbery, 30/1/51 (PRO-DSIR17/559).
46. Chapters 12 and 13 deal specifically with the United Kingdom. What we have here is therefore nothing more than a brief note setting out the prevailing state of mind in a few lines.
47. Skinner's reaction is reported in Gowing (1974,2), 227; letter Blackett to Cockcroft, 13/2/51 (PRO-AB6/612).
48. See document NP 54 (Nuclear Physics Committee), 31/1/51 (PRO-AB/912). The answers appear in the letters Halford to Finch, 15/1/51, and Crivon to McMillan, 26/1/51 (CE-O695).
49. We have consulted *Le Monde* (19/12/50; 26/12/50); *Combat* (19/12/50; 26/12/50); *L'Humanité* (22/12/50); *L'Observateur* (28/12/50); *La Tribune des Nations* (29/12/50).
50. *Le Monde* and *Combat* were alone among the newspapers quoted in limiting themselves to a report of the Geneva meeting without extrapolating to the problem of nuclear energy.
51. The four expressions which we have emphasized show a rather serious misunderstanding of what was decided on the 12th. Firstly, there was not just one such machine being built in the world but at least three: two in the United States (Brookhaven, Berkeley), and one in England (Birmingham). Their energies were respectively 3, 6, and 1 GeV. Moreover, the term 'pool' did not appear in the resolution and was never used by UNESCO; the resolution spoke of a *laboratory* based on an accelerator and Rabi spoke of setting up a *centre* or *laboratory*. Finally, these machines were not for 'the study and peaceful development of atomic energy'. Apart from the confusion that this communiqué created in the press in general and in Great Britain in particular, one question arises: who could have written such a communiqué? In note 38 we quoted an extract from a letter which de Rougemont sent to Auger in February 1951, and in which he said that the communiqué had been drafted 'jointly'. The tenor of the communiqué makes this unlikely: only a non-scientist could have written such a text. Moreover, neither Auger nor the other scientists present on the 12 December spoke of atomic energy—which opened up 'a whole range of politico-economic questions beyond the assessment and competence of scientists' (as Ferretti put it), and which had been rejected for this reason—and Auger would not have put his foot in it by writing that the committee would 'immediately establish' the study office. His knowledge of UNESCO would have made him act with greater care and go back to the terms used in the resolution, which had indeed been drafted jointly.
52. Hartland and Gibbons, unpublished article written in November 1972 and entitled *Britain joins CERN: an analysis of the decision process, 1951–1953*.
53. As was argued by Hartland and Gibbons.
54. For Auger, see, among other things, the British document IOC (50) 95, 26/6/50 (PRO-CAB134/405); for Dautry see McMillan to Awbery, 31/1/51 (PRO-DSIR17/559); on both see section 12.1, note 14 and sections 14.3 and 14.4.
55. For France and Italy, see section 4.1: for Britain, section 13.5.1; for more general remarks, section 14.7.

CHAPTER 4

The period of informed optimism[1]
December 1950–August 1951

Dominique PESTRE

Contents

4.1 The implementation of the resolution of 12 December: French, Italian, and Belgian roles in the organization of the project, December 1950–May 1951   124
    4.1.1 The Franco-Italian front   125
    4.1.2 Belgium   126
    4.1.3 UNESCO and the Centre Européen de la Culture (CEC)   126
    4.1.4 Auger and other European countries   127
    4.1.5 Situation and difficulties at the end of April   128
4.2 Meeting of consultants at UNESCO in Paris, May 1951   130
    4.2.1 The outcome of the consultants' meeting   131
    4.2.2 A remark on the functioning of the envisaged laboratory   133
4.3 Discussions arising out of the consultants' report, May–July 1951   134
    4.3.1 First reactions   135
    4.3.2 The seventh General Assembly of the International Union of Pure and Applied Physics (IUPAP), Copenhagen, 11–13 July 1951   137
4.4 The intergovernmental meeting is called   140
4.5 A marriage of convenience   141
    Notes   142

The most innovative aspect of the meeting held at the Centre Européen de la Culture on 12 December 1950 was that, for the first time, a medium-term scientific and political project was drawn up in rather clear terms: western European states who so wished could share in the common task of setting up a laboratory around a very large particle accelerator. As far as its proponents were concerned, a framework had been proposed which marked the end of the phase of consultation and ushered in that of implementation. As the resolution passed at Geneva on 12 December stated, 1951 should be devoted to making preparatory technical and scientific studies and to obtaining the official acceptance of the project by governments, while in 1952 the construction work on the big machine would start. This meant that, in the immediate future, one had to find money to finance the preparatory study, to secure UNESCO's approval for the programme proposed in December, to set up a study office to define the machine's scientific parameters and, finally, to prepare an intergovernmental meeting.

In fact, it soon became clear that this timetable was somewhat optimistic and that it would have to be reconsidered. In this chapter we shall concentrate on the first eight months of 1951 when those leading the project came up against the *inertia* of scientific circles for the first time. While there was no open opposition at this stage, doubts would soon be voiced. This did not stop a number of decisive steps being taken, notably the decision to call a meeting at governmental level for the end of the year: the letter of invitation was sent out on 31 August 1951 by the Director-General of UNESCO.

4.1 The implementation of the resolution of 12 December: French, Italian, and Belgian roles in the organization of the project, December 1950–May 1951

In chapters 2 and 3 we attempted to describe the particular reasons that prompted some scientists and administrators of certain European countries to make such an effort in 1950 to fill out the idea of a common nuclear physics laboratory. We showed that, while ideological factors played some role—for example the conviction that science could help develop an internationalist spirit, the desire to circumvent traditional conflicts by uniting Europe, and so on—national weaknesses in physics, particularly in the new field of accelerators, often seemed to have been an essential determinant. Faced with the real backwardness of their countries some people realized that radical measures were called for.[2]

We noted also that some scientists and administrators were receptive to these ideas because the situation was similar in several European countries, more particularly in

France and Italy, and because there were good arguments in favour of such projects. Confronted with the American and, to a lesser extent, the British giants, and with the new status of science in the economic and political life of the post-war world, national responses were, in the long run, surely doomed to fail in Europe: only a European science could meet these challenges.[3]

4.1.1 THE FRANCO-ITALIAN FRONT

The thing that struck us when studying the period between December 1950 and May 1951 was that the project's backbone was made up of only a few individuals: Amaldi, Auger, Colonnetti, Dautry and, to a lesser degree, Cacciapuoti, de Rose, Ferretti, Perrin — all French or Italian.

In both countries the Geneva resolution of 12 December had been taken very seriously. Chapters 9 and 10 look at this in greater detail; for the present we simply note that Colonnetti contacted the Italian government on 18 December, that the Consiglio Nazionale delle Ricerche, referring to the timeliness of the Geneva decision, offered 2 million lire for the preparatory studies, and that in France, Auger (via de Rose) contacted the Ministry of Foreign Affairs on 19 December.[4]

At the same time Auger, Colonnetti, and Dautry remained closely in touch with one another, dining together on 11 January in Paris, while Amaldi, who happened to meet Francis Perrin in Bombay in December, was invited by Pierre Auger to come to Paris on his return from India. The idea of establishing a study office 'to draw up detailed plans for the laboratory' was confirmed. Its aim was essentially technical: it was to consider the equipment needed, and in particular the accelerator for which it had to make provisional designs. The office also had to make a study of the buildings required and to prepare a budget, both for the machine-building years and beyond. In Auger's view, the office should consist of three or four people working together for about eight months including, if possible, a Frenchman and an Italian. Apart from the office, Amaldi and Auger would be responsible for the overall organization. They would form a sort of two-man management in which Amaldi would be a 'superconsultant on the physics side', to quote Cacciapuoti.[5]

Does this imply that representatives of other countries were not to be invited during the preparatory studies? Of course not. Indeed the idea was to invite in turn the top European accelerator specialists to help the study office in its day-to-day work. Nonetheless, overall responsiblity for the project was felt to rest principally with the French and Italians. After all, as Cacciapuoti said in a letter to Amaldi, only France and Italy had put up any money; it was therefore natural that 'at least in the organizational phase, the two countries [had] a priority which, with Auger's agreement, [they intended] to take full advantage of as quickly as possible'.[6]

Notes: p. 143

The work programme that was drawn up in broad outline in February 1951 was not put into practice until two months later, apparently for administrative reasons. The Italian CNR had to agree that some of the money offered on 18 December could be used to finance Amaldi's trip. This it did at the beginning of April, and Amaldi arrived in Paris at the end of the month. He and Auger had a first discussion on 26 April. In the ensuing days they met a number of French physicists including Francis Perrin and Louis Leprince-Ringuet, who Amaldi reported to be in favour of the project, and Frédéric Joliot, who they met on the same occasion and was said to be against. Finally on 2 May a working meeting was held with Francis Perrin who had recently been appointed High Commissioner of the Commissariat à l'Energie Atomique.[7]

4.1.2 BELGIUM

Auger's contacts with Belgium, though not as close as those he had with Italy, were nonetheless special. On 6 January Willems wrote to de Dadelsen at the Centre Européen de la Culture congratulating him on the good work done on 12 December. A month later Verhaeghe, chairman of the Institut Interuniversitaire de Physique Nucléaire (IIPN) informed Auger of 'the [Institute's] agreement in principle with the European collaboration project on the cosmotron'. He said that he would recommend to the appropriate authorities that Belgium should take part and that he wished to maintain close links with the study office in Paris. Verhaeghe also mentioned his readiness to go to Paris with Paul Capron. After Auger had expressed his satisfaction in letters to Verhaeghe, dated 16 February, and Willems, dated 21 February, the Management Committee of IIPN, through its president Jean Willems, informed Auger that it had decided 'to make 50,000 Belgian francs available to the study office which was to design the European bevatron'. The money arrived at UNESCO within the next few weeks and Verhaeghe was invited to go to Paris (before leaving for the United States) to discuss the information he should try to obtain on American accelerators. By the beginning of March, therefore, three countries had built up a rather special relationship amongst one another, and felt it their duty to bring the project to fruition.[8]

4.1.3 UNESCO AND THE CENTRE EUROPÉEN DE LA CULTURE (CEC)

In parallel with his contacts with Italy, France, and Belgium, Pierre Auger had to clarify the relationship of the study office to the CEC and to UNESCO. In the days following the meeting at the CEC's headquarters in Geneva the Centre had sent the resolution then voted to different European organizations and personalities. All the same on 3 January Auger wrote to de Dadelsen reminding him that it had also been agreed at Geneva that 'the Centre Européen de la Culture would send a letter to the Director-General of UNESCO asking him to submit to the Executive Council [which was to meet from 15 January onwards], if he considered it appropriate, the Centre's offer of collaboration'. Auger remarked that this had not been done and that it seemed all the more urgent 'since articles of varying degrees of accuracy have started to appear in the press on the subject'. By return of post Denis de Rougemont, Director of the CEC, sent a letter to the Director-General of UNESCO which complied with Auger's request.[9] Unfortunately difficulties arose which neither Auger nor de Rougemont seem to have anticipated. As Denis de Rougemont noted, the Director-General of UNESCO '[said] that he [could not] accept that the Centre and UNESCO be joint patrons of the study office until we [CEC] had signed a convention with UNESCO'. This implied that the Centre would have the status of an official body and that it would be recognized, for example, by the Council of Europe. In addition, the French government hesitated to transfer the money it was considering advancing 'to a private body like the Centre which may be unable to guarantee reimbursement if need be'.[10] Auger therefore proposed that the Executive Committee of UNESCO be the only body authorized to handle the money intended to finance the study office. It was to be paid into a special account to which Auger would have direct access and, until the Centre became an officially recognized body, the study office would be set up under the auspices of UNESCO alone. Pierre Auger submitted this proposal to the 14th meeting (on 27 January 1951) of the 25th session of UNESCO's Executive Committee. Having thanked the Italian Research Council and any future donors, it authorized the Director-General to accept the donations 'for the execution of resolution 2.21 of the programme for 1951' (the 'Rabi' resolution).[11]

4.1.4 AUGER AND OTHER EUROPEAN COUNTRIES

Once the general outline decided in Geneva in December had been worked out in a little more detail between France and Italy, Pierre Auger sent a circular letter to a number of European scientists on 21 February. He informed them that the study office would soon start work, and he asked his correspondents to send an expert to Paris for consultations for about ten days so that they could give the office 'their views on the implementation of the project in question'. Mentioning that UNESCO had only a little money available—two million lire and two million French francs promised so far—Auger said he hoped that the experts' expenses would be borne by their own countries. In his letters to Kramers, Randers, and Scherrer he proposed Cornelius Bakker, Odd Dahl, and Peter Preiswerk as possible consultants, but left the decision to the various national institutes concerned. In the case of Sweden (in a letter to Waller and Svedberg) he asked the Swedish scientific authorities to decide whom to send.[12]

A few days later, on 2 March 1951, Auger wrote to Sir John Cockcroft. Significantly this letter was slightly different to those he sent on 21 February to the Netherlands, Norway, Sweden, and Switzerland. Firstly, although Auger said he

Notes: p. 143

would like to have experts in Paris and especially a British expert, he added: 'even if the U.K. is not interested in being full partner in the laboratory when organized'. Auger also asked if it would be possible for Cockcroft 'to give permission to one of [his] young men, especially competent perhaps in the high-frequency part of the accelerating instrument, to help with our Planning Bureau'. If so, he said, '[he] would arrange for his expenses here for any period of time [Cockcroft] could find suitable (say from 2 to 6 months)[...]'.[13] Put more bluntly this was tantamount to asking the British to provide technical know-how, know-how which they alone in Europe possessed to any extent, while France and Italy assumed the costs. Clearly then these two countries were 'petitioners' for the project whose success was so important to them.

In the following weeks the replies came in. On 13 March Ivar Waller said he welcomed the developments in Paris, reported the interest which the project had aroused among young physicists, but noted that nothing had yet been done with the government. Cockcroft replied on 19 March that there would be no difficulty in sending an expert for about ten days although it might be more difficult to assign someone for a longer period 'owing to pressure of work here'. Three days later he proposed that Auger have a more general discussion of the whole scheme with people like Chadwick, Fry, and himself. He added: 'There is, I know, some feeling that the actual energy of the accelerator requires careful consideration in relation to the effort which will be available. We have a good deal of experience here on the effort required to the big jobs' (sic). At the beginning of April Scherrer replied that he would ask Huber or Preiswerk to go to Paris for two weeks 'to collaborate with your group'. Finally Bakker, on the 10th, and Kramers on 16 April, replied for the Netherlands. In principle, wrote Bakker, 'it has not yet been decided to what extent the Netherlands can share in the setting up of a European institute for nuclear physics' but he would be coming to Paris at Kramers' request. For his part, Kramers sent a very cordial letter emphasizing that the collaboration between the Netherlands and Norway pointed the way forward for the more ambitious UNESCO project, that all the same the problem of cost should not be forgotten ('the main problem is the enormous amounts being spent on rearmament' he noted), and that he eagerly awaited the technical results 'of the small committee which is preparing the future discussions in Paris'. He also confirmed that Bakker for the Netherlands and Odd Dahl for Norway would be able to spend some time in Paris.[14]

4.1.5 SITUATION AND DIFFICULTIES AT THE END OF APRIL

Now that the main facts have been established, a tentative assessment can be made of the situation prevailing on the eve of Amaldi's arrival in Paris for a week's stay.

The first thing to note is that most European countries were contacted, eight to be precise. However, some of the states on the list drawn up in Geneva had not been informed. It is probably not very significant that Greece, Luxemburg, and Yugoslavia were left out — they were small countries scientifically speaking and their participation at this stage was therefore not crucial. It is also highly likely that Auger did not know anyone to contact in them — as seems also to have been the case for Austria. There remained the Federal Republic of Germany and Denmark. For the first, the political reasons we mentioned earlier probably still applied, but they certainly did not apply to Denmark. Why no contact with Bohr? This remains an open question for us.

The second point to try to establish is that of the 'atmosphere' as Auger or Amaldi might have felt it. Obviously, the situation was clear for Belgium, France, and Italy: in these countries influential people had furnished proof of their support. For the other countries, at least in the short term, there were encouraging signs. Admittedly Bakker had said that nothing substantive had been decided about Dutch participation, admittedly Great Britain was not prepared to send a physicist to Paris for any length of time (and we will see that, despite what they said, the French and Italians were in a similar position), admittedly no *government* had yet been contacted save those of France and Italy. Nonetheless, each country was prepared to send a well-known specialist for a week or two to help the UNESCO study office while, in the case of Britain, there was talk of a very high-level meeting with the project's leading lights.

The grounds for an 'informed optimism' were, however, somewhat offset by the disquieting fact that it was proving difficult to set up a real study office. In Geneva Ferretti had been proposed and it had been hoped to find a Frenchman or a Norwegian. Unfortunately, despite the time which had passed, nothing had happened: the Ferretti idea was dropped without the Italians being able to find a replacement, and the hoped-for Norwegian never materialized. Auger's efforts to find a British specialist in March also failed, as we have seen.[15] That left the French. Auger probably discussed the possibility of their releasing one or more young research scientists for a long period with various laboratory directors in Paris. It seems that this was no easy task though, since it was only at the end of April that, thanks to Pierre Grivet, professor of electronics at the Sorbonne, he could recruit Edouard Regenstreif, an engineer of the Ecole Supérieure d'Electricité in Paris, and a specialist in electronic optics. He was, therefore, by no manner of means an accelerator specialist. More seriously no one else was recruited in 1951 and in fact the study office was never actually set up — at least in the form initially envisaged.[16]

The sources of these difficulties seem relatively easy to understand: each country needed all its specialists and none was prepared to 'give up' even one, particularly at this stage of the project when there was no question of doing real research or of constructing a machine. This completely disrupted the initial plan. Obviously Edouard Regenstreif alone could not 'speedily make concrete and detailed plans for the laboratory', nor could he alone make the 'highly technical study [...] for the laboratory's main machine' as Auger had first hoped. Accordingly the way in which the whole preparatory phase was structured had to be rethought. It was to this that Amaldi and Auger applied themselves when the former visited Paris.[17]

A last remark on this turn of events. We have hinted at the more or less predictable impossibility of setting up the study office. Nevertheless, even if it had been established, it certainly could not have done everything expected of it in six or eight months. As we now know, that would require several years' work by some of the top European specialists in the years to come.

It must not be thought, however, that we are making a blanket criticism—all too easy from the vantage point of the present![18] As we have said we know that the initiators of the project were not accelerator specialists and that they viewed matters from a more general standpoint. Their goal was not merely to construct a medium-sized accelerator, it was to awaken Europe and, through the construction of a giant accelerator, to make her states understand the urgency and necessity of developing fundamental scientific research on a large scale as had happened in the United States since the war. To do that they had to be ambitious, to take risks, to arouse the enthusiasm and interest of the younger generation. Understandably then the initial conception of the intermediate stage of the project was liable to be vague, and when their aspirations were put to the test their over-optimistic aspects were revealed.

4.2 Meeting of consultants at UNESCO in Paris, May 1951

The most conspicuous result of Amaldi's visit to Paris was the redirection of the plan of work. While Auger could still write, in a letter to Perrin on 30 April, that, 'thanks to these special funds [from Belgium, France and Italy], UNESCO [was] now in a position to set up a study office', and that Regenstreif '[had] been appointed to carry out the technical work' under the supervision of European experts, he also added—and this was new—that 'a first meeting [of these] experts [would] probably be held from 23 to 25 May 1951 at UNESCO headquarters so as to have an exchange of views on the work to be done by the said study office'. In other words, faced with the difficulties of having a preliminary report drawn up by a nonexistant study office, Auger considered holding a meeting of all those whom he had previously invited separately as consultants. Their task would be to prepare the ground for the report needed for the intergovernmental conference.[19]

Auger therefore sent another series of letters to Bakker, Cockcroft, Kramers, Randers, Scherrer, Waller, and Willems, asking them to send consultants to the meeting of 23–25 May. They all agreed to do so and the meeting was attended by Alfvén (standing in for Waller, who was ill), Amaldi, Capron, Dahl, Goward (on Cockcroft's suggestion), Heyn (proposed by Bakker, who was in the United States), Kowarski, Perrin, and Preiswerk.[20]

4.2.1 THE OUTCOME OF THE CONSULTANTS' MEETING

We have two reports on this meeting—one probably drafted by Jean Mussard, Auger's right hand man on the administrative side of the project, and a second by F. A. Heyn prepared for the Netherlands Institute for Fundamental Research into Matter (FOM).[21]

The first of these two reports, which was to be submitted to UNESCO's General Conference for information and approval, begins with a statement of the reasons justifying the project. Arguments of a general kind such as we have already encountered are first given: the cost exceeded that which individual countries alone could afford; there was a need to slow down the brain drain to the USA by making available to European scientists machines equivalent in quality to the American ones; it was important to deliver a psychological blow to European science, and to support its technological and scientific development. On the scientific level it was stressed that 'it [was] hardly possible to overestimate the importance of a modern and powerful cosmotron': theorists would have first-hand knowledge at their disposal, other areas of research would benefit from the laboratory's work (chemistry, biology etc.), and scientific collaboration would be strengthened. In other words, the laboratory was vital if Europe was not to lose touch with fundamental research and, by a domino effect, in other areas of science.

The envisaged equipment was then described. At the outset it was again stressed that a nuclear reactor should not be constructed, *at least for the time being.* Scientifically and politically a powerful accelerator was to be preferred, a project for which it should be possible to have the backing of governments. However, as there were already a number of small accelerators in Europe, the authors of the report noted that, to interest *all* the countries, a *big* accelerator should be planned. Its energy range should be between 3 and 6 BeV because 'the creation of heavy particles by protons [was] possible from about 2.5 BeV upwards.' Mentioning then the financial and industrial constraints and the time factor, the group proposed to *copy,* generally speaking, the machine at Berkeley, and to abandon the idea mooted at Geneva to construct the biggest machine in the world. Heyn observed that, in this way, the machine plans could be submitted immediately to UNESCO without having first to establish a provisional organization.[22]

The cost of such a project was estimated, on the basis of figures for the Berkeley machine, at about 12 million 1951 dollars. About the same sum would be needed to equip the laboratory. Between 20 and 25 million dollars were therefore required. Spread over five years—the estimated time for the construction of the accelerator—the cost would be about 5 million per year, distributed between the participating states. As in Geneva, this distribution was to be according to the rules in force in UNESCO or the United Nations, no state having to pay more than one-third of the total.

The next issue that the reports raised was that of the site. As in December, Geneva was already considered to be the best place. This was made clear in Heyn's unofficial report, though in the official UNESCO report the approach adopted in Geneva in December was again followed: criteria were defined without mentioning a particular preference.[23]

Before going any further, one remark could perhaps be made. The proposals at this meeting seem to be rather similar to those made five months earlier in Geneva. This is easily understood since no more work on the technical aspects of the project had been done since December. The organizers, who had not been able to set up their study office, had nothing new to offer. This accounts for the dissatisfaction felt by some of the participants. We know of only one example of this, but we suspect that others shared it. When Professor Alfvén reported to the Swedish *Atomic Energy Committee* in August 1951, he said that 'the conference was badly prepared. UNESCO had let it be understood that detailed plans would be available but nothing of the kind was ready'.[24]

All the same there were some differences between the proposals made in December 1950 and those made in May 1951. They concerned the rate at which the various stages were to be carried out, and the resources that were thought necessary to get the project off the ground. For the first time it was suggested that a *provisional* organization, financed by the participating states, would be needed, 'since the funds available for the first stage [would] not be adequate to carry out detailed studies'. A four-stage timetable was therefore drawn up instead of the three-stage one put forward in Geneva, and a fairly clear distinction was made between questions relating to the technical characteristics of the accelerator and those relating to the way the new body would be organized and financed.

The *first stage* covered the year 1951 and so included the work in hand. The 'study office' was made responsible for collecting the documentation necessary for the preparation of a copy of the Berkeley Bevatron, and for cost estimates related to that and to the other equipment which would be vital for the laboratory. For its part the group of consultants was to make a study of the administrative structure as well as of the organization and financing of later stages. This work was not intended to be definitive but merely to prepare the way for an intergovernmental conference to decide both on the project's practicability and on the setting up of a provisional organization funded with 200,000 dollars.

The establishment of this body would initiate the *second stage* of the project, to be devoted to the detailed technical study of the accelerator. This time the study was to be exhaustive, since the third stage was to start with the construction of the accelerator. A design office of at least ten people would draw up the plans and it was expected that at least a year would be needed. In Heyn's opinion, in view of the major decisions which would have to be taken, the most delicate problem would be to choose a scientist to lead the project. To assist him the intergovernmental conference would appoint a board of experts to whom the design office would report. The board would also study the administrative and legal aspects of the project.

The *third stage*—the building of the laboratory—would start as soon as the technical study had been approved and the administrative and financial decisions had been taken by the countries concerned. The *final stage*—that of operation—would then begin, in the course of which 75 (sic) 'scientists and engineers' would work in the new laboratory with an annual budget of about one million dollars.

4.2.2 A REMARK ON THE FUNCTIONING OF THE ENVISAGED LABORATORY

The idea which the consultants formed in May of the relationship between the laboratory and the participating states merits a closer look. Heyn's report gives us perhaps the best insight into this aspect, since it probably reflected the spirit prevailing at the meeting more faithfully than the other report. Heyn had no official function and he was therefore able to be more direct and frank. We have already pointed out that he said quite clearly that Geneva was considered by those present to be the most attractive site and that agreement had been reached that, by and large, the Berkeley machine should be copied. The other report was more diplomatic and was carefully vague on both these points. In addition Heyn, who did not have a pre-eminent position in the Dutch physics community, needed to report as precisely as possible to those who had sent him to Paris, namely Bakker and Kramers. For all these reasons it seems useful to discuss his description of what was decided about the relationship between the laboratory and its member states.

Briefly, Heyn saw the laboratory as operating like any national laboratory or like Brookhaven in the United States, i.e. as an institution created and funded by a government but where the decisions and subsequent options were the sole concern of scientists. Thus in *stage 2,* once the intergovernmental conference had voted the 200,000 dollars and approved the setting up of the *provisional* laboratory, the latter was to be managed solely by the board of experts made up of high-energy physicists. Similarly, in *stage 4* (and probably in stage 3, though that was not expressly stated) the 'permanent governing board' which was to define the programmes and manage the day-to-day affairs of the laboratory was to work 'under the authority' of a sort of council called the 'new board of experts'. This was also to be made up solely of nuclear-physics specialists, though it was made clear that they should come from various member states. The structural chart therefore did not allow for any non-scientists, which at first sight may seem a little surprising for an intergovernmental body.

Nevertheless, government representatives had a role to play, but not in the running of the laboratory as such. For example, *at the beginning* of the second stage they would have to assess the practicability of the project and its importance and make available the 200,000 dollars; at the *beginning* of the third stage they would have to approve the technical study, the administrative structure (rules governing appointments, ...), and grant the 5 millions needed annually over 5 years; finally, *before* the last stage started they would have to adopt definitive regulations and undertake to pay the one million dollars annually which were essential to the smooth running of the laboratory. Once that had been done, however, *political circles would have no further role* and the laboratory would be exclusively in the hands of scientists. In all likelihood, there would be reports on the work done—as in any national laboratory—but no mention was made of this.

We would suggest four main reasons why these 'scientific experts' conceived the project in this way. Firstly, all the existing models operated thus. We must not forget that we have here the first major international laboratory to be set up. Secondly, the model was consistent with then-current attitudes among scientists: a scientific organization, albeit international, managed even partially by 'politicians' seemed somewhat strange. The state should provide a scientific institution with an overall budget, in this case for a particular project, but everything else should be left to the scientists. Otherwise conflicts between states, or unnecessary bureaucratic obstacles, would surely be introduced. What is more—and this brings us to the third reason—since the laboratory's expenditure was roughly known, there would be no need to renegotiate with states as such. For minor modifications (a little more than one million per year was spent, for example), the scientists representing the states on the new board of experts could take the necessary steps. Lastly, the laboratory would only carry out fundamental research in a field in which no significant industrial application was thought imminent. The member states therefore had no need to be involved directly in the administration of such a laboratory so as to protect their own industrial or commercial interests. The structure proposed therefore did not lack for arguments.

On the other hand, the official report drafted by Mussard was couched in more cautious terms. Heyn's board of experts (or the new board of experts) was called the 'Committee [or board] made up of representatives of governments'. Was that merely semantic cautiousness which had no bearing on the real problem? Or was it a hint that governments would try to place 'their' men on the committee, that they would want to have 'their' say? We do not know with certainty since Mussard was not very explicit on this point. However, it is not unreasonable to think that he had those doubts in mind when writing his official report for the UNESCO conference, and that he probably considered it a little naive to visualize a multinational body without permanent political representation.

4.3 Discussions arising out of the consultants' report, May–July 1951

The first task of Auger and Mussard at UNESCO at the beginning of June 1951 was to submit to the organization's annual conference the proposals advanced by the group of consultants. Another aspect of their work was to complete the preparatory studies (which the consultants called *stage 1*) for submission at the end of the year to the intergovernmental conference which UNESCO was to convene. In other words while the official procedure for convening the conference had to be started by obtaining the UNESCO conference approval for the consultants' report, scientific preparations also had to continue, which was the business of the physicists and scientific administrators working with them (Willems, Colonnetti, ...). In this section we shall study the latter.

4.3.1 FIRST REACTIONS

Having drafted a report summarizing the discussions held in Paris from 23 to 25 May Jean Mussard sent, between 8 and 14 June, a circular letter to those who had taken part. Replies were received between 15 and 30 June.[25] In his letters Auger's assistant put four questions. The first concerned his report: do you find it satisfactory? do you think other points should be added? In the main the consultants agreed with Mussard's record and only Preiswerk, Dahl, and Amaldi proposed that further details be added to justify better the need for a European laboratory to governments. Odd Dahl thus wrote:

> It may also be pointed out that a modern nuclear physics laboratory is in its scope a *universal* laboratory, even if the final aim is for highly specialised knowledge. The projected laboratory will thus serve as a *training centre* in coordinated research, covering physics, bio-physics, chemistry, biochemistry, technology and medicine in its basic and applied forms, creating a type of research worker *adaptable to industrial research* in the respective home-countries[26] (our emphasis).

The second point raised by Mussard concerned that of the documentation needed for the study office: accelerators under construction or operating in each country, bibliographies which might be useful to the study office, industrial resources or qualified personnel available nationally. Although in some cases the answers were rather brief, all the consultants sent the information requested, giving Auger and Regenstreif a better picture of the European situation.

This inquiry merely prefaced a more detailed study that Regenstreif was to undertake between June and September, the consultants' meeting having agreed that he should spend a few days in each country so that he could learn more about the possibilities for accelerator construction on the spot. Mussard raised this third issue in his letters, and the trips were accordingly arranged. Regenstreif went to Switzerland at the end of June, to Belgium at the beginning of July, to Norway in mid-July. In September he went to the Netherlands and to the United Kingdom where he spent a week.[27]

The last question asked by Mussard in his letters was: to whom should we send the report to make it known and to obtain support? In reply some lists of names were sent comprising, in the main, nuclear physicists and administrators in big institutions to whom the report was then distributed.

Notes: p. 145

One thing is clear from the correspondence exchanged between Mussard and the consultants in June: everyone was interested in the project and did his best to help the study office set up under the auspices of UNESCO. However, the level of commitment varied from one person to another. Some were satisfied simply to reply to the letters, others showed a far greater determination.

Amaldi, Preiswerk, and Verhaeghe were typical of the latter. Amaldi wrote to Regenstreif on 11 June and then to Auger to warn the latter of unwelcome developments in the United States;[28] the following day he wrote to Mussard to ask what might be done in relation to Austria and ... Ireland; on 21 June he replied to Mussard's letter and sent a copy of a letter from Collins of Brookhaven along with the Cosmotron plans; on 30 June he sent Mussard details of Italian accelerators ... The same could be said of Preiswerk who gradually became one of the project's most ardent supporters. On 12 July, for example, he proposed to Auger to take advantage of his visit to Lucerne in September (the annual meeting of the Swiss Society of Natural Sciences) to give a 'brief summary of the projects for the European laboratory'; on 17 August he asked when the intergovernmental conference was scheduled, remarking that it was fairly pressing and that a meeting of consultants should be held to define the project in scientific terms; at the end of that month he published an article in the Swiss press ... As for Verhaeghe, on his return from the United States he wrote to Auger to say that he was prepared to come to Paris to discuss the project as a whole, he reiterated that the machine at Berkeley was superior to that at Brookhaven, he put forward a number of technical variants for the European accelerator, he proposed that the injector be built by Philips in order to ensure Dutch support ...[29]

Very positive responses, therefore, seemed to be forthcoming from some Italian, Swiss, and Belgian scientists.[30] Can we then infer overall support from the scientific communities in these countries? Not necessarily, particularly in the case of Switzerland. When reporting on his trip to Switzerland Regenstreif had said that 'not all physicists agreed to recommend to the Federal Council that it vote the funds needed to build the cosmotron'. If Preiswerk, Staub, and Wöffler were in favour, Huber and above all Paul Scherrer were hostile to the project. Accordingly 'a change in Mr Scherrer's negative attitude [seemed] vital', Regenstreif wrote, since it could influence the position of other countries like the Netherlands and Sweden.[31] On the other hand, a week later Regenstreif wrote on his return from Belgium that 'in contrast to Switzerland, the general feeling in Belgium [was] very much in favour [...] There [was] no opposition at all', and the Belgians were 'even surprised at the project's slow rate of progress'. They were, nevertheless, aware that there was opposition: they knew about 'Britain's hostility', they were concerned about 'Mr Scherrer's coolness' particularly since, in Bakker's view, 'it would have repercussions on the reservations in the Netherlands'. The Belgians therefore proposed that 'France started negotiations with Britain on a "pool of accelerators" since the sharing of "accelerator resources" would be likely [...] to encourage

Britain to join the project [...]'. For his part Regenstreif suggested that Professor Weisskopf of MIT should be asked to talk to Scherrer since he was 'very much in favour of the cosmotron' and had, it seemed, 'already taken the first steps in this direction on his way through Zurich'.[32]

It can be seen then that there was a wide spectrum of attitudes, from Amaldi and Auger who were the architects of the project on the one hand, to Scherrer and the British on the other, with a large grey area in between. In Auger's view Bohr fell into this last bracket. In a letter to Weisskopf of 5 July Auger asked him to try and convince Scherrer of the wisdom of the project and he added: 'It would be very useful if you could also talk to Bohr. Bohr is interested, it was his suggestion to build the laboratory around a cosmotron, but it would be very useful to know what he thinks about the situation in general'.[33]

This mention of Weisskopf brings us to the question of the attitude of the consultants and of Auger towards the Americans. As John Krige has already explained, the pioneers of CERN looked upon the United States as 'a model to be emulated and a resource to be tapped'. And indeed during June and July a number of elements justifying this description can be identified. The machines under construction at Berkeley and Brookhaven were considered as models to be imitated and the consultants asked American scientists to send them any documentation available on the subject—Amaldi contacted Collins, Preiswerk got in touch with Weisskopf and Bishop, Auger wrote to Rabi and Brobeck.... But the United States was also seen as the country to which those who would be in charge of the technical studies should be sent on a study tour. In February it had been thought that Amaldi should make another trans-Atlantic trip, and in June it was decided to send Regenstreif. To do this two important American laboratories were asked for help and the visit was organized between them, the American government and UNESCO. Finally, American physicists in favour of the project (Rabi, Weisskopf) were contacted so that, in so far as they were able, they could use their influence to convince the less enthusiastic Europeans. This Weisskopf did, and Rabi proposed to do.[34]

4.3.2 THE SEVENTH GENERAL ASSEMBLY OF THE INTERNATIONAL UNION OF PURE AND APPLIED PHYSICS (IUPAP), COPENHAGEN, 11–13 JULY 1951

Although probably not intended, the General Assembly of IUPAP was to play a not insignificant role in the definition of the European laboratory's programme. And not without reason: for the first time since December 1950 a meeting of many of the most eminent physicists from ten western European countries, the United States, and Canada took place *and* the project had begun to take shape. It was no longer a question 'of one of the high-flown and crazy ideas which emanate from UNESCO' but a scheme for which some governments had put up money and which would probably soon be discussed officially at an intergovernmental meeting. Nothing could be more normal than that the subject would be raised. All the more so since Amaldi, in accordance with the decision taken at the previous meeting of the IUPAP Executive Committee held in September 1950, intended to make a report on the lines set out by the consultants in May supplemented by Regenstreif's latest observations.[35]

Two days before the Assembly met the matter was debated by some of those attending the colloquium on *Problems of Quantum Physics* organized by Bohr and Rozental from 6 to 10 July, after which Bohr and Kramers sent the following telegram to Auger: '[...] PROJECT NUCLEAR CENTER RAISES INTRICATE PROBLEMS DEBATED INFORMALLY BY PARTICIPANTS COPENHAGEN CONFERENCE AND BE FURTHER DISCUSSED DURING INTERNATIONAL UNION ASSEMBLY STOP DESIRABLE YOU COULD ATTEND PERSONALLY'.[36]

As far as we know there is no very detailed record of this meeting. However, indirectly we can reconstruct most of the decisions taken using the many letters written afterwards. The documents we have are listed in the notes.[37]

The first impression one has is that it was on this occasion that criticisms were first voiced about the overall conception of the project. To quote Kramers in September 1951, although 'the idea of the project met with considerable sympathy' (that idea being 'to provide opportunities for a strong international group of older and younger physicists to perform fruitful research on a level and a scale which no local European scientific laboratory would have the means to attain'), 'doubts were expressed from various sides as regards some points in the contemplated programme'.[38]

These objections concerned the very principles on which the construction of the laboratory was based i.e., in Kramers' words, the setting up of a 'costly new institution the immediate purpose of which [...] would be to build a big accelerator of a type now under construction in the USA'. In the opinion of Kramers and Bohr this was unrealistic. It seemed to them to be more prudent, both scientifically and financially, to proceed step-wise and to build, in the first instance, a smaller machine. In that way 'the effort and expenditures in the initial stages would not be solely devoted to preparatory activities which [would] only bear fruit after a considerable lapse of time', and the laboratory could quickly prove its worth. Furthermore this would justify demands for an increase in aid from governments which at the time were perhaps 'hesitant to commit themselves to obligations towards an enterprise the success of which would seem rather far off and difficult to assess'.[39]

This argument reflected the concerns which had been expressed in the preceding months, both by those who were very much in favour of the project and by those who were less convinced. For example, at the meeting between Perrin and Amaldi in Paris on 2 May, Perrin had already 'suggested that the laboratory might start by building a less powerful machine in the first instance [...]. A group of physicists could thereby start work on a definite and immediate task'. Similarly Wideröe, whom Regenstreif met in June, had proposed 'to proceed in stages' building 'one or more intermediate machines before starting on the bevatron proper'. At the beginning of July Verhaeghe had also recommended this approach, adding that 'a 200 MeV machine would greatly facilitate the training of the cosmotron's scientific and technical personnel'. It was possible therefore to agree without undue difficulty that a smaller machine should also be built.[40]

This did not solve all the problems, however. For although everyone could accept that in the long term a big synchrotron was needed, the time when one should start its construction remained an open question. For reasons of cost ('Bohr and Kramers were suddenly afraid of the consequences (especially financial)' wrote Auger to Rabi on 21 August), but probably also for scientific reasons, a number of people were in favour of delaying the construction of the big machine. Conversely, Auger and his colleagues emphasized that the attraction of the project was precisely to build a machine that no country could build on its own. If construction was postponed some states might lose interest in the whole scheme. It is difficult to be more precise about the discussions held at the time but the document drafted after the conference, which altered the work programme, seems to justify our assertions. This document also proves that an agreement was more or less reached. It foresaw that *stage 2* (1952) would still be devoted to the design of the accelerators, but that the small machine would be constructed in *stage 3* (1953–1954), while the bigger accelerator would only be started in *stage 4* (1955– ...). In the course of the latter the small machine would come into operation.[41]

'The still more complex question of organization and effects — favourable and unfavourable — on the functioning of the universities, on the training of students, and on the availability of scientific manpower'[42] still had to be discussed. These questions were not studied at the Copenhagen meeting but it was agreed that an *Institute for Advanced Studies in Nuclear Physics* be set up, the planning for which would be done in 1952, and which would organize courses, colloquia, and a summer school in *stage 3*. At the same time it would demonstrate to governments putting up money that progress was being made and that their investment would not only show returns in the far distant future.[43]

The result of all these dealings was that at least Auger and his friends had the impression that the difficulties were being overcome. Of course everything was perhaps not yet completely clear, but some hurdles had been crossed. Despite Kramers' doubts, Bakker seemed determined; Cockcroft, during a long conversation with Amaldi and Bakker, is said to have promised his support with the reservation that he could not say what the attitude of the British authorities would be; finally and most significantly for Auger and Amaldi, Scherrer no longer seemed against. In a letter to Auger dated 27 July 1951 Preiswerk wrote: 'I have just received a very optimistic card from Mr Scherrer. On his return Mr Weisskopf told me that important changes to the project had been decided in Copenhagen. I would

Notes: pp. 145 ff.

be very pleased to know in detail about the decisions which were taken'. And in August Auger himself concluded in a letter to Rabi: 'After two meetings, and lots of help from younger physicists (like Bakker and Preiswerk), we came to a satisfactory agreement. I hope to be able to call the first constitutive meeting this fall'.[44] Apparently all that remained to be done was to organize the intergovernmental meeting.

4.4 The intergovernmental meeting is called

We have seen that the document drafted by Mussard and sent to the consultants between 8 and 14 June for their approval was of an official nature, its aim being to help the Director-General draft his paper for the sixth session of the UNESCO General Assembly *(Report on preliminary studies regarding the establishment of a European Regional Laboratory for Nuclear Physics)*. This report, published on 19 June 1951 and numbered 6C/PRG/25, started with an introduction setting out the origins of the idea, from the 'Rabi' resolution approved in Florence in June 1950 to the most recent meeting of consultants in May 1951, the record of which was attached as an annex. On the basis of the conclusions reached in May, the Director-General proposed that 'a conference of governmental representatives be held in the autumn of 1951', with the aim of deciding on 'definite proposals concerning the organization and financing of the detailed study of the project' for 1952.[45]

One sentence of the report deserves special mention since it shows that in June 1951 Auger—and the Director-General—had no inkling of the criticisms that would be made at Copenhagen. According to the report the meeting in May 'provided an opportunity of putting forward suggestions relating to the chief points to be elucidated by the preliminary study [...]. As a result, the secretariat [was] now in possession of sufficient information to enable it to arrive at the conclusion that the preliminary survey called for by resolution 2.21 [would] be completed before the end of 1951' so that *stage 2,* which would be the responsibility of a provisional organization, could be launched.

The report was submitted to a working group set up by the Programme Committee of the General Assembly at its first meeting. The working group met four times before publishing its conclusions (document 6C/PRG/26) on 28 June. The group said that it was satisfied with the results obtained so far and recommended that work be continued along the lines set out in 6C/PRG/25, provided that 'interested governments [would] assume the financial [...] responsibility'. It added that in its view consideration should also be given to letting other states participate in the project, to the conditions on which students and scientists from non-participating countries could have access to the laboratory's facilities, and to the means by which a link could be maintained with UNESCO. In short, the UNESCO working group gave its blessing to 6C/PRG/25, merely emphasizing that it wished the laboratory to be as open as possible.[46]

These findings were accepted by the Programme Committee and the General Assembly entrusted UNESCO's Director-General with the task of continuing the studies along the lines set out in documents 6C/PRG/25 and 26. After some hesitation as to the date, the official letter of invitation to the member states of UNESCO was prepared, proposing to hold a 'conference on the organization of studies relating to the possible establishment of a European Nuclear Physics Laboratory'. The letter was sent to the various governments on 31 August 1951.[47]

This letter has several noteworthy features. In the first place, it referred to a laboratory 'built around *one* or *more*' powerful accelerators. The feelings expressed in Copenhagen in July had therefore been taken into account even if the proposed changes were never discussed in detail. It was also clearly stated that 'member states willing to associate themselves [with stage 2 in 1952] [would] not thereby commit themselves to taking part later in setting up the laboratory'. This stressed the provisional character of the body responsible for the detailed studies, as well as the fact that a step-wise approach would be adopted. No large financial commitment was therefore envisaged at this stage. Thirdly, in order not to put off those who might be afraid of being caught up in an uncontrollable sequence of events, it was stated that observers could play a full part in drawing up the programme 'should it become apparent in the course of the discussion that the government in question would be prepared to participate'. Lastly, it was stressed that, given the nature of the subjects to be discussed, it was desirable that 'each delegation should consist of two members, one a scientist and the other an administrative and financial expert'. The Director-General — in actual fact probably Auger — thus adopted a more supple position than the scientists at their May meeting. Not that he implied that 'administrative and financial' specialists should subsequently be members of the Council of the provisional organization. He left that to the conference to decide. But when officially convening the conference he chose to emphasize that both sorts of specialists would be useful.[48]

4.5 A marriage of convenience

To conclude we would like to emphasize the changes which took place between December 1950 and August 1951. The consultants' report of May represented a first move. By proposing a new agenda, a different timing, it recognized the value of an intermediate stage. Such a stage was worthwhile for a number of reasons. Scientifically it meant that the design of a giant accelerator would require time and a certain number of qualified people. Financially it showed that the cost of the design studies would exceed the initial estimates, or at least the resources of UNESCO, even though they would be supplemented by Belgian, French, and Italian help.

Politically it demonstrated the wish to move forward more gradually so as not to scare governments with demands which might be considered not only burdensome but above all hasty and uncontrollable.

The trend towards more 'caution' was reinforced by the decisions taken in Copenhagen. Participation in the provisional stage (1952) would not imply any subsequent commitment. Reliable designs and estimates would be available in due course and only then would countries have to take a firm decision. Furthermore staggering the construction programme entailed that governments would only become progressively involved. The sums of money requested would therefore not be enormous to start with. The idea of building a smaller machine and of setting up an Institute for Advanced Studies would enable enough specialists to be trained to use the big machine to its full potential and while it was being built to let them 'do some physics'. A certain 'realism', a certain 'pragmatism' had indubitably crept in.

Did this mean that no problems remained and that there was now unanimous agreement? It would be hazardous to assert this because two quite separate concerns persisted despite the common wish to build a European organization. For some the big machine was the main preoccupation; for others it was precisely that which posed a problem. In this sense the 'agreement' of Copenhagen can be seen as a kind of *modus vivendi* which placed these two concerns on an equal footing without really choosing between them. However, if it was possible to place them side by side in one document it was because everyone expected that enough time would elapse before major decisions had to be taken and because no substantial financial commitment was envisaged in the short term (1952). It was therefore possible to start to work together. Put differently though we can speak here of a marriage of convenience, it was one which seemed to be viable, at least in the near future.

And what of the immediate tasks, those to be accomplished before the intergovernmental conference scheduled for December? There were two. Firstly, the scientific parameters of the machines had to be fixed. This would not be easy since no study office existed. A means would therefore have to be found to carry out the Director-General's promise in his letter to governments that 'a scientific and technical report [would be] submitted to the Conference'. The second task was to draw up the financial estimates and internal regulations for the provisional organization: national representatives in December would not only be scientists, and they were likely to be quite interested in these matters. In any event, the 'specialists', the 'consultants' had to meet together again very soon.

Notes

1. The documents consulted for this chapter are to be found mainly in the following archives:
    - UNESCO archives, Paris, *CERN, Official Correspondence* file.
    - Jean Mussard's papers, deposited at CERN, Geneva.

Notes

2. See chapter 2, especially 2.1 on the international laboratories under the auspices of the UN and UNESCO; 2.3 and 2.4 on the importance of the European idea; 2.4, on scientific requirements. For more systematic studies on those points, see section 9.2 on France, chapter 10 on Italy.
3. In the particular case of the debates surrounding the birth of CERN around 1950, we return to a more general conclusion put forward by Jean-Jacques Salomon (1970a), 336: 'The international character of the intergovernmental institutions which pursue scientific research in Europe is less the result of a particularly internationalist or even European attitude amongst researchers, *than the result of the lack of national scientific institutions capable of undertaking these same activities on the same scale*' (our emphasis). See also sections 14.2, 14.3 and 14.7.
4. Minutes of the CNR Council meeting, 18/12/50 (CNR, 154-159); letter Colonnetti to de Gasperi, 18/12/50 (CHS, Colonnetti file, CERN); letter Auger to de Rose, 19/12/50 (Mussard file, CERN). Regarding the funds promised by the French in December 1950 and the delayed payment of the two million francs, see section 3.4. See also section 9.3.
5. Letters Dautry to Colonnetti, 15/12/50; Colonnetti to Dautry and to Auger, 18/12/50; Dautry to Auger, 21/12/50 (Mussard file, CERN); Dautry to Auger, 9/1/51 (CHS, Auger file, CERN) concerning a meal on the 11th with Colonnetti and Montel. It was Cacciapuoti who got in touch with Amaldi: letters Cacciapuoti to Amaldi, 7/2/51; 20/2/51; Amaldi to Cacciapuoti, 10/2/51 (CHS, Cacciapuoti file, CERN). The term 'super-consultant on the physics side' comes from Cacciapuoti's letter of 20 February. The description of the study office was given by Auger in letter Auger to Colonnetti, 8/2/51 (Mussard file, CERN). Additional details on the study office were given in an unsigned text (probably by Auger) entitled: *Request for permission to receive funds for the study of a regional laboratory,* which must have been prepared for the 25th Session of the Executive Board of UNESCO held from 15 to 27 January 1951 (CERN-CHIP/27/II). The quotations regarding the study office are taken from Auger's letter to Colonnetti dated 8 February.
6. Letter Cacciapuoti to Amaldi, 7/2/51 (CHS, Cacciapuoti file, CERN).
7. See correspondence between the Italian delegation to UNESCO, the Italian Embassy in Paris and Pierre Auger, particularly letters Auger to Colonnetti, 19/3/51; Auger to Morelli, Secretary-General of the Italian CNR, 27/3/51; Morelli to Auger, 5/4/51; Cacciapuoti to Amaldi, 12/4/51; Mussard to Morelli, 12/4/51 (UNESCO). The agreement of the CNR was obtained early in April, but since Auger was away until 23 April Amaldi's arrival was postponed until the 26th. Regarding Amaldi's stay, see also Edoardo Amaldi's diary (CHS, Amaldi file, CERN).
8. Letters Willems to de Dadelsen, 6/1/51 (CEC); Verhaeghe to Auger, 10/2/51; Auger to Verhaeghe, 16/2/51; Auger to Willems 21/2/51; Willems to Auger 2/3/51; Auger to Willems, 15/3/51; Auger to Verhaeghe, 22/3/51; Auger to Willems, 23/3/51 (UNESCO). On 27 April 1951, Auger informed Willems that UNESCO had received the money (UNESCO).
9. See section 3.4; letters Auger to de Dadelsen 3/1/51; de Rougemont to the Director-General of UNESCO, 8/1/51; de Rougemont to Auger, 8/1/51; Silva to Dautry, 9/1/51 (CEC).
10. *Diary* of Denis de Rougemont, *Denis de Rougemont's journey to Paris from 16 to 20 January* (CEC).
11. Document entitled: *Request for permission..., op. cit.* note 5; *25th session of the Executive Board of UNESCO, Minutes, 14th sitting, 27 January, item 12.7 of the agenda,* doc.25 EX/SR 1-15, 127, 28/2/51; *Resolutions and Decisions adopted by the Executive Board at its twenty-fifth session, 15 to 27 January 1951,* doc. 25 EX/Decisions, 34, 9/2/51 (UNESCO).
12. Identical letters Auger to Kramers, Randers, Scherrer, 21/2/51; to Waller, almost identical, 21/2/51 (UNESCO).
13. Letter Auger to Cockcroft, 2/3/51 (PRO-AB6/912). Auger concludes by adding: 'I am very glad to see that a real interest is shown by the countries constituting the 'region' and if some interest is shown concretely by the U.K. it will be even more gratifying'.
14. These letters are all in Mussard file, CERN. For the Dutch-Norwegian plan, see for example the introduction to Dahl & Randers (1951).

15. If the truth be told, the proposal of Ferretti never resurfaced in the correspondence between Auger and Amaldi or Colonnetti. His name only appears in connection with the Geneva meeting of December 1950. The question of a replacement is raised in several letters, such as Auger to Colonnetti, 8/2/51, or Mussard to Amaldi, 14/6/51 (Mussard file, CERN). Norwegian participation appears in a report which Waterfield sent to London following a conversation with Auger in January 1951; see letter McMillan to Awbery, 30/1/51 (PRO-DSIR17/559).

16. Letter Auger to Perrin, 30/4/51 (Mussard file, CERN); *Notice sur les Titres et Travaux* by Mr Edouard Regenstreif, Geneva, July 1959. In the report of the *Interview of 2 May (?) with Mr Perrin, High Commissioner for Atomic Energy* (Mussard file, CERN), which was probably written by Mussard and to which Amaldi seems to have contributed, we find the following:

    '3) I asked him [Perrin] if he would nominate a young physicist to take part in the weekly meetings which were to take place at the Study office. He appointed Mr Abragam, who works at Châtillon [...].

    5) [...] He was greatly in favour of the plan to invite young physicists representing the various French laboratories interested in nuclear physics to the weekly meetings, and as often as possible foreign physicists as well. He thought it would be a good idea to ask Mr Joliot-Curie to nominate one of his associates, and to ask the same of Mr Leprince-Ringuet and possibly also of Mr Rocard. The number of people attending the weekly meetings should not exceed ten'.

    Unfortunately, we do not know whether these meetings took place or not. Since we have not come across other references to them we think they probably did not.

    One final comment: Lew Kowarski (1961), 4, talked about Jean Mussard as some kind of secretary to this office. It would seem more accurate to say that, being a staff member of UNESCO (according to UNESCO terminology in 1951 he was 'Programme Specialist' in the Department of Exact and Natural Sciences), he was chosen by Auger to be his assistant on the European laboratory project.

17. The quotations are taken from the report of January 1950 quoted in note 5: *Request for permission ...* .

18. It is easy, of course, for the historian to make such comments, knowing, as he always does, how events subsequently transpired.

19. The quotations are taken from letter Auger to Perrin, 30/4/51 (Mussard file, CERN). On 21 February Auger had written:

    '[...] I hope that a number of experts from various countries will accept to come as consultants for short periods of about 10 days in order to give us their views concerning the implementation of this project. These experts will thus be able to make the acquaintance of the persons responsible for designing the Laboratory and to give advice regarding its tasks and its structure.' He added: 'I hope that it will be possible to organize these visits [...] in such a way that the consultants from the various interested countries do not all come at the same time' (circular letter, 21/2/51, UNESCO).

20. We have no clear proof that this redirection of work was the *result* of Amaldi's arrival in Paris. It is our impression that during the Italian physicist's stay the question of the study office was raised and that it was then that the decision was taken to call the meeting of 23-25 May. The only indication is of a chronological nature: the change took place while Amaldi was in Paris. Moreover, the letters of invitation were dated 2 May (the seven letters are in the UNESCO archives, along with the replies).

21. The report drawn up by Mussard formed the annex *(Possible establishment of a European Regional Nuclear Physics Laboratory)* of the Director-General's Report 6C/PRG/25 dated 19/6/51 (UNESCO). The report drawn up by Heyn is entitled *European Nuclear Research Laboratory,* 5/6/51 (Mussard file, CERN).

22. Although this is made clear in Heyn's report, it is more ambiguous in Mussard's report.

23. In letter Mussard to Kramers, 4/7/51, one reads: 'Please note also that in chapter III of the Consultants' report the surroundings of Geneva were explicitly mentioned as the best place for the future Laboratory. This mention has been deleted by the Director-General in document 6C/PRG/25'.

24. Auger had written on 2 May: 'This meeting will be devoted to a preliminary exchange of views on the organisation of the preliminary studies and on the work of the Study office [...]' (letter Auger to Kramers, 2/5/51, Mussard file, CERN). Alfvén's statement is given in Nyberg & Zetterberg (1977), 13.

25. This correspondence (several dozen letters) is in the UNESCO archives.
26. Letters Preiswerk to Mussard, 15/6/51; Dahl to Mussard, 24/6/51; Amaldi to Mussard, 30/6/51 (UNESCO). For an analysis of this aspect of high-energy laboratories as 'centres of expertise', see section 14.9.
27. It was never intended that Regenstreif should travel to France or to Italy, either because there was little value in doing so, or because the particular role of these two countries in organizing the project made such visits unnecessary. The trip to Sweden was postponed several times because of timetable difficulties. It apparently took place from 5 to 8 December 1951. See letter Alfvén to Mussard, 29/11/51 (UNESCO).
28. Amaldi wrote to Auger, 11/6/51 (Mussard file, CERN): 'I have just heard from Bernardini that the United States delegate to the next UNESCO conference is probably going to put forward a proposal for the creation of other European Laboratories [...]. Some of our American friends think that this would be the best way of *killing* the Nuclear Physics Laboratory [...]'.
29. This correspondence is in the UNESCO archives. See also Regenstreif's report dated 11/7/51 and numbered UNESCO/NS/MEMO/240168: *'Report on my trip to Belgium (4-7 July 1952)'*. (The date is obviously a mistake, it should read 1951.) We should add that Goward (UK) and Bakker (NL) showed considerable sympathy, even if it was not so marked as in the case of Amaldi, Preiswerk or Verhaeghe. As regards Goward, see letter Auger to Rabi, 21/8/51 (UNESCO).
30. The same applies to several Frenchmen such as Perrin and Kowarski. However, this does not appear in the correspondence, and their contacts with Auger must have been established by personal meetings.
31. Regenstreif's report dated 3/7/51 and numbered UNESCO/NS/MEMO/7984: *'Report on my trip to Switzerland, 28-30 June 1951'*. It should be noted that, despite persistent searching, we have only located the reports on Switzerland and Belgium. This is a pity, because these reports give a very detailed picture of the situation in the middle of 1951: the state of mind of scientists and administrators, accelerators which had been or were being built, plans, industrial potential, etc.
32. Reports quoted in notes 29 and 31. The Belgians' only real worry was the cost: 'they [in Belgium] agree with the figure of 12 million dollars for the machine itself, but they think [...] that the overall cost should not exceed 18-19 million dollars'.
33. Letter Auger to Weisskopf, 5/7/51 (Mussard file, CERN). Regarding Scherrer, Auger wrote: 'One of the people we *must* have with us on this project is Scherrer. You know that he is not in favour of the plan [...] but we must convince him [...]. Without putting yourself out too much, do you think you could try to persuade him?'.
34. See Krige (1983). On 21/6/51, Amaldi sent Mussard a letter from Collins of Brookhaven dated 1/5/51, along with the plans of the Brookhaven machine and a study of its cost which Collins had attached to his letter. On 15/6/51, Preiswerk put Bishop from Berkeley in touch with Auger and suggested that during Regenstreif's visit to Switzerland he get in touch with Weisskopf in an attempt to persuade Scherrer. On 2/7/51 Auger wrote to Rabi and on 10/7/51 he wrote to Brobeck of the Radiation Laboratory at Berkeley. Rabi on 15/7/51 and Brobeck on 15/8/51 gave extremely positive replies to Auger's requests for assistance. On 21/8/51 it was Haworth of Brookhaven who offered his services to organize Regenstreif's trip to the United States, Haworth having been contacted by Rabi. To show the state of mind of these American physicists, Rabi's letter of 15/8/51 can be quoted: 'I very greatly hope [...] that the important project will go ahead full steam. If I can be of any help to you in this connection, please let me know'. These documents are all in the UNESCO archives.
35. The first quotation is from Skinner, see Gowing (1974), vol. 2, 227. Regarding Amaldi's plan, see letter Amaldi to Regenstreif, 11/6/51 (UNESCO).
36. Telegram, 9/7/51 (UNESCO). See also interview with Amaldi by Gowing and Kowarski, 25/7/73 (CERN).
37. International Union of Pure and Applied Physics, *Position at 1st November 1951, Report of the Seventh General Assembly (1951)*, UIP.4, November 1951, 26; International Union of Pure and Applied Physics, Doc. SG.51.8 (July 1951), *Summary Record of the Seventh General Assembly,* typewritten document;

Copenhagen General Assembly of the International Union of Pure and Applied Physics, July 11, 12, 13, 1951, report prepared and submitted by J.A. Wheeler, chairman of the USA delegation, typewritten document (these three documents are in the Archives of the National Academy of Sciences, Washington D.C.). Letters Preiswerk to Auger, 27/7/51; Rabi to Auger, 15/8/51; Auger to Rabi, 21/8/51; Kramers to the Director-General of UNESCO, 24/9/51 (UNESCO). See also letters from the Director-General of UNESCO, 31/8/51, sent to the governments and entitled: *Conference on the organization of an enquiry into the possible establishment of a European Laboratory for Nuclear Physics,* NS 240185; document UNESCO/NS/PN/1, 19/9/51 and entitled: *Report of the result of the enquiries carried out since the meeting of consultants held from 23 to 25 May 1951* (UNESCO). Finally, see Nyberg & Zetterberg (1977), 13.

38. Quotations from letter Kramers to the Director-General of UNESCO, *op. cit.* note 37. Regarding these objections Rabi wrote in his letter to Auger on 15/8/51: 'I have heard rumors that the plan for the high energy laboratory has run into very strong opposition from some important personalities' (UNESCO).
39. Quotations from letter Kramers to the Director-General of UNESCO, *op. cit.* note 37. Our mention of Bohr as being on the same side as Kramers during this discussion is inferred from their later correspondence (see section 5.1).
40. See *Interview of 2 May ..., op. cit.* note 16, for Perrin; Regenstreif's report on his trip to Switzerland, *op. cit.* note 31, for Wideröe; his report on his trip to Belgium, *op. cit.* note 29, for Verhaeghe. In Kowarski (1961), the author confuses the results of this meeting in Copenhagen with the results of the consultants' meeting in May. In fact he said that it was in May that the construction of the intermediate machine was proposed. Although it is possible (or even probable) that this possibility was *discussed* in May, it was certainly not *decided to proceed* with it. That only occurred in July after pressure from Kramers and Bohr. A letter from Kowarski to Chadwick (15/12/51, CC –CHADI,1/11) confirms this fact beyond doubt. For an analysis of this letter see section 5.4.
41. Letter Auger to Rabi 21/8/51; document UNESCO/NS/PN/1, 19/9/51 (UNESCO).
42. Quotation taken from Wheeler's report, *op. cit.* note 37.
43. UNESCO/NS/PN/1, 19/9/51.
44. Letters Preiswerk to Auger, 27/7/51; Auger to Rabi, 21/8/51 (UNESCO). The conversation with Cockcroft is reported by Amaldi in the interview quoted in note 36.
45. Document 6C/PRG/25, 19/6/51 (UNESCO).
46. Document 6C/PRG/26, 28/6/51 (UNESCO).
47. This letter was sent to all member states of UNESCO. UNESCO had received 27 replies by the end of the year.
48. The statement that observers would be able to take part in the debate if their governments were likely to join the project later was undoubtedly intended to leave the door open for possible British delegates.

CHAPTER 5

The period of conflict[1]
August–December 1951

Dominique PESTRE

Contents

5.1 The gradual emergence of an alternative programme, August–October 1951   148
    5.1.1 The rise of the opposition around Kramers and Bohr   148
    5.1.2 The proposals put to Auger by Kramers and Bohr and the arguments advanced at the time   151
    5.1.3 Auger's reactions   153
    5.1.4 An attempt at a wider interpretation   156
5.2 Second meeting of consultants, Paris, 26 and 27 October 1951   157
    5.2.1 Preparations for the meeting, 19 August–25 October   157
    5.2.2 The meeting   160
5.3 Tension peaks, late November–early December 1951   163
    5.3.1 The document drawn up at the third meeting of the consultants (17–18 November) and submitted to the conference as the working document   163
    5.3.2 The official replies to the letters from Kramers, Bohr, and the Netherlands Minister of Education (20 November)   166
    5.3.3 The Oslo crisis   167
5.4 The roots of the division   169
    Notes   174

The previous chapter ended with a reference to the IUPAP General Assembly held in Copenhagen in July 1951. Here two groups confronted each other, the one putting forward a fairly detailed plan to build a very large accelerator, the other more sceptical, fearing that the project was overambitious. In fact both were surprised by the turn of events. Amaldi and Auger were faced with unexpected criticism of the very item they considered to be essential, i.e. the machine of energy about 6 GeV. For their part Bohr and Kramers were alarmed by what they took to be the unrealistic nature of such an expensive project.

As we have suggested, once the initial shock had worn off a kind of *modus vivendi* emerged between these two schools of thought, inspired by a spirit of mutual goodwill. At the time this was given no formal expression; deeper reflection was needed. And it was only to be expected that, once each group had reconsidered the situation, the original desires or fears re-emerged. Indeed, a gap began to reappear at the end of August.

It was to widen over the next few months. Actual hostility gradually took the place of what, in July, had been only a difference in conception. As the date of the intergovernmental conference drew nearer, the need to frame the specific decisions to be taken became ever more pressing. Tension mounted, and the differences between the groups condensed into two radically opposed projects. It will be the purpose of this chapter to describe the development of the dispute and to try to understand its roots.

5.1 The gradual emergence of an alternative programme, August–October 1951

Let us begin by giving a chronological survey of events, showing how they evolved and who the main protagonists were. We shall then take a closer look at the arguments put forward by both sides in support of or criticizing the projects presented. Finally, we shall attempt to explain the reasons for the attitudes described.

5.1.1 THE RISE OF THE OPPOSITION AROUND KRAMERS AND BOHR

In about the middle of August Kramers was in Copenhagen. After giving further thought to the plan discussed at the IUPAP meeting in July, he arrived at a fresh proposal that he put to Bohr: Why not consider setting up the European laboratory in Copenhagen, using the Bohr Institute as a core around which it could develop? This proposal was discussed with Chadwick, then on a visit to Copenhagen, who immediately made his support known to Bohr. Summarizing his feelings in a letter to Thomson, he later wrote:

> As you know, I was strongly opposed to the proposals which Auger made some time ago, for I thought they were quite impracticable [...]. Kramers' suggestion appeals to me very much. It is certainly practicable, and it is based on facilities, both in men and apparatus, which already exist.[2]

On being informed of Chadwick's attitude, Kramers decided to write to Auger on 23 August and to put his idea before him. The next day he told Bohr, Mott (then chairman of IUPAP), and Chadwick, whom he asked to warn Darwin. On 30 August Chadwick replied to Kramers, reiterating his complete confidence.[3]

Kramers wrote to Auger again on 31 August. Three days previously he had met the Director of the ZWO (the Netherlands Organization for the Advancement of Pure Research) who had told him of his reservations concerning the European laboratory project, at least in the form it had taken within UNESCO. The May consultants' report 'would hardly be considered by our government', wrote Kramers, 'as a satisfactory working paper on the basis of which it would be willing to send a government representative to the planned meeting of November 1951'. He added that the situation appeared to be similar in Denmark, Norway, and Sweden, 'the main point—just as in Holland—being that they [the governments] are afraid to commit themselves to plans of which they cannot oversee the consequences, about the financial implications of which they have certain misgivings'. This is why Kramers suggested that the envisaged programme be reconsidered using, for instance, the channel of states' official delegates to UNESCO.[4]

The same day Auger replied to the letter of 23 August. He thanked Kramers for having thought again about the European laboratory project. Auger also told him that he was willing to submit the proposal to use Copenhagen as the nucleus for the European laboratory to the scientists he would be meeting in Switzerland and in Chicago on the occasion of the inauguration of the new Fermi institute for nuclear studies.[5]

One week later, around 8 September, Ronald Fraser, the representative of the International Council of Scientific Unions (ICSU) at UNESCO, was in Copenhagen, where he was to meet Bohr to organize a symposium on epistemological problems in physics. As Auger had shown him the letter from Kramers before his departure from Paris, he and Bohr discussed the new situation. Both agreed that it would be a good idea to look more closely at how to put Kramers' suggestion into practice and they decided to meet Jakob Nielsen, Denmark's official delegate to the UNESCO General Conference. Two ideas arose from this meeting: that the Danish Foreign Minister had to be approached, and that there had to be a meeting in Paris with the Director-General of UNESCO, James Torrès-Bodet, in order to make an official announcement of

Notes: p. 174

the Danish desire to have Kramers' proposal considered. On 11 September, in a letter to Auger, Fraser gave an account of his visit to Copenhagen, stressing Bohr's enthusiasm and the advantages of Copenhagen as a site, but adding that such a proposal might not please the French. By the same post he informed Bohr, Kramers, and Chadwick of what he had done. Chadwick told Darwin and Thomson what he knew.[6]

On 14 September Bohr informed Chadwick of Fraser's visit. He emphasized Fraser's concern over the project drawn up in May[7] and informed Chadwick that Fraser had arranged a meeting for him (Bohr) and Nielsen with the Director-General of UNESCO for 27 September. Bohr then asked what the attitude of British physicists in this discussion might be—the information could be useful to him in his talks at UNESCO. Before receiving an answer, however, he again wrote to Chadwick (on 20 September) explaining that the time did not yet seem ripe to have discussions with Torrès-Bodet and Auger. Before doing so he would prefer to have a better idea of reactions in Europe to Kramers' proposal. Chadwick's reply arrived a few days later: Darwin's attitude to the idea for a European laboratory was in any event lukewarm, even if Thomson was 'definitely attracted by the Copenhagen scheme'.[8]

For his part Kramers seemed to think that the situation was fairly clear, for he sent an official letter to the Director-General of UNESCO on 24 September which he circulated in Europe. Of course he sent a copy to Auger with a covering note explaining the reasons for his action. As part of the preparations for the intergovernmental meeting, he said, it was his duty to warn the Director-General of the reluctance in some governmental circles, to inform him of the need to prepare a different working scheme from document 6C/PRG/25 drawn up on the basis of the May consultants' report, and to tell him of his offer to use Copenhagen as a 'pilot centre'.[9]

Auger had spent September in the United States and did not reply to Kramers until 11 October. His letter encompassed all the correspondence which had accumulated since 31 August. The reply is complex and we shall examine it shortly in greater detail. Before doing so it must be added that, on 12 October, Malte Jacobsson, Chairman of the Swedish Atomic Energy Commission, wrote to the Director-General of UNESCO, mentioning Kramers' proposal; on 24 October the Netherlands Minister of Education, Arts, and Sciences wrote to Torrès-Bodet expressing the Dutch government's fears; while on 26 October Bohr wrote to Auger, sending him a memorandum on the procedure to be followed if Kramers' proposal were to be accepted. He stated that this memorandum had been drawn up together with Swedish 'colleagues', who had also been in contact with Bakker in the Netherlands.[10]

The purpose of the foregoing account, which may seem somewhat elliptical, was simply to describe the network of contacts that Kramers and Bohr built up. Let us now turn to a more systematic examination of their project.

5.1.2 THE PROPOSALS PUT TO AUGER BY KRAMERS AND BOHR AND THE ARGUMENTS ADVANCED AT THE TIME

It is no easy matter to provide an overall account of what Bohr and Kramers had in mind in August, September, and October 1951 for a fairly simple reason: it seems that their thoughts had not, like Minerva, sprung fully armed from Jupiter's head, but had rather been gradually clarified between 20 August and 26 October. Thus, in his first letter to Auger (23 August) Kramers concentrated on his idea of using Copenhagen as a nucleus for building up the European laboratory. No mention was made of the need to rethink the entire project proposed by UNESCO. By contrast, two months later Bohr proposed to dismiss completely everything that had been done during the previous eighteen months: there was no longer to be any question of accelerators, at least for the near future; there was no longer to be any question of important commitments on the part of the states. There was thus a perceptible evolution towards a project increasingly at odds with Auger's.

We have decided to present these 'counter-proposals' by looking at the problems in the order in which they arose to those who had drawn up the project during the preceding months. This will lead us to examine:

1. the proposal for having Copenhagen as a site;

2. the criticisms levelled at the May consultants' report, judged to be an inadequate basis for the intergovernmental conference;

3. Bohr's suggestion for a different work programme for the following months or even years.

1. The proposal to have Copenhagen as a site[11]

Kramers' first idea was that 'the only way to ensure such an enterprise [the building of the European laboratory] a degree of success which would justify the big efforts demanded from the participating countries [was] to have the contemplated institution grow and develop in intimate connection with an existing active centre of research', using it as a 'pilot institution' from which the other could develop with all 'the necessary flexibility' and in 'close contact with the newest developments in theoretical and experimental physics'. According to Kramers, an immediate corollary of this proposal was that considerable budget savings could be made. His second idea (not separated from the first in the letter of 23 August) was that the Bohr Institute in Copenhagen 'could really act as a most effective nucleus for an international European laboratory'. There were many reasons for this conviction: firstly, there was Bohr's personality ('a wonderful security' wrote Kramers in August, adding a month later: 'the confidence he enjoys in all circles would be an inestimable asset'); then there was the institute itself, 'an efficient theoretical group with a long tradition in international co-operation'; finally, in Copenhagen one had 'a hard working nucleus of scientists and engineers who already [had] experience in

Notes: pp. 174 ff.

building large machines', and the institute had 'just undergone a major enlargement [...] and very good facilities [were] available for starting work on the preparations for the high energy machines [...]'.[12] For his part Fraser added two arguments, one of a geographical nature ('the very great economic and topological advantage of a site at sea level'), the other more important and of a more political kind: the installation of the laboratory in Copenhagen would imply 'the automatic full collaboration of Holland', and 'supremely important the backing of the British group through Chadwick'.[13]

2. Criticism of the May consultants' report

All the same, starting with his letter of 31 August, Kramers began to go beyond the simple question of a suitable site and raised the issue of whether or not the project drawn up by UNESCO was realistic. His conclusion was that the May report could not provide a serious foundation for the intergovernmental conference. The main point in the criticism was apparently one of cost. We have already quoted an extract from Kramers' letter of 31 August ('governments are afraid to commit themselves to plans [...] about the financial implications of which they have certain misgivings'), a sentiment echoed in the letter from the Netherlands Minister of Education in October, which focused on this point. He wrote: 'if one reads in your document 6C/PRG/25, Annex 1, that the costs of the project [...] are estimated during five years to be twice the annual contribution of the participating countries to the UNESCO budget, it is clear that the financing of the project will bring almost insurmountable difficulties for the post-war Netherlands and for many other European countries which are interested in this project'. He added that it would seem preferable to him to establish the laboratory in association with an existing institution, as should be the groups responsible for the design studies in 1952. Only thus could the laboratory function 'usefully and without unnecessary costs' and a guarantee of scientific seriousness be given.[14]

3. The proposal for another working programme

This new basis for discussion was certainly most explicitly expressed in Bohr's letter to Auger dated 26 October. Some hints were already given in some of Fraser's letters. Nevertheless, the first coherent and clearly stated presentation of the new proposal which Auger read was that spelt out in Bohr's letter.

In making his suggestion Bohr started from Kramers' line, this being already a step towards a programme alleviating fears 'as regards commitments to plans and expenditure before everyone has had an opportunity of convincing himself of the soundness of the programme'. He, however, felt it necessary to go further: the project in its existing form should be abandoned and the whole matter rethought. To do so he put forward a new procedure:

– firstly, agreement should be reached on the idea of associating the laboratory 'in its initial stages with an existing centre of research';

- should such an agreement be reached on the suitability of the Institute of Theoretical Physics of the University of Copenhagen, a board of directors would have to be set up, within which each state would be represented by a delegate 'appointed by the National Academies' (modelled on ICSU, for example);
- this board, widely representative of European science, would appoint a group of experts to consider 'thoroughly how the new institution could best stimulate inter-European co-operation in the field of atomic physics [...]'; proposals would be made leading in the short term to an outline of research work to be done, and in the long term to a specification of the facilities that the laboratory should be equipped with;
- on the basis of these proposals the directorate would draw up a programme to be submitted to the governments 'regarding grants of the necessary financial support and the distribution of expenditure';
- the construction phase would then commence.

To sum up, Bohr strongly insisted that 'in the first stage of the project, there should be no question of any government committing itself to expenditures other than those connected with the participation in the direction board [...]' since the Copenhagen Institute would provide all the necessary infrastructure, and any decision on 'the erection, equipment and management [...] would thus be taken only after most thorough considerations'.[15]

This was, then, a comprehensive counter-proposal which argued that everything done so far was pointless. It was based on two *a priori* conceptions: on the one hand, it was deemed inadvisable to make major financial commitments in the short term—and the argument here was that one should not put off governments. On the other hand, from the scientific angle, one had to proceed methodically. Only after a systematic survey of the state of the art of nuclear research, and with an accurate knowledge of the type of equipment and the experiments that could be carried out with it, would it be possible to make rational choices. One could not, relying merely on impressions, decide that the most important thing to do was to build a giant accelerator.

Here then was the alternative project. As Auger only became aware of it piecemeal, it seems as well to put forward in succession the answers given by the Director of UNESCO's Department of Exact and Natural Sciences.

5.1.3 AUGER'S REACTIONS

Naturally his first reactions concerned the idea of setting up the European laboratory in association with Bohr's Institute in Copenhagen. A handwritten note, definitely written by Regenstreif for Auger at the end of August and marked 'confidential', gives an initial impression of the feelings aroused at UNESCO by Kramers' letter of 23 August.[16]

Notes: p. 175

Regarding the argument that there was a group of important theoreticians in Copenhagen and that Bohr himself would be a guarantee for the future of the European laboratory, Regenstreif pointed out that 'while Bohr's scientific personality is beyond dispute, confidentially there may be reservations about his spirit of collaboration, his organizational abilities, and that modicum of dynamism which is essential to bring a project of the size considered to fruition'. Auger himself said to Alexander King at the beginning of October that 'Bohr, in spite of his vast experience and activity, is a little too old to undertake this international work', and a few days later he told Waterfield, the British Scientific Attaché in Paris, that the Bohr Institute was 'an old one and one which is now past its prime': he knew of no young and brilliant physicists around Bohr.[17]

Concerning the 'hard working nucleus of scientists and engineers' mentioned by Kramers, Regenstreif noted that the Institute's equipment consisted only of a rather antiquated cyclotron dating from 1939 and a classical Van de Graaff, and that only three people were allocated to this equipment, two of them being physicists 'working at one and the same time as physicists, engineers, draughtsmen, technicians, turners, fitters, etc.'. He added that Kramers' phrase could therefore be applied to almost every European country and that, even when enlarged, the premises would still be 'totally inadequate to house a machine like a synchrotron'.

With regard to the political advantages, Regenstreif remarked that if the aim was to ensure Dutch or British participation, the wisest course would be to install the laboratory in Great Britain; it would then also be possible to take advantage of British scientific know-how. For his part Auger pointed out to Kramers that if the laboratory were to be set up in Copenhagen, there might be less support from France or Italy—the countries which were the main contributors.[18]

Finally, Auger repeatedly stressed the language problem which would probably arise with lower-grade staff (a problem which did not exist in Geneva), Copenhagen's rather remote geographical position, its climate ('Auger told me', noted Waterfield, 'how surprised he was at how strongly this point had been put to him'), and finally the industrial weakness of the country. In this last respect, Switzerland or Belgium were much better. The impression given, then, was that Auger and his associates were not convinced by the arguments put forward: insofar as the main problem was one of location, Copenhagen had little to offer.[19]

The second set of arguments—the excessive cost, the apprehension expressed in some governmental circles concerning an ambitious and uncontrollable project—was taken more seriously by Auger. In his letter to Kramers of 11 October for instance, he recognized that, in itself, document 6C/PRG/25 was inadequate. But he added: 'we are therefore preparing a more elaborate report on technical as well as on administrative, financial and other questions. We have also issued a small additional document (UNESCO/NS/PN/1, enclosed), containing new proposals as resulting from the discussion in Copenhagen'. Let us remember that this document staggered accelerator construction in time, work on the bevatron not starting until two years after the beginning of work on the small accelerator. Auger thus proposed to take account of the fears expressed by proceeding more slowly and by cutting costs. This was confirmed at the second meeting of consultants on 26 and 27 October.[20]

Finally, there remained the proposal to begin the matter again from scratch abandoning, at least for the time being, any idea of a large accelerator. No reply to this appears in UNESCO documents before November. Did Auger do this deliberately? Had he understood the extent of the proposed revisions concerning both the procedure to be followed and the consequences for the bevatron before he received Bohr's letter of 26 October? These questions are difficult to answer despite the plethora of correspondence in our possession.

In fact, Auger's attitude seems to reveal two aspects. On the one hand, he behaved as if no radical doubts had been raised.[21] He went on preparing the next consultants' meeting according to the May report as amended in Copenhagen in July, answering only the first criticisms (those concerning the site or dealing with the rate or cost of building the accelerators). This he did in the first letters he wrote on his return from the United States, the one to Blackett dated 4 October in which he raised the problem of location, as well as that to Verhaeghe on 10 October which he tactfully concluded by saying: 'you will note that we [UNESCO] have so far refrained from making any concrete proposal concerning the site of the future laboratory. For the time being I think it best for UNESCO to remain neutral in this matter'. He maintained this attitude, moreover, until the end of October in *all* his and Mussard's *letters* to Kramers (11 October), Preiswerk (12 October), Blackett (12 October), Alfvén (17 October), and Amaldi (19 October).[22]

On the other hand, a certain anxiety is perceptible, particularly concerning the British situation. Having met Fraser on his return from Chicago,[23] Auger quickly wrote to his best friend in Britain, Patrick Blackett, telling him that Fraser was to meet Chadwick 'but not on my behalf, on his own'; a few days later he spoke to Alexander King, chairman of the Committee on Overseas Scientific Relations, which had considerable influence in Britain with respect to the European laboratory; finally, around 15 October, he asked Waterfield to come and see him, explaining that 'the ultimate goal *must be* a machine, otherwise several of the interested states [would] withdraw their support' (our emphasis).[24] However, Auger never went so far as to alert his friends (we are thinking of Amaldi, Dahl, or Preiswerk) that he thought the bevatron was actually threatened. It was only five weeks later and in Auger's absence, once it was clear that there were two diametrically opposed projects and that the conference would end in deadlock, that Auger's deputy, Jean Mussard, took it upon himself to do so.[25]

It seems then possible to suggest that *if* Auger had an idea of the ultimate goals of Bohr and Kramers, that idea remained imprecise. He probably had more of an

Notes: p. 175

impression than an explicit knowledge of the situation before receiving Bohr's letter. This would explain why he made no reference to such proposals in his letters, why at the same time he was anxious and tried to find out what was happening in Britain, but also why he kept on preparing the intergovernmental conference according to the initial plans.[26]

5.1.4 AN ATTEMPT AT A WIDER INTERPRETATION[27]

To conclude, can we suggest which factors shaped the conduct of those involved or, more modestly, what induced the protagonists to take up the positions which they did? To do this we must go back a little.

Let us begin with Auger. For a year he had been busy with Rabi's idea: to establish in Europe a large fundamental research centre in a field where the essential equipment was expensive, so making the continent once more scientifically competitive. Following informal consultations between June and October 1950 he and others had realized that the main wish was to build a laboratory centred on a giant accelerator, the largest one possible in fact. The support had been there and had been enthusiastic, the scheme had expanded rapidly, and had culminated in the consultants' report of May 1951.

In July 1951 Bohr and Kramers, two of the leading scientific statesmen of the time, were confronted with an already well-advanced project. They were surprised by the cost and by the amount of scientific manpower which the establishment would drain off from national laboratories. They expressed their anxieties, but the spirit of collaboration was strong on all sides. Everyone sincerely wanted to see European science assert itself, concessions were made, and a kind of compromise deal was struck in Copenhagen in July. All went away satisfied.

What was this compromise? It was: we shall build a laboratory equipped initially with a small machine; we shall set up an institute for advanced nuclear studies; we shall design a large accelerator. There may nevertheless have been some ambiguity. What was the meaning of the sentence: we shall *design* a big machine? For Auger it implied: we shall *build* the bevatron two years after the start of work on the small machine. For their part Bohr and Kramers probably interpreted it as: after a few years we shall *take a decision* on whether to build it or not. This can be inferred from a letter Kowarski sent to Chadwick on 15 December. He wrote: 'Let us build as quickly as possible a meson-producing machine and let us make at the same time detailed plans for a bevatron; our politicians, after a glance at these plans, may or may not vote the money, but in the meantime there will be the tangible benefits of a serviceable machine and a concrete co-operative endeavour. This proposal, originally put forward by Alfvén and myself (I used the slogan 'mesons in a hurry') was at first tactfully soft-pedalled by UNESCO, but *reasserted itself at Copenhagen in July.*' (our emphasis).[28]

With that they parted, Auger to draw up a document taking account of these new suggestions and to prepare the intergovernmental meeting, Bohr and Kramers to think over the matter again—let us not forget that it was new to them. It was then that Kramers proposed the idea of setting up the laboratory in Copenhagen under Bohr's leadership. In the short term unnecessary expenditure and the dispersal of human resources would be avoided. In the longer term events would be subject to a minimum of control—the possibly rash enthusiasm of Auger and his colleagues would be tempered by Bohr's wisdom.

Continuing his enquiry, Kramers came face to face with northern European administrators. They emphasized one main point: if such a project was to be carried out, it would be better to be clear about what one wanted to build. Such sums could not be laid out very often, if ever. Anxiety therefore increased in Amsterdam and Copenhagen, and it was around this time that the idea arose that it would be better to begin all over again, scientifically and diplomatically, and to abandon the bevatron.

This does not mean that Kramers or Bohr wanted to stop all efforts at setting up a European laboratory. It was simply that they had other priorities and ones that were, moreover, not easy to explain to the others, to the fervent 'big-brookhavenists', as Kowarski called them.[29] They therefore tried to make them realize the utopian nature of their project in a very roundabout way. Unfortunately the others did not understand, or did not want to. Were they not talking about an even more powerful accelerator (6 to 10 GeV) on their return from the United States? There remained only one thing to do: to explain clearly that the project had taken a wrong turning, and that everything had to be cast back into the melting pot. This was what Bohr explained to Randers, and even more clearly to Auger on 26 October.[30]

5.2 Second meeting of consultants, Paris, 26 and 27 October 1951

5.2.1 PREPARATIONS FOR THE MEETING, 19 AUGUST–25 OCTOBER

Preparations for this second meeting of consultants were spread over two months, so in parallel with the debate we have just been following. Its origins lay in the proposals made in May and in the resolution adopted by the sixth General Conference of UNESCO held in June and July 1951, which authorized the Director-General to continue investigations 'aimed at setting up a European nuclear physics laboratory'. In the light of the 'consensus' reached in Copenhagen, the decision was taken in August to call an intergovernmental conference for the end of the year. We have seen that the letter of invitation was sent to UNESCO member states on 31 August 1951.[31]

Nevertheless, UNESCO still had to settle the details of the administrative structure to be established in 1952 and the parameters of the 6 GeV proton

Notes: pp. 175 ff.

synchrotron. Unfortunately the design office theoretically responsible for these tasks still did not exist. Some interim measure was therefore needed and Auger proposed to set up 'a small scientific committee consisting of people whose competence and authority would ensure that this work was suitably performed'. Although open to all interested scientific institutions, the core of this committee was to consist of only four or five people so that it could meet quickly, and several times if necessary, before the intergovernmental conference. In a series of letters dated 19 August to Alfvén (Sweden), Bohr (Denmark), Cockcroft (United Kingdom), Colonnetti (Italy), Kramers (Netherlands), Perrin (France), Randers (Norway), Scherrer (Switzerland), Hans Thirring (Austria), and Verhaeghe (Belgium), Auger suggested that Bakker or Heyn (Netherlands), Dahl (Norway), Perrin (France), and Preiswerk (Switzerland) form it.[32]

On 23 and 24 August Cockcroft, Scherrer, and Thirring replied, approving both the idea of the small scientific committee and the names proposed by Auger. Alfvén wrote on 27 August. For him and for Waller two types of committee were needed, one consisting of physicists 'with long experience of accelerator construction', and the other of people 'who [had] a general background of the field and at the same time much initiative and a vivid interest for the project'. For the first group they agreed to the choice of Bakker, Heyn, and Preiswerk, but also proposed the participation of a British member such as Goward, and noted that Wernholm of Stockholm could be useful. For the group which was to discuss 'the more general problems', they mentioned Dahl, Perrin, and Kowarski. For his part Kramers replied in his letter dated 31 August. Having made the comments reported above concerning official Dutch hesitations, he went on to say that the idea of the small committee and the list of proposed names were acceptable. Colonnetti replied on 1 September that he hoped that, 'although no Italian [was] taking part in the preparatory work, we [would] be given the opportunity of obtaining information at the November conference, which [he would] be glad to attend'. On 13 September Verhaeghe approved Auger's proposals on Belgium's behalf, pointing out that Widerøe could provide valuable help to the committee and that American specialists could offer their services in the future. Finally, Randers expressed agreement with Auger, adding that Amaldi should be a member of the committee. 'As a matter of fact', he said, 'I would consider that most important'.[33]

By the time he returned from the United States Auger had therefore obtained agreement in principle from scientists from eight countries (Austria, Belgium, France, Italy, Norway, the United Kingdom, Sweden, and Switzerland), some of them adding fresh suggestions. The Netherlands (via Kramers) had replied with reservations, and Bohr had not replied at all. There remained the Federal Republic of Germany, which had become a member of UNESCO on 11 July 1951 and had therefore received an official letter of invitation to the intergovernmental conference. Auger had not written to a German scientist on 19 August but had taken advantage of the Fermi birthday celebrations to have discussions with Wolfgang

Gentner, who had suggested that he contact Otto Hahn, chairman of the Max Planck Gesellschaft.[34]

All the same as time went on two major problems—already implicit in some of the replies—emerged. Firstly, there was the need to reach agreement on the exact terms of reference of the committee. For Auger priority had to be given to examining 'the state of progress of Regenstreif's work on the bevatron', though 'we could also exchange views on the question of the possible construction of a medium-sized accelerator' and to discussing the institute of advanced nuclear studies 'which [seemed] to be favourably received in all circles'. For Kramers, on the contrary, the first thing to do was to discuss the idea of a 'pilot institute' and to decide on the location of the laboratory. Only then could other matters be tackled.[35]

The second problem was to know who should take part in the committee meetings and who should decide on the participants. It seems that every country wanted to be represented in the group. This explains the rather cold tone of Colonnetti's reply ('I have taken note of your decisions' was all he said) and Alfvén's suggestion that thought should be given to his colleague Wernholm for the committee. This kind of requirement is not surprising of course in the context of international collaboration. However, the full extent of the problem was revealed in a scathing letter from Kramers dated 9 October. Mussard having sent a telegram to Heyn a few days before asking if he was free on 26 October, Kramers replied: 'what is the meaning of the telegram to Mr Heyn? Has Mr Auger appointed Mr Heyn [...]?' If so, he said, I suppose that his expenses will be paid by UNESCO and 'we Dutchmen and our government will have nothing to do with your settlement with Mr Heyn, and it will be he who replies to you'. By contrast, he continued, if you ask Dutch physicists to appoint one of their number, we shall send Bakker. But only if our government agrees, and only to explain our attitudes from the strictly scientific viewpoint.[36]

Faced with these two kinds of reaction Auger had to refine his thinking and partly modify it. This he did in a series of letters to Verhaeghe (10 October), Kramers (11 October), Bakker and Preiswerk (12 October), Alfvén and Dahl (17 October), Bakker, Bohr, Colonnetti, and Goward (18 October), and Amaldi (19 October). He maintained that the main task should be to discuss the bevatron, but he left all other possibilities open. In fact when it met the committee dealt with the entire problem. Auger also defended his idea of a small technical committee but stressed that 'all those wishing to give us their advice [would] be welcome'.[37] There is even reason to believe that by mid-October he had become converted to the idea that it would be better to enlarge the committee, so to confer greater authority on the proposals it might make. Thus Mussard wrote to Goward: 'We shall be happy to have your comments and of course also your visit, in case you wish to participate in the discussion on 26 October', and to Amaldi: 'It is possible that Mr Bakker, who, it seems, shares Mr Kramers' views, will be present next Friday; this is one reason why we are hoping that you will be able to come too'.[38]

The result was that, except for Alfvén who was absent on 26 and 27 October, the committee which then met was surprisingly like the May one: Amaldi, Auger, Dahl, Goward, Kowarski, Mussard, Perrin, Preiswerk, and Regenstreif were there, while Bakker and Verhaeghe respectively replaced Heyn and Capron.

5.2.2 THE MEETING

We have two accounts of this meeting, one written by Bakker and the other by Goward. UNESCO itself summarized the proposals in a document used as a working paper for the third consultants' meeting in November. This document, as amended on 17 and 18 November, was called the *Working Document* for the December intergovernmental conference. Unfortunately we do not have the original version and it seems most unwise to use the working document itself since it was revised and corrected in the interim. We shall therefore restrict ourselves to the two first-mentioned reports. More often than not they give the result of the debates without going into the details of alternative proposals.[39]

The first item concerned the equipment for the future laboratory. Two machines were proposed: a 5 BeV proton synchrotron and a synchro-cyclotron of about 0.5 BeV. The envisaged time of construction of the two machines was respectively six and from two to three years. According to Goward 'the design and construction of the two machines [had] to proceed in parallel'. Note here that, by comparison with the proposals made in Copenhagen, there had been an important change, and the reaffirmation of an earlier position. The new idea, as proposed by Cockcroft in his letter to Auger of 24 October, was to have a synchro-cyclotron instead of an electron synchrotron. At the same time the plan to build the two machines *in parallel* was reaffirmed, though in Copenhagen it had been provisionally accepted to build them one after the other. As Goward said, some participants might have agreed to construct only a synchro-cyclotron, but most of them were unwilling to give up the big machine. For them, whatever the interest of the small device, there was no question of postponing the start-up date for constructing the bevatron.[40]

From American data on the Brookhaven Cosmotron and the Berkeley Bevatron, the costs for the two machines were estimated at three and ten million dollars respectively. The consultants concluded that three million dollars per year would therefore have to be spent during the five or six years of construction. It will be remembered that in May the cost of the 6 GeV bevatron was assessed at twelve million dollars—roughly corresponding to the ten million proposed here for a 5 GeV synchrotron—but that at the time the same figure was given as an estimate for the 'other expenditure needed for the establishment of a large design office, the erection of buildings, the purchase or construction of auxiliary equipment, etc.'.[41] Here, with no explanation, and although research was to be scheduled and performed with two machines, the amount of the 'other' expenditure was reduced to a few million dollars—to take account of the oft-levelled criticism that the costs of the project were excessive. This 'cost-minimizing process' was however patently artificial.

Indeed, who can have failed to ask the delicate question as to how the cost of a more grandiose project could be 30 to 40% lower than that of a more modest one and how the change could come about in five months, as if by magic.

The third matter discussed was the organization of work in the preliminary phase when no machine construction was envisaged. Four groups were specified: two to examine the two accelerators, one to arrange the organization of the institute of advanced studies, and the last to examine how the laboratory, the administrative structure, the buildings, etc. were to be organized. The first two would be made up of about fifteen people each, including four physicists and six engineers. Names were put forward: Dahl and Heyn for the design of the bevatron—assisted by Wideröe (Switzerland), Fry (United Kingdom), and de Braine (France); Bakker and Preiswerk for the synchro-cyclotron—assisted by Pickavance (United Kingdom), Wernholm (Sweden), and Surdin (France); Fierz (Switzerland), Møller (Denmark), de Groot (Netherlands), Ferretti (Italy), Perrin and Proca (France), and Géhéniau (Belgium) for the third group; and Kowarski for the 'organization' group—assisted by Staub (Switzerland), Koch (Denmark), and Siegbahn (Sweden). Finally, a directorate-general was also planned, to be headed by Edoardo Amaldi.

Concerning these lists, Goward commented in his letter to Cockcroft of 29 October: 'They seemed to have some confidence that Britain would take part in this project and, for example, that Harwell might release some of its accelerator staff. I discouraged them, for my part, but they wrote down the names Fry and Pickavance'. It will be noted in this connection that the October reports were much more precise and ambitious than the May one, which merely said that some ten specialists would be needed in 1952. This time the participation of about 40 or 50 people was envisaged, including seven half-time physicists and four working full-time (plus the unpaid ones in the 'theory' group), 14 full-time engineers, 13 technicians, and a few administrative staff.

The fourth item examined was the financing of the provisional phase. A figure of 200,000 dollars had been quoted in May, but no details had been given. Here we find the same figure but with a host of details. Salary scales were first drawn up, ranging from 30–35,000 SF for the director to 12,000 SF for the technicians. Goward added: 'It was considered that the scales should be such that they would not upset European university people and it was considered that Americans, as an example, would be willing to come with salary drops'. A calculation of what would be paid out in salaries, together with about 150,000 SF for overheads (travel, telephone, etc.), gave 819,000 SF, i.e. 200,000 dollars. No one will be surprised to find the May figure again. Since the figure had already been published in official UNESCO documents, it *had to* reappear. Perhaps the salary scales had to be reduced, the travelling expenses to be nibbled at but the result was there: the figure was precisely substantiated, as promised by UNESCO.

Thereafter the distribution of this budget among the European states was dealt with. No precise scale was drawn up as it depended upon the participating countries.

Notes: p. 176

Nonetheless the principles first put forward in December 1950 were reaffirmed: a scale close to that used by UNESCO would be adopted, i.e. in proportion to national resources, with a ceiling of 30%. According to Goward, 'most of the consultants were a little optimistic of government backing but had no assurances'. For the November meeting UNESCO was to draw up various schedules by way of example, and to calculate what each state would have to pay in each case and according to various indices.

The fifth item concerned the site. Once again, as in December 1950 and in May 1951, the consultants first drew up a list of criteria as a basis for site selection: ease of access, presence of qualified labour, proximity to a university and a large city, the use in the area of a reasonably widely-spoken language, and the availability of a cheap electricity supply. Once again, as in December and May, the conclusion was that Geneva was the best place even if there was an awareness 'that other locations were being, and were likely to be, suggested'.

In conclusion, we would like to stress a few points about this October meeting, the most important being *the complete lack of explicit or implicit reference to the proposals of Bohr and Kramers*. What is more, as we pointed out, there had even been a reaffirmation of positions held prior to the Copenhagen conference. In other words this meeting was a direct continuation of the one held in May, and was conducted as though nothing important had happened meanwhile.

Why were Bohr and Kramers' proposals ignored? Firstly, because Auger had probably screened the consultants from part of the dispute—most of them did not realize the nature or extent of the opposition to 'their' project.[42] Secondly, and quite apart from this, because the consultants had considerable confidence in their abilities and did not doubt that they would receive the support they needed. As Goward reported, 'The main points I asked them to consider carefully were:

> a – Whether their technical and industrial resources were sufficient [...]. They expressed confidence in their sufficiency [...].
> b – Whether they could make machines more cheaply than the Americans. They thought they could, despite our experience [...]'.[43]

There were many reasons for this confidence. On the one hand several American scientists had expressed warm support at the Chicago conference: 'everybody seems to be ready to help us with technical and scientific advice' wrote Auger on 11 October, adding that many physicists had stated that with American help and experience it would be possible to build an even more powerful machine. Moreover, recognized European specialists working for industrial firms, men like Casimir (Philips) and Wideröe (Brown-Boveri), had told Auger of their interest and had offered to collaborate. Finally, the project as conceived by Auger had received the support of several scientific administrators, which led one to hope that the support of the countries themselves would be forthcoming: the Belgian, French, and Italian cases were fresh in everyone's memory.[44]

5.3 Tension peaks: late November–early December 1951

In the light of the two previous sections, no-one will be surprised to learn that November was a month of considerable tension. The record of the meeting of 26 and 27 October, for example, could not but infuriate Bohr and Kramers. To understand what follows we first need a few chronological markers. Here they are.

From the end of October to mid-November each of the two groups was working on its own and there was virtually no correspondence between them. At UNESCO the document to be submitted to the conference was being prepared, and was discussed one last time on 17 and 18 November.

This meeting was attended by the October consultants together with J.C. Jacobsen and Torsten Gustafson for Denmark and Sweden. The atmosphere was, apparently, already somewhat strained and Amaldi noted in his diary concerning Jacobsen and Gustafson: 'Strana opposizione di questi due ultimi'.[45] Nevertheless it adopted a document which UNESCO was to put into final shape for December: this was the conference's working document.

Two days later, on 20 November, the Director-General of UNESCO replied to the official letters from Kramers (24 September) and from the Netherlands Minister of Education (24 October), while Auger replied to Bohr's letter of 26 October. In essence they repeated the conclusions reached at the meeting of 17 and 18 November.

One week later, on 28 and 29 November, the Joint [the Netherlands and Norway] Establishment for Nuclear Energy Research was inaugurated. Mussard, who was representing Auger, was shocked by the violent opposition to the UNESCO project, particularly from Bohr and Kramers. On his return to Paris he therefore called a last-minute meeting of the consultants for the Friday before the conference (14 December) to reconcile viewpoints, if possible.

The meeting was duly held and the conference opened on Monday, 17 December 1951.

5.3.1 THE DOCUMENT DRAWN UP AT THE THIRD MEETING OF THE CONSULTANTS (17–18 NOVEMBER) AND SUBMITTED TO THE CONFERENCE AS THE WORKING DOCUMENT

The version we are using here is dated 19 December and numbered UNESCO/NS/NUC/1. We assume that no amendments were made to it between the end of November, when a start was made on drawing it up, and 19 December. Only the meeting of 14 December could have made any changes to it, but we do not think that it did—the text is completely in line with the letters sent by UNESCO on 20 November, the time difference apparently being due merely to delays in UNESCO.[46]

The document opens with a two-page introduction which gives a brief historical survey of the studies and a glimpse of the purposes of the conference. There follows the body of the text which comprises two parts: *General outline of the project,* and *Items for discussion*. Finally, there are seven annexes (23 pages). Annex I reproduces the previous texts or letters showing 'the interest pertaining to the construction of a [large] particle accelerator in Europe'; annexes II and III describe in detail the studies to be undertaken; annex IV presents the draft 'organization of the second stage of studies [1952]'; annex V gives 'cost estimates for the second stage'; annex VI provides possible lists 'of States willing to participate' and some calculations 'of their respective financial contributions'; finally, annex VII lists the criteria for selecting a site and reproduces three letters to the Director-General of UNESCO, one from Eugène Bujard, Vice Chancellor of the University of Geneva, dated 9 June 1951 and proposing Geneva as a location, one from Malte Jacobsson and Gösta W. Funke dated 12 October 1951, and finally Kramers' letter of 24 September.

1. Two important preliminary remarks

One point was underlined (both literally and figuratively) from the very start: it was to be the 'interested governments [which] will assume the financial and technical responsibility necessary for carrying out stage 2 of the project', and not UNESCO; nevertheless, 'it is reminded that a final decision regarding the construction of the Laboratory is not expected from the Conference'. That was to come *at the end* of phase 2, at the beginning of 1953. One conclusion of the May consultants was thus emphatically reaffirmed: participation in phase 2 did not imply any subsequent commitment. We have already emphasized the reasons for this proposal—to reassure governments, to move ahead gradually—and there is no need to repeat them here.

Another point was also heavily stressed, recurring three times: in the first lines of the body of the text, in annex I, and in the introduction to the important annex II: *object of studies,* where we read: 'The following report is based on the assumption that the equipment of the Laboratory would be chiefly intended for the acceleration of particles of high-energy'. Indeed the document clearly stated that the *only* working assumption was that of establishing a laboratory equipped principally with a very high-energy bevatron (3 to 10 GeV). The insistence on this is obviously not coincidental: faced with the proposals of Bohr (and Kramers), and despite the presence of Jacobsen and Gustafson, the November consultants stated once more that the fundamental choice had been made, that there was to be no going back to square one. Even if the cost was high, the need for a powerful bevatron was—to quote the text—*'maintained'*.[47]

2. The laboratory equipment

There is nothing to surprise us here. Presenting the various possible types of accelerator—synchro-cyclotron, electron synchrotron, proton synchrotron, linear accelerator—and giving a summary of the existing problems in high-energy physics, the report concluded in favour of a 0.5 GeV proton synchro-cyclotron, 'a well-engineered machine which works in a safe, satisfactory way', for 'studying the formation of light mesons and for a number of other problems [...]'; and of a proton synchrotron of at least 3 GeV, which would make it possible 'to elucidate a certain number of phenomena such as cosmic showers', and to study 'the artificial creation of pairs of heavy mesons'. If possible, however, its energy should be 6 to 10 GeV, especially for the study of 'the conversion of kinetic energy into protons and neutrons'. The 'performance [sic] of the European synchrotron' would thus have to be decided (3 to 6 or 6 to 10 GeV).[48]

Nothing new emerged concerning the cost of the machines either. As the Chicago synchro-cyclotron's costs had been estimated at 2.7 million dollars, a figure of 3.5 million dollars was considered likely for its European equivalent. Based on the Berkeley machine, but also taking American experience into account (the document pointed out, for instance, that the weight of the Berkeley magnet had been overestimated), a 6 GeV bevatron was regarded as likely to cost about 8 million dollars. This, of course, did not allow for the 'buildings and auxiliary equipment'.

3. Setting up an institute for advanced nuclear studies

In November, stress was laid on the proposed institute's methods (organizing a summer school, holding seminars and meetings, etc.), which were to ensure the collaboration of non-European specialists. It was thought that the institute could be formally established in the summer of 1952.

4. Organization of studies in phase 2

As in October, the document proposed four study groups and a directorate-general, the whole to be under the responsibility of an advisory committee consisting of 'representatives from the contributing countries'. This body was to call the conferences which would take 'final decisions regarding the creation of the Institute [...] and the construction of the Laboratory'. There was only one really new item which had not been covered in the May and October proposals: 'it [was] suggested to *decentralize* the studies as much as possible' (our emphasis). This meant that the four groups could be set up in different places in order to make best use of the facilities available at existing national institutions. The document put no names forward for group leaders and gave no list of host institutions, either; these were to be discussed at the conference.

This procedure calls for some comments. First of all it shows that, tactically speaking, Kramers' suggestion had been taken into consideration, and in a rather clever way: the problem was not to select *the* pilot institute for the laboratory—a thorny and sensitive question that was likely to split the conference down the middle—but to choose four group leaders plus a director. Accordingly one could expect there to be groups in Norway (around Dahl), in the Netherlands (around Bakker), in France (around Kowarski), and probably in Copenhagen (around Bohr),

Notes: p. 176

while Amaldi would set up the directorate in Rome. One could even envisage having sub-groups, so soothing all national susceptibilities.

Apart from its tactical merits, this approach had other real advantages over the May suggestions, both in terms of the manpower available in Europe and budget savings: it allowed those who were prepared to be members of the various groups (Amaldi, Bakker, Dahl, ...) to continue their own work in their own institutes without having to leave them immediately. In addition they only needed to be paid part-time.

5. The participating states and the scale(s) of contributions

The document pointed out that the scale of contributions could not be drawn up since it depended upon the number of contributing states. By way of illustration it gave only the two extreme cases (all European countries or only one) and one intermediate case (where it was 'arbitrarily' (sic) assumed that the participating countries would be Belgium, France, Italy, Switzerland, and Yugoslavia). It proposed that the conference should set a deadline, 'for example January 31 1952, before which all states willing to participate in the second stage of the project should officially inform the Director-General of UNESCO'.

6. Site for the future laboratory

No final answer was expected from the conference on this item. The text merely recommended that selection criteria be drawn up and gave information on the offers already received. Hence the publication of the three letters in annex VII.

Our conclusion will be brief: the document drawn up on 17 and 18 November obviously restated the proposals made by the consultants in May, sometimes adding some of the steps proposed in Copenhagen in July and included by the consultants in their October report. The only new feature was that concerning the decentralization of the study groups in 1952.

5.3.2 THE OFFICIAL REPLIES TO THE LETTERS FROM KRAMERS, BOHR, AND THE NETHERLANDS MINISTER OF EDUCATION (20 NOVEMBER)[49]

These three letters were important as they really came to grips with the main opponents with whom UNESCO was in contact. They were directly inspired by the work done at UNESCO during the previous three days and thus did not contain anything really new. What does deserve mention, however, is the diplomatic attitude adopted by Auger who was undoubtedly the author of the three letters (even though two were signed by the Director-General).

This attitude had three distinctive features. Firstly, UNESCO's authority was affirmed: for example, the Director-General wrote to Kramers that it was 'the

UNESCO Secretariat [which was] responsible for drawing up the Working Document which [would] be submitted to the Conference'. This meant that the working framework was established and that Bohr's proposal to reconsider everything was not acceptable.

Secondly, it was emphasized that 'the wording of this document has been most carefully studied in order to take into account all views expressed by the experts who took part in the preparatory consultations'. In his replies to Kramers and the Dutch minister the Director-General went even further, since he tried to show that *all* their suggestions had been met.

However, Auger was aware that a solution would not be found simply by stating that it was UNESCO which decided in the final instance, or by stylistic devices intended to convince readers that all points of disagreement had been overcome. He also knew—and this is the third aspect of his attitude—that *without a consensus among the physicists* ('The success of the Conference of Governmental delegates in December', he wrote to Bohr, 'will very much depend upon the unanimity of opinion among scientists'), the European laboratory project, in this case *his* project, the most expensive and ambitious project, might not be adopted. Accordingly he again left a door open for possible discussions between physicists. Hence the conclusion of his letter to Bohr: 'I therefore very much hope, in case any question still requires clarifying among scientists, that it will be possible to do so *before* the Conference' (our emphasis).

Can we be more precise about Auger's motivations for this last move? We think so. What he apparently understood was that the 'non-scientific' governmental representatives would quickly become suspicious if it were to appear that some key scientists at the conference were not persuaded of the sound basis of, and need for, the bevatron. The only way of convincing those who could not judge for themselves on the importance of the scientific proposals was to present a united front. This was all the more important as considerable sums of money were being asked for and as, at the time, while some circles in the United States were beginning to realize the important role that science and research would play in the post-war world, this had not yet been fully recognized by continental politicians. Or at least, if there was some awareness, it had not yet been expressed in budgetary terms.[50]

5.3.3 THE OSLO CRISIS

We have said that the Joint Establishment for Nuclear Energy Research in Kjeller was inaugurated on 28 and 29 November. The first day of festivities was taken up with official ceremonies in the presence of the King of Norway, and the second with a scientific conference in Oslo. In the morning Kowarski presented a report on the design and use of nuclear reactors. In the afternoon Cockcroft opened a discussion on future prospects for atomic energy.[51] However, other matters were on people's minds, especially the intergovernmental conference which was to be held very shortly. Many delegates soon realized that Bohr and Kramers were still completely opposed to UNESCO's plans, and they welcomed Bohr's suggestion for a further exchange of views on this matter. Bohr repeated that 'quite frankly [...] a European Institute of Advanced Nuclear Studies must come first, and any new machines only as an afterthought. [For him] this sort of co-operation [would] require very little money'. Some of those present were clearly shocked, not to say disillusioned and despairing. Kowarski, for example, knowing the importance of Chadwick in Britain and aware of his close links with Bohr and Kramers, considered it worth writing a long letter to him on 15 December in order to confront him with the various points of view. Likewise, on his return from Norway Mussard, aghast at the prospect of a deadlock at the conference, immediately took action.[52]

His first step was to write to Bohr to explain that 'the views expressed [in Oslo] did not exclude each other'. Then he sent a series of rather anxious letters to Dahl (4 December), Preiswerk, Bakker, and Amaldi (6 December), in other words to his 'political' associates.

In his letter to Bohr, Mussard began by justifying his silence in Oslo: 'I thought', he said, 'it was your intention on this occasion to have a free exchange of views [...] without intervention from a UNESCO representative'. In fact, he had felt that he was being deliberately marginalized and that, as he told Amaldi, 'most of those in favour of the project found it virtually impossible to express their opinions'. He then tried to reassure Bohr—'No commitment [was] expected from any country [...] for the construction of the laboratory itself'—concluding with the proposal we have already mentioned for another consultants' meeting on 14 December 'to discuss the questions which may need a more precise formulation'.

In his letters to Amaldi, Bakker, Dahl, and Preiswerk Mussard was slightly more explicit about his feelings and the events at Oslo. He sent them a copy of his letter to Bohr, adding that he thought that they would understand his attitude. To Dahl he once again spelt out the official standpoint of UNESCO: 'We shall in any case maintain the principle that the purpose of the Conference is to reach an agreement between those states who are willing to collaborate *on the basis of the plan submitted by UNESCO and not to discuss on fundamentally different proposals*' (our emphasis). In other words Mussard reaffirmed that the purpose of the studies was to establish a laboratory 'and not merely the creation of a nuclear physics committee'.[53]

Bakker answered Mussard's letter on 7 December, saying that he agreed with what Mussard had explained to Bohr. In any event, he pointed out, since we would have to wait for the governments' decision concerning their contributions, and therefore hold another conference within a few months, why not use the time to meet one of Bohr's requirements, and gather a wide commission of experts just after the December conference with instructions to review all the problems. It would thus be possible for the final decision to be taken between March and May. He ended by adding two details: the British could offer the Liverpool machine—which should also satisfy Bohr and Kramers—and Kramers was now convinced 'that there were many drawbacks to choosing Copenhagen as the location of the European laboratory'.[54]

Somewhat relieved, Mussard commented in a letter to Kowarski on 11 December: 'This morning I received a relatively optimistic letter from Bakker. He [...] hopes that it will be possible to reach general agreement by allowing a further three-month period for exchanges of opinions on the types of accelerator which should be investigated. As in any case we need three months to obtain official governmental agreement and the funds, such a compromise would make virtually no difference to the schedule'.[55]

5.4 The roots of the division

To conclude this chapter, we should like to return to the nub, the focal point of the events of these six months, that is to say the conflict between different proposals. According to Kowarski,[56] one had, firstly, 'the "original Rabi" or "Big Brookhaven" school [...]. For them, Rabi's definition could mean only a bevatron "possibly bigger than Berkeley" '. This was the group represented by Amaldi and Auger since 1950. Alongside them, the 'modest Brookhaven school', i.e. those who wanted first to build a smaller machine. Here Kowarski was thinking of Alfvén, Bakker, and himself. However, these two viewpoints '[had] been provisionally reconciled in the UNESCO working paper' under the pressure of 'the common threat coming from view (3)'. For that, i.e. the view associated with the names of Bohr and Kramers 'the main emphasis was on the co-ordination and promotion of existing European activities, and on growth rather than creation'.

The interesting problem to tackle, once this classification is accepted, is of course to understand its roots. Some have already appeared, and in section 5.1.4 we have tried to give an idea of the *process* by which the opposition could have formed. What was done there, however, remains incomplete and we must try to go a little deeper, bearing in mind that our analysis is based on a small sample (30 to 35 people).

Let us begin with what Kowarski called 'view (3)'. Those supporting it were Kramers in the Netherlands (but not the other Dutch scientists concerned), Bohr in Denmark and, with him, the Danish and Swedish scientists (but not Randers and Dahl, the Norwegians), and finally some of the leading personalities in British physics like Chadwick and Thomson. A certain difference is perceptible between the attitudes of Bohr and Kramers, and Chadwick and Thomson. Its basis was that the former pair, like Amaldi or Auger, wanted to set up effective inter-European forms of collaboration, whereas the latter were much more reserved on this point. It must not in fact be forgotten that Kramers had taken part in the scientific commission of

Notes: p. 177

the Centre Européen de la Culture in December 1950, that he had shown himself in sympathy with Auger's initiatives in April 1951, that he, like Bohr, had always taken an active part in the discussions, and that Bohr was to participate actively in subsequent years.[57] Conversely, Chadwick or Thomson never intervened in discussions with other Europeans, they always regarded the idea of a European laboratory as a continental affair and, though supporting Bohr, they did so while letting him draw up his counter-proposals on his own.

All the same one finds certain common characteristics in this group. Firstly, they belonged to the scientific 'establishment', to those who were comfortably settled, to those who held power. This was quite obvious in the cases of Bohr, undisputed father of European theoretical physics, of Kramers, chairman of IUPAP from 1947 to 1951, and of Chadwick, administrative and political senior statesman of British physics. Moreover, these men were elderly and formed the old guard of the heroic period of physics of the twenties and thirties, and had now more or less retired from active research work. To have a clearer idea of this remember that, while Randers was 37, and Amaldi, Kowarski, and Preiswerk 43 or 44 years old, Bohr was 66, Darwin 64, Chadwick 60, Thomson 59, and Kramers 57. Finally, most of them were not really involved in big science (unlike Cockcroft or Perrin), and their interests, or at least those of Bohr and Kramers, lay primarily in theoretical physics.

Is it possible then to 'explain' some of their attitudes? We think so. It might, for instance, be said that their spirit of enterprise or adventure was perhaps less strong than that of people like Randers or Amaldi, that it had diminished with age; or that it was now a little late for them to break the habits they had acquired, habits which may have proved themselves in the past but which were now somewhat obsolete—the habit of working in small teams, or even alone, the habit of working with limited resources within the well-established framework of universities, and not as part of larger collaborations using heavy equipment. This may help understand the reluctance of Bohr and of Kramers to spend so much money on a single fundamental research project, or their wariness in the face of the challenge embodied in the proposal of the 'big-brookhavenists': could one be sure of the importance and urgency of such a large accelerator, could one be sure that it would be successfully built, could one be sure that a body independent of the conventional research establishments and comprising people of entirely different backgrounds would be viable? Or indeed, under these circumstances, would it be capable of what was expected of it?

If we add that it may have been hard for a theoretician of the old school to believe in 1950–51 that Europe's priority should be to build a 'bevatron'—there were so many other problems to be solved, especially in theoretical physics—if we remember that the most urgent thing for some was to train young people and to use the money to reconstruct national physics—and everyone knew that this was needed—finally, if we bear in mind that there was a flourishing high-energy physics programme in the United Kingdom in which a great deal—some said too much—had already been invested, and from whence came the desire to 'do physics again', one can understand some of the attitudes we have described.

These factors, which might be termed sociological, were partially perceived at the December conference. Waterfield, the British Scientific Attaché, spoke of two groups: on the one hand the old friends of Sir George Thomson, the British delegate, among them Bohr with whom Thomson spent most of his time; on the other, 'a group of younger men' whom he described as being interested in the *construction* of machines but which 'while admiring Bohr [...] [was] clearly jealous for its own independence. [...] As Kowarski put it', Waterfield went on, 'they [were] prepared to have Bohr president of a democracy but not an absolute monarch'.[58]

It would nevertheless be one-sided to regard the older group solely as a brake on a necessary evolution. Admittedly their reaction was somewhat negative, and certainly their intentions were not purely innocent, but in their remarks and in their criticisms there lay a certain wisdom, a certain realism. As we have already said, this was understood by their opponents who took up some of the ideas put forward by Bohr and Kramers (albeit in modified form), and were to adopt others in the future. In this respect the part played by Bohr and Kramers was also a positive one. It forced the younger group to 'come down to earth' and to face up to Europe's limitations—in money, in manpower, in awareness too, for one had to take account of the way of thinking of the administrators and politicians concerning the financing of research, and to adapt the project accordingly.

The characteristic features of the other group—even though it was sometimes divided over the order of priorities in the construction of the machines—was that it had been formed in a wave of enthusiasm, that it now espoused a fairly mature project, and that it was strongly determined to re-enter into *competition* with the USA in the very field which the United States had opened up.[59] Their reasons were rooted in a number of concerns (increasing emigration, the fact that scientifically Europe was lagging further behind, etc.), and in the desire to overcome them by shaking an old continent which found it hard to understand what was happening at the leading edge of research and at the interface between science and politics. It is true that this group comprised a younger generation of physicists, men who were more inclined towards experimental nuclear physics or cosmic rays, more involved in large research centres, often more interested in the construction of heavy equipment. They were also still fully active at the forefront of research, and thus probably more in touch with what was, and would be, at stake in their discipline.

However, it is not enough to stop at these oppositions: theoreticians versus experimentalists or engineers; young people versus the 'establishment'; old working methods versus 'big science'. Superimposed on this split was another split on national lines. For Waterfield or Thomson attending the conference, for Kowarski presenting the situation to Chadwick, for many others, two poles had emerged:

Belgium, France, and Italy on the side of the 'big-brookhavenists'; Denmark, Sweden, and Britain on the other. Between them and more divided were Switzerland, the Netherlands (even though it seemed that by and large, under the influence of Preiswerk and Bakker, these two leaned towards the first pole), and Norway, where the scientists (Dahl and Randers) backed the UNESCO project though the 'politicians' seemed to prefer the Scandinavian front.[60]

Drawing on a few ideas already mentioned we can try to understand the reasons for this new split. The first striking feature of this rift between Belgium–France–Italy and Britain–Denmark–Sweden was that it mapped onto an overall difference in quality in engineering and perhaps scientific output. To keep matters simple, we have on the one hand prestigious teams or men (the Institute for Theoretical Physics of the University of Copenhagen, Kramers) and groups with considerable technical achievements behind them, especially in the accelerator field (we are thinking here of Harwell around Cockcroft, of Oliphant's team at Birmingham, Skinner's at Liverpool, Dee's at Glasgow, and of the Swedish 200 MeV synchro-cyclotron which was almost completed at the end of the year). On the other hand we have countries which were lagging somewhat behind, *especially in the construction of accelerators,* or were actually somewhat isolated (France). Which leads one to interpret the rapid commitment of these countries (Belgium, France, and Italy) to Amaldi's and Auger's projects as a way of dealing with the situation.[61]

This kind of motivation, which is easy for us to grasp on looking at the extent of scientific or technical development in the various countries, was also quickly perceived by some of the protagonists themselves. We have recalled elsewhere the comments made by several British scientists in the spring of 1951; we could add that the Swedish Committee for Atomic Affairs asked in March 1952 'if it would not be better to create not one, but several centres [in Europe]; one for an intermediate accelerator in southern Europe and another centre aimed at advanced research in northern Europe, where intermediate accelerators already [existed]'. Thus the uneven level of development between European countries had certainly been noticed, and the idea that it could have affected the positions adopted is not without foundation.[62]

This suggestion is reinforced by the fact that we know *who* organized the 'political' support in Belgium, France, and Italy: Willems, Dautry, and Colonnetti. These men were at the intersection between the political world and the big scientific institutions financing research or organizing it in the nuclear domain, they were well acquainted with the balance of forces and the distribution of knowledge in Europe, especially in the high-energy field. And if, for over a year, they had been in contact with their governments, often at the highest level, if, since 1951, they had found the money to help Auger and had promised that more financial support would be forthcoming, it was because they were aware of the urgent needs in their respective countries.

We nevertheless think that there was another reason for the split down national lines. We have drawn attention to the commitment of Colonnetti, Dautry, and Willems to the European Movement, and we know that they were first active via the Centre Européen de la Culture. And we cannot but notice that this reflected a more general trend differentiating Belgium, France, and Italy from Britain on the one hand, and from Sweden and Denmark on the other, the degree of involvement of these various countries in European affairs being the determining factor. What will, of course, spring to everyone's mind is the dedication of the former three countries, together with the Netherlands, the Federal Republic of Germany, and Luxemburg, to the pro-European projects of 1950 and 1951: the Schuman Plan, the founding of the European Coal and Steel Community, and the attempt to create a European Defence Community. Conversely, the reluctance of the Scandinavian countries to accept European ideas is as well known as the deep reservations of Britain, which hinged on the view that she had of her own position on the world political scene, namely as a bridge between Europe and the United States.[63]

We therefore consider it justified, at least from the factual viewpoint, to place side by side two sets of data relating to these six countries: on the one hand, the attitudes of their physicists and scientific administrators towards the integrative project of Amaldi and Auger; on the other, the political and diplomatic attitudes of the same countries to the European ideal, as expressed for example in the Schuman plan. To this list we could even add the cases of Germany, Norway, and the Netherlands. Were we not in fact to see, during the succeeding two months, Bakker the scientist and Bannier the administrator take the Netherlands into the camp of Auger and Amaldi? Were we not to see during the conferences of December 1951 and particularly of February 1952 the German delegates do the same thing, whereas the Norwegian delegates (though not Randers or Dahl) brought their country into line with the rest of Scandinavia? The correlations between the two sets of facts thus seem very systematic, indeed too regular to be due to coincidence alone. They certainly reflect states of mind and national approaches that more or less permeate all the citizens of a country and in particular—which is hardly surprising—its scientists and administrators.

Our model could be refined even further. We have chiefly shown *individual motivations* due to age, training patterns, or different scientific practices, and *national motivations* correlated with levels of scientific/technological development and with attitudes towards the European ideal. However, these by no means exhaust all possibilities, and more particular ones can be envisaged—the fact that the laboratory was likely to be installed in Geneva, which undoubtedly informed the positive attitude of Swiss officials, or the fact that the 'brain drain' appeared to be especially marked in Italy, or Auger's deep internationalist convictions and those of his great friend Laugier at the UNO. In all these cases we can identify more specific 'reasons' contributing to, or in some instances weakening, the interplay of the more general motivations described above.[64]

Notes: p. 177

Finally, to come down to individual cases, we could show how the various explanatory factors were articulated with each other, and how the weights they had varied from one person to another. For example, whereas overall the same factors were at work in the cases of Amaldi, Auger, Colonnetti, and Dautry, not necessarily the same were decisive. And when motivations were contradictory, at times one side, at times the other, was dominant. We need merely compare the attitudes of Bakker and of Kramers, the two Dutch scientists, or the attitudes of Randers and Dahl with those of the official representatives of Norway at the intergovernmental conference of February 1952.

We cannot claim, however, to have given an exhaustive account of the factors which governed the attitudes adopted. Our more modest aim has been to try to identify the regular features, the correlations, and to note those we considered essential. In this way we hope to have set out the main considerations shaping various people's behaviour. For further detail studies going beyond the scope of our work would be essential.

Notes

1. The UNESCO viewpoint—and that of Amaldi, Dahl, Preiswerk, Randers, ... —is well known to us from:
   - UNESCO archives, Paris, *CERN, Official Correspondence* file
   - Jean Mussard's papers, deposited at CERN, Geneva.

   The 'opposition' viewpoint—that of Bohr, Kramers, Chadwick, ... —is reasonably well known to us from British documents:
   - Chadwick's papers, Churchill College, Cambridge
   - DSIR, Ministry of Education and Science, United Kingdom Atomic Energy Authority files, Public Record Office, Kew, London.

   Consultation of the Niels Bohr archives in Copenhagen revealed very little that was not already known. We consulted neither the Dutch nor the Swedish archives. However, see Nyberg & Zetterberg (1977).

2. Account based on letters from Kramers to Auger, 23/8/51 (UNESCO); Kramers to Chadwick, 24/8/51 (CC–CHADI,1/13); Chadwick to Thomson, 14/9/51 (CC–CHADI,1/9). The quotation is taken from the last letter.

3. Letters Kramers to Chadwick, 24/8/51 (CC–CHADI,1/13); Chadwick to Darwin, 13/9/51 (CC–CHADI,1/13).

4. Letter Kramers to Auger, 31/8/51 (UNESCO).

5. Letter Auger to Kramers, 31/8/51 (UNESCO).

6. Account based on letters from Fraser to Auger, 11/9/51; Fraser to Chadwick, 11/9/51; Bohr to Chadwick, 14/9/51; Chadwick to Fraser, 19/9/51; the draft of a letter from Chadwick to Bohr, probably dated 19 or 20/9/51 (CC–CHADI,1/3;1/9).

7. This is how Bohr describes Fraser's feelings (letter Bohr to Chadwick, 14/9/51, CC–CHADI,1/3):

   'He had himself been worried over the plan to have a meeting in November of government representatives to decide upon extensive plans involving considerable expenditures, and he felt that the only proper procedure would be first to agree upon the location of the project and on the composition of an international board to be entrusted with the organization at the locality in question of an expert group of experimental and theoretical physicists to study and make recommendations as to the best way in which such a European centre could fulfil its task'.

8. Letter Bohr to Chadwick, 20/9/51; draft of a letter Chadwick to Bohr, 19 or 20/9/51 (CC-CHADI,1/3). Referring to Darwin, Chadwick said: 'but, assuming that such a laboratory is to be established, he would strongly agree that Copenhagen is by far the best place for it'.
9. Letters Kramers to Auger, 24/9/51; Kramers to the Director-General of UNESCO, 24/9/51 (UNESCO).
10. These four letters are in the UNESCO archives.
11. Letters Kramers to Auger, 23/8/51 (UNESCO); Fraser to Auger, 11/9/51 (CC -CHADI,1/3), Kramers to the Director-General of UNESCO, 24/9/51 (UNESCO).
12. All quotations are taken from the letters Kramers to Auger, 23/8/51, and Kramers to the Director-General of UNESCO, 24/9/51 (UNESCO).
13. Perhaps we should remind the reader that as early as December 1950 Kramers had expressed the fear that the European laboratory might deprive the national laboratories of their best brains (see section 3.3). His proposal that the European laboratory be developed out of an existing institution was certainly an attempt to overcome this problem.
14. Letter from the Dutch Minister for Education, Arts and Sciences to the Director-General of UNESCO, 24/10/51 (UNESCO).
15. Letter Bohr to Auger, 26/10/51 (UNESCO). This letter was accompanied by a proposal drafted after discussion with Swedish scientists. It is principally the latter which we have summarized here.
16. Handwritten note on three sheets entitled: *Comments on Kramers' letter dated 23 August 1951* (UNESCO). These comments probably reflected more the opinion of certain physicists whom Regenstreif met on his travels than his own personal opinion, since he was still very young at the time.
17. Letter King to Chadwick, 11/10/51 (PRO-ED157/302); report Waterfield to Verry, 18/10/51, on his conversation with Auger (PRO-DISR17/559).
18. Letter Auger to Kramers, 11/10/51 (UNESCO). According to Alexander King *(op. cit.* note 17), Auger had developed this latter argument at length.
19. For these arguments, see Waterfield's report quoted note 17, and Regenstreif's note quoted note 16.
20. Regarding documents 6C/PRG/25 and UNESCO/NS/PN/1, see sections 4.2 and 4.3.
21. Let us not forget that the first *letter* which was explicit in this regard was Bohr's letter (to Auger) dated 26 October.
22. All these letters are in the UNESCO archives.
23. This information is taken from a letter Fraser to Chadwick, 24/9/51 (CC -CHADI,1/9). The quotation in note 7 shows that Fraser had rather precise ideas on how to proceed in such situations.
24. Letters Auger to Blackett, 4/10/51 (UNESCO); King to Chadwick, 11/10/51; Waterfield to Verry, 18/10/51 (PRO-ED157/302 and PRO-DSIR17/559).
25. See section 5.3.3 below.
26. On 22 October, Auger received a letter from Randers dated 18/10/51 (UNESCO) which, when read closely, was quite clear about Kramers' and Bohr's ultimate aims. Randers wrote: 'I have also read the remarks of Professor Kramers with the suggestion that the job be approached gradually by beginning with appointing a few physicists to work in Copenhagen and then developing this into a European laboratory with a smaller instrument than the Bevatron originally suggested for the European laboratory [...]. *By giving up the idea of building a very big machine* I also am afraid that the project loses much of its appeal and that it would consequently be even more difficult to raise money for such an undertaking' (our emphasis). However, it is likely that it was in his discussions with Kramers—and not on reading Kramers' letter—that Randers understood what Kramers was after. Let us not forget that both of them were at the head of the Dutch-Norwegian collaboration for a nuclear reactor at Kjeller.
27. For a more sociological kind of analysis, see section 5.4 below.
28. Letter Kowarski to Chadwick, 15/12/51 (CC-CHADI,1/11).
29. In his letter to Chadwick, 15/12/51 (CC-CHAD I,1/11).
30. Auger wrote to Kramers on 11 October (UNESCO): 'I have found much encouragement in the USA, where everybody seems to be ready to help us [...]. Some people however asked why the European group

did not contemplate the construction of a stronger machine. As we shall have the benefit of American experience [...] it would be useful if we, Europeans, try to go further, for example up to 10 BeV'.
31. See section 4.4.
32. These letters are all in the UNESCO archives. Quotations from letter Auger to Verhaeghe, 19/8/51 (UNESCO).
33. These replies are all in the UNESCO archives. Randers first gave Auger an oral reply at the Chicago meeting, which he subsequently confirmed in his letter dated 18/10/51 (UNESCO). The quotation is taken from this letter. On 8 November Auger replied (UNESCO): 'Amaldi did not wish to be a regular member of this Committee but he assured me that he would attend meetings whenever possible'.
34. Letter Gentner to Auger, 15/11/51 (UNESCO). In this letter Gentner proposed that Auger get in touch with Walther Bothe of Heidelberg instead of Otto Hahn, the former being both the most important experimental nuclear physicist in Germany, and at the same time a supporter of the plan. For further details, see chapter 11.
35. For Auger, see for example his letter to Preiswerk dated 12/10/51 (UNESCO), or his letter to Kramers, 11/10/51 (UNESCO). For Kramers, see his letter dated 24/9/51 to the Director-General of UNESCO (UNESCO).
36. Letter Kramers to Mussard, 9/10/51 (UNESCO).
37. The quotation is taken from letter Auger to Verhaeghe, 10/10/51 (UNESCO), but a similar phrase is to be found in all the letters sent between 10 and 19 October. In reply to Kramers' letter dated 9 October and quoted in note 36, Mussard was to say that the members of the committee would be 'appointed by UNESCO', that their 'travelling and accommodation expenses would be refunded by UNESCO', but that they would be invited 'with the agreement of the relevant circles in the countries concerned'. This is why Mussard told Bakker that if he was nominated by the Dutch scientists, then he would be invited. See letter Mussard to Bakker, 12/10/51 (UNESCO).
38. Letter Mussard to Goward, 18/10/51 and Mussard to Amaldi, 19/10/51 (UNESCO).
39. Bakker, *Meeting of consultants to UNESCO, 25-27 October 1951, UNESCO-building, Paris* (Mussard file, CERN); Goward report attached to a memo to Sir John Cockcroft, 29/10/51 (PRO-AB6/912). The report drafted by UNESCO (which we do not have) was mentioned in a circular which Mussard sent to the consultants and dated 13/11/51 (UNESCO).
40. Quotations from the reports cited in note 39. Letter Cockcroft to Auger, 24/10/51 (UNESCO).
41. Report 6C/PRG/25 drawn up after the May meeting of the consultants.
42. This has only recently become clear to us. When reading a first version of this chapter, Amaldi remained dubious about what we had called 'the opposition'. However, after reading the correspondence we had gathered, he told us that he never realized at the time the extent of this opposition to the consultants' project. Apparently, Auger took it upon himself to handle Bohr and Kramers, leaving the consultants free to refine the project at the technical level.
43. Goward doubtless is referring here to the difficulties facing Oliphant's group in Birmingham and Skinnner's group in Liverpool.
44. Letter Casimir to Auger, 13/9/51; letter Wideröe to Auger, 26/9/51 (UNESCO).
45. 'Strange opposition of these last two', *Amaldi's diary* (copy in CERN).
46. Thus Mussard wrote to Bakker on 5 December 1951: 'the official document will not be ready before the Conference, but it will not contain any substantial changes to the provisional text' (UNESCO).
47. All quotations in section 5.3.1 are taken from document UNESCO/NS/NUC/1, unless otherwise stated.
48. The choice of an energy greater than 3 GeV would allow for the production and investigation of particles then grouped under the heading 'heavy mesons'. The choice of an energy greater than 6 GeV would allow antiprotons to be produced. There was nothing new about these arguments in 1951, as they had already been used in the United States with regard to the decision on the 6 GeV bevatron. For further details on these points, see chapter 1.
49. Letter Director-General of UNESCO to Kramers (in reply to Kramers' official letter of 24 September);

letter Director-General of UNESCO to the Dutch Minister of Education, Arts and Sciences (in reply to the Minister's official letter of 24 October); letter Auger to Bohr (in reply to Bohr's personal letter of 26 October); all are dated 20/11/51 (UNESCO).

50. If we may give just one example, it can be said that in France, outside the field of nuclear energy proper, this did not become translated into reality until the government of Mendès-France. See for example Châtelet (1960) and section 9.2 of this volume.
51. *Program for the inauguration of the Joint Establishment for Nuclear Energy Research,* document attached to a letter from J.E.N.E.R. to Mussard, 15/11/51 (UNESCO).
52. Letter Kowarski to Chadwick, 15/12/51 (CC-CHADI,1/11), from which the quotation was taken; see also interview with Jean Mussard, 1/9/73 (CERN); letters Mussard to Dahl, 4/12/51, Bohr, 5/12/51, Amaldi, Bakker, and Preiswerk, 6/12/51 (UNESCO). Note that Mussard took these initiatives because Auger was in Asia at the time.
53. He did not send Dahl a copy of his letter to Bohr, because Dahl was present at the Oslo meeting, and because he had written to him the day before he wrote to Bohr. It should be noted that, just as Auger had done on 20 November, Mussard used an argument appealing to authority (that the decision was up to UNESCO) to refuse to alter the text submitted to the conference.
54. Letter Bakker to Mussard, 7/12/51 (UNESCO). Kramers' 'conversion' should be compared with the following remark by Nyberg and Zetterberg: 'Nearly all of them [the scientists Funke met in November 1951 in Europe] had also taken a clearly negative attitude towards a location in Copenhagen [...]. After his round trip, Dr Funke began to doubt the possibilities of realizing the Copenhagen project, "however nice it would be from our point of view" ' (Nyberg & Zetterberg (1977), 14).
55. Letter Mussard to Kowarski, 11/12/51 (UNESCO). One thing has struck us about Bakker and Mussard: neither of them appeared to understand fully what Bohr was proposing; they both thought that if he was allowed to set up his 'committee of experts' the problem would be resolved in three months.
56. Here we have followed the sequence of events as depicted by Kowarski in his letter to Chadwick quoted in note 52.
57. Kramers, who was already ill in 1951, died early in 1952.
58. For his part, Heisenberg would also observe that the UNESCO plan was essentially supported by young researchers. See for example the account of the third meeting of the Senat der Deutschen Forschungsgemeinschaft, Sitzung am 29 February 1952 (DFG-AZ.46/2/52).
59. In his report quoted in note 58, Heisenberg remarked that one of the motivations of the younger group was that Europe should catch up with the United States.
60. It must be noted that even in the most extreme cases —France and the United Kingdom, for example—we are not suggesting that there was unanimity. We are aware of various exceptions, various shades of opinion. However, the fundamental trends were clear, and it is these that we have described.
61. On these points, see part IV of this volume.
62. See section 4.3.2. The quotation about the Swedish Committee for Atomic Affairs is taken from Nyberg & Zetterberg (1977), 17.
63. See for example Grosser (1972).
64. Let us simply describe the Swiss case. We said that in June 1951 Scherrer (who was a friend of Bohr's) was opposed to the plan for a giant accelerator. In August, after the Copenhagen conference, Preiswerk described him as being more enthusiastic. In December, at the conference, Switzerland—and Scherrer was head of the Swiss delegation—adopted a position very much in favour of the UNESCO plan. This did not prevent Skinner from writing in January that Scherrer was sceptical about the bevatron. 'It seems', he added, 'that he was pressed by the Swiss Government into being a delegate more or less against his wish [...]'. Here then we have the strange case of a government supporting the most ambitious plan against the opinion of its most illustrious scientist. See letter Skinner to Chadwick, 14/1/52 (CC-CHADI,1/8).

CHAPTER 6

The establishment of a Council of Representatives of European States[1] December 1951–February 1952

Dominique PESTRE

Contents

6.1 The intergovernmental conference in Paris, 17–20 December 1951   180
    6.1.1 The delegations   180
    6.1.2 Opening statements by the delegations   181
    6.1.3 Adoption of a resolution to 'save' the conference   183
    6.1.4 Final matters discussed by the conference   186
6.2 Negotiations between December and February   188
    6.2.1 First controversial issue: a joint laboratory or 'simple' co-ordination?   189
    6.2.2 Second controversial issue: priorities in the allocation of funds   190
    6.2.3 Third controversial issue: who was or could become a member of the Council?   191
    6.2.4 Fourth controversial issue: division of powers in the provisional organization   192
6.3 The second session of the conference, Geneva, 12–15 February 1952   194
    6.3.1 Once again: laboratory or co-ordination?   195
    6.3.2 The proposal for a six-week international conference   196
    6.3.3 Acceptance of last remaining matters   196
6.4 Concluding remarks on the respective roles of scientists and politicians in the process of establishment of the Council   198
    Notes   202

The previous chapter ended on a rather gloomy note. Two fundamentally opposed proposals had been put forward, one based in Copenhagen around Bohr, and supported by Kramers and several leading British, Danish, and Swedish physicists; the other based in Paris around Auger, and backed by scientists and administrators from Belgium, France, Italy, Switzerland, and certain leading figures from the Netherlands and Norway. In the weeks leading up to the intergovernmental conference tension heightened to the extent that Mussard of UNESCO considered it essential that the consultants come together one last time three days before the meeting opened.

The atmosphere was therefore rather tense on the morning of 17 December 1951. The aim of this chapter is to understand how a compromise was reached. It should be noted immediately that it took a long time. After meeting for four days the conference adjourned without achieving its objectives but agreed to reconvene on 12 February 1952. In the interim there was intense diplomatic activity and new documents were drawn up. On 12 February the conference resumed, hoping to terminate its business with a meeting of the Council of Representatives of States which it wanted to set up. However, when it ended on 15 February 1952 the Council had not met: it had been formally constituted but not enough states were willing to commit themselves immediately. A month and a half later, however, five states having ratified the Agreement, the Council could be convened, and it held its first session from 5 to 8 May 1952.

6.1 The intergovernmental conference in Paris, 17–20 December 1951 [2]

It is not easy to describe what happened during the four days of the conference as everything has to be inferred from a small number of reports and resolutions. In particular it is difficult to retrace the process of informal negotiations though we all know how important this is in diplomatic gatherings. Nevertheless we believe that our knowledge of the situation immediately before 17 December, which is based on a large volume of correspondence and is therefore quite accurate, allowed us to 'read between the lines' of the resolutions.

6.1.1 THE DELEGATIONS

Although in August the Director-General of UNESCO had invited more than 30 countries to participate in the conference, only thirteen European countries sent delegations: Belgium, Denmark, France, Greece, Italy, Norway, the Netherlands, the United Kingdom, Sweden, Switzerland, as well as the Federal Republic of Germany, Turkey, and Yugoslavia. For their part Australia, Brazil, China, Cuba, India, Israel, and Japan were represented by observers. Without listing the members of all the delegations, the composition of some of them is worth noting. Germany nominated Werner Heisenberg and Alexander Hocker, though the latter did not attend because the French refused to grant him a visa. Denmark's official delegates were H.M. Hansen and Jakob Nielsen, Niels Bohr and J.C. Jacobsen being only advisers. Italy sent Senator Alessandro Casati and Gustavo Colonnetti, Edoardo Amaldi not even being an adviser. The reason for this seems to have been political as de Clementi, Italy's permanent delegate at UNESCO, had gone as far as to say to Auger that he would leave the conference hall if Amaldi set foot in it. This was something most of the European scientists and Colonnetti were not prepared to accept. Verde from the Institute of Physics at the University of Turin called on Amaldi at his hotel early in the afternoon on the day the conference opened; at six o'clock he returned with Colonnetti; half an hour later Bakker telephoned, and that evening Amaldi dined with several delegates including Heisenberg. The following day he came to the conference where he was greeted demonstratively by Thomson, the British delegate — and nothing more was said.[3]

Norway's delegate was Professor E. Hylleraas from the University of Oslo; Odd Dahl was only an adviser. Amaldi noted in his diary on 17 December that Dahl's position with regard to the Norwegian delegation seemed to be rather similar to his own. Lastly, the Netherlands' delegation was composed of Cornelis J. Bakker and J. Hendrik Bannier, director of the Netherlands Organization for the Advancement of Pure Research (ZWO), Kramers being absent for reasons of health.

6.1.2 OPENING STATEMENTS BY THE DELEGATIONS

After listening to Auger's welcoming speech, appointing its officers, adopting rules of procedure, and studying the credentials conferred by the different states, the conference heard the official statements of the delegations. Only Norway and Turkey did not speak.[4]

As can be imagined, the statements were of two distinct types. On the one hand Denmark, Sweden, and the United Kingdom put forward similar ideas; on the other Belgium, France, Italy, and Switzerland made identical proposals concerning the working programme of the conference. For their part Germany, Greece, the Netherlands, and Yugoslavia adopted more diverse positions.

The Scandinavian countries and the United Kingdom began their statements by reaffirming that they were 'very positive to a European collaboration in the field of nuclear physics', which is readily understandable as it was *they* who appeared to be *opposed* to the UNESCO project. They then stated that the procedure which should be followed was the one proposed in Kramers' letter, i.e. to make use of existing institutions at least to begin with. That granted, Thomson added that the United

Kingdom envisaged giving the European centre certain rights over Liverpool's machine, even if 'one necessary condition' was that 'the Centre should be associated with an existing research organization'. Similarly, Malte Jacobsson declared that 'Sweden [was] willing to make available for European collaboration' the 200 MeV synchro-cyclotron then starting up in Uppsala. They concluded by observing that their project did not 'exclude planning the building of experimental devices on a big scale'. Jakob Nielsen from Denmark, however, stated that 'instead of pointing to experimental facilities which would only come into operation in four or six years from now on', it seemed 'essential to us that we should start as soon as possible using the available means of experimental research [...]'. He also expressed satisfaction at having heard Auger say in his opening speech that 'the plan laid before us in the working paper [was] not to be regarded as a rigid one'.[5]

Belgium, France, Italy, and Switzerland began by emphasizing that the working document '[provided] a satisfactory basis for discussion', which meant that these countries considered it 'essential to build some kind of a big machine on which research would be centred' and that they wanted a 'truly international laboratory', 'an institution which [would] be their own [the European countries], and not an already existing national organization'. However they added that the 'proposals such as the one submitted by the delegation of the United Kingdom, which would permit to begin, in the very near future, joint scientific work even on a limited scale', were very important. In plain language this meant that they accepted the offers of the use of specific existing institutions or equipment, but that this did not in any way authorize the conference to overlook the very reason for its existence: the study of plans for a bevatron. Finally, during these opening statements Belgium, France, and Switzerland announced their contributions for phase 2. They would respectively amount to 20,000, 71,000, and 25,000 dollars. In the days thereafter Italy and Yugoslavia offered 25,000 and 10,000 dollars for the planning stage.[6]

Now for the uncommitted cases. Yugoslavia declared that she was 'in favour of making the large accelerator of more than 3 BeV' but also 'fully [supported] the proposal to increase the co-operation by the use of existing facilities made available in various countries'. Germany, through Heisenberg, began by saying that if the European centre was located in Geneva 'this plan [would] probably find much support in our country, on account of the old tradition of neutrality and friendly European cooperation in Switzerland'. Then, stressing the critical economic situation in Germany, Heisenberg said that he attached great importance to the offers of the immediate use of the Liverpool machine and of Professor Bohr's Institute in Copenhagen. As far as the bevatron was concerned, he emphasized that in his opinion an attempt should be made to construct more advanced machines than the American ones, and in any event not to be restricted to copying one of them. Finally there was the statement made by J.H. Bannier of the Netherlands. Remaining strictly neutral in the debate, he only said that as the Netherlands delegation was not 'empowered to commit itself on any definite decision', it wanted to 'express the willingness of the Netherlands to collaborate' and to do 'its utmost to help the conference in improving the plans, which at the moment may leave way to some doubts [...]'.[7] Note here that the newcomers—Yugoslavia and Germany—had apparently not fully realized what was at stake. Their main wish was to be part of the new organization—at the limit, no matter what its precise aims. In the case of the Netherlands, however, no such innocence was possible. Bannier knew the situation too well to declare that he was in favour of everything. Cautiously, he only offered to help find an intermediate solution.

Before proceeding any further it should be added that several countries mentioned the problem of financing the project *in the long term*: 'in the present state of financial stringency further large expenditure by Britain on nuclear physics would not be justified', said Thomson; 'the Greek delegation expresses its regret at being unable to participate in the financing of the nuclear laboratory, on account of the conditions prevailing in their country [...]', declared Professor D. Hondros; 'it seems to me very doubtful whether our country could in a near future consider expenses of the order of magnitude necessary for carrying out the third stage', noted Heisenberg; and if Jakob Nielsen supported Kramers' plan, one reason for doing so was that 'this is in no way a costly plan'.[8]

6.1.3 ADOPTION OF A RESOLUTION TO 'SAVE' THE CONFERENCE

On reading the previous statements, one realizes that the positions had not seriously altered for two months. To paraphrase the closing speech of chairman de Rose,[9] at the end of the first day none of the delegates was unaware that their points of view were not identical. Unless an approach could be found that would cut across the existing proposals, the conference would have to be adjourned.

A temporary solution was provided by the Netherlands delegation, who put forward a resolution towards the end of the the first day. It contained five articles, two trying to meet the wishes of the British, the Danes, and the Swedes, and three to satisfy those of the supporters of the UNESCO working document. The Netherlands delegation 'proposed to start with':[10]
- 1) Creating the Institute for Advanced Studies in Nuclear Research in close liaison with an existing centre, that in Copenhagen;
- 2) 'Accepting the offer made by the United Kingdom representative' to use Liverpool's machine 'on a European basis'. To satisfy Thomson's condition it was stated that 'the theoretical guidance should be provided by the Institute for Advanced Studies in Nuclear Research' located in Copenhagen;
- 3), 4) and 5): Creating three study groups, one 'for an intermediate machine', the second 'for a large machine', and the third 'responsible for organizing the central European Laboratory for Nuclear Research'.

In short the Netherlands delegation proposed not to choose between the two existing projects, but to *juxtapose* them, to have them develop *in parallel*. On the

Notes: p. 203

one hand, Bohr in Copenhagen would be able to organize a general survey of the state of the art in nuclear physics and of European needs, a survey which he had said was necessary and which he thought should be done above all else; on the other hand, Auger and his associates were given an agreement in principle for both machines and for the central laboratory they wanted, as well as for the creation of three groups in charge of studying them. Nevertheless, there was no mention of where the three groups would be located, who would lead them, or how they would be financed. Moreover no mention was made of any need to establish a Council of Representatives of States, this being an issue of secondary importance for Bohr and his supporters (after all it was not an urgent matter to establish such a Council to use the Liverpool cyclotron), but a crucial one for the others who had to recruit a large full-time staff and who needed the money to do so. It was therefore unlikely that the Netherlands proposal would be adopted without debate.

In fact renegotiation of the proposal took up the best part of the following two days, the third and fourth versions of the Dutch resolution being dated 19 December. One thing must be said before studying them. Since the first version of the resolution took into account the Danish and British offers (a possibility not envisaged in the initial UNESCO working document), annexes IV and V of this document (which described the four study groups and the distribution of funds between them) obviously could not be adopted in that form: a new balance had to be struck. This was, of course, a sensitive issue and the feeling developed very early on that it could not be resolved quickly and that a second session of the conference would accordingly have to be held a few weeks later.[11] If we have given these details now, it is to draw attention to the fact that the precise wording of the successive versions of the Dutch resolution seems to have implied a wider 'bargain' concerning the whole of what was accepted (or rejected) by the end of the December conference.

What are the third and fourth versions of the resolution? That the conference recommends:

- Point 1: 'Setting up a Board of Representatives from the participating countries, with headquarters in Geneva' to direct the implementation of the programme in phase 2. Later the conference would decide to form a working group to prepare a draft agreement for creating this Council, to be placed before the second session of the conference;[12]
- Point 2: Accepting the offer of the British delegation;[13]
- Point 3: Accepting the offer of the Danish delegation to use Bohr's Institute in Copenhagen 'to assemble a *study group* for theoretical research on a European basis' (our emphasis). The resolution goes on: '[This group] should provide the theoretical guidance for experimental work, to be carried out with the machines'. One point must be noted here. The intention was to create a study group to prepare the work to be done with the various machines. This put the group set up in Copenhagen on a par with the other three groups, and the tasks assigned to it were those which the UNESCO working document had envisaged from the outset;

– Points 4, 5, and 6: Creating the three study groups for the two machines and the laboratory. Here the formulation was as vague as in the first version: no detail concerning the machines, the group leaders, or the locations was given. However — and this seems to have been part of the 'bargain' which led to the adoption of this version of the Dutch resolution — after passing this resolution the conference adopted annexes II and III of the working document, which were entitled: *Object: main equipment of laboratory* (the 0.5 BeV synchro-cyclotron and the 3 to 10 BeV proton synchrotron) and *Creation of an Institute for Advanced Studies in Nuclear Research*. In other words the type of machine to be built seems to have been agreed on. Indeed, during the discussion Auger said that 'the working groups, referred to under items 4 and 5 of the resolution previously adopted [...], would be responsible for designing two instruments [as defined in Annex II], the construction of which could, from an engineering point of view, be started immediately'.[14]

To sum up the terms forming the basis of the compromise reached at the first session of the conference, we can say that the following were conceded to the Bohr–Jacobsson–Thomson group:

1 – The principle of beginning collaborative work using existing institutions and equipment (the institute in Copenhagen, the Liverpool machine and also apparently the Uppsala machine, the Swedish delegate having confirmed this offer during the discussion on the Netherlands resolution);

2 – The possibility of undertaking theoretical studies in Copenhagen 'without delay'. The group responsible for this was clearly described, unlike those for the 'intermediate machine', the 'large machine' and the 'laboratory'. During the debate the Swedish delegation even asked for it to be specified that the conference 'agreed' that only the Council of Representatives would have the authority to create the latter three groups, and this was accepted;

3 – The principle of creating an institute for advanced studies in nuclear research. Although the Copenhagen group was responsible for working out the scheme in detail, this was not to be regarded as a decision on the future location or on the work of the theoretical study group, which was independent of it.

The supporters of the UNESCO project were granted:

1 – The constitution of the Council of Representatives. As we have said, this was crucial for them as their long-term project required an inter-state agreement, if only for financial reasons. And it was precisely the question of the Council which caused the British not to sign the agreement in February. On the 12th of that month the following message would arrive: 'His Majesty's Government [...] regard the International Governmental Organization envisaged in the draft Agreement as too elaborate a mechanism' ... too elaborate for the aims Britain thought it should have; consequently the British delegation would not sign the proposed agreement.[15]

2 – The principle of creating the three study groups they wanted. However, since annexes IV and V were to be debated again, everything concerning personnel and

funds allocated to these groups depended on the agreement to be reached during the second session of the conference. We have not drawn attention to this purely for convenience: it was in fact the main topic of discussion at the second session.

6.1.4 FINAL MATTERS DISCUSSED BY THE CONFERENCE

Regarding the site a *modus vivendi* was soon achieved. The Belgian and Italian delegations having proposed that the future laboratory be located in Geneva, the chairman observed that 'it was not the task of the conference to take such a decision'. He was immediately backed by the Danish and Swiss delegations, whose countries were the two main candidates for the site, the former stating that it wished to 'reserve its right to propose, in due course, Copenhagen as a possible location for the laboratory'. In fact as the issue was too far removed from the preoccupations of the moment and also much too sensitive, the conference decided not to discuss it.

With regard to those who wished to participate in phase 2, 'the majority of the delegations stated that their governments were willing to participate in the work'. However, a definitive list could not be drawn up since 'some delegations' did not have the authority to commit their governments. All the same the financial offers made by Belgium, France, Italy, Switzerland, and Yugoslavia were confirmed and Denmark announced 'that a financial contribution, which would, under no circumstances, be inferior to the figures resulting from the preliminary calculations made by UNESCO, could possibly be expected'. For their part the United Kingdom and Sweden repeated their proposals concerning Liverpool and Uppsala. At this point Auger publicly expressed his satisfaction, not to say jubilation, at seeing such rapid progress and he remarked that 'the availability of funds necessary for the completion of the second stage [as he saw it] was practically certain'—after all more than 151,000 dollars were already offered.[16]

It remained only to organize the second session of the conference. A working group was created, composed of M. de Hemptinne (Belgium), J. Nielsen (Denmark), F. de Rose (France), G. Colonnetti (Italy), J.H. Bannier (Netherlands), A.H. Waterfield (United Kingdom), and A. Picot (Switzerland). It was to prepare the draft agreement creating the Council of Representatives and defining its powers and functions. Its task was to be completed by 20 January 1952 so that the text could be sent to the different countries. The conference would reconvene on 12 February in Geneva under the chairmanship of Paul Scherrer, and it was hoped that by the end of the second session the agreement would be ready for signature by the states. It was specified that the agreement would include a binding commitment to make contributions in money or in kind.

Secondly, the conference asked the UNESCO secretariat—that is, the group of scientific consultants—to prepare a revised version of annexes IV and V of the working document in the light of the resolution proposed by the Netherlands and adopted by the conference.

Finally, Paul Scherrer and Niels Bohr were asked to begin negotiations with the British authorities responsible for the Liverpool machine in order to determine ' the terms under which the synchro-cyclotron presently under construction at Liverpool could be temporarily made available for the European group'.

To conclude, we would like to try to answer one question: how was it possible to reach an 'agreement' given what we know of the wide differences of opinion that separated the various delegations at the start of the session, of the disaster of the last joint meeting in Oslo the month before, and of the grave doubts entertained (by Kowarski and Mussard for example) over the chances for success at the conference?

We think different types of argument are needed to answer this question. The first, which we have already mentioned, is that the issues dealt with by the conference were of *limited enough scope* to allow for a juxtaposition of the basic ideas contained in the two projects. By 'limited enough' we mean that only short-term decisions were made and that the preparatory studies did not imply any decision regarding construction and therefore no major financial commitment. This made it possible *not to choose* and to go on studying both projects in parallel. The corollary of this was that the competition for funds between the two would start all over again.

A second element explaining why an agreement was possible is that the Netherlands resolution was couched in terms sufficiently vague for all to read into it a victory for themselves, or at least to hope that their point of view would prevail *in the long run*. The supporters of the UNESCO project did indeed have reason to think that, having an agreement in principle on the bevatron and the study groups for 1952, they would be in an excellent position during the February session: they would provide the money themselves and would therefore be able to determine how it was spent; what they had conceded was of no consequence. For his part Bohr was prepared to agree that there be a continuation of studies for '[the] big new apparatus, the construction and placing of which is left to further decisions', but the important thing as far as he was concerned was that it had been decided to use existing machines and to set up immediately in Copenhagen 'a group of experts to explore the general situation as regards high energy research [...]'. And Thomson could write that 'the French and Italians [...] will have to be allowed to try to raise the money', which, he added, 'I don't think they can do'. In other words the agreement was possible because each party felt that it would come out on top in the end.[17]

However, we need a third type of argument to grasp why a minimum of agreement was possible in December but not a month earlier in Oslo. To appreciate its importance, we must firstly recall that, even if the physicists disagreed about the best way to proceed, most of them shared a common aim: to strengthen European co-operation in science. There was therefore a genuine desire not to put at risk the first European conference called in order to study such a possibility. The second

Notes: p. 203

element to recall is of a more structural nature. It characterizes the relationship between the scientists and the outside world and can be described as follows: scientists planning a large-scale project generally tend to paper over their divisions and present a united front to those to whom they apply for funds. For the good of their community as a whole, they behave as a pressure group projecting an image of unity: they know that if they appear to be divided, they will get nothing.[18] Now, for the first time, the European scientists found themselves in the context of an official conference, a diplomatic conference from which *decisions* were expected by the *states* which were represented. Everyone was aware that if the conference ended in failure, it would tarnish the image of the nuclear physicists and weaken the whole community. For this reason the moral pressure on the participants, particularly on the more hesitant delegations, was particularly intense (Thomson wrote to Chadwick on 22 December that the offer to use the Liverpool synchro-cyclotron on a European basis 'has prevented us having to take a purely negative attitude, which would have been unpopular'[19]). Hence the necessity, during that official gathering, to ensure some kind of agreement. In brief, the fact that diplomats were participating in the conference (which was not the case at Oslo) and that its results would come under public scrutiny served to contain centrifugal tendencies.[20]

6.2 Negotiations between December and February

The conference held in December 1951 asked three working groups to prepare the second session to be held in February. The first, referred to as 'the working group' in texts of that period, was to prepare the draft agreement creating and defining the powers and functions of the Council of Representatives of States. It met three times (21 December, 3 and 4 January, 14, 15, 16 January) and the draft was sent to the states on 24 January 1952. The second group established by the conference was to reconsider the documents submitted in December concerning the organization of studies in phase 2. This was the group of UNESCO consultants. It met on 18 and 19 January 1952 in the presence of Amaldi, Bakker, Dahl, Goward, Kowarski, Preiswerk, and Verhaeghe, Auger and Mussard representing UNESCO. It should be noted that the Danish and Swedish experts were unable (or did not want) to be there. However J. Nielsen, the Danish representative at UNESCO, was present as an observer. Lastly the third group, composed of Bohr and Scherrer, had the task of negotiating with the British authorities over the use of the Liverpool machine. This they did during a visit to Britain from 5 to 12 January 1952. The detailed results of this journey are examined by John Krige in section 12.4.1.[21]

It must be added that there was intense diplomatic activity surrounding these 'official' meetings, and a substantial number of amendments were proposed by direct agreement between national delegations. However some issues remained too controversial to be resolved in this manner. Thus Denmark officially submitted a series of amendments to the conference office on 5 February. Sweden did the same a few days later. As the proposed amendments caused some concern in the pro-UNESCO camp, on 11 February Italy submitted one at variance with the previous ones.[22]

Before examining the main controversial issues, it is perhaps worth giving a brief outline of the most important document submitted to the February session of the conference: the draft agreement constituting the Council. Article II described the *composition* of the Council and the conditions under which new member states could be admitted; Article III the *functions* of the organization to be established; Article IV the *method of work* of the Council and of the committee which would meet between the Council sessions; Article V the powers of the *Director-General;* Article VI the financial resources and the method of drawing up the *budget* of the organization; Article VIII the *duration* of the agreement (18 months or longer if the Council so decided); and finally, Article IX described the conditions which had to be fulfilled for the agreement to come into force: that it had been signed without reservation by five states and that 150,000 dollars had been contributed.

6.2.1 FIRST CONTROVERSIAL ISSUE: A JOINT LABORATORY OR 'SIMPLE' CO-ORDINATION?

In the final version of their 'draft Agreement constituting a Council of Representatives of States', the members of the working group specified that the function of this Council was the 'Establishment of a European Nuclear Research Centre', the word centre having been chosen in order 'to include all activities of the Council, as mentioned in the first Resolution of the Paris Conference [the Netherlands resolution]'.[23]

On learning of this formulation, very determined opposition again arose in Copenhagen, Stockholm, and Cambridge. The Danes even proposed a change of title, feeling that the Council should only be created 'for the establishment and preparation of European co-operation in Nuclear Research'. In their view this formulation was closer to the wishes expressed in December where the dominant idea was co-operation, a co-operation, which '*in the course of time may lead* to a continuous pooling of experimental facilities and possibly more than one special laboratory [...]' (our emphasis). For their part the Swedes insisted that before making plans one should discuss 'the problems to be attacked and the types of apparatus suitable for these purposes' and that instead of deciding to establish a centre it was more important to encourage 'new developments and inventions', and to stimulate 'the enterprising spirit of European scientists in this field [of nuclear research]'. This line of thinking was directly connected with Bohr's proposal of 26 October taken up in December. The adoption of annex II (definition of the machines to be constructed) was apparently forgotten.

The Danes and the Swedes also protested, logically, against one of the corollaries

Notes: pp. 203 ff.

of this definition of the aims of the Council. The working group had said (in Article III of the draft) that the Council was to prepare for the establishment of the centre by carrying out technical investigations and by making all the necessary arrangements for the use of the installations which had been offered it. They declared that this definition of functions was incomplete. Certainly the Council was to carry out 'technical investigations relating to new experimental equipment', but it had first to go ahead with the 'study group for scientific research'. This statement, which referred to the theory group located in Copenhagen, served to distinguish it from the three others and to underline its special nature: it would not be merely a 'technical' group, but would have the more immediate and general function of making proposals. For the Swedes the relationship of the group with the Council should therefore be clearly specified in the text of the agreement itself.[24]

6.2.2 SECOND CONTROVERSIAL ISSUE: PRIORITIES IN THE ALLOCATION OF FUNDS

As can be imagined, the difference which we have just described led to two different orders of priority with regard to the money collected. For Bohr, for Jacobsson, for Chadwick little money would be needed in the short term and 'a considerable part of the total expenditure in the second stage of the project should be used for putting into effect the objects defined in points 2 [use of the Liverpool machine] and 3 [creation of a study group in Copenhagen] of resolution I [the Dutch resolution]'. In other words to provide stipends covering the travel and accommodation expenses of those sent to the two cities. Consequently 'the number of full-time posts with fixed salaries [to be created for the 'technical' groups] should be as low as possible'.[25]

Of course this view was not shared by the advocates of the UNESCO project. For example, the consultants who met on 18 and 19 January spent most of their time defining the tasks of the study groups, discussing salary levels and qualifications for the different posts, and working out the best recruitment procedure. As far as they were concerned the proposals they had made in November for the December conference (annexes IV and V of the working document) were still valid, and they intended to recruit dozens of people, mainly for the SC (synchro-cyclotron) group and the PS (proton synchrotron) group. The working group had dealt with the same matters during its meeting on 3 and 4 January and they were referred to in the letter sent to the states on behalf of the Director-General of UNESCO on 24 January.[26]

Similarly, Auger reaffirmed that 'the 150,000 U.S. dollars would virtually all be used in the planning studies'. 'Should there be any small sum left over from the plannning studies' wrote a British official, 'it is Auger's hope that it might be used for posting research physicists either to Liverpool or Copenhagen'. As can be seen, the position was quite unambiguous and it was the Italians who took on the task of drafting the counter-amendment to the Danish proposals. On 11 February they asked that following Article III.2, which read that the Council should 'take measures appropriate for utilizing the equipment [...] put at its disposal', the rider be added: 'provided that the financial obligations arising out of such agreements shall not prejudice the achievement of the principal aims set out in Paragraph 1' — which stated that the Council should carry out 'technical investigations relating to the necessary experimental equipment'.[27]

6.2.3 THIRD CONTROVERSIAL ISSUE: WHO WAS OR COULD BECOME A MEMBER OF THE COUNCIL?

For the members of the working group the answer to the first part of this question was simple: 'The States which took part in the regional Conference [...] and which undertake to contribute in money or in kind to the Council [...] should be members of the Council'. This definition of the body directing the provisional organization posed problems only in the case of Great Britain. It was not a matter of finance, although Sir George Thomson had said in December that it would probably be difficult for the United Kingdom to allocate additional funds to nuclear research. After all, the continental countries were ready to consider the offer of partial use of the Liverpool machine as a contribution in kind, being only too happy thereby to involve a reluctant United Kingdom in their undertaking. It was at the political level that the problem arose.

The 'affair' began at the working group meeting on 3 and 4 January, when the British delegate asked whether provision could be made for the possibility that 'the U.K. might be represented by such a body as the Royal Society' on the Council and not by an official representative of the government. The other participants replied that this was unthinkable since an *intergovernmental* organization was being created, and the matter was not raised again in the working group.[28]

However the issue came up again at the highest level during a Cabinet Steering Committee meeting held in Britain on 8 February, attended by representatives of the Foreign Office, the Treasury and other government departments. In its conclusions this committee noted firstly that 'standing policy of Her Majesty's Government [was] against joining or promoting the creation of new international organizations'; it then stated that whatever the aim of the proposed centre nothing, at this stage of preliminary studies, required 'the setting up of a special Council of States [...]'; consequently 'the proposed machinery was unnecessarily cumbersome for the task in hand' and Her Majesty's Government would not sign the proposed agreement.[29]

We would like to make an observation at this point. The attitude we have just described is perfectly 'logical' from the point of view of the Treasury or the Foreign Office. For them, the government had a standing policy which had to be followed unless good arguments were given against it. As none was given, *notably by the scientists,* they refused to let Britain be part of the Council. Why did the Foreign Office or the Treasury think the scientists had no good reasons? Because they

Notes: p. 204

rejected the UNESCO project and the need for an accelerator 'which no European Country could build alone'. By contrast, they supported Bohr's scheme, Bohr who (rightly) felt in October that a group of leading scientific institutions (*and not states*) would be enough for the kind of co-operation he proposed. In the light of the Bohr–Kramers–Chadwick project, the Committee's position was a 'natural' one — as were those of France or Italy in the light of the Amaldi–Auger scheme. For them, an interstate Council was probably the best solution.[30]

The second part of the question was: who could become a member of the centre in the future? Here the problem was the possible candidacy of eastern European countries, something which at least neither France nor Great Britain wanted. Concern about this question resurfaced in January. On the 31st Waterfield wrote to Rackham that 'the possibility of an application from [eastern European] countries to join the Council seems to be in the air'. His only reason was that the project had been far less attacked in the Communist press during the last month. To 'solve' this problem it was decided that applications should be examined by the Council (the French specifying a two-thirds majority), and that every candidate would have to declare in advance its readiness to 'co-operate in the work of the Council on a footing of the free reciprocal exchange of persons and scientific information and technique', which in the view of Waterfield (and others) would serve to 'exclude eastern European countries'.[31]

6.2.4 FOURTH CONTROVERSIAL ISSUE: DIVISION OF POWERS IN THE PROVISIONAL ORGANIZATION

For the working group the provisional organization would have to be managed by a Director-General because, as Waterfield said, 'somebody must "run the show" '. Therefore the draft agreement stated: 'The Council shall commit to a Director-General the task of carrying out its decisions and its technical work'. For their part the UNESCO group of scientific consultants were more concerned about the co-ordination of scientific work carried out by the study groups. For this reason they thought that a Director-General (or a 'top Executive Officer') should be appointed to call and to chair regular meetings of group leaders. In this way continuity between present activities and the work already done would be ensured, and a link with the Council established.

However Bohr, Chadwick, and some states (the United Kingdom, Denmark, and Sweden) reacted negatively to these proposals. Their basic idea was that responsibility for all important decisions, including appointments, should lie with the Council or its chairman, rather than with a too-powerful Director-General who would be independent of the states. France, who shared some of their anxieties, emphasized however that this person should be given sufficient authority 'to act for the Council in carrying out any necessary co-ordination in the intervals between the meetings of the Council'.[32]

In conclusion we should like to mention a point which might seem obvious, if not trivial: the December conference did not seem to have resolved anything, and at the beginning of the February session the situation seemed to be little different from the one which had prevailed after the Oslo meeting in November 1951. The projects were still conflicting ones, the terms of the debate had not evolved and the resolutions adopted in December carried little weight: they had been accepted because the delegates did not wish the conference to end in failure and needed time to strengthen their own positions.

Indeed, between December and February the two opposing groups behaved like well-organized clans communicating only by amendments. On the one hand the activity of the supporters of the UNESCO project (represented by the consultants) was co-ordinated by Auger's secretariat; on the other Bohr, Jacobsson, Kramers, and Chadwick corresponded and met in Copenhagen to 'concentrate their fire' on vital issues. And although each group claimed to be faithful to the spirit and the letter of the Dutch resolution adopted in December, no-one was fooled. Certainly not the British officials who met on 8 February: they commented in their text that in fact 'the draft agreement [seemed] to embody two entirely separate proposals' — an assessment which only served to increase their own scepticism.[33]

Nevertheless, by the 12 February one aspect of the situation had changed, namely the balance of forces. Four points are to be borne in mind here:

a – Firstly, three texts existed which had been adopted in December and were therefore 'binding' on the delegations.[34] These texts clearly defined the laboratory, the two machines to be constructed, and the four groups instructed to carry out studies in 1952. The Danes and the Swedes might have believed that the main task of the Copenhagen group was to reconsider the entire situation, but it would be easy to confront them with the texts they had signed.

b – Moreover the Swedish and Danish camp, whilst it could claim possible support from a divided Norway, had lost the opportunity for blackmail provided by the threat of Britain's withdrawal, since the United Kingdom had already announced that she would not participate in the centre. It had also lost the benevolent neutrality of the Netherlands which decided at the beginning of February to offer 10,000 dollars and to participate in stage 2, adopting the UNESCO position.[35]

c – The UNESCO camp, for its part, had gained in strength with the addition not only of the Netherlands but also of Germany. This is clear from the minutes of the February conference where, for example, Waterfield spoke of a 'strong group France–Germany–Belgium–Holland–Italy–Switzerland', Denmark being relatively isolated.[36]

d – Lastly those putting up money supported the UNESCO project: they were able to provide the 200,000 dollars needed for phase 2 on their own. This would be a powerful argument when deciding how to distribute financial resources.

Notes: p. 204

6.3 The second session of the conference, Geneva, 12–15 February 1952[37]

We know quite a lot about the February session of the intergovernmental conference because the three British delegates, who did not take part simultaneously in the proceedings, wrote interim reports which we have retrieved. In addition the Geneva press showed more interest in the debates than the French press had done in December.

We would first like to describe how the work of the conference unfolded, as this will allow us to identify clearly the issues which provoked the most heated debates. On 12 February Albert Picot, member of the Swiss Council of State, opened the session. After the usual formalities each delegation addressed the conference, outlining its wishes and putting its amendments. It was apparent from this first exchange of views that two issues would not impede the work: each delegation said that it would sign the draft agreement and authorize the United Kingdom 'to send observers to meetings of the Council and to maintain contacts with it through unofficial bodies such as the Royal Society' — even if they found this British request 'thoroughly illogical'. It also seemed that it would be easy to reach an agreement concerning the powers of the Director-General.[38]

When discussion of the draft agreement was again taken up in the afternoon, it appeared that everything could not be resolved in the plenary session. It was therefore decided to form two working groups, one to redraft Article V on the subject of the Director-General and the other, which was broader and composed of Amaldi, Bannier, Heisenberg, Jacobsson, Kowarski, Nielsen, Preiswerk, and Verhaeghe, to reconsider the Preamble and Articles I, II, and III. The issue here was the aims of the provisional organization: should it set up a laboratory or limit itself to co-ordination?

The two groups met that evening and the first soon reached agreement on Article V. The second met inconclusively for three and a half hours. It came together again on the morning of 13 February, but still could not reach a satisfactory conclusion. This led to a third meeting, apparently on the morning of 14 February. This time the text was adopted in plenary session.[39]

In the meantime the conference had adopted the remaining articles of the draft agreement without significant change and had discussed Bohr's and Scherrer's visit to the United Kingdom. The broad outlines of an arrangement for collaborating with Liverpool were accordingly made.

Articles I, II, and III having been redrafted, on the 14th the conference approved the agreement, the Council's rules of procedure, and a text providing for collaboration between it and UNESCO. In the afternoon there was a debate on Bohr's proposal to organize a six-week conference in Copenhagen at the end of the spring, followed by a preliminary discussion 'about training staff for the study group'.[40]

A press conference and a reception closed the day on 14 February, and 15 February was devoted to signing the agreement and its annex.

6.3.1 ONCE AGAIN: LABORATORY OR CO-ORDINATION?

The opening statements all concentrated on this issue. On the one hand Denmark, Sweden, Norway, and the United Kingdom reaffirmed that the main purpose of the Council was to develop all possible forms of collaboration, that 'the Copenhagen group, in particular, would have a very important task as long as the plans for a European Laboratory were not more definite', that the funds collected for phase 2 'should be primarily used for stipends' for those posted to Liverpool and Copenhagen, that the Council would not need to recruit a large staff, and so on.[41]

For their part, Switzerland, France, Belgium, and Italy observed that some points had already been agreed on in Paris, that the working groups should get started as soon as possible and 'without waiting until results had been obtained by the Theoretical Group at Copenhagen' and that 'their financial contributions were offered mainly in order to permit the planning of the Laboratory's equipment and not for other purposes, however useful these might be'.[42]

As for the German and Dutch delegations, whilst their statements were less categorical, they nevertheless believed 'that the creation of a joint laboratory was the final aim of the Council', and that 'the necessity and the possibility of building a big machine should be studied by the Council'.[43]

There remains the hard-won agreement reached by the working group instructed to clarify this matter during the conference. It can be summed up as follows. Firstly, the word *Centre* (in European Nuclear Research Centre) was abandoned. Chosen initially for its neutrality and vagueness, it was overtaken by the severity of the ensuing polemic. It was therefore decided to be explicit: the Council would examine plans for a laboratory and *simultaneously* organize other forms of collaboration. It remained to determine the order of priority ('There was considerable argument about the order of these two items' remarked Waterfield). On this point the advocates of the laboratory won the day and the title of the agreement became: 'Agreement Constituting a Council of Representatives of European States for Planning an International Laboratory and Organizing Other Forms of Co-operation in Nuclear Research', one of the longest titles in the history of titles, and one which reflected the asperity of the debate.[44]

Secondly, Article III, which defined the Functions of the Council, was completely redrafted. The original wording: 'the Council [...] shall prepare the constitution of a [...] Centre' was far too vague and was replaced by the definition of three functions:

1) The Council 'shall make plans for the establishment of a [...] laboratory [...]'.

2) The Council 'shall take measures appropriate for utilizing the equipment and facilities' made available to it (Liverpool, Copenhagen). However, the Italian amendment was incorporated and the text added: 'provided that the financial obligations arising out of the agreements [should] not prejudice the achievement of the purposes of the Council set out in paragraph 1) of this section'.

Notes: pp. 204 ff.

3) The Council shall 'undertake theoretical research in connection with the work described in paragraphs 1) and 2) of this section'.

It seems therefore that all in all, as Waterfield said on 14 February, 'the laboratory group won'. They had gained a reaffirmation that the principal aim was to study the establishment of a laboratory, of which the plans would be drawn up immediately; they had secured financial priority for the 'synchro-cyclotron', 'proton synchrotron', and 'laboratory' groups agreed to in December; and although they had formally granted a rather special status to the 'theory' group (subparagraph 3 above), they had rejected Bohr's definition of the task of this group ('surveying the entire field of atomic research in Europe') and substituted for it a more limited function: assisting in the work described in paragraphs 1 and 2.[45]

6.3.2 THE PROPOSAL FOR A SIX-WEEK INTERNATIONAL CONFERENCE

It would be an illusion to imagine that Niels Bohr could drop an idea which he considered decisive: if he was determined to go ahead with 'a survey of the present situation in atomic physics', of 'high-energy accelerators' and of 'prospects of future developments in this field', it was because he wanted the European initiative to have the best possible scientific base from the outset. He therefore proposed a six-week conference to meet in Copenhagen some months later. The results obtained would be sent to the Council to help it in its work and its decisions.[46]

While the Netherlands and the Federal Republic of Germany were apparently favourably disposed towards this proposal, it was the cause of some anxiety in the 'pro-laboratory' camp. The French, for example, raised the question of financing the conference (that is, they expressed the fear that it was a way of circumventing the decision to allocate most of the funds to studying plans for the laboratory), and the Swiss expressed doubts that 'such a conference could yield immediate and tangible results'. Finally, after a long debate, it was '*agreed* that the [first] Council [meeting] should discuss Prof. Bohr's proposal and especially its financial aspects'.[47]

What we find noteworthy here is that the scientific delegates (who were in a large majority) were probably all interested in Bohr's proposal. The idea of holding this type of conference before deciding on research to which Europe would be committed for ten or twenty years could not fail to attract them. However the atmosphere of the conference was such that it was impossible not to be basically distrustful and suspicious. Hence the hesitation. Three months later at the first Council meeting, when sufficient money was available, the problem was solved and a three-week conference was arranged.[48]

6.3.3 ACCEPTANCE OF LAST REMAINING MATTERS

Regarding the Director-General, the working group responsible for redrafting Article V finally proposed that the Council 'shall appoint a Secretary [...] and shall commit to him the task of carrying out its decisions under the authority of the Chairman [of the Council]'. To satisfy the consultants it was added that 'The Secretary of the Council shall keep in close contact with the study groups as provided for in Section 2'. In other words a formulation had been devised which gave fewer grounds for thinking that the Secretary would have considerable autonomy *vis-à-vis* the Council. This satisfied the British and the Scandinavians, and was sufficiently vague to allow the other delegations to hope that the director (now called Secretary of the Council) would have wide enough powers to 'run the show'.

As far as the admission of possible new members was concerned, the French amendment requiring agreement by a two-thirds majority was not adopted, as the Swiss wanted to leave the door open to all European countries, including those of the eastern bloc. However subparagraph 2 of Article II was altered 'to avoid a fight between Swiss [...] and French'. The text reads: 'Any European State which has not taken part in the above-mentioned Conference, which undertakes to co-operate [...] on a footing of the free exchange of persons and information [...] and to make an adequate contribution [...] is eligible for membership of the Council'.[49]

Lastly, the commitments of the different countries were discussed. Three delegations were empowered to sign the Agreement without reserve: the Federal Republic of Germany, the Netherlands, and Yugoslavia. Six were authorized to sign subject to ratification: Denmark, France, Greece, Italy, Sweden, and Switzerland. The Belgian and Norwegian delegations declared that 'their governments would participate', although they were not able to sign that day. They signed, subject to ratification, on 2 April (Belgium) and 5 May (Norway). In other words eleven of the twelve States that had participated in the conference declared themselves parties to the Agreement.

The following contributions in money or in kind were promised:

Federal Republic of Germany, 35,000 dollars (152,000 Swiss francs);

Denmark, use of the Institute of Theoretical Physics at the University of Copenhagen;

France, 25 million French francs (312,000 Swiss francs);

Italy, 25,000 dollars (109,000 Swiss francs);

Netherlands, 10,000 dollars (43,000 Swiss francs);

Sweden, 57,000 Swedish kroner (48,000 Swiss francs);

Yugoslavia, 10,000 dollars (43,000 Swiss francs).

In addition *Belgium* announced that its contribution would be 20,000 dollars (it actually offered a million Belgian francs, i.e. 87,000 Swiss francs); *Norway* announced that it would contribute 5,000 dollars (22,000 Swiss francs, which it did); lastly *Switzerland* confirmed its December announcement that it would contribute 100,000 francs.

Adding these together gives a total of 917,500 Swiss francs (211,000 dollars) which would quickly be made available—that is, slightly more than the 200,000 dollars originally calculated.[50]

6.4 Concluding remarks on the respective roles of scientists and politicians in the process of establishment of the Council

The first question that springs to mind after reading this chapter is whether the Agreement could now be considered as providing a real opportunity to carry out joint work. After the number of developments we have described, the question would seem to be a reasonable one.

Generally speaking it is always tricky to answer this type of question, since the way one implements an agreement, the state of mind in which one sets out to put it into practice, are crucially important. In many respects the question can only be answered in retrospect. It must be noted, however, that a text binding the leading physicists and administrators of European science had been signed for the first time, that a Council had been established, and established at an intergovernmental level, and that states had offered financial support. A hitherto tacit contract had therefore been made *official,* which leads one to think that joint work should be possible *at least for the next year or eighteen months.* This is not to deny that the proposed programme was ambiguous, resulting as it did from the juxtaposition of two projects. Even though the relative weights of the two had been roughly defined (less of the money for the Bohr project, and the majority of it for the UNESCO scheme), there was nothing to suggest that the competition between them would end.[51] In this respect everything depended on the Copenhagen group: according to the conclusions it would draw from its general survey of nuclear physics, the conflict might—or might not—resurface in a year or so.

This could only feed the doubts and anxieties which already existed about the situation in the *medium term.* There was no guarantee at the beginning of 1952 that the states would actually be prepared to offer the fifteen, twenty, or thirty million dollars that the main project would call for mid-1953. Until now the debate had dealt only with what was to be done in the short term; but this only made sense (for the UNESCO project) if some hundred times as much money could be provided in a not-too-distant future. And in that respect the signs were anything but auspicious, as we have seen.

This way of posing the problem is, however, very static and in this sense misleading. Two major unknowns overshadow our remarks. The first is the *future evolution* of the field of high-energy physics. If, as seemed to have been the case since 1948, the physics community found this new field more and more promising; if the Brookhaven Cosmotron and the Berkeley Bevatron fulfilled the hopes which had been placed in them; above all, if significant improvements could be made to accelerator technology—and the proposal of a European giant could not but stimulate thought in this direction—it was quite likely that the way in which the problem was posed would change, so overcoming the remaining doubts of certain scientists.

On the other hand this type of development would not be sufficient in itself.

Whilst it could make scientists and their administrative colleagues more determined and strengthen their belief in the choice they had made, governments would not necessarily think it sufficient reason to give priority to the project and to provide the money asked for. We must not forget the economic situation of countries devastated by the war, particularly Germany but also France and Italy, three of the principal contributors; we must not forget too the still very hesitant attitude of European politicians towards science in general. But again we must be careful not to present a static picture: if the economic situation became more favourable, then considerable changes in attitude would be possible.

We would now like to raise a more important issue. We have seen that the negotiations were long, difficult, tense, and sometimes on the point of breaking down. Nevertheless an agreement was reached. So far we have advanced three arguments to explain this. In the first place, the determination of several key people to achieve closer collaboration between European countries. These are not empty words when one thinks of the many initiatives which came from people like Colonnetti and Dautry, or of the attitudes of Auger and Bohr during the debate — firm but conciliatory, ready to compromise if failure seemed likely. Secondly, we pointed out that the financial commitment was rather small: an interim stage requiring 200,000 dollars. As in a poker game, it was then possible to put up a stake and play just 'to see what happened'. There was no compulsion to follow higher bids. Thirdly, we said that the game was being played out in a context in which it was difficult to be the one responsible for sabotaging it, the context of an official intergovernmental conference.[52]

While all this is quite plausible, we still cannot overlook a remark often made about CERN, *namely that it succeeded because the entire project remained in the hands of the scientists.* Of course this assertion does not only apply to CERN's prehistory, but it is perhaps possible to test it in relation to this period.

The first thing to note is that it is not easy to say whether or not this assertion is true as the real problem is to know exactly what is meant by it. If it means that the *process* leading up to the agreement signed on 15 February 1952 was successful because only (or mainly) scientists were involved, then the assertion is false in spite of the fact that the scientists were in the majority. Three points suffice to justify this view.

To begin with, this was a *big science* project. And where there is 'big science' there is inclusion in a budget, and so the involvement of politicians, or at least of administrators from the higher echelons of politics. We have only to think here of Dautry and de Rose, and of their vital role in obtaining support for the project in French political circles, of Bannier in the Netherlands, of Colonnetti in Italy, of Willems in Belgium, etc.

Moreover, whenever *international* big science is at issue, there is an agreement between states, which implies the involvement of diplomats and non-scientists. And

Notes: p. 205

the importance of the latter in the negotiating process itself (between December 1951 and February 1952) was not negligible. We have said that their role was partly an indirect one, being to personify public awareness and sense of propriety, and so to 'compel' the scientists to reach a compromise. But they also played a more direct role when it came to drawing up texts: the working group which drafted the agreement was composed mainly of diplomats and administrators; some of the amendments resulted from close co-operation, on an equal footing, between diplomats and scientists; and it was the political circles who finally decided whether or not to sign. The British case is the best example of this.

Lastly, we must mention the role played by the members of the European Movement, and by the Centre Européen de la Culture in particular. They helped to spread the initial idea of European scientific centres in Belgian, French, and Italian scientific circles through Colonnetti, Dautry, and Willems; they helped to find the necessary finance (directly for 1951; indirectly for 1952); finally, the CEC provided hospitality and firm support between October and December 1950.

We are convinced, nevertheless, that the assertion we are dealing with contains a vital element of truth in that it emphasizes one of CERN's special characteristics: governments and their departments had virtually no role in the *conception* of the CERN project, in determining the *aims* of the future organization, or in drawing up the *rules* governing its existence. Put differently, the CERN project was not consciously situated within the diplomacy of any state, unlike Euratom five or six years later, for example. On the contrary, governments merely served as channels for advancing money for the next stage, channels which were contacted, which were to be persuaded and seduced by all possible means (Colonnetti writing to Gasperi, Dautry to the Ouai d'Orsay, and so on), but which the scientists and their administrative friends kept as far away as possible from the heart of the project.

The *initiators* of the project were thus not to be found in the world of politics and diplomacy, nor were the initial *roots* and *motivations,* in spite of the existence of an active European ideal. It is in the scientific *need* felt by scientists and scientific administrators, the need to equip Europe with machines beyond the means of individual countries, that the foundation of the CERN project is to be found. This is crucial because from the outset centrifugal forces of national origin were reduced to a minimum. This was made easier, though, by the fact that the scientific field in which CERN was involved was not associated with direct military or commercial interests so that political circles never felt the need to interfere in the affair.[53]

It is possible to go further, however, and acknowledge that the leading scientists involved in the project showed a strong determination to let the physics community plan the *whole* project itself, including the best ways of 'selling' it to the politicians. They felt there should be two distinct phases: *planning,* which should be left to them, and *approval* by the politicians, which should be postponed until the last possible moment. To substantiate this claim we shall describe how Auger and Mussard organized their work at UNESCO throughout 1951.

Auger received money for 'his' project (from Italy, Belgium, and later France) from the beginning of 1951. He placed it in a UNESCO account, but one which was independent in the sense that he alone had access to it. In April he needed to call a meeting of specialists ('experts' in UNESCO terminology) nominated by the leading figures in European physics. This was a complicated matter because *officially* he had to obtain 'the authorization of the UNESCO Executive Council' and reply in advance to an 'avalanche of questions which would inevitably be asked by the relevant UNESCO departments (financial, legal, political, etc.)'. Consequently Auger preferred the affair to remain outside the official sphere of UNESCO. After discussing it with some friends like Amaldi and Perrin, as we have seen, he brought together what he called a ' "Group of Consultants", which was a notion unknown to the administration and therefore not covered by the regulations'. Only the scientists with whom Auger was in contact were therefore implicated in the process and, as Jean Mussard remarked, the system worked perfectly up to the end of November 1951.[54]

But there are more spicy examples. In order to circulate quickly documents prepared by the 'consultants' (officially 'UNESCO documents had to go through a series of checks before arriving, complete with all the necessary endorsements, at the official copying and distribution service' (Mussard)), but also with the idea of keeping the UNESCO bureaucracy and the state delegates out of the negotiations, Auger and Mussard set up '—in an unused bathroom in the former Hôtel Majestic [where their offices were located]—a small private document copying service', which was, as can be imagined, 'strictly prohibited'. In this way, with the complicity of several people including René Maheu, the UNESCO Director-General's Chef de Cabinet, and with the help of some colleagues in other departments, they succeeded in organizing an intergovernmental conference called by UNESCO without the UNESCO hierarchy ever coming to hear of it. The 'affair' was only discovered a fortnight before the conference—whereupon the Director-General threatened to cancel it completely! He gave way under pressure from Auger and Maheu, but refused to chair the inaugural meeting on 17 December.[55]

Leaving anecdotes aside, these men who showed so great an involvement in the European laboratory project, believed that the recipe for success was to keep scientists in sole charge of the project. It is probably true that this was what stopped the states' delegates to UNESCO from looking too closely at the way the project was taking shape or at the progress of negotiations, which prevented the scheme from getting bogged down in the UNESCO bureaucracy, but which also prevented *scientific* opponents from using all the resources of this bureaucracy to throttle the 'adventurous' ideas of Amaldi and Auger.

This last comment leads us to point out two of the limitations of this tactic, a tactic which left the scientists free to study matters that they considered to be in their

Notes: p. 206

domain. Firstly, since at some stage or other the politicians were bound to be brought in, at least at national level, it was effective only when scientists in the country concerned were *united*. When they were hesitant or divided, they raised doubts among those they were negotiating with. In other words a pressure group such as this could not obtain results unless it was determined and convincing as a group. The case of the United Kingdom illustrates this: the fact that the scientists themselves were not convinced of the need for the large (UNESCO) project served not only as a pretext, but was one of the basic reasons for the refusal in political circles to participate in the preliminary work.

A second limitation of the tactic is that, however well conceived a project may be among scientists themselves, it will not be readily approved by politicians unless it fits into, or at least does not run counter to, the main lines of national policy and diplomacy. Otherwise, regardless of the extent of 'scientific control' in the elaboration of the project, failure is likely. This is why, although the initial autonomy of the CERN project from government decisions and national policies allowed the text of the agreement to be completed quite rapidly in February 1952, and although this independence prevented additional constraints being imposed on the scheme by recognized state representatives, it is nevertheless true to say that it was able to make such rapid progress (with the subsequent blessing of the governments of the principal western European nations) only because it was in harmony with the main lines of diplomacy of those countries. After all, without the agreement on the Schuman plan and the signature of the treaty establishing the European Coal and Steel Community (18 April 1951), without the Pleven plan and the imminent signature of the treaty establishing the European Defence Community (27 May 1952), the climate would probably not have been the same, and Amaldi and Auger's project might not have been as readily accepted.

Notes

1. The main archival sources used in this chapter are:
    - UNESCO archives, Paris, *CERN, Official Correspondence* file
    - Jean Mussard's papers, deposited at CERN, Geneva.
2. The main documents referred to in this part include the *Working Document* tabled by UNESCO (UNESCO/NS/NUC/1, dated 19/12/51); the *Draft Final Report* drafted by J.H. Bannier of the Netherlands (UNESCO/NS/NUC/9(Prov.), dated 21/12/51); some of the documents produced at the conference including document UNESCO/NS/NUC/6, *Draft Resolution proposed by the Netherlands Delegation,* of which we have the first, third, and fourth versions; Edoardo Amaldi's notes in his diary (CERN); the documents sent by Waterfield, British Scientific Attaché in Paris, to Rackham, Secretary to the British UNESCO Committee (PRO-ED157/302); a letter from Thomson to Chadwick dated 22/12/51, and one from Bohr to Chadwick dated 29/12/51 (CC-CHADI,1/10 and 1/9).

3. This information is from Edoardo Amaldi's diary. It is also to be noted that in his preliminary intervention Colonnetti, on behalf of Italy, praised Amaldi's role in the preparatory phase of the Conference. See UNESCO/NS/NUC/9 (Prov.), 7.
4. For Norway, Hylleraas remarked the following February that when he had attended the Paris conference it had been 'to some extent' as an observer (UNESCO/NS/NUC/16(Prov.), 3). Also worth noting is the balance among the office-holders at the conference: Chairman, de Rose (France); Vice-chairmen, Thomson (United Kingdom), Scherrer (Switzerland), Nielsen (Denmark). A 'neutral' held the fifth post (Bannier, rapporteur).
5. Quotations taken respectively from the opening statements of Malte Jacobsson (Sweden), George Thomson (United Kingdom), Malte Jacobsson, Jakob Nielsen (Denmark), Jakob Nielsen and Nielsen again (UNESCO/NS/NUC/9(Prov.), annex IV).
6. Quotations taken respectively from statements of Francis Perrin (France), Paul Scherrer (Switzerland), Francis Perrin, Paul Scherrer, Francis Perrin (UNESCO/NS/NUC/9(Prov.), annex IV).
7. UNESCO/NUC/9(Prov.), annex IV.
8. UNESCO/NUC/9(Prov.), annex IV.
9. De Rose's speech is reproduced in UNESCO/NUC/9(Prov.), annex IX.
10. First version of the Netherlands resolution referred to in note 2. It is surely no coincidence that the resolution was put forward by a divided delegation that was wavering between the two proposals that had been tabled.
11. It was officially voted for on 20 December.
12. It seems to have been considered diplomatically wise to select Geneva as the location for the Council because Copenhagen had been named as the location for the 'theoretical' group. However, this clause in no way foreshadowed the selection of the future site for the laboratory.
13. No further mention is made here of the fact that the Copenhagen group was to be in charge of the theoretical aspects of the work to be done by European physicists in Liverpool. More details on this are given in 'Point 3' of the new resolution.
14. The fourth version of the Netherlands resolution was the one finally adopted by the conference. It is annexed to UNESCO/NUC/9(Prov.), which is the source for the other information and for Auger's quotation. In connection with annex II of the *Working Document,* Bannier's minutes contain the statement 'The proposals embodied in annex 2 of the Working Document were *accepted*'.
15. The Foreign Office message is in the UNESCO archives.
16. All this information, and that which follows, is taken from UNESCO/NUC/9(Prov.).
17. The quotations from Bohr and Thomson are taken from the letters referred to in note 2.
18. For this aspect, see Gilpin (1962), Salomon (1970a), Greenberg (1971).
19. Letter Thomson to Chadwick, 22/12/51 (CC-CHADI,1/10).
20. François de Rose, conference chairman and career diplomat, was to give his own version of the situation in his closing speech to the conference. 'If you had failed', he said, 'the repercussions of this failure would have been felt far beyond the field of nuclear physics. On the contrary, this first success will represent for the governments and public opinion of our countries, a new evidence of the vitality of our old continent [...]' (UNESCO/NUC/9(Prov.)).
21. On the so-called working group, see letter Waterfield to Rackham, 8/1/52, 5 pp. (PRO-ED157/302); letters Director-General of UNESCO to governments, 22/1/52 (UNESCO-NS269950); the text of the *Draft Agreement constituting a Council of Representatives of States for the Establishment of a European Nuclear Research Centre,* attached to the above letter; letter Waterfield to Rackham, 31/1/52 (PRO-ED157/302); letters Skinner to Chadwick, 14/1/52; Chadwick to Bohr, 19/1/52; Bohr to Chadwick, 1/2/52 (CC-CHAD I,1/3 and 1/8). For the consultants' meeting, see UNESCO/NUC/10, 4/2/52, drawn up by Mussard.
22. The amendments are reproduced in UNESCO/NS/NUC/16(Prov.), entitled *Draft Report* (for the second session of the Conference, 12-15 February 1952).

23. The information relating to the selection of the term *Centre* was given by de Rose at the February conference (see UNESCO/NUC/16(Prov.), 4). For his part, Waterfield noted on 8 January (letter to Rackham, 8/1/52, PRO –ED157/302) that the term was selected 'as the laboratory in the previous title may never be achieved'.
24. Taken from the Swedish and Danish amendments referred to in note 22.
25. Taken from the Swedish and Danish amendments. Denmark also proposed replacing the words 'one hundred and fifty thousand dollars' (the amount considered necessary for creating the Council) by 'fifty thousand dollars', which is logical but a pure formality since 150,000 dollars had already been pledged. The figure settled on was to be 100,000 in the final version of the text. See Article IX in both the draft and final agreements.
26. Thus the consultants again put forward the salary scales drafted in December and, proposing a nine-point recruitment procedure for candidates, compiled 'a draft application form'. Details are to be found in letter Waterfield to Rackham, 8/1/52 (PRO–ED157/302), letter D-G to member states of UNESCO, 24/1/52 (UNESCO-NS269950).
27. Auger's statement is given in a letter signed 'Assistant Secretary' and sent to Chadwick, 25/1/52 (PRO–ED157/302); the Italian amendment is reproduced in UNESCO/NUC/16(Prov.).
28. Letter Waterfield to Rackham, 8/1/52 (PRO–ED157/302).
29. Cabinet Steering Committee on International Organisations, meeting of 8/2/52 (PRO–CAB134/943).
30. See section 12.4 and chapter 13, appendix 2.
31. Letter Waterfield to Rackham, 31/1/52 (PRO–ED157/302). We know only the French and British attitudes on this matter. The same position may have been true of other countries, while that of Switzerland appears to have been peculiar to it. The last type of clause (demanding on the part of would-be members a kind of moral commitment to freedom of exchange of persons and knowledge) was probably a very formal one. This is the view, incidentally, of a British official who said on 8 February: 'This [clause] had no doubt been included in order to safeguard the position of other countries should Iron Curtain countries apply for membership of the Council. It might, however, prove to be a double-edged weapon, since it was to be expected that while other countries would observe their obligation, Iron Curtain countries would not, and this point should be carefully watched' (Document referred to in note 29).
32. See in particular the reports by Waterfield and the consultants referred to in note 21. The Scandinavian and British concern was that, as they would be in a minority, the Director-General might well not be the one of their choice. Conversely, if power was to lie within the Council, they had more hope of being in a position to control events.
33. To appreciate the intimacy of the links and the extent to which activities were co-ordinated, see exchange of letters between Bohr and Chadwick referred to in note 21; the quotation is drawn from the document referred to in note 29.
34. We have in mind the Netherlands resolution and annexes II and III of the December working document (see section 6.1.3 above).
35. The opportunity for blackmail provided by a threat of British withdrawal was based on the wish of the continentals to have Britain in the European Council. For the Netherlands, see the long letter from Bannier, Secretary, and Wagenvoort, President of the ZWO, to the Netherlands Ministry of Education, Arts and Science, 1/2/52, 4 pp. (CHS, Bannier file, CERN). In the Netherlands the position proposed in this letter appears to have been the one that prevailed.
36. For details see below, section 6.3.
37. Document UNESCO/NS/NUC/16(Prov.), entitled *Draft Report* and written by Bannier; minute Murray to Verry, 12/2/52, ref. 969-4-2; Outward Savingram from U.K. Delegation, Geneva, to Foreign Office, 14/2/52; letter Waterfield to Murray, 14/2/52 (PRO–DSIR17/559 and ED157/302). Articles from the *Continental Daily Mail,* 13/2/52 and *La Suisse,* 15/2/52.
38. Minute Murray to Verry referred to in note 37.
39. Document UNESCO/NS/NUC/16(Prov.), 4–5 and letter Waterfield to Murray referred to in note 37.

40. Document UNESCO/NS/NUC/16(Prov.), 6–8 and *La Suisse* referred to in note 37. The quotation is taken from the letter from Waterfield to Murray (note 37).
41. Quotations taken from the statement by Malte Jacobsson (UNESCO/NS/NUC/16(Prov.), 2). According to the *Continental Daily Mail,* from the first day Bohr proposed the convening of a large scientific conference in Copenhagen 'to make a new start on efforts to promote the co-ordination of atomic research in Europe'. This matter was discussed at some length on the 14th.
42. Quotations from French and Italo-Belgian statements, the latter being grouped together in UNESCO/NS/NUC/16(Prov.), 3. According to the *Continental Daily Mail,* 'France, Belgium, Switzerland and Italy urged the maintenance of the original plan and accused Denmark and her supporters of failing to stand by agreements reached at the Paris meeting'.
43. Quotation from the statement by the German delegation as given in UNESCO/NS/NUC/16(Prov.), 3.
44. The Waterfield quotation is taken from the letter to Murray quoted in note 37. The word 'international' appearing in the title was the fruit of a long discussion. Initially, the working group had proposed the term 'European', which was opposed by the British and the Scandinavians who suggested 'regional'. Their official reason was that there was nothing to indicate that *all* the states signing the agreement had necessarily to participate in establishing the laboratory. In their view, the countries of southern Europe could build it by themselves since it was their project. They finally settled on the word 'international'.
45. The quotation defining Bohr's proposal is taken from the *Continental Daily Mail,* the Waterfield quote coming from his letter given in note 37.
46. The quotes are taken from the 'Plan proposed by the Danish delegation for the activities and organization of the study group at the Copenhagen Institute for theoretical physics' (UNESCO/NS/NUC/16(Prov.), Annex IX).
47. Our only source for this discussion is *La Suisse,* 15/2/52. It is therefore important to be cautious about the actual terms of the debate. In particular, we know nothing about Belgium and Italy. Nonetheless the discussion was probably bitter, since no decision was taken on the Danish proposal. The last quotation is taken from UNESCO/NS/NUC/16(Prov.), 7.
48. See Minutes of the Session, European Council for Nuclear Research, First Session, Paris, 5–8 May 1952, document CERN/GEN/1,15/10/52.
49. The only source here is Waterfield (see note 37). This is why we provide no further details as to the attitudes of other delegations. It will be noted that the final phrasing does not presuppose any obligation to admit new members, even Europeans.
50. Information taken from UNESCO/NS/NUC/16(Prov.), annex V and the report referred to in note 48. Note the strange position of Greece and Switzerland, which signed the agreement subject to ratification without a definitive signature of the annex setting forth the contributions, something which by rights should not have been allowed. After Bannier had expressed his concern over the situation to Mussard, the latter replied to him on 20/2/52: 'Miss Thorneycroft [of the UNESCO Legal Service] has explained to me that such a signature was not sufficient for these countries to belong to the Council [...] That might seem illogical but Miss Thorneycroft thought it would be useful for them to be allowed to sign the agreement now, so as to set in motion the ratification machinery in the two countries' (letter Mussard to Bannier, 20/2/52, UNESCO).
51. The problem arose again at the first meeting of the Council. In a letter to Chadwick dated 6/3/52 (CC-CHADI,1/3), Bohr explained that he was hoping that the Council would grant 50,000 to 60,000 dollars to the Copenhagen group. At the first meeting of the Council he received 80,000 Swiss francs, i.e. less than 20,000 dollars.
52. We could also suggest a fourth argument: the players also had the Americans looking over their shoulders, as it were. While it is true that there was no delegate from the United States present at the conference, nonetheless their shadow hung over the entire proceedings. In addition to the information already provided on Rabi and Weisskopf in chapter 4, we would like to quote the letter from Lawrence himself to Bohr dated 21/12/51 (AIP-Bohr correspondence): 'There was further discussion of the UNESCO

proposal in Stockholm and when I got back here I had an opportunity to say something in Washington and found a very receptive ear, so that I should not be surprised that some help may be forthcoming in the normal course of events. I trust you will keep this in confidence'. On the same day, moreover, he wrote to Dahl: 'Wherever I went the UNESCO project was discussed and invariably it was said that if the project materialized efforts should be made to obtain Odd Dahl's services. I assured them that if you took on the job things would turn out all right' (Berkeley: Bancroft Library, Lawrence papers, box 10, folder 2). In our view this American 'presence' seems to have played the same role as the wish not to be 'against' or 'unpopular', the desire not to be the one responsible for failure. After all it was obviously no accident that Auger had a letter from Livingston of MIT distributed to the delegates at the February session, a letter encouraging the Europeans to build a large accelerator that would be 'a step forward in size and energy' (copy in UNESCO/NS/NUC/17, original dated 5/2/52).

53. All this is discussed at greater length in chapter 14. The secondary literature is given there.
54. Quotations taken from Mussard (1974) and Mussard (1984). With regard to the 'Consultants Group', Mussard adds: 'The trick may have been a little crude, but it has worked well for six months [May to November 1951]'.
55. Articles by Mussard quoted in note 54. Even if Jean Mussard has humorously dramatized events, we have little reason to doubt the truth of what he says.

PART III

The provisional CERN
February 1952–October 1954

CHAPTER 7

Survey of developments[1]

John KRIGE

Contents

7.1 Establishing the study groups and the secretariat   211
7.2 Fixing the energies of the accelerators   211
7.3 The discovery of the strong-focusing principle   213
7.4 The choice of a site for the new laboratory   213
7.5 Britain takes the plunge   215
7.6 Consolidating the scientific work   216
    7.6.1. The accelerator groups   216
    7.6.2. The Theory Group   217
    7.6.3. Other forms of co-operation   218
7.7 The Convention and its signature   219
7.8 The new mood in CERN   222
7.9 Putting down roots in Geneva   223
7.10 The nomination of the first Director-General   225
7.11 The ratification of the Convention   226
7.12 The philosophy of the organization   228
Appendices
    7.1 Scale of percentage contributions to the permanent organization applicable during the period to 31 December 1956   231
    7.2 Rates of exchange used for accounting purposes (a) as from 1 December 1952 and (b) as from 1 January 1954   231
    7.3 Dates of deposit of instruments of ratification of the Convention at UNESCO House in Paris   232
    7.4 Official delegates to the nine Council sessions   232
Notes   234

With the signature of the Agreement constituting a 'Council of Representatives of European States for Planning an International Laboratory and Organizing Other Forms of Co-operation in Nuclear Research' on 15 February 1952, the prehistory of CERN effectively drew to a close. The financial means for the European laboratory project had been secured; it was only a matter of time before the 'treaty' establishing the body entered into force. To celebrate what they called 'THE OFFICIAL BIRTH OF THE PROJECT [RABI] FATHERED IN FLORENCE', some two dozen delegates telegrammed the good news that same day to Rabi himself. 'MOTHER AND CHILD [WERE] DOING WELL', they added—so well, in fact, that on 5 May 1952 the new Council met for the first time in Paris.[2] Ten European states were officially represented, most of them by two delegates, one a leading scientist, the other an influential science administrator. Paul Scherrer was again in the chair.

The immediate task of the Council and its executive was to draw up plans for the new laboratory and its equipment, and to draft an intergovernmental convention to place the organization on a permanent footing. This was to occupy them for the next fourteen months. From 1 July 1953, when the Convention was signed, another fifteen months were needed before it entered into force, fifteen months during which increasingly heavy commitments to the future laboratory were made.

This chapter and the next deal with the history of these two phases in the Council's life. To study them we use two different kinds of approach. In contrast to the earlier stage of development, during this period the several distinct activities involved in launching the new research centre became both progressively institutionalized and relatively autonomous of one another. With that the main stream divided into a number of different tributaries, each advancing at its own distinctive pace and leaving its own distinctive trace on the landscape of the organization. To deal with this situation we present both a chronological overview of the general flow of events (in this chapter), and more analytical case studies of particular developments (in the next). In this way, and at the expense of some repetition, we hope both to orientate our readers in the period and to give them some insight into the deeper forces at work in the establishment of CERN.

Before we begin a brief word about terminology. It is customary, if somewhat anachronistic, to refer to the organism whose life spanned our period—the period, that is, from early May 1952 to late September 1954—as the 'provisional CERN', so distinguishing it from the 'permanent CERN', the European Organization for Nuclear Research. The provisional period, in turn, is usually split into two stages or phases: the planning stage up until the Convention was signed (i.e. 1 July 1953), and the interim stage, the period of waiting until it was ratified on 29 September 1954. Conforming to custom, we shall adopt this terminology in what follows.

7.1 Establishing the study groups and the secretariat

The main business at the first session of the provisional Council was to appoint the heads of the four study groups and a Secretary-General who were to carry out its programme, and to allocate funds to each. Edoardo Amaldi was appointed Secretary to the Council. Cornelis Bakker and Odd Dahl were put in charge of the groups to study the accelerators, respectively labelled the Synchro-cyclotron (SC) Group and the Cosmotron Group. Theoretical studies, and arrangements for using the cyclotrons in Liverpool and Uppsala, were Niels Bohr's responsibility, while Lew Kowarski was appointed head of the Laboratory Group. Its aim was to plan the infrastructural context into which the accelerators would be embedded. At the second session of the Council it was decided that the Secretary and the group directors would be collectively known as the Executive Group.[3]

It was generally understood that, at least initially, the group leaders would remain at their home stations — Amaldi in Rome, Bakker in Amsterdam, Bohr in Copenhagen, Dahl in Bergen, and Kowarski in Paris. Contact between them was maintained by regular meetings of the Executive Group, and by weekly reports in which Amaldi informed his colleagues of recent activities at the secretariat. Each group was relatively autonomous — it had its own bank account and directors had considerable control over how the funds in 'their' account were spent, for example.

As the scientific work picked up momentum there was further decentralization of the organization, notably within the accelerator groups themselves.[4] Bakker and Dahl established a division of labour and distributed tasks among institutes all over Europe. By the end of 1952 there were subsections of both groups working in Harwell and in laboratories in Paris. The radio-frequency system for the SC was being developed at Philips in Eindhoven, that for the bigger machine was being studied at Walther Bothe's institute in Heidelberg. Bakker also had a magnet group in Sweden and some theoretical work was being done for Dahl in Copenhagen.

This dispersal of the groups obviously restricted the possibilities for communication between their members. On the other hand, there was the notable advantage that work on the design of the accelerators could proceed without any of the personnel involved having to move from their home stations. Apart from providing indirect financial support to the provisional organization, this arrangement enabled the subgroups to draw on the expertise and technical resources available in their institutes for the European laboratory project.

7.2 Fixing the energies of the accelerators

Having established its technical teams, the next major task facing the new organization was to settle the design energies of the two accelerators. Officially

Notes: pp. 234 ff.

this was done at the second Council session held from 20–21 June 1952 in Copenhagen.

The meeting took place immediately after an international conference of theoretical and experimental physicists organized by Bohr, and held at his Institute for Theoretical Physics from 3–17 June. The idea of holding such a conference had been raised by Bohr in February, but had been greeted with suspicion by the delegates to the UNESCO-sponsored meeting, who had left the detailed arrangements to the new Council (see section 6.3). At its first session the Council agreed to sponsor and to subsidize Bohr's conference, stipulating that part of it 'be devoted to a survey of the present stage of knowledge regarding very high energy particles and, in particular, the experimental equipment necessary for improving such knowledge'.

Werner Heisenberg reported on the proceedings at the second Council session. Having surveyed the relative merits of various electron and proton accelerators in terms of energy, beam intensity, and cost/GeV, Heisenberg suggested that Dahl's group make a feasibility study of a powerful proton synchrotron. He also proposed that Bakker's group design a 600 MeV (proton) synchro-cyclotron, and investigate its financial and industrial requirements. The Council approved Heisenberg's report and its recommendations.

Two factors in particular prompted these decisions: the scientific interest of the devices, notably the opportunities they provided for doing meson physics, and the success of the Brookhaven 3 GeV Cosmotron. This machine was designed to be almost an order of magnitude more powerful than any existing, working accelerator. When it gave its first beam in March 1952 doubts about the technical feasibility of a device of this size were dispelled. Indeed Heisenberg suggested that Dahl's group consider constructing their machine 'along the lines of the Brookhaven Cosmotron', and to operate in the energy range 10–20 GeV.

At this session the name of the Council was also agreed on. This was no mere formality. In January, between the two UNESCO-sponsored conferences, the project had officially come to be called the 'European Nuclear Research Centre', and abbreviated CERN (the acronym deriving from the French equivalent). 'Centre' was chosen as being sufficiently vague as to accommodate the wishes of both those who wanted to build a new laboratory and those who favoured only 'simple' co-ordination of existing facilities. In fact the ensuing debate was so heated that the term was dropped in the title of the Agreement constituting the provisional Council. The Council thus inherited an acronym but not its reference. Something had to be done about this, and at the second Council session the issue of what the 'C' stood for came up *again*. Apparently overcome by tedium Auger compiled an enormous list of names to distract himself—names like *Cabale, Casino, Concubinage, Cirque,* ...[5] Ignoring these suggestive alternatives, 'The Council *decided* that its name shall be henceforth "European Council for Nuclear Research"'.

7.3 The discovery of the strong-focusing principle

Now that the types of accelerator had been settled, it was time to take up the generous offers of help made by the American machine groups. Accordingly, soon after the Copenhagen conference Odd Dahl, his deputy Frank Goward, and the accelerator expert Rolf Widerøe, visited the United States. They spent the week beginning 4 August 1952 at Brookhaven, and then joined Cornelis Bakker in California, where they visited Berkeley and Stanford.

While at Brookhaven their American hosts explained to the European visitors a new idea they had just had for revolutionizing the design of high-energy accelerators.[6] It was based on a principle which came to be known as the alternating-gradient (AG) or strong-focusing principle. In essence this amounted to a way of arranging the magnets guiding the charged particle beam so as to confine it more narrowly. In effect applying the AG principle meant that the cost/GeV of a proton synchrotron could be considerably reduced.

Dahl returned to Europe determined to explore the feasibility of the new concept. Brookhaven promised to co-operate as much as possible, to the extent of sending some of its own personnel to join Dahl's team. The idea was discussed at a meeting of his renamed Proton Synchrotron (PS) Group at the end of August. Although a set of machine parameters for a scaled-up version of the Cosmotron had already been produced, the group decided to proceed forthwith with the design of an accelerator embodying the new principle.

Dahl presented his proposals to the third Council session which met in Amsterdam from 4–7 October 1952. The design energy of the envisaged accelerator had been raised from 10 GeV to 30 GeV, the construction time was reduced by one year to about six years, and the cost was to be 'not more than the other machine', i.e. $14 million. The Council sanctioned Dahl's decision to abandon the original project, and entrusted his group with the task of designing a strong-focusing proton synchrotron.

The importance of these developments cannot be over-estimated. Scientifically speaking an entirely new energy range would be probed for the first time, promising to give results as exciting as they were difficult to predict. Technically there was the challenge posed by breaking new ground and of making 'new contributions to the art of experimental physics'. Politically there was the opportunity for Europe to leap from a relatively backward position to a world leader in this field of nuclear physics research. From the point of view of the proponents of the laboratory, the discovery of strong focusing could not have come at a more propitious time.

7.4 The choice of a site for the new laboratory

The choice of a site was one of the most difficult and potentially divisive decisions which the Council had to take. At the time of the UNESCO meetings in Paris in

Notes: p. 235

December 1951 and February 1952 two strong candidates had emerged, Geneva and Copenhagen.[7] By the end of July two further governments had offered locations for the European laboratory — France (Paris) and the Netherlands (Arnhem).

The merits of these four candidates were eloquently pleaded by their respective official delegations to the third Council session. In the subsequent discussion first the Danes then the French withdrew their offers. This left Arnhem and Geneva. The former was withdrawn after a protracted battle, so enabling the Council unanimously to agree that CERN should be located in the Swiss canton.

The circumstances surrounding this 'decision' are complex, and we analyze them at some length in section 8.1. Briefly, the following factors were involved in it:

- on technical grounds (size, accessibility, availability and cost of electricity and water, accommodation and educational facilities in the region, etc.) there was little to choose between the four sites;

- Copenhagen stood little chance of success. The younger members of the Executive Group were opposed to it, fearing that Bohr would dominate the laboratory if it was located there, it was too far north for the more southern Europeans, notably the French and the Italians, and even the British were unenthusiastic about it;

- the major objection to Paris was that it was located in a large country which, it was thought, would inevitably come to play a dominant role in the laboratory's life;

- the candidature of Arnhem was strongly supported in the Council and was apparently only withdrawn to preserve unanimity in the face of vigorous opposition;

- Geneva had the backing of some members of the Executive Group, and of France (for whom it was second to Paris), Italy, and Belgium — precisely the three states who had taken the lead in advocating a collaborative European laboratory and who had made substantial financial contributions to the project. Also in Geneva's favour was the fact that it was in a small, politically neutral country, with an established tradition of hosting international organizations, and that it had a particularly beautiful natural environment.

The Council decision to locate CERN in Geneva was only accepted begrudgingly by the Scandinavians, who resolved to be compensated in some way for the loss of Copenhagen (see later). It was also opposed in the canton itself. A spirited campaign spearheaded by the local Communist Party, but enjoying relatively widespread support, objected that radiation from the accelerators would be a health hazard, and that the presence of a western European nuclear research facility on her soil would endanger Switzerland's neutrality. The matter was settled in a characteristically Swiss way. At a referendum held in the canton on 27 and 28 June 1953, a proposal prohibiting the location of any nuclear physics institute there was defeated by a two-to-one majority.

7.5 Britain takes the plunge

The February 1952 Agreement establishing the provisional CERN was not signed by the British. Her Majesty's Government judged the envisaged machinery for scientific collaboration to be unduly complex—all the more so considering that some of her most senior physicists were not convinced of the need to build a very big accelerator in Europe. No official British delegate was thus present when the Council met for the first time in May 1952.

Whatever her formal status, both the Council and the United Kingdom had a mutual interest in keeping contact with each other. Observers from the Royal Society attended the second and third sessions. At the same time scientific opinion in the country changed steadily in favour of Britain joining CERN. Influential members of the government took up the cause, notably Lord Cherwell, Churchill's closest scientific advisor, and Sir Ben Lockspeiser, the Secretary of the Department of Scientific and Industrial Research (DSIR). Yielding to the pressure exerted by scientists and senior ministers, on 29 December 1952 the Chancellor of the Exchequer agreed that Britain should join CERN. She duly signed the Convention establishing the permanent organization on 1 July 1953.

Although Britain was never an official member of the provisional organization, from January 1953 onwards she was treated, and behaved, like a fully-fledged member of the Council. Her delegates were observers representing their government. She made two 'gifts' of £5000 each during the planning stage, and paid her share of the monies raised during the interim period. And she provided both the chairman and the secretariat of the Interim Finance Committee (IFC) which first met on 1 and 2 July 1953, and which was responsible for making the financial arrangements for the transitional period, and for laying the administrative foundations of the new organization.[8]

Why did Britain decide to join CERN, and why was the Council so keen to welcome her into its ranks? We study these questions at length in chapters 12 and 13, and merely wish to summarize some of the salient points here.

The key factor determining Britain's change in policy was the growing conviction among a group of younger nuclear physicists based at Harwell that their scientific interests were best served by membership of the European laboratory. There were many reasons for this—they included the success of the Cosmotron, the discovery of AG focusing, the recognition that unless Britain's physicists had access to a very big accelerator they would lose their position among the world-leaders in high-energy physics, and the savings in cost and manpower which the collaborative venture offered over a purely domestic initiative. Added to this there was the belief that if she did not join CERN quickly Britain would lose all opportunity of filling senior posts in the organization and of shaping its overall development in ways which she thought fit.[9]

From the Council's point of view, Britain had much to offer. Scientifically, she

Notes: p. 235

could bring a wealth of experience to the construction of the accelerators, along with the advantages of her 'special relationship' with the United States. Financially, she was among the wealthiest countries in Europe, and would be a major contributor to the organization's budget. Administratively, in the DSIR she had a great deal of experience in the management of a basic scientific research institute, experience which was particularly useful when it came to drawing up salary scales, making arrangements for social security, establishing a contract policy for industry, and so on.

7.6 Consolidating the scientific work

In this section we wish to provide a quick overview of developments within the three scientific study groups during the life of the provisional CERN. It is divided into three sections. The first deals with the two accelerator groups. Then we look at Bohr's theory group and, thereafter, describe some of the steps taken to extend 'other forms of co-operation' in nuclear research. Our treatment is very much in the nature of an introduction to these topics. As we mentioned in the preface, they have histories which transcend the limits of our period, and will be handled more comprehensively in our study of the permanent organization in volume II.

7.6.1. THE ACCELERATOR GROUPS

The SC Group made steady progress during this period. By October 1952 a division of labour into five main sectors had been made, and the responsibility distributed between laboratories in several member states. Six months later estimates of the cost of all major items were at hand, and by the end of 1953 the group was in a position to call for tenders for the magnet. Only the radio-frequency system, which was being developed in association with Philips in Eindhoven, was causing trouble. It continued to do so throughout the construction period of the synchro-cyclotron.[10]

By contrast, the PS Group had considerable difficulty during this early period. John Adams, at the time one of the accelerator experts from Harwell who had recently become attached to the group, has described the situation as follows:

> By the time of the fifth Council meeting, held in Rome at the beginning of April 1953, the honeymoon period with the new idea was coming to an end. What had been thought to be a very simple machine design using very strong alternating gradient focussing was found on closer examination to have serious drawbacks in that the machine design was very sensitive to irregularities in the field of its magnet.[11]

In effect this meant that if one took 'too great advantage' of the strong-focusing principle, as Dahl put it, the machine would be 'rather difficult to line up and adjust for operation'.[12] These technical constraints forced the group to revise the design parameters, as a result of which earlier estimates of the cost/GeV of the new-style PS began to look distinctly optimistic.

At its fifth session at the end of March 1953 the Council agreed that the group could be joined by John and Hildred Blewett from Brookhaven. By October that year most of the problems surrounding the application of the AG principle seemed to be in hand. The group presented its findings at an international symposium organized for that purpose in Geneva, and received high praise for their work to date. Costs were escalating, however, and at the seventh Council session which started on 29 October 1953, the group was asked to cut the design energy of the PS from 30 to 25 GeV to save money. A revised set of machine parameters satisfying this condition was presented in March 1954.

There is a striking contrast between the smooth, almost unhesitant progress achieved by the SC Group in our period, and the halting, less certain development of the PS team. This contrast was first and foremost a consequence of the different kinds of project each was involved in. The task of the SC Group was to scale-up an existing machine already built in the United States. In deciding to apply the strong-focusing concept, on the other hand, the PS Group took on a research and development project, and had to work out 'a completely different design [to a scaled-up Cosmotron] without even a model to guide it let alone a large working machine'.[13] We discuss the implications of this difference at greater length in the next chapter.

7.6.2. THE THEORY GROUP

The organizational structure and rhythm of activities of the Theory Group were essentially those of an academic institution devoted, in this case, to the advanced training of physicists from the member states. The nucleus of the group was a four-man team comprising N. Bohr, C. Møller, J. Jacobsen, and S. Rozental. Bohr was the nominal director until September 1954, when he handed over to Møller; Rozental was responsible for much of the administrative work. They were assisted in their task by other staff members of Bohr's Institute for Theoretical Physics in Copenhagen, and by visiting lecturers from all over the world—people like Hendrik Casimir, Ernest Courant, Robert Marshak, Wolfgang Pauli, and Cecil Powell.

In its role as a training ground for theoreticians the group concentrated on studying fundamental problems in high-energy physics. Main areas of interest included nuclear constitution, quantum electrodynamics, and mesons and the field theory of nuclear forces. Some group members investigated beam-focusing problems in AG accelerators, liaising with the PS group who were similarly preoccupied.

The Theory Group was unusual among the four study groups in that it had a quasi-separate identity of its own. Its key staff were already centralized around an existing institute of international repute when the group was set up, and were simply called on to incorporate activities of concern to CERN into an ongoing, independently developed research and training programme. Inevitably, then, the decision to locate CERN in Geneva raised questions about the future of the Theory Group, and its relationship with the central laboratory in particular.

The Swedish delegation opened the discussion on this at the fourth Council session in Brussels in January 1953. They proposed that the Council make a firm commitment to support theoretical studies in Copenhagen for five years, and then take a decision about the future status of Bohr's group. The Swedes emphasized that they recognized that a theoretical group would also be needed at the central laboratory, but they felt that, at the same time, one should take advantage of the traditions and international contacts built up at Copenhagen to further the development of European theoretical physics.

This proposal met with considerable hostility in some quarters at the meeting. The 'evident disagreement' among the delegates, wrote one British observer, was 'a cloud casting a dark shadow over the meetings'.[14] We shall explore the roots of these divisions at greater length in section 8.1.3. Briefly, and too simply, the Scandinavians were concerned to preserve the future of Bohr's institute as one focal point in Europe for international scientific collaboration. Their opponents saw this initiative as divisive, and as another indication of the reluctance of Bohr and of some of the northern countries to devote themselves wholeheartedly to the main laboratory project. A compromise of sorts was reached at the sixth Council session in June 1953, where it was resolved that theoreticians could be offered long-term contracts to work at Copenhagen, but only on the understanding that they be prepared to move to Geneva if the Council called on them to do so.

7.6.3. OTHER FORMS OF CO-OPERATION

In addition to planning an international laboratory, the provisional Council was responsible for encouraging other forms of co-operation in nuclear research.[15] In the event this came down to discussing the use of the cyclotrons made available to the Council at Liverpool university and in Uppsala, and to sponsoring cosmic-ray research.

Bohr played an active role in fostering the links with the accelerators. His group discussed the experimental programme of the Swedish machine, and interacted frequently with those of its members who were posted there. And a special committee, chaired by Bohr, was set up at the third Council session to select candidates to work around the 400 GeV Liverpool accelerator, which served as a training ground for members of the SC group.[16]

Amaldi was the main driving force behind the cosmic-ray activities. At the third session of the Council he reported on the results obtained in a two-month long expedition to the central Mediterranean organized by several European laboratories.

A year later a similar expedition on a larger scale was held, and its results reported by Amaldi to the seventh session at the end of October 1953.[17]

There was considerable support for cosmic-ray studies in the Council. At the second session in June 1952 the Danes successfully proposed that CERN 'give its patronage', but no financial support, to ongoing Europeanwide collaborations in cosmic-ray physics, particularly those using the photographic method 'in the extreme high energy region'. The importance of cosmic-ray work was further stressed by British physicists early in 1953. Cockcroft and Powell, in particular, felt that at least in its early days, the permanent organization should do research in this area, both 'to get techniques useful for nuclear work established', and to create a pocket of scientific research in an otherwise machine-dominated environment.[18] And indeed the Convention establishing the permanent CERN made provision for cosmic-ray research in its basic programme (see below), and 300,000 SF were put aside for such activities in the budget for the first financial period of the new organization.[19]

The decision to do high-energy physics with both cosmic rays and accelerators in the new laboratory reflects the continuing interest in both techniques as sources of particles at this time. Their relative merits were clearly elaborated by Heisenberg in his report to the second session. The high energies available in cosmic rays made them 'superior to the use of big machines [...] for qualitative observations (discovery of new particles, rough determination of their masses and interaction cross-sections)'. For quantitative work, however, the 10^{12}-fold increase in intensity obtainable in accelerators made them 'an indispensable tool' for elementary-particle studies.

Whatever the merits of cosmic-ray research, the Council was only prepared to make a limited effort to support it. It always ensured that the bulk if not all of the funds at its disposal for experimental work were spent on the PS and the SC. This was simply another symptom of the gradual eclipse of cosmic-ray investigations by accelerators which was more or less total by 1955.[20]

7.7 The Convention and its signature

The Agreement constituting the provisional Council instructed it to produce the draft of a convention establishing the permanent organization. A first version was prepared by Kowarski with the help of UNESCO officials, and briefly discussed at the third Council session in October 1952. It was circulated soon thereafter along with a so-called 'First report to member states on the organizational and financial implications of future European co-operation in nuclear research'. In May 1953 a considerably revised version of the Convention was again circulated, this time along with a 'second report', which superseded the first and which gave a good deal of general and technical information on the two accelerators and the planned

Notes: p. 235

laboratory, including the estimated costs of building and running it: 120 MSF for the first seven years, and 8.6 MSF per annum thereafter.[21]

The text of the Convention was finally agreed at the sixth Council session in June 1953. It had some twenty articles dealing, *inter alia,* with the purposes of the laboratory, the conditions of membership, the structure of the Council, the role of the Director, and the financial demands on member states. Several articles detailed the formal mechanisms for amending, ratifying, etc. the Convention, and for handling problems with or between member states e.g. disputes and withdrawal. Attached to the Convention was a protocol which made provision for the financial administration of CERN. In what follows we shall briefly discuss three of the Convention's most important clauses—that distinguishing between the 'basic' and 'supplementary' programmes of research, that concerning the admission of new member states, and the scale of contributions to the main budget as fixed in the financial protocol. The first two of these topics are analyzed at greater length in section 8.2.

(i) Article II of the Convention distinguished between the 'basic' research programme of the laboratory and any additional 'supplementary' programme. In signing the Convention a member state agreed to participate in the former. It involved the construction and operation of the laboratory, with its accelerators and ancillary apparatus, as well as cosmic-ray work and the organization of other forms of co-operation in nuclear research. Any activities undertaken over and above these constituted a 'supplementary' programme to which no one was obliged to contribute, and which required the approval of a two-thirds majority of the member states.

It was the United Kingdom which first advocated this distinction, and support for it was widespread and spontaneous. Essentially Britain's aims were two. From a scientific point of view, the British wanted to restrict the activities of the laboratory as much as possible to work directly related to the two accelerators. They also wanted to introduce some mechanism for limiting the financial obligations imposed on member states, at least once the 'basic' facilities had been provided.

(ii) The most controversial and divisive clause in the Convention concerned the admission of new states to the organization. In the original draft it was laid down that 'any other European state' could join CERN provided that it was invited to do so by a two-thirds majority of the Council. Early in 1953 the British objected strongly to this provision. On the one hand, they did not want membership of the laboratory to be restricted to European states. On the other hand, they wanted to exclude the Soviet Union and eastern Europe from CERN. Neither of these demands won universal assent. In particular, the French preferred to limit membership to western European states and were not happy about widening membership to include the Commonwealth and the USA, while the Swiss were opposed to any clause which could be interpreted as deliberately excluding the eastern bloc. In the event a compromise was reached whereby the reference to Europe was dropped from the clause, and the unanimous assent of the Council was required for the admission of a new state. In principle, then, membership of CERN was open to *any* state. In this way the appearance of openness was saved; in reality, any existing state could veto the application of a new candidate if it was in its interests to do so.

(iii) Granted the magnitude of CERN's budget and, more importantly, of the relatively large slice of national scientific income that it would consume, the scale of national contributions to the organization is of some interest.

Various possible ways of determining this scale were discussed in the Council in 1953. While it was generally agreed that it should be related to national wealth, there were of course different ways of assessing that. On the one hand, Italy and Yugoslavia favoured taking *per capita* income into account, a suggestion most unattractive to Sweden and Switzerland, for example. Their income per head in 1951 was the highest of all the member states, and some 50% above that of Britain and France. On the other hand, if the scale was based on national income, the small 'rich' countries in general, and Sweden and Switzerland in particular, would be relatively lightly assessed.

According to Jean Mussard, this question 'provoked an endless and confused discussion and Ben Lockspeiser, who was an impatient man, suddenly exploded. He was fed up with this "shameful horsedealing" '. Offended, a Swedish delegate broke down and wept openly.[22] But to no avail. The minutes of the fourth session record that 'It was finally [sic] agreed that the national income should be taken as a basis, with the exception that France and the United Kingdom should bear an equal percentage' — an arrangement which slightly favoured the latter. The resulting percentages, as specified in the financial protocol to the Convention, and applying for the period up to the end of 1956, are given in appendix 7.1.

On 1 July 1953 the representatives of the eleven member states who were party to the Agreement constituting the provisional CERN, plus the United Kingdom, met at the Ministry of Foreign Affairs in Paris. They approved the text of the Convention establishing the European Organization for Nuclear Research, and its associated financial protocol. Both documents were open for signature until the 31 December 1953.

Delegates from nine states signed forthwith — Belgium, the Federal Republic of Germany, France, Greece, Italy, the Netherlands, Sweden, the United Kingdom, and Yugoslavia. A Swiss delegate signed in mid-July, and only after his authorities had formally agreed to the final wording of the clause admitting new member states. Denmark and Norway signed at the end of December.

The Convention was a kind of international treaty, and signature of it had to be ratified in each member state. It came into force when the instruments of ratification of seven member states were deposited at UNESCO House in Paris. The host state, Switzerland, had to be one of the seven, and the total of their financial contributions to the organization's budget had to be at least 75%.

Notes: p. 235

7.8 The new mood in CERN

The proceedings on 1 July 1953 took place in a mood of confidence and optimism about the future of the organization, a mood which contrasted starkly with the doubts and uncertainties which had prevailed during the UNESCO-sponsored conferences some eighteen months before. There were two marked symptoms of this change of heart, namely,

- around the end of 1951 and into 1952 many had feared that the European laboratory project was too ambitious and costly for the member states. The Council's role was seen to be essentially exploratory, and it was felt that a firm commitment to the scheme could only be made after calling a special intergovernmental conference. This conference was still envisaged in the first report to member states at the end of October 1952. In fact it was never held. The Convention was signed at what amounted to an extention of the sixth Council session, and by regular Council delegates—people like Willems (B), Perrin (F), Heisenberg (FRG), Colonnetti (I), Bannier (NL), Waller (S), and Lockspeiser (UK);
- no one signing the Convention had the least doubt that it would be ratified and that the permanent organization would come into being. Indeed it was thought at the time that the 'interim' CERN would last no more than six or seven months, and specific financial provision was made in a so-called 'Supplementary Agreement' which authorized the provisional Council to proceed 'with a number of tasks in view of the construction of the Laboratory [...]'.[23]

What had caused this change in outlook? More specifically, why had it turned out to be relatively easy to secure the commitment of the member states to the project? Of course circumstances differed somewhat between countries, but the following factors appear to be of general importance.

First and foremost, there was the discovery of alternating-gradient focusing, and the opportunity it provided for building accelerators far more powerful and at no extra cost than those originally envisaged.

Then there was the growing conviction that nuclear physics research as a whole was important, and that European governments should establish appropriate mechanisms to support it. For example, in 1950/51 the French reorganized their atomic energy 'Commissariat', expanding its scope with a view to bridging the gap between France and other 'atomic nations'. In July 1952 a national committee for nuclear research was set up in Italy. In 1952/53 the Greek government took steps to establish a national commission for atomic energy. And on 1 August 1954 the U.K. Atomic Energy Authority was formed, thus significantly changing the way in which many nuclear activities were organized and funded in Britain.[24]

Finally, these developments occurred against a background of steady improvement in the overall economic situation. Almost $10 billion of Marshall aid were injected into the European economies between the end of 1947 and June 1950.

A period of sustained prosperity was soon under way, inaugurating a decade in continental Europe in which 'growth of output and consumption, investment and employment [surpassed] any recorded historical experience [...]'.[25] Therewith the means for financing the construction of CERN became available.

7.9 Putting down roots in Geneva

As we mentioned above, when signing the Convention the Council agreed that preparations could be made during the interim period for the construction of the laboratory. The most urgent matters needing attention were spelt out by the Executive Group at a meeting in Amsterdam in mid-June 1953. They wanted the PS Group centralized at Geneva, a supporting administrative nucleus established, and an architect commissioned to draw up the plans for the laboratory buildings.[26] We shall discuss each of these initiatives in turn.

Scientifically speaking the timing of the PS move to Geneva was set more or less by the progress made on the development of the strong-focusing principle. Originally it was scheduled for the spring of 1953, but due to difficulties encountered in applying the new concept it was decided to postpone centralization until the general specifications of the machine had been settled, and a programme of detailed engineering studies could begin. In the event it was decided that about half the total squad (numbering about 20 persons) should move to Geneva early in October 1953, to be followed soon thereafter by the rest. Accommodation had been offered by Prof. R. Extermann, director of the Institute of Physics of the University of Geneva.

On 5 October 1953 the advance guard of the PS Group arrived in Geneva—Frank Goward, John Adams, and Mervyn Hine from Harwell, John and Hildred Blewett temporarily released from Brookhaven, Kjell Johnsen from the headquarters in Bergen, and Edouard Regenstreif from Paris. In anticipation of the move Dahl had written to Amaldi remarking that he assumed that the group could 'just move into Extermann's Institute', despite some 'formal difficulties'.[27] Things did not quite work out that way, however. To their dismay, when the group arrived at the university they 'were not even permitted into the building'. Dahl recalls that, before allowing them to enter his institute, Extermann 'demanded that [they] should be cleared by high Geneva officials'. While Goward tried to clarify the situation the rest of the group were left 'sitting on the steps' outside!'[28]

Apparently this confrontation was triggered by Extermann's desire to ensure that the PS Group did not act as a more or less autonomous unit alongside his own team of researchers. Once measures had been taken to reassure him on this score, relationships between Extermann and the group improved considerably. All the same the CERN staff never really felt at home in the institute itself, and were pleased to move into an adjacent annex and four nearby huts which were put at the organization's disposal during the course of 1954.[29]

Notes: pp. 235 ff.

These were not the only difficulties encountered by the group in these early months. There were the problems of adjusting to the new environment, felt particularly acutely by the British some of whom thought that living in Geneva was 'something like living in a desert'.[30] There was insecurity, for the Council had no authority to offer the group members contracts with the permanent organization, and when they moved the IFC had not yet settled a system of social security benefits. There were financial problems, for while steps were taken to ensure that no one who was transferred to Geneva incurred financial losses, their salaries were soon found to be inadequate for life in Switzerland. Finally there was the unexpected death of Goward from a brain tumour on 10 March 1954—less than a month after he had had 'some form of short blackout', as he put it to Amaldi.[31]

Despite these difficulties the work of actually designing the PS was under way in Geneva by the end of 1953. The laboratories were gradually equipped with experimental apparatus, stores, and a medium-sized workshop. By September 1954 the PS Group occupied some 1200 m^2 of space and its staff numbered a little over 50 persons.[32]

To support the PS group the nucleus of an administrative structure began to be built up in Geneva in the closing months of 1953. A Dutch purchasing officer (N.E. Groeneveld Meijer) moved there in mid-October to secure the equipment it needed, to be followed shortly by a personnel officer (R. Penney) temporarily released from the UK Ministry of Supply. Early in 1954 accommodation was provided for the administration, and for the offices of the Director-General and part of the Laboratory Group in a villa near Geneva airport—the so-called 'Château de Cointrin'. At the same time the cantonal authorities offered CERN premises in one of the main hangars on the airport. This was used for the SC Group which began to move to Geneva late in 1954.[33]

All of this accommodation was of course temporary, and intended only to house the personnel and their equipment until such time as the buildings at the permanent site in Meyrin, on the Franco–Swiss border, were ready. The task of designing these was entrusted to a Zurich architect, R. Steiger, who worked in close collaboration with the Laboratory Group.

Steiger's original commission was to prepare the plans for the SC and administration buildings, and some ancillary facilities. It was clearly understood that the provisional Council could not authorize steps beyond 'reasonable progress with the planning'. However, early in 1954 it was confronted with a dilemma arising because the start of the building programme had to be timed so that facilities would be available as soon as the major components of the SC were delivered. This meant that excavations on the site had to commence in spring 1954 so as to have made sufficient progress by the time the Swiss winter set in. A delay of a few months, it was claimed, could set the building programme back by as much as a year. But the Convention was not ratified early in 1954, as had been expected when the move to Geneva was agreed to, and the first contract signed with Steiger. Indeed it was clear that the permanent organization would not come into being before the summer at the earliest. We have described the measures taken by the Council and the IFC to deal with this situation in section 8.3.

On 17 May 1954 the first earth was turned in the rolling fields just outside Geneva where CERN was to be built. The ceremony, which was not official, was attended by local members of the CERN staff, by cantonal authorities, and by Council chairman Robert Valeur. No speeches were made, and no journalists reported the historic event.[34]

7.10 The nomination of the first Director-General

At its sixth session in June 1953 the Council set up a Nominations Committee to put forward candidates for the 'principal officers' of the permanent organization. The committee, which was essentially composed of physicists, was convened by Francis Perrin. Its other members were Nobel Laureates Niels Bohr, John Cockcroft, and Werner Heisenberg, plus Gustavo Colonnetti and Paul Scherrer. At the ninth meeting of the Council in April 1954 the committee presented its recommendations — that the Council designate 'Mr. Felix Bloch, Nobel Prize, Professor at Stanford University, as Director [...]', and that his appointment be coupled with those, 'as other members of the Directorate of Mr. Edoardo Amaldi [...] in the capacity of Deputy-Director and of Mr. C.J. Bakker [...] in the capacity of representative of the Scientific Group Leaders'.

The complex negotiations surrounding this recommendation will not detain us here, and we wish merely to comment briefly on the attitudes of the three men involved in the arrangement, bringing out the somewhat remarkable fact that none of them was particularly happy with it.

Bloch was never enthusiastic about the idea of leading CERN. He was not particularly interested in constructing accelerators, and he feared that he would be overburdened with administrative duties. From the start he specified that he could not bind himself for more than two years (at most), that he wanted to be free to continue with his work on nuclear magnetic resonance (which meant that CERN had to bring over his equipment and two assistants from Stanford, and put the latter on its payroll), that his salary be the same as what he was earning at Stanford (so almost double the figure allowed for in the CERN salary scales), and that his nomination have unanimous Council support (though formally only a two-thirds majority was required). With all these conditions satisfied he accepted the post, though even then with some reluctance.

Amaldi too had his doubts. He realised that, as Bloch's deputy, he would play at best a secondary role in organizing CERN's scientific life, and that his task would be mainly administrative. This did not appeal to him, and he made it clear that he wanted to spend a good deal of his time in Rome and working around an American

Notes: p. 236

accelerator. If he accepted to serve on the Directorate it was mainly because of the considerable pressure put on him to do so by Council chairman Robert Valeur, who realized that unless some arrangement could be made to release Bloch of his administrative duties the Stanford professor would never be lured to CERN.

Bakker's dealings with the Nominations Committee constantly went awry. They had offered him a 'temporary' directorship (probably with reduced powers) in October 1953, not out of the conviction that he was the best man for the job, but because they could not find a really eminent physicist to take it on at short notice. Resenting their attitude, he turned the offer down. Now, to placate the Dutch, and the (bulk of the) Executive Group, he was being offered a post of somewhat ambiguous status in the Directorate, to which was later added the directorship of the SC division. Bakker's pride was again hurt, all the more so since he claimed that he had been led to believe that someone else would be put in charge of the smaller accelerator, and that he, Bakker, would be on the same level as Amaldi, and senior to the heads of the machine divisions.

In the light of these circumstances it is not surprising that the Directorate of the permanent organization was in danger of crumbling to pieces even before the body officially came into being. No sooner had Bloch arrived in Geneva in mid-September than he learnt that Amaldi wished to return to Rome as soon as possible, and that Bakker was unwilling to come to Geneva full-time. A brief period in office was enough to convince Bloch that he had made a mistake. In mid-November 1954 he informed Council president Sir Ben Lockspeiser that he wished to be relieved of his post as Director-General by autumn 1955.

7.11 The ratification of the Convention

On 29 September 1954 the instruments of ratification of both Germany and France were deposited at UNESCO House in Paris. With that the conditions necessary for the ratification of the Convention had been met, and the provisional Council officially ceased to exist. For a few days Secretary-General Amaldi was the sole owner of all the assets of CERN. They passed to the permanent organization when it first met in Geneva from 7 to 8 October 1954. In this section we first comment on the rate of the process of ratification, and then describe the situation inherited by the permanent Council.

There are two noteworthy features about the rate of ratification (see appendix 7.3). Firstly, by 31 January 1954, when it was originally assumed that the Convention would come into force, in fact only one member state had completed the formalities. Obviously the Council delegates had seriously miscalculated the time taken for the process. One reason for this was their lack of experience with the procedures involved: as Verry put it to Mussard, 'it is evident that many members of the council have an inadequate understanding of what is meant by ratification

[...]'.³⁵ The immediate consequence of this was that the IFC had its time cut out raising funds to cover an interim period that lasted fifteen months instead of the anticipated six or seven.

A glance at our list in appendix 7.3 also reveals that, somewhat ironically, the United Kingdom, who had been more or less indifferent to the project in its early stages was the first to ratify, whereas France (and Italy) who were two of its most enthusiastic promoters were among the last. The main reason for this was that Britain alone among the countries involved did not join the provisional Council. For her the question of membership always concerned membership of the permanent CERN. As a result the various ministries involved in the decision, including the all-important Treasury and Foreign Office, were in frequent contact long before the Convention was signed; in giving the green light for signature they were simultaneously committing Britain to the permanent organization. This gave the United Kingdom a considerable headstart over, say, France, where policy was articulated to meet successive phases in the evolution of the project. Membership of the provisional CERN was one thing, that of the permanent organization quite another. In practice this meant that some key officials in the bureaucracy were only brought into the picture towards the end of 1953. This inevitably delayed matters a little.³⁶ On top of this the French (and German) national assemblies had questions of far greater urgency to attend to early in 1954 (e.g. the war in Indochina for France), and CERN tended to slip down the parliamentary agenda.

What state of development had CERN reached when the permanent Council first met? The list of achievements of the provisional body was impressive. Work on the site was well under way. The access road to the SC and PS buildings was finished, the first electricity and telephone cables connecting the site to the city had been laid, the order for the excavation and foundation works for the SC building had been placed, and tenders from firms interested in constructing its upper structure had been received.³⁷

At the same time some 2500 m² of accommodation were occupied in various parts of Geneva. Personnel numbered 112 full-time and 8 part-time — almost half of them in the PS division. There was an unspent balance of over 4.4 MSF, of which about half was already committed for the building programme alone.³⁸

On the organizational side CERN comprised six divisions and a directorate. We have already described the unstable situation prevailing in the latter. Here we shall simply list the divisions and the directors recommended by the Nominations Committee and provisionally confirmed by the Council when it first met.³⁹

John Adams and Cornelis Bakker were put in charge of building the proton synchrotron and synchro-cyclotron, respectively. The Laboratory Group was split in two for the construction period. Lew Kowarski was director of a newly-formed Scientific and Technical Services division, responsible for instrumentation, the workshops, documentation, and so on.⁴⁰ Peter Preiswerk headed Site and Buildings division, and was the main link between CERN and the architect Steiger. Administration was temporarily in the hands of Sam Dakin. Finally the directorship of the Theory Group in Copenhagen passed from Niels Bohr to Chris Møller, who had done much to help the aging doyen of European physicists during the provisional period.

This catalogue of information does not really capture the full extent of the achievement of the provisional organization. It is better expressed in the Christmas message from some of the senior staff to Council president Sir Ben Lockspeiser at the end of 1954. As they put it,

> Less than three months after its official birth, CERN finds itself in possession of an active programme of research and building in full progress, adequate accommodation and considerable staff. The stage of teething troubles is behind us; our approaching adolescence will bring difficulties of its own but we can look ahead with confidence.[41]

7.12 The philosophy of the organization

Among the scientists and administrators who launched CERN there were several who had a particular conception of the kind of organization that it should be. Though of course idealized, this conception of the nature and aims of the laboratory had a determinate effect on their behaviour. To conclude this chapter we shall first list five related elements of what we call their organizational philosophy, alluding to the contexts in which each was brought into play, e.g. the choice of a director and the formulation of certain aspects of staff and financial policy. Having identified the philosophy we go on to discuss some of its implications, which are developed at greater length in chapter 14.

The five most prominent constituents of CERN's organizational philosophy as articulated during the provisional period were the following:

(i) CERN should be imbued with a scientific spirit. The roots of this concern lay in the fear that the intellectual life of the laboratory would be dominated by technical problems surrounding the accelerators, particularly in its infancy when the machines were being constructed. To counteract this eventuality several concrete measures were taken: an eminent physicist was appointed Director-General, a small theoretical studies group was established in Geneva, regular seminars on scientific topics were arranged, and experimental work on cosmic rays was encouraged.

(ii) In its research characteristics CERN should be like a university.[42] This had several related implications. It meant that the laboratory should be devoted to work 'of a pure scientific and fundamental character, and in research essentially related thereto', and should have 'no concern with work for military requirements' (Article II of the Convention). It also meant that the work should be openly published and freely discussed—a Swiss proposal to limit the disclosure of information by staff members was vehemently opposed by Amaldi with the support of Lockspeiser in the

IFC.[43] Finally, the notion that CERN should be like a university was seen to entail that permanent posts should be discouraged so that there would be a 'flow of scientists through the Laboratory so necessary to avoid stagnation'. This would also increase the accessibility of the laboratory's facilities to scientists in the member states.[44]

(iii) CERN should encourage idealists, people who did science for its own sake.[45] What it wanted was people with 'a pioneering and adventurous bent of mind', whose attitude was 'in harmony with the attempt to create in Europe a new front-line of advanced physical research'. In Kowarski's view this was another reason for keeping permanent contracts to the minimum — 'the promise of a secure position carrying a combined academic and diplomatic prestige' would attract the 'wrong sort of candidate'. So too would unreasonably high salaries, typically those paid in other international organizations. As Kowarski remarked twenty years later, in drawing up the salary scales 'the idea in [his] mind was to take the level of salaries which existed at UNESCO, at the Geneva office of the United Nations and so on, and to knock off a rather sizeable percentage'.[46]

(iv) Merit should be the main criterion in the allocation of resources. This criterion had an important impact in two areas of policy. Firstly, in recruiting staff, the selection 'of persons of the highest ability and integrity ... [was] of first importance'.[47] Secondly, in furnishing the laboratory with plant, equipment, etc. at least three competitive tenders would normally be obtained, and the contract would be awarded to the firm whose tender was the lowest which satisfactorily complied with technical and delivery requirements.[48]

Granted that merit was the *primary* criterion for distributing these rewards, it was accepted that in recruiting staff CERN should also seek to achieve 'a wide distribution of posts among the nationals of Member States'. Pressure to have a similar provision ensuring that 'no one Member State shall receive an unduly high proportion of the total value of the orders placed competitively' was resisted on the grounds that it would produce 'inextricable political difficulties'.[49]

(v) CERN should not compete with research facilities in its member states. As Kowarski put it, 'CERN is a co-operative organization rather than a supranational one. It aims at adding to the existing network of research institutes rather than standing outside or above that network'.[50] He used this as an additional argument for keeping salaries well below those of other international organizations, and for limiting the number of permanent contracts — people thus employed would be 'automatically removed from the national academic scene'.

Like any ideals, those which inspired the launching of CERN had their limitations. In the first place they were based on a romanticized and outdated conception of what kind of people scientists were. As our account in a previous section showed (at least) the contenders for the top post of Director-General were not the self-sacrificing idealists that Kowarski thought they should be. Like the members of any other profession — for that is what these scientists were — they were concerned in varying degrees with prestige, financial reward, and their power within the organization. This is not to judge them, merely to dispel a myth.

What is more, it was naive to hope that CERN would not compete with member states for personnel and keep them at the Geneva laboratory. This was an inevitable consequence of the uneven level of development of its member states. Quite apart from its facilities, which were likely to be enormously superior to anything that some of the participants could afford, there was the question of salaries. Kowarski's original scales were revised upwards by Penney, whose main concern was to ensure that personnel who moved to Geneva would have a salary effectively equivalent to that of a Swiss employee of comparable status.[51] To take an extreme case from Penney's data, this meant that an unmarried middle-income employee from Italy would earn two to three times more at the Geneva laboratory than at home — a difference considerably greater than the difference in the cost of living between the two countries. As a result many who came to CERN were increasingly reluctant to leave it.

Some of the other ideals we have mentioned are vulnerable in more subtle and indirect ways. Consider, for example, the question of openness. Consistent with its Convention CERN was an international laboratory welcoming researchers from all over the world who wished to do 'pure' research. However, like all scientific activities in the modern world, and in particular the activities of a front-line physics laboratory equipped with highly-sophisticated technology, some work done at CERN could not but be of interest from an applied angle. This was doubtless in the minds of the Swiss delegates when they suggested to the IFC that some provision be made to limit the disclosure of information by CERN staff. In the event they were overruled, but as a result the laboratory was placed in an 'asymmetrical' relationship *vis-à-vis* its users: being open, CERN could not exclude visitors from research centres which were not — and who could exploit the results of their research behind closed doors for purposes at variance with CERN's ideals.[52]

Appendices

Appendix 7.1

Scale of percentage contributions to the permanent organization applicable during the period to 31 December 1956

Member state	Contribution (%)
Belgium	4.88
Denmark	2.48
Federal Republic of Germany	17.70
France	23.84
Greece	0.97
Italy	10.20
Netherlands	3.68
Norway	1.79
Sweden	4.98
Switzerland	3.71
United Kingdom	23.84
Yugoslavia	1.93

Appendix 7.2

Rates of exchange used for accounting purposes (a) as from 1 December 1952 and (b) as from 1 January 1954

Country	Currency	(a)	(b)
Belgium	Sw. Franc 1 = Belgian francs	11.40	11.46
Denmark	Sw. Franc 1 = Danish crowns	1.58	1.58
Federal Rep. of Germany	Sw. Franc 1 = Mark	0.96	0.96
France	Sw. Franc 1 = French francs	79.37	80.16
Italy	Sw. Franc 1 = Lire	142.35	142.86
Netherlands	Sw. Franc 1 = Florin	0.87	0.87
Norway	Sw. Franc 1 = Norw. crowns	1.63	1.63
Sweden	Sw. Franc 1 = Swed. crowns	1.18	1.18
Yugoslavia	Sw. Franc 1 = Dinars	69.12	69.12
United Kingdom	Sw. Franc 1 = Pound	0.0817	0.0818
United States	Sw. Franc 1 = Dollar	0.233	0.233

These figures were revised every three months or in case of fluctuations of more than 5%.

The data in columns (a) and (b) are from CERN/Memo/5(Rev.2) and CERN/Memo/5 (Rev.4), respectively (CERN–CHIP10031).

Survey of developments

Appendix 7.3
Dates of deposit of instruments of ratification
of the Convention at UNESCO House in Paris

Country	Date of deposit
United Kingdom	30 December 1953
Switzerland	12 February 1954
Denmark	5 April 1954
Netherlands	15 June 1954
Greece	7 July 1954
Sweden	15 July 1954
Belgium	19 July 1954
France	29 September 1954
Federal Republic of Germany	29 September 1954
Norway	4 October 1954
Yugoslavia	9 February 1955
Italy	24 February 1955

Appendix 7.4
Official delegates to the nine Council sessions[1]

	Official National Representatives[2]	UNESCO[3]		Council Session								
		A	B	1	2	3	4	5	6	7	8	9
B	J.L. Verhaege	X	X	X	X	X	X	X	X	X	X	X
	J. Willems		X	X	X	X	X	X	X	X	X	X
D	H.M. Hansen	X			a	X	X	X	X	X	X	X
	J.C. Jacobsen			X								
	J. Nielsen	X	X	X	X	X	X	X	X	X	X	a
F	F. de Rose	X	X	a								
	F. Perrin	X	X	X	X	X	X	X	X	X	X	X
	R. Valeur				X	X	X	X	X	X	X	X

1. This list has been compiled from the official minutes of the meetings, which are not always accurate, e.g. Alexander Hocker was mistakenly recorded as present at the first UNESCO conference.
2. Nominated substitutes for the two official representatives at each meeting are not recorded.
3. Columns A and B refer respectively to the UNESCO-sponsored preparatory conferences in Paris (17–20/12/51) and Geneva (12–15/2/52). A cross in the column means that the CERN Council delegate was also a member of his country's official delegation to that conference.

Appendices 233

	Official National Representatives[2]	UNESCO[3]		Council Session								
		A	B	1	2	3	4	5	6	7	8	9
FRG	W. Heisenberg	X	X	X	X	X	X	X	X	X	X	X
	A. Hocker	X	X	X	X	X	X	X	X	X	X	X
G[4]	Th.G. Kouyoumzelis									X		X
	J. Papayannis										X	
	N. Hadji Vassiliou									X	X	X
I	A. Casati	X	X	X								
	G. Colonnetti	X	X	X	X	X	X	X	X	X	X	a
	A. Pennetta				X	X	X	X	X	X	X	X
NL	C.J. Bakker	X	X	X								
	J.H. Bannier	X	X	X	X	X	X	X	X	X	X	X
	S.R. de Groot					X	X	X	X	X	X	
N	J. Holtsmark			X	X	X				X	X	X
	E. Hylleraas	X	X		X		X	X		X		
	B. Trumpy						X	X		X		X
	H. Wergeland		X			X						X
S	H. Alfvén	X	X	X			X					
	T. Gustafson	X	X				X		X	X	X	X
	M. Jacobsson	X	X		X			X				
	I. Waller	X	X	X	X	X	X	X	X	X	X	X
CH	A. Girardet				X							
	A. Picot	X	X	X		X	X	X	X	X	X	X
	P. Scherrer	X	X	X	X	X	X	X	X	X	X	X

4. Greek observers to the 2nd and 6th Council Sessions are not recorded.

	Official National Representatives[2]	UNESCO[3] A	B	Council Session 1	2	3	4	5	6	7	8	9
UK[5]	P.M.S. Blackett					RS						
	D.W. Fry				RS	RS			X			X
	J. Cockcroft						X	X		X	X	
	B. Lockspeiser						X	X	X	X	X	X
Y	S. Dedijer	X	X	X	X	X	X	X	X	X	X	
	S. Nakicenovic											X
	A. Peterlin							X				
	P. Savic	X	X	X					X			
	I. Supek						X		X		X	X
	R. Walen									X		

5. The UK delegates were always observers, either on behalf of the Royal Society, otherwise on behalf of H.M. Government. A.H. Waterfield was unofficially present at the first Council session. He also attended both UNESCO conferences.

Key
X, a = Respectively present/absent as member of the delegation
RS = Royal Society (UK)

Notes

1. The basic sources for this chapter were the official proceedings of the provisional Council as recorded in its minutes, and the more detailed case studies given in chapter 8. The minutes of the Council are part of a series of CERN/GEN/... documents, among which one also finds the three reports to member states dealing with the scientific, financial, and organizational aspects of the European laboratory, and the texts of the Agreement setting up the provisional CERN, that of the Supplementary Agreement prolonging it, and that of the Convention establishing the permanent organization. All are gathered together in (CERN–CHIP10014).
If the context makes it clear no detailed references to these official documents are given in the body of the text.
2. Telegram from some two dozen conference delegates to Rabi, 15/2/52 (UNESCO).
3. Lew Kowarski's deputy, Peter Preiswerk, and Odd Dahl's deputy, Frank Goward, also frequently attended meetings of the Executive Group.

4. For information on the dispersal of the PS Group, see J.B. Adams, 'Introductory talk on the alternating gradient proton sychrotron', 27/7/54 (CERN-DG20551). Bakker's progress report appended to the minutes of the third Council session describes the situation in his group. The Laboratory Group had two offices, one in Paris (Kowarski), the other in Zurich (Preiswerk). Although the senior nucleus of Bohr's group was in Copenhagen, fellows on stipends also worked in Skinner's laboratory at Liverpool university, and at the Gustaf Werner Institute for Nuclear Chemistry in Uppsala.
5. Auger's full list is available in (CERN-Mussard file). For more detail on the conflict around the word 'centre' see sections 6.2.1 and 6.3.1.
6. For details of the events surrounding the communication of the new focusing principle to the CERN visitors, see Dahl's trip report, CERN-PS/54, 25/8/52 (CERN-DG20551).
7. The circumstances surrounding the emergence of these two candidates is complex, and have been discussed at length by Dominique Pestre in chapter 6.
8. The Interim Finance Committee was a direct descendant of an administrative and financial working group set up by the Council at its fifth session to make financial arrangements for the transitional period, and to draw up financial regulations, salary scales, staff regulations, social security provisions, etc. for the permanent organization. It comprised delegates from Belgium, France, Italy, Sweden, Switzerland, and the United Kingdom. Sir Ben Lockspeiser was its chairman, and the DSIR provided the secretariat. The second meeting of the IFC held on 27-28/8/53 was also the third and last meeting of the working group. The new committee retained its basic structure and terms of reference, though when Valeur was Council chairman, for example, and so attended IFC meetings in that capacity, France was replaced by the Netherlands in the list of six officially represented countries. The working group minutes and papers are in (CERN-DG20534), those of the IFC in (CERN-CHIP10025).
9. For a more detailed analysis see particularly section 13.4.2.
10. For more detail on the progress of the SC see section 8.5 and Mersits (1984).
11. Adams (1968), 89.
12. In the second report to member states, CERN/GEN/5, on page 12.
13. Adams (1968), 87.
14. The phrase was used by Waterfield, the British Scientific Attaché in Paris, in a report sent to Lockspeiser, which he forwarded to Cockcroft on 23/1/53 (SERC-NP24).
15. This was included in the Council's work as part of an effort to accommodate the position held by Bohr and his allies.
16. This 'Committee for other forms of co-operation' was chaired by Niels Bohr. Its other members were Edoardo Amaldi, Stefan Dedijer, Werner Heisenberg, and Francis Perrin. The minutes of its four meetings are in CERN-DG20795.
17. The reports are CERN/16, 30/9/52, and CERN/GEN/11, both in (CERN-CHIP10014).
18. Letters, Powell to Lockspeiser, 10/2/53 (SERC-NP24) and Cockcroft to Amaldi, 8/5/53 (PRO -AB6/1076). The quotation is from the latter.
19. See CERN/IFC/37, 25/3/54 (CERN-CHIP10025). This was about 3% of the estimated total budget.
20. This has been analyzed at greater length by Ulrike Mersits in chapter 1 of this volume.
21. The two reports are in CERN-CHIP 10014.
22. Letter Mussard to the author, 2/9/85 (private communication). The event is also mentioned in letter Waterfield to Verry, 23/1/53 (PRO-DSIR17/561). Waterfield claimed that 'The Swedes had made themselves rather unpopular, blackmailing the Council [...] by taking the line that they expressed the authentic voice of Britain, and it was very upsetting to them when Secretary came along and said other things'.
23. See 'Note to the Supplementary Agreement', CERN/44, 1/6/52.
24. For the French aspect, see section 9.1 of this book. The Italian initiative is mentioned in Amaldi's weekly report no.7, 21-25/7/52 (CERN-CHIP10029). Developments in Greece are described in CERN/37, 31/3/53. For Britain, see for example Jay (1955).

25. Laqueur (1982), 181–182.
26. Their views were spelt out in CERN/44, Addendum I, 11/6/53.
27. Letter Dahl to Amaldi, 15/9/53 (CERN-DG20551).
28. Interview of Dahl and Johnsen by Gowing, Bergen, 5/7/74 (CERN). The story is also told by John Blewett (private communication).
29. Letters Picot to Amaldi, 23/9/53 (CERN-CHIP10018), and 9/10/53 (CERN-DG20829) explain Extermann's position. Some reasons for the PS Group's subsequent discontent are described in letter Amaldi to Bloch, 13/6/54 (CERN-DG20512).
30. Interview of H. Blewett, Hine, and Johnsen, with Gowing, Kowarski, and Newman at CERN, 3/7/75 (CERN), 32.
31. Letter, Goward to Amaldi, 12/2/54 (CERN-DG20551).
32. See the final report to member states, CERN/GEN/15.
33. See CERN/73, 14/1/54.
34. For information about the proceedings at this ceremony see letter Picot to Micheli, 20/5/54 (CERN-CHIP10016). We are not sure about the precise date of the first *coup de pioche*. Lew Kowarski in his 'An account of the origin and beginnings of CERN', i.e. Kowarski (1961), gives the date as 17/5. Picot's letter, of which we only have an extract made by Monique Senft, puts it a day later.
35. Letter Verry to Mussard, 2/7/54 (CERN-DG20805). He pointed out that, for ratification, the instrument had to be signed by the head of state or someone under his authority. Most delegates confused it with approval for membership being given by the national legislature.
36. The position in France has been discussed by Dominique Pestre in section 9.5.2. For developments in Germany see, for example, letter Hocker to Amaldi, 17/2/54 (CERN-DG20822).
37. Memo, Preiswerk to the Executive Group, CERN/L, 24/9/54 (CERN-DG20600).
38. See final report to member states, CERN/GEN/15, and Preiswerk's memo, note 37.
39. The proposals made by the Nominations Committee after a meeting in London on 7/7/54 are in a file of papers concerning its activities temporarily shelved on row H in the CERN archives.
40. The dispute surrounding the role of Kowarski's division is touched on in section 8.6.
41. Letter on behalf of the Director-General to Lockspeiser, 23/12/54 (CERN-DG20798).
42. In using this analogy we are drawing on a document produced by Kowarski, CERN/L/WP.2, 22/5/53 (CERN-DG20534).
43. See minutes of the second meeting of the IFC, CERN (I.F.C.)(1953) 2nd Meeting, 3/9/53 (CERN-CHIP10025).
44. A point particularly stressed by the British, see CERN/57, 15/10/53.
45. This conception of the scientist was uppermost in the discussion of the salary scales at the seventh Council session—see its minutes, CERN/GEN/12. For the quotations which follow, see Kowarski, *op. cit.*, note 42.
46. Interview of Penney by Kowarski, Boston, 13/10/75 (CERN), 9.
47. See article 7 of the provisional staff regulations, addendum V to CERN/GEN/12.
48. In this respect CERN differs from other European organizations established later like ESRO and ELDO, where a principle of just return applied, i.e. the principle that contracts should be awarded to member states proportionately to their contributions to the organization's budget. For some of the problems this policy created in the two space organizations see Schwarz (1979).
49. The discussion was originally held at the second meeting of the IFC (minutes CERN(I.F.C.)(1953) 2nd Meeting, 3/9/53 (CERN-CHIP10025)), with reference to the draft financial rules, appended to CERN(I.F.C.)1, 27/7/53 (CERN-CHIP10025). The political problems engendered by retreating from a policy of competitive tendering were stressed by Lockspeiser at the seventh Council session—see CERN/GEN/12.
50. Kowarski, *op. cit.*, note 42.
51. Penney's report is CERN PS/S19, 8/1/54 (CERN-CHIP10025).
52. On the notion of asymmetry see Grinevald et al. (1984), 33.

CHAPTER 8

Case studies of some important decisions[1]

John KRIGE

Contents

8.1 The choice of a site   238
    8.1.1 The decision to locate CERN in Geneva   239
    8.1.2 The local opposition to the site in Geneva   241
    8.1.3 Winning something for Copenhagen   243
8.2 The Convention. Two key clauses   246
    8.2.1 The basic and the supplementary programmes   246
    8.2.2 The admission of new states   250
8.3 Financing the interim organization   252
    8.3.1 The first request for funds   253
    8.3.2 Raising further funds 'to continue reasonable progress with the planning work'   254
    8.3.3 Taking further measures to avoid 'serious delays in the orderly and economical progress of the work'   257
    8.3.4 Concluding remarks   260
8.4 The nomination of the first Director-General and its aftermath   261
    8.4.1 The first initiatives   262
    8.4.2 A second main candidate emerges   264
    8.4.3 Political realignments   267
    8.4.4 A decision is taken   268
    8.4.5 The director is appointed   271
    8.4.6 Concluding remarks   272
8.5 The strong-focusing principle: the decision and its early consequences   273
    8.5.1 The proton synchrotron   274
    8.5.2 The synchro-cyclotron   276
    8.5.3 Some problems surrounding the building of an alternating-gradient synchrotron   278
8.6 Planning the future laboratory   282
    Notes   285

In the previous chapter we presented an overview of the historical development of the provisional CERN. We described the activities of the various study groups, of the Council, and of the Interim Finance Committee (IFC). We also depicted the first steps taken to establish the laboratory in Geneva, and to prepare for the birth of the permanent organization. In this chapter we complement that discussion with more comprehensive, in-depth analyses of some of the important policy decisions taken during the period in question.

As we have mentioned before the reasons for this division lie in the nature of the provisional CERN itself. The study of this transitional organism has posed a number of problems for us, for it bridges two very different periods in the organization's life. The steady increase in the size and complexity of the organization alone, not to speak of the growing diversity of its activities, make a 'comprehensive history' of it impossible. At the same time one would like to preserve the richness of understanding which an in-depth analysis can provide. To satisfy these potentially conflicting demands, we have decided to complement a general overview such as that in chapter 7 with more intensive studies of particular topics or key events.

The events we have selected for closer analysis highlight different aspects of CERN itself. In the first three sections the bias is towards political and administrative matters—we discuss the debate surrounding the choice of Geneva as the site for CERN, two clauses in the Convention (one specifying the research activities of the laboratory, the other the conditions for admitting new states) and, in section 8.3, the steps taken to finance the interim organization. Thereafter we analyze three issues intimately associated with the scientific life of CERN, namely, the nomination of the first Director-General, the decision to adopt the strong-focusing principle and its early consequences, and the preliminary steps taken to plan the laboratory in which the two accelerators would be housed.

8.1 The choice of a site

One of the most pressing tasks facing the new Council was to choose a site for the European laboratory. As early as December 1950 a set of criteria for selecting such a site, drawn up very much with Geneva in mind, had been formulated at a meeting at the Centre Européen de la Culture held in that city. Subsequently Bohr and Kramers argued forcibly that the laboratory should be associated with the Institute of Theoretical Physics in Copenhagen. To deflect them, Auger successfully insisted that the matter of location be settled after the new Council had been set up. By the time the Council first met in May 1952 further procrastination of this delicate and contentious issue was no longer possible or desirable. In what follows we shall first discuss the circumstances surrounding the decision to locate CERN in Geneva, and then describe the subsequent reactions to it, both in Geneva itself and in some Scandinavian circles.

8.1.1 THE DECISION TO LOCATE CERN IN GENEVA

When the Council met in May only one official offer of a site had been made—that of Geneva. Other governments wishing to propose a location were given until 31 July 1952 to do so, by which time three more offers had been made—Copenhagen by Denmark, Arnhem by the Netherlands and Longjumeau, just outside Paris, by France. Each proposal was accompanied by a reply to a detailed technical questionnaire drawn up by the Laboratory Group.[2]

The decision on the site was taken at the third Council session in Amsterdam held from 4-7 October 1952. The debate was opened on the first morning by Lew Kowarski, who presented the results of his questionnaire. That afternoon the representatives from each of the four countries involved were invited to put their cases, and on the following day all of the delegates visited the site at Arnhem. On the morning of 6 October the debate began in earnest.

Though we do not have detailed information on the course of the proceedings, one participant has described the choice of location as involving an 'amazing fight'.[3] Patrick Blackett and Donald Fry, who attended as observers of the Royal Society (UK), reported that the delegates from the four countries offering a site 'had clearly been officially briefed to make a stiff fight for locating the laboratory in their country. The scientific prestige attached to it was clearly rated very high; also perhaps the expectation of an appreciable net financial gain' through industrial contracts and technological spin-off.[4] In the event, according to the minutes of the third Council session, 'the French, Danish and Dutch delegations successively withdrew their offers in order to permit the Council to take a unanimous decision' in favour of Geneva.

There is little doubt that this 'decision' was primarily the outcome of hard bargaining; certainly the findings of the questionnaire played little or no role in it. Kowarski himself concluded his report on the survey of the various candidates with the observation that 'every one of the four proposed locations is acceptable on narrowly technical grounds'.[5] This left considerable scope for the interplay of political forces. In what follows, it is these that we shall first explore, discussing the reasons behind the successive withdrawal of Copenhagen, Paris, and Arnhem. For this study, our written sources have been supplemented by observations made by some of the participants at a meeting of pioneers held at CERN in September 1984.[6]

It appears that *Copenhagen* was rather quickly eliminated from consideration as a site for the new laboratory. For one thing it was deemed too far north, too remote from the centre of gravity of western Europe as it was put. For another the younger members of the Executive Group were strongly opposed to having a close association with Bohr, fearing that he would play a dominant role if the laboratory were in Copenhagen. They were also unhappy about having to move with their families to Denmark. On the other hand, they were not against using Bohr's institute as a temporary centre for European theoretical work. Herein lay the basis for a compromise. Though the main laboratory was not to be in Copenhagen one of CERN's important activities would be located there for the time being. This helped to defuse the debate around Copenhagen and removed it from serious consideration as a possible site.

Although Britain took no official part in the debate, her attitude may also have played a role in discouraging the Scandinavians from pressing their case. An official brief prepared for Blackett after discussion with Lord Cherwell, the Prime Minister's closest scientific advisor, stated that, if he were asked about the site, he 'should say that he is not informed of any views which H.M. Government may have on this'.[7] At the same time, on 3 October, the eve of the third Council session, Chadwick wrote to Bohr to say that Cherwell was 'strongly against location in Denmark on the grounds that Denmark was too vulnerable to Russia'.[8] This view, Chadwick felt, was likely to be shared by the British government as a whole. Granted the psychological and material importance which British support had played in Bohr's campaign, this information may well have discouraged him and his delegation even more.

The complex political circumstances surrounding the offer, and subsequent withdrawal of *Paris* will be discussed by Dominique Pestre in section 9.4.2. He shows that, while of course there were some French officials who were determined to locate the laboratory near Paris, those more in touch with the actual forces at work in the Council recognized that this would be difficult to achieve. There was a strong feeling among a majority of states that the laboratory should be located in a small country, for fear that a larger, more powerful host could take an unfair advantage of a facility of this type on its soil. Thus it was argued that, though the French should plead the cause of Paris, they should be prepared to withdraw it if the climate of opinion seemed hostile. As an alternative France could then support Geneva, with whom she shared a border, a language, and a culture.

With Copenhagen and Paris out of the reckoning, only *Geneva* and *Arnhem* remained. It was on the choice between these two sites in particular that feelings ran high, sparked by deep-seated political alliances. Italy had long since declared her support for Switzerland, and remained unwavering. France had also 'retreated' to defend Geneva. For Arnhem, the growing links between the Scandinavians and the Dutch in the nuclear field, coupled with geographic proximity, lead one to suspect that they would have supported it as the next best thing to Copenhagen. French sources suggest that both Germany and Britain also favoured Arnhem, the latter with enthusiasm, though their claims were certainly exaggerated.[9]

Surveying the above account, we see that the alliances surrounding the choice of a site reflect that regional pattern we have encountered before. On the one hand, we have the southern European 'latin' countries, France, Italy and, for this purpose, Switzerland, who favoured a location which was both culturally and geographically close to themselves. These were also three of the earliest and most enthusiastic proponents of the entire venture. Indeed France, Italy, and Belgium (which also had strong French ties) provided the money which enabled UNESCO to launch a feasibility study in the first place. On the other hand, we have the Scandinavians and possibly the Germans and the British preferring a more northern location. In the event a divisive and ultimately paralyzing rift between these two regional factions was averted. As reported by Bannier 30 years later it became clear to him during the debate 'that if the fight went on between Arnhem and Geneva at least the French delegation would retire from the whole project [...]. If I insisted on Arnhem it might have invalidated the whole project'.[10] Placing the interests of 'the whole project' first, Bannier withdrew Arnhem and the Council unanimously 'decided' in favour of Geneva.

Apart from these political considerations, there were arguments of a general kind in Geneva's favour.[11] There was the fact, already mentioned, that it was situated in a small country. There was the advantage of Swiss neutrality. This was particularly important for a lay public still sensitive to collaboration with Germany, and doubly so in the nuclear field. There was the Genevan tradition of hosting international organizations, going back to the time of the League of Nations. The proponents of the laboratory were keen to have it granted international status. While all the states offering sites could no doubt have arranged for this, the experience and familiarity of the Swiss with the requisite procedures probably attracted the delegates. There was the centrality of the city, more or less equidistant from Brussels and Frankfurt, from London and Rome, from Belgrade and Copenhagen. There was the ready availability of many sophisticated mechanical workshops and of a skilled labour force able to speak one of the more common European languages. Finally, there were the living conditions in Geneva—the convenience of good rail and air connections, the presence of an international school and a university with a physics institute keen to co-operate, and the natural beauty of the environment. While such considerations certainly did not determine the choice of Geneva as site, they undoubtedly helped to tip the scales in its favour.

8.1.2 THE LOCAL OPPOSITION TO THE SITE IN GENEVA

No sooner had the decision been taken to locate the laboratory in Geneva, when a vigorous campaign against it was launched in the canton itself.[12] It was spearheaded by the Communist Party, though supported by a far more representative section of the population, including big business and the Red Cross! In what follows we shall restrict ourselves to the objections raised through the columns of the local press and

Notes: p. 285

in debates in the *Grand Conseil*. The Communists quickly decided to sponsor an 'Initiative prohibiting the establishment of any nuclear physics institute in the canton of Geneva', and by early February 1953 had mustered a sufficient number of signatures (well over 7,000) to force a referendum on the issue.

The Communists based their opposition on two main arguments. Firstly, they suggested that the accelerator would endanger the health of the local population. Secondly, they argued that the presence of the laboratory in Switzerland would undermine her neutrality and embroil her in war. Drawing attention to the recent history of the discipline, the Party pointed out that the nuclear research to be done at CERN might have 'indirect' military applications. They mentioned the important role which the American Nobel prizewinner Isidor I. Rabi had played in the birth of the organization, and noted that many of the member states were part of the Atlantic Pact. The Communists also made much of a report in Britain's *Economist* which remarked that 'Switzerland's role [in accepting the laboratory] may be a new departure in that country's attitude towards European integration and defence'.[13]

The first of these arguments was dealt with by pointing out that many accelerators were already located in urban areas including one in Zurich hospital, without any adverse effects. The second argument was less easily disposed of. Indeed in suggesting that Swiss neutrality was endangered the Communists had touched a raw nerve.

The issues were put rather well in a confidential letter to his Foreign Office from an official in the British Legation in Berne.[14] He admitted that, 'in spite of all the explanations offered, I have not been able entirely to clear my own mind of doubts as to whether Switzerland's collaboration with a number of powers of whom only one, Sweden, can be considered as neutral *vis-à-vis* the Soviet Union will not ultimately (if quite unreasonably) prejudice her neutrality'. And would the laboratory at Geneva not contribute, if only indirectly, to the production of atomic bombs? Would the existence of such a valuable plant on Swiss soil not provoke a Russian invasion of her territory in time of war?

In the light of such fundamental concerns, it is not surprising that the savants and university intellectuals who led the counter-attack, along with officials in the canton (one of whom was a Swiss delegate to the CERN Council), found themselves 'soir après soir' addressing public meetings 'which raised passionate discussions'.[15] Their main arguments were two. They stressed that the laboratory would be devoted to pure research. Its results would be openly published, and were of no military concern. And they pointed out that the organization was an essentially European one, that its conception had developed under the auspices of an international body, UNESCO, and that it was intended precisely to *free* the continent from its dependence on the United States in this field of research.

Voting on the referendum took place on the 27-28 June 1953, just before the Convention setting up CERN was due to be signed. Some 24,000 people, almost 40% of those eligible to do so, cast their vote. About two-thirds of them were in favour of locating the laboratory in Geneva.[16]

The choice of a site

However one assesses the arguments in this dispute, the anxieties they revealed certainly had an impact on the drafting of the Convention. The first Swiss amendment submitted on 30 December 1952 insisted that it be specified that the laboratory only undertake pure scientific research, and asked that a new paragraph be inserted in the first version of the Convention stating that 'discoveries made at the Laboratory shall be published'. She also asked that the reference to it in Article II as a 'regional European' laboratory be deleted, for this 'may give the impression that it concerns only a part of Europe'.[17] As we shall see in section 8.2.2, this was simply the first of several determined Swiss attempts to protect her neutrality by ensuring that no clauses in the Convention could be interpreted as selectively excluding eastern European countries from CERN membership.

8.1.3 WINNING SOMETHING FOR COPENHAGEN

The decision to locate CERN in Geneva was not only given a hostile reception by several local groups. It also provoked some of the Scandinavian countries who had been angling to have the laboratory for about a year, and were now determined to retrieve something for Copenhagen from the ashes of defeat.

To this end Funke, the secretary of the Swedish Atomic Energy Committee, and Swedish CERN Council delegate Jacobsson approached Bohr and Amaldi around Christmas 1952.[18] They were informed that, before taking a final decision on whether or not to join CERN, Sweden would like more information about 'the future development in Copenhagen of the work of the Theoretical Study Group'. It was of the 'utmost importance' to her that 'plans are prepared for the coming organization of the work which should secure it the necessary stability and continuity'.

At the fourth Council session in January 1953, the Swedes proposed that there be two theory groups, one in Geneva and one in Copenhagen. The Council, they felt, should make a firm commitment to support the latter for five years, and should then take a definite decision about the long-term status of Bohr's group.

The Council minutes record that this proposal was supported by the German, Norwegian, and Yugoslavian delegations. (The original draft of the minutes also recorded Danish support for the idea, but this was removed at the request of Nielsen, who claimed that his delegation had 'taken no action in connection with the Swedish proposal'.[19]) Most other delegations were less enthusiastic. Indeed it seems that the disagreement on this issue was both divisive and emotive, and no decision on it was taken at the fourth session. In what follows we shall analyze the roots of this division, though our documents limit us to discussing the complexities of the Danish attitude to the scheme, and the opposition to it in the Executive Group and in some quarters in Britain and France.

The attitudes in the Danish camp were of course strongly influenced by Bohr's position during 1952. Needless to say Bohr's wish to have the new laboratory closely

Notes: p. 286

associated with his Institute for Theoretical Physics in Copenhagen persisted despite the forces ranged against him in the provisional organization. To this end he took a number of steps designed to secure, at the very least, an important role for Copenhagen in the life of the new centre. To mention five:

- in June 1952 Bohr unsuccessfully suggested changing his group's name to 'Group for Studies and Co-operation' — a move which Kowarski for one saw as opening the way to locating *all* forms of co-operation, including accelerator construction, in Copenhagen;[20]
- early in September Bohr told Council chairman Paul Scherrer that he would like to 'have the organization center [for] cooperation in european physics' in Copenhagen, and that he would seek an 'official mandate' to do so at the third Council session in October.[21] He elaborated on the idea in a letter written to Chadwick around the same time. It would, wrote Bohr, probably 'be necessary to set up a more permanent organization especially entrusted with the promotion of co-operation'. Its work would 'in some way be a natural continuation' of the activities of his Theoretical Studies Group, and 'though an integral part of the common endeavour', 'it ought perhaps rather be considered as a supplement to the Laboratory than as a department of it';[22]
- throughout 1952 Bohr tried to get Chadwick — at once the most prominent British nuclear physicist and (for much of the year) the most hostile to the idea of equipping Europe with a big bevatron — to play an active role in the provisional organization. He hoped that Sir James would 'act as a main link in our co-operation with the British'. Just before the first Council session he telegrammed 'YOUR PARTICIPATION PARIS EVEN SHORT PERIOD MOST VALUABLE'. A month before the third Council session he wrote at length, explaining the state of play, stressing that it was 'very important that [Chadwick] take part in the beginning of the meeting where surely opportunities for expressing general views will be given', and asking him to come to Amsterdam for a tête-à-tête a few days before the Council meeting;[23]
- at the end of September 1952 Bohr had a long discussion with Alexander King, a high-ranking official in the British Department of Scientific and Industrial Research who was one of those responsible for defining his country's official line on the European laboratory project. Expressing his views 'very forcibly', Bohr pointed out that the new organization's 'general form and shape' were still not settled, and encouraged the British to persist with the idea that, rather than having the contributions to the laboratory's budget go to a common fund, each project be financed separately, 'only those countries interested in the project being expected to pay for it'.[24] Was Bohr suggesting that British influence — and money — be directed towards setting up the hoped-for 'supplement' to the main laboratory in Copenhagen?
- with the discovery of the strong-focusing principle a member of Bohr's group immediately built a simple model to test the principle. On its own this means very

little of course. However in November 1952 the British physicist Jim Cassels visited Bohr's institute. He wrote that the people there 'were talking about building a 600 MeV P.S.', adding that what he called 'this private project' explained 'the interest being taken here in strong focusing'. Was Bohr's group seriously thinking of building a small strong-focusing machine of the same energy as the synchro-cyclotron to be built in the Geneva laboratory?[25]

Taken together these several elements suggest that, while Bohr and his group may not have had a particularly clear view of what they wanted to do in 1952, they continued to harbour the idea of developing Copenhagen as a major centre for European co-operation in nuclear physics loosely affiliated with, but relatively autonomous from, the main European laboratory project. According to Cassels, they were supported in this by Danish officials. The Danes were particularly concerned about the consequences of Bohr's imminent retirement. His personal influence and prestige attracted the world's best to Copenhagen. It also enabled him to secure considerable funds from the Carlsberg Foundation and the Danish government. No one then at Copenhagen had Bohr's charisma, however. To ensure the long-term future of the Danish capital as a leading centre for theoretical studies the Danes hoped to set up a new institute, administratively distinct from Bohr's and led by another eminent European physicist, for example, Hendrik Casimir or Victor Weisskopf.[26]

We turn now to the opponents of the scheme, beginning with the *Executive Group* (excluding Bohr, of course). According to a confidential French memo, they believed that the main theory group should be shifted to Geneva as soon as accommodation on the site was available for them, and no later than two or at most three years after building had started.[27] In this way they would have at least a year to plan experiments to be done with the synchro-cyclotron, due to give its first beam in year four. If the theory group remained at Copenhagen much after year three it would become difficult to attract its members to Geneva. There was also the danger of division and rivalry between the two centres, with one being devoted to theory, the other to experiment or, more generally, one being the center for 'science' (Copenhagen), and the other for 'machines' (Geneva).

The *French* saw much to recommend the Executive Group's position.[28] *Britain's* attitude was complicated by the history of her solidarity with Bohr, which had to be weighed against the recognition that he had lost some of his leadership qualities, and that it would be divisive to establish a part of CERN permanently in Copenhagen. Thus at the fourth session her two delegates, Sir Ben Lockspeiser and Sir John Cockcroft suggested that the theoretical group 'should stay in Copenhagen for a limited period only which would, in fact, relate to the period of Professor Bohr's active participation, and that subsequently the group should move to Geneva'.[29]

The French and British delegations embodied their ideas in two papers prepared for the sixth Council session.[30] Following the arguments of the Executive Group, the former proposed that to enable the theory group to be in Geneva during the year before the synchro-cyclotron operated, theoreticians in Copenhagen should be given contracts of no more than three years. The British were prepared to support a centre located in the Danish capital for five years, but were careful to say that new premises should not be built for it in this period. The agreement reached at the sixth session in June 1953 fused these two proposals. It was resolved that theoreticians could be offered five-year contracts if that was necessary to attract outstanding personnel, though they had to be prepared to move to Geneva after three years if the Council so decided.

Meanwhile other steps were being taken to secure the future of Bohr's institute. Disappointed by the opposition to their proposal at the fourth session in January, and stripped of effective British support, CERN Council delegates from Norway, Sweden, and Denmark met in Gothenburg on 17 February 1953, with representatives from Iceland and Finland.[31] Their aim was to discuss the possibility of establishing a Scandinavian theoretical institute in Copenhagen to be independent of CERN and funded with Scandinavian money. With that the idea of what was to become NORDITA—the 'Nordisk Institut for Teoretisk Atomfysik'—was born.

8.2 The Convention. Two key clauses

The Agreement constituting the Council of the provisional CERN instructed it to submit to the governments of its members 'the draft of a convention for the establishment of an international laboratory and for the organization of other forms of co-operation for nuclear research'. The first draft of this Convention (document CERN/19) was laid before the third session of the Council in October 1953. In the following months it was subjected to meticulous scrutiny in Britain, in particular, from which emerged a 'United Kingdom Provisional Re-Draft of the Draft Convention', dated 24 December 1952. This document, in fact, provided the basis for the Convention that was finally signed by delegates on 1 July 1953.

Two items in this Convention are of particular interest to us: the distinction between the basic and supplementary scientific research programmes, which forms part of Article II of the Convention dealing with the purposes of the organization, and the procedure for admitting new member states, specifically Article III.2.

8.2.1 THE BASIC AND THE SUPPLEMENTARY PROGRAMMES

The suggestion that one should distinguish between two *kinds* of research programme in the envisaged European laboratory emerged 'officially' from a meeting held at the British Department of Scientific and Industrial Research on 11 December 1952.[32] At this meeting a number of government officials and science administrators discussed some of Britain's reservations about the existing version of the draft Convention (CERN/19) with Edoardo Amaldi, who had been specially invited to London for the purpose. The results of the discussion were summarized in an important memorandum circulated to all member states.[33] Here it was proposed that a distinction be drawn between research projects initially agreed on, and further projects and initiatives to be decided on in the future. Whereas all member states were financially responsible for the former, those who wished to do so could opt out of any subsequent extension of the laboratory's activities.

These ideas were embodied in a number of clauses in Britain's provisional re-draft of 24 December 1952. They were incorporated more or less unchanged in the draft agreement prepared for the Brussels session (January 1953), at which 'our [i.e. British] anxiety about further financial commitments was found to be shared by other countries'.[34] There was thus little controversy around the incorporation of the British idea, and in its subsequent refinement.

The Convention signed on 1 July 1953 distinguished between a basic programme — called an initial programme in the earliest drafts — and a supplementary programme of research.[35] The basic programme made provision for two kinds of activities. Firstly, there was to be an 'International Laboratory [...] for research on high energy particles, including work in the field of cosmic rays'. It was to be equipped with the proton synchrotron and the synchro-cyclotron then being planned, as well as the requisite ancillary apparatus and buildings. Secondly, CERN would be responsible for the 'organization and sponsoring of international cooperation outside the Laboratory'. This covered essentially the activities of the Theoretical Studies Group. Provision was again made for work on cosmic rays.

Any extension of this basic programme, i.e. any supplementary programme of research, required the approval of a two-thirds majority of member states. On the other hand, no member state had an obligation to contribute to a supplementary programme, even if it agreed in principle that such a programme was acceptable. The scale of contributions of those that did participate was calculated in the same way as that used for assessing contributions to the basic programme. However, the proviso that no state should pay more than 25% of the costs of the programme was lifted. A state which was not party to a supplementary programme lost the right to participate in it, though could vote on matters relating to 'facilities to the cost of which it [had] contributed'.

Distinguishing between the basic and supplementary programmes was, and has continued to be, of some importance in the development of CERN. By giving no state the power of veto over an extension of CERN's activities it ensured that new initiatives favoured by a majority did not run aground simply because one member did not wish to take part in a particular supplementary programme. At the same time it protected the interests of such members by freeing them from the need to contribute financially towards such a programme. It thus provided for the future expansion of the organization, while allowing for the uneven level of development of national scientific research effort in its member states.

Notes: p. 286

The British had two specific concerns in launching the initiative which resulted in the distinction we are discussing. They wanted to limit the objectives of the laboratory to research on high-energy particles, and they wanted to ensure that they only paid for what they actually needed at CERN and could not build themselves. Let us elaborate.

(i) Concern in British circles that the laboratory was becoming involved in activities only loosely connected with accelerators was sparked by Cassels during his European trip,[36] and subsequently reinforced by the second report to member states. Cassels visited Kowarski on his way to Copenhagen in November 1952. In Paris he learnt that, in addition to having scientific divisions devoted to the two accelerators and to theory, provision was being made for instrumentation, general physics, and chemistry divisions. More specifically, according to Cassels 'some people in Paris' were talking of equipping the Geneva laboratory with two or three Van de Graaff accelerators, an electron microscope, and an X-ray machine. 'These suggestions for "diversionary activities" ', he wrote to Cockcroft and to Chadwick, 'underline the importance of getting a firm programme into any agreements we sign'.

(ii) The second British concern combined the anxieties just expressed with rather more narrow dictates of national interest. As her physicists would soon have access at home to machines powerful enough to do meson physics, it was the opportunity of working with the proton synchrotron that particularly attracted the British to the collaborative venture. To translate this into policy Alexander King suggested, and his departmental Secretary Lockspeiser agreed, that Britain press the CERN Council to adopt the OEEC procedure whereby countries only contributed to the cost of projects in which they were directly interested themselves—an arrangement which also appealed to Bohr, as we mentioned a short while ago. Hence the idea that 'we should from the outset wish to contract out of certain work in the initial programme—by which, in fact, we meant that we did not want to contribute towards the cost of the 600 MeV machine'.[37] Waterfield went further. In a letter to Mussard the British Scientific Attaché in Paris wrote that, in his view, 'the U.K. is nervous of having to pay its share of what one might call standard low power equipment (e.g. Van de Graaff, small cyclotron, etc.) which would not interest us, though it might seem very important to a country like Yugoslavia which, I believe, has almost nothing at present'.[38]

Both of these matters were discussed with Amaldi at the meeting in London on 11 December 1952. On the first, he confirmed that there were two trends of thought among the Council members, some of them wanting to limit the aims of the organization rather precisely, others 'more in favour of a fairly elastic set of objectives with a prospect of subsequent extension of the Organization's field of interests'. Personally Amaldi sympathized with British fears about the possible 'waste of effort' entailed by the more flexible approach, and thought that many Council members would welcome a more specific and limited definition of the organization's tasks.

On the second point both Amaldi and Cockcroft, who was present at the meeting, argued that it would be unwise for Britain not to share the cost of the smaller accelerator. An alternative measure was proposed to meet British fears that they would be making an open-ended financial commitment by joining CERN: while all signatories to the Convention would have to participate in the initially-agreed programme of the laboratory, individual member states could be allowed to contract out of *subsequent* projects agreed to by the Council.

The British took this advice to heart. Their provisional re-draft of the Convention specified that the laboratory should be for 'research on high energy particles' and, while distinguishing between its initial and supplementary programmes, included both accelerators in the former. As mentioned earlier both proposals met with general support in the Council.

Britain's attitudes, and particularly her wish to pay only for what she really needed, were indicative of her leading position in accelerator physics in Europe, and of her lack of ideological commitment to the collaborative venture. Unlike many participants she did not see the laboratory *both* as satisfying a national scientific need and as contributing to the reconstruction of Europe, and of European physics in particular. Relatively unwilling to make the sacrifices that the latter might entail, her initial approach was one dictated primarily by the desire to give her physicists access to the strong-focusing accelerator which they did not have at home, and did not wish to build themselves.

There is a rather interesting difference to note here between intention and effect. Britain's aim was to opt out of contributing to the costs of relatively standard inexpensive equipment of no particular interest to her. In the subsequent development of CERN, however, the distinction between the 'basic' and the 'supplementary' programmes has played another role.[39] It has allowed states to remain members of the organization without having to contribute financially to large and expensive projects involving qualitative advances in the laboratory's research facilities.

We should like to make one further comment. We have often remarked on the tendency to think anachronistically of CERN as if the scientific aims and equipment of the laboratory were clear from as early as June 1950, and that the main task henceforth was to win political support for the scheme. The material in this section has again emphasized that this was not so. As late as the end of 1952, and indeed well beyond (see section 8.6), there were those who wanted the organization—as its name implied—to devote itself to nuclear physics in the wide sense of the term (even to atomic physics), and not only to high-energy particle physics. This diffuseness of its scientific aims was at least partly a consequence of the diverse scientific needs of the participants, who saw in the laboratory the means for advancing their capability in a range of nuclear-related fields. If the British were able to take the lead in getting the aims of the laboratory precisely specified it was because they were especially aware of what they wanted from it, and of what particular gap in their own research activities it filled.

Notes: p. 287

8.2.2 THE ADMISSION OF NEW STATES

One of the most controversial clauses in the Convention, and the one whose underlying principles were in dispute until just before it was signed, concerned the admission of new member states to the organization, foreseen in Article III.2. In what follows we shall discuss the more important revisions to this article and explore the reasons behind them.

In the first version of the draft Convention (CERN/19) Article III.2 read:

> Other European States may join the Organization, provided that their applications for membership are approved by the Council and that they become parties to this Convention.

It was the British who first questioned this in their provisional re-draft of 24 December 1952,[40] in which they proposed that

> The Council may, by a two-thirds majority of Member States, decide to invite any other European State to join the Organization and any State so invited may thereupon become a member by becoming a party to this Convention.

This version differed from the original in two substantial ways. Whereas in CERN/19 a prospective European member *applied* for admission, now the initiative lay with the Council, which had to *invite* such a state. And whereas before the Council's approval of an application required only a simple majority—as specified in Article V.6 of CERN/19—now a two-thirds majority was required before an invitation could be issued.

At this stage Britain's main concern was to exclude eastern Europe from the organization. Of course the two-thirds majority rule did not guarantee that invitations would not be issued to eastern European countries. On the other hand, Britain could count on French support for this position and it was hoped that their combined influence on the other delegates would be sufficient to muster the required majority.[41]

It was realized that 'the U.K. phraseology [was] likely to upset the Swiss'. They felt that their neutrality would be impugned unless *any* European state was free to apply for membership. There was concern in some quarters about this, but the Foreign Office was emphatic, and 'likely to be extremely difficult—up to Ministerial level' on the principle of invitation.[42]

The British proposal was accepted, apparently without any controversy or much discussion, at the fourth Council session in Brussels on 12–14 January 1953, and was incorporated with one minor technical addition in the draft Convention which emerged from it.[43] Sir Ben Lockspeiser returned home satisfied that the document contained the 'safeguards we were instructed to secure'. And for the time being at least the Swiss, if disturbed, were not showing it.[44]

The British Foreign Office was still not satisfied, however. Already in January they had let it be known that they would like the word 'European' deleted from

Article III.2, thus leaving the Council free to invite *any* state to join if there was a two-thirds majority in favour. At the time of the Brussels session it was felt that 'this need not be made a sticking point'. However, on the eve of the fifth session of the Council, held in Rome from 30 March to 2 April 1953, a new set of British amendments to the draft Convention were produced. These included the proposal that the word 'European' be dropped from Article III.2.[45]

The British had two reasons for this move.[46] The one, stated quite openly, was that they wanted the door to be left open for Commonwealth membership. The second reason why Britain wanted to delete the word 'European' from Article III.2, and one about which she was somewhat more discreet, was the question of the United States.

Britain's main concern here was to protect her 'special relationship' with America. To do so 'we should be able to show to the world that the organization had in fact no political significance as a European body'. A convention which limited membership to European states left just the opposite impression. 'I cannot help feeling', wrote A. Dudley in the Foreign Office, 'that in the minds of some of our own scientists as well as those of some Europeans there is a definitely anti-American bias, and that one of the objectives of establishing a *European* (or in fact Western European) organization is to ensure that work is undertaken on a large scale in Western Europe from which the Americans are excluded. I imagine that we do not want to encourage exclusiveness of this kind', Dudley said. Although it was unlikely that America would, in fact, want to join the organization, he went on, it was against the government's political interests to be party to a convention which *a priori* excluded the United States from membership.

The proposed amendment was strongly opposed by the Swiss in Rome, who tried unsuccessfully to reintroduce the principle of application, and who stressed that the British initiative played into the hands of the Communists and 'neutralists' then waging a campaign against siting the laboratory in Geneva.[47] When it came to the vote, however, the amendment was adopted by seven votes to one (Switzerland) with France and West Germany abstaining.

It was not long before the Swiss fought back. Towards the end of May they circulated an *aide-mémoire* to all member states insisting that there should be no equivocation on the essentially European character of the new organization, and proposing an amendment which again allowed any European state to apply for membership.[48] Among other arguments, they stressed — and the French agreed with them — that to delete the word 'European' from Article III.2 was to vitiate the whole character of the organization. It had been launched by a group of European states who were at roughly the same level of scientific, industrial, and economic development after the war, and who saw in the collaborative venture a means of collectively rebuilding European physics. If the United States, so far ahead of her European allies in these respects, were allowed to join fully in the project, the equilibrium between the other member states would be upset by the 'preponderance of American influence and finance in the organization', as Verry put it.

Notes: p. 287

As the date planned for the signing of the Convention drew closer Britain put forward yet another amendment.[49] This time she suggested that Article III.2 be dropped altogether from the Convention. In effect this meant that the admission of new members required an amendment to the Convention and so unanimity among the member states. This proposal was approved at the sixth Council session in Paris on 29–30 June 1957.[50] To get around the technical problem of having to amend the Convention each time a new member joined, France then advocated including a clause without the word 'European' but enshrining the principle of unanimity. Thus the final version of (the first part of) Article III.2, accepted unanimously, was

> Other States may be admitted to the Organization by the Council [...] by a unanimous decision of Member States.

In a concession to Switzerland Article III.2(b), which described the mechanism for admission, left it to an aspirant to 'notify the Director' if it wished to join.[51]

Article III.2(a) was the best compromise one could hope for under the circumstances. By not excluding *a priori any* state from admission to CERN it enabled Switzerland to preserve her neutrality—for she could always say that eastern European states were free to apply—and it enabled Britain to avoid 'offending' America. At the same time, in its demand for unanimity, the article meant that, in practice, existing members had a veto power over who was allowed to join. In this way the article preserved the appearance of openess while masking the reality of exclusivity.

There were losses too, though. The founders of the laboratory had seen it as being built primarily by Europeans for Europeans. Though they had always kept the concept of Europe deliberately vague,[52] one of their main aims was to restore their continent to its erstwhile glory in physics. For that reason they wanted the laboratory to be a regional one. And while they welcomed the participation of non-European states in its scientific life—after all it was an *international* laboratory—they did not want them to shape its overall policy. For them the deletion of 'European' from Article III.2 was a defeat, and a sad recognition that there was little place for idealism in the world of power politics. As Amaldi wearily remarked early in June 1953 to a correspondent who wrote of his regret about omitting the adjective 'European', he had 'already done all [he] could to try to maintain the original wording of this Article...'. However, Amaldi went on, he did 'not entertain any too serious hope of a draw-back to the original idea'.[53]

8.3 Financing the interim organization

Our aim in this section is to explore the measures taken by the representatives of the member states to ensure the uninterrupted development of the European laboratory project while waiting for the Convention to be ratified. What makes the topic particularly interesting is that the provisional Council delegates were faced with an unexpected dilemma as 1954 wore on. Just before signing the Convention on 1 July 1953, the delegates adopted a Supplementary Agreement prolonging the life of the provisional CERN, and making financial arrangements for its continued wellbeing.[54] Enough money was set aside for some six or seven months, the time they estimated that it would take for the Convention to come into force. However, it soon became clear that this estimate was hopelessly optimistic. And, as time passed, the financial demands of the interim organization steadily escalated. The task of the Council and the Interim Finance Committee (IFC) was to raise funds from member countries using an agreement which had been signed on the assumption that it would be short-lived, that it entailed a limited commitment to the permanent CERN, and that it involved no heavy financial engagements. The steps they took to deal with this situation throw considerable light on how the decision-making mechanisms of the interim CERN worked, on the attitudes adopted by the representatives of member states to the scientists involved in the day-to-day management of CERN, and on the determination of all to push ahead with the European laboratory project come what may.

8.3.1 THE FIRST REQUEST FOR FUNDS

Article II.1 of the Supplementary Agreement provided for the funding of the interim body 'until the thirty-first of January 1954'. The precise amount called for was specified in an annex to the Agreement. Here the signatories were asked to contribute a further 756,500 SF to the Council's work. This sum was distributed proportionately between the ten member states who had signed, using the scale provided in the Convention as a basis. The figure was fixed in the light of the needs of the organization as spelt out by the Executive Group, and was adjusted to meet the peculiarities of the British position.

Essentially the Executive Group gave the Council the choice of calling for 0.5 MSF or for 1 MSF to fund the interim CERN for the first six months of its life.[55] The former was 'the minimum sum necessary for CERN to remain in existence until the 31st of December, 1953'. 1 MSF, however, was 'the sum necessary to make an adequate use of the possibilities offered by CERN'.

Here the Executive Group had two projects in mind. They wanted the Proton Synchrotron (PS) Group to be transferred to Geneva as soon as possible, and a supporting administrative nucleus to be established there. This was essential if the work of the group was to proceed in the most efficient and rapid manner possible. In addition they wanted to employ an architect in August 1953 to start on the detailed architectural planning and structural engineering design. This, they felt, was necessary to provide accommodation in time for the scientific activity required before and after the start-up of the synchro-cyclotron.

The difference between the 1 MSF required for these projects and the 756,500 SF actually stipulated in the annex to the Supplementary Agreement is accounted for by the special circumstances of the British.[56] The new agreement could only be signed by states party to the original agreement of February 1952, which established the provisional CERN. The United Kingdom was thus formally excluded from it. However, Britain let it be known that she would voluntarily pay her share of the total funds raised if the other signatories accepted their financial responsibilities. For the initial period this amounted to 243,500 SF, the same as the French share; 243,500 SF + 756,500 SF = 1 MSF.

The Executive Group advised the Council that the choice between raising 0.5 MSF and 1 MSF for the period until January 1954 was not simply a financial one; it also involved a decision of principle. If it was agreed that the PS Group, then dispersed among the home stations of its members, be moved to Geneva, its staff would have to be given some prospect of job security. If an architect was employed in August 1953, his designs would obviously influence the architectural policy of the permanent CERN. In short, as the Executive Group pointed out, in approving its proposals and the 1 MSF required to implement them, the Council would to some extent commit the future organization. As it was generally understood that steps of this kind were needed during the latter half of 1953, the delegates who signed the Supplementary Agreement were not unduly disturbed by this at the time.

8.3.2 RAISING FURTHER FUNDS 'TO CONTINUE REASONABLE PROGRESS WITH THE PLANNING WORK'

Our aim in this section is to discuss the measures taken at the eighth and ninth Council sessions to raise additional funds under the Supplementary Agreement to secure the 'reasonable progress' of the interim organization. Perhaps the most striking feature of the process is the *relative ease* with which the additional resources called for were agreed to. In what follows we shall first briefly discuss the steps taken to raise the funds, and then give some reasons why they were obtained with little difficulty.

The *eighth session* (14 January 1954) was the first at which the delegates had to face the fact that the Convention would not come into force by the end of January 1954. The ground for raising the additional funds required by the interim organization was laid at a meeting of the IFC on 16 December 1953. All member states, and not simply those officially represented on the committee, were invited to send delegates. The principal business of the meeting was 'to consider the present financial position, the estimated expenditure during the life of the interim organization, and the method of finding the necessary funds'.[57]

The IFC had before it detailed estimates of group expenditures for December 1953 to February 1954. It appeared that there were adequate resources in hand until the end of January to cover 'the reasonable continuance of the work at the general level already agreed by the Council'. Thereafter some 880,000 SF would be required for

February–April 1954. This included a special request for 100,000 SF by the PS group for test apparatus and essential stores.

Article II.2 of the Supplementary Agreement provided the method of finding the necessary funds. It stipulated that, in the event of the interim Council continuing in existence after 31 January 1954, it 'may from time to time recommend further financial measures to enable it to carry on its work prior to the entry into force of the Convention [...]'. Delegates were asked to secure the necessary resources from their governments, and at the next meeting of the IFC, on the eve of the eighth session, all twelve states who had signed the Convention agreed to accept the new burden.[58] Several of them were willing to pay their shares at the beginning of February. The Netherlands was even ready 'to pay, if necessary, forthwith'.

The spadework was now done, and when the eighth session of the Council met on 14 January 1954, it unanimously accepted the recommendations of the IFC. In the light of the ratification situation it was thought prudent to assume that the Convention would not come into force before 30 April. To cover February, March, and April 1954, 670,000 SF were raised under Article II.2 of the Supplementary Agreement. Britain agreed to contribute her share of 214,300 SF making the total levied for the three months 884,300 SF.

The rate of development of the ratification situation during the first three months of 1954 was such that additional funds had to be raised at the *ninth session* of Council in April 1954. When it met to review matters on 9 March the IFC was confronted with a request for a further increase in the three-monthly budget.[59] This was primarily due to 'necessary increases in staff and equipment for the PS Group to the extent of 100,000 SF per month' and, related to this, an increase of 20,000 SF per month in the Laboratory Group's budget 'to meet the extra administrative load due to the increased P.S. budget'. Thus the total amount needed was estimated to be about 880,000 SF + 3 × 120,000 SF = 1.24 MSF.

The IFC duly recommended that the Council approve a resolution calling for 945,000 SF for May–July 1954. The Council obliged at its ninth session. Britain thereupon agreed to add her share of 304,900 SF, bringing the total for the three months in question to 1.25 MSF.

The only difficulty raised by these requests for money in 1954 was caused by the special situation in France.[60] Although her contributions to CERN were provided for in the national budget presented in 1953, it was stipulated that these funds be frozen until the Convention was ratified by the French parliament. In the event this only occurred around mid-1954. To cover all the French contributions promised under the Supplementary Agreement two loans of 250,000 SF each, guaranteed by the Swiss government, were arranged.[61]

Apart from this, all apparently went smoothly. Informed in advance by the IFC of the additional financial needs of the interim organization, delegates to the eighth and ninth sessions were empowered by their governments to accept their share of the burden.[62] This authority was independent of whether or not the country in question had ratified the Convention. It was also not affected by the fact that the amounts called for were higher than a figure of about 800,000 SF per three months originally estimated by the Executive Group as needed for the period after January 1954.[63]

We would suggest several reasons why these additional funds, totalling some 2 MSF, were raised so easily:

(i) The delegates had *little option* but to agree to the additional money. In accepting a budget of 1 MSF for the first six or seven months of the interim CERN they had given the go-ahead for the PS Group to move to Geneva, for an architect to be employed to start designing the laboratory, and so on. Having once accepted that it was within their powers to authorize such moves, they could now hardly take steps to jeopardize them. The chairman of the IFC warned delegates on the eve of the eighth Council session that unless they were prepared to pay the additional funds called for, and to pay them rather quickly, 'it might well be necessary to stop all preparatory work until funds were available'.[64] Given the prevailing definition of 'preparatory work', to ensure its 'reasonable progress' one could not but accept the new financial burden.

(ii) The funds called for were *far less* than delegates and their governments had originally expected to have to pay in 1954. When they signed the Convention they assumed that, during the following year, they would have to pay their share of the annual budget of the permanent organization, which was some 17 MSF. Forewarned, they had little difficulty in accepting that only 2 MSF were needed to keep the organization going until July 1954.

(iii) Another factor facilitating the payment of additonal funds under the Supplementary Agreement was the *source of money* in some countries. It is difficult to generalize here. However, we do know that in the case of Italy, the Netherlands, and Sweden the funds were disbursed by scientific bodies—by the National Committee for Nuclear Research in Italy, by the Dutch Organization for the Advancement of Pure Research (at least in 1952 and 1953), and by the Atomkommittén in Sweden. It was thus a relatively easy matter for their delegates to be empowered to pay their shares demanded at the eighth and ninth sessions: the procedure was purely administrative and not subject to the vagaries and delays of the ongoing political process.

The situation in France provides a useful contrast here. As it happened her contributions to CERN had to be paid by the Foreign Ministry. Thus they were subject to a rather more formal set of controls than if they had been in the hands of purely a scientific committee. In particular, as we have seen, it was stipulated that France's contribution only be released after her parliament had ratified the Convention.

(iv) There was the *possibility* that some or all of the money paid under the Supplementary Agreement could be *deducted* from the member states' contribution to the permanent organization. The situation is confused here, because the policy on this underwent several changes, and was still not clear in mid-1954.

The early official position, as stated by Amaldi in his letter of invitation to the sixth Council session, was that sums paid under the Supplementary Agreement could be regarded as if they were contributions to the new organization.[65] 'They represent therefore no additional charge on member states', he said. Subsequently he wrote that this proposal was not approved by the Council at its sixth session, so that a payment made to the interim CERN was to be 'considered as a payment completely independent from the contribution to be paid under the Convention'.

The Swedes, in particular, were not convinced by this, and continued to pursue the issue in 1954. The correspondence that we have suggests that there was, in fact, no clear official position on it for some time.[66] Finally early in November 1954 the Chief Administrative and Finance Officer of the permanent organization agreed that all the contributions Sweden had made hitherto under the Supplementary Agreement could be regarded as contributions to 'the definite CERN'.[67]

8.3.3 TAKING FURTHER MEASURES TO AVOID 'SERIOUS DELAYS IN THE ORDERLY AND ECONOMICAL PROGRESS OF THE WORK'

In addition to dealing with the more or less straightforward request for 1.25 MSF to cover the period up to the end of July, the ninth session of the Council had to confront a crucial question of policy concerning the future rate of development of its work. Specifically it was called on to decide whether foundation and excavation work on the site in Meyrin could commence before the Convention was ratified.

The official proposal to follow this course of action emerged from an Executive Group meeting held just before the ninth session. It was communicated by Edoardo Amaldi to Sir Ben Lockspeiser, the chairman of the IFC, in a letter dated 7 April 1954.[68] The letter asked that the IFC, which met that day, 'consider the financial possibilities of starting the construction on the site next May'. At the meeting the chairman felt that 'this proposal raised a most important matter of policy which went quite beyond the powers of the Interim Finance Committee to make recommendations [...]. The whole matter must therefore be discussed in the Council on the question of principle'.[69] This it was at the session which opened the next day.

The basic argument was that, if one wanted to have the buildings ready for the big machines at the beginning of 1957, one should start work on the site well before the Swiss winter set in. It was claimed that a delay of a few months at this stage could result in the whole construction programme being thrown back by as much as a year due to bad weather conditions. All group activities would be delayed accordingly, with serious consequences both for the technical programmes and for staff recruitment and morale.

The Laboratory Group had recommended that work be started on the roads, the synchro-cyclotron building, the power station, and the workshop.[70] At the ninth session Lew Kowarski reported that this programme would cost some 200,000 SF for the first three months, and much larger sums thereafter. The whole programme

Notes: p. 288

involved a commitment of some 5.5 MSF. Lockspeiser thought that the 'big financial risk' involved in this 'most hazardous course' was 'difficult to justify at the moment'. However, he felt that a solution should be found, and undertook to have a personal interview with the architect, R. Steiger, forthwith. He reported on the results of his deliberations on the second day of the Council session.

Steiger confirmed that if foundation work did not start shortly bad weather could well force a postponement of everything to the following spring. He was however prepared to accept a limited contract for 100,000 SF up to the end of June. Thereafter the situation would have to be reviewed by the IFC, since further progress would require spending some 0.5 MSF by the end of September.

The Council accepted these proposals. The initial amounts requested by Steiger were relatively small, and on a par with what he had been paid previously. Clearly, then, he was not being authorized to make a qualitative change in his level of activities. At the same time, to avoid reconvening before the Convention entered into force, the Council granted 'full powers to the IFC in order to enable them to take such decisions as they might deem necessary about work on the site'.

All member states were represented at the next meeting of the IFC held in London on 23-24 June 1954.[71] A survey of the ratification situation led the committee to consider what steps had to be taken to finance the organization for a further three months. Some 2.1 MSF were needed to keep it going at the 'present level' from August to October. The delegates readily agreed to provide this money. To proceed with work on the site, however, which had started in a modest way on 17 May, an additional 2.9 MSF were required. On this there were 'strong differences of opinion about how far it was legitimate or wise to incur expenditure which much exceeded interim planning, particularly while the date of entry into force was still problematical'.

Given her somewhat embarrassing position, France apparently did not take part in this debate. However, the other two countries who had strongly backed the project from the start—Belgium and Italy—along with Greece, declared themselves willing to proceed forthwith. They were in the minority, however. Most other delegates felt that they did not have the authority to commit their governments to expenditure beyond that required for 'preliminary work', and that any such authority was conditional on full ratification of the Convention. At the same time all felt the urgency of pressing on with construction work. In the light of the ratification position an ingenious compromise was devised, and embodied in a resolution adopted by the IFC after an overnight adjournment.

The resolution distinguished carefully between two contributions, the one distributed between the member states as shown in its 'Column A', and the other as given in 'Column B'. Contributions made under Column A were towards the 2.1 MSF required to continue with preparatory work. They were called for in the usual manner. Column B distributed 'an additional sum of 2,900,000 SF necessary to meet the cost of excavation and work on the foundations during the months of August, September and October [...]'. This money was needed 'to avoid serious delays in the orderly and economical progress of the work'. Delegates were asked to agree to provide these resources if the following conditions were met:

(a) on ratification by all states other than France, or

(b) on ratification by Germany and on a vote in favour of ratification by the National Assembly of France, and

(c) provided that by 15 July no formal objection had been made by any state.

What was the reasoning behind points (a) and (b) of this formula? It will be remembered that the financial requirement for the Convention to come into force was that it be ratified by seven states whose combined contribution amounted to at least 75% of the CERN budget. Conditions (a) and (b) represented two ways of effectively achieving this requirement.

Condition (a) is easily understood. It was a direct consequence of the fact that France's share was 23.84% of the total. Condition (b) was devised in the light of the ratification situation prevailing at the time. Five states—the United Kingdom, Switzerland, Denmark, the Netherlands, and Greece—had ratified when the IFC met. Their total contribution was 34.68%. If France and Germany's shares were added one had seven states—including the three major financial contributors—whose combined share was 76.22%.

The date of 15 July was chosen in the light of reports by the delegates of France and Germany on the state of the ratification process. Robert Valeur, the Council chairman, ascertained that the matter was on the agenda of the French National Assembly for 6 July, where a favourable vote was expected.[72] Alexander Hocker, for Germany, reported that his country would almost certainly ratify between 15 and 25 July.

On 15 July the chairman of Council and the secretary of the IFC met to review matters.[73] None of the conditions laid down by the IFC three weeks before had been fully satisfied. As for (a) only six of the dozen member states had formally ratified the Convention. Concerning (b) the condition governing the French vote had been satisfied on 6 July, but (full) ratification by Germany was some way off yet; the Bundestag had only just adopted the bill of ratification (on 8 July). There were also uncertainties surrounding condition (c). In a letter dated 8 July 1954 the Swiss objected to financing important construction work before the go-ahead had been given by the Council, though let it be known the following week that they would not press the matter.[74] Undeterred, on 20 July 1954 Amaldi wrote to the member states asking that the funds called for in Column B be transferred to CERN's account 'at the earliest opportunity after the beginning of August, thus enabling CERN to proceed with important excavation and foundation work before winter'.[75] By the end of September 1.8 MSF of the additional 2.9 MSF were in the bank.

Notes: p. 288

8.3.4 CONCLUDING REMARKS

Table 8.1 surveys the funds raised by the Council and spent by three of its study groups from May 1952 until the permanent organization came into being. It reveals the extent of the transformation in the rate of expenditure on CERN during 1954 — first doubling, then increasing by a factor of five, so that we have a ten-fold increase between January and September. Indeed for the last three months of the interim period the monthly amounts called for (about 1.7 MSF) were more or less those budgetted for the permanent body (17 MSF per year, so about 1.4 MSF per month).

Table 8.1
Summary of (a) total funds pledged to the Council from May 1952 to September 1954, and (b) the amounts actually spent by three of its groups.

(a) Council's budgetary measures[76] Time period	Total (SF)		Average expenditure per month (× 1000 SF)
Pledged to cover May '52–June '53	1,075,399		77
Called for to cover July '53–Jan. '54	1,000,000		143
Called for to cover Feb. '54–Apr. '54	884,300		295
Called for to cover May '54–July '54	1,250,000		417
Called for to cover Aug. '54–Oct. '54	2,100,000		700
	2,900,000		967

(b) Amounts actually spent (SF)[77] Time period	PS	SC	Lab	Total = PS + SC + Lab
From May 1952 – May 1953	146,191	127,617	96,115	369,923
From June – December 1953	336,930	119,062	260,555	716,547
From January – September 1954	998,000	165,000	756,000	1,919,000
Total	1,481,121	411,679	1,112,670	3,005,470

The difference between the amounts raised and spent is largely absorbed in forward commitments from the period 1/10/54 and in the budgets of the Theory Group and the Directorate. The amounts actually spent from January – September 1954 include an estimate for 1/7/54 – 30/9/54.

In every respect then — from the financial angle, and from the point of the development of the scientific and construction programmes — the interim CERN merged imperceptibly with the permanent organization. The credit for achieving this smooth transition lies with the delegates to the provisional Council and the Interim Finance Committee. Always responsive to the needs of the scientists, never doubting that the Convention would be ratified, they were prepared to stretch to the limit what was meant by ensuring 'reasonable progress with the planning work'. Called on to finance a transitional phase with an ever-receding terminus, they found ways of satisfying the escalating demands made of them without seriously impeding the work of the study groups or losing the confidence of the governments of the member states. Their success in doing so was partly due to structural reasons: for one thing most European countries had not yet institutionalized science policy, so that high-ranking officials in the state apparatus had considerable room for manœuvre. But it was also due to their own determination, to their conviction that CERN was worth fighting for, to the hopes that the project engendered, to the possibilities that it embodied. Together with the scientists these delegates presented a united front to the member states. Together they launched the permanent CERN.[78]

8.4 The nomination of the first Director-General and its aftermath

The Convention signed in Paris on 1 July 1953 empowered the Council of (the permanent) CERN to appoint a director to be 'the chief executive officer of the organization and its legal representative'. The Convention did not specify how long his term of office should be. This was left to the discretion of the Council, which could also decide to postpone his appointment 'for such period as it considers necessary'.

Despite this latter provision it was generally agreed that it was preferable for a director to be appointed as soon as possible after the Convention came into force. The post was a crucial one, both administratively within the organization, and in the relationship between CERN and its member states. In addition there were those, like Sir John Cockcroft, who thought that without a director to manage the laboratory from the start the Executive Group would rapidly take control of its overall development.[79]

It was initially suggested that nominations for the directorship be called for from the governments of member states. The IFC, however, felt that this would politicize matters unnecessarily. The appointment was a scientific one. It was therefore preferable that it be handled by the Nominations Committee. This was a predominantly scientific body, chaired by Francis Perrin, the other members being Niels Bohr, John Cockcroft, Gustavo Colonnetti, Werner Heisenberg, and Paul Scherrer.

After a long and complex process of negotiation the Nominations Committee put forward the name of a candidate for the post of director. This was approved by the interim Council at its ninth session on 8-9 April 1954, and confirmed by the Council of the permanent organization when it met in October. Our aim in this section is to describe the steps which led up to his appointment, to discuss some of its consequences, and to explain why it was such an unhappy choice.[80]

Notes: p. 288

8.4.1 THE FIRST INITIATIVES

As early as May 1953 three candidates were already being seriously considered for the top post at CERN—Edoardo Amaldi, the Secretary-General of the provisional organization, Patrick Blackett, British cosmic-ray physicist and Nobel prizewinner, and Hendrik Casimir, then a co-director of the Philips Research Laboratory in the Netherlands. Around September at least two other names were put forward—those of Pierre Auger, who was still closely associated with CERN, and Cornelis Bakker, the Director of the SC Group.[81]

Auger soon dropped out of the picture. He let it be known that he was not keen on the post, and his countryman Perrin saw no point in pressing the matter, particularly since he felt that Auger's candidacy would be strongly opposed by Britain and the Nordic countries.[82] Blackett's chances were also negligible. Not only did Cockcroft himself prefer Casimir. There was also the 'problem' of Blackett's left-wing political views. In this connection the Foreign Office let it be known that they 'would most strongly deprecate' Britain's sponsoring Blackett as a candidate or even giving active support to his candidature—whatever his scientific merits, he 'could not possibly hope to get close co-operation from Brookhaven', and he would be 'likely to attract' 'an increasing number of politically like-minded workers' to CERN, which, as 'we know only too well from experience in other places' 'does not make for a happy staff'.[83]

What of Casimir? Casimir visited Copenhagen in late July or early August 1953, and in discussion with Bohr was not unwilling to consider the directorship of CERN. On 5 October he was approached by Cockcroft on behalf of the Nominations Committee. Casimir confirmed his interest in what he called 'in many ways a very attractive and fascinating task and also a rather difficult one', but felt that he could not give a definite answer until he had thoroughly discussed the issue with his employers and colleagues at Philips. After further consideration of the matter he decided early in December not to accept the position, despite further approaches from Bohr and from Heisenberg.[84]

This brings us to the two members of the Executive Group, Amaldi and Bakker. Amaldi was officially offered the directorship by Scherrer early in October 1953. Replying on the 16 October, he said that 'for many reasons that [he had] explained to various members of the Committee during private conversations, [he had] decided quite firmly not to accept' the proposal.[85]

One of the 'many reasons' for Amaldi's refusal at this point—and the one that he himself stresses—was to allay Scandinavian suspicions about the motives of the members of the Executive Group. In their view the group was far too concerned to further its own interests, seeking to gain control of European physics and to secure their own positions in the new laboratory. According to Amaldi, it was precisely to counteract such criticism that he turned down the offer of the directorship.[86] Indeed he went further, stating publicly at the seventh Council session at the end of October that he would no longer be available to the organization once the Convention entered into force.

A second reason for Amaldi's refusal was his desire to return to a more active scientific life. As Secretary-General of the provisional organization he had laboured long and hard with administrative matters, many of them somewhat trivial. Despite the load, he had found time to report to the Council on two cosmic-ray expeditions in the Mediterranean. He wanted to develop such interests, returning to his institute in Rome, and spending some time working around a big American accelerator. As Amaldi realized only too well, it would be difficult if not impossible to combine these activities with the post of director of CERN.[87]

When the Nominations Committee met during the seventh Council session in Geneva from 29 to 31 October 1953, they had then only one strong candidate for Director-General, Cornelis Bakker. Though we have little direct evidence for the circumstances surrounding their offer and Bakker's reaction to it, our documents permit us to reconstruct a plausible account of them.

It was generally felt in 1953 that the development of the new laboratory would pass through two distinct phases, each making somewhat different demands on the director. During the first three years the main activity of CERN would be one of construction, when both the laboratory itself and the machines would be built. Around the fourth year the scientific work as such would start (around the SC). This would gradually increase thereafter, and would be in full swing around 1960/61 with the commissioning of the PS.

In the light of these considerations the Executive Group, in particular, favoured a man like Bakker. As far as they were concerned the main task of the first director was pragmatic: it was to set up the new laboratory. Bakker had both the organizational and administrative skills required for this task, and an intimate knowledge of what was involved in designing and building an accelerator.

Of course the members of the Nominations Committee could see the merits of this argument. On the other hand many of them felt that it was desirable to have a really prestigious scientist as director from the start. In this way a scientific atmosphere could be generated within the organization at an early stage. Furthermore the scientific respectability of the laboratory would be ensured. With an eminent physicist as director the laboratory would start off with real *éclat*, other leading physicists would be attracted to it, and CERN would gradually become one of the most important research centres in the world.

Granted the difficulty of reconciling these demands the Nominations Committee decided early in October that its first choice was an eminent physicist as director. Then, if a suitable candidate could not be found, they would elect a temporary director for the period of machine construction. It was in this spirit that, in October 1953, Bakker was offered the post of interim director for three years by the Nominations Committee.[88] He refused.

Reflecting somewhat bitterly on Bakker's reaction a year later, Valeur remarked

Notes: p. 289

on his 'intransigence' and concern for 'prestige', adding that if he had accepted he would almost certainly have become director when his temporary appointment expired.[89] Let us consider each of these points in turn.

Prestige certainly was involved. Formally speaking it seems that Bakker may only have had 'such power and responsibilities as the Council may direct', to quote the Convention, i.e. not necessarily those of a 'full' director. Furthermore, it was quite clear that, as a competent though not really *outstanding* physicist, he was a second-best choice for the committee. This was hardly a compliment, and doubtless he resented the snub.

Furthermore, the chances of Bakker's staying in office once the SC started may have been less secure than Valeur believed. Our documents indicate that, in approaching physicists at this time, the Nominations Committee asked them whether they would be willing to serve as director forthwith, *or in about three years time*. Thus Bakker may have been offered the post on the understanding that his appointment would be coupled with that of his successor.[90]

In the light of the above it is perhaps not surprising that Bakker was 'intransigent'. He was being asked to take on a post which demanded the organizational and administrative role of a director, without the associated status and authority—and that because the Nominations Committee had not found a suitably eminent physicist to fill the directorship. He was not prepared to leave Holland and to relinquish his leadership of the SC Group for that. On the other hand, if he was offered the post of director of CERN...

8.4.2 A SECOND MAIN CANDIDATE EMERGES

The advantages of having a physicist experienced in machine building as CERN's first director, and the support he enjoyed in the Executive Group, ensured that Bakker would remain a main candidate for the post. At the same time the Nominations Committee continued to search for a prestigious physicist who would take on the job. On 26 November 1953 its chairman Francis Perrin approached another candidate—Felix Bloch, European *emigré,* professor at Stanford University since 1934, and joint winner of the Nobel Prize in 1952 for his work on nuclear magnetic resonance. The tone of the letter was exploratory: Would Bloch consider being the director of the laboratory during the period of construction? Failing that, could he envisage filling the post in three years time? If he was interested, the directorship would be offered to him for at least five years. Would his acceptance imply his intention to return to work in Europe for a longer period? And if he did not want the post would he consider coming to spend a year in Geneva in the not too distant future as a 'grand Conseiller du C.E.R.N.'?[91]

Before receiving Perrin's letter Bloch had been contacted by Bohr and Heisenberg. Bohr suggested that he come over to Europe at the University of Copenhagen's expense to discuss matters in more detail. For his part Heisenberg went to considerable lengths to explain to Bloch what the post involved, warmly recommending his American colleague to accept it.[92] On 18 November Bloch replied to Bohr. It would be very difficult for him to get away from Stanford just then, and hardly justifiable in the light of his 'lack of eagerness' for the post. 'My tendencies', wrote Bloch, 'have actually always been to stay free from administrative duties as well as from the construction of big machines to the extent that this was possible'. Yet it was just these tasks that were to be the main concern of the director in the early years. All the same Bloch was tempted by the idea of coming to CERN. Casimir, for example, was a man eminently suited for the post, and if he were appointed Bloch 'might be very well inclined to work at the CERN either temporarily or permanently'. Alternatively, 'if the committee', wrote Bloch, 'should find that the chances for success of the laboratory might be so materially improved by my choice as to justify a calculated risk I would reconsider the directorship'.[93] After learning in mid-December that Casimir had decided against candidature Bloch yielded to the temptation.[94] On 26 December 1953 he wrote again to Bohr, spelling out the terms on which he would consider being Director-General, and apologizing for not accepting the offer 'more enthusiastically and whole-heartedly'. 'It is unthinkable to me', wrote Bloch, 'at this stage, to break off my connections with the U.S.A. in general and with Stanford in particular [...]'. However, he would be willing to ask the university to grant him a limited leave of absence for two years, while CERN, for its part, would need to grant him 'the right to leave even sooner if it should take less than two years to see that [he] was not suited'.[95]

The Nominations Committee considered Bloch's letter of 26 December at a meeting in Paris early in January 1954. On 4 January they telegrammed him with the news that they 'WOULD LIKE TO RECOMMEND TO THE C.E.R.N. COUNCIL MEETING ON JANUARY 13 YOUR APPOINTMENT AS DIRECTOR UNDER THE TERMS OF YOUR LETTER OF DECEMBER 26 TO PROFESSOR BOHR'. A few days later Perrin received Bloch's reply—Stanford was willing to grant him up to two years leave of absence, and he could be in Geneva around 1 October 1954.[96] Accordingly, when the Council met on 13 and 14 January, the Nominations Committee announced that it hoped to secure agreement from a well-known physicist in the near future. And arrangements were made for Bloch to visit Europe for further discussions at CERN's expense.

On 25 February 1954 Bloch arrived in Copenhagen to visit Bohr. He was in Zurich on 1 March and he went to Geneva on 3 March, where he met Amaldi. Together they travelled to Paris the following day. On 5 March Bloch met with (at least) the Executive Group, one member of the Nominations Committee (Bohr), Council chairman Valeur, and IFC secretary Verry.[97]

The discussions with Bloch in Paris did not go particularly smoothly. Three issues in particular caused some unpleasantness.[98]

Predictably, there was Bloch's determination not to let administrative duties impede his scientific research. In Bloch's view his main task, if appointed, would be to create a scientific spirit in CERN—a view shared by Bohr—and he saw this as

Notes: p. 289

involving three main activities: the development of theory 'in close contact with Copenhagen', cosmic-ray research, and studies in nuclear magnetism, for which he proposed 'to bring complete equipment and an assistant from Stanford'.[99]

Then there was the question of money, and of Bloch's salary in particular. The CERN salary scales granted the director an annual taxfree salary of 36,000 SF plus. Bloch wanted to earn the same as he was getting in Stanford — over 60,000 SF per annum. Valeur, in particular, feared that in paying him this much above the agreed levels one would create an unhealthy precedent. To make matters worse Bloch wanted in fact two assistants put on the CERN payroll, and he expected the organization to bear the cost of their (return) travel and moving expenses as well as that of his scientific equipment at Stanford.

Finally, there was the problem of Bloch's citizenship. Although he was born in Switzerland, Bloch had also taken out American nationality in 1939. This could have unfortunate repercussions, particularly in France and Italy. Neither the French nor the Italian parliaments had yet ratified the Convention. Both had rather strong Communist parties. It was certain that their delegates would be highly critical of the appointment of an 'American' to direct CERN, interpreting his presence as yet another symptom of American influence on the organization. And although it was most unlikely that they could stop parliamentary ratification of the Convention in these two countries, they could doubtless delay it for quite some time.

No final decision was reached at the Paris meeting. Instead two alternative solutions to the problem of the directorship were suggested. In the first Bloch was director, Amaldi his deputy. In the second Amaldi was director, Bloch chairman of the envisaged Scientific Policy Committee (SPC). As Amaldi refused the second solution outright, only the first stood any chance of success. Both he and Bloch were willing to consider it, but under certain circumstances.[100]

For his part, Amaldi made it clear that he would only be in Geneva to pass on his administrative experience in the initial period of Bloch's term of office, and to ensure continuity with the new person who would replace him. In between times he wanted to concentrate on his scientific activities in Rome and in the USA, though this did not preclude his getting 'some fun' from collaborating with Bloch to create a scientific spirit at CERN.

What were Bloch's demands? He was not prepared to take on a heavy administrative load. He would bind himself for two years at most. And although in principle only a two-thirds Council majority was required to appoint the director, he insisted on unanimity. He did not want to accept such an important office knowing that even a minority of member states were opposed to it on political grounds.

Considering the difficulties in the 'latin' member states surrounding Bloch's American citizenship, this last demand would be difficult to satisfy. The Paris meeting ended with Bloch being asked to reconsider his demand for unanimity. For his part, Bloch asked for a decision no later than the time Council next met. That left a month.

Before going on to describe the circumstances leading up to the Council's recommendation, we would like to focus briefly on Amaldi's position. Why did Amaldi refuse the directorship, but accept to be considered as Bloch's deputy? Undoubtedly one crucial reason was that the former arrangement would leave him administering CERN while Bloch, as he himself saw it, would be 'put in charge of the scientific activities'. The prospect of playing a predominantly administrative role had as little appeal to the one as to the other. On the other hand Amaldi was prepared to help Bloch temporarily with his administrative duties, both out of concern for a scientific colleague, and for the health of the infant organization.

But there were also tactical considerations. These had been important when Amaldi refused the directorship the previous October. They were equally significant now. Bloch was being extraordinarily difficult. Time was running short, and it was rather unlikely that any other candidate acceptable to the Nominations Committee would be found. Perhaps if Amaldi agreed to serve as Bloch's deputy and to help him a little at the beginning, Bloch would be persuaded to accept the post, and the tiresome task of finding the first director for CERN would be settled, at least for the immediate future.

8.4.3 POLITICAL REALIGNMENTS

Although the choice of the director was, first and foremost, the task of the Nominations Committee, it was inevitable that political considerations should also be involved in it. After all it was the most senior post in a European organization. Furthermore, in the particular case of Bloch, there were the two additional factors which touched the nerves of the political process—his American citizenship and his wish that his appointment have the unanimous support of the Council.

What political support did the two main candidates have? Around the time of the Paris meeting it was clear that neither Bloch nor Bakker would easily be unanimously recommended by the Council.[101] The former had the 'anglo-nordic' countries behind him, notably Britain and Germany, but also the Scandinavians and Yugoslavia. The 'latin' countries, particularly France and Italy, and of course the Netherlands preferred Bakker. Their position was weak, however, for although the Executive Group also wanted the first director to be a man experienced in machine building, the Nominations Committee as a whole supported Bloch. Indeed any suggestion that Bakker be appointed raised, to quote Valeur, 'objections as obstinate as they were unjust on the part of [its] important members [...]'.[102] And it was their wishes that carried the greatest weight.

Immediately after the Paris meeting two letters were sent to Bloch.[103] One, from Amaldi, summarized the solutions discussed there (Bloch/Amaldi (deputy), Amaldi/Bloch (chairman of the SPC)) and added a third, Bakker (director)/Bloch (chairman of the SPC), which Amaldi warmly recommended. The other letter, from Cockcroft on behalf of the Nominations Committee, officially offered the first

Notes: pp. 289 ff.

solution. Cockcroft added that he did not think it important that Bloch have unanimous Council support. In his view it was sufficient that his candidature be backed by France. With the most influential of the 'latin' countries behind him, he need not fear that his term of office would be marred by major disagreements in the Council.

Cockcroft conveyed this news to Valeur when they met at Goward's funeral on 13 March 1954. The French delegate and Council chairman took stock of the situation. It was clear that the forces arraigned against Bakker were considerable, and unyielding. It was also clear that, without France's support, the Council would remain seriously divided over Bloch, and that he would not accept the post of director anyway. In the light of these considerations Valeur and Perrin concluded in mid-March that the French delegation had no choice but to support Bloch's candidature.[104]

In Valeur's view France could only safely adopt this stance if two conditions were met:[105] it was essential that Amaldi accept to be Bloch's deputy, and, it was imperative that the Belgian and, more especially, the Italian governments fall in line with France. In this way unanimity would be secured in the Council and likely criticisms in the French press and parliament could be more easily dealt with.

These concerns dominated Valeur's activities during the last fortnight of March. He put the case for a change in policy to higher authorities of his government. He asked, even implored, Amaldi to agree to be Bloch's deputy. And he persuaded Council delegates Willems (B) and Pennetta (I) to recommend that their governments support his position.

On 28 March Valeur informed Lockspeiser that the French government had instructed its delegation to support the team Bloch/Amaldi. About a week later on 5 April 1954, that is only three days before the ninth Council session was due to start in Geneva, Switzerland agreed to back France. By that time ten of the twelve member states were in favour of the Bloch/Amaldi pairing. The positions of Belgium and Italy were still not known.[106]

What of Bloch himself? In a letter to Amaldi of 20 March 1954 he rejected the third solution. He was not prepared to be nominated as a deputy to Bakker 'in one package'.[107] If the Council wanted Bakker they should appoint him, and leave him free to choose someone to run the scientific activities. As for his own candidature, Bloch would be willing to accept to be director with Amaldi as his deputy, as proposed by Cockcroft—but, he insisted, only if the Council *unanimously* supported him.

8.4.4 A DECISION IS TAKEN

On 8 April the Nominations Committee presented its proposal to the Council at its ninth Session. It recommended Felix Bloch as director, Edoardo Amaldi as his deputy, and Cornelis Bakker as representative of the scientific group leaders, with particular responsibility for the construction of the machines. The proposal was unanimously approved by the Council.

The most striking feature of this proposal is the inclusion of Bakker in the package. As far as we know up until the end of March it had never been suggested that CERN be managed by a 'troika' — at most a directorship 'in parallel' had been considered. Indeed, it is almost certain that the decision to complement the Bloch/Amaldi team with the name of Bakker was taken at the last minute, and in the light of pressure put on the Council from two distinct sources — the Dutch delegation and the PS group leaders.[108]

The Dutch were never particularly enthusiastic about Bloch as director, all the more so as they had their own rival candidates, Bakker and Casimir. Indeed early in February Bannier wrote to Valeur putting the latter's name forward again. It had emerged from earlier exchanges between the two that both Casimir and Bloch were interested in a team Casimir (director)/Bloch (leader of the theory group in Geneva). Nothing came of this idea. But when the Bloch/Amaldi combination won French support at the end of March, Bannier made it clear to Valeur that the Dutch were not too happy and wanted Bakker included in some sort of scheme.

The pressure from Adams and Dahl was expressed in a strongly-worded internal report written on 2 April and plainly directed at the Council which was to meet in less than a week. The authors suggested that the importance of the big machines, notably the PS, 'may not be fully recognised at Council level'. The reputation and success of CERN, even its very existence, depended almost entirely on them. Their progress and success would be the main attraction for good physicists. That granted, the best director during the construction period would be the person who could 'forward the machine projects'.

Although a decisive step was taken by the Council at its ninth session, a considerable amount of work remained to be done. In particular the consent of the three members of the 'troika' had to be obtained and their roles clarified. In what follows we shall discuss the activities surrounding each member in the months leading up to the foundation of the permanent CERN, starting with Bakker.

Bakker's position in the organization was discussed at an Executive Group meeting held immediately after the ninth Council session.[109] Surveying the distribution of senior posts in the permanent CERN, the group considered that he could either continue as director of the SC division, or that he could lead the PS division, with Adams and Dahl as his collaborators. The latter was a full-time appointment, and Bakker was reluctant to accept it, stating that the Dutch authorities would probably not allow him to devote more than 50% of his time to CERN business. Accordingly at a meeting of the Nominations Committee in London on 7 July 1954, Bakker was proposed as director of the SC division.[110] In addition the rather loose triplet born at the ninth Council session was baptized 'The Directorate'. Bloch was to be its 'president', Amaldi and Bakker its

Notes: p. 290

'members', the latter with the responsibilities bestowed on him by the Council in April.

Bakker was most unhappy with this arrangement. He had been led to believe otherwise, he said.[111] In particular he had expected to have a position as leader of the machine groups which would place him above Adams and whoever was nominated as director of the SC division. As this was not the case, he was not willing to move to Geneva or, at best, to do so only part-time.

Once again, as in the offer of the 'interim' directorship, it appears that Bakker's pride had been hurt.[112] Neither the leadership of the PS Group, nor a post on the directorship below Amaldi's in status would satisfy him. Indeed there is evidence to suggest that his claim that the Dutch authorities would not release him on more than a half-time basis was an excuse, a polite way of refusing to fill the gap left by Goward's death. Similarly his unwillingness to come full-time to Geneva to serve on the directorate jeopardized that appointment too, and was tantamount to a refusal. Apparently what Bakker wanted was a post at least as prestigious as that of deputy director. Anything less than that could not persuade him to commit himself fully to CERN.

Let us now turn to Bloch.

Three cables were sent to Bloch on 9 April 1954.[113] One spelt out the recommendations of the Nominations Committee. Another, from Lockspeiser and Valeur, assured Bloch that 'the unanimity of the Council was spontaneous and genuine'. A third, from Valeur alone, informed Bloch that his salary would be that agreed in Paris.

Bloch's response to Valeur was positive, though unenthusiastic.[114] He was 'strongly inclined to accept the directorship', but there were 'still a few points, concerning payments by CERN' which first needed clarifying. These concerned the return costs of transporting his family, his scientific equipment, and his assistants from Stanford to Geneva, the latter's salaries, contributions to Bloch's retirement fund, his income tax position.... The list seemed endless, and CERN had no option but to accede to Bloch's demands. Since the Council wanted him as director, wrote Valeur to Lockspeiser, they would have to take him 'lock, stock, and barrel'.[115]

In parallel with these activities Bloch began to prepare the ground for his scientific activities in Geneva.[116] He discussed with Amaldi space requirements for the theoretical work and for his own research facilities. He explored the possibility of recruiting personnel to start work on cosmic rays and to form the nucleus of a theory group in Geneva. Amaldi undertook to handle the former in consultation with leading European cosmic-ray physicists. Bloch himself wrote to B. d'Espagnat and A. Abragam in July asking them if they would consider joining CERN as theoretical physicists, and help to create 'a truly scientific atmosphere' in the laboratory.[117]

On 11 September 1954 Bloch arrived at CERN. To his shock and disappointment Amaldi immediately informed him that he would soon be leaving Geneva.[118]

Bloch undoubtedly felt that Amaldi had left him in the lurch. He had expected Amaldi to devote considerable time to easing him into his post and to dealing with administrative affairs. Now it seemed that this would not be the case.

Several factors need to be taken into account in assessing Amaldi's action. Amaldi had always made it clear that his main interest now was in scientific work, and that he would be only temporarily available to assist Bloch in Geneva. More to the point, when he had accepted to bear an administrative load for some months as Bloch's deputy, it was generally believed that the Convention would come into force in July 1954. It was now September, Amaldi had spent the summer as Secretary-General of the provisional CERN, and he was itching to get back to Rome.

Moreover, Amaldi had hoped that somebody else would be appointed specifically to assist Bloch with administrative duties.[119] This idea had been raised as early as the Paris meeting in March. In July the Nominations Committee had confirmed its wish to have such a post in addition to that of the head of administration, and recommended Jean Mussard for the job. In the event Mussard, who was in Auger's office at UNESCO, did not take it, and the chances of Amaldi being sucked into the vacuum thus left increased accordingly. This was a risk he was simply not prepared to take.

Finally, and perhaps most important of all, Amaldi had never really wanted the post of Deputy-Director anyway. It gave him no special authority in the organization—the machine groups were primarily responsible to their directors Adams and Bakker, and Bloch clearly wanted to take charge himself of the scientific policy of the organization. The pressure on Amaldi to accept the nomination had been enormous, and he had done so partly for tactical reasons. He had enjoyed the opportunity he had had as Secretary-General to run the provisional CERN. The post of Deputy-Director of the permanent organization left him far less scope to take the initiative.

8.4.5 THE DIRECTOR IS APPOINTED

With less than a week to go before the first session of the permanent Council, scheduled for 7–8 October 1954, the situation surrounding the directorate was desperate. Amaldi had announced his intention to leave in the near future. Bakker had effectively withdrawn from the troika, and was most reluctant to come to Geneva. And to cap it all, Valeur had just received a copy of the letter from Bloch to Bohr of 26 December 1953 which he probably had not seen before, the letter in which Bloch did not simply say that he would only bind himself as director for a maximum of two years, but added that he 'would have to ask for the right to leave even sooner' if need be. Apparently the 'ridiculous' arrangement (Valeur) agreed on in April was in danger of crumbling to pieces before it had even been sanctioned by the first permanent Council.[120]

The Council duly appointed Felix Bloch as Director-General. However, all the other members of the directorate and the divisional directors were only appointed provisionally, and on the understanding that this should in no way prejudice the final appointments.

Bloch's first six weeks were not happy ones. As he put it to Bohr, 'my idea in accepting the position was that for at least half a year it would be a more or less joint enterprise in close and constant cooperation with Amaldi [...]'. The latter had been 'pretty steadily' in Geneva during the second half of September and in October 'and he gave me [Bloch] during that time all possible help in getting initiated in my work'. Come November, however, Amaldi was not prepared to spend more than one week a month at CERN, and he had 'remained unshakeably firm in this decision'.[121] Bloch turned to Sam Dakin, his Chief Administrative and Finance Officer, who recalls that the physicist 'gave only 30/60 minutes a day to managing the place (by talking to me), and who claimed that even that gave him a headache'. When Bloch learned that Dakin was only 'on loan' from Britain for one year his mind was made up.[122]

On 25 November 1954 Bloch went to London to see Sir John Cockcroft and the president of the Council, Sir Ben Lockspeiser.[123] He explained to them that when he had decided to come to Geneva in the spring he had assumed that his activity there 'would be primarily of a scientific character'. The situation he had found on arriving at CERN 'sharply contrasted' with that, however. He had found that the post entailed 'heavy administrative duties and responsibilities', and that he was 'effectively without a Deputy', and with only a temporary director of administration (i.e. Dakin). Realizing that the directorship demanded 'an almost total sacrifice of [his] scientific work', Bloch informed Cockcroft and Lockspeiser that he was not prepared to retain the post beyond the autumn of 1955.

Felix Bloch's resignation was accepted 'with great regret' by the Council at its second session on 24 February 1955. At the same session it invited Cornelis Bakker to replace Amaldi as Bloch's deputy until his term of office expired on 31 August 1955. The Council also 'unanimously invited' Bakker 'to accept the appointment as Director-General, as from 1 September, 1955'.[124]

8.4.6 CONCLUDING REMARKS

The appointment of Felix Bloch as the first Director-General of CERN was obviously a mistake. How did it happen that a man so reluctant to take the post, and so obviously uninterested in constructing an accelerator laboratory, came to Geneva in this capacity when, as Amaldi likes to put it, the bulldozers had just started to break the ground?

While the answer to this question is obviously complex, the role of the Nominations Committee in Bloch's appointment was obviously crucial, even determinant. It was they who had the authority to nominate the leading staff of the new organization and to secure their approval. And it was they who, on the whole, believed that the first director should be an eminent physicist who would bestow scientific respectability on CERN.

The links between Bloch and (some) members of the Nominations Committee were intimate. On the one hand, there were the historical associations: Bloch had studied physics under Scherrer in Zurich, had done his Ph.D. in Leipzig under Heisenberg, and had been a fellow of Bohr's Institute in Copenhagen. Thus no less than half the members of the committee had played a role in Bloch's intellectual formation. Then there was Bloch's Nobel Prize. This bound him tightly to the Nominations Committee's most important members, Bohr, Cockcroft, and Heisenberg, his fellow Nobel Laureates. Between these three there existed 'une grande solidarité professionnelle', as one Swiss Council delegate put it, a solidarity which spilt over into an arrogance which was shared by Bloch, and which irritated at least Council chairman Valeur, who referred to them collectively as 'les Bramines'.[125]

There is no doubt that Bloch's appointment owes much to the influence of these three men.[126] It was they who persuaded, some would say 'bribed' him, to come to Geneva against his better judgement. It was they, or at least Heisenberg, who was 'violently opposed' to Bakker's appointment as director, believing that he was not sufficiently eminent and fearing that he would be in the thrall of the CERN executive.[127] In the event they got what they wanted, but at great personal cost to Bloch himself, and to the organization whose interests they were trying to serve.

8.5 The strong-focusing principle: the decision and its early consequences

In the summer of 1952 it was decided to equip CERN with a proton synchrotron (PS) embodying the newly discovered strong-focusing principle. This step had far-reaching consequences, not the least of which was that for a while continental Europe was equipped with the biggest accelerator in the world — and that less than a decade after she had been lagging far behind America in this regard. In this section we want to discuss the strong-focusing concept in a little more detail and to assess the early impact of the decision on the PS Group and on the embryonic organization to which it belonged.

The main thread of our argument is determined by the recognition that, in adopting the AG principle, the PS underwent 'a strange metamorphosis' from the initial project, which involved scaling up the 3 GeV Cosmotron to 10–15 GeV.[128] It is the nature and effects of this metamorphosis that interest us here.

To study it we shall compare the development of the PS during our period with that of the smaller synchro-cyclotron (SC). The SC's design and construction were 'based on installations of the same type and comparable energy as already proved elsewhere'.[129] It was thus a different *kind* of project from the PS, which demanded a substantial amount of new research and development. The SC thus provides a useful

Notes: pp. 290 ff.

point of reference for grasping the specific effects of embarking on a project of this kind, and what we say about it is chosen primarily to throw the activities of the PS group into sharper relief. We focus on these in the third section below, having first described the progress made on the two accelerators up to about September 1954.

8.5.1 THE PROTON SYNCHROTRON

In its initial conception the PS was to be an extrapolated version of the Brookhaven Cosmotron. To find out more about the American machine three members of the group, Odd Dahl, Frank Goward (Harwell) and Rolf Wideröe (Brown-Boveri, CH) visited the laboratory early in August 1952.[130] While discussing the CERN project in anticipation of their visit, the Brookhaven people had come up with a new concept in machine design which was presented to their European visitors by 'Drs Courant, Livingston, Blewett, Sneyder, White and more'. This concept came to be known as the alternating-gradient (AG) or strong-focusing principle.[131] Its implications for the dimensions, cost, and engineering of a high-energy accelerator emerge particularly clearly if one compares the first three columns of table 8.2, which presents the historical evolution (up to about mid-1954) of some of the less technical parameters of the CERN PS. Project 1 was the scaled up Cosmotron scheme, project 2 the new machine proposed to the third Council session after the visit to Brookhaven. It was clearly based on the tentative design worked out for the visitors by Livingston (column 2).

Comparing projects 1 and 2 with each other, it emerges that one can, in principle, build a much bigger (cf. orbit radius), more powerful accelerator for the same cost using the strong-focusing concept. The reason is that, by using magnets with high alternating field gradients (high field index n), the beam becomes more tightly focused. It is then possible drastically to reduce the dimensions of the vacuum tank without losing beam intensity due to collisions with the walls of the vacuum chamber. This means that the size of the magnets surrounding the tank can be reduced, with a corresponding reduction in the iron required for them (about a factor ten in our case). The considerable savings made allow one to achieve much higher energies at a fixed cost.

While there were no obvious objections to the AG principle, it proved astonishingly difficult to apply successfully in practice. Indeed although the machine parameters presented at the fourth session of the Council in January 1953 (see table 8.2) showed little change from the original set, serious doubts had already been raised about the technical feasibility of implementing the concept.[134] In particular it had been discovered that the more one tried to confine the beam (by increasing n) the more sensitive its trajectory was to magnet inhomogeneities, which scattered the particles to the walls of the vacuum chamber. This meant that one had to balance the benefits of a large n against the subsequent demands in the precision to which the magnets had to be manufactured and aligned, and a corresponding 'touchiness' in the operation of the device. In other words, from a technical point of view, the optimum value of n was far lower than the original theoretical calculations had led one to believe.

The resulting uncertainty in the parameter situation in April 1953 (column 5, table 8.2) had been resolved six months later (column 6, table 8.2). At a technical symposium in Geneva attended by about a hundred European scientists and several experienced guests from America, a new set of design characteristics for a 30 GeV machine was presented by the PS group. To compensate for magnet inhomogeneities the injection energy was increased considerably, and n was reduced by a factor of nine or ten, with corresponding increases in the dimensions of the vacuum chamber, the weight of the magnets and, of course, the cost.

The new proposal was discussed at considerable length during the seventh session of the Council, held immediately after the symposium. Of particular concern to the delegates was the new cost estimate of some 70 MSF—some 10 MSF more than Dahl's initial estimate, and indeed some 15 MSF more than the Council had allowed

Table 8.2
The change in the design parameters of the PS

	Project 1 approx. July '52	Livingston's proposal Aug. '52	Project 2 Sep. '52	4th Session Jan. '53	2nd Report to M/S April '53	Technical Sympos. Oct. '53	9th Session March '54
Maximum energy (GeV)	10–15	30	30	30	30	30	25
Magnet weight (tons)	6000	below 500	700	800	up to 10,000	4000	3300*
Equilibrium orbit radius (m)	29	90	100	100	?	112	72
Vacuum chamber inside dimensions (cm)	45 × 15	20 cm^2	5 × 4	4 × 5	increased by a factor 3 or 4	8 × 12	8 × 12
Field index, n	?	3600	?	4000	900	392	278
Injection energy (MeV)	30	3.5	3.5	10–15	50	50	50
Cost (incl. buildings) (MSF)	60	-	60	60	to be estimated	69	68*

The data in this table are derived from the following sources:
Columns 1 and 3 - the minutes of the third Council session
Column 2 - trip report by Dahl covering his visit to the USA (see note 130)
Columns 4 and 7 - the minutes of the fourth and ninth Council sessions, respectively
Column 5 - the second report to member states
Column 6 - presented at the conference on the alternating-gradient PS (see note 132)
The data at * in column 7 are from a report written by H. Blewett in December 1953 (see note 133)
The symbol '?' means that detailed information is not provided in the source used

for in its budget. Put starkly, the choice which now confronted the delegates was whether to keep within the existing budget and to reduce the energy of the machine, or to hold the energy at 30 GeV and to ask member states for more money. In the event the Council decided that the scientific disadvantages of reducing the energy a little would be more than compensated for by the possible savings in time and money. Dahl's group was instructed to restudy the machine for an energy of 25 GeV, and at the ninth Council session a revised set of parameters satisfying this requirement was presented (column 7, table 8.2). Expected costs were not given, but a preliminary estimate made some months before in the light of a new appraisal of the expense of some items, suggested that the machine was still going to be far more expensive than the figure of 55 MSF initially approved by the Council, and mentioned in the second report to member states.

As we pointed out earlier, the decision to incorporate the new strong-focusing principle into the CERN PS meant that the group's task was changed 'from a fairly straight forward engineering study with cost estimate to a difficult development programme [...]'.[135] It was an exciting and challenging project in which Europeans could make an important contribution to the art of experimental physics. But it was also a hazardous one, for when it was decided to opt for project 2 no one knew quite what was involved in translating a basically simple idea into a technically feasible device. Before discussing some of the consequences of this 'adventurous high risk — high gain course of action',[136] we shall briefly describe the parallel development of work on the SC.

8.5.2 THE SYNCHRO-CYCLOTRON

In the first few months of our period SC group leader Bakker's main tasks were to settle the energy of the machine, to allocate the work on its design to various subgroups scattered across Europe, and to make enquiries about manufacturing possibilities, delivery times, and costs. By the time the third session of the Council met in Amsterdam he had established a division of labour with five main components (see table 8.3). The planning and general layout of the machine, and the design of the vacuum chamber, were entrusted to groups based at Harwell and at Liverpool university, the latter including 'apprentices' from the continent, the magnet design was the responsibility of a team located at Uppsala, and the shielding and control system was being developed at a laboratory in Paris.

The most demanding main component in the SC was the radio-frequency system. Although this was not the only novel design feature in the accelerator,[137] it was the one which required a considerable amount of both research and development work. This was undertaken in Holland by a sub-group which worked in close collaboration with Philips of Eindhoven.

Table 8.3 reveals the steady progress achieved by the SC Group during the first two years of its life, during which it reaped the benefits of designing a scaled up

The strong-focusing principle

Table 8.3
The development of the SC—a simplified overview

(a) General layout	
By 1 October 1952:	Preliminary design given
By 15 June 1953:	Gradually reaching its final stage in consultation with the Laboratory Group
By 15 March 1954:	Being slightly modified in discussion with the architect
(b) Vacuum chamber and pumping layout	
By 1 October 1952:	Preliminary design given
By 15 June 1953:	Design ready and manufacturers asked for quotations
By 15 March 1954:	New drawings incorporating minor modifications, and detailed specifications, prepared
(c) Magnet	
By 1 October 1952:	Pole base diameter settled at 5 m. Pole shape and type of coil under investigation. Enquiries re manufacturing possibilities made
By 15 June 1953:	Design reaching final stage. Al preferred to Cu for coils. Quotations for manufacture of magnet and coils collected
By 15 March 1954:	Magnet design frozen. Details of shimming being studied. Final invitations for tender being prepared ready for despatch
(d) Radio-frequency system	
By 1 October 1952:	Three alternative modulating systems being considered; the 'vibrating reed' seems the most promising
By 15 June 1953:	Two models of the 'vibrating reed' system in operation
By 15 March 1954:	A visit to Berkeley has confirmed the soundness of the reed's design. Tests continue
(e) Shielding and control	
By 1 October 1952:	Agreement reached on basic features of the organization of the control system
By 15 June 1953:	Drawings on the control system elaborated, and quotations for parts obtained
By 15 March 1954:	Minor changes made

The data at the dates given were obtained from Bakker's progress reports presented to the third, sixth, and ninth Council sessions, respectively. They are reproduced as annexes to the minutes. For more detail on the rate of development of the SC see Mersits (1984).

version of an existing accelerator (the Pittsburgh machine). It is noteworthy how quickly a basic parameter like the magnet pole diameter was settled, how soon contact was established with people who would actually be involved in the building of the machine (the architect, commercial firms), and how rapidly agreement could be reached between the several parties on what was scientifically necessary and industrially possible. Speaking generally, by November 1953, some 15 months after the group had been formed, the design of the machine was almost complete.

Notes: p. 291

It is difficult to draw meaningful comparisons between the SC and the PS on the basis of table 8.3, since it was always realized that it would take much longer to build the bigger machine, even without the complications introduced by AG focusing. On the other hand, table 8.3 amply confirms that the synchro-cyclotron lived up to the requirement that it be 'easy and quick to build'. And herein lay its strength. Although the CERN machine was the biggest of its type under construction in Europe at the time, it was never intended to be a pioneering project. It had other purposes: to do useful meson physics, to provide a training ground for young Europeans inexperienced in accelerator technology, and to serve as a nucleus around which the laboratory in Geneva could grow while the PS was being built. In the event the solid advances achieved by the SC group must also have reassured Council delegates and their governments immersed in the anxieties surrounding the early development of the PS. In fulfilling these objectives the SC made an important if unspectacular contribution to the early history of CERN.

8.5.3 SOME PROBLEMS SURROUNDING THE BUILDING OF AN ALTERNATING-GRADIENT SYNCHROTRON

Our aim in this section is to identify some of the non-technical demands consequent on the decision to build the CERN PS using the strong-focusing principle. The choice that we have made is not arbitrary. For reasons given earlier, we often contrast the element identified with the parallel situation in the construction of the SC. By doing so we want to suggest that the stresses and difficulties we mention were peculiar to the PS *as innovative project.*

(a) Although, with the wisdom of hindsight, the decision to build the 30 GeV accelerator is seen as 'outstandingly courageous', at the time there was considerable anxiety about whether the technical problems which arose in implementing the new principle could be solved. John Adams, for example, has spoken of the 'Jeremiah-like prognostications' made by himself and others at Harwell about the defocusing effects of inhomogeneities in the magnetic field, 'which became a depressing feature of the group meetings'.[138]

Coupled with this was the fear, particularly strong in the early days, that the group lacked the competence to overcome these difficulties. Early in 1953 senior German physicists doubted whether Goward or Dahl had mastered the new focusing principle.[139] The British were similarly sceptical around this time. Pickavance (Harwell) spoke of a 'weakness at the top, and near the top' of the group, and argued that it should be strengthened by Adams and Hine, who had already done some valuable work on the inhomogeneity problem. Adams and Hine, for their part, did not have 'any good opinion of Goward', and when they first got interested in the new principle, 'it was clear to us that clearly Goward did not understand it'. Cassels (Harwell) was also alarmed. After attending a group meeting in Paris in November 1952 he wrote to Cockcroft and Chadwick that 'It was obvious that the technical discussion had been at a high level, but that the practical approach of the group was alarmingly unrealistic'.[140]

The criticism of Dahl may well have been due to his particular way of working. Dahl is said to have been more intuitive than rational, better at sensing the importance of a new idea than at working out its detailed implications. Correlatively his real strength from the point of view of the PS project lay in his determination to explore the new concept of AG focusing and in his ability to hold the team working on it together. Certainly his immediate decision that CERN should build a strong-focusing machine was something of a leap of faith, 'rationally' guided by little more than his conversations at Brookhaven and a set of preliminary accelerator parameters drawn up by Livingston. What is more Dahl saw himself as complementing Goward, the latter being responsible for the more scientific and technical sides of the project, while he took care of the administrative and policy-making angles.[141] In short those who criticized Dahl for being impractical may well have identified what was a weakness from a technical point of view, but a strength when it came to launching and organizing a daring and innovative project.

Not everyone doubted the professional competence of Dahl and Goward in the early days. The Americans, in particular, were rather impressed. Thus after a short visit to the group in the Autumn of 1952 John Blewett wrote to Dahl that 'your progress thus far is quite remarkable, your group is indeed a talented one, and their spirit of cooperation is very good'.[142] This leads one to suspect that anxieties about their ability were mingled with other factors, like personal or national rivalries, or simply caution in the face of the unknown. Indeed Hildred Blewett has recalled that at the time she called John Adams and Mervyn Hine 'the miserable English', and that the 'Harwell twins' 'didn't think that anybody ever understood anything'. For his part, Hine has suggested that his distrust of Goward, and the 'extremely negative and critical attitude' he and Adams took may be partly due to the fact that 'we were running the kind of counteraction to the official line... whatever Goward's people were saying, we were agin it'.[143]

Wherever the truth lies, the fact remains that as ways were found of dealing with the technical problems associated with the strong-focusing principle, confidence within, and of, the group increased. By May 1953 Dahl felt that his group (which now included a considerable Harwell-linked component, including Adams and Hine) 'has developed the strong-focusing principle further than any of the American groups'.[144] And by the time of the technical symposium held in October that year Cockcroft spoke openly at the seventh Council session of the 'excellent work' that it had done.

(b) Initially both the PS and the SC Groups were dispersed through a number of centres, and were run 'by remote control' as Dahl put it.[145] However, whereas the SC Group remained decentralized until late in 1954, it was felt from as early as November 1952 that the PS Group should be drawn together in one place. An investigation of the facilities available in Geneva convinced Dahl that the team should move there as soon as possible, a view confirmed by John Blewett.[146]

Drawing on his experience at Brookhaven Blewett argued that the PS members be brought together 'as soon as possible, integrating the group effort and establishing an esprit de corps and a real working cooperation. We feel that only when this is done do any possible weak points or bottlenecks become evident'. The importance of this was generally accepted and, after some delay, the first nucleus of the PS Group arrived in Geneva in October 1953.

(c) Related to this was the question of the composition of the group, notably the role in it of part-time members. Pickavance felt that the group leaders and their deputies in both the PS and SC Groups should work full-time on the project once it got under way. 'An international project, also involving industry in several countries, is difficult enough without this additional source of inefficiency', he wrote. Cockcroft and Chadwick tended to agree with him, at least as far as the head of the PS group was concerned.[147] In fact Cockcroft felt so strongly about this that he told Amaldi that before the Harwell scientists 'finally made recommendation to our government' to join CERN, they wanted assurances that the PS group leader would be full-time, at least after the move to Geneva. The proposal was put to Dahl who, while understanding the arguments for it, felt that he could not leave his home station permanently. Goward nominally headed the squad based in Geneva, until his sudden and untimely death on 10 March 1954. Dahl handed over the directorship to Adams in October of that year.

Some people also questioned the value of employing consultants. John Blewett, in particular, argued that, having 'only a partial grasp of the problems involved, [they] produced information which was completely useless'. Blewett believed that only full-time workers could do the 'real work of design and the initiation of fabrication'.[148]

Whether or not Blewett was right, the fact remains that when the group got down to the 'real work of design' as he called it, the balance swung heavily in favour of employing full-time members. The change occurred with the move to Geneva in October 1953. At this point the project entered a qualitatively new phase, the phase of 'engineering studies in the practical sense' as Dahl put it. Table 8.4 shows that a

Table 8.4
Staff build-up and composition in the PS and SC groups

Date	PS group				SC group			
	FT	PT	Con	Tot	FT	PT	Con	Tot
1 Oct. '52	3	1	4	8		6	8	14
15 Mar. '53	6	3	5	14	5	3	8	16
1 Oct. '54	54	0	1	55	10	2	6	18

Key: FT = Full-time, PT = Part-Time, Con = Consultant, Tot = Total.
Data at the cited dates were derived from the minutes of the third Council session and the second and final reports to member states, respectively. The figures are not strictly comparable across time periods. The first set includes only scientists and engineers, the second includes one full-time clerical worker in each group in addition, and it is not known what categories of staff are included in the third set.

year later there was only one consultant left in the group (Dahl himself) and no part-time workers. By contrast the SC group remained decentralized, and roughly unchanged both in size and composition during the period. In short to make progress in its early stages the innovative PS project required a different *form of organization* to that needed for the planning and design of the SC.

(d) One factor of great importance for the PS during these early years was the support given by America.[149] Livingston's role was particularly crucial during 1952, for not only did he strongly encourage the Europeans to be ambitious, and to build a PS much larger than anything in the United States,[150] he also revolutionized its design and shared his first ideas about strong-focusing with Dahl and the others when they visited Brookhaven. In fact both this laboratory and Berkeley supported the PS group, going so far as to release staff for six months to work with it (John and Hildred Blewett for the second half of 1953, and Hugh Bradner from Berkeley in 1954).

It must be stressed that this link with America went beyond the normal international traffic in scientific personnel. The Blewett's did not simply visit the PS Group. They were fully-fledged members of it, engaged in all its aspects, from theoretical research to drawing up cost estimates. Behind this involvement lay the expectation that the future of high-energy accelerators lay in exploiting the AG principle. Brookhaven and a consortium of large mid-western universities wanted to base big new machines on it, and indeed in January 1954 the U.S. Atomic Energy Commission announced that a 25 GeV AG proton synchrotron would be built at the former laboratory. Cornell went as far as to offer to direct the PS Group.[151] In short it was because the Americans and the Europeans had a joint interest in being in on the ground floor of a new development in accelerator technology that the collaboration between them was so intimate and of considerable importance to both.

Though the AG principle was developed jointly by groups at CERN and at Brookhaven, there were times when the former had to rely exclusively on their own resources. During the first half of 1954—a rather crucial period—the US Department of Commerce imposed restrictions on the transfer of technical knowledge, and the group in Geneva had no official news of developments across the Atlantic. Then a 'very welcome parcel' arrived, to quote John Adams,[152] containing reports and minutes of the meetings of the Brookhaven specialists. During this period at least, Hildred Blewett felt that 'in many ways Brookhaven got more from the cooperation than CERN did'.[153]

(e) When Dahl first introduced 'Project 2' to the Council he optimistically remarked that the machine embodying the AG principle 'may be constructed faster and cheaper than a Project 1 machine, and will give three times the particle energy'.[154] However, whereas the estimated cost of the SC remained firm during our period, it rapidly became clear that it would be impossible to build a 30 GeV PS within the limits of the budget originally specified.

Notes: pp. 291 ff.

This simply confirms that major research and development projects, like the construction of the PS, involve spiralling costs which appear to be as unavoidable as they are difficult to control. These dangers were recognized and accepted in some quarters as early as 1953. The British had seen the costs of their post-war accelerator programme escalate steadily, and the Department of Scientific and Industrial Research, which was responsible for them, was repeatedly asked for additional resources. Indeed it was the Department's Secretary, Sir Ben Lockspeiser, who stressed at the seventh Council session that one could not know the cost of the PS until the tenders were received. Governments he said, would 'certainly understand' that 'the Council's task was an entirely new one, for which there were no precedents which could serve as a basis'. In similar vein, Jan Willems (B) felt that if after a couple of years the original estimates of the cost of the PS 'appeared to be financially unrealistic, Governments would certainly understand the necessity to raise their contributions'. As the next section shows, a considerable amount of 'understanding' was demanded of governments, not only regarding the costs of the PS, but indeed of the construction and exploitation of the laboratory as a whole.

8.6 Planning the future laboratory

So far in this volume our aim has been to discuss as best we could the birth of CERN, the context in which it emerged, the central actors who shaped its broad outlines, the institutional factors which played a role in its development. In chapter 14 we draw together the several strands of our story, and try to give a global interpretation of how and why the laboratory was launched. Here, to conclude parts II and III we want, instead, to look ahead. More specifically we want to try and capture something of the image of the future laboratory which its protagonists had in 1953/54, to ask what they thought the laboratory would look like around 1960.

The first striking feature of the prevailing mentality in the early 50s is that there was *no idea of intrinsic growth* of the basic nucleus of the laboratory. The dominant image drew a sharp distinction between constructing the centre (to take seven years and to cost 120 MSF) and exploiting it (to cost about 8.6 MSF per annum thereafter).[155] The bulk of the money during the first period was for capital expenditure, that during the operating period for salaries, administration, and maintenance. Thus the idea that one had of CERN was similar to that of any other major capital construction project, say, a ship or an aeroplane: one built it, and then used it.

Related to this was the assumption that by the late fifties the staff of the organization would have grown to about 300 (from some 120 in September 1954), *and would stay there*. It was realized that this number would already be reached around the time that the SC came into operation, i.e. about 1957. But it was assumed that 'most of the people engaged in machine building [would] leave after the completion of the machines'.[156] What would change, then, as one moved from construction to exploitation of the accelerators was not the total number of staff, but their distribution between, say, engineers and technicians on the one hand, and experimental physicists on the other.

In the light of these findings it is interesting to look a little more closely at the provisions made for the exploitation of the accelerators. Again this can be approached from the financial and the staff angles. Concerning the former the long-term budget drawn up in 1953 set aside 72 MSF for the two accelerators (without their buildings), and 5 MSF for scientific material to exploit them. Of this, some two-thirds was for detectors—counters and electronics (1.7 MSF), cloud chambers, nuclear emulsions—and for beam handling and shielding material. In fixing this sum it was realized that 'radical changes' in experimental technique were likely during the seven-year construction period, the trend being towards more complex, difficult, and costly experiments. To accommodate this, 30% of the 5 MSF was for 'unforeseen equipment'.[157]

The question of staff is more complex because the Laboratory Group's ideas about the numbers needed, the rate of their recruitment, and the nature of their activities were hotly disputed by some British physicists.

The first concrete steps taken to ensure that trained physicists would be ready with suitable equipment to exploit the accelerators when they gave their first beams were taken by Yves Goldschmidt-Clermont and Peter Preiswerk, two members of Lew Kowarski's Laboratory Group. In a report distributed early in April 1953 they assumed that the SC would produce its first usable beam around 1957.[158] To prepare for its use they divided the intervening period into two main blocks—a 'farming' period and a 'warming up' period.

During the farming period, which would last about a year and a half, the authors proposed that about ten physicists be recruited to study, design, select, and order the basic scientific equipment for the laboratory and its workshops. During the ensuing warming up period sizeable experimental teams would be built up at a number of European centres. Originally their concern would be less to do research of scientific importance, than to test their instruments and to gain experience in working together in teams. As the time for the commissioning of the SC drew closer the teams would move to Geneva. 'Armed with suitable paraphernalia', as the report put it, they would be 'waiting for the beam at year 4', and would be capable of using it 'immediately [...] for significant experiments'.

The British delegation vigorously criticised these proposals in a note circulated in mid-December 1953.[159] The scheme outlined by the Laboratory Group, they said, 'would be a bad policy which would lead to a waste of physicists and money'. As an alternative they proposed that the SC division recruit 6–8 physicists in 1956 who had had experience on 'large' cyclotrons in England or Sweden. These would then accumulate the necessary stock of scientific equipment they needed for the experimental programme, and set to work on the accelerator when the first beams emerged in 1957.

Notes: p. 292

There were two distinct reasons for the British move. They argued that there was no point in recruiting physicists immediately to build equipment which would be used initially for cosmic-ray studies, for example. 'Nuclear research equipment gets out of date extremely rapidly', and anyway apparatus 'devised and built for one purpose is hardly ever suitable for another purpose', they said.[160] They also saw the move as part of an attempt by Kowarski to build up a team of physicists outside the accelerator groups and situated in his own division.[161] This they deemed intolerable: already in May 1953 Cockcroft had told Lockspeiser of his 'fear that Kowarski [would] get more and more control' which, he went on, 'would be very dangerous'.[162]

Goldschmidt-Clermont was not moved by these arguments.[163] He agreed that some of the experimental groups would be in the PS and SC divisions, but added that others needed to be free from the duties of accelerator construction, so concentrating all their attention on 'the conception of instruments for experimentation'. And he insisted that if experimentation was to begin with the beams as soon as they became available, it was necessary to recruit personnel and to start building apparatus immediately. If this was not done, Goldschmidt warned, the 'somewhat unique opportunity' offered by CERN would be lost, that is, the opportunity 'of allowing ample time to develop instruments for definite purposes in accelerator research'.[164]

By and large Goldschmidt-Clermont's arguments were overlooked. Kowarski was at pains to point out that they were not the 'official' point of view of the Laboratory Group.[165] And when the Executive Group met in April 1954 they drastically cut down the number of scientists in what was to be Kowarski's division, in line with British ideas.[166] As a result, for the first two years of its life the only major piece of scientific equipment built at CERN was a double cloud chamber for K meson experiments using cosmic rays.

To what extent did the CERN of 1960 correspond to the laboratory as envisaged by those who launched it in the early 50s? A few numbers are helpful here. The total expenditure for the five years 1955–59 inclusive was over 230 MSF—to be compared with about 100 MSF foreseen by the UNESCO consultants in May 1951, and the 120–130 MSF envisaged in CERN/GEN/5 in May 1953.[167] The number of staff on 31 December 1959 was 886, of which 155 were classified as scientists and engineers—this latter being double the number anticipated in May 1951, and the total being nearly triple the size allowed for in CERN/GEN/5. As for the expenditure on the experimental programme with the existing PS and SC, in 1959 *alone* 20 MSF was spent—four times the amount put aside for the first seven years.[168] And as early as 1955 anxieties were being raised about the lack of preparation for the SC experimental programme, and the serious shortage of physicists at CERN experienced in machine work who were in a position to take on the task...[169]

No one who pioneered CERN could have imagined that it would mushroom in this way. No one could have anticipated the qualitative transformation that high-energy physics was to undergo at the end of the decade. All, the scientists, the Council delegates, the governments in the member states, had to find ways of dealing with the new situations created, the new opportunities opened up, the new demands made. How they did so is another story, a story we will tell in volume II.

Notes

1. A comprehensive range of sources have been used for this chapter, including the minutes of the provisional Council and Interim Finance Committee meetings (in CERN–CHIP10014 and 10025 respectively), and other CERN official documents, many files in the CERN–DG (Director-General's Office) and CERN–LK (Lew Kowarski) series, and documents in the French Ministry for Foreign Affairs and in the Public Record Office in London (notably the AB6 series).
2. See minutes of the third session, CERN/GEN/4. The detailed replies to Kowarski's questionnaire, and his technical report on them, comprise annex V of this document. Detail on the order of proceedings is from a Swiss document prepared for a meeting of Geneva's *Grand Conseil*. It can be consulted in the *Mémorial des séances du Grand Conseil, session ordinaire, cinquième séance, samedi 8 novembre 1952 à 15 heures* at p. 1452. See (CERN-CHIP10017).
3. Bannier (private communication), September 1984.
4. See Blackett and Fry, *Report to the Royal Society on the 3rd Session [...]* undated, but received by Cockcroft on 16 October 1952 (PRO-AB6/1074).
5. See annex V of CERN/GEN/4 (note 2).
6. The occasion was a seminar on CERN history held at CERN on 20 September 1984. The seminar was attended, *inter alia*, by E. Amaldi, P. Auger, J.H. Bannier, D. de Rougemont, A. Mercier, J. Mussard, I.I. Rabi and V. Weisskopf. The proceedings were recorded, and are in the CERN archives.
7. This is from an undated *Note for Professor Blackett's use at Amsterdam* (PRO-DSIR17/551).
8. Letter Chadwick to Bohr, 3/10/52 (NBI).
9. The German Foreign Office told Heisenberg that, as far as they were concerned, the choice of the site 'should primarily be based on practical considerations' (in a letter dated 26/9/52 (AA-PA287)). At the Amsterdam session Heisenberg took no strong position on the site, though he recognized that Arnhem had geographical advantages for his country (see chapter 11). The only hint we have found of British preference for Arnhem is a comment by Verry that 'Probably we should regard Holland as slightly more favourable than Copenhagen or Switzerland' (in a memo to Lockspeiser, 1/10/52 (PRO-DSIR17/551)). All the same, five days later Verry proposed that a request from the Dutch Chargé d'Affaires for British support for Arnhem (the request is made in an *aide-mémoire* dated 23/9/52 (PRO-ED157/302)) should be treated coolly. He should be told 'that note has been made of the Netherland's proposal, but that at present the U.K. find it premature to reach a decision about which site they would regard most favourably owing to the fact that there are a number of technical and other factors still unresolved'. See letter Verry to Butler, 6/10/52 (PRO-DSIR17/551).
10. Bannier, at the seminar, see note 6.
11. These were, for example, some of the arguments presented by Councillor Picot at a meeting of the Geneva *Grand Conseil* on 8/11/52—see note 2.
12. The documents on which this account is based were collected together by Monique Senft (CERN-CHIP10017). They include, in particular, the *Mémoire des séances du Grand Conseil* of Geneva for 8/11/52, 5/12/52, 15/3/53, and 16/5/52. See also a document dated 19 November 1952 from the French Ambassador in Switzerland to his Minister of Foreign Affairs (CERN-CHIP10022). See also the unpublished thesis of C. Mathys, *L'Etablissement du CERN à Genève et l'initiative du Parti du Travail*, Faculté des Lettres Mémoires d'Histoire Contemporaine, Geneva University, July 1971 (CERN).

13. The article, entitled 'Europe's own Harwell,' was in the *Economist,* 8/10/52, 194.
14. Letter Macdermot to Somers Cocks, 18/11/52 (PRO-DSIR17/560).
15. Letter Picot to Amaldi, 19/6/53 (CERN-DG20829).
16. The detailed figures are in a folder 'Notes Chancellerie d'Etat, Genève, 1953', in (CERN–CHIP10017).
17. See *Amendments Proposed by the Swiss Government,* CERN/19, Addendum IV, 30/12/52.
18. Letters Jacobsson and Funke to Bohr, 23/12/52, and Jacobsson to Amaldi, 30/12/52 (CERN-DG20828).
19. See Nielsen's amendment of the draft minutes, CERN/26, Addendum I, 27/3/53.
20. Letter Bohr to Amaldi, 30/6/52 (CERN-DG20611) in which he proposes and actually uses the new name, and Kowarski's reaction, in his letter to Amaldi 7/7/52 (CERN-DG20593). Kowarski feared that this move would re-establish the unfortunate divisions of the Geneva session of the UNESCO Conference held in February 1952, and would dampen the enthusiasm of member states to found a new laboratory in 1953, for they would feel that a kind of European laboratory already existed in Copenhagen.
21. Bohr's views as reported by Scherrer are in a letter Scherrer to Amaldi, 12/9/52 (CERN-DG20798). For Amaldi's immediate reaction see his reply, 17/9/52, *ibid.* Amaldi remarked that 'the Secretariat of CERN is in itself the organization center of European nuclear collaboration [...]'.
22. Letter Bohr to Chadwick, 9/9/52 (CC–CHADI-1/3).
23. For Bohr's attempts to have Chadwick serve as the main link with the British, see letters Bohr to Chadwick, 6/3/52 and 26/3/52 (CC–CHADI,1/3). Also on this file are the telegram, Bohr to Chadwick, 2/5/52, and letter Bohr to Chadwick, 9/9/52, recommending the tête-à-tête.
24. Letter King to Verry, 29/9/52 (PRO-DSIR17/560).
25. Letters Cassels to Cockcroft, 11/11/52 (PRO-AB6/1074), and to Peierls 27/11/52 (PRO-AB6/1014). The enthusiasm with which the Bohr group responded to the discovery of strong focusing is described in a report from the British Scientific Attaché in Stockholm, who visited Bohr's institute with Alexander King on 29/9/53 (PRO–DSIR17/561).
26. Cassels' views are given in the letters quoted in note 25. For the names of Bohr's possible successors see, for example, letter Cockcroft to Lockspeiser, 4/5/53 (PRO-AB6/1076).
27. *Note confidentielle sur l'installation des études théoriques du CERN à Copenhague,* undated, but later than 28/4/53 (AEF-77).
28. For France's attitude, see previous note.
29. See *Notes on Draft Report to the Royal Society,* undated, (probably) written by Lockspeiser, and discussing the proceedings of the fourth Council session (PRO-AB6/1076).
30. The British and French proposals are, respectively, CERN/46, 20/6/53, and CERN/47, 27/6/53.
31. The meeting was mentioned by Dahl in a letter to Amaldi, 16/2/53 (CERN-DG20551). He reported on it in a subsequent letter dated 2/3/53 (CERN-DG20551). Here Dahl listed the delegates from Norway, Sweden, and Denmark who attended—Bohr's name was amongst them. In Dahl's view the basis for the decision was that it was 'judged unlikely that Europe [would] support the idea of a European Institute in Copenhagen'. This should not be seen as a 'demonstration', Dahl went on, 'but rather as a second best choice' for the Scandinavians. Bohr also insisted that the move was not to be seen as a threat to CERN, in his letter to Amaldi, 2/3/53 (CERN-DG20551). Alfvén informed someone in the British Foreign Office of the meeting, and the reasons for it, on 14/2/53—see *Note on Foreign Office letter of 14th February, 1953, reference USE 1241/23* (PRO-AB6/1076).
32. For the minutes of this meeting see memo Verry, *European Nuclear Research Organization—C.E.R.N.,* 11/12/52 (PRO-DSIR17/551). Its proceedings are also discussed in section 13.5.1.
33. Letter Amaldi to all Member States, 20/12/52 (CERN-DG20924). The memorandum was document CERN/19, Addendum II, 12/12/52.
34. Lockspeiser (?), in his report on the fourth session, see note 29.
35. It was changed in CERN/38 (Rev.2), 1/6/53, because the drafting committee felt that the adjective 'initial' could be misleading 'since it seems to imply that this programme is for an initial period, whereas it was really intended to describe the programme initially agreed upon'. They thus suggested that the phrase 'initial programme' be replaced by 'basic programme'.

36. Letter Cassels to Cockcroft, copied to Chadwick, 11/11/52 (PRO-AB6/1074).
37. See letter King to Verry, 29/9/52 (PRO-DSIR17/560). The quotation is from Verry's minute on the meeting with Amaldi on 11/12/52—see note 32.
38. Letter Waterfield to Mussard, 19/12/52 (PRO-DSIR17/551).
39. See, for example, Pestre (1984c).
40. Document CERN/19, Addendum V, 24/12/52.
41. For the other ways of securing British interests, see letter Lockspeiser to Dudley, 4/3/53. On the French attitude see enclosure to *Confidential brief for U.K. Observers for use at Council Meeting at Brussels, on the 12th January, 1953,* 8/1/53, both in (SERC-NP24).
42. See enclosure to *Confidential brief...,* note 41.
43. For the proceedings of this session see document CERN/GEN/6. The *Draft Convention Establishing a European Organization for Nuclear Research* which emerged from it was document CERN/28, 9/2/53.
44. For the quotation see the report quoted in note 29. Lockspeiser's report is typical of several we have from the British observers in making no reference to Swiss opposition at the meeting. And indeed at the end of March in a list of *Amendments Suggested by the Swiss Delegation* (document CERN/28, Addendum III, Corrigendum, 31/3/53), Article III.2 was that drafted after the Brussels session six weeks before.
45. *Further comments by the United Kingdom and Proposals for Additional Amendments to the Draft Convention,* CERN/28, Addendum VII, 29/3/53. Warning of this move was given in an *aide-mémoire* sent to governments of all member states on 19/3/53 (CERN-CHIP10022).
46. For material in these two paragraphs see British Foreign Office document USE1241/83, signed by Dudley, 22/12/52 (PRO-FO371/101517), and memos from Verry entitled *European Organization for Nuclear Research (E.O.N.R.),* 31/12/52, and to Lockspeiser, 16/4/53, headed *Confidential* (SERC-NP24). See also Despatch No. 35 from the Secretary of State (F.O.) to the British government's representatives at the fourth Council session 14/2/53 (PRO-DSIR17/552).
47. See minutes of fifth session, CERN/GEN/9, and report on it by Bartlett, 8/4/53 (SERC-NP24).
48. The Swiss *aide-mémoire* was sent with a covering letter by Thevenaz to Valeur, 28/5/53 (AEF-77). The amendment is CERN/38, Revised, Addendum III, 2/6/53. Typical of the French attitude is a *Note pour l'Ambassadeur de France à Londres,* 29/4/53 (AEF-77).
49. See CERN/38, Rev.2, Addendum III, 19/6/53. For Lockspeiser's anxieties about the unanimity provision see his letter to Dudley, 4/3/53 (SERC-NP24).
50. The French played an important role in securing the smooth passage of the British amendment. In particular they pointed out to the Swiss that if they insisted on explicitly leaving the door open for eastern European countries, Britain would insist on leaving it open for non-European states—see the telegram French Foreign Office to Ambafrance Berne, June 1953 (AEF-77). On 20/6/53 French ambassadors in all the member states were informed that France intended to support the British amendment. For some of France's preliminary versions of the new Article III.2 see (AEF-77). For the behind-the-scenes negotiations between Britain and France, see memo Verry, *CERN,* 22/6/53 (PRO-DSIR17/554).
51. The Swiss delegation, while supporting the new article, did not feel itself empowered to sign the Convention on behalf of its government. In fact Switzerland signed some two weeks later, on 17 July 1953—see *Déclaration Suisse à la conférence,* undated (CERN-DG20829).
52. Letter Verry to Dudley, 13/12/52, where he reports on a communication from Amaldi that Israel had applied for membership (PRO-DSIR17/560).
53. Letter Amaldi to Kronig, 8/6/53 (CERN-DG20925).
54. 'Supplementary Agreement Prolonging the Agreement Constituting a Council of Representatives of European States for Planning an International Laboratory and Organizing Other Forms of Co-operation in Nuclear Research', 30/6/53, CERN/GEN/7. The signatories were delegates from Belgium, Denmark, France, Germany, Italy, The Netherlands, Norway, Sweden, Switzerland, and Yugoslavia.
55. See CERN/44, 1/6/53 and CERN/44, Addendum I, 11/6/53.
56. See minutes of the sixth Council session, CERN/GEN/10,9.
57. For the minutes of the fourth meeting of the IFC, held on 16–17 December 1953, see CERN/IFC/18

(CERN-CHIP10025). An important supplementary document on the financial position is CERN/IFC/19 (CERN-CHIP10025).
58. This was the fifth meeting of the IFC. Its minutes are CERN/IFC/23 (CERN-CHIP10025).
59. This was the sixth meeting of the IFC. Its minutes are CERN/IFC/33. See also CERN/IFC/32, *Finance Required for the Interim Organization during May, June and July, 1954* (CERN-CHIP10025).
60. France's position has been described at some length by Dominique Pestre in section 9.5.
61. The first loan was applied for by Amaldi on 21/12/53, and confirmed in a letter to him of 8/1/54. It was interest-free for the first six weeks, and at a rate of 3% per annum thereafter. The second loan was officially confirmed in a letter from the Swiss Federal Department of Finance and Customs, Berne to Amaldi of 3/5/54. All letters in (CERN-DG20829).
62. In fact delegates were specifically asked to arrive at the Council meetings empowered by their governments to accept the payments called for by the IFC. See for example the letters of invitation from the Secretary-General to delegates to the ninth session, 16/3/54 (CERN-DG20927).
63. See CERN/44, Addendum I, 11/6/53.
64. At the fifth meeting of the IFC—see note 58.
65. The letter of invitation is letter CERN/810, of 6/5/53 (CERN-DG20925). The clear reversal of position is in a letter from Amaldi to Funke, 5/9/53 (CERN-DG20828). For an intermediate position adopted by the Executive Group see CERN/44, Addendum I, 11/6/53. It stated that money paid after 1/2/54 would 'reduce the amount necessary for the initial budget of the new Organization'. A difference was therefore drawn between the first 1 MSF raised under the Supplementary Agreement and any more money that might be called for in the interim period.
66. In a marginal note on a letter from Funke to Amaldi dated 27/7/54 (CERN-DG20828) Verry wrote that 'there seems no objection' to the proposal that the contributions paid under the resolution voted at the eighth meeting of the IFC be deducted from contributions to the permanent CERN. The vagueness of this remark attests to the absence of a clear official policy on the matter.
67. See letters Funke to European Organization for Nuclear Research, 1/11/54, and Dakin to Funke, 4/11/54 (CERN-DG20828).
68. The letter is from Amaldi to 'Sir Ben', 7/4/54. It is included as an annex to the minutes of the Executive Group's meeting held in Geneva from 6–7 April 1954, CERN/EX/10, 23/4/54 (CERN-CHIP10031).
69. The minutes of this (seventh) meeting of the IFC are document CERN/IFC/40 (CERN-CHIP10025).
70. See CERN/L/Z, Zurich, 12/12/53 (CERN-DG20593).
71. Its minutes are CERN/IFC/46. The budget estimate for expenditure for August–December, 1954 is CERN/IFC/45. Both are in (CERN-CHIP10025).
72. A favourable vote in the Assembly was tantamount to ratification by France. Formally the matter had to be agreed to by the Senate, but this body rarely went against the wishes of the National Assembly. Even then it could only delay a final parliamentary decision; it did not have the authority to overrule it.
73. This is reported in Amaldi's weekly report No. 76, 5–16/7/54 (CERN-CHIP10029).
74. Letters (Swiss) Departement Politique Fédéral to Amaldi (CERN-DG20829), and Verry to Eliane (Bertrand), 15/7/54 (CERN -DG20805). For the ratification process in France and Germany see section 9.5 and chapter 11.
75. Letter Amaldi to Member States, 20/7/54 (CERN-DG20928).
76. The figures are derived respectively from annex IV, minutes of the sixth Council session, CERN/GEN/10; annex to the Supplementary Agreement, CERN/GEN/7 (see note 54); annex IV, minutes of the eighth Council session, CERN/GEN/13; annex IV(c), minutes of the ninth Council session, CERN/GEN/14; annex II, minutes of the eighth IFC meeting, CERN/IFC/46 (CERN-CHIP10025).
77. The first row of figures is from annex IV, minutes of the sixth Council session, CERN/GEN/10. The second row has been calculated by subtracting the first from the figures given for 1952/53 on p. 9 of the final report to member states, CERN/GEN/15. The third row is taken from the same source.
78. These issues are dealt with at greater length in chapter 14.
79. Letter Cockcroft to Lockspeiser, 4/5/53 (PRO-AB6/1076).
80. What follows is based on a variety of sources. At CERN we have used (CERN-DG20512), which contains

correspondence with Amaldi, and a Nominations Committee folder stored in a DG file, referred to as (CERN-DGNomCt). Another main source was the Quai d'Orsay. Boxes 104 and 105 dealing with 'affaires atomiques' contain correspondence with Valeur in particular. Additional correspondence with Lockspeiser (and Verry) is to be found on (SERC-NP38). For Bloch's own papers see (SU). Dr. Casimir has kindly shown us relevant parts of his correspondence dealing with the matter, for which we should like to thank him here.

81. For information in this paragraph see letter Cockcroft to Lockspeiser, 8/5/53 (PRO-AB6/1076), and letters Valeur (?) to Perrin, 29/8/53, and Perrin to Valeur, 22/9/53 (AEF-104).
82. See the exchange of correspondence between Valeur and Perrin referred to in note 81.
83. Letter Cockcroft to Lockspeiser, 8/5/53 (PRO-AB6/1076), and Memo Verry to Lockspeiser, 7/12/54 (SERC-NP38). This memo was written when Blackett's name came up again, this time as a possible successor to Bloch. All quotations are taken from it.
84. This paragraph is based on letter Mrs. Bohr to Rozental, 17/8/53 (AIP-Bohr Library), and letter Casimir to the author, 20/11/85, and enclosed letters Cockcroft to Casimir, 5/10/53, Bohr to Casimir, 4/12/53, and Bloch to Casimir, 11/5/54.
85. Letters Scherrer to Amaldi, 8/10/53, and reply, 16/10/53 (CERN-DGNomCt).
86. In an informal discussion with the author at CERN on 18/3/85.
87. This emerges particularly in a letter Amaldi to Bloch, 12/3/54 (CERN-DG20512).
88. For the different kinds of demands placed on the first director, see the report of the Nominations Committee to the eighth Council session, 14/1/54, CERN/GEN/13, letter Perrin to Bloch, 26/11/53 (CERN-DGNomCt) and letter Verry to Valeur, 8/1/54 (CERN-CHIP10022). For the difference of opinion between the Executive Group and the Nominations Committee on the best kind of candidate, see the report of a Swiss Council delegate, Guggenheim, on the eighth Council session (CERN-CHIP10016). For the idea that if an eminent physicist could not be found one could appoint a 'temporary' director, see letters Scherrer to Amaldi, 8/10/53 (CERN-DGNomCt), and (indirectly) Cockcroft to Bohr, 7/7/53 (AIP-Bohr Library).
89. Letter Valeur to Cabonat, 8/3/55 (AEF-104).
90. We infer this from the last two letters referred to in note 88 and, more concretely, from the letter Perrin to Bloch, 26/11/53 (CERN-DGNomCt), and letter Cockcroft to Lockspeiser, 2/12/53 (PRO-DSIR17/554), where it is specifically said that 'it would be a mistake to make [Bakker] a *permanent* Director of the organization'.
91. Letter Perrin to Bloch, 26/11/53 (CERN-DGNomCt). This was a follow-up to an earlier letter from Scherrer—see letter Bloch to Bohr, undated, but around late October (SU).
92. Letters Heisenberg to Bloch, 1/11/53, Bohr to Bloch, 2/11/53 (SU).
93. Letter Bloch to Bohr, 18/11/53 (SU), from which all the quotations are taken.
94. It should be added that he was also being pressured by Perrin to take a decision, though this was probably of secondary importance - see letter Perrin to Bloch, 11/12/53 (SU).
95. Letter Bloch to Bohr, 26/12/53 (SU).
96. Both telegrams are in (SU).
97. For these arrangements see letter Bloch to Amaldi, 16/2/54, and the reply 23/2/54, and cable Bloch to Amaldi, 26/2/54, all in (CERN-DG20512). Those present at the Paris meeting may be deduced from this correspondence, and letter Valeur to Amaldi, 26/2/54 (CERN-DGNomCt) (for Verry), and letter Bloch to Valeur, 10/4/54 (CERN-DG20512) (for Bohr).
98. What follows is reconstructed from letter Bloch to Bohr, 1/3/54 (SU), and from letter Amaldi to Bloch, 12/3/54, and reply, 20/3/54, cable Valeur to Bloch, 9/4/54, letter Bloch to Valeur, 10/4/54, and reply, 16/4/54 (CERN-DGNomCt), letter Amaldi to Bloch, 10/4/54, and reply, 21/4/54, letter Amaldi to Bloch, 24/4/54. All of these are in (CERN-DG20512), except for Valeur's letter indicated. For a rather alarmist rendering of the political difficulties surrounding Bloch's American citizenship, see Guggenheim's report, note 88. See also a Swiss note on a conversation with Valeur in Geneva on 17/2/54 (CERN-CHIP10016). We return to this matter again in the main text.
99. The quotations are from letter Bloch to Bohr, 1/3/54 (CERN-DG205 12)
100. See letter Amaldi to Bloch, 12/3/54, and reply, 20/3/54 (CERN-DG25 12).

101. See report on a conversation with Valeur in Geneva, 17/2/54 (CERN-CHIP10016). See also the exchange of letters between Bannier and Valeur, 6/2/54, and 22/2/54 (CERN-DGNomCt).
102. Letter Valeur to Amaldi, 31/3/54 (CERN-DGNomCt).
103. Letter Amaldi to Bloch, 12/3/54 (CERN-DG20512). For Cockcroft's letter, see reference in letter Valeur to Amaldi, 15/3/54 (CERN-DGNomCt).
104. Letters Valeur to Amaldi, 15/3/54, 31/3/54 (CERN-DGNomCt).
105. The material that follows is from letters Valeur to Amaldi (cf. note 104) and to Lockspeiser, 28/3/54 (CERN-DGNomCt).
106. Letter Petitpierre to Scherrer, 5/4/54 (CERN-CHIP10016).
107. Letter Bloch to Amaldi, 20/3/54 (CERN-DG20512).
108. For the Dutch initiatives see letters Bannier to Valeur, 6/2/54, and reply, 22/2/54 (CERN-DGNomCt). The Dutch wish to have Bakker included along with the Bloch/Amaldi team in a three-man group is reported in a letter Valeur to Bannier, 21/4/54 (AEF-104). The report by Dahl and Adams is CERN/PS/Admin.6/OD.JBA, 2/4/54 (CERN-CHIP10031).
109. See letters Valeur to Bannier (note 108), and Amaldi to Bloch, 10/4/54 (CERN-DG20512).
110. For the conclusions of this meeting see F. Perrin, *Propositions présentées par le comité des nominations*, undated (CERN-DGNomCt).
111. As reported in a letter Valeur to Verry, 17/9/54 (AEF-105).
112. After the Executive Group meeting on 10/4/54 Valeur wrote to Bannier asking him why the Dutch authorities were not willing to release Bakker for more than 50% of his time, when only the week before Bannier had said that the Dutch wanted him included on the directorate (letter, 21/4/54 (AEF-104)). Bannier replied somewhat ironically that Valeur was putting too much weight on what Bakker had said—he would be largely available—letter Bannier to Valeur, 6/5/54 (AEF-104). For Bakker's attitudes after the meeting of the Nominations Committee in July see letters Valeur to Verry, 17/9/54, and reply, 22/9/54 (AEF-105), and letter Cockcroft to Bloch, 28/9/54 (AEF-105).
113. They are in (CERN-DG20512).
114. Letter Bloch to Valeur, 10/4/54 (CERN-DG20512).
115. Letter Valeur to Lockspeiser, 17/4/54 (AEF-104). For other aspects of Valeur's reactions to this letter, see his reply to Bloch, 16/4/54 (CERN-DGNomCt).
116. See letters Amaldi to Bloch 21/5/54, 13/6/54, 20/7/54 (CERN-DG20512) and Bloch to Amaldi, 29/5/54, 29/6/54 (CERN-DG20512).
117. Letters Bloch to Abragam and to d'Espagnat, 10/7/54 (CERN-DG20611).
118. Letter Cockcroft to Bloch, 28/9/54 (AEF-105), and interview of Bloch with Kowarski, Stanford, 21/11/74 (CERN). See also Amaldi's contribution to 'Felix Bloch, 1905-1983', CERN/DOC, August 1984.
119. See letter Amaldi to Bloch, 24/4/54 (CERN-DG20512), and Perrin's report on the July meeting of the Nominations Committee (note 110).
120. Letter Bloch to Bohr, 26/12/53 (SU). It was passed on to Valeur on 30/9/54. Valeur's remark about a 'ridiculous' arrangement is in his letter to Carbonat, 8/3/55 (AEF-104).
121. Letter Bloch to Bohr, 22/12/54 (SU).
122. Letter Dakin to Senft, 27/3/75 (CERN-CHIP10037). Amaldi's version of events is that he said he would stay long enough to transmit to Bloch 'all the information necessary for a smooth and successful continuation of CERN leadership', and that he asked to be relieved of his post of Deputy-Director 'after a few months'—see his contribution to 'Felix Bloch 1905-1983', CERN/DOC, August 1984.
123. Letter Bloch to Lockspeiser, 14/1/55 (CERN-CHIP10038)
124. We do not wish to enter again into the details surrounding these decisions. Suffice it to say that all did not go smoothly. Several candidates were approached for the directorship—Blackett, Amaldi, Dahl, Auger, Bakker, Leprince-Ringuet... The last two emerged as the strongest candidates. The Dutch were initially reticent about Bakker but, somewhat to the surprise of the French, arrived on the eve of the Council session in February 1955 with his candidature more or less sewn up. In the light of this, and their

candidate's hesitation, the French decided not to press for Leprince-Ringuet. Apparently Bohr and Heisenberg tried behind the scenes to have Bloch kept on as director of the SC. The idea was dropped in the face of strong opposition from Lockspeiser. For more details, see telex Valeur to de Bourbon-Busset, 14/2/55, letter de Bourbon-Busset to Valeur, undated, and reply, 21/2/55, and report on second Council session, sent by Cabonat to Valeur, 25/2/55, all in (AEF-104).

125. See Guggenheim report, note 88, and letter Valeur to Cabonat, 8/3/55 (AEF-104).
126. See interview of Bloch by Kowarski, note 118, of Bloch by Weiner (AIP), at page 52, Guggenheim report, note 88, and letters Valeur to Amaldi, 31/3/54 (CERN-DGNomCt), and Bloch to Casimir, 11/5/54 (SU).
127. For Heisenberg's conviction that Bakker was not sufficiently eminent, see chapter 11. The suggestion that Heisenberg feared that Bakker would not be able to control the executive arm of CERN was made to us by Jean Mussard (letter to the author, 2/9/85), and is indirectly confirmed in letter Heisenberg to Scherrer, 15/9/52 (CERN-DG20798), where Heisenberg complains that the Executive Group (including Bakker) was 'a group of young activists [who tried] each time to take the conclusive decisions before [the Council met], and the Council could do nothing but say yes and amen afterwards'.
128. The phrase is John Adams'. See Adams (1968).
129. As described by Gentner (1960/1).
130. For details of this visit see Dahl's 'trip report', CERN-PS/S4, 25/8/52 (CERN-DG20551).
131. For technical details see Livingston (1966), section IV.
132. M.H. Blewett, *Notes from the conference on the alternating-gradient proton synchrotron held at the Institute of Physics of the University of Geneva, Geneva, Switzerland, October 26-27-28, 1953.* The lectures presented were published separately: *Lectures on the theory and design of an alternating-gradient proton synchrotron.* Both available in the CERN archives.
133. M.H. Blewett, *Tentative schedules of cost, manpower and construction, for CERN 25-GeV proton synchrotron,* Report CERN-PS/MHB3, 1/12/53.
134. See J.D. Lawson, *The Effect of Magnet Inhomogeneities on the Performance of the Strong Focussing Synchrotron,* Report CERN/PS/JDL1, 2/12/52, and the revised version, 10/12/52, both on (CC-CHADI,1/2). See also, for example, Adams (1968). Apart from the problem mentioned, it was also found that 'the accelerating particles had to go through a transition energy at which their phase focussing suddenly disappeared', to quote from Adams, p. 89.
135. The quote is from Dahl's progress report to the sixth Council session, annex V of the minutes.
136. Adams (1968), 87.
137. For example it also made use of a new method of beam extraction incorporated in the Liverpool SC. See Gentner (1960/1) and Mersits (1984).
138. All material in this paragraph is from Adams (1968).
139. See draft minutes of the fourth meeting of the German 'Kommission für Atomphysik', 28/2/53 (DFG-A721-16,H.1)
140. For Pickavance, see his letter to Cockcroft, 18/12/52 (PRO-AB6/1076). For the opinions attributed to Adams and Hine, see the interview of H. Blewett, Hine, and Johnsen with Gowing, Kowarski, and Newman on 3/7/75 (CERN). For Cassels's attitude, see his letter to Cockcroft, 11/11/52 (PRO-AB6/1074).
141. Letter Dahl to Amaldi, 4/2/53 (CERN-DG20551).
142. Letter J. Blewett to Dahl, 25/11/52 (CERN-DG20551).
143. See interview, *op. cit.,* note 140.
144. Letter Dahl to Amaldi, 27/5/53 (CERN-DG20551).
145. Letter Dahl to Amaldi, 26/7/52 (CERN-DG20551).
146. See note 142.
147. See Pickavance, note 140. For Cockcroft, see his letter to Amaldi, 23/1/53 (PRO-AB6/1076), for Chadwick, see the enclosure to his letter to Martin, 18/2/53 (CC-CHADI, 26/1).
148. See note 142.
149. For the importance of this see Krige (1983).

150. Letter Livingston to Auger, 5/2/52 (UNESCO).
151. Letter Dahl to Amaldi, 2/3/53 (CERN-DG20551).
152. Letter Adams to Haworth, 3/6/54 (CERN-DG20551). For information on what caused the hold-up, see section 14.9.
153. See interview, note 140.
154. See Dahl's progress report, annex III of the minutes of the third Council session, CERN/GEN/4.
155. See CERN/GEN/5, 5/5/53, a copy of which is on (CERN-CHIP10014).
156. For British estimates of CERN staff distribution after three and six years, see Fry's figures given in CERN/L/Z, 12/11/53 (CERN-LK22448). The quotation is from a paper prepared by Adams, and circulated as annex II to Amaldi's weekly report No 62, 13-19/2/54 (CERN-CHIP10029).
157. A first list of the laboratory's nuclear instruments was made at a three-day meeting of members of the Laboratory Group in Zurich starting on 30 January 1953—their report is on (CERN-LK22448). An extended report on nuclear equipment, including cost, appeared as CERN/L, 10/2/53 (CERN-LK22448). See also CERN/GEN/5.
158. See CERN/L/Rep. 10, 8/4/53 (CERN-LK22449).
159. See CERN/67, 15/12/53. For the background to the debate in Britain see, for example, letter Skinner to Cockcroft, 10/10/53 and Skinner's attached *Detailed Criticism of Draft Report for CERN* (PRO-AB6/1076).
160. *Ibid.*
161. British fears were not unfounded. CERN/GEN/5 included an 'organizational scheme' which, using Brookhaven as a model, broke down the internal structure of the completed laboratory into seven divisions overseen by a directorate. These were administration, theory, proton synchrotron, synchro-cyclotron, and three that came to be grouped together under the heading of general services – instrumentation, general physics, and chemistry. The report placed about 50 staff in each of the machine divisions and about 120 in general services, 75 of these being in the instrumentation division. 'General services' was thus bigger than the two machine divisions *combined,* and swallowed up *40%* of the staff. It was a foregone conclusion that Kowarski assisted by Preiswerk would be the division leader. See also note 165 below.
162. Letter Cockcroft to Lockspeiser, 4/5/53 (PRO-AB6/1076).
163. See CERN/L/Z, 12/12/53 (CERN-LK22448).
164. Goldschmidt-Clermont proposed an interesting parallel. 'The situation', he wrote, 'can perhaps best be visualised by drawing an analogy to say fighter planes. They also have a tendency to be rapidly outmoded and they can only be built for a definite purpose. Yet no nation would postpone the equipment and training of its air force because the planes built today would be outmoded at the eventual outbreak of battle'—see CERN/L/Z, 12/12/53 (CERN-LK22448).
165. In a note added by Kowarski to CERN/L/Z on 18/12/53. Indeed right from the start Kowarski had tended to preoccupy himself with the organizational implications of the proposals made by Goldschmidt-Clermont and Preiswerk—see his CERN/L/WP.4, 22/5/53 (CERN-LK22449). In this document Kowarski also foresees the nucleus for the detailed planning of scientific equipment as being created in the Laboratory Group. To repeat, it is not surprising that the British saw him as wanting to be in charge of a division actively engaged in scientific research.
166. The Executive Group's proposals are CERN/86 and CERN/88, both of 7/4/54.
167. The total figure is obtained by adding the expenditure given in the CERN annual reports for 1955 to 1959 inclusive. For information on the May 1951 consultants meeting see section 4.2.1. Of course data in CERN/GEN/5 were soon revised upwards; an adjustment for inflation (probably no more than 15%), also has to be made. If we choose CERN/GEN/5 as a reference point it is because it was *the* official document on the basis of which governments had to decide on CERN membership.
168. The data for 1959 in this and the previous sentence are obtained from the CERN annual report for 1959.
169. See remarks by Skinner in document NPSC94, 28/11/55 (SERC-NP34). Ironically Skinner was one of the most virulent critics of the organizational scheme originally proposed in CERN/GEN/5—see material quoted in note 159, and Skinner's memo *C.E.R.N. Organization,* 7/12/53 (PRO-AB6/1076).

Bibliography for Parts II and III

John KRIGE and Dominique PESTRE

Accelerators (1954)	F.E. Frost and J.M. Putman, *Particle Accelerators,* University of California, UCRL-2672, November 1954.
Adams (1955)	J.B. Adams, 'The Alternating Gradient Proton Synchrotron', *Nuovo Cimento,* Suppl. Vol. 2, Series 10 (1955), 355-374.
Adams (1957)	J.B. Adams, 'Some Engineering Problems of the CERN Proton Synchrotron', *Discovery* (July 1957), 286-292.
Adams (1964)	J.B. Adams, 'Nuclear Particle Accelerators', *Proceedings of the Royal Society A,* **278** (1964), 303-322.
Adams (1965)	J.B. Adams, 'CERN: The European Organization for Nuclear Research', J. Cockcroft (ed.), *The Organization of Research Establishments* (Cambridge: Cambridge University Press, 1965), 236-261.
Adams (1968)	J.B. Adams, 'Odd Dahl and the Machine at CERN', *Festskrift til Odd Dahl* (Bergen: A.S. John Griegs Boktrykkeri, 1968), 83-92.
AIP (1978)	M.L. Perl (ed.), *Physics Careers, Employment and Education* (New York: AIP, 1978).
Alexander (1975)	K.J.W. Alexander (ed.), *The Political Economy of Change* (Oxford: Basil Blackwell, 1975).
Allison (1971)	G.T. Allison, *Essence of Decision: Explaining the Cuban Missile Crisis* (Boston: Little, Brown & Co., 1971).
Allison & Morris (1975)	G.T. Allison and F.A. Morris, 'Armaments and Arms Control: Exploring the Determinants of Military Weapons', *Daedalus,* **104** (summer 1975), 99-130.
Amaldi (1955a)	E. Amaldi, *The Scope and Activities of CERN 1950-1954* (Geneva: CERN 55-2, 1955).
Amaldi (1955b)	E. Amaldi, 'CERN, the European Council for Nuclear Research', *Nuovo Cimento,* Suppl. Vol. 2, Series 10 (1955), 339-354.

Bibliography for Parts II and III

Amaldi (1977)	E. Amaldi, 'Personal Notes on Neutron Work in Rome in the 30's and Post-war European Collaboration in High Energy Physics', *Scuola Fermi (1977)*, 294–351.
Amaldi (1979)	E. Amaldi, 'The Years of Reconstruction', C. Schaerf (ed.), *Perspectives of Fundamental Physics* (London: Harwood Academic Publishers, 1979), 379–461.
Armacost (1969)	M.H. Armacost, *The Politics of Weapons Innovation: The Thor-Jupiter Controversy* (New York: Columbia University Press, 1969).
Atomic Energy (1946)	Special Committee on Atomic Energy, United States Senate, *Essential Information on Atomic Energy, Committee Monograph No. 1* (Washington: US Government Printing Office, 1946).
Auger (1956)	P. Auger, 'Science as a Force for Unity among Men', *Bulletin of Atomic Scientists*, **12** (1956), 208–210.
Auger (1963)	P. Auger, 'Scientific Cooperation in Western Europe', *Minerva*, **1** (1963), 428–438.
Auger (1973)	P. Auger, 'Henri Laugier', *Cahiers Rationalistes*, **300** (1973), 309–319.
Auger (1975)	P. Auger, 'Discours de M. Pierre Auger', *CEC (1975)*, 30–35.
Baracca & Bergia (1975)	A. Baracca and S. Bergia, *La spirale delle alte energie* (Milano: Studi Bompiani, 1975).
Belloni (1986)	L. Belloni, 'La Preistoria del CERN', *Il Nuovo Saggiatore*, **2** (1986), 72–78.
Berkner report (1951)	'Science and Foreign Relations: Berkner Report to the US Department of State', *Bulletin of Atomic Scientists*, **6** (1950), 293–298.
Bernstein (1975)	J. Bernstein, 'Profiles, Physicist' [on Isidor I. Rabi], *The New Yorker*, 13/10/75 and 20/10/75.
Brouland (1970)	P. Brouland, 'Le CERN et la coordination des activités européennes en matière de physique des hautes énergies', *Cadres Juridiques (1970)*, 353–362.
Brown & Hoddeson (1983)	L.M. Brown and L. Hoddeson (eds.), *The Birth of Particle Physics* (Cambridge: Cambridge University Press, 1983).
Cadres Juridiques (1970)	*Les cadres juridiques de la coopération internationale en matière scientifique et le problème européen,* Actes des colloques d'Aix-en-Provence et Nice (Bruxelles: Commission des Communautés Européennes, 1970).
CEA (1958)	*Commissariat à l'Energie Atomique, 1945–1958* (Paris, undated).
CEC (1975)	*Deux initiatives du CEC, Documents sur les origines du CERN et de la fondation européenne de la culture* (Geneva: Centre Européen de la Culture, XIV, 4, hiver 1975).
CERN (1955)	*Laying of Foundation Stone. Signature of the Agreement with the Swiss Federal Council, 10th and 11th June 1955* (Geneva: CERN, Nov. 1955).
CERN (1964)	'L'Organisation européenne pour la Recherche nucléaire célèbre son dixième anniversaire', *Courrier CERN*, **4** (novembre 1964), 151–155.
CERN (1979a)	*25 CERN, CERN 1954–1979* (Geneva: CERN, 1979).
CERN (1979b)	*25 CERN, 1956–1979. A Photographic Record* (Geneva: CERN, 1979).
CERN (1979c)	*25 CERN, Allocutions* (Geneva: CERN, 1979).
CERN (1979d)	'25 CERN', *CERN Courier*, **19** (September 1979), special issue.
CERN (1984)	*30ème anniversaire CERN, Exhibition brochure* (Geneva: CERN, 1984).
Châtelet (1960)	A. Châtelet, *La France devant les problèmes de la science* (Paris: Notes et Etudes Documentaires, La Documentation Française, No. 2552 (20/6/59), No. 2580 (20/10/59), No. 2671 (28/5/60), No. 2721 (30/11/60)).
Colloque Caen (1957)	'Comptes-rendus du Colloque National de Caen (1–3 novembre 1956), Numéro spécial sur l'enseignement et la recherche scientifique', *Les Cahiers de la République*, 2ème année, 5 (1957).

Colloquium-Paris (1982)	*International Colloquium on the History of Particle Physics: Some Discoveries, Concepts, Institutions from the Thirties to Fifties*, 21-23/7/1982 (Paris: Editions de physique, supplément *Journal de Physique,* Tome 43, décembre 1982).
Colonnetti (no date)	*A Ricordo di Gustavo Colonnetti* (Torino: CNR-IMGC, no date).
Conant (1950)	J.B. Conant, 'Science and politics in the XXth Century', *Foreign Affairs,* **28** (1950), 189-202.
Coulam (1977)	R.F. Coulam, *Illusions of Choice. The F-111 and the Problem of Weapons Acquisition Reform* (Princeton: Princeton University Press, 1977).
Cushing (1982)	J.T. Cushing, 'Models and Methodologies in Current Theoretical High-Energy Physics', *Synthese,* **50** (1982), 5-101.
Dahl & Randers (1951)	O. Dahl and G. Randers, 'Heavy-water Reactor at Kjeller, Norway', *Nucleonics,* **9** (1951), 5-17.
Dautry (1952)	*Raoul Dautry, 1880-1951* (Paris: Cité Universitaire, 15 septembre 1952).
Day, Krisch & Ratner (1980)	J.S. Day, A.D. Krisch, and L.G. Ratner (eds.), *History of the ZGS,* AIP Conference Proceedings No. 60 (New York: American Institute of Physics, 1980).
Delmas (1981)	C. Delmas, *L'O.T.A.N.* (Paris: PUF, 1981).
Drechsler (1951)	G.J. Drechsler, 'The US State Department and World Science', *Bulletin of Atomic Scientists,* **7** (1951), 121-122.
Dufour (1970)	J. Dufour, 'Le CERN et la physique européenne des hautes énergies', *Cadres Juridiques (1970),* 343-352.
Gentner (1960/1)	W. Gentner, 'The CERN 600 MeV Synchrocyclotron at Geneva. I. Object and Design', *Philips Technical Review,* **22** (1960/1), 141-149.
Gilpin (1962)	R. Gilpin, *American Scientists and Nuclear Weapons Policy* (Princeton: Princeton University Press, 1962).
Ging-Hsi (1950)	W. Ging-Hsi, 'UNESCO and International Scientific Organization', *Bulletin of Atomic Scientists,* **6** (1950), 283-285.
Glantz (1978)	S.A. Glantz, 'How the Department of Defence Shaped Academic Research and Graduate Education', *AIP (1978),* 109-122.
Godement (1978-79)	R. Godement, 'Aux sources du modèle scientifique américain', *La Pensée,* (oct. 1978), 33-69; (fév. 1979), 95-122; (avr. 1979), 86-110.
Gowing (1974)	M. Gowing, *Independence and Deterrence, Britain and Atomic Energy, 1945-52,* vol. I, *Policy Making,* vol. II, *Policy Execution* (London: MacMillan, 1974).
Greenberg (1971)	D.S. Greenberg, *The Politics of Pure Science,* (New York: Plume Books, 1971).
Grinevald et al. (1984)	J. Grinevald, A. Gsponer, L. Hanouz, and P. Lehmann, *La Quadrature du CERN* (Lausanne: Editions d'en Bas, 1984).
Grodzins & Rabinowitch (1963)	M. Grodzins and E. Rabinowitch (eds.), *The Atomic Age* (New York: Simon and Schuster, 1963).
Grosser (1972)	A. Grosser, *La IVe République et sa politique extérieure* (Paris: Armand Colin, 1972).
Harwell (1952)	*Harwell, The British Atomic Energy Research Establishment, 1946-1951* (London: Her Majesty's Stationery Office, 1952).
Haskins (1962)	C.P. Haskins, 'Technology, Science and American Foreign Policy', *Foreign Affairs,* **40** (1962), 224-243.
Heilbron, Seidel & Wheaton (1981)	J.L. Heilbron, R.W. Seidel, and B.R. Wheaton, 'Lawrence and his Laboratory: Nuclear Science at Berkeley', *LBL News Magazine,* **6** (Fall 1981), special issue.

Herzog (1977)	Y. Herzog, *Le CERN, structure et fonctionnement,* Thèse pour le doctorat d'Etat, 1977, Université de Droit, d'Economie et de Sciences Sociales de Paris (Paris II).
Hirsch (1959)	E. Hirsch, 'A guide to Euratom', *Bulletin of Atomic Scientists,* **15** (1959), 250–252, 265.
Huxley (1970, 1973)	J. Huxley, *Memories* (London: Allen and Unwin, 1970); *Memories II* (London: Allen and Unwin, 1973).
Jay (1955)	K.E.B. Jay, *Atomic Energy Research at Harwell* (London: Butterworths, 1955).
Jungk (1968)	R. Jungk, *The Big Machine* (New York: Scribners, 1968).
Kaufmann (1980)	W. Kaufmann (ed.), *Particles and Fields* (San Fransisco: W.H. Freeman, 1980).
King (1953)	A. King, 'International Scientific Co-operation; its Possibilities and Limitations', *Impact of Science on Society,* special number on international scientific co-operation, **4** (1953) 189–218.
Kowarski (1955)	L. Kowarski, 'The Making of CERN. An Experiment in Cooperation', *Bulletin of Atomic Scientists,* **11** (1955), 354–357.
Kowarski (1961)	L. Kowarski, *An Account of the Origin and Beginnings of CERN* (Geneva: CERN 61-10, 1961).
Kowarski (1967)	L. Kowarski, *The Making of CERN, Memories and Conclusions,* unpublished, 17 pp., March 1967 (Kowarski interview file, CERN).
Kowarski (1973)	L. Kowarski, 'Le CERN, source d'esprit européen', *Courrier CERN,* **13** (1973), 251–254.
Kowarski (1977a)	L. Kowarski, 'New Forms of Organization in Physical Research after 1945', *Scuola Fermi (1977),* 370–401.
Kowarski (1977b)	L. Kowarski, 'Some Conclusions from CERN's History', A. Zichichi (ed.), *New Phenomena in Subnuclear Physics* (New York: Plenum Publishing Corporation, 1977), 1201–1211.
Krige (1983)	J. Krige, *The Influence of Developments in American Nuclear Science on the Pioneers of CERN* (Geneva: CERN, CHS-1, 1983).
Laboratoire Lawrence (1981)	'Le Laboratoire Lawrence de Berkeley à 50 ans', *Courrier CERN,* **21** (1981), 335–346.
Laqueur (1982)	W. Laqueur, *Europe since Hitler. The Rebirth of Europe* (Harmondsworth: Pelican Books, 1982).
Laves & Thomson (1957)	W.H.C. Laves and C.A. Thomson, *UNESCO, Purpose, Progress, Prospects* (Bloomington: Indiana University Press, 1957).
Lecerf (1965)	J. Lecerf, *Histoire de l'unité européenne* (Paris: NRF, Gallimard, 1965).
Livingston (1966)	M.S. Livingston, *The Development of High Energy Accelerators* (New York: Dover Publications, 1966).
Long & Wright (1975)	T.D. Long and C. Wright (eds.), *Science Policies of Industrial Nations* (New York: Praeger Publishers, 1975).
Malina (1950)	F.J. Malina, 'International Cooperation in Science: the work of UNESCO', *Bulletin of Atomic Scientists,* **6** (1950), 121–125.
Marrou (1961)	H.I. Marrou, 'Comment comprendre le métier d'historien', *L'Histoire et ses méthodes* (Paris: Gallimard, Encyclopédie de la Pleiade, 1961), 1465–1540
Masclet (1973)	J.C. Masclet, *L'union politique de l'Europe* (Paris: PUF, 1973).
Melanson (1973)	P.H. Melanson, *Knowledge, Politics and Public Policy. Introductory Readings in American Politics* (Cambridge: Winthrop Publishers, 1973).

Mercer (1978)	R.A. Mercer, 'The Reflections of a Former High Energy Physicist doing Industrial Applied Mathematics', *AIP (1978)*, 193-196.
Mersits (1984)	U. Mersits, *Construction of the CERN Synchro-cyclotron (1952-1957)*, (Geneva: CERN, CHS-13, 1984).
Moonman (1968)	E. Moonman (ed.), *Science and Technology in Europe* (Harmondsworth: Penguin Books, 1968).
Mussard (1974)	J. Mussard, 'Des idées et des hommes', *Revue Polytechnique,* **1329** (1974), 1025.
Mussard (1979)	J. Mussard, 'Le CERN enfant naturel de l'Europe', *Revue Polytechnique,* **1386** (1979), 573.
NATO (1971)	*NATO, Facts and Figures* (Brussels: NATO Information Service, 1971).
Needell (1983)	A.A. Needell, 'Nuclear Reactors and the Founding of BNL', *Historical Studies in the Physical Sciences,* **14** (1983), 93-122.
Needham (1976)	Interview with J. Needham, 'La coopération scientifique de la guerre à la paix', *Proton, CERN staff association,* **110** (1976), 12-16.
Niebuhr (1949)	R. Niebuhr, 'The Illusion of World Government', *Bulletin of Atomic Scientists,* **5** (1949), 289-292.
Nimrod (1979)	J. Litt (ed.), *Nimrod. The 7 GeV Proton Synchrotron,* Proceedings of a Nimrod Commemoration Evening, Rutherford Laboratory, 27 June 1978 (Didcot: Science Research Council, 1979).
Noyes (1946)	W.A. Noyes, Jr., 'The United Nations Educational, Scientific and Cultural Organization', *Bulletin of Atomic Scientists,* **2** (1946), 16-17.
Noyes (1947)	W.A. Noyes, Jr., 'UNESCO holds first session in Paris', *Bulletin of Atomic Scientists,* **3** (1947), 92-93.
Noyes (1950)	W.A. Noyes, Jr., 'International Cooperation in Science: the Work of UNESCO', *Bulletin of Atomic Scientists,* **6** (1950), 121-25.
Nyberg & Zetterberg (1977)	S. Nyberg and K. Zetterberg, *Sweden and CERN, the Decision-Making Process, 1949-1964* (Stockholm: FEK, report 9, 1977).
OECE (1956)	*L'OECE au service de l'Europe* (Paris: Château de la Muette, 1956 (3e éd.)).
OEEC (1956)	*OEEC at Work* (Paris: OEEC, 1956).
OTAN (1953)	*OTAN* (Paris: Organisation du Traité de l'Atlantique Nord, 1953 (2e éd.)).
Penney (1964)	R.W. Penney, 'Le onzième anniversaire. Quelques souvenirs peu scientifiques', *Courrier CERN,* **4** (dec. 1964), 169-170.
Pestre (1984a)	D. Pestre, *Physique et Physiciens en France, 1918-1940* (Paris: Editions des archives contemporaines, 1984).
Pestre (1984b)	D. Pestre, 'Aux origines du CERN: Politiques scientifiques et relations internationales', *Gesnerus,* **41** (1984), 279-289.
Pestre (1984c)	D. Pestre, 'L'Organisation européenne pour la recherche nucléaire (CERN): un succès politique et scientifique?' *Vingtième Siècle, revue d'histoire,* **4** (1984), 65-76.
Petrucci (1959)	G. Petrucci, 'Il CERN Organizzazione Europea per le ricerche nucleari', *Scientia,* **94** (1959), 209-217.
Pickering (1985)	A. Pickering, 'Pragmatism in Particle Physics. Scientific and Military Interests in the Postwar United States', paper presented at the History of Science Society Annual Meeting, Bloomington, Indiana, 31 Oct.-3 Nov. 1985.
Polach (1964)	J.G. Polach, *EURATOM, its Background, Issues and Economic Implications* (New York: Oceana Publications, 1964).
Puppi (1968)	G. Puppi (ed.), *Old and New Problems in Elementary Particles. A volume dedicated to Gilberto Bernardini* (New York: Academic Press, 1968).

Ramsey (1966)	N.F. Ramsey, *Early History of Associated Universities and Brookhaven National Laboratory,* Brookhaven Lecture Series, 55 (Brookhaven: BNL 992 (T-421), 1966).
Rapport CEA (1950)	*Rapport d'activité du Commissariat à l'Energie Atomique du 1er janvier 1946 au 31 décembre 1950* (Paris: Imprimerie Nationale, 1952).
Rapport CEA (annuel)	*Rapport sur l'activité et la gestion du CEA, année 19..* (Paris: Imprimerie Nationale, annuel après 1951).
Ronayne (1984)	J. Ronayne, *Science in Government* (Victoria, Australia: Edward Arnold, 1984).
Rougemont (1948)	D. de Rougemont, *L'Europe en jeu* (Neuchatel: Editions de la Bacconnière, 1948).
Rudwick (1985)	M.J.S. Rudwick, *The Great Devonian Controversy* (Chicago: University of Chicago Press, 1985).
Salomon (1964a)	J.J. Salomon, 'International Scientific Policy', *Minerva,* **2** (1964), 411–434.
Salomon (1964b)	J.J. Salomon (ed.), *Organisations scientifiques internationales* (catalogue), (Paris: OCDE, 1964).
Salomon (1968)	J.J. Salomon, 'European Scientific Organizations', *Moonman (1968),* 65–86.
Salomon (1970a)	J.J. Salomon, *Science et Politique* (Paris: Seuil, 1970).
Salomon (1970b)	J.J. Salomon, 'Les bases nationales d'une politique européenne de la science', *Cadres Juridiques (1970),* 254–264.
Salomon (1977)	J.J. Salomon, 'Science Policy Studies and the Development of Science Policy', in *Spiegel-Rösing & de Solla Price (1977),* 43–70.
Salomon (1981)	J.J. Salomon, 'Science Policy Studies and Science Policy-Making–The Principle of Serendipity', *Fundamenta Scientiae,* **2** (1981), 401–411.
Schilling (1961)	W.R. Schilling, 'The H-Bomb Decision. How to Decide Without Actually Choosing', *Political Science Quarterly,* **76** (1961), 24–46.
Schwarz (1979)	M. Schwarz, 'European Policies on Space Science and Technology 1960-1978', *Research Policy,* **8** (1979), 204–243.
Schweber (1985)	S.S. Schweber, 'Some Reflections on the History of Particle Physics in the 1950's'. Talk given at *International Symposium on Particle Physics in the 1950's: Pions to Quarks,* Fermilab, 1–4 May, 1985, (to appear in proceedings).
Scuola Fermi (1977)	*Rendiconti della Scuola Internazionale di Fisica Enrico Fermi, LVII Corso* (New York: Academic Press, 1977).
Seidel (1983)	R.W. Seidel, 'Accelerating Science: the Postwar Transformation of the Lawrence Radiation Laboratory', *Historical Studies in the Physical Sciences,* **13** (1983), 375–400.
Seidel (1986)	R.W. Seidel, 'The Political Economy of High Energy Physics', in L.Brown, M. Dresden, and L. Hoddeson (eds.), *Elementary Particle Physics in the 1950's* (Cambridge: Cambridge University Press, forthcoming).
Shils (1948)	E.A. Shils, 'The Failure of the UN AEC: an Interpretation', *Bulletin of Atomic Scientists,* **4** (1948), 205–210.
Simpson (1983)	J. Simpson, *The Independent Nuclear State: the United States, Britain and the Military Atom* (London: MacMillan, 1983).
Skolnikoff (1967)	E.B. Skolnikoff, *Science, Technology, and American Foreign Policy* (Cambridge: MIT Press, 1967).
Spiegel-Rösing & de Solla Price (1977)	I. Spiegel-Rösing and D. de Solla Price (eds.), *Science, Technology and Society* (London: Sage, 1977).

Teitgen (1963)	P.H. Teitgen, *Les étapes de l'idée européenne* (Paris: Etudes et Documents, Documentation Française, 17, 1963).
Urey (1949)	H.C. Urey, 'The Paramount Problem of 1949', *Bulletin of Atomic Scientists,* **5** (1949), 283-288.
Vitale (1974)	B. Vitale, 'Quelques considérations sur le rôle de la science dans le monde capitaliste', *Fundamenta Scientiae,* **14** (1974), 1-30.
Vogt (1985)	E. Vogt, 'Accelerators—Instruments and Symbols for Power', *IEEE Transactions on Nuclear Science, NS-32* (1985), 3834-3837.
Wade (1977)	N. Wade, 'Particle Beams as ABM Weapons: General and Physicists Differ', *Science,* **196** (1977), 407-408.
Wentzel (1953)	A.L. Wentzel, 'International Laboratory for Nuclear Research', *Bulletin of Atomic Scientists,* **9** (1953), 233-234.
Willems (no date)	*Jean Willems (1895-1970), Recueil de souvenirs* (Bruxelles: Imp. Lielens, no date).
Williams (1973)	R. Williams, *European Technology. The Politics of Collaboration* (London: Croon Helm, 1973).

CONFIDENTIEL

NOTE SUR LA CREATION D'UN ORGANISME COOPERATIF
DE RECHERCHE ATOMIQUE EN EUROPE OCCIDENTALE

I. DISCUSSION DES MOTIFS.

Le spectaculaire succès des recherches atomiques américaines effectuées <u>pendant la guerre</u> était dû à la transposition sur le plan industriel des découvertes faites durant la période précédente dans des laboratoires purement scientifiques, dont la plupart était située en Europe. Le personnel dirigeant de l'effort scientifique américain était, pour une très grande partie, d'origine ou de formation européenne.

Si, dans <u>l'après-guerre</u>, l'Europe pouvait garder ce rôle d'initiateur, laissant à l'Amérique le soin de parachever les applications les plus coûteuses, on pourrait envisager l'avenir sans trop d'inquiétude. En fait, cette division des rôles s'avère impossible à maintenir. L'étude des forces fondamentales de la nature exige des moyens de plus en plus éloignés de l'échelle de l'organisme humain non équipé; la découverte résulte de la mise en jeu de mécanismes de plus en plus variés. L'Amérique, servie à la fois par sa richesse, par la multiplicité de vos ressources, et par sa tradition propre de l'effort producteur coordonné, prend aujourd'hui la première place non seulement quant aux applications, mais aussi quant aux découvertes fondamentales elles-mêmes.

L'Angleterre, partenaire de l'Amérique durant la guerre, parvient aujourd'hui à maintenir un effort indépendant qui lui permet de se tenir au courant du progrès de la science atomique. La situation internationale l'oblige à se tenir à l'écart; à défaut de guider l'Europe Continentale, elle lui montre cependant le chemin. Si l'Europe n'arrive pas à s'équiper sur une échelle pour le moins analogue, elle condamnera ses savants atomiques et corpusculaires d'abord à l'inefficacité, puis à la tentation de s'exiler, et finalement à l'impossibilité de recruter des élèves ou des successeurs. Le nombre total des centres de recherche ou de mécanismes puissants ne doit nécessairement pas être conçu sur l'échelle américaine; mais aucune économie n'est plus possible sur le coût <u>d'un</u> centre ou <u>d'un</u> appareil efficace.

Quelques récents succès (la pile française, le synchrocyclotron hollandais) montrent que la construction <u>d'un appareil</u> relativement moderne est encore à l'échelle des possibilités <u>des nations</u> européennes prises séparément. La construction d'un ensemble adéquat de ces appareils ne l'est manifestement plus. Dès lors, pour sauver la situation, les nations doivent combiner

First page of Kowarski's Note on the creation of a western European atomic research centre, dated April/May 1950, and sent to Auger, Dautry, Perrin, de Rose, and several French officials. (Copyright CEA, Paris)

Not for Release before　　　　　　Press Release 311
11 a.m. Friday, 9 June　　　　　　PARIS, 9 June 1950

UNITED NATIONS EDUCATIONAL,
SCIENTIFIC AND CULTURAL ORGANIZATION

AMERICAN SCIENTIST EXPRESSES VIEWS ON UNESCO-SPONSORED
SCIENCE RESEARCH CENTRES

(Florence, 9 June) The desire of leading American scientists to have the strongest competition of comparable western European scientists in creative research on behalf of peace lies behind the United States proposal that UNESCO help organize research centres in Western Europe and elsewhere in the world into newer knowledge in physical and other sciences.

This is the view of Professor Isidor I. Rabi, Nobel prize winner in physics, who presented on behalf of the United States the proposal which was approved by the Programme and Budget Commission of UNESCO's General Conference.

"We scientists in the United States, and to a lesser extent, the scientists in the United Kingdom", Professor Rabi said, "possess instruments of research which do not now exist in Western Europe and elsewhere in the world, nor can they exist there under present circumstances because they are too expensive. We propose that UNESCO use its good offices to get nations together on a regional basis in order to make possible the setting up of creative research facilities compared to those in the United States.

"The purpose we have in mind is to get the most vigorous competition of our fellow-scientists in Europe and elsewhere in the world in creative work on behalf of peace. After all, Science had its birth in Europe, and there are many men of the greatest ability in Europe who are being prevented from fulfilling their parts in the great European scientific tradition only because of lack of the instruments so necessary in modern research.

"We want to preserve the international fellowship of Science, to keep the light of Science burning brightly in Western Europe. Moreover, we want very much to help remove a sense of frustration which, very understandably, is growing among scientists of countries which do not have the material means that we have in the United States. So far as I am concerned, these centres which UNESCO is now to help set up, are one of the best ways of saving western civilization."

Professor Rabi said he hoped the centre in Western Europe would be able to afford a large Cyclotron of the type which exists in the United States in such places as the Massachusetts Institute of Technology, the University of Chicago, Columbia University in New York, and Berkeley, California. He thought it should also have a synchroton: material for expensive biology such as exists at Woods Hole, Massachusetts, and elaborate computing laboratories such as the one at Harvard University.

"UNESCO", the American scientist declared, "should be the catalyst for the science of the world. I do not mean that UNESCO should run the

.../...

Press Release 311 - page 2
9 June 1950

Research Centres but it should make the initial plans, get them started and see that they are kept going. It seems to me that Western Europe could have a Centre as good as the best in my country if France, Italy, Holland, Belgium, Switzerland, and West Germany got together under UNESCO auspices to form it.

"UNESCO already, I believe, goes far toward carrying out the role which I believe to be a major part of its task, that is to act as a kind of Ministry of Education for the United Nations, but it is even more important that UNESCO should provide facilities for making the scientific achievements today that will be taught in the schools and Universities tomorrow. Therefore, it should take concrete steps actually to encourage achievements on the creative side of science."

Professor Rabi's proposal made clear that UNESCO should not contribute to the cost of construction of the Research Centres or of maintaining them out of its regular budget. The Funds, he said, should come from the Governments and from private sources in the Nations of each region involved.

He expressed the opinion that some of the money could come from the United States, and the point was made by the United Kingdom delegate to the Committee, in the course of the discussion of the proposal, that steps might be taken to get financial help from the United States Economic Cooperation Administration.

Press release after Isidor I. Rabi's intervention in Florence in June 1950, in which he proposes that UNESCO help set up a western European research centre. (Copyright UNESCO, Paris) le 14 Octobre 1950

S.P.N° 3207

Cher ami,

    Le Mouvement Européen vient, la semaine dernière, dans la réunion de son Centre Culturel tenue à Genève, de mettre au premier rang de ses préoccupations la question des Centres Scientifiques européens (énergie nucléaire, astrophysique mécanique des fluides, etc...). Une autre réunion se tiendra le 12 Décembre.

    J'aimerais bien savoir d'ici là ce que l'U.N.E.S.C.O. veut faire et ne pas contrarier son action. Si un jour vous avez quelques minutes de libres, pouvez-vous passer rue de Varenne pour que nous en parlions ?

    Cordialement.

R. DAUTRY

Monsieur Pierre AUGER
Directeur de la Section des Sciences
Exactes et Naturelles à l'U.N.E.S.C.O.
19, avenue Kléber
PARIS

Letter from Dautry to Auger dated 14 October 1950 offering the co-operation of the Centre Européen de la Culture. (Courtesy P. Auger, Paris)

Consiglio Nazionale delle Ricerche
IL PRESIDENTE

Pollone 18 dic 1950

A Sua Eccellenza Alcide De Gasperi
Presidente del Consiglio dei Ministri   ROMA

Caro Presidente,

Come Io Ti avevo preannunciato nel mio "Pro memoria riservato" su gli studii di fisica nucleare, questi si stanno sempre più rapidamente sviluppando in tutti i paesi; e siamo ormai alla vigilia della creazione di un "Laboratorio Europeo di fisica nucleare" a cui l'Italia non potrà assolutamente restare estranea.

La Commissione per la cooperazione scientifica del Centro Europeo di Cultura – in cui l'Italia è rappresentata dal Sen. Casati e dal sottoscritto, nonchè da due esperti, i professori Rollier del Politecnico di Milano e Ferretti dell'Università di Roma – riunita il 12 dicembre a Ginevra, ha votato alla unanimità la seguente deliberazione:

La Commission de Coopération Scientifique du Centre Européen de la Culture, patronné par l'Assemblée Consultive Européenne,

réunie le 12 décembre 1950 à Genève, au Centre Européen de la Culture,

vue la résolution n. 2.2 de la Conférence Générale de l'Unesco à Florence en juin 1950

vue la résolution du Conseil Economique et Social des Nations Unies adoptée à Genève le 14 août 1950

recommande:

a) la création, conformément aux voeux de l'Unesco et du Centre Européen de la Culture, et en relation avec leurs secrétariats, d'un Laboratoire européen de physique nucléaire, centré sur la construction d'un grand instrument d'accélération des particules élémentaires. La puissance de cet instrument devra être supérieure à celle prévue pour les appareils actuellement en construction.

b) la constitution d'un fonds européen pour la construction et le fonctionnement de ce Laboratoire. Ce fonds serait alimenté annuelment par les cotisations des Etats fondateurs, d'après un barème établi selon la même formule que celle sur laquelle sont calculées les cotisations nationales aux Nations Unies. Le total annuel de ces cotisations pouvant être évalué à 5 millions de dollars pendant les cinq premières années.

c) le choix d'un emplacement qui satisfasse divers critères tels que: 1) proximité d'un centre important de recherches et d'enseignement; 2) ravitaillement facile en main d'oeuvre spécialisée et proximité de sources d'énergie; 3) commodité d'accés pour tous

...glio Nazionale delle Ricerche
IL PRESIDENTE

les pays fondateurs; 4) facilité d'accorder à cet emplacement un statut d'extraterritorialité. Après examen la Commission suggère que soient étudiées plus spécialement les possibilités de la zone française qui entoure Genève, de la région entre Bâle et Moulouse, et des environs de Copenhague.

d) l'exécution de ce projet selon les étapes suivantes: 1951 études préparatoires par un Bureau d'études; 1952-55 construction du grand appareil; 1953 mise en place de l'équipement auxiliare.

e) la création immédiate à Paris, en relation avec l'Unesco, d'un Bureau d'études chargé de preparer les plans de construction, le futur programme de travail et l'organisation technique et administrative du Laboratoire.

f) la création dès à présent d'un Centre de formation de physiciens théoriciens, destinés à constituer la section technique indispensable au Laboratoire.

La partecipazione dell'Italia è stata prevista dalla Commissione nella misura del 12 % (di fronte al 30 % assunto dalla Francia) La quota potrà naturalmente essere discussa in sede della stipulazione della convenzione tra gli Stati interessati; ma non potrà variar di molto ed importerà una spesa che si potrà presumibilmente aggirare tra i 300 ed i 400 milioni annui.

Se tieni presente che perchè la nostra partecipazione sia effettiva e proficua è assolutamente indispensabile che una somma almeno eguale venga annualmente spesa in Italia per la preparazione scientifica del personale e la costruzione di installazioni sperimentali in cui il personale stesso possa sviluppare in modo autonomo gli studii e le ricerche a cui parteciperà in sede internazionale converrai che non era ingiustificato il preventivo, che io ti ho comunicato a titolo di prima informazione, di un miliardo annuo per i soli studii atomici.

Per il momento, ed in attesa delle decisioni che il Governo crederà di prendere, era urgente che l'Italia partecipasse ai lavori preparatorii di cui alle lettere e) ed f) della deliberazione di Ginevra. E poichè in assenza di una Commissione atomica la responsabilità di questa partecipazione ricade evidentemente sul Consiglio delle Ricerche, io ho senz'altro previsto lo stanziamento, sul bilancio del Consiglio, di un milione di franchi (di fronte a 50 milioni già stanziati dal Governo francese. Ma questo non è evidentemente altro che un punto di partenza, che resterebbe forzatamente senza seguito se le mie istanze in tema di finanziamento del C.N.R. non venissero sollecitamente accolte.

E a questo proposito non posso che richiamare la Tua benevola attenzione su quanto mi sono permesso di farti presente con mio telegramma del 13 u.s.; che cioè l'argomento delle necessarie economie con cui il Ministero del Tesoro cercherà presumibilmente di eludere le mie istanze, non è in questo caso assolutamente accettabile; poichè sono proprio le attuali condizioni di emergenza che impongono

Consiglio Nazionale delle Ricerche
IL PRESIDENTE

imperiosamente ed improrogabilmente una vera e propria mobilitazione della scienza e degli scienziati ai fini della difesa nazionale, ciò che implica che il finanziamento del C.N.R. venga in questo momento considerato come parte integrante delle spese ~~delle spese~~ per la difesa.

Ciò non avviene soltanto negli Stati Uniti che su questa via si sono messi da gran tempo; avviene in tutti i paesi che si accingono a fronteggiare la grave situazione internazionale. Nè ciò riguarda soltanto il settore della fisica nucleare a cui si riferisce particolarmente questa mia lettera, ma tutti quei settori della scienza e della tecnica in cui l'ottima preparazione dei nostri studiosi e dei nostri ricercatori rischia di restare inutilizzata.

La mia esperienza delle due passate guerre mi ha permesso di valutare i gravi danni che sono venuti al paese dalla mancata tempestiva mobilitazione degli scienziati e dei tecnici.

La mia attuale carica di Presidente del Consiglio delle Ricerche mi fa sentire imperioso il dovere di segnalarti il pericolo di una eventuale ripetizione del medesimo errore, e mi induce a dirti ancora una volta che il Consiglio delle Ricerche, i suoi organi e gli studiosi tutti che ad essi fanno capo non chiedono che di essere utilizzati al servizio del Paese.

Gradisci l'espressione della mia immutata devozione e credimi

sempre dev.mo

G. COLONNETTI

Letter from Colonnetti to Alcide de Gasperi dated 18 December 1950, drawing his attention to the importance for Italy of a European laboratory for nuclear physics. It also reproduces the entire text of the resolution passed in Geneva on 12 December 1950. (Copyright State Archive, Turin)

Le 5 juillet 1951

Professor Weisskopf
c/o Professor Niels Bohr
Blegdamveg 15
Copenhague

PERSONNELLE

Cher ami,

    Je me permets de vous déranger au milieu de vos travaux, pour vous demander si vous ne pourriez pas rendre encore un service important au Laboratoire Européen de Physique nucléaire.

    Parmi les personnalités qu'il faut que nous ayons avec nous dans cette entreprise, il y a Scherrer. Vous savez peut-être qu'il n'est pas favorable au projet, peut-être parce qu'il en a d'autres et ne voudrait pas voir l'argent aller ailleurs, peut-être pour d'autres raisons. Mais il faut le convaincre, parce qu'il détermine l'attitude du Gouvernement fédéral, et parce que l'opinion hollandaise et peut-être suédoise en dépendent un peu.

    Pouvez-vous, sans que cela vous coûte trop d'efforts, tâcher de l'influencer? Je ne peux guère le faire directement, car c'est un peu trop évident, et une action indirecte par des savants qu'il estime hautement serait sans doute plus efficace. Également si vous pouvez en dire un mot à Bohr, ce serait très utile. Bohr s'y intéresse, c'est lui qui a suggéré de centrer le Laboratoire sur un cosmotron, mais il serait très utile de connaître sa réaction à la situation générale.

    Mes meilleurs souhaits à vous et votre femme.

    Bien cordialement,

P. Auger.

Letter from Auger to Weisskopf dated 5 July 1951 asking him to convince Scherrer and Bohr of the soundness of the European Laboratory project. (Courtesy P. Auger and V.F. Weisskopf)

I did not reply to your letter because I was wishing to have a discussion with Charles Darwin. He came to dine with me last night and we had a long talk.

I think it would be fair to say that Darwin is not quite convinced that a European Laboratory is necessary or really desirable; but, assuming that such a laboratory is to be established, he would strongly agree that Copenhagen is by far the best place for it. Darwin will write to Kramers in this sense, probably adding some reservations.

I have also written to George Thomson, who replied that he was "definitely attracted" by the Copenhagen scheme and would like to discuss it with me. I have not yet seen him and it will be some days before we can meet.

Fraser sent me a copy of his letter to Auger. I thought that this letter betrayed a misunderstanding of the present position in this matter and that it might have unfortunate effects. I therefore wrote to Fraser to point out that the suggestion about Copenhagen came from Kramers and that your attitude, as I understood it, was that you were willing to accept the burden and responsibility, not that you were seeking it.

I added a remark to the effect that there

First page of a draft letter from Chadwick to Bohr, around 20 September 1951, expressing his support for Kramers' idea. (Courtesy Lady Chadwick, Cambridge and the Master, Fellows and Scholars of Churchill College in the University of Cambridge)

UNESCO/NS/NUC/6 (Rev. 3)

Paris, 19 December 1951.

UNITED NATIONS EDUCATIONAL, SCIENTIFIC AND CULTURAL ORGANIZATION

Conference on the organization of studies relating to
the establishment of a European Nuclear Research Laboratory

Paris, 17 - 21 December 1951

Draft Resolution proposed by the Netherlands Delegation

The Conference recommends :

1 - Setting up a Board of Representatives from the participating countries, with headquarters in Geneva, to supervise the programme embodied in items 1 to 5.

2 - Accepting the offer made by the United Kingdom representative to use the Liverpool synchro-cyclotron for 400 MeV protons as an instrument to be operated on a European basis.

3 - Accepting the offer made by the Danish representative to use the Institute of Theoretical Physics in Copenhagen to assemble a study group for theoretical research on a European basis. It should provide theoretical guidance for experimental work, to be carried out with the machines.

4 - Establishing a planning group for an intermediate machine.

5 - Establishing a planning group for a big machine.

6 - Establishing a planning group for the organization of the European Laboratory for Nuclear Research. In this Laboratory, the machines should be installed and advanced studies should be carried out.

UNESCO/NS/NUC/9 (Prov.)
Annexe IX
Original : Français.

ORGANISATION DES NATIONS UNIES POUR L'EDUCATION, LA SCIENCE ET LA CULTURE

Conférence pour l'Organisation des Etudes concernant la création
d'un Laboratoire Européen de Recherches Nucléaires,

Paris, 17 - 21 décembre 1951.

Discours de clôture prononcé par le Président, M. François de Rose,
le 20 décembre 1951 (matin).

Au moment où va prendre fin la première partie de la Conférence pour l'organisation des études concernant la création d'un laboratoire européen de recherche nucléaire, je voudrais vous féliciter des résultats auxquels vous êtes parvenus. Je ne veux pas revenir sur le détail des décisions et des recommandations que vous avez adoptées mais je ne puis m'empêcher de rendre hommage à l'esprit dans lequel vous les avez adoptées.

Nul d'entre nous n'ignore que les points de vue n'étaient pas identiques au début de nos travaux. Mais vous aviez tous en commun la volonté d'établir une collaboration scientifique entre vos pays. Pour atteindre ce but vous avez fait les concessions suffisantes pour qu'un programme commun de travail soit tracé et pour en confier l'exécution à un organisme où tous vos gouvernements seront représentés.

Vous n'avez pas seulement servi ainsi la cause de la coopération sur le plan scientifique. Si vous aviez échoué, si des hommes de science n'avaient pu se mettre d'accord sur un programme d'action commun, les répercussions de cet échec eussent été ressenties bien au delà du domaine de la physique nucléaire. Au contraire, ce premier succès sera pour les gouvernements et pour l'opinion publique de nos pays, et pas seulement de nos pays, une preuve nouvelle de la vitalité de notre vieux continent et de sa volonté de garder son rôle dans le développement d'une civilisation de plus en plus marquée par la pensée scientifique.

Peut-être me permettrez-vous d'ajouter une remarque tirée de mon expérience professionnelle. Il est très facile de présider vos débats. Lorsque les choses paraissent un peu difficiles, il n'y a qu'à laisser se créer une certaine confusion et à lever la séance. Vous vous réunissez alors entre vous et, à la séance suivante, quelqu'un apporte une résolution sur laquelle tout le monde est d'accord. Ce qui prouve qu'alors que l'on chercherait en vain un physicien parmi les diplomates il se trouve beaucoup de diplomates parmi les physiciens.

ACCORD

PORTANT CREATION D'UN CONSEIL DE REPRESENTANTS D'ETATS
EUROPEENS POUR L'ETUDE DES PLANS D'UN LABORATOIRE INTERNATIONAL
ET L'ORGANISATION D'AUTRE FORMES DE COOPERATION
DANS LA RECHERCHE NUCLEAIRE

GENÈVE, LE 15 FÉVRIER 1952

AGREEMENT

CONSTITUTING A COUNCIL OF REPRESENTATIVES OF EUROPEAN
STATES FOR PLANNING AN INTERNATIONAL LABORATORY AND ORGANIZING
OTHER FORMS OF CO-OPERATION IN NUCLEAR RESEARCH

GENEVA, 15 TH FEBRUARY 1952

Cover page of the official version of the Agreement signed on 15 February 1952. (Copyright CERN, Geneva) of the Cosmotron had attracted a very representative group.

Prior to our visit, Brookhaven had our tentative specifications for a 10 GeV machine, and our project had caught the interest of the permanent and visiting staff.

During discussions of our magnet designs and n value preservation, the thought came up that much could be gained in vacuum chamber cross section and n value control by arranging the magnet blocks with alternately opposing gaps of very high n value.

This idea with elaborations was presented to us jointly by Dr's Courant, Livingston, Blewett, Sneyder, White and more.

As a result of discussions over several days between the Brookhaven group, Goward and Wideröe, it became clear that this approach must be looked into very thoroughly.

During the course of the days a tentative design was worked out by Livingston in order to demonstrate the advantages in dimensions, cost and engineering as compared to the cosmotron. His specifications are as follows:

Energy: 30 GeV
Injection: 3.5 MeV
Orbit radius: 300 feet
Number of alternate magnet sections: 240
Length pr. magnet section: 7 feet
Total magnet cross section: 12" on the radius by 20"
Magnet gap on orbit: 2"
n: 3600
Vacuum chamber cross section: about 2" diam.
Iron weight: not more than 500 tons
Copper weight: about 300 tons
Cost: same as 3 GeV Cosmotron
Power: Same as 3 GeV Cosmotron
Cooling: Radiation from coils to surroundings.

The Brookhaven group felt quite confident about this new idea. The laboratory is, however, not in position to follow up with plans for a new machine in the near future, but there is an understanding between Brookhaven and Berkeley to the effect that if a larger machine is to be built some time, Brookhaven is to have the

Extract from Dahl's trip report to the United States (CERN-PS/54) dated 25 August 1952 announcing the discovery of the alternating-gradient principle. (Copyright CERN, Geneva)

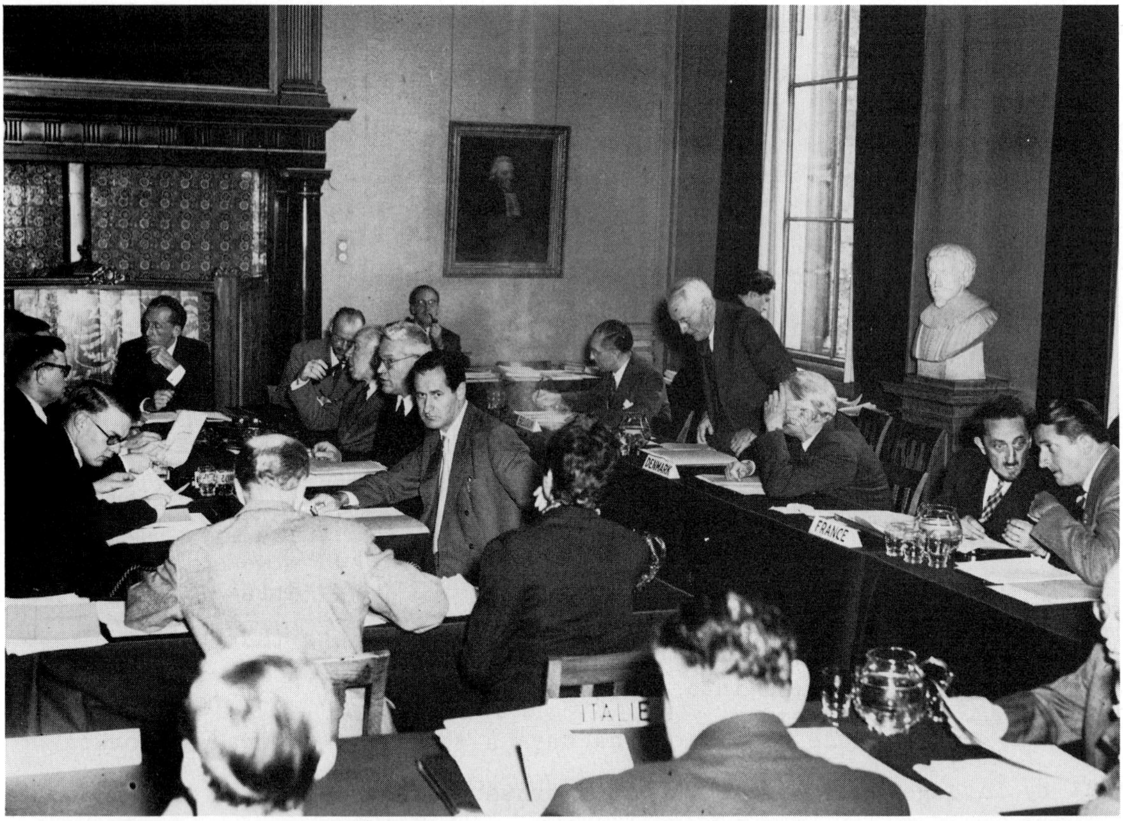
DENMARK
FRANCE
ITALIE

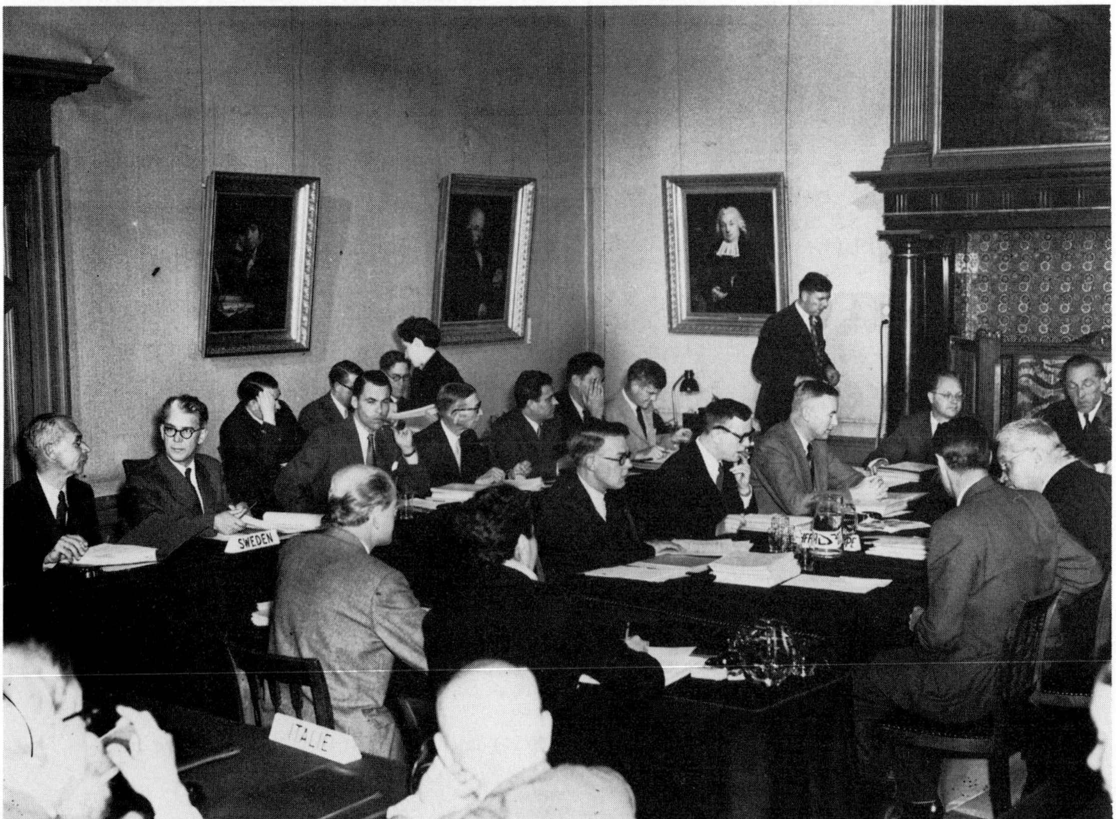
SWEDEN
ITALIE

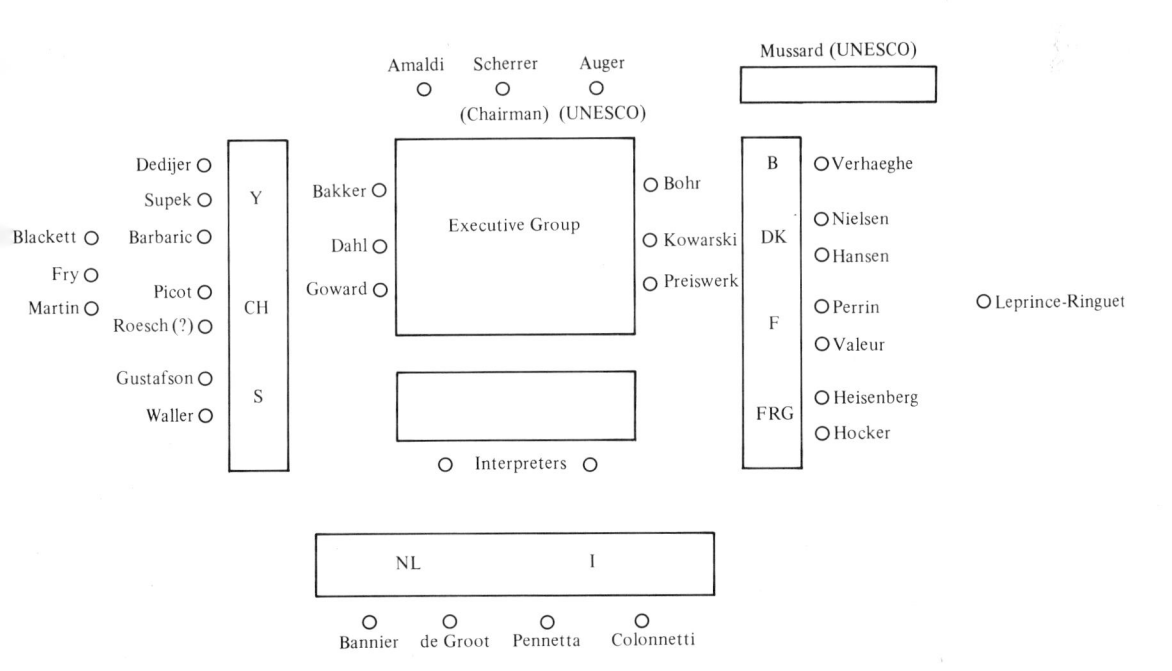

```
                    Amaldi    Scherrer   Auger           Mussard (UNESCO)
                      O         O         O            ┌─────────────────┐
                              (Chairman) (UNESCO)       │                 │
                                                        └─────────────────┘
                                               ┌──┐
               Dedijer O                       │ B│  O Verhaeghe
                Supek O  ┌──┐                  │  │
                         │ Y│  Bakker O        ├──┤  O Nielsen
   Blackett O  Barbaric O│  │         ┌──────────┐ │DK│ O Hansen
         Fry O           │  │  Dahl O │Executive │ O Bohr
                 Picot O │  │         │  Group   │ ├──┤
       Martin O          │CH│ Goward O│          │ O Kowarski  O Perrin       O Leprince-Ringuet
             Roesch(?) O │  │         └──────────┘ │ F│ O Valeur
                         │  │                    O Preiswerk
            Gustafson O  │ S│                     ├──┤
              Waller O   │  │                     │  │ O Heisenberg
                         └──┘                     │FRG│ O Hocker
                                                  └──┘
                               ┌──────────────────┐
                               │                  │
                               └──────────────────┘
                                    O Interpreters O


                        ┌─────────────────────────┐
                        │   NL            I       │
                        └─────────────────────────┘
                            O        O        O        O
                         Bannier de Groot Pennetta Colonnetti
```

Third Session of the provisional CERN Council, Amsterdam, 4–7 October 1952. (Photographs: Lindeman, Amsterdam)

alle 15.30.
Alle 16 parto in macchina con
Adams per Harwell ove arrivo
verso le 19. Cena e conversazione
con Fry, Goward, Adams, Laws
Walkinshaw, Wilson.
Dopo cena a Abingdon Hotel
Crown & Thistle

12 Parto in macchina da Abingdon
alle 8.20 e sono alle 10.30 al
D.S.I.R. (Regen street 5   ) discuss
con King e Verry dello schema di
convenzione studiato dagli inglesi.
Pranzo con King e Waterfield al
Athenaeum club.
Di nuovo al D.S.I.R. per scrivere
varie lettere. Alle 4 vado a
Imperial College ove incontro
Geroge, Wynn-Williams, Klein, poi
Hodson ecc. A cena dai
George.

13 Vado al D.S.I.R. a firmare le

Extracts from Amaldi's diary, Quaderno No. 1, from December 1952 dealing with his trip to Britain and the inauguration of the Cosmotron. (Courtesy E. Amaldi, Rome)

varie lettere e poi passeggio per Londra.
Alle 8.30 p.m. parto per New York

14 Giungo alle 8.30 all' Aeroporto
internazionale di New York. Con una
macchina vado a Brook Haven Nat.
Lab. Nel pomeriggio arrivano Dahl
e Randers. Siamo a Cena con
Johnsen dai Green. Ci sono gli Haworth
e Kruger.

15 Il cosmotrone è dedicato. Alla sera
Cena e mio discorso.

16 e 17 Congresso on High energy accelerator
Alla sera voliamo a Rochester

18 Inizia il congresso on High energy
physics
Prosegue il congresso e alla sera faccio
il mio secondo discorso sul CERN.
Finisce il congresso e vado a Ithaca
in macchina con Cocconi Silverman
Morrison. Sono ospite dai Cocconi.
È domenica e passo la giornata
con i Cocconi

Copy to Sir James Chadwick

Baltic Hotel,
Copenhagen,
Denmark.

11th November, 1952

Dear Sir John,

I thought it might be useful to send back some impressions of the state of things in C.E.R.N., as seen in Paris and from two days in Copenhagen. There are some points which need careful attention, though I expect you know about most of them already.

1) There is some pressure to use the proposed laboratory at Geneva not only for high energy physics but also for conventional nuclear and even atomic physics. Some people in Paris, for example, were talking of having 'two or three Van de Graaffs', an electron microscope, and x-ray machines. At the last Council meeting there was a motion to change the wording in a certain agreement from 'high energy physics' to 'nuclear physics'. The vote was evenly divided and there was a procedural muddle, but fortunately in the end no change was made.

The provisional plan for the laboratory envisages an administrative block with six scientific divisions, with roughly equal accommodation; theory, proton synchrotron (P.S.), synchro-cyclotron (S.C.), chemistry, techniques and general physics. It is obvious that the last three of these might easily get out of hand unless their functions are closely defined from the outset. Personally I do not see that a general physics division is necessary. Kowarski said that it might, for example, be used to develop nuclear emulsions, but the techniques division should surely do this. A chemist (whose name I do not know) was suggesting an electron microscope to study changes in materials

---

First page of a letter from Cassels (in Copenhagen) to Cockcroft dated 11 November 1952, expressing anxieties about the range of activities envisaged for CERN. (Crown copyright U.K. Public Record Office. Doc. AB6/1074)

zu verwenden (Ballonversuche, Arbeiten an den Maschinen in Liverpool, Uppsala usw.). Eine Zusammenarbeit auf diesen Gebieten sei im Entwurf der Konvention in Aussicht genommen. Wenn es für die experimentellen Arbeitsgruppen sinnvoll sei, gleich nach Genf zu gehen, so treffe das nicht in gleicher Weise für die übrigen an den "anderen Formen der Zusammenarbeit" beteiligten Forscher zu.

Herr G e n t n e r befürchtet, dass es im Rat noch Krisen geben werde, wenn konkret verlangt würde, den Wohnsitz nach Genf zu verlegen. Das gelte vor allem für die Zeit, in der das Laboratorium gebaut werde. Um später Mitarbeiter für Genf parat zu haben, habe er jetzt schon zwei Nachwuchskräfte zur Ausbildung nach Cambridge und Liverpool geschickt.

> Nach weiterer Aussprache beschliesst die Kommission, die Deutsche Forschungsgemeinschaft zu bitten, beim Auswärtigen Amt eine Ergänzung der Konvention dahin zu beantragen, dass 10% des Budgets für andere Formen der Zusammenarbeit ausserhalb von Genf verwendet würden.

Herr H e i s e n b e r g rechnet damit, dass ein solcher Vorschlag von den drei skandinavischen Staaten, aber auch von England und Italien unterstützt würde.

Herr H o c k e r bittet die Kommission noch um eine Aeusserung, ob eine deutsche Beteiligung an dem Genfer Projekt, die auf sechs bis sieben Jahre hinaus jährlich 2,5 bis 3 Mill.DM erfordern würde, unter wissenschaftlichem Gesichtspunkt verantwortet werden könnte.

Er gibt den auf der letzten Sitzung des Rats ausgearbeiteten Vorschlag über die prozentuale Beteiligung der Mitgliedsstaaten an den Gesamtkosten bekannt.

Herr H a x e l betont das wissenschaftliche Interesse an dem Projekt, glaubt aber, dass das politische Interesse an einer deutschen Beteiligung noch stärker ins Gewicht falle.

Auch Herr H e i s e n b e r g ist der Ansicht, dass man die Aufwendungen sowohl wegen der wissenschaftlichen Bedeutung des Projekts als auch insbesondere unter dem Gesichtspunkt der europäischen Zusammenarbeit vertreten könne.

Herr H o c k e r fragt noch, ob die Kommission ein wirtschaftliches Interesse der Bundesrepublik an der Beteiligung für gegeben erachte.

Herr G e n t n e r sagt hierzu, es sei noch nicht entschieden, ob das Synchrozyklotron von Schneider-Creusot oder im Ruhrgebiet gebaut würde.

> Die Kommission ist der Ansicht, dass eine Beteiligung der Bundesrepublik an der Konvention mit jährlich 2½ bis 3 Mill.DM, die sich nach der Fertigstellung des Laboratoriums auf knapp die Hälfte verringerten, angemessen sei.

For the German Federal Republic Pour la République Fédérale d'Allemagne

[signature]
subject to ratification

For the Kingdom of Belgium Pour le Royaume de Belgique

[signature]
sous réserve de ratification

For the Kingdom of Denmark Pour le Royaume de Danemark

[signature] Elvaerum
sous réserve de ratification

23.12.53

For the French Republic Pour la République Française

[signatures] Alexandre Parodi, Robert Valeur
sous reserve de ratification

For the Kingdom of Greece Pour le Royaume de Grèce

[signature] N. Kmbirsicos
sous reserve de ratification.

For Italy Pour l'Italie

[signatures] Gustavo Colonnetti, Antonio Pennetta
sous reserve de ratification

For the Kingdom of Norway   Pour le Royaume de Norvège

Subject to ratification
31/12/1953.
[signature]

For the Kingdom of the Netherlands   Pour le Royaume des Pays-Bas

[signature]
Subject to ratification.

For the United Kingdom of Great Britain and Northern Ireland   Pour le Royaume-Uni de la Grande-Bretagne et de l'Irlande du Nord

[signature]
Subject to ratification.

For the Kingdom of Sweden   Pour le Royaume de Suède

Ivar Waller
Torsten Gustafson
Subject to ratification

For the Confederation of Switzerland   Pour la Confédération Suisse

[signature]
sous réserve de ratification.

For the Federal People's Republic of Yugoslavia   Pour la République Fédérative Populaire de Yougoslavie

Pavle Savić
sous réserve de ratification

Signatures of the Convention establishing a European Organization for Nuclear Research in Paris on 1 July 1953, and thereafter by the authorized representatives of twelve European governments. (Original deposited at UNESCO, Paris)

Pierre Auger, Gustavo Colonnetti and Miss Thorneycroft (UNESCO) at the meeting in Geneva on 15 February 1952 when the Agreement establishing the provisional CERN was signed.

Eliane Bertrand (de Modzelewska) and H.L. Verry (U.K.) at a session of the provisional CERN Council in 1953.

Villa de Cointrin at Geneva Airport, preliminary headquarters of the newly established laboratory (1954). (Photo CERN)

A delegation of Geneva officials visit the site on 17 May 1954 when the first earth was moved. (Photo CERN)

Seminar organized by the authors on CERN's history and held on the occasion of the laboratory's thirtieth anniversary in September 1984.

Bannier and de Rougemont at the coffee break.

Front row, left to right,
Amaldi – Mercier – Bannier – Pestre
Second row, left to right,
Auger's daughter – de Rougemont – Mussard – Rabi – Weisskopf.

PART IV

National decisions to join CERN

CHAPTER 9

French attitudes to the European laboratory[1]
1949-1954

Dominique PESTRE

Contents

9.1 The French political and diplomatic context, and nuclear policy   305
    9.1.1 The Quai d'Orsay and France's European policy   305
    9.1.2 The Commissariat à l'Energie Atomique and French nuclear policy   307
9.2 The French scientific context, 1945-1955   309
    9.2.1 Aspects of the material infrastructure for scientific research in France   310
    9.2.2 Essential characteristics of physics in France around 1950, with particular reference to high-energy physics   315
9.3 French initiatives regarding the UNESCO project, the decision-making process, 1949-May 1952   317
    9.3.1 The French project, November 1949-December 1950   318
    9.3.2 The project for a laboratory centred around a bevatron: first encounters of the initiators of the project with the governmental machinery, December 1950-May 1951   319
    9.3.3 The interministerial meeting of 18 May 1951   321
    9.3.4 The French decision to participate in phase 2 of the UNESCO project, October 1951-May 1952   323
    9.3.5 A few remarks on the decision-making process in France   325
9.4 Towards the signing of the Convention, February 1952-July 1953   328
    9.4.1 Financing the first year of CERN   328
    9.4.2 France and the problem of the site of the laboratory   329
    9.4.3 French requirements regarding the Convention   331
9.5 The debate surrounding French ratification, July 1953-September 1954   334
    9.5.1 Communist opposition to the CERN project   335
    9.5.2 The ratification process   339
    9.5.3 The debate in the National Assembly   340
9.6 Concluding remarks   342
    Notes   344
    Bibliography   350

Those who have read the chapters devoted to the prehistory of CERN and to the period of the provisional organisation may feel that a study of French attitudes is superfluous: the importance of France's participation, and the role of men such as Dautry, Auger, Perrin, de Rose, and Kowarski have already been considered. However, there are at least two considerations justifying such a study.[2]

The question arises first as to whether the attitudes we have described were representative. We have seen that a section of the scientific community responded favourably and was even partially responsible for initiating the European project. What, however, were the reactions of the rest of the physics community, of the administrators, and in political and diplomatic circles? These aspects have not yet been considered. In fact, it should be realized that the physics community was hardly consulted before 1952, that a significant proportion expressed its hostility to the project, and that in certain respects it was the Ministry of Foreign Affairs which played one of the decisive roles in France.

We must also explore the motivations of the various parties. We mentioned two obvious motives in our earlier description of the process which led to the establishment of CERN. We stated that one reason why the French were so keen to co-operate in a European project was that French science, and especially high-energy physics, were in a relatively mediocre state around 1950–52, and that such co-operation offered the opportunity to emerge from this critical situation and rapidly to rejoin the world leaders. We added that at the beginning of the fifties French diplomatic circles were *a priori* in favour of all forms of European integration—which would have facilitated the approval of the project by the Ministry of Foreign Affairs. *However, we never explored other avenues* in a systematic way and there is nothing to say that these two motives were the only ones, nor even the most common. This is something we would like to look at now.[3]

A few words on how we approach this study are called for. We begin by reconstructing the French 'context', analyzing it on two levels. The first focuses on the politico–diplomatic situation—in particular on the projects for European integration drawn up by the Ministry of Foreign Affairs—and on nuclear policy. Here we consider the role played by the Commissariat à l'Energie Atomique (CEA) in the conception of the European laboratory project (section 9.1). At the second level of analysis we discuss the resources of manpower and of finance available to French science, and the quality and quantity of equipment available for fundamental particle physics research (section 9.2).

Only once we have done this do we more specifically consider French behaviours and moves—namely, the way the decision-making process unfolded inside the country from 1949 to early 1952 (section 9.3), the way the Ministry of Foreign Affairs handled the key questions of the site and of the Convention during the 'planning' period (section 9.4), finally, the way the whole governmental machinery, the political parties, and the parliament dealt with the matter between the summer of 1953 and that of 1954 (section 9.5). The reader should therefore not be surprised to encounter remarks of a relatively general nature in the first pages, remarks which we nevertheless deem essential for a full appreciation of the French case.

9.1 The French political and diplomatic context, and nuclear policy

9.1.1 THE QUAI D'ORSAY AND FRANCE'S EUROPEAN POLICY [4]

We do not propose to give here a detailed account of France's European policy during the decade after the war. However, as we have just explained, it would be difficult not to make passing reference to it in view of its apparently determining influence on French attitudes to the European laboratory project, at least at governmental level.

To understand what led France to its pro-European stance at the start of the fifties, two observations must be made: that France, which had emerged from the war in a very weakened state (be it in terms of population, industrial capacity, or military power, and with respect to the United States, the Soviet Union as well as the United Kingdom), had become a second-rate world power. On the other hand, French leaders refused to recognize this fact. They dreamt nostalgically of France's former greatness and tried to make her 'once more into one of the great world powers and to ensure her a haughty independence from the other powers'.[5]

These pretensions naturally led the politicians of the Fourth Republic into a series of cul-de-sacs caused by the all-too-evident contradiction between their stated aim of playing a leading role and the means actually at their disposal. In this connection one may simply cite, in the economic field, the example of the Marshall Plan, the negotiations for which began in 1947 and which led, on 16 April 1948, to the convention establishing the Organization for European Economic Co-operation (OEEC).[6] Similarly, in the military field, it was necessary for France to join NATO (1949) under the *de facto* leadership of the United States and to call upon American assistance in her war in Indochina. Thus, in the words of one of the foremost French sociologists of the post-war period, it can be said that 'General de Gaulle ordered France to play a role commensurate not with her means, but with his own conceptions of her grandeur. The parliamentarians who followed him [...] took over without abandoning the policy of greatness' (Raymond Aron).[7]

The consequences of the impossibility of adjusting ends to means are easily imagined. French politicians who had defined objectives which the country could not achieve alone—and which her allies were obliged to contest in the name of

Notes: pp. 344 ff.

realpolitik—suffered repeated setbacks, and were obliged in the end to accept positions which had originally been considered unthinkable.[8]

In the case of European policy, this desire for greatness initially expressed itself in the aim, announced in 1945, of preventing by all means 'the revival of German power and influence on the Continent'.[9] Unfortunately for France, from 1947, the cold war led the Anglo-Americans to dispute this French priority: they wanted Germany to become again a power acting in the interests of the 'free world'. Subjected to increasing pressure to adopt a more flexible approach, France's political leaders resisted, but could not prevent the European policy advocated by Washington and London from gradually becoming a reality. At the beginning of 1950 an alternative emerged in Paris. Since the initial objectives were no longer tenable, 'why not combine [France's] weakness with that of others to create a common force capable of exorcizing the threat of one of the great powers and of inspiring the respect of the other'?[10]

Thus, in May 1950, Robert Schuman, the French Minister of Foreign Affairs, proposed the plan which bears his name, a plan which aimed at the establishment of a European Coal and Steel Community (ECSC), this to be the first step towards a more complete integration of the nations of the old continent. The new policy, 'drawn up under the banner of Franco-German reconciliation', had two related advantages. On the one hand, it made it possible not to abandon totally the initial objectives and to achieve, through integration, a degree of control over the evolution and the development of the new Germany. On the other hand, it opened up for Europe (and thus for France which hoped to play a key role there) a positive alternative—the opportunity to mobilize greater resources and so to act as a buffer between the two super-powers. Once the idea was launched this second advantage, namely to make a genuine contribution to the development of Europe, came to be more important than the first.[11] Thus, on 13 December 1951, the French parliament ratified the agreement signed on 18 April by Belgium, Germany, Italy, Luxemburg, the Netherlands, and France for the establishment of the ECSC.

The 'German problem' was not, however, totally resolved for France. The outbreak of the Korean War in June 1950 immediately highlighted another issue, that of the German contribution to the war effort of the Atlantic camp. Urged by the Americans to find a solution, but still refusing to allow the remilitarization of the Federal Republic of Germany, the head of the French government, René Pleven, thought of re-applying the 'Schuman model'. On 23 October 1950 he proposed the formation of a European army in which German contingents would be integrated. On 27 March 1952 the Six signed an agreement embodying proposals of this kind. All the same in 1953 and 1954 there was greater hesitancy than in the case of the ECSC. The question of national sovereignty posed by the 'dissolution' of the national armies was at stake. Accordingly on 30 August 1954, after several years of interminable discussions, the French parliament refused to ratify the agreement.[12]

What conclusion is one to draw regarding the French attitude to Europe and its reconstruction? That between 1950 and 1953 (a key period for this study) it was both offensive and positive, regardless of the complexity of the motives and the paths which led to it. It seems to have enjoyed public support and to have been consciously based on a Paris–Bonn axis, the United Kingdom and the Scandinavian countries showing no interest in its development. The projects drawn up by the scientists for a European laboratory for nuclear research in 1951 and 1952 thus *resonated* with those of French diplomacy. They therefore stood a good chance of obtaining support—if only for this reason—from the Ministry of Foreign Affairs, even if they had not been initiated there. We will see that this is indeed what happened. By contrast, in the British case, projects of this type were *a priori out of phase* with the Foreign Office's policy—which could only complicate their path through the various government departments.

9.1.2 THE COMMISSARIAT À L'ENERGIE ATOMIQUE AND FRENCH NUCLEAR POLICY [13]

The interest of describing quickly French nuclear policy and so that of the CEA in the decade 1945–1955, lies in the fact that it was part of the senior staff of the CEA, in collaboration with a few highly-placed civil servants in the Quai d'Orsay, who organized support in France for the projected European laboratory for nuclear research. This is no mere coincidence, being due rather to the quality of these men, the way the CEA was organized, and the importance of the financial means at the Commissariat's disposal.

A preliminary general remark: just as we have stated that one of the mainsprings of French diplomacy in the post-war period was a strong will to re-establish a '*grande politique* which would maximize France's primary goals of security and independence', so it may be said that the very rapid establishment of the Commissariat à l'Energie Atomique was a response to the desire to prevent France from being left out of a field without which no state could claim to be a great power. As the statute setting out the motives for the establishment of the CEA put it: 'Pressing necessities at the national and international levels demand the taking of measures necessary so that France *can take its place* in the field of atomic energy research' (our emphasis).[14]

The CEA was set up by statute 45-2563 of 18 October 1945, a statute promulgated by the provisional government of the French Republic presided over by General de Gaulle, and directly inspired by two men, the scientist Frédéric Joliot and the administrator Raoul Dautry, recently appointed Minister of Reconstruction and Town Planning.

The Commissariat had numerous purposes, the most important being to pursue 'scientific and technical research with a view to the utilization of atomic energy in the various fields of *science, industry* and *national defence*' (our emphasis). It was also to develop 'on an industrial scale devices generating energy from an atomic source' and to take 'all necessary measures to place France in a position to benefit from the development of this branch of science'. Its role was therefore twofold—to organize both fundamental research and applied and industrial research.[15]

Amongst the panoply of French research institutes, it had a unique structure. Coming under the direct control of the Prime Minister, it enjoyed considerable administrative autonomy (Articles 5 to 8 of the statute), and was protected from conflicts over the distribution of financial resources between ministries. Its expenditure, in particular, was not *a priori* subject to financial control and, being considered vital to the national interest, it received important sums of money compared to other scientific establishments.[16] Clearly these three factors endowed it with wide material possibilities for action and considerable freedom in its scientific options, notably with respect to fundamental research.

It was managed jointly by an Administrator-General (responsible for administrative and policy matters) and a High Commissioner (with responsibility for scientific matters). Raoul Dautry and Frédéric Joliot held these posts for the first five years. An *Atomic Energy Committee* monitored the programmes and approved the draft budgets. Until it was reorganized in 1951, the committee had a membership of six, the High Commissioner and the Administrator-General, a representative of the Ministry of National Defence, and three scientists (Irène Joliot-Curie, Francis Perrin, Pierre Auger)—thus four of the six members were scientists.[17]

At the time of its establishment, the CEA had four trump cards: competent scientists (i.e. Joliot and members of the Anglo-French team in Canada, notably Kowarski, Guéron, and Goldschmidt), a stock of uranium oxide hidden during the occupation, the possibility of buying heavy water in Norway and, finally, deposits of uranium-bearing minerals on national territory and in Madagascar. In the light of this situation, four priorities were defined in 1946: to train people in large numbers, to create new industrial techniques which were essential for an atomic programme but which were non-existent in France, to organize prospecting for mineral deposits, and to construct a first research reactor. In 1947 it was decided to construct the simplest possible reactor (using uranium oxide and heavy water). This went critical on 15 December 1948 and was called ZOE.

In August 1949 a project for a more powerful and sophisticated reactor was approved. Meanwhile, it had become possible to set up formally an *accelerator division* (summer 1947), a division of chemical physics, a division of nuclear physics and one of pile management (1948), a mathematical physics division (beginning of 1949), as well as divisions of technology (which was strengthened markedly during the summer of 1949), of electrical construction, and of mechanical construction. As a result the CEA became the main French supplier of laboratory equipment. It also became a major centre for fundamental research at the instigation of the scientists, who seem to have been more influential than the administrators from 1946 to 1950.[18]

Nevertheless, around 1949–1950 major tensions arose, particularly between Dautry (with the backing of an increasing number of politicians) and Joliot. Several associated factors account for this: the fact that Joliot (and a considerable proportion of the CEA's staff) were Communists, and had proclaimed this loud and clear at a time when the cold war was at its height; the latent conflict, mainly among scientists on the one hand and administrators and politicians on the other, regarding the possible military applications of the work carried out at the CEA; finally, the ever more clearly expressed desire of Dautry and his colleagues to orientate the Commissariat more and more specifically towards industrial research. The result of this was the sacking of Joliot by the government on 28 April 1950 and the inauguration of a phase of reorganization and reorientation of the CEA's activities.

The nature of the reorganization was spelt out on 3 January 1951. It was characterized by a reinforcement of the powers of the Administrator-General at the expense of those of the High Commissioner, and an increase in the size of the Atomic Energy Committee from six to ten, of whom only four were scientific members. A reorientation of activities followed suit. A five-year development plan was introduced which could be described briefly, in Lawrence Scheinman's words, as 'an industrial production decision' as opposed to a 'research decision'. It was decided to opt for plutonium production—which in general explains the increase in governmental scrutiny of the CEA's activities—though not at the expense of fundamental research.

If one wished to explain in a few words the significance of the events of 1950–51, it could be said that for the preliminary period 'the primary goal was to bridge the gap between France and other atomic nations'. This accounts for the importance of the scientists. But after five years 'a new plateau had been reached which could serve as a springboard for industrial development and the eventual exploitation of atomic energy as a source of power'. Hence the reorientation towards applied research, which reflects this change in the technical and material level of France's atomic capacity.[19]

9.2 The French scientific context, 1945–1955 [20]

As we mentioned earlier, we would now like to present a brief study of the scientific situation in France, that very poor situation being apparently one of the major concerns of the French *scientists* who initiated the European laboratory project. Unfortunately we cannot refer the reader to studies already published, since none exists as far as we know. Unfortunately too, the documentation with regard to physics between 1945 and 1960 is not very good. Accordingly we can do little more than give a rough idea of some of the relevant aspects of French science, namely, the size of the French effort compared to that of the other major developed countries, the level of the research budget (in order, for example, to assess the relative importance of the finance France was prepared to offer for CERN), the rate of reconstruction of science in France, and finally, a few words more specifically on the quality of physics with particular reference to cosmic-ray physics and accelerator technology.

9.2.1 ASPECTS OF THE MATERIAL INFRASTRUCTURE FOR SCIENTIFIC RESEARCH IN FRANCE

The simplest way to give an impression of the order of magnitude of France's investment in science compared with that of other developed countries is to choose two examples from before and after the period we are considering. In February 1948, speaking at Brookhaven on the French nuclear effort, Kowarski said that this effort 'can be summarized in a simple formula: France stands in about the same ratio to Britain as Britain to the US. The numerical value of this ratio is, roughly, of the order of 10, if we consider such criteria as the annual budget [...] or the total manpower employed'. He added: 'the state of our industry imposes on us [the CEA] the necessity of making almost everything in our workshops with our own manpower. For example [...] amplifiers, scalers, surveying instruments, beta and gamma Geiger counters [...]'.[21] Eight years later the national colloquium held at Caen on scientific research and education drew conclusions which were hardly more optimistic. It stated that to catch up with other developed countries, it would be necessary '*to double* the number of students *within 10 years*', '*to triple* the number of engineers *over the same period*' and '*to increase tenfold in 10 years* the personnel engaged in scientific research and teaching'. If one adds that at that time the number of scientific diplomas awarded annually in France was still rather small (4,750 in 1954 as against 7,900 in the Federal Republic of Germany, 8 to 13,000 in the United Kingdom, 66,000 in the USA, and 72,000 in the USSR), that the number of researchers as a proportion of the working population was also small (1 in 4,750 in France in 1954 as opposed to 1 in 650 in the USA) and that the rate of increase of science graduates was one of the lowest in Europe (1% for engineers, for instance, in 1956 as against 2.5% in Sweden, 3.6% in the United Kingdom, and 4.2% in the Netherlands), one readily concludes that there was a persistent structural weakness in French science between 1945 and 1955.[22]

This does not mean that there were no developments. To grasp their rhythm we give some statistical information in the tables and figures presented in this section. Our data are not exhaustive, but they are sufficient to highlight some essential aspects. The numbers of physics teaching staff ('professors' and 'senior lecturers') in science faculties in France are given in table 9.1. They show that outside Paris there was a perceptible rise from 1946, but that the pace only increased in the fifties. In Paris, for which more data are available, there was hardly any growth from 1945 to 1950, while the average annual increase rose to 9% between 1950 and 1956 and 22% between 1957 and 1959.[23]

Table 9.2 and figures 9.1 and 9.2 give the numbers of staff and the budgets of the Centre National de la Recherche Scientifique (CNRS) and of the CEA. Relatively

Table 9.1
Number of physics teaching staff (professeurs et maîtres de conférences) in science faculties in France

Year	In Paris	Outside Paris	Total
1934	30	72	102
1938	31	74	105
1943	31	74	105
1946	32	78	110
1948	34	86	120
1950	35		
1951		109	147
1952	41		
1954	47		
1956	54		
1959	90	242	332

Sources: *Annuaire général de l'Université et de l'enseignement français,* published annually, 1929–1943. *Annuaire général de l'Education Nationale,* 1946, 1947, 1949, 1952, 1960. These yearbooks provided the information for the regions outside Paris. For Paris, *Livret de l'étudiant,* published annually by the Société des amis de l'université de Paris.

Table 9.2
CNRS and CEA budgets (in millions of francs)

Year	Budget (CNRS)	Budget expressed in 1945 francs	Budget (CEA)	Budget expressed in 1945 francs
1945		172		
1946	559	341	500	305
1947	895	363	600	243
1948	1 024	251	1 600	394
1949	2 006	435	3 400	739
1950	2 440	483	4 700	930
1951	3 028	497	3 800	630
1952	3 644	539	8 600	1 272
1953	4 380	683	9 900	1 544
1954	5 257	831	10 700	1 690
1955	7 135	1 113	35 800	5 580
1956	7 989	1 206	55 400	8 360
1957	9 593	1 391	77 900	11 300
1958	11 883	1 699	81 200	11 590
1959	17 504	2 171	84 700	10 510
1960	25 018	3 102	151 900	18 850

Sources: For the CNRS, information was derived from Encyclopédie (1960), 253. For the CEA, Rapport CEA (1961), 156 (approximate figures).

Notes: p. 345

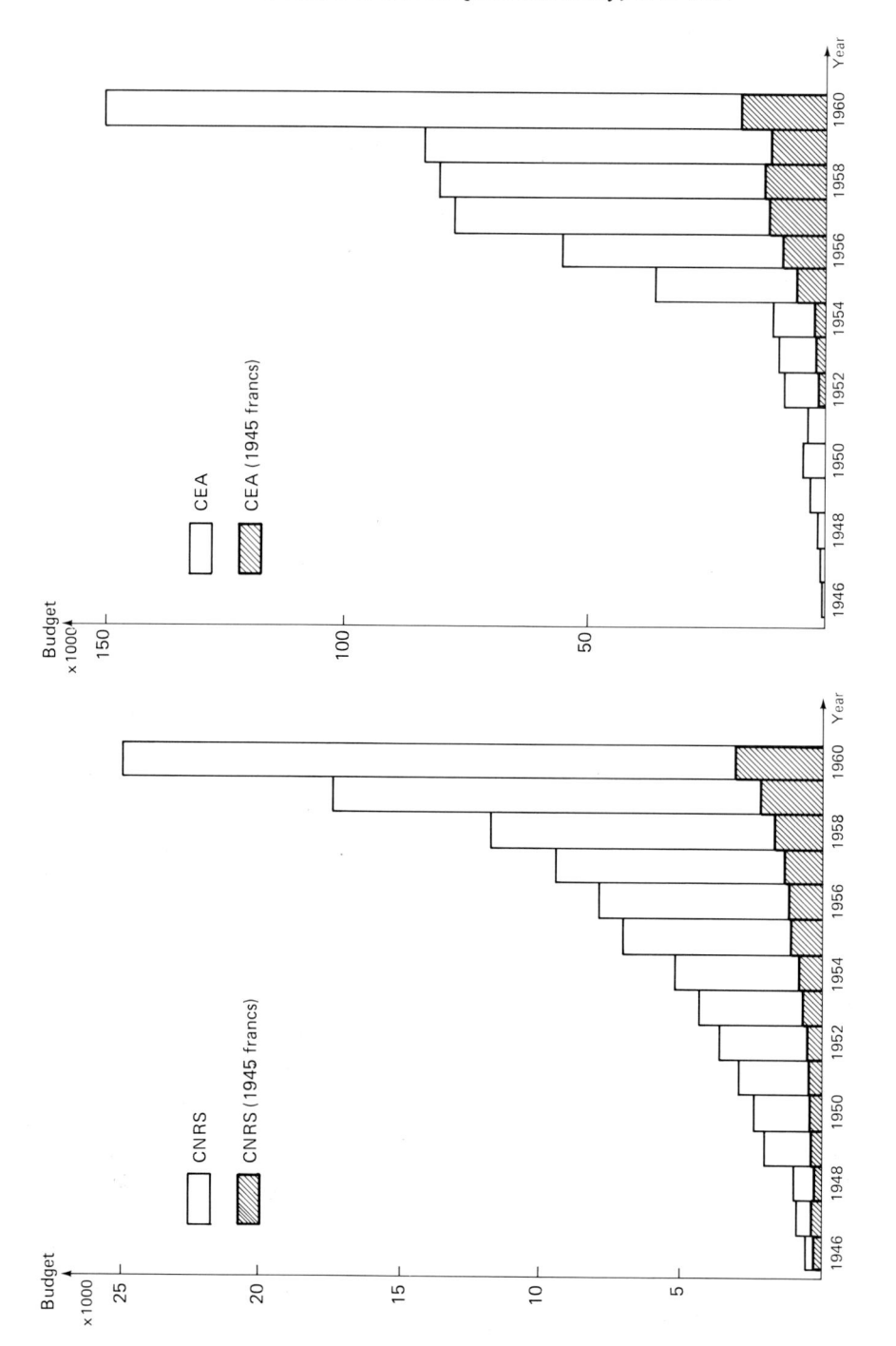

Figure 9.1.
CNRS and CEA budgets (in millions of francs)

The French scientific context 313

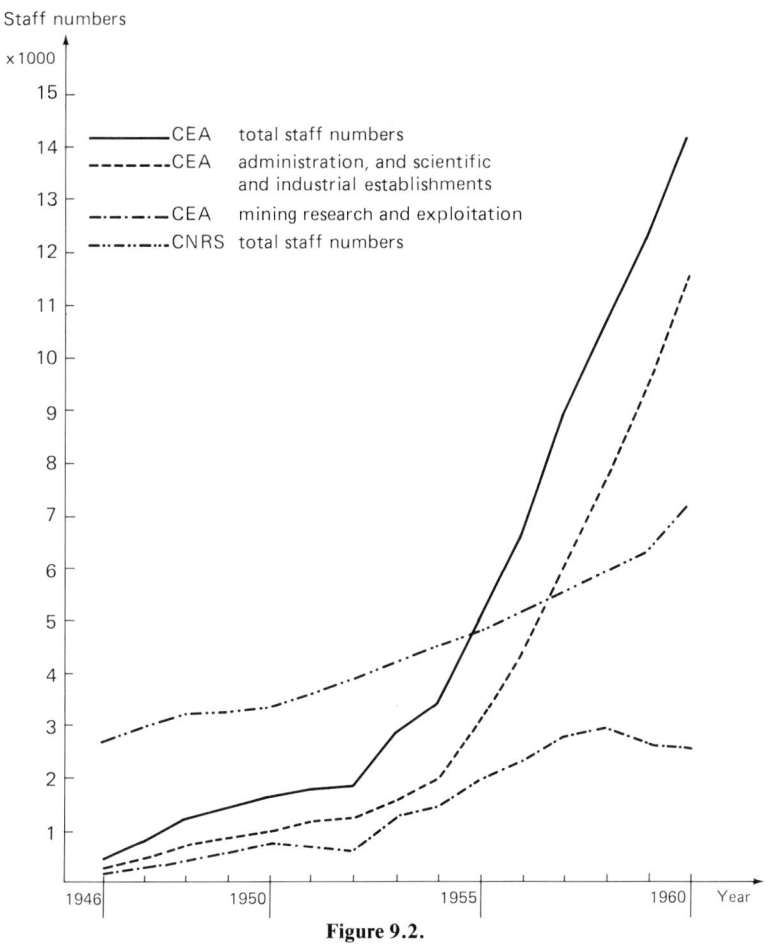

Figure 9.2.
CNRS and CEA staff numbers [Sources as for table 9.2. For CEA staff numbers, see Rapport CEA (1961), 12]

more advantaged than higher education, the CNRS increased its staff between 1945 and 1948. Numbers remained static in 1949 and 1950; thereafter there was an average annual increase of some 8% during the fifties. Obviously these figures conceal the fact that in real terms the annual increase was some 270 per year in 1951 and 1952, about 310 from 1953 to 1956, but 410 from 1957 to 1959. As for the budget, there were wide fluctuations (at constant prices) between 1946 and 1949, the average rate of increase between 1946 and 1952 being approximately 10% per annum. Thereafter the CNRS's budget increased annually by 150 million francs (1953 and 1954), by 190 millions during 1955, 1956 and 1957, by 390 millions in 1958 and 1959, and by a billion francs in 1960.

Turning now to the budget figures for the CEA, we find that though they differed somewhat in magnitude, they followed the same temporal pattern—slow and irregular growth between 1946 and 1951, more sustained growth from 1952 to 1954, a considerable increase during the years 1955–1959, and an explosive increase around 1960–1961.

In the light of these statistics we can say that a somewhat mediocre situation prevailed at the end of the forties and the early fifties, the increase in the material means allocated to science having been slow and uneven since the war. Thereafter, although matters improved gradually from 1951–1952 to 1956–1957, growth only really took off at the very end of the fifties. Every indicator therefore seems to confirm the image of a French science weakly supported around the time when the proposals for a European laboratory for nuclear research began to emerge.

An assessment of the material resources available for physics, in particular, is difficult for lack of easily accessible documentation. However, if one can believe the numerous articles devoted to this subject between 1948 and 1956,[24] and if one adds the partial information we have given in tables 9.3 and 9.4, we can reasonably

Table 9.3
Physics teaching staff numbers (professeurs et maîtres de conférences) in the Paris science faculty

Year	Total	Theoretical physics	Nuclear and corpuscular physics	Electronics
1945	31	4	2	0
1948	34			
1950	35	4	2	1
1952	41			
1954	47			
1956	54	9	2	4
1959	90	12	6	7

Sources as for table 9.1. For 'nuclear and corpuscular physics', we considered the courses entitled: *Radioactivité* and *Physique nucléaire;* for electronics, *Diffraction électronique* and *Electronique.* No course titles incorporating the terms *Physique corpusculaire* or *Physique des hautes énergies* appeared until 1960.

Table 9.4
Rough estimate of physics research staff numbers (excluding CEA staff)

Year	In Paris	Outside Paris	Total
1939	200– 230	180	380– 410
1956	700– 800	725	1425–1525
1960	1100–1200	1200–1250	2300–2450

Sources: For 1939, we have used the estimates we arrived at in Pestre (1984a). For 1956 and 1961, we have a summary account based on the information in *Laboratoires Scientifiques. Répertoire de l'Office National des Universités et Ecoles Françaises,* 1956 and 1961. The figures are, however, only indicative, as the lists are not always either very exact or complete. Up until the early sixties in France there were no reliable statistical data on scientific research. For 1956, Châtelet (1960) may also be used, which gives figures that are lower by 10 to 12% on average.

conclude that the position was hardly different. Regarding table 9.3, it should be noted that the universities in France were far from being the spearhead of modernization between 1945 and 1960. This is why the figures for nuclear and particle physics, like those for electronics, are so desperately low. Efforts were made by the CNRS, the CEA, the Collège de France, and the Ecole Polytechnique to compensate for these deficiencies. In the case of the CNRS, for instance, one is reminded of the proposal formulated by the corpuscular physics section which met in November 1950. Before spelling out its wishes for the CNRS five-year plan for 1951–1956, the meeting called for the following:

1) *Allocation for equipment:* a further 200 million for 5 years, of which 50 million is for the first year.
2) *Allocation of research scientists* (all categories): 30 new research scientists per annum for 5 years.
3) *Allocation of technicians* (all categories): 15 new technicians per annum for 5 years.
4) *Special allocation* for the construction of a generator for the acceleration of particles to 500 million electronvolts including premises and specialized staff. Cost: *1 billion 200 million* in three years.[25]

This post-war situation was not spared harsh criticism, the best known being that of the scientists. However, it was only in April 1953 that the question was seriously considered by the public authorities. In that month a decree was passed constituting a Scientific and Technical Research Commission under the auspices of the Commissariat au Plan. The main conclusion of the Commission's work was that it was imperative to create a responsible authority at *governmental level*. The government of Pierre Mendès-France did so in June 1954 in the form of a Secretariat of State. The Secretariat soon formed a Conseil Supérieur de la Recherche Scientifique et du Progrès Technique, but owing to the slow rate of progress made, a parliamentary committee was set up which organized two very important colloquia in 1956 and 1957. The conclusions reached there, along with those of the Conseil Supérieur which was preparing the third modernization and equipment plan, brought about an important, albeit gradual, change in the state of mind of French society. Helped by a euphoric economic atmosphere, important financial and structural changes were introduced at the end of the fifties which affected the very organization of research — as is all too evident from the statistical data we have given.[26]

9.2.2 ESSENTIAL CHARACTERISTICS OF PHYSICS IN FRANCE AROUND 1950, WITH PARTICULAR REFERENCE TO HIGH-ENERGY PHYSICS

In order to be brief, we describe the characteristics of French physics by relying mainly on an article which appeared in the November 1951 issue of *Physics Today*. We have chosen this article because we thought it particularly pertinent and accurate

Notes: p. 345

on the basis of what we know about physics in France at the time, because it was written by someone who was in Paris for the academic year 1950-1951, which is a key year for our purposes, and finally, because the author was Victor Weisskopf, a high-energy theoretical physicist.[27]

Weisskopf begins his article by emphasizing 'the very thorough background of classical physics and mathematics' provided by the French university system. However, he adds that in France one frequently finds 'an overemphasis on obsolete subjects, which are kept in the curriculum purely for historical reasons'. Moreover, 'there are in France very few courses in advanced subjects on the graduate level which are in direct relation to actual research work'. Consequently, apprentice researchers who had to educate themselves wasted an inordinate amount of time. For our part we would add that during this period there were virtually no courses for students having a first degree, specialized doctorate courses only being introduced in 1954, that seminars were not yet in general use, and that, as table 9.3 shows, the most contemporary subjects (particle physics, electronics, and even quantum physics) were hardly taught at university.

Weisskopf goes on to say that in spite of the considerable number of laboratories, 'the results are not as impressive as they once were', that 'the work is based upon ideas which were conceived elsewhere, and to which only a detail or a refinement is added here and there', and that 'the same work is done more thoroughly and reliably at some other places in the world'. The reasons he gives for this situation are inadequate financial support, the destruction and disruption caused by the war, the small number of 'experienced senior physicists able to guide the work', and the absence of co-ordination or even contact between the various research centres, all factors which had been criticized before the war but which still held true at the beginning of the fifties.[28]

Lastly, Weisskopf reviews the various laboratories. We will only mention those which are either directly or indirectly related to high-energy physics. Speaking of theoretical physics, he emphasizes 'the rather abstract character and perhaps too much of the formal mathematical type' of this research which was only loosely connected to experimental work—with the exception of the theorists of the CEA mathematical physics group, of whom he writes: 'many problems became clearer to us during our meetings'.[29]

Regarding cosmic rays, he mentions Louis Leprince-Ringuet's group at the Ecole Polytechnique. At this time the laboratory was of European stature studying 'the mass of the meson, the decay products of mesons and the nuclear effects of high energy cosmic rays' and preparing the large installations at the Pic du Midi. Apart from this team, only the members of Pierre Auger's small group were working on cosmic rays, studying mainly extensive showers.[30]

In nuclear physics and chemistry, the main laboratories were Irène Joliot-Curie's laboratory at the Institute of Radium, Frédéric Joliot's at the Collège de France and at Ivry (in the suburbs of Paris), the Rosenblum team at Bellevue (near Paris) working on α-radiation and, above all, the CEA's teams. Restricting ourselves only to the work and the equipment directly related to particle physics, there were a few of Joliot's students working at the Collège de France on the 'cyclotron producing a deuteron beam of 7 MeV', a group at Maurice de Broglie's laboratory building a 3 MeV linear accelerator and the CEA's accelerator division which was completing the installation of a 3 MeV Van de Graaff and of a classical 25 MeV cyclotron at Châtillon near Paris. According to Kowarski this division, which was the most important group building accelerators in France around 1950–1951, comprised 'some thirty engineers and technicians well versed in all the necessary techniques'. Unfortunately we do not know the proportion of engineers to technicians and, what is more, Kowarski does not mention if there were any physicists working in this division.[31]

In the light of the above we can conclude that the existing potential was rather low, not only in terms of manpower but also in terms of equipment, and that, taken as a whole, French output was of somewhat average quality. In the high-energy field, with the exception of Leprince-Ringuet's group, the situation was worse even when compared to that of smaller countries like Sweden or the Netherlands.[32] Of course no meaningful comparison was possible with the United States or the United Kingdom, the latter having already commissioned, amongst other things, a 175 MeV synchro-cyclotron in 1949, and now building a 300 MeV electron synchrotron, a 400 MeV synchro-cyclotron, and a proton synchrotron of about 1000 MeV. On a lesser scale, the Netherlands had commissioned a 27 MeV synchro-cyclotron (deuterons) in 1949, while Sweden completed a 200 MeV synchro-cyclotron (protons) in 1952. By contrast, France would not complete its 25 MeV cyclotron (deuterons) for some time.[33] However, the leading figures in physics in France knew of the seriousness of the situation, as is clear, for example, from the proposal drawn up by the CNRS's corpuscular physics section in November 1950. In requesting that design studies be rapidly begun for an accelerator for heavy particles of some 500 MeV (and in hoping that it would be commissioned by the CEA), the section showed that it was aware of the urgency and necessity of such a development.[34] It is probable that it was the same kind of considerations which lay behind the vigorous support by Auger, Perrin, Kowarski, and Dautry for the various projects for a European research laboratory, and in particular their immediate support for the 'bevatron' project which was first clearly formulated in December 1950.

9.3 French initiatives regarding the UNESCO project, the decision-making process, 1949–May 1952

Clearly it is not necessary to describe in detail the role of the French in the developments between 1949 and 1952 which led to the setting-up of the provisional CERN. This has been done elsewhere and does not bear repetition here. Instead, we

Notes: pp. 345 ff.

will concentrate our attention on the negotiations between French scientists, administrators, and politicians.

9.3.1 THE FRENCH PROJECT, NOVEMBER 1949–DECEMBER 1950

The origins of the French project, whose main proponents were Raoul Dautry and the heads of the scientific divisions of the CEA (Perrin, Kowarski, Guéron, and Goldschmidt), are to be found in the preparation for the European Cultural Conference convened by the European Movement in Lausanne between 8 and 12 December 1949. Invited by Dautry, who was also chairman of the French Executive Committee of the European Movement, to come up with an appropriate proposal for this meeting, Kowarski and his colleagues sketched out a project. The central idea was to establish a 'super-Brookhaven' for Europe, if possible at Saclay where the CEA was then building its research laboratories. As they envisaged it, there would be a multinational centre mainly devoted to nuclear physics and equipped with at least one reactor, an accelerator division, a department of nuclear chemistry, and probably several peripheral units (metallurgy, medicine ...) associated with nuclear activities.[35]

Various factors lay behind the support for this proposal: firstly, and in the case of Raoul Dautry in particular, the political will to push for the construction of a united Europe; secondly, the desire to create a locus of nuclear research able to compete with the Americans and the British who were shielded behind their barriers of secrecy; thirdly, the opportunities for economies of scale which the establishment of such a centre would provide (obvious markets for Norwegian heavy water, for pure graphite produced by the CEA ...); finally and more generally, the need for science in Europe (and by implication in France, even if this was not spelt out in black and white in the documents of that period) to unite in order to face the challenges imposed by big science.[36]

In December 1949 Dautry forwarded the Lausanne resolutions to the Ministry of Foreign Affairs, hoping apparently that something could be started there.[37] In fact he did not press the matter and nothing was done. By April 1950, the CEA's group of directors had met and drawn up a more precise text than the one written by Kowarski in November 1949. This text was sent in May to a good friend they had in the Quai d'Orsay, François de Rose. They had known him in the UN Atomic Energy Commission in New York as early as 1946 where he was a member of the French delegation.[38] Back in Paris early in 1950, de Rose was nominated Deputy Director of the Department of Economic Affairs of the Ministry of Foreign Affairs, which did not prevent his retaining a keen (and personal) interest in nuclear matters. Sometime in May 1950, then, de Rose and the scientists of the CEA devised several schemes to establish some kind of official contact with other European countries—but without arriving at a simple and applicable solution. As a matter of fact, they decided first to take unofficial action and to test the reactions of two or three key European scientists. For this reason Francis Perrin and François de Rose paid a visit to their former colleague on the UN AEC, Hendrik Kramers. This visit was a failure. Kramers showed no interest and, disappointed, the two French officials dropped the idea.[39]

On 7 June 1950 the 'project' was revived by the resolution submitted by Isidor I. Rabi at the Fifth General Assembly of UNESCO in Florence. His proposal for a European regional physics centre was interpreted in Paris as not incompatible with the French scheme.[40] Diplomatic activity therefore resumed. De Rose drafted a letter for the Prime Minister on 19 June 1950, so only 12 days after the vote on the Rabi resolution. This letter was used in July by the Minister of Finance at a meeting of the Council of the OEEC. On 20 November, through his Secretary of State, the Prime Minister requested the Minister of Foreign Affairs to arrange that 'the representatives of [his] department, of the Ministry of Finance and of the Commissariat à l'Energie Atomique meet shortly with a view to submitting to the government the main outline of the policy to be pursued in order to achieve this objective [the setting-up of European research centres, with particular reference to nuclear physics]'.[41]

In France, then, around the end of 1950 a project had already been communicated to some government departments and the state machinery had started moving — even if slowly. Nothing very precise had yet been drawn up, but one had reason to be optimistic — and early in December Dautry could tell the Director of the Centre Européen de la Culture that 50 million francs would be available for the European project.[42] More precisely the French government, via the intermediary of the CEA and the Ministry of Foreign Affairs, seemed ready to study a working proposal around which to negotiate and which could be transmitted to other European countries through diplomatic channels. Apparently the procedure was to be modelled on the very recent Schuman plan (submitted in May) and Pleven plan (submitted in October 1950).[43]

9.3.2 THE PROJECT FOR A LABORATORY CENTRED AROUND A BEVATRON: FIRST ENCOUNTERS OF THE INITIATORS OF THE PROJECT WITH THE GOVERNMENTAL MACHINERY, DECEMBER 1950–MAY 1951

This situation changed drastically with the convening of the meeting on 12 December 1950 at the Centre Européen de la Culture in Geneva, a meeting at which eight European scientists first proposed setting-up a joint laboratory centred around a proton synchrotron of some 6 GeV, thereby abandoning all other ideas such as a reactor or 'computing laboratories'. One consequence of this meeting was that the scientific goals of the laboratory were narrowed. More importantly, however, were the changes the meeting brought about in the context of negotiation — *the French as a national group were no longer in a position to take a diplomatic initiative, since the project had passed into the hands of a supranational group, namely the physicists.* At the instigation of Auger and the consultants (among them some French physicists) a more informal negotiating mechanism was proposed: the scientists were to define the scientific, organizational, and financial aspects of the project and it was only when they had reached an agreement among themselves that their proposal was to be sent to governments for approval. Clearly the ground would have to be prepared for this approval, but the politicians would no longer play a major role in the elaboration of the project. The initiative, the control, had passed into other hands.[44]

The French 'network' gathered at the CEA immediately after the Geneva meeting. On 19 December 1950, Dautry, Auger, the three leaders of the scientific departments (Kowarski, Guéron, and Goldschmidt), and the CEA's Secretary-General, Lescop, met to discuss the best method of helping Auger to set up his 'study office'. As money was the first concern, a request was made to de Rose at the Ministry of Foreign Affairs that 'France advance funds'. Whereupon protracted negotiations started between the Ministry of Foreign Affairs and the Ministry of Finance.[45]

These negotiations comprise an exchange of letters which was started on 2 February 1951 with the request of the Department for Cultural Relations of the Ministry of Foreign Affairs (which was responsible for relations with UNESCO) to the Department of Accounts and Personnel of the same ministry (which negotiated with the Ministry for the Budget), to grant two million francs for the study office to be set up by Pierre Auger at UNESCO. Although we lack precise information in this respect, it appears that the request was designed to be of the same order of magnitude as the amount offered by Italy (two million lire, or approximately one million francs). This exchange of letters ended on 11 May with the letter from the Minister of Foreign Affairs to the Minister for the Budget, the Prime Minister, the Minister of Defence, the Minister of National Education, and to the Directors of the CEA and the CNRS, convening an interministerial meeting on 18 May. The meeting was needed because the Minister for the Budget had refused to release the funds requested.[46]

Three reasons were given for this refusal. It was noted that this amount 'would be superfluous if an immediate, even if very approximate, assessment led us to the conclusion that the required expenditure on the construction of the international laboratory exceeded the contributions which the different interested governments could pay'. The Ministry for the Budget therefore feared that a process was being initiated which would be unviable in the long term. It was also pointed out that 'the payment to UNESCO of special contributions to fund a kind of supplementary budget to the budget of this institution would be tantamount to an increase in that organization's budget'. This would be in conflict with the undertaking to keep 'to a minimum the financial burdens, often considered too great, of the participating states'. Finally, the Ministry for the Budget remarked that 'the very question of the setting-up of a laboratory' seemed to pose a fundamental problem which should be thoroughly examined by the various ministerial departments.[47]

Defending its request, the Ministry of Foreign Affairs confirmed that the Prime Minister was in favour of this move (letter of 20 November 1950). It was also pointed out that Italy had offered money, though France had taken the initiative, and if she wished to keep it she must do likewise. This would also be the price she would need to pay if she wanted to ensure that the laboratory could be situated on French territory. In addition the Director of Cultural Relations noted that 'as this enterprise is an intergovernmental one, the sum it will call for from each government will not, by definition, exceed the contributions they are able to pay'. In his view, this invalidated the first of the Ministry of Finance's arguments.[48]

Two brief remarks should be made at this point. This exchange of letters proves that no precise negotiations had taken place within the French government before December 1950. Dautry had spoken for himself or on behalf of the CEA when he mentioned the 50 MFF and not on behalf of the government. Secondly, if the money was not offered immediately to Auger, it was because the request for the funds was channelled through the government machinery, and was not sent directly to bodies such as the CEA or the CNRS (unlike Italy where it was the Consiglio Nazionale delle Ricerche that offered money).

9.3.3 THE INTERMINISTERIAL MEETING OF 18 MAY 1951

The meeting was attended by four civil servants from the Ministry of Foreign Affairs, including François de Rose and Louis Joxe, Director of the Department for Cultural Relations, representatives of the other Ministries invited, including General Bergeron, a member of the Atomic Energy Committee, Donzelot, Director of Higher Education, Dupouy, Director of the CNRS, and Perrin, High Commissioner of the CEA. Raoul Dautry was absent, but at the request of de Rose he had the following message communicated to Louis Joxe: 'Sorry that I cannot be there for the meeting on 18 May. Tell Joxe that it is my project and that I recommend it to him'.[49]

In its deliberations the group began by replying to the Minister for the Budget. It emphasized that the amount of two million francs 'would in no sense constitute an undertaking which would be binding [on France] in the future'. It added that if France ultimately wished to take part in the European project, this initial expenditure would place her in a strong position during subsequent negotiations. However, as these arguments alone were not convincing, the group turned its attention to the crux of the problem.[50]

The first point it stressed was that 'the scientists present at this meeting [had] given their complete support to the principle of the construction' of a European laboratory equipped with a 6 GeV accelerator, since, in their view, this was the only way to maintain 'science in our countries at the level it [had reached] in North

Notes: p. 346

America, in Great Britain and certainly in the Soviet Union'. As this argument was accepted by the representatives of the other departments, the group asked whether France could build the laboratory alone. It concluded that she could not. The resulting financial burden would be beyond her means, and since she had insufficient qualified manpower at her disposal, 'these machines could not be built without draining off useful men from other fields of activity in the country'. In other words, since France could not do it alone and since the planned laboratory '[would] in no sense be duplicating work already being done [...] and on the contrary [offered] the cheapest and most profitable means of equipping Western Europe with a research tool equal to the needs described above', the group's advice was to take part in the UNESCO project. The scientific weakness of France—with explicit reference to accelerators—was therefore put forward as the primary reason for giving firm support to the European project.[51]

That granted, the group immediately emphasized that another important argument should be added to 'these scientific and technical motives': 'if this project was realized in a Western European framework, it would fall in line with the policies followed by all French governments' and in this respect 'the Department of Foreign Affairs cannot but approve the principle'. On this occasion then, a reference to French diplomacy was explicitly put forward as an argument for supporting the project. Moreover, the group pointed out, the results obtained at the European laboratory would be 'sufficiently remote from military applications' and would consequently not 'alter the balance of power in Europe'.

Having put forward these fundamental arguments, the working group raised three associated questions. For one thing, it wished to establish what attitude to adopt with regard to the possible participation of Germany. Its conclusion was that 'the question must be borne in mind' but that no major obstacle should arise. With regard to the Soviet bloc countries, by contrast, the group's reply was clearly negative: 'The French delegation should oppose' any attempt to have those countries participate. This possibility led the group to raise a third question, namely whether UNESCO was the best channel for negotiation. The reply was yes and no, no because there was the problem of the eastern European countries, yes because the UNESCO initiative had been triggered by an American, Isidor I. Rabi, which offered two advantages: 'active support of American scientific circles' was guaranteed and fears of small countries such as Belgium, Norway, and Switzerland that they would be invited only 'to come and fit into a French scheme' would be somewhat allayed. Even if the disproportion of resources between these countries and France remained large, and even if the preparatory studies were being directed by a Frenchman, Pierre Auger, it was reassuring to them that consultations would be conducted under the auspices of a neutral organization like UNESCO. The 'UNESCO procedure' therefore offered the maximum benefits for the minimum drawbacks. For this set of reasons, the interministerial group recommended to the Minister for the Budget that the two millions requested be granted.

At this point the reader might conclude that the decision was motivated by a clear awareness of French interests and by the fact that this project was primarily French and would be under French control. To this we would say that the first observation is substantially true, and that it reflects what was at stake in the debate, while the second remark is only true incidentally and was not a decisive ground for the choice that was made. It is only to be expected that it should be entirely in France's interest for the group to propose to allocate the two million francs. If Italy granted two million lire and Belgium offered 50,000 Belgian francs, it was because it was also in those countries' interest to do so, and because they hoped to get more out of the enterprise than they had put into it. Indeed, it is always this question of *national* interest which is first asked of the scientists of a country when its official representatives are trying to decide whether to lend their support to an international project.[52]

As we have said, the second argument—we support the project because it is a French project under French control—was not decisive. Certainly Auger's presence was mentioned as an advantage and was used as an argument, and certainly nobody claimed that attempts would not be made to protect as best one could the interests of French scientists while the laboratory was being built. But the overriding feeling was that the preliminary studies were being prepared under the auspices of UNESCO and that concessions would be necessary. And anyway, the group which met on 18 May recognized that, after Rabi's proposition, no alternative channel for advancing the project could be seriously proposed.[53]

9.3.4 THE FRENCH DECISION TO PARTICIPATE IN PHASE 2 OF THE UNESCO PROJECT, OCTOBER 1951–MAY 1952

Once the main lines of argument leading France to support the UNESCO project to construct an accelerator of some 6 GeV had been spelt out at the meeting of 18 May, the Ministry for the Budget granted the two million francs. Responsibility for continuing negotiations on the scientific programme of the laboratory now returned to the CEA scientists, principally Francis Perrin and Lew Kowarski who attended the meetings of UNESCO's consultants in May, October, and November 1951, the Copenhagen conference in July, and the meeting held on the occasion of the inauguration of the joint Dutch and Norwegian reactor at Kjeller in November.

Dautry died on 21 August 1951. De Rose for his part continued to follow up the matter. At the end of October he had a private meeting with the Secretary of State to the Prime Minister, Félix Gaillard, to inform him of the results of the consultants' meeting of 26 and 27 October. Gaillard instructed him to contact the Ministry for the Budget regarding the 23 to 25 million francs which France would have to spend for her participation in phase 2 of the project ('the planning stage'). By the beginning of December de Rose was concerned by the Ministry's tardy and ambiguous reaction. On 5 December he once more alerted Gaillard to the situation.

Notes: p. 346

On 6 December he wrote a note to his minister and drafted a letter to the Minister for the Budget. Here de Rose informed him that he would include on the agenda of the interministerial committee meeting to be held on 8 December a discussion of the stance to be adopted by the French delegates at UNESCO's intergovernmental conference on 17 December. We do not know whether the question of the European laboratory was considered at this particular meeting (or elsewhere) but a decision was taken at governmental level, since a delegation was appointed and Francis Perrin could publicly state on 17 December that the French delegation supported the UNESCO project, and was prepared to contribute 25 million francs (so some 30% of the 200,000 dollars considered necessary for phase 2).[54]

The arguments used by François de Rose in the letters of 5 and 6 December were those that had come to the fore after the meeting of 18 May, namely that scientifically the project was crucial for France, while politically it could not but be approved by the Minister of Foreign Affairs. In addition de Rose emphasized two particular points: that the importance of the project had increased since May because 'as a result of the contractual agreements to be signed', the Federal Republic of Germany would again have complete freedom in the field of scientific research; and that France must make a clear commitment at the start of the intergovernmental conference, as Copenhagen had been proposed as a site for the laboratory. Again, as in May, he stressed that it would be in France's interest to have the laboratory set up on its soil or at least on that of a French-speaking country. France would then play host to European research scientists, which would foster 'the development of French in these circles' and contribute to her intellectual influence.

During these negotiations one question seems to have caused some difficulty, namely, the source of the French contribution. Was it to be the CEA or the Ministry of Foreign Affairs? In the view of the Ministry for the Budget, it should be the CEA. In the opinion of de Rose and the scientists involved in the negotiations, it should be the Ministry of Foreign Affairs. They reaffirmed that certain countries were 'afraid that the project would come under the influence of the French Commissariat à l'Energie Atomique and would thereby lose its international character'—an impression that would be reinforced if the 25 millions were offered by the Commissariat. It is difficult to assess whether this was the only or even the most important argument, but one thing is certain: there was no opposition to the European project from the Ministry of Foreign Affairs, whereas attitudes were less clear within the French scientific community. A section of it was reticent, even hostile, as many scientists were afraid that the national research budget would be reduced to finance CERN. We therefore think that de Rose's argument possibly concealed another consideration, namely, that if the money for CERN was not taken out of the funds allocated to the CNRS or to the CEA, it would be clear that it posed no threat to national research expenditure. Moreover, as the money would be paid by the Ministry of Foreign Affairs, it would subsequently be its *task* to set up the group which would determine France's stance on the CERN project, thereby making it easier to keep hostile elements at bay. Wherever the truth lay, de Rose's argument prevailed and the money was paid by the Ministry of Foreign Affairs.

To conclude our account, we shall give what little information we have on the period up to the first meeting of the CERN Council early in May 1952. As for the negotiations in the working group set up by the UNESCO conference of December 1951, there is no indication that the governmental machinery was consulted on the line to be taken—apart from a meeting of the Atomic Energy Committee held on 3 January 1952. This seems to have been left to the initiative of de Rose, Perrin, Kowarski and several of their colleagues, while the other 'departments' were only informed of the outcome after the negotiations had taken place. On the other hand an interministerial meeting was held on 27 January—so after the working group had submitted its findings—to define the French position with regard to the new working documents drafted for the February session of the UNESCO Conference. Judging by the attitudes adopted by the French delegation during that session, the meeting approved the work done by the working group.[55] Since the Ministry of Foreign Affairs judged that 'in the circumstances parliamentary ratification was [not] needed', the text of the Geneva Agreement was forwarded directly to the President of the Republic for ratification ten days after the February session of the conference. As this procedure was both flexible and rapid, the Council of the provisional organization was officially informed of the ratification on 11 April 1952. The agreement having also been signed without reserve by Germany, the Netherlands, and Yugoslavia on 15 February, and ratified by Sweden on 2 May, the Council met on 5 May.[56]

9.3.5 A FEW REMARKS ON THE DECISION-MAKING PROCESS IN FRANCE

In chapter 13 John Krige presents a model for the decision-making process in the United Kingdom in 1951 and 1952 with regard to the European laboratory project. In what follows we wish to see to what extent it is useful for an understanding of the French case.[57]

The model starts by affirming that one should not regard the government and the scientists as two discrete entities: the 'government' is an articulated collection of departments each having the right to put its case and with authority in its own allotted domain; the group of scientists for its part ought also to be considered as one (or several) of these 'departments'. The value of this first level of conceptualization is apparent if one considers the preparations for, and the proceedings of, the meeting of 18 May. Indeed one does not see 'scientists' debating with the 'government', but the representatives of various departments (Budget, Defence, Foreign Affairs, CNRS, CEA) expressing their opinions in their areas of competence, and reacting according to their own logics, i.e. in accordance with the stances they would generally adopt under similar circumstances.

Thus we see the Ministry for the Budget first restating the position of his department (and so that of the government) with regard to UNESCO's budget, and refusing the grant of two million francs on the grounds that he did not want to increase that budget. To reverse this decision would be tantamount to the department's reconsidering its established policy, and particularly strong arguments would be needed for this. Then again one sees the representatives of the Quai d'Orsay supporting the project, in particular because it was consistent with the department's policy, or one finds Dupouy and Perrin in favour, emphasizing the importance of a 6 GeV accelerator for science in France.

Having established this first point, John Krige specifies that one may reasonably expect there to be 'border wars' between departments. We have no clear evidence that any took place in France in 1951, except perhaps with regard to the source of the funds. However, for two months at the end of 1953 there was a serious battle of this nature,[58] whose *only* effect was to cause a further delay in the ratification procedure. Further refinement is needed since the model is still too limited: it assumes that the various departments have rationally defined positions. In fact the arguments used by the various representatives of a department to determine the position they feel they should adopt vary from one person to another and with time. One thinks here of the line of argument used by the Ministry of Foreign Affairs. In 1950 no reference is made to France's pro-European policy, perhaps because it was too recent; in 1951 the argument appeared; in 1952, it became crucial, when in an internal note to the Secretary-General one could read: 'it seems only appropriate that France, as the principal champion of European unity, should play a prominent role in the establishment of this laboratory'.[59] The model is also limited, as John Krige notes, in that it does not take account of changes of view resulting from the existence of a hierarchy within departments or within the government as a whole. Stances adopted at a certain level can be 'appealed against', either at a higher level in the same department or through convening an interministerial committee. To illustrate, this is what the Ministry of Foreign Affairs did in May 1951 when it was unable on its own to get the Ministry for the Budget to reverse its decision.

One must nevertheless distinguish two moments in the analysis. On the one hand there is the situation in which a proposal having been submitted, is studied by the various departments so that a common position may be worked out. At this stage of the negotiations, the model we are considering represents the unfolding process very closely. But there are also moments when matters are more informal, in general *before* a specific proposal is submitted, or *after* if the decision is negative. In such situations it is the activities of pressure groups, which often cut across departmental boundaries, that dominate events. In France such a group existed, comprising scientists (Perrin, Kowarski, Auger), Dautry, and de Rose. Strongly united, they lobbied for a positive decision. Typically, they contacted the President of the

Council in 1950 to obtain a statement of principle which they used, unsuccessfully as it turned out, between February and May 1951 to get the Ministry for the Budget to change its mind. Similarly, de Rose asked Dautry to send a word to Louis Joxe who was to chair the meeting on 18 May; or again de Rose made a further appeal to Félix Gaillard in December 1951 in the light of the inertia of the Ministry for the Budget. This type of pressure group uses the authority of some of its members to get a project off the ground, to draw up appeal procedures, to by-pass government quarrels, or above all to speed up decision-making processes constantly slowed down by these conflicts. The relative weights of the formal (bureaucratic) procedures of negotiation and of the lobby activities may also vary from one national tradition to another. In this sense, notable differences may exist between, say, Britain and France.

To improve the model the place and role of the scientists should also be more precisely defined. We have stated that during the decision-making process they are to be considered as one (or several) of the departments which interacted with the aim of defining the official position of the country to which they belonged. This is true but must be qualified by two remarks already made in chapter 13.

1. Firstly, their strength resided in a *knowledge* which they alone possessed. The power of their arguments therefore depended largely on their presenting a united front. If they affirmed together that a certain option was vital for the future of science in their country, the matter was given serious consideration by the other departments. If, on the contrary, there was no unanimity, doubts arose. This is why Louis Joxe emphasized the unanimity of the scientists at the *beginning* of the report he sent to the Minister for the Budget on 19 May.

2. The second remark is that the terms 'scientists' and 'unanimity of the scientists' are too vague. In May, for example, Dupouy spoke on his own behalf and Perrin on behalf of the scientific directors of the CEA, and little more. At that stage, other scientific institutions or key figures in French nuclear physics such as Joliot and Leprince-Ringuet were not officially consulted.[60] Thus for 'the scientists' one should read 'the scientists who were recognized as leaders of a key scientific institution *and* who were consulted'. This distinction is not merely academic since, at a more refined level of analysis, the scientists do not constitute a 'department' organized like the others. Individual views and passions are more in evidence. More importantly the distinction is crucial for us since a certain opposition to the CERN project arose in France between 1951 and 1954, an opposition which was largely spearheaded by the scientists close to the Communist Party. Since they were Communists, these physicists were perceived as having no right, in the eyes of the officials of other departments, to 'represent' the scientific community in France and so to be 'consulted' in an affair such as this. Neither their opinion nor their opposition were pertinent. In other words, scientific excellence is not enough, it has to be coupled with a political outlook which is not too radically at variance with that of 'the government'. We do not wish to suggest that if there are such divergences

there is automatically a lack of consultation, but only that this *can* happen in some contexts. Ours is a case in point: an international affair, the cold war, Joliot's recent dismissal from the Directorate of the CEA.

We see then that any impression one might have gained from reading part II of this volume of a France that had unanimously supported the Auger project from the beginning, and of a France that, along with Italy and Belgium, had been one of the three champions of the project, is too simple, even if it is not false. Viewed from without, this may have appeared true, but seen from within France the matter was much more complex. There was no straightforward encounter between the 'unanimous wishes of the scientists' and the no less unanimous aims of 'the government'. There was, rather, a process of negotiation, marked sometimes by conflict, often following the roundabout paths of an effective pressure group, but leading all the same to a positive official attitude in favour of the UNESCO project.

9.4 Towards the signing of the Convention, February 1952–July 1953

The broad outlines of the history of the provisional CERN are presented in chapter 7.[61] As we are only concerned here with the specificities of the French position, three problems seem to us to be worthy of attention: the decisions relating to the financing of CERN, the question of the site of the laboratory (the French government having officially offered a location near Paris), and French demands regarding the drafting of the Convention. This choice requires comment. It was made on the basis of the documentation at our disposal, namely that of the Ministry of Foreign Affairs. As we have said, there is virtually no material in the CNRS archives relating to the provisional CERN, and we were not allowed access to the CEA's archives. Could these limitations have distorted our perception of what was essential for France during this period? We do not know but in the light of what we have found in British and Swiss archives, and from those of CERN, there seem not to have been other major issues, and *especially no other issues of a scientific kind,* on which France's position deserves separate treatment.[62]

9.4.1 FINANCING THE FIRST YEAR OF CERN

In December 1951 France had offered 25 million francs for phase 2, a sum which she had paid before the second session of the intergovernmental conference was held in February 1952.[63] On returning from the second meeting of the Council of Representatives of European States (the provisional CERN Council) held in Copenhagen on 20–21 June 1952, the French delegation submitted a report to the Secretary-General of the Ministry of Foreign Affairs. Having emphasized how 'remarkable it [was] to note that the scientists of these twelve European states [had] reached a unanimous agreement in this regard [the type of machine to be built]', and that this project '[constituted] an impressive achievement at the level of European co-operation', Francis Perrin and Robert Valeur (who had replaced de Rose on the Council) drew attention to the fact that the construction of the laboratory could begin in September 1953. In accordance with the proposals made by the CERN Council, they asked that steps be taken immediately to budget in 1953 for the French contribution of about 100 million francs.[64] This proposal was accepted by the various ministries concerned, and 75 million francs were earmarked for CERN in the budget of the Ministry of Foreign Affairs and voted by parliament. It was, however, not to be released until the forthcoming agreement constituting the final or 'permanent' CERN had been ratified by the French parliament. At the beginning of 1953 the same procedure was implemented for 1954. Following the CERN Council session in Brussels (12-14 January 1953), the Ministry of Foreign Affairs 'asked the Ministry of Finance to approve in principle, in the budget estimates for 1954, an appropriation of 336 million francs', corresponding to the French contribution. In the summer, when it appeared that construction work in Meyrin (Geneva) would not begin in 1953 as planned, the government adjusted the Ministry of Foreign Affairs budget which was then being submitted to parliament. Only 261 million francs were requested for 1954, the 75 millions voted for 1953 being still available. The vote took place in parliament on 4 November 1953. 285 millions were finally set aside in the 1954 budget, to be released when the convention was ratified.[65]

9.4.2 FRANCE AND THE PROBLEM OF THE SITE OF THE LABORATORY

The problem of the site proved to be much more complex than the issue we have just discussed, to which there was a straightforward solution since French policy with regard to CERN had largely been defined in May and December 1951. Here there were two complicating factors. For one thing, stakes were high—the question of the site roused national feelings, for which reason Auger had always carefully managed to postpone discussion of the issue until the provisional CERN was set up.[66] Then, on the French side, there was the intransigence of two key figures, the Secretary-General of the Ministry of Foreign Affairs, Alexandre Parodi, and the Secretary of State to the Prime Minister, Félix Gaillard, who both believed until July or August 1952 that the laboratory would be built in France. Summing up the situation on 20 August, Kowarski wrote: 'Lucet and Valeur [...] have informed me of the positions adopted by the government, either publicly (Council of Ministers at the end of July, Gaillard's statement to the press) or in conversation especially with Parodi. They feel that, in line with the government's wishes, the French delegation to Amsterdam should adopt a firm, even inflexible position in favour of a site in France'.[67] And it is true that the studies putting the French case in support of her candidacy, which are collected together in a file at the Quai d'Orsay, were prepared

Notes: p. 347

with considerable care and thoroughness. A chief engineer of 'Ponts et Chaussées' was given responsibility for co-ordinating the work and several ministries and major departments—Aménagement du Territoire (Regional Development), Douanes et Droits Indirects (Customs and Indirect Duties), Electricité de France...—contributed to it. We are therefore led to ask what made the government—Kowarski noted that 'the focal point in the government seems to be Gaillard'—take this matter so seriously.[68]

The first factor was the narcissistic and self-interested desire to have the international laboratory on French territory. This was of course also true of the other candidates as shown, for example, by the diplomatic efforts of Switzerland and the Netherlands to secure the support of other Council members in the vote. The same forces were at work later when, in 1969 and 1970, Austria, Belgium, France, Germany, Italy, and Switzerland tried to set up a new joint laboratory with a 300 GeV accelerator. The desire of most of them to have the laboratory on their own territory made agreement on a site impossible to achieve.[69]

There were also reasons of a more specific kind. As François de Rose recently pointed out to us, at that time it was relatively common to set up European-inclined bodies in Paris.[70] Even if this was not done systematically, the rate at which it did take place was not negligible. Moreover, as we have explained, the overwhelming impression in France was that the UNESCO project had naturally grown out of the 1949/1950 project (whose principal architect had been Dautry). This led people to think that France deserved preferential treatment. Indeed, the case put by Dautry, de Rose and their colleagues had always been constructed on this model. In their notes to the Prime Minister or to the Ministry of Foreign Affairs, they always let it be thought that France could have the laboratory situated on her soil—if she adhered to the course she had followed since 1950. The rhetoric used by the project's advocates can thus be considered as partly responsible for the hopes and 'intransigent' attitudes of Félix Gaillard and Alexandre Parodi. Probably for de Rose and Perrin these arguments were primarily tactical, their aim being to force the governmental authorities into taking a decision which would result in a firm French commitment. Unfortunately for them they were taken literally![71]

The French delegates to the Council were therefore in a delicate position in the summer of 1952. There were four proposals for a site—Arnhem in the Netherlands, Copenhagen, Geneva, and Paris, and there was no indication that Paris would be chosen. According to Robert Valeur,[72] the small countries, which were in the majority in the Council, were reluctant to set up the laboratory on the territory of one of the 'big' states, especially France which seemed to want 'to extract undue advantage from a collaborative venture'. In addition Switzerland had offered a site, which deprived France of the support of that country and of Italy who, we know, had officially supported Geneva since 1951. Lastly and above all, there was the Dutch proposal which, according to information given by Valeur on 26 August, seemed to have the best chance of obtaining a majority of the votes of the participating states. Although we do not know how reliable this 'information' was,[73] Robert Valeur spelt out the factors in Arnhem's favour: the site was rather well situated in Europe, it had good communication links, it would receive, for obvious reasons, the support of the Scandinavian countries, and Germany and the United Kingdom would support it against Paris so as to maintain a balance among the larger states.

This is why, in the hope of softening the attitudes of Félix Gaillard and Alexandre Parodi, Perrin, de Rose, and Valeur used a three-tiered argument throughout August and September. They first tried to convince them that it seemed impossible to have the site in Paris, and that to adopt an extreme position in this regard would be disastrous, either because France 'would lose face in painful circumstances' if the vote went against her, or because 'the laboratory would be set up in a climate of ill-will' if the vote went in favour of the French proposal. Next, they pointed out that there was an alternative which was almost as satisfactory as Paris: Geneva, because French was spoken there, and because the city was situated at the very entrance to France. Lastly, they showed how the inevitable defeat which would result if the cause of Paris was pursued to the bitter end could be transformed into victory: one could support Paris strongly, and when the situation appeared to reach a stalemate make a grand gesture of compromise by withdrawing the French proposal and asking the Danes and the Dutch to do the same to the benefit of... Geneva.[74]

Did this line of argument, often used in writing and during meetings with Gaillard and Parodi, convince the two French politicians? One cannot be sure, but it seems likely that it did, since the course of events at the third session of the Council in Amsterdam in October 1952 suggests that the French delegates had not been given instructions to fight for Paris to the last ditch. By contrast, as far as Arnhem was concerned (Copenhagen apparently not being a serious rival), de Rose always described Parodi as remaining 'extremely inflexible', going so far as to say that if Arnhem were chosen 'it would be better not to participate'.[75]

The outcome? Geneva was 'unanimously chosen' after the other proposals had been withdrawn. Unfortunately we do not have the detailed information needed to describe with any subtlety the unravelling of the 'drama'. Both the Council minutes and the French state archives are astonishingly discreet in this regard. This is understandable, since the urge for reconciliation surely prevailed after these excesses of national pride—CERN had to continue on its way.[76]

9.4.3 FRENCH REQUIREMENTS REGARDING THE CONVENTION

The drafting of the Convention raised various problems which have been described in section 8.2 of this volume. The French took part in the debate along with all the other national groups, but came down heavily on one side only in the

Notes: pp. 347 ff.

controversy surrounding the admission of new member states which, they felt, touched on the very nature of the project itself.

Article III, paragraph 2 of the Convention as approved by the various delegations until March 1953 specified that

> The Council, referred to in Article V, may, by a two-thirds majority of all Member States, decide to invite any other European State to join the Organization, and any State so invited may thereupon become a member by becoming a party to this Convention.[77]

On 19 March, in an aide-memoire sent to all the states, the British government proposed that the word European be deleted from Article III, paragraph 2. The stated purpose of this amendment was to leave the door open 'for the later inclusion of non-European and specifically of Commonwealth Countries'. On 29 March the committee responsible for drafting the convention decided not to delete the word, arguing that it was not 'a pure drafting matter'. During the Council session in Rome of 30 March to 2 April 1953, the British amendment was accepted by 7 votes to 1 (Switzerland) with two abstentions, the French delegates stating that they had insufficient instructions to vote against it and the Germans wondering 'whether there [was] anything behind the British proposal'.[78]

During April the Quai d'Orsay also tried to determine the exact motives of the Foreign Office. As it had seemed unlikely at the Council session that the desire to leave the door open for Commonwealth countries was the only reason behind the British proposal, the Quai d'Orsay assumed that the Foreign Office 'would not object to the possibility of the United States becoming a member of the Organization'.[79]

The French reaction was therefore strong and the French ambassador in London was invited to act. In a memo sent to him on 29 April, Paris noted that the British amendment modified the spirit in which the laboratory had been established, and would jeopardize its aims and even possibly its very character. It was pointed out that the laboratory was initially a response to a particular situation in a group of countries (those of western Europe), countries with roughly equivalent scientific and economic capacities, and which, confronted with a specific need, wished to unite to improve their situation. This of course in no way precluded close scientific co-operation with other countries (including the United States), and such collaboration had indeed already begun in the Proton Synchrotron Group. But the *decision-making* power, particularly with respect to the programmes, should remain in the hands of those for whom the laboratory was intended—otherwise the mass of different interests would become so unwieldy that the organization would be unable to operate. The French therefore requested that Article III be renegotiated.[80]

However, the re-drafting of this article was complicated by another consideration, namely the possibility of applications for admission from eastern European countries. On this point, the British (like the French) were adamant, whereas the

Swiss wished the organization to remain strictly neutral. As the French ambassador in London noted, 'the British difficulty arises from a certain reluctance to agree to statutes, which while preventing the full and complete accession of western democracies outside Europe to the organization in question would leave the door open to European states from the Communist bloc'. Therefore, after various attempts, the British proposed, with French agreement, that paragraph 2 be deleted. There would be no clause taking account of the possibility of the accession of new states. Instead, as and when the problem arose, it would be necessary, under the terms of Article X, to modify the Convention. Since this would require unanimity, the Swiss (and the French) could take action to ensure that CERN retained its European character, while the British (and the French) could refuse to admit countries from eastern Europe. The proposal was not accepted in this form and a version of paragraph 2 expressing the same idea was retained. It reads:

> 2. (a) Other States may be admitted to the Organization by the Council referred to in Article IV by a unanimous decision of Member States.[81]

A remark is called for here. The drafting of the Convention involved hundreds of amendments but only this item, on the accession of new member states, was the subject of heated controversy. It went to the very heart of the foreign policies of the European countries, all of which wanted to adopt a position on CERN which was consistent with their general diplomatic stance. It was for this reason that the United Kingdom wanted CERN to *appear* to be part of the Atlantic sphere, to be open to it, and to be different from existing European institutions, e.g. the European Coal and Steel Community or the European Defence Community. As Verry wrote in a confidential note, 'The other idea behind our proposal had been that we should be able to show the world that the organization had in fact no political significance as a European body'.[82] Similarly, if Switzerland was so opposed to the British amendment, it was because its first requirement was that of absolute neutrality with respect to all European countries, and in the East-West debate. As a British document stated, 'the Swiss opposed it, [...] arguing that the amendment would *be interpreted* as aiming at American membership and hence the utilization of the proposed Organization for the Cold War'.[83] Of course if during a vote certain countries refused the membership of a Communist state, no great harm would be done, but formally nothing should forbid this or even hint at it. Finally for France, and particularly for the Quai d'Orsay, if the laboratory was to be in the West, its character as a European project was essential. A French note dated April 1953 states: 'One of the main reasons for our participation has always been, apart from the scientific interest, precisely the fact that it [the organization] was a joint European venture'. In other words, it was to be consistent with France's wish to restore her erstwhile glory by *constructing an integrated Europe* —an objective which did not, of course, conflict with the defence of the 'free world' of which that Europe was clearly a part.[84]

Notes: p. 348

What can one say of the solution finally arrived at? That it seemed the least undesirable in the light of these three types of requirement. And by introducing the right of veto, further consideration of the matter could be postponed.

9.5 The debate surrounding French ratification, July 1953–September 1954

The text of the Convention which precipitated the debates we have just described was finally adopted by the CERN Council at its sixth session on 29 and 30 June 1953. It was ceremoniously signed by nine of the twelve delegations on 1 July in the Salon de l'Horloge at the Ministry of Foreign Affairs in Paris.[85] It now had to be ratified. The members of the Council thought that this would be a rapid procedure and that the Convention could come into force within six or seven months, that is, before the end of January 1954.[86] Unfortunately this view was over-optimistic, as the dates on which the instruments of ratification were deposited at UNESCO show:

United Kingdom	30 December 1953
Switzerland	12 February 1954
Denmark	5 April 1954
Netherlands	15 June 1954
Greece	7 July 1954
Sweden	15 July 1954
Belgium	19 July 1954
Germany and France	29 September 1954
Norway	4 October 1954
Yugoslavia	9 February 1955
Italy	24 February 1955.[87]

There was particular concern in the Council with regard to the ratification procedure in France.[88] Some people, like Guggenheim, an adviser to the Swiss delegation, had the impression 'that the situation [was] much more serious than it [appeared]' and that something was going wrong in France.[89] Four considerations of a somewhat different kind lay behind these uncertainties.

Firstly, there was the contrast between the relative slowness of the ratification process and the rather positive attitude adopted by France in previous years. In 1950 France had helped to promote the idea of European laboratories. In 1951 she had offered money. At the beginning of 1952 she was among the first to ratify the Agreement establishing the provisional organization and to offer her contribution. Now suddenly there was a lengthy delay and this was worrying. Secondly, the French contribution represented about 24% of the CERN budget. Since 75% of the total had to be paid before the Convention could come into force, French ratification was crucial. Without it and that of any other country no further progress could be made. Thirdly, France was the only country not to pay its share of the budget contributions regularly requested from the member states to enable CERN to continue its activities during the transitional period (July 1953–September 1954). In fact parliament had made payment of the French contribution conditional on ratification of the Convention. There is nothing surprising in this but the psychological effects can be imagined. Finally, 'the French and Italian situations have one thing in common, namely the entrenched opposition of the Communist parties [...] and the very serious hesitation on the part of certain non-Communist elements which regard CERN as an American institution in disguise'.[90]

This latter consideration, which probably caused most concern, was strengthened by the lack of progress in the debate on the ratification of the treaty establishing the European Defence Community (EDC). Opposition to it had been increasing daily since the end of 1953. Thus, if CERN was seen during the parliamentary debate as another example of European integration, there was a serious risk that ratification of the Convention be refused.[91]

In fact, as we will see below, the fears appear to have had little foundation. Only the Communists voted against the project in parliament while the other members hostile to the EDC disassociated it from the CERN case. This, however, should not be allowed to conceal two things, the political opposition to the project, which we shall study forthwith, and the fact that France's ratification was rather slow, if not by contrast with other countries (remember Germany and Italy, for example), at least by contrast with her earlier alacrity in committing herself to the project. We need, then, to pay some attention to the ratification process. Finally, we consider the debate in the National Assembly.

9.5.1 COMMUNIST OPPOSITION TO THE CERN PROJECT

Before discussing this general opposition to the project, we must look rapidly at the hesitation regarding the synchro-cyclotron which was not specific to the Communists, and which arose in France early in 1952. In October 1950 the Corpuscular Physics Commission of the CNRS proposed that the CEA construct an accelerator of about 500 MeV, and this proposal had been included in the CEA's programme in November 1951. In January 1952 there arose the obvious question of duplication with the planned CERN synchro-cyclotron. On 3 January, on Perrin's suggestion, the French Atomic Energy Committee proposed that 'the Commissariat offer to construct the 400 MeV accelerator and to make it available to the European centre'. It is difficult to say what became of this offer, as the French delegates appear not to have *officially* formulated this proposal in Geneva in February 1952 or during the first CERN Council sessions, though the matter was unofficially discussed in May. Apart from that there is no further mention of the idea in our documents. The French delegates apparently thought that the proposal would not meet with the approval of the other CERN member states.[92]

This episode was only peripheral to the criticism of the entire project launched

Notes: p. 348

after 1950 by scientists and politicians close to the Communist Party. The arguments they used were to remain unchanged until July 1954. We present them thematically, proceeding from the more general to the more particular.[93]

At the heart of the opposition there was CERN's lack of neutrality in the East–West conflict. It was considered to be an Atlantic organization inspired by the Americans, similar to NATO, the CEA, or the European Defence Community. The proof of this was that like the other organizations CERN excluded the countries of eastern Europe and that, as everybody agreed, an American had been responsible for launching it. More than that: an article in *Lettres Nouvelles* dated June 1953 and signed Nucleus, revealed that Isidor I. Rabi had not acted alone. Early in 1950 he had belonged to a commission of the American State Department responsible for carrying out a study on 'Science and Foreign Relations'. In its report, according to Nucleus, the commission explained that 'in our countries there [was] "a strong intellectual potential" and a "reservoir of scientific talent" ' which remained under-exploited due to a lack of material resources. It went on to deplore the fact 'that for its own "security" the United States [had] not yet exploited this reservoir "in an intelligent way" nor [had] she yet "tapped this stock of new ideas" '. 'In these circumstances', the article concluded, 'the establishment of a European atomic pool appears in a different light'. It fitted in with American designs, and as such was unacceptable.[94]

The second type of argument was that, while the idea of setting up a body whose purpose would be 'to combine the efforts of several countries to build extremely costly equipment' was sound, its implementation would be unacceptable if it ultimately sabotaged French research. Unfortunately, the Communists argued, this was already happening—consider the 400–500 MeV accelerator, or the amounts set aside for French science and which were not being increased. As Irène Joliot-Curie put it in June 1954, CERN was being offered a great deal of money at a time when 'the financial effort which would have been necessary [for French research and in particular for the implementation of the CNRS's five-year plan] was not agreed: in nuclear physics, the accelerator requirements [had] not even begun to be met'.[95]

Thirdly, it was noted that, compared to her financial commitment, France seemed to have little influence within the organization. She had only one vote out of twelve in the Council, she had not managed to get the laboratory built on her soil, her scientists were not responsible for the construction of the machines or for leading the theory group, there would only be a handful of French physicists at CERN, and so on. More maliciously, Nucleus wrote that through Kowarski, France would be responsible for the 'dimensions of doors and windows', the 'height of ceilings', the 'choice of floor tiling and paint colours', the 'number and location of the lavatories'... He concluded: 'In short, France, which gave the world the first studies in radioactivity, finds herself having to pay one quarter of the CERN budget and being given a bookkeeper's post by way of compensation. Moreover the bookkeeper is expected to be a member of the orthodoxy ...'.[96]

This last remark led to a fourth criticism: who was consulted and who took the decision on behalf of France? Nucleus claimed that it was largely the Ministry of Foreign Affairs, Francis Perrin having been merely consulted for the sake of form. 'Has the opinion of the highest French scientific authority, the Academy of Sciences, ever once been sought?' Nucleus asked. 'Over the last two years has the matter ever once been discussed in scientific circles?' The answer was no. Moreover, 'Mr. travelling-salesman Auger failed to consult the three French Nobel prize winners [Louis de Broglie and Irène and Frédéric Joliot-Curie], who were specialists in the matter and directly concerned'. Which only went to show that the scientists were not trusted. 'If the project is so irreproachable,' Nucleus added, 'what is there to hide?'[97]

The conclusion? According to Irène Joliot-Curie, 'French scientists would not understand why commitments [were] being made to the CERN project before reliable guarantees [had] been given concerning the construction of accelerators in France'. She added that an emergency plan had just been drawn up to this end and that it was 'absolutely vital that it be implemented'. However, she did not explicitly recommend that France withdraw from CERN, as did the Communist deputy Georges Cogniot during the debate in the National Assembly.[98]

These arguments were naturally countered on several occasions. The various notes drafted for the Ministry of Foreign Affairs dealt mainly with technical matters or with 'the facts', while the more political responses were left to members of parliament.

The counter-arguments were as follows:[99]

1 – The purpose of a European laboratory was first and foremost to make up for 'the lack of well equipped laboratories' in Europe and to curtail the resulting 'brain-drain';

2 – It was the French who had originally had the idea, not the Americans. Dautry had been the first to submit the proposal officially in Lausanne in December 1949;

3 – France's contribution to the CERN budget came from the Ministry of Foreign Affairs and could not therefore be the cause of any restrictions imposed on the national research budget;

4 – The machines built in Geneva, in particular the proton synchrotron, would allow the CEA to concentrate its efforts on other construction projects. Duplication was therefore ruled out;

5 – France had no leading posts in the technical and scientific groups 'because she [had] no physicists of sufficient experience' to whom such responsibilities could be given;

6 – France was not, however, in a weak position. Both the chairman of the Nominations Committee and the president of the Council were French. In any case, compromise was inevitable in an international body;

Notes: pp. 348 ff.

7 – Kowarski's role had covered more than just architectural details. He was in charge of the general physics services, of the preparation of experimental equipment, etc.

8 – It was wrong to claim that de Broglie had not been consulted. CERN was also his idea (Lausanne, December 1949) and he was invited to Copenhagen for the European conference of June 1952. If he did not attend, that was his affair;

9 – However, it was true 'that the Joliots [had] not [been] consulted. This [was] a political fact and should be considered as such'. Anyway Frédéric Joliot had publicly opposed the project as early as January 1952;

10 – All the same, in a spirit of reconciliation, 'the Ministry of Foreign Affairs has decided [January 1954] to set up a scientific advisory committee responsible for advising our official representatives to the Council'. The 'Academy of Sciences, the Commissariat à l'Energie Atomique, the CNRS and the higher education authorities will be represented on it'.[100]

Without wishing to comment on all the controversial points, three remarks are worth making. Concerning the *origin* of the project, we have already explained that the quest for paternity is completely futile. The UNESCO project was the outcome of the meeting of many currents which emerged between 1948 and 1950, and responded to a number of different interests.[101] If there was a dispute in France over the 'father of CERN', it was because people thought, by this expedient, to condemn or to justify the project *politically*.

As for the fact that the French scientific community was hardly consulted, there is no simple answer. However, we can say that from the outset, all the important figures in French physics (i.e. some 40 to 50 physicists) were *informed* about the UNESCO project since Auger had described it when the CNRS five-year plan was being prepared between October 1950 and February 1951.[102] On the other hand the only people *formally associated* with the decision-making process were the Director of the CNRS (Dupouy, a physicist), the scientific directors of the CEA (including Perrin and Kowarski), and, after 1952, the Atomic Energy Committee (including Leprince-Ringuet).

Negotiations were therefore not conducted in the greatest secrecy, and the Ministry of Foreign Affairs did nothing without specifically consulting certain scientists—even if far fewer were *consulted* than were merely *informed*. However and more generally, it is important to note that the launching of any new project always seems to require a leader or a group of individuals who make the scheme their own and who try to achieve rapid success using a wide variety of means. An ideal, more 'democratic' or more 'rational' procedure never seems to occur. This is partly because there cannot be neutral and objective criteria which allow one to choose definitively. The other reason is that it is difficult to strike a balance between efficiency and holding a wide preliminary inquiry, between pragmatism and the 'democratic' ideal. Consequently certain people are always in a position to complain that they have not really been consulted (Joliot in France, but also Bohr with regard to the Amaldi and Auger project). In the case of the Joliots, however, this 'technical' argument should not be allowed to conceal the fact that there was considerable political opposition to them.

Our last remark. Can it be said that CERN deprived French science of funds between 1950 and 1955? The answer seems to be not, since the money for CERN came from a source which, on the insistence of de Rose and Perrin, and against the advice of the Ministry for the Budget, was entirely separate from that of other sources of expenditure on science. However, this fear was not specifically Communist—the Ministry of National Education and the governing bodies of the CEA had also expressed it in 1953—nor indeed specifically French—we have come across it in virtually every country.[103]

9.5.2 THE RATIFICATION PROCESS

A year passed between the signing of the treaty (1 July 1953) and ratification by the French parliament (6 July 1954). By way of comparison, it took eight months for the ratification of the agreement establishing the European Coal and Steel Community (treaty signed on 18 April 1951, parliamentary vote on 13 December 1951).

In the case of the Convention for the establishment of CERN, the ratification procedure can be divided into three stages: the drawing-up of the draft law for submission to parliament, the consultation with other ministerial departments, and the parliamentary decision.[104]

The *drawing-up of the draft law* by the Ministry of Foreign Affairs began early in July. The text was ready by 3 August. On the eighth it was forwarded to the Minister's office from which it emerged duly signed not later than the beginning of September.

The period of *consultation with the other ministries concerned* (Ministry of Education, Secretariat of State to the Prime Minister, Secretariat of State for Economic Affairs, Atomic Energy Committee, and the CEA) lasted from mid-September to mid-December 1953. The agreement of the various departments was obtained without any problem between 5 November and 16 December. On 19 December the completed file was forwarded to the office of the Prime Minister for 'examination by the Council of State [to assess the constitutionality of the text] and the Council of Ministers'. Thereafter, from mid-December to the beginning of February, one of those 'border wars' between ministerial departments which we described in section 3.3 took place. The Secretariat of State for the Budget did not consider that it had been adequately consulted, and suddenly announced that it would be obliged 'to oppose examination of the draft law by the government'. In a series of notes and conversations, the Secretary raised several questions which in his view required further study. In the event, since this problem seemed to derive solely from one department's over-sensitivity, a solution was quickly found.[105]

Notes: p. 349

A period of three months for the consultation procedure (from September to December 1953 if one disregards the conflict with the Secretary of State for the Budget) is easily accounted for. The project did not have priority and its consequences had to be reconsidered. Indeed, the decision taken in France in favour of CERN between December 1951 and January 1952 applied only to phase 2 of the project. This had been clearly stipulated in the preliminary document for the Paris conference (December 1951) which states: 'it is reminded [sic] that a final decision regarding the construction of the laboratory [phase 3] is not expected from the Conference [...] Such a decision can only be taken on the basis of the data to be obtained during the second stage'. For this reason the French parliament made payment of the contribution for phase 3 conditional on the ratification of the Convention, that is, on its giving official approval to France's participation in the *construction* of the laboratory.[106]

The project having been finally approved by the government, it was *sent to parliament* which reconvened on 9 February 1954. However, it was not until June that it was submitted to the Foreign Affairs Commission of the National Assembly for study. On 11 June that Commission studied the text, on the 25th the National Education Commission did likewise, on 6 July the Assembly voted the text, on 5 August the Council of the Republic approved it and it was promulgated eight days later on 13 August 1954.

Why was there something of a delay between February and June? Simply because parliament seems to have been preoccupied with more important matters than CERN, namely the Tunisian and Moroccan questions, the debate on the EDC (Marshal Juin resigned on 31 March), and above all the conflict in Indochina. At the end of January the siege of Dien Bien Phu by the Vietnamese began, and between the 5th and the 9th of March and the 4th and the 6th of May this was the only matter dealt with by the Assembly.[107]

9.5.3 THE DEBATE IN THE NATIONAL ASSEMBLY [108]

The debate took place on Tuesday, 6 July 1954. After parliament had heard the rapporteurs of the Commissions of Foreign Affairs, National Education, and of Finance, the Communist MP Georges Cogniot submitted a counter-project proposing that the 700 million francs for CERN specified in the financial protocol should be used to set up a 'French institute for fundamental nuclear physics research'. This proposal was rejected, and the ratification of the CERN Convention was then approved, only the Communist group voting against. An amendment moved by the Commission of National Education was incorporated into the text of the law. It specified: 'The Government shall take all measures to provide particle accelerators for French nuclear research, covering the range of various energies required for modern research and to train research scientists in sufficient numbers'.

During the debate most of the arguments already encountered were used. Only two points are worthy of special mention. Three people defended the project against Georges Cogniot's criticisms: Maurice Naegelen (rapporteur of the Foreign Affairs Commission), Jules Moch (on behalf of the Socialists) and Jacques Soustelle (on behalf of the Gaullists). Interestingly enough—and was it just by chance that these men were chosen?—all three were *opposed* to the European Defence Community. This led them to defend the project in ways which differed from those used by the departments of the Quai d'Orsay in 1952–1953, at least on one crucial point: they played down the *European* character of the project. Thus Maurice Naegelen could state: 'if a man like me [...] can be in favour of a project like this, it is precisely because it is of interest to many states which did not support the European Defence Community project, notably the United Kingdom and Switzerland, a neutral country'. Or Jacques Soustelle: 'the planned nuclear research centre seems to me precisely to be free from a number of vices which I find particularly regrettable in other [European] projects'. For his part, Jules Moch insisted that 'what [was] at issue [was] neither a movement for Europe nor a political movement, not the cold war, not the European Defence Community, not the European Coal and Steel Community, still less the production of atomic bombs, but simply the construction of an important laboratory which France would be hard put to build on her own'. The advantages of having men like this defend the project in this way were clear, particularly with respect to the Communist opposition: the scheme, for which there were no political motives, even European ones, was approved on purely national scientific grounds, and it was the Communists who seemed to be the unpatriotic ones.

This leads us to our second remark. If the interest of the project was primarily scientific, it had to be part and parcel of 'a serious, co-ordinated, coherent, and considerable improvement in the budget for fundamental research in France' (Jacques Soustelle). This demand, reiterated by all the speakers, led to the ready acceptance of the amendment proposed by the National Education Commission. Could it then be said that CERN not only did not deprive French research of funds, but in fact enabled it to obtain more? This would be an exaggerated claim. Although the CERN debate clearly highlighted France's scientific weaknesses, the realization in political circles that something had to be done about the disastrous state of French scientific research cannot be attributed to this alone. Let us remember that from the day it took office (on 18 June 1954) the government of Pierre Mendès-France made scientific research one of its priorities. For the second time in the history of France, a 'Secretariat of State for scientific research and technical progress' was created under the direction of a scientist, Henri Longchambon. Between 18 June and 6 July, the date of the CERN debate, the Secretariat had already taken decisions in this field, notably with regard to two accelerators for which funds were granted on the morning of the debate. One sees then that it would be too much to say that CERN lay behind this new science policy—even if it served as one of the catalysts.[109]

To conclude this study of France's ratification, we will consider again the anxiety expressed by some CERN Council members: was there a serious risk of the Convention not being accepted by the French parliament? In the light of the Assembly debate, we do not think so, though we recognize that such fears were more than understandable. It must not be forgotten that the debate on the EDC, in the words of Jean-Pierre Rioux, '[had raised] two major questions over which France [had] no intention of admitting she [had] had no control since 1944: American hegemony and the resurgence of Germany'. Hence the return of the bogey of nationalism, with the Communists and the Gaullists 'at the heart of the "resistance" against any potential "Vichystes"'.[110] Add to this the revival of the 'neutralists' and the 'progressives' (whose anti-Americanism was rekindled by their opposition to the colonial wars in Indochina and Morocco), and the hesitations of the traditional parties *vis-à-vis* the construction of Europe (the Socialists, for example)—and the fears expressed by CERN Council delegate Guggenheim are readily understood. All the same parliament dissociated the CERN project from the others and kept it within the initial framework which Auger, Amaldi, and a few associates had wished to give it from the outset, treating it as a strictly scientific project which could easily be realized within the framework of western Europe. And since the new government had made research a priority, and had already taken concrete decisions to this end, a positive vote was obtained without difficulty.

9.6 Concluding remarks

Among the ensemble of French attitudes which we have described, three should be singled out, each corresponding to different men, to different patterns of behaviour, and to different motives.

We have first the active defenders of the European laboratory project. Among them Dautry was the most long-standing elder statesman, a man who was aware of modern scientific requirements, who knew that small nations would be increasingly unable to meet them, and who saw himself as the architect of European solutions. Alongside him, civil servants at the Quai d'Orsay such as François de Rose and Robert Valeur spring to mind. Atypical of diplomats, who were usually unable to understand the demands imposed by big science, they owed their awareness in this regard to their contacts with scientists: in New York in the United Nations Atomic Energy Commission in the case of de Rose, in Paris at the Quai d'Orsay for Valeur, where he was appointed head of the department responsible for relations with UNESCO. Rapidly convinced of the importance for Europe (and for France) of Dautry's and Auger's projects, they made them their own, even though the issues were somewhat outside their usual scope (de Rose's tasks in the Quai d'Orsay, for example, covered neither scientific affairs, for which no one was responsible, nor relations with UNESCO.) Finally, a few scientists who were associated both with the

French CEA and science administration, such as Auger and Perrin, also fall into this group. Their commitment to the project grew from their awareness of the structural weaknesses of French science, and out of a left-wing tradition of universalism which their fathers had stood for in the pre-war period.

Underpinning the cohesion of the group was the idea that there was a scientific need which could not be satisfied in the national context, and that the inertia of political circles should be combatted by all possible means. Thanks to them, a first guiding thread appeared in the French position: from the outset, France worked for the establishment of supranational laboratories without ever reneging on her commitments, often adopting a flexible stance during negotiations. This is why at the end of 1950 the French delegates abandoned their original ideas, accepting instead a laboratory dedicated only to high-energy physics; why, in 1951, they managed to secure an unqualified French engagement in phase 2, the planning stage of provisional CERN; why, in 1952, without having really defended the idea, they withdrew the proposal to have the French CEA construct the synchro-cyclotron and to offer CERN the use of it; why, six months later, they managed to persuade Parodi and Gaillard that CERN could not be set up near Paris without doing serious harm to the spirit in which it had been born; and finally why, in 1953, when the Convention was being drawn up, they made a minimum of demands and, in the opinion of many people and of the British in particular, played a vital mediating role. In other words, it was to Dautry, de Rose, Valeur, Auger, and Perrin that between 1950 and 1954 France owed its image as one of the first states to have championed unwaveringly the cause of the nascent organization.

But it is not only they who deserve mention. After all they only represented their country by proxy. Beyond them, the Ministry of Foreign Affairs was the final authority. Here two men should be mentioned: Louis Joxe, Director of the Department of Cultural Relations and Robert Valeur's superior, and Alexandre Parodi, whom de Rose knew well and who was Secretary-General of the Ministry and second in the hierarchy after the Minister. What is typical of them is the support and the confidence they had in their subordinates, although not directly involved themselves. There were two independent factors behind their backing for the project—firstly the scientific considerations which Auger, Dupouy, and Perrin had continually emphasized and, secondly, the political ones, since this scientific co-operation was in line with the general foreign policy which their Minister, Robert Schuman, had introduced in 1950, and of which they were supporters.

Naturally they had their own demands, and hoped that France's interests would be catered for. However, they too remained rather flexible and provided the support that de Rose, Perrin, and the others needed. It was they who were responsible for a second guiding thread in French attitudes, namely the repeated wish to see France play a leading role in all European affairs. This is why France prided herself on having been the initiator of the project through Dautry, on the one hand, and Auger on the other, on having committed two million francs in 1951 for preliminary studies, and on having supported the bevatron project from the time of the Paris and Geneva conferences, offering in January 1952 one third of the money required for the provisional CERN; and why she insisted, finally, on paying as much as the United Kingdom to the CERN budget, though the latter ought, on the basis of its gross national product, to have paid more. These attitudes reinforced the image which the Quai d'Orsay wanted to give of France, namely, of being magnanimous in diplomatic affairs, of being the founder of the European spirit, and of having lost none of its greatness.

Finally, a study of French attitudes would not be complete without considering the role of the political world or at least, in this context, of parliamentary circles. It is difficult to treat this comprehensively, since parliament only appears briefly in our story, and in a very specific political situation. However, one point is to be stressed. The attitude of the members of parliament was far more pragmatic than that of the Ministry of Foreign Affairs, and it was their pragmatism which enabled them—the Communists apart for obvious political reasons—unanimously to ratify the Convention establishing CERN. Many of them were able to do so by differentiating CERN from the ECSC and the EDC; it was by stressing that its interest was primarily scientific that they were able to justify their stance. And if there was also a European dimension, they could compliment themselves on this too, always being quick to add that national sovereignty was not jeopardized in this case, and that for once countries like Switzerland, the United Kingdom, and Scandinavia were taking part.

Notes

1. The main archival source used in this chapter is that of the Ministry of Foreign Affairs (frequently referred to as the *Quai d'Orsay* after the street where it is situated in Paris). The series of files is entitled *Affaires Atomiques*. For the period 1949-1954, the most important files are numbers 75 to 80. We have also used *Raoul Dautry's personal archives* which are lodged with the *Archives de France,* series 307AP, boxes 209-214 entitled *Mouvement Européen*. Unfortunately, we did not find the papers of Francis Perrin (apparently destroyed) and the Commissariat à l'Energie Atomique categorically refused to let us consult their archives. Finally, we surveyed the CNRS's and Ministry of Education's archives. The net result was rather disappointing. Indeed the remarks we made for the 20s and 30s in Pestre (1984a), 327-328, apply at least for the late 40s/early 50s. For bibliographical information on the diplomatic side, see *Annuaire Diplomatique et Consulaire de la République Française,* published annually. In these notes the archives of the Ministry of Foreign Affairs will be abbreviated to (AEF), those of Raoul Dautry to (AF-307AP), those of CNRS to (AF-CNRS).

    For an overview of the French situation after World War II, see Carré, Dubois & Malinvaud (1972), Earle (1951), Gilpin (1968), Gilpin (1975), Goldschmidt (1964), Hoffmann et al. (1963), and Kuisel (1984). On more specific problems, see note 4 (on the politico-diplomatic context), note 13 (on French nuclear policy), and note 20 (on the scientific situation).

2. See parts II and III of this volume.
3. See for example section 5.4.

4. For this study, see Grosser (1972), Rioux (1980), Scheinman (1965), and Vincent (1977). It should be remembered that the Fourth Republic was set up at the end of WWII and came to an end in 1958 with General de Gaulle's accession to power.
5. Grosser (1972), 33.
6. Grosser points out that the interim aid granted to France during the winter of 1947-1948 amounted to 66% of the ration of bread officially distributed, 60% of the oil requirement, and 20% of coal needs. Grosser (1972), 220.
7. Quoted in Scheinman (1965), xviii.
8. See Grosser's *conclusion* on what was considered to be 'unthinkable'.
9. Scheinman (1965), xx.
10. Grosser (1972).
11. Here we follow Grosser (1972), 231-238; Rioux (1980), 199-204 paints a slightly different picture.
12. See Grosser or Rioux on this very complex European Defence Community initiative.
13. In addition to Scheinman (1965), see Weart (1979), Rapport CEA (1950, 1951...), Dév. Nucl. France (1965), *Atomic Energy Developments in France,* talk given by Dr. Kowarski at Brookhaven, Feb. 6th 1948 (Brookhaven National Laboratory Library), *Aperçu de l'activité des Services de physique et de technologie du CEA,* unsigned and undated (the document is typed and was probably written by Kowarski around 1950, Kowarski interview file, CERN). Naturally there is much more information available on the CEA—the titles run to dozens.
14. The first quotation is from Scheinman (1965), xviii; the second is from the statute of 18 October 1945 reproduced in Rapport CEA (1950), 179-183.
15. Text of the statute cited in note 14.
16. CEA and CNRS budgets are compared in table 9.2. Whereas the ratio is roughly 1 to 1 in 1946-1947, it rises to 2 to 1 in 1950, 4 to 1 in 1955, and 6 to 1 at the end of the fifties.
17. All this information is culled from the documents cited in note 13. It should be noted that Pierre Auger resigned from all his posts in France when he became Director of UNESCO's Department of Exact and Natural Sciences in 1948. He was not replaced on the Atomic Energy Committee.
18. Rapport CEA (1950) and texts by Kowarski cited in note 13.
19. The last three paragraphs are directly inspired by Scheinman (1965), 1-89. The quotations are from pp. 85 and 88.
20. Some references are given in the notes to tables 9.1 to 9.4. The following should be added: Châtelet (1960), Colloque Caen (1957), CSRSPT (1957), Weisskopf (1951).
21. Kowarski, *Atomic Energy...,* op. cit. note 13, 1.
22. Colloque Caen (1957), CSRSPT (1957). The reader should note that the situation was perhaps not as catastrophic as the figures may lead one to believe. These two reports *wanted to demonstrate* that France was *particularly* late.
23. The percentages have been calculated as follows (the example shows the annual 9% increase from 1950 to 1956):

$$\frac{\text{No. for 1956} - \text{No. for 1950}}{\text{No. for 1950} \times 6} = \frac{54 - 35}{35 \times 6} = 9\%$$

24. By way of example from among many possible sources, see Rech. Phys. (1948), Barrabé (1951).
25. Document entitled 'Groupe III, Physique', no written date, 11 pp. (AF -CNRS/102).
26. Châtelet (1960), No. 2580, sets out these initiatives in detail. For the economic recovery, see Carré, Dubois & Malinvaud (1972).
27. Weisskopf (1951).
28. See Pestre (1984a).
29. He lists the following as members of the CEA group: Yvon, Abragam, Horowitz, Mercier, Trocheris, and Meyer.
30. Most of this information was confirmed in a series of interviews carried out by the author and K. Pomian between July 1979 and June 1980. In particular, K. Pomian met Pierre Auger, Georges Charpak, Mme

Dewitt-Morette, Louis Leprince-Ringuet, Albert Messiah, Francis Perrin, Charles Peyrou, and Jacques Prentki.
31. Kowarski, *Aperçu..., op. cit.* note 13, 5.
32. Out of curiosity we also consulted the articles published in the *Journal de Physique,* the most important physics journal in France along with the *Comptes-rendus de l'Académie des Sciences.* From 1950 to 1954, whereas several articles deal with cosmic rays or detectors, *only one* deals with accelerators [Michel-Yves Bernard, 'Un modèle théorique simple pour l'étude du mouvement des ions dans un accélérateur linéaire' *Journal de Physique,* **15** (October 1954), 121A–132A]. Moreover, no mention is made of accelerators in contemporary editions of the *Revue Scientifique,* a popular science review.
33. Accelerators (1952–1954). See also section 1.3.
34. This proposal was written into the five-year plan of the CEA drawn up and approved in November 1951. The project was subsequently reconsidered.
35. For a more precise analysis, see sections 2.2 and 2.4.
36. See the notes sent by Kowarski and Guéron to Dautry in November–December 1949 (AF-307AP/212).
37. Copy of the file Dautry sent to the Quai d'Orsay is in (AF-307AP/212).
38. See section 2.1 and appendix 1 of this volume.
39. See section 2.4.
40. See sections 2.5, 3.1, and 3.2.
41. Draft letter Bidault to Ministre des Affaires Etrangères, 19/6/50 (AF-307AP/212); letter Président du Conseil to Ministre des Affaires Etrangères, 20/11/50 (AEF-76); *Extrait du discours prononcé par M. Petsche, ministre des finances et de l'économie nationale au Conseil de l'OECE le 7 juillet,* in letter de Rose to Auger, 8/7/50 (Mussard file, CERN). The *Président du Conseil* was the head of the government. He presided over the Council of Ministers and was the equivalent of a British Prime Minister. In the text, we have called him the Prime Minister. The *Secrétaires d'Etat* (called Secretaries of State in the text) were attached to a minister or to the *President du Conseil.* They were a kind of second-rank minister. For details, see for example Suleiman (1974).
42. See section 3.4.
43. 50 million francs represented approximately 1% of the CEA budget and 2% of the CNRS budget for 1950; the amount was thus not negligible.
44. For the role of the scientists in the Auger project see sections 6.4 and 14.5.
45. See letter Dautry to Lescop, 19/12/50 (AN-307AP/211) and letter Auger to de Rose, 19/12/50 (UNESCO). The quotation comes from Auger's letter.
46. Eleven letters and reports go to make up the exchange of correspondence which runs from 1 February to 11 May 1951 (AEF-76).
47. *Note pour la Direction Générale des Relations culturelles,* 7/2/51 (AEF-76) for the first quotation; Letter Ministre du Budget to Ministre des Affaires Etrangères, 2/4/51 (AEF-76) for the following quotations.
48. *Note no. 485, Direction des Affaires Economiques (à l'attention de Monsieur de Rose),* 7/3/51 (AEF-76).
49. Quoted in letter Mme Lestringuez, Dautry's secretary, to Louis Joxe, Director-General of the Cultural Relations Department, 16/5/5 (AEF-76).
50. Our knowledge of this meeting is based on a long report by Louis Joxe sent on behalf of the Minister of Foreign Affairs to the Minister for the Budget, no typed date (AEF-76). The quotations are taken from this report.
51. Details of the French situation as regards accelerator construction are given in report cited note 50, 5.
52. See, for example, the way the problem was presented at the Steering Committee on International Organizations meeting held in London, 8/2/52 (PRO-CAB134/943).
53. For his part Pierre Auger steered his own course, without referring to the CEA or the Ministry of Foreign Affairs.
54. See in particular letter de Rose to Gaillard, 5/12/51 (AF-307AP/211), *Note pour le Ministre,* 6/12/51 and *Projet de lettre au Ministre du Budget,* undated, probably of 6/12/51 (AEF-76). It is impossible to be more precise on the decision-making process in France: the quality of the archival material does not permit

it. However, it *seems* that the governmental machinery for handling matters of international science policy was far less formal than in the UK, for example—see chapter 12 for a contrast.

55. See *Extrait du Procès-Verbal de la 104ème réunion du Comité de l'Energie Atomique tenue le 3 janvier 1952* (AF-307AP/211); and the note from the Ministry of Foreign Affairs (17/1/52) sent to the other ministerial departments, the CNRS, and the CEA, containing a summary record of the December intergovernmental meeting in Paris, and convening a interministerial meeting on 28 January to give a ruling on the new UNESCO documents (AEF-76). The meeting was held on the 27th but we have no minutes of it.
56. See *Note de la Direction des Relations Culturelles au Service du Protocole,* 26/2/52 (AEF-76).
57. Chapter 13, appendix 2.
58. See section 5.2 below.
59. *Note,* drawn up by Robert Valeur for the Secretary-General of the Ministry of Foreign Affairs, 9/7/52 (AEF-76). Robert Valeur was head of the Cultural Exchanges Service (which came under the Department of Cultural Relations), and which dealt with matters relating to CERN. He occupied this post from 23 January 1952 and gradually replaced François de Rose, who was appointed Ambassador to Madrid.
60. Regarding Joliot, see section 5.1 below.
61. Part III of this volume.
62. As far as we know, the only scientific point discussed among the French delegation was the need for the synchro-cyclotron. This point, however, does not seem to have been discussed officially by the Council (see section 5.2 below). We should also add that the question of French financial support differs from the other two points in that it was an *internal* matter, whereas the question of the site or the drafting of the convention concerned France's relations with other countries. The reader will forgive us for having nevertheless raised them together.
63. See note Mussard to Miss Thorneycroft entitled: *Lettre du 25 janvier 1952, du Ministère des Affaires Etrangères, annonçant le versement de 25 millions de francs, pour le Conseil de Représentants CERN,* undated (UNESCO).
64. *Note pour le Secrétaire Général,* 9/7/1952, drafted by Robert Valeur from an initial report by Francis Perrin (AEF-76). Quotations from pp. 3 and 7.
65. See, for example, *Journal Officiel de la République Française,* 5/11/53 and 4/12/53; letter Valeur to Verry, 7/12/53 (CERN-DG20821). The documentation is very limited on all these points. The way the negotiations were carried out with other ministries is unknown to us.
66. It should be remembered that in December 1950 and in May, October, and December of 1951, the matter of the site had always been deferred. It had always been found preferable to consider only the *criteria* for the choice.
67. Note from Kowarski entitled: *Emplacement du Laboratoire International, Etat de la question au 20 août* [1952] (Kowarski interview file, CERN).
68. This file entitled: *Proposition Française, mai-octobre 1952* (AEF-79) comprises dozens of letters, reports and plans. P. Alix was responsible for the study which he began with a preliminary note dated 27/5/52.
69. Switzerland entered into consultation with the other countries at ambassadorial level from December 1951 onwards. See Archives fédérales, Département politique fédéral, Berne, Dossier 2001 (W) 3 Band 39, F.11.3.51.; the Netherlands took similar steps in September 1952 (PRO-ED157/302; DSIR17/551). For 1969-1970, see the minutes of the CERN Council and the CERN Committee of Council (CERN/C and CC/E, 901-950). Finally, see Pestre (1984b).
70. Private conversation with François de Rose, Paris, 19/10/83.
71. See for example the letters and notes sent by de Rose in December 1951 and referred to in note 54.
72. The best example of this line of reasoning is in the *Note pour le Secrétaire Général,* 26/8/52 (10 pages), very probably written by Valeur (AEF-79). The quotations are taken from this report.
73. For example, according to the report cited in note 72, it was certain that the United Kingdom would campaign and even exert pressure behind the scenes to ensure that the Dutch proposal would be supported by the Scandinavian countries, by Germany, and by the Yugoslav delegation. According to the British archives (PRO-DSIR17/551), this was untrue.

74. The quotations are taken from the report cited in note 72. We cannot go into detail here, but the arguments used in the report of 26 August are masterpieces of rhetoric applied to the theme: how does one bring one's minister to reach the conclusion one wants. Regarding the United Kingdom for example, Valeur juggles adroitly with the truth (the importance of British high-energy physics, or British tardiness in joining the project), resorts to righteous indignation (how dishonourable it would be for Britain to snatch our pre-eminent position from us at the last minute!), plays on latent French anglophobia, uses flattery, etc.
75. Letter de Rose to Perrin, 22/9/52 (AEF-79). Here is another reason to be careful with Valeur's 'information' that Arnhem was likely to be chosen (note 73). Arguing like this was the best way to make Parodi and Gaillard more 'flexible' on Geneva.
76. On this matter, the documentation in the French archives is very abundant up to the beginning of October, and then stops suddenly. There is only one letter from Perrin to Bouchard at the Ministry of Foreign Affairs informing him, without giving any details, that on 4 October the CERN Council had decided to select the Geneva site.
77. CERN/28, 9/2/53.
78. *Aide-mémoire,* British Embassy, Paris, 19/3/53 (AEF-76); CERN/34, 30/3/53; *Minutes of the Session,* Fifth session, Rome, CERN/GEN/9, 5/8/53, 5–6. On the German attitude, see letter to P.J.E. Male signature illegible, 17/4/53 (SERC-NP24). On the French attitude, see *Note pour l'ambassadeur de France à Londres,* 29/4/53 (AEF-76). It should be noted that the British version differs at this point. Bartlett wrote in an internal British note (8/4/53, SERC-NP24) that: 'Although M. Valeur had been instructed by his Government to oppose our amendment he abstained'.
79. *Note pour l'ambassadeur...,* op. cit. note 78. Here we find: 'When our delegates suggested [...] introducing a clause that would "invite any other European State or, in the case of non-European States, such States as have constitutional links with European States, etc." into Article III (2), the British observers clearly indicated that they considered this formulation too restrictive'.
80. *Note...,* op. cit. note 78.
81. The note from the French Ambassador to London is *Télégramme à l'arrivée,* no.2652/2653, 15/6/53 (AEF-76). Article III(2) quoted in the text is taken from the *Convention for the Establishment of a European Organization for Nuclear Research,* CERN/GEN/8. For the negotiations as a whole, see sections 7.7 and 8.2.
82. See Bartlett's note cited in note 78 and Verry's note, 16/4/53 (SERC-NP 24).
83. Bartlett's note cited in note 78 (our emphasis).
84. Quotation drawn from the *Note...* cited in note 78.
85. Switzerland signed on 17 July, Denmark on 23 December, and Norway on 31 December 1953.
86. See for example *Minutes of the Session,* Sixth Session, Paris 29–30/6/53, CERN/GEN/10, 9.
87. Kowarski (1961), Appendix IV, 12.
88. Some of the remarks which follow apply to Italy. See *Exposé du prof. P. Guggenheim sur la 8e session du Conseil,* 14/1/54, Archives fédérales, Département politique fédéral, Berne, Dossier 2001 (f) 1970/1 Band 208 F.11.3.51. The notes on these documents were taken in the Berne archives by Monique Senft.
89. Quotation taken from note Campiche to Micheli, 1/2/54 (Berne–*id.* note 88).
90. *Exposé du prof. Guggenheim...,* op. cit. note 88.
91. See Rioux (1983), 18–29.
92. See *Extrait du Procès-Verbal de la 104e réunion du Comité de l'Energie Atomique tenue le 3 janvier 1952* (AF-307 AP/211). That the matter was informally debated in May is established in a letter from Gentner to Scherrer, 12/5/52 (MPI/K), in which Gentner informed Scherrer that he had discussed the problem with Joliot. Here, the problem for the French was that the national interest was in conflict with the collective interest of the member states. This was also true for the British who, for the same reasons, did not want to have to 'pay' for the synchro-cyclotron. See the memorandum bearing Verry's signature *European Nuclear Research Organisation—CERN,* 11/12/52 (PRO-DISR 17/551), and sections 13.3 and 13.5.

93. For this section, Communist newspapers were consulted. We retained four articles. 'Des savants nazis participeront aux recherches atomiques à Paris', *L'Humanité,* 22/12/50; 'Le pool atomique européen: une grave menace pour notre indépendance', *Energies Nouvelles,* Jan 52 (copy AF-307AP); 'La science française sacrifiée sur l'autel du CERN?', *Les lettres françaises,* 28/5-4/6/53; opinion column by Irène Joliot-Curie, 'Le projet de Centre européen de recherche nucléaire', *Le Monde,* 11/6/54.
94. In his speech to the Assemblée Nationale on 6 July 1954, the Communist MP Georges Cogniot also used the report from the American International Science Policy Survey Group, entitled *Science and Foreign Relations* (Department of State, 3860, May 1950). This anti-Americanism was coupled with violent, anti-German nationalism. The following statement by Georges Cogniot himself is typical: 'If the French proposal [for a site near Paris] was rejected, it was at the instigation of the West German delegation, and nominally of the physicist Heisenberg, who as well as being a famous physicist and currently on intimate terms with Chancellor Adenauer, was also previously responsible for organizing atomic research in Nazi Germany', *Journal Officiel de la République Française, débats parlementaires, Assemblée Nationale,* 7/7/54, 3233. One final remark: the Irène Joliot-Curie article is the only one from amongst those cited in note 93 not to put forward any anti-German arguments.
95. First quotation taken from *Energies Nouvelles;* second quotation from *Le Monde.*
96. This line of reasoning is explicit in *Energies Nouvelles* and *Lettres françaises.*
97. Quotations taken from *Lettres françaises.*
98. On the subject of the emergency plan, see section 5.3 below.
99. The information here is taken from the *Note pour le Secrétaire d'Etat* [au Budget], 25/1/54 (AEF-77) and from the *Aide-mémoire pour la séance du 16 juin* [1954], prepared at the Quai d'Orsay, and commenting on Irène Joliot-Curie's article (AEF-77). These documents seem to have been written by Valeur.
100. All quotations are from the articles cited in note 99.
101. See section 2.6.
102. *Op. cit.* note 25.
103. *Avis donné par le Conseil scientifique* [of the CEA], 15/10/53; *Avis donné par le Comité de l'Energie Atomique,* 5/11/53; letter 'Secrétaire d'Etat auprès du Président du Conseil', to Minister of Foreign Affairs, 9/12/53; letter Minister of Education to Minister of Foreign Affairs, 16/12/53 (AEF-77).
104. The documentation detailing these stages is in (AEF-76 to 80).
105. The file on this episode is to be found in (AEF-77). We do not have the space here to describe this polemic in detail, a polemic which reveals much about the functioning of the state apparatus. First quotation taken from letter Minister of Foreign Affairs to Merveilleux du Vignaux (at the office of the Prime Minister) 19/12/53; second quotation in letter from the Secretary of State for the Budget to the Minister of Foreign Affairs, 19/1/54 (AEF-77).
106. Quotation taken from the *Working paper,* 18/12/51 (UNESCO/ NS/NUC/1), 2. See also sections 6.2 and 6.3 for more details. It should perhaps be emphasized that if the United Kingdom was the only country in a position to ratify the convention in less than six months, it was probably because at that time it was not necessary for her to renegotiate the project with all the ministerial departments. In fact, the United Kingdom's decision to take part in the project dates back to the end of 1952, when she agreed to join the permanent organization without first becoming a member of the provisional CERN. See section 13.5.3.
107. To appreciate the intensity of French political activity between November 1953 and August 1954, see for example Rioux (1983), 18-40 and Vincent (1977), 73-77.
108. *Journal Officiel..., op. cit.* note 94, 3225-3234. All the quotations in this section are drawn from this source.
109. *Journal Officiel..., op. cit.* note 94.
110. Rioux (1983), 19, 23.

Bibliography for Chapter 9

Annuaire (annuel)	*Annuaire diplomatique et consulaire de la République Française* (Paris: Imprimerie Nationale)
Barrabé (1951)	L. Barrabé, 'La crise de l'enseignement supérieur', *Cahiers Laïques,* **2** (1951).
Bloch-Lainé & Bouvier (1986)	F. Bloch-Lainé and J. Bouvier, *La France restaurée* (Paris: Fayard, 1986).
Carré, Dubois & Malinvaud (1972)	J.J. Carré, P. Dubois, and E. Malinvaud, *La croissance française* (Paris: Seuil, 1972).
Carter, Forster & Moody (1976)	E.C. Carter (II), R. Forster, and J.N. Moody (eds), *Entrepreneurs in Nineteenth- and Twentieth-Century France* (Baltimore: The Johns Hopkins University Press, 1976).
Châtelet (1960)	A. Châtelet, *La France devant les problèmes de la Science* (Paris: Notes et Etudes Documentaires, La Documentation Française, No. 2552 (20/6/59), No. 2580 (20/10/59), No. 2671 (28/5/60), No. 2721 (30/11/60).
Colloque Caen (1957)	'Comptes-Rendus du Colloque Nationale de Caen (1–3 novembre 1956), Numero spécial sur l'enseignement et la recherche scientifique', *Les Cahiers de la République,* 2ème année, **5** (1957).
CSRSPT (1957)	Conseil Supérieur de la Recherche Scientifique et du Progrès Technique, Rapport, *Pour assurer l'avenir, investir en hommes,* 2nd trimestre 1957.
Debiesse (1960)	J. Debiesse 'Enseignement nucléaire et Commissariat à l'Energie Atomique', *Revue Politique et Parlementaire,* **702** (1960), 474–480.
Dév. Nucl. France (1965)	*Le développement nucléaire français depuis 1945* (Paris: Notes et Etudes Documentaires, La Documentation Française, No. 3246, 18/12/65).
Earle (1951)	E.M. Earle (ed.), *Modern France: Problems of the 3rd and 4th Republics* (Princeton: Princeton University Press, 1951).
Encyclopédie (1960)	*Encyclopédie pratique de l'Education en France* (Paris: IPN, 1960).
French Atom (1950)	'French atomic scientists report on their work in 1949', *Bulletin of Atomic Scientists,* **6** (1950), 299–302.
Gilpin (1968)	R. Gilpin, *France in the Age of the Scientific State* (Princeton: Princeton University Press, 1968).
Gilpin (1975)	R. Gilpin, 'Science, Technology, and French Independence', *Long & Wright (1975),* 110–132.
Goldschmidt (1964)	B. Goldschmidt, *The Atomic Adventure* (Elmsford: Pergamon Press, 1964).
Goldschmidt (1985)	B. Goldschmidt, 'Les débuts du Commissariat à l'Energie Atomique', *Revue du Palais de la Découverte,* **14** (1985), 41–51.
Grosser (1972)	A. Grosser, *La IVe République et sa politique extérieure* (Paris: Armand Colin, 1972).
Grosser (1984)	A. Grosser, *Affaires extérieures. La politique de la France, 1944–1984* (Paris: Flammarion, 1984).
Guerlac (1951)	H. Guerlac, 'Science and French National Strength', *Earle (1951),* 81–105.
Hoffmann et al. (1963)	S. Hoffmann, CH.P. Kindleberger, L. Wylie, J.-R. Pitts, J.-B. Duroselle, and F. Goguel, *In Search of France* (Cambridge: Harvard University Press, 1963).
Julliard (1968)	J. Julliard, *La Quatrième République (1947–1958)* (Paris: Calmann-Lévy, 1968).
Kowarski (1961)	L. Kowarski, *Origine et débuts du CERN* (Genève: CERN 61-20, 1961).
Kuisel (1984)	R.F. Kuisel, *Le capitalisme et l'Etat en France: modernisation et dirigisme au XXe siècle* (Paris: Gallimard–NRF, 1984).

Leprince-Ringuet (1985)	L. Leprince-Ringuet, 'La Renaissance de la recherche à l'X', *La Jaune et la Rouge,* **408** (1985), 7-35.
Long & Wright (1975)	T. Dixon Long, C. Wright (eds.), *Science Policies of Industrial Nations* (New York: Praeger, 1975).
Pestre (1984a)	D. Pestre, *Physique et Physiciens en France (1918-1940)* (Paris: Editions des Archives Contemporaines, 1984).
Pestre (1984b)	D. Pestre, 'L'Organisation Européenne pour la Recherche Nucléaire (CERN): un succès politique et scientifique?', *Vingtième Siècle,* **4** (1984), 65-76.
Rapport CEA (1950)	*Rapport d'activité du Commissariat à l'Energie Atomique du 1er janvier 1946 au 31 décembre 1950* (Paris: Imprimerie Nationale, 1952).
Rapport CEA (annuel)	*Rapport sur l'activité et la gestion du CEA, année 19..* (Paris: Imprimerie Nationale, annuel après 1951).
Rech. Phys. (1948)	*Les recherches physiques en France* (Paris: La Documentation Française Illustrée, No. 18, juin 1948).
Rioux (1980, 1983)	J.P. Rioux, *La France de la Quatrième République* (Paris: Seuil, tome I, 1980; tome II, 1983).
Scheinman (1965)	L. Scheinman, *Atomic Energy Policy in France under the Fourth Republic* (Princeton: Princeton University Press, 1965).
Suleiman (1974)	E.N. Suleiman, *Politics, Power, and Bureaucracy in France, the Administrative Elite* (Princeton: Princeton University Press, 1974).
Vincent (1977)	G. Vincent, *Les Français, 1945-1975, chronologie et structures d'une société* (Paris: Masson, 1977).
Weart (1979)	S. Weart, *Scientists in Power* (Harvard: Harvard University Press, 1979).
Weisskopf (1951)	V. Weisskopf, 'Physics in France', *Physics Today,* **4** (1951), 6-11.

CHAPTER 10

The Italian scenario

Lanfranco BELLONI

Contents

10.1 CNR and the 'Years of Reconstruction' 354
10.2 Rome physicists' involvement in the European project 359
10.3 CNR and Italian support of the European laboratory project 369
    Notes 376
    Bibliography for Chapter 10 380

10.1 CNR and the 'Years of Reconstruction'

The following pages provide a brief sketch of the history of Italian scientific institutions in the post-war years, with the aim of setting the frame for the events dealt with in the second part of this chapter, where Italian participation to the European initiative will be discussed. In order to begin from the beginning, we have to go back to the end of the glorious but all too short 'Fermi era' of Italian physics.

As is well known, Italian science had suffered grievous losses before the outbreak of World War II due to Mussolini's racial laws of 14 July 1938. Besides Fermi, Emilio Segré and Bruno Rossi were forced to leave the country.[1] No one could possibly hope for their return: in these post-war years Italy had very few attractions both from the scientific and the political point of view.

Already before the war Edoardo Amaldi[2] was practically left alone and as a consequence tried to concentrate Italian talents and resources in nuclear physics around Rome's Physics Institute.[3] Thus Amaldi arranged the move to Rome of several physicists, who were scattered in Italian universities and were left without an adequate support and a proper guide. Among them was Bernardo Nestore Cacciapuoti,[4] who in the immediate post-war years opted for a career at UNESCO and played a role in CERN's prehistory as the link between two of the Founding Fathers, Amaldi and Pierre Auger.

Before the war, Consiglio Nazionale delle Ricerche (CNR) was the only government agency dealing with the problems of scientific and technological research[5]. After its foundation in 1923, CNR had managed to make its presence felt amidst many difficulties trying to act as a link between University research and industry.[6] Just before the end of hostilities, CNR was reorganized to meet the challenges of 'the years of reconstruction'.[7] As an autonomous body of the state, directly dependent on the head of government, the main task of CNR remained 'promotion, coordination and regulation of scientific research'. Among CNR's primary functions, the 'decreto' of March 1945 explicitly stated 'consulting for the State in scientific and technical matters', while post-war emergency further recommended 'investigation of scientific and technical problems bearing on the reconstruction of the country'.

In order to fulfil its mandate, CNR was entitled a) to coordinate national activities in the various branches of science, b) to promote the establishment and transformation of scientific laboratories providing financial support within its budgetary limits, possibly in association with other administrations, c) to implement and support research programs of national interest, d) to assist and help scientific institutes and individual scholars and researchers through grants, prizes etc., e) to do bibliographical, documentary and editorial work, f) to take care of Italian participation in international scientific and technical organizations in agreement with the Ministry for Foreign Affairs. Such a list of 'attribuzioni' provided the juridical framework of the CNR President's future action in support of the European laboratory project.

The main innovation of newly elected (in December 1944) CNR President Gustavo Colonnetti[8] was the idea of CNR funded research 'centers'. Colonnetti's idea was duly embodied in the March 1945 Decreto. In order to fulfil CNR's mandate, specialized research institutes would have been the ideal solution, but the situation of the immediate post-war years was extremely difficult. An alternative and viable solution, suggested by Colonnetti, was to help already existing and particularly distinguished research groups. By establishing its 'centers', CNR could provide funds to University-based groups on a regular basis.

As far as physics was concerned, Amaldi's group was no doubt prominent. Thus already on 1 October 1945, the 'Centro di studio per la fisica nucleare' was established in Rome University under Amaldi's directorship. Personnel and funding were provided by Rome University, while CNR provided funds for equipment and research expenses. The initial endowment of Rome Centro was 5 Million Lire for research expenses.[9] The convention between CNR and Rome University establishing the Centro was supposed to last for five years.[10] Rome Centro was the first one in Italy as far as nuclear physics was concerned and provided a useful tool for Amaldi's policy of concentrating resources and man-power in the Italian capital. In the following years, two more centers for nuclear physics were established in Northern Italian universities — at Padua (1947) and Turin (1951).

Besides CNR Centro in Rome, another important initiative in nuclear physics developed in industrial Northern Italy right after the war. In November 1946, a new laboratory, named Centro Informazioni Studi ed Esperienze (CISE), was established in Milan with financial support from several Northern Italian private firms. The laboratory's name was purposely kept vague, since at the time of its inception peace talks were still going on in Paris, and it was not clear yet whether Italy would be allowed to develop a nuclear energy program. In fact, CISE's basic goal was the design and construction of a low power experimental reactor.[11]

Until 1963, production and distribution of electric energy in Italy was in private hands.[12] The interest of electrical and non electrical firms in the new nuclear-energy technology was stimulated by engineer Mario Silvestri, who was working for a private electric-utility company in the years immediately following the war. He joined forces with Milan University physicist Giuseppe Bolla[13] and managed to get support from industrial circles. It is not possible to do justice here to CISE's important and seminal role, since a comprehensive history of the Milanese laboratory is still not available. Suffice it to say that Bolla and Silvestri gathered a group of brilliant young physicists and engineers and started a vigorous research program. CISE's young researchers had the benefit of frequent visits from Rome

Notes: pp. 376 ff.

Centro physicists, namely Amaldi, Gilberto Bernardini and Bruno Ferretti, as well as Enrico Persico from Turin.[14]

The CISE physics group started doing measurements of uranium nuclear constants, measurements of neutron diffusion and the like, thus concentrating mainly on fundamental research relevant to reactor technology. Not surprisingly the physicists were more interested in such fundamental work rather than in technological puzzles themselves. One might possibly add in this connection that the consultants from Rome, with the possible exception of Ferretti, never showed more than a passing interest in reactor physics and technology. It seemed in fact that the Rome physicists were rather inclined to carry on the tradition of the 'Italian Fermi', so to speak, and thus never became deeply involved in the kind of applied nuclear physics to which Fermi himself had turned after his move to the United States for well known reasons. An interesting assessment of the state of the art in nuclear physics in the post war period is given in a brief preamble to a conference Amaldi gave in 1949 on the subject ot the latest developments in nuclear reactors. The acknowledged leader of the Italian nuclear 'tribe' explained that after 1945 nuclear research branched in three different directions.[15] The first one was 'straight' nuclear physics, that is, the study of nuclear structure, nuclear energy levels and reactions. The second line of development of nuclear physics was represented by the study of elementary particles' interactions, that is the kind of interactions ultimately determining nuclear structure. Such investigations were carried out either through the study of cosmic radiation or through the study of nuclear processes artificially produced by big accelerating machines. The third branch of post-war nuclear research was nuclear engineering.

Amaldi stated that in his opinion, nuclear technology, whose main goal was reactor construction, would soon cut its umbilical chord connecting it with 'mother' physics to become a straight engineering subject. He further drew another distinction between civilian and military applications of 'nucleotecnica'.

Amaldi remarked that the three sectors of nuclear physics were not fully independent from each other. As he put it, 'it is deemed certain that the solution of fundamental problems of elementary particles' interactions will have a decisive influence, even if not an immediate one, on all researches of nuclear physics and engineering'. There is little doubt that Amaldi's preference went to the 'fundamental' problems of elementary particles. It is perhaps worthy to stress here that Amaldi's characterization of the different branches of nuclear physics given at the beginning of his 1949 conference was apparently but one example of his relentless efforts aimed at clarifying the different directions taken by nuclear research. It seems that the authoritative Roman professor lost no opportunity to explain that several trends had developed pointing either toward military and energetic applications or toward purely scientific goals, and we will return to this briefly.

To go back to the Rome Centro, reports of research activities published once a year in the CNR bulletin[16] provide a rather direct evidence of Amaldi's group preferences regarding the three sectors in which nuclear physics had branched out after 1945. Without providing here a quantitative or in-depth analysis of the research work carried out in Rome from 1945 to 1951 one can roughly state in approximate terms that reactor physics was definitely not a favourite subject among Rome Centro physicists. Furthermore, interest in 'straight' nuclear physics went through a steady decline, while increasingly more effort went into particle physics. In this connection it is perhaps to be noted that Rome Centro officially extended its activities to particle physics as far back as in 1947, when its denomination underwent a significant change from 'Centro di studio per la fisica nucleare' to 'Centro di studio per la fisica nucleare e delle particelle elementari'. In fact, the second subject of research, that is particle physics, started prevailing on the first and original one at an overwhelming rate. Another example of Amaldi's characterization of the different activities one could label as belonging to the wide realm of nuclear research is provided by a memo sent off by Roman physicists to CISE's director, Ingegner De Biasi.[17] The circumstances of the writing of this memo do not need to be spelled out here, being just an episode in the somehow tormented history of CISE.[18] The memo itself, which was actually written by Bruno Ferretti, provides a clear outline of Rome physicists political and institutional program and was even prophetic under some respect. Amaldi and Ferretti basically argued in favor of 'fundamental' or pure physics and stressed the need of separate institutions for 'fundamental' nuclear physics on one side, and applied nuclear research on the other. They also pleaded for establishing a sort of authority for nuclear affairs modeled on the US Atomic Energy Commission. However it took a relatively long time before Italian politicians and the government started realizing the social and cultural importance of new developments in nuclear research.

In the present context it is probably worth noting that in Amaldi's and Ferretti's memo to De Biasi there is but another punctilious characterization of the different 'branches' of nuclear physics. As stated above, there are several traces of Amaldi's relentless 'preaching' on this very point[19] and we will see later that Amaldi's 'distinctions' eventually made their way to the 'relazione' of the Ministry for Foreign Affairs on the occasion of the ratification of the treatises establishing CERN by the Italian Senate and parliament.

There is little doubt that Amaldi's repeated sermons on the different branches of nuclear research served a rather clear purpose. Namely Amaldi wished to dispel the general impression that nuclear research was a rather monolithic body of knowledge and the associate feeling that the nuclear 'business' was altogether off limits for a country like Italy. By pointing out that several trends had developed in the field, Amaldi tried successfully to convey the idea that Italy ought to engage in an effort in the purely scientific line of nuclear research, given her physicists' particular expertise and the politico-economical boundary conditions of the country.

Continuing our rough sketch of the history of Italian scientific institutions, an important event is to be recorded as far as 'fundamental' nuclear physics was

Notes: p. 377

concerned. In the summer of 1951, the embryo of Istituto Nazionale di Fisica Nucleare (INFN) was formed within CNR. As mentioned before, besides Rome Centro, a second CNR Centro for nuclear research was established at Padua in 1947 and a third Centro was set up in Turin in the summer of 1951. These three Centri joined in coordinating their activities thus giving rise to a new CNR institute, the above mentioned Istituto Nazionale di Fisica Nucleare.[20] The financial situation of the country had considerably improved since the immediate post-war years and the idea of having a specialized research institute was no more just a forbidden dream. Support for nuclear physics had gone up from the few Million Lire granted to Rome Centro in 1946 to a projected budget of 200 Million Lire for the first year of operation of INFN.[21] A first step was thus made toward the creation of a specialized institute for fundamental particle physics.[22]

There is yet no comprehensive history of INFN written from the historian's point of view, although several reports exist which emphasize the difficult situation INFN pioneers had to face in the years 1951–53.[23] Financial means provided by CNR soon turned out to be inadequate for supporting the development of the new Institute.

The ambitious project of a national electron synchrotron of 1000 MeV had thus to be budgeted outside CNR. As far as this project was concerned, INFN physicists took advantage of the establishment of a Comitato Nazionale per la Ricerche Nucleari (CNRN) within CNR itself.[24] Again there is no comprehensive history of the developments that led to the creation of the new Comitato, neither have we a full historical account of CNRN activities, with the exception of official reports.[25]

The task of CNRN was allegedly to supervise and promote 'studies and researches concerning nuclear energy and its applications for industrial purposes'. To fulfil its mandate, CNRN was entitled a) to carry out studies, research and experimentation in nuclear physics, b) to promote and encourage development of industrial applications of nuclear energy, c) to keep international connections and develop collaboration with international organizations and foreign institutions dealing with nuclear research.[26] As far as financing was concerned, CNRN received a rather substantial start up sum of 1 Billion Lire for its first year of operation.[27]

On the other side of the coin, CNRN was born as a branch of CNR and depended on the old Consiglio for financial and administrative support and even for secretarial services. Besides CNR was entitled to have a say on the general research policy of the new Committee, being one of the major sponsors of it. For the same reason, the Ministry of Industry was entitled to a supervision of research projects of industrial interest advocated by the Committee.[28]

Briefly stated, the juridical and administrative status of CNRN was far from being well defined in the first years of operation, hence persistent and repeated efforts had to be made in order to reach a better definition of the power and the responsibility of the new Committee.[29]

As just stated, CNRN was not an autonomous institution from many points of view, being connected to CNR by strong ties. However, since it had to be considered the sole authority responsible for nuclear research, starting from its inception CNRN took over, at least nominally, also the administration and management of INFN.[30] Thus from the summer of 1952, INFN started being financed indirectly by CNR, that is via CNRN.

The institutional scenario as far as physics was concerned, was further complicated by the fact that CNR Comitato per la Matematica e la Fisica still existed as representative of both nuclear and non-nuclear physicists within CNR. A further episode might be worth mentioning in this connection. After having rather generously sponsored nuclear physics developments and just on the eve of being formally deprived of their control, CNR decided to lend an ear also to the claims of those physicists who had been less favoured by fortune. On CNR's initiative, an informal meeting of both nuclear and non-nuclear physicists was held in Rome in June 1952.[31] A compromise was then reached on an informal basis to the effect that the ratio of expenditures for 'classical' and nuclear physics respectively had to be kept at 1 to 2.5. Under the denomination of 'classical' physics fell also solid state physics, then in a nascent state in Italy.[32] During the following years INFN provided support for solid state physics as well, somehow bypassing its statute limitations. However, the ratio agreed upon informally in Rome always remained rather wishful thinking. A remarkable unbalance between nuclear and particle physics on one side and the other branches of physics on the other, namely solid state physics, was to become a basic feature of the Italian scenario for the years to come.[33] In conclusion, one can state that for the period under consideration, that is from 1950 roughly to 1955, CNR was the only established body supervising the whole of government supported research in Italy, while both INFN and CNRN were, so to speak, infant institutions struggling their way toward maturity, that is toward full recognition of their power and responsibility and toward financial independence.[34]

10.2 Rome physicists' involvement in the European project

Italian involvement in the European laboratory project was primarily connected with the initiatives of two scientists and science organizers, namely Gustavo Colonnetti and Edoardo Amaldi. Colonnetti gave his high patronage to Italian participation in the European initiative that eventually culminated in CERN, while Amaldi's specific competence was of paramount importance in directing Italian (and possibly also non-Italian) efforts toward the creation of a European laboratory for particle physics.

Both Amaldi and Colonnetti were deeply convinced of the need for wide international collaboration especially at the European scale. Besides both were involved in the Italian initiatives to join UNESCO.[35] It was through UNESCO channels that Colonnetti got the invitation to attend the Lausanne meeting sponsored by the European Movement from 8 to 12 December 1949. Back in Rome,

Notes: p. 377

Colonnetti reported to CNR's Scientific Board on the recently held 'Conferenza Europea della Cultura'.[36] His report dealt with general science administration problems, and centered upon the need for further integration among national research councils beyond the level of mere bilateral agreements. In order to establish multilateral agreements among national organizations, a first step was to reduce and possibly to eliminate structural differences existing among them.[37]

Colonnetti was well aware of the importance of nuclear physics developments[38] but his report did not touch upon the issue of European collaboration in nuclear research raised at the Lausanne meeting.[39]

However, copies of both De Broglie's 'message' and Dautry's 'Discourse' are held among Colonnetti's papers.[40] Relevant passages of both documents bear marks on the margin, indicating that Colonnetti's attention was possibly attracted by the first vague proposal of a European collaboration in 'big science'. A more direct interest was obviously to be expected from professional nuclear scientists, namely Rome Centro physicists, who had already discussed among themselves the idea of a European collaboration in their own special field.[41] It is not clear from Amaldi's recollections the exact date at which he 'became aware that similar problems were discussed in other European countries, in particular in France [...]'[42] neither when he heard of the European Cultural Conference held in Lausanne in December 1949. No written evidence exists as to how and when he came to know of French discussions. Unfortunately there remains no written evidence either of discussions and exchanges among Rome physicists 'on the idea of setting up a European laboratory for high energy physics',[43] although Amaldi is quite positive about them. This brings us to a brief discussion of Amaldi's own contributions to CERN's historiography. Amaldi wrote articles and gave interviews on the subject of CERN history on several occasions.[44] His recollections, in both oral and written form, follow a regular pattern, since they are based on the sketchy notes of his diary supplemented by his legendary memory. Before giving an example of the procedure on which Amaldi's recollections are based, a description of his Diario is necessary.[45] Amaldi's Quaderno No 1 is a typical Italian schoolboy notebook of the '40s and '50s. It is filled partly with Ginestra Amaldi's neat handwriting and partly with Edoardo's more scrambled calligraphy. It contains notes of special encounters Amaldi had, of important visitors to his institute and of significant meetings he had either in Rome or during his frequent trips abroad.

Periodically, notes taken on paper slips were assembled and copied down in the Quaderno either by him or by his wife Ginestra. This explains the uniform handwriting of entire pages in Amaldi's diary.

Amaldi took notes for his Quaderno for several reasons. No doubt, those notes were intended primarily for his personal documentation, in accordance with a well-known trait of his personality, namely his concern for accurate annotation. However, there is more to it than that. In an interview,[46] Amaldi stated that he decided to keep a diary after he heard of Hiroshima and Nagasaki. Allegedly he was so impressed by the devastating results of a certain kind of nuclear research that he wanted to leave written evidence of his complete lack of involvement in any sort of military enterprise. Accordingly, Amaldi's Quaderno shows no evidence of even consulting for military projects.

As stated above, Amaldi's recollections are based on his diary supplemented by his memory. For an example of his procedure, let us consider page one of Quaderno No 1. There we find sketchy notes concerning a visit of Frédéric Joliot to Rome:

> October 1949 27 (Thursday) I meet
> F. Joliot-Curie at his hotel
>
> November 1st Small party at CNR in
> honour of Joliot and Bernal.[47]

In a set of 'Notes for CERN History—from E. Amaldi Diary', compiled later by Amaldi himself, he further elaborated on the episode contributing personal memories of it: 'The 27th of October 1949 I had a conversation with Joliot who was spending a few days in Rome and I raised the problem of the creation of a European laboratory. He had completely different ideas. First he did not consider high-energy physics as a very important field: he was much more interested in low-energy nuclear physics. Second he understood the European collaboration as an organization which had the task to send European physicists to work in Paris, essentially under his leadership'.[48]

In a subsequent interview, he added that Joliot wanted to take advantage of his, that is Amaldi's expertise, in neutron physics by having him as director of a division of a future 'European' laboratory to be set up in Paris.[49] Apparently Amaldi rejected Joliot's offer for both scientific and political reasons. Allegedly Amaldi, together with Bernardini, was in those years a follower of the European movements, thus did not quite like Joliot's particular brand of 'Europeanism'. On hindsight, he went as far as to say that Joliot was a sort of Gaullist 'avant la lettre' as far as European science policy was concerned.[50] These might well have been the political reasons of Amaldi's dislike of Joliot's offer. From the scientific point of view, Joliot was also disappointing because he wanted somehow to push Amaldi into a field, namely low energy nuclear physics, that in Amaldi's opinion was not exactly at the forefront of nuclear research.

By 1949 the frontier of 'fundamental' research had moved to high-energy physics and it had become also clear that the principal tools of advanced experimental physics were particle accelerators. Besides nuclear reactors, big American laboratories, like Berkeley and Brookhaven, were already endowed with accelerators while particle physics was claiming the limelight on the American scene, just as reactor physics was fading into the background.[51]

In order to meet 'le défi américain' in advanced experimental physics, European physicists needed to find possible ways of building their own big accelerator. In fact the British were trying a foray into big science on an independent basis albeit with limited success. Oliphant's ambitious project of building a 1 GeV proton synchrotron in Birmingham was proceeding at a very slow pace amidst great difficulties.[52]

The prospects for European progress in high-energy physics were improved considerably by the intervention of an American, Isidor Rabi, at UNESCO Conference in Florence in June 1950.[53] Amaldi and Rabi have divergent recollections of this crucial episode, which is discussed at length in other chapters of this book. Namely Amaldi claims that he was informed in advance of Rabi's speech by Gilberto Bernardini, who was commuting from Rome University to Columbia University in New York. Amaldi remembered that 'during the General Assembly of UNESCO in Florence Rabi went to spend two days in Rome during which views were exchanged about his proposal of 'creating regional laboratories''.[54] On the other hand, Rabi sharply denies having been influenced by anybody before giving his speech in Florence, least of all by a European.[55]

Unfortunately Amaldi did not take notes of Rabi's visit or they were not copied down in his Quaderno, so there is no written evidence of his exchange with Rabi, if there was one. Amaldi also recalled the meeting of the Executive Committee of IUPAP on 7-8 September 1950: 'A l'occasion de la réunion du Comité Exécutif de l'Union Internationale de Physique Pure et Appliquée qui a eu lieu à Cambridge Massachusetts (USA) les jours 7-8-9 septembre, on a parlé de la proposition présentée par le Prof. I.I. Rabi à l'Assemblée Générale de l'UNESCO à Florence, au sujet de la construction d'un Laboratoire Européen de Physique Nucléaire. Après quelques discussions, l'Exécutif a décidé de faire préparer deux rapports sur cet argument: l'un devrait être rédigé par Rabi, qui devrait préciser aussi bien que possible sa pensée, l'autre par moi [...]'[56] Those reports, if they have ever been written, have not yet surfaced. It is highly probable that they have never been written.

On 3 October 1950, Amaldi wrote to Auger informing him of how things were developing in IUPAP and asking for an exchange of views on the envisaged European laboratory. Among other things, Amaldi wrote to Auger: 'A présent je viens d'apprendre de Ferretti qu'à l'occasion du Congrès d'Oxford[57] vous avez provoqué et dirigé une discussion intéressante à ce sujet [...]. J'ai appris de Ferretti les lignes générales que vous avez données à la discussion à Oxford et je suis complètement d'accord avec vous'.[58] 'Les lignes générales' of the Oxford discussions were summarized by Ferretti in a report to the Scientific Board of CNR a few months later.

To introduce the discussion of this report, we have to mention that the Centre Européen de la Culture called a meeting of its Groupe d'Etude des Recherches Scientifiques for 12 December 1950 in Geneva. Four Italian personalities were invited, namely Colonnetti, Amaldi, Mario A. Rollier[59] and Alessandro Casati.[60] Upon receiving the invitation, Amaldi wrote to Raymond Silva, CEC Secretary General, that owing to a previous commitment he was unable to attend the meeting and suggested that his close collaborator Bruno Ferretti could take part in the conference in his place.[61] Rollier's invitation to Geneva resulted from contacts with Casati, Raoul Dautry and Denis de Rougemont. Apparently Dautry had informed Casati that he wanted to have a meeting with Italian members of Parliament or representatives of responsible circles in order to discuss possibilities of a closer Italian-French collaboration in nuclear research. At that time, Rollier happened to be a private consultant on nuclear affairs to the Minister of Foreign Trade, Social Democrat Ivan Matteo Lombardo. Casati alerted Rollier of Dautry's wish and urged him to write in order to arrange a meeting. Accordingly Rollier wrote to Dautry suggesting that they could meet and discuss together with Casati in Strasbourg from 20 to 23 November on the occasion of the meeting of the 'Conseil Européen de vigilance', to which both Dautry and Rollier belonged.[62] Dautry answered suggesting they meet in Geneva instead, on the occasion of the forthcoming conference called by the CEC. Accordingly Dautry wrote to Silva recommending that he invite both Casati and Rollier to Geneva.[63] Upon receiving the invitation, Rollier wrote to the Minister of Foreign Trade informing him of these events.[64] Moreover, shortly before the conference, Rollier wrote a preliminary report to the Minister in which he informed him that both de Rougemont and Silva had the impression that Dautry considered him, that is Rollier a 'porteparole' of the Italian government in matters of international scientific cooperation.[65] That was hardly the case, since Rollier had no official position, being simply a consultant to a relatively minor Ministry and he could not speak on behalf of the Italian government.

Casati did not go to Geneva and Colonnetti did not attend the conference due to illness. Thus only Ferretti and Rollier went to the CEC meeting and both reported on it. Ferretti made a report to the CNR Scientific Board,[66] while Rollier sent off a written report to both Colonnetti and Casati.[67] Both their reports provide source material for reconstructing the events together with the Compte-Rendu by Jean-Paul de Dadelsen of the Centre Européen de la Culture.[68]

Ferretti was supposed to report at the CNR Scientific Board meeting of 18 December 1950. Colonnetti was still ill, yet he wrote a letter to Guido Castelnuovo, President of CNR Committee for Mathematics and Physics, on 13 December and sent a telegram to the CNR on 15 December. Both the letter and the telegram were read at the CNR meeting before Ferretti's report, so that the letter's content is summarized in the minutes and the text of the telegram is reproduced in full. In his letter to Castelnuovo, Colonnetti referred to the Geneva Conference and its decision to establish an office in UNESCO headquarters in Paris for a preliminary study of a European laboratory. He wrote that a commitment had already been made for Italian participation in such an office and stressed that it was necessary to provide an initial modest contribution. Colonnetti asked his CNR colleagues to vote in favour of such a modest contribution, waiting to find other sources for future expenses. In the telegram, Colonnetti announced that Ferretti had

Notes: p. 378

the task of presenting his, that is Colonnetti's proposal to provide 1 Million French Francs for Italian participation in the Paris office from the CNR budget. Colonnetti urged both the Scientific and the Administrative Board of CNR to approve his proposal. After Colonnetti's letter and telegram were read, Ferretti started his report with a brief history of the project for an 'International Atomic Center' and a summary of previous discussions on the nature of the future laboratory. Ferretti remarked that two different 'tendencies' had emerged. According to the first one, which he ascribed to the French, the future laboratory ought to be equipped not only with big apparatus for nuclear physics, like accelerating machines, but also with atomic piles. The second tendency, allegedly championed by Bohr, was in favour of accelerating machines, 'con astensione dalle pile atomiche'. Apparently Bohr's dislike of atomic piles was due to the fact that their 'construction implies a complex of economical and political problems escaping scientists direct evaluation and lying outside their competences'.

As explained before, Ferretti had assisted in the discussion at the Oxford Conference in September 1950 on the nature of the future European laboratory and had reported on it verbally to Amaldi. In his report to the CNR of December 1950, Ferretti did not mention explicitly the discussion held in Oxford a few months before. Apparently he only referred to 'previous discussions' in general terms, although he must have had the Oxford discussion on his mind.

On the other hand, Pierre Auger as well mentioned 'two tendencies' just like Ferretti did in his own summary of the Oxford discussions concerning the nature of the future laboratory. At page 3 of the 'Compte-Rendu analytique de la réunion du 12 décembre 1950' compiled by Jean-Paul de Dadelsen of the Centre Européen de la Culture, one reads that 'Pour le programme du Laboratoire, M. Auger signale qu'à une récente réunion de physiciens à Oxford, deux tendances se sont manifestées:
1. ne pas trop limiter au début le domaine des recherches;
2. comme le proposait M. Niels Bohr, commencer par créer un grand instrument d'accélération de particules (d'1 milliard de volts) et se grouper autour.'

Clearly enough the first 'tendance' as indicated by Auger ('ne pas trop limiter au début le domaine des recherches') coincides with the 'tendenza' ascribed by Ferretti to the French, that is the tendency not to limit the equipment of the future laboratory to particle accelerators. On the other hand, Bohr is indicated by both Auger and Ferretti as the major figure among the advocates of the 'second tendency', that is the tendency favouring the construction of a big accelerator with the more or less temporary exclusion of atomic piles.

Auger's and Ferretti's summaries of the Oxford discussions of September 1950 were substantially in agreement with each other. The only difference between them stemmed directly from the different attitudes of Professor Ferretti and diplomat Auger. Namely, Ferretti provided a neat characterization of the opposite 'tendencies' that emerged during the Oxford discussions and clearly emphasized their differences, carefully distinguishing black from white, so to speak. While diplomat Auger had perforce to take a different course. No doubt Auger was aware just as Ferretti that the two 'tendencies' that emerged during the Oxford discussions were at least partially at odds with each other. Furthermore as it will be recalled, the 'first' tendency was ascribed by Ferretti to the French, that is to Auger's own countrymen. Hence, Auger's rendering of the Oxford discussions had to have a rather more diplomatic turn than Ferretti's. Apparently Auger tried somehow to reconcile the two opposed tendencies, by simply listing them one after the other as if they were perfectly compatible, thus smoothing out their differences in a purely fictitious succession of different research programmes. In fact, on reading Auger's summary of the Oxford discussions, one does not have at first sight the impression of a basic disagreement between the two 'tendencies'. As just stated, Auger simply lists them so that Bohr's proposal of building a big particle accelerator appears to be merely the first step toward the creation of a laboratory whose research programme needed not to be too much limited 'au début', thus leaving room for the hopes and expectations of those who were longing for big reactors. However, once a big accelerator had been built, the orientation of the laboratory toward high-energy physics was likely to be irreversible and rather exclusive as well.

Another conclusion from this analysis of Ferretti's and Auger's summaries of Oxford discussions might well be that Auger himself was not a supporter of the tendency ascribed by Ferretti to the French. This fact may also explain why Amaldi, himself relatively uninterested in reactors, could write to Auger that he was 'complètement d'accord' with the French scientist-diplomat. Apparently one thing on which Amaldi could be 'complètement d'accord' with Auger was not the French tendency as characterized by Ferretti, that is the tendency in favour of reactor development along with accelerators. It is much more likely that Auger and Amaldi were 'complètement d'accord' on the point of view attributed by both Auger and Ferretti to Niels Bohr.

In fact, Ferretti continued his report to the CNR Scientific Board by recalling that IUPAP also took an interest in the project of a European laboratory for advanced nuclear physics and had given Amaldi the task of formulating a detailed plan. Together with Ferretti, Amaldi had prepared a summary project that was 'essentially in agreement with Bohr's point of view'. Ferretti told CNR officials that the plan he concocted with Amaldi was discussed at the Geneva meeting called by CEC.

Ferretti's report then went on with a rough sketch of Geneva deliberations indicating that the Italian share of expenses for the project would be 12.5%. He further stated that he had closely considered the problems of Italian participation together with Colonnetti and that they both agreed on the convenience in participating in the Paris office expenses as well as on the necessity 'to take a position with respect to the European center for Atomic Studies, so that our physicists can play the role they deserve'.

Thus apparently Ferretti failed to report to the CNR Scientific Board that he himself had been offered a leading position in the UNESCO office in Paris which was in charge of a preliminary study of the European laboratory project. In fact, from Rollier's report one learns that following a suggestion by the Belgian delegate Paul Capron, Ferretti had been 'appointed' to the Paris office together with a 'French member'. Ferretti did not mention that offer in his report to CNR. Concerning the Paris office, he simply informed that the French government had already decided to give a contribution of 50 Million French Francs, adding as a remark that he considered such an amount of money to be exceedingly high for the kind of preliminary study to be undertaken. Apparently Ferretti did not emphasize the offer made to him in Geneva because he was reluctant to accept it. In fact he was going to turn that offer down and never joined the Paris office. One can only speculate on possible reasons for Ferretti's refusal to work for the Paris office. Most likely he was not willing to take up what seemed to be a very heavy organizational duty. It is also probable that Ferretti was concerned about the danger of having to abandon active research work, as it had been the case with Cacciapuoti's previous appointment with UNESCO.

In the discussion following Ferretti's report, Guido Castelnuovo declared himself in favour of the project just presented and recommended Italian participation in the European enterprise right from the start. Francesco Giordani, President of the CNR Committee for Chemistry, briefly remarked on the extraordinary amount of money that was needed for nuclear research. Giordani said that support for nuclear physics was becoming a problem in every country and that it was impossible to cope with the financial needs of nuclear research with ordinary means. Given the extraordinary situation connected with the nuclear sector, Giordani stated that a new ad hoc committee had to be established for the promotion and coordination of nuclear research.[69] He also declared, that he was convinced that the establishment of a European laboratory was an excellent opportunity that had to be seized and fully endorsed Colonnetti's proposal to provide 2 Million Italian Lire, equivalent to 1 Million French Francs from the CNR budget to cover the Italian share of expenses for the Paris UNESCO office operations. Giordani added that it should be made clear to the Italian government that the CNR contribution had only a 'symbolic' character and that it represented the maximum effort CNR could make given its budgetary limitations. Other members of the CNR Scientific Board agreed with Colonnetti's proposal and voted in favour of it. On the same day, the CNR Administrative Board approved the transfer of 2 Million Italian Lire to the Paris Office.[70]

Unlike Ferretti's report, Rollier's had less of an official impact and was addressed personally to both Casati and Colonnetti. Accordingly, Rollier's report lacked any official tone and was written in his typically lively style. Rollier briefly summarized discussions and speeches given by de Rougemont and Auger at the Geneva meeting. He reported that 'Auger had recalled the Harwell [Oxford] conference, the acknowledged need of a laboratory for high energy elementary particles for Western

Europe, and Niels Bohr's interest for the creation of such a laboratory'. When discussions turned to the future laboratory's apparatus and research programme, Rollier reported that 'the Commission unanimously adopted Niels Bohr's suggestion', namely that the future laboratory should be equipped for the production of high-energy particles and should be centered around a big instrument like a 'cosmotron' of no less than 6 BeV. Rollier added that, following discussions of the tentative plans and budget presented by Amaldi and Ferretti, and rough estimates by Kowarski, a conclusion was reached over a start-up sum of 5 Million Dollars per year for five consecutive years.[71]

Rollier wrote that a discussion followed concerning the opportunity of publicizing such a figure. Apparently he himself had insisted that national governments be made aware immediately of the order of magnitude of expenses to be incurred for the support of the European project. Rollier claimed that 'his' point of view was eventually accepted by the participants to the Conference.

Even before Rollier's report had reached Colonnetti, the CNR President had informed the head of the Italian government on the European initiative and its costs for Italy.[72] On 18 December 1950, Colonnetti wrote to Alcide De Gasperi:

> [...] we are on the eve of the creation of a European Laboratory for nuclear physics to which Italy must absolutely participate (...siamo ormai alla vigilia della creazione di un 'Laboratorio Europeo di fisica nucleare' a cui l'Italia non potrà assolutamente restare estranea).

In his letter Colonnetti transcribed the text of the Geneva conference deliberations he had received before. Then he indicated that the Italian share of expenses was 12%, which meant a contribution of roughly 300–400 Million Lire per year for a five year period. In order to make Italian participation more effective, it was necessary to spend at least an equal amount of money for the construction of local facilities in Italy for the training of personnel. Colonnetti stated that Italy had to engage in the preliminary stage of the project, that is the establishment of the Paris UNESCO office, while waiting for government decision regarding the whole matter. Since there was no 'Commissione Atomica', Colonnetti felt that responsibility for Italian participation fell upon the CNR. Therefore he had decided 'without hesitation' to provide 1 Million French Francs from the CNR budget to cover expenses for Italian participation in the Paris office.

'This is clearly just the beginning [...]', Colonnetti explained, warning that there would be no sequel if his requests of an increase in funds for CNR were to be rejected. Here reference is made to a previous exchange between Colonnetti and De Gasperi concerning CNR's financial situation, and the need of major funding especially for nuclear research.

The idea of asking money for both domestic and European nuclear physics facilities was almost certainly inspired to Colonnetti by Amaldi. Allegedly Amaldi's strategy was to use the European project as a booster for domestic development as

Notes: pp. 378 ff.

well. One of the advantages of this strategy was that the supporters of the European programme could not possibly be accused of damping Italian initiatives in order to favour the European effort. Needless to say, the advantages of Amaldi's strategy were not merely of a dialectical nature, as far as the development of Italian nuclear physics was concerned.[73]

Besides this point, Colonnetti's letter to De Gasperi deserves further comments. Anticipating difficulties for his request of an increase in funds from the Treasury, Colonnetti stated that such difficulties were not acceptable, since the 'emergency' of the times made absolutely necessary a true 'mobilization of science and scientists for national defence purposes, which means that funds for CNR should be considered in this moment an integral part of defence expenses'. According to Colonnetti, 'this does not happen only in the United States, [...] it happens in all countries that are going to face the serious international situation of today'.

Colonnetti's recommendation did not apply only to nuclear physics, which was the specific topic of his letter to De Gasperi. It also extended to all branches of science and technology, since the 'excellent capabilities' of Italian scientists and researchers were in danger of being underemployed. Colonnetti wrote that his personal experience during past wars had taught him that the country had suffered severe damages from the lack of 'mobilization' of scientists and technicians, so that he felt it his duty, as CNR President, to warn against a repetition of the same mistakes.

The CNR President was writing this letter at a time of great international political tension, so that he could hardly avoid emphasizing one of his favourite arguments in favour of science. In fact, Colonnetti was convinced that one of the main purposes of scientific research was to serve the country in war time. His idea was that in order to help war effort, research need not necessarily be confined to strictly military projects, like the development of new weapons. On the contrary, the country could be better served in case of war by pointing exactly in the opposite direction, that is by developing and supporting all branches of science independently of their immediate applications. Past experience had taught that even scientific investigation having initially no connection whatsoever with war needs later bore results of tremendous military importance. Furthermore industry could only benefit from a strong research effort, thus putting the country in a better position in case of war.[74]

Colonnetti's call for a 'mobilization' of scientists for national defence was no doubt influenced by analogous calls made at about the same time by several American scientists and politicians in favour of scientists' involvement in defence programmes.[75] Apparently Italian government circles did not take seriously Colonnetti's offer of putting science at work for national defence.[76] Furthermore, Italian scientists like Amaldi or Rollier were strongly opposed to the idea of working for military projects.[77] Besides, Colonnetti himself was definitely no 'hawk', as he might appear to have been from this letter to De Gasperi.[78] As for the European laboratory project, Italian politicians were probably more inclined to be convinced of the need to engage in a very expensive research programme by its actual European character, rather than by its remote military returns. In fact, Italian government circles were extremely sensitive to the issue of European collaboration 'per se', so that by resorting to his science mobilization proposal, Colonnetti was indeed just trying to force an already open door.

10.3 CNR and Italian support of the European laboratory project

The following section contains a rough description of the Italian involvement in the European laboratory project from December 1950 to the ratification of the Paris Convention by Italian Parliament in February 1955. This schematic account is based on Colonnetti's papers held at Turin State Archive and on the minutes of the meetings of CNR Consiglio di Presidenza. No documents pertaining to CERN prehistory were deposited in the Archivio Storico-Diplomatico of the Italian Ministry for Foreign Affaires in Rome, thus the reaction of the Ministry officials can be gauged only from the extant Colonnetti's letters and his reports to the CNR Scientific Board.

On 24 October 1951, Colonnetti informed his CNR colleagues that the first or preliminary phase of the project was over although it really was not, and remembered that Italian participation was made possible by CNR appropriation of 2 Million Lire.[79] The next step was the actual design of the laboratory's equipment that had to be completed by 1952. The estimated amount of money needed for the second phase was 200,000 US Dollars and Italy's share of expenses was around 11 Million Italian Lire. Colonnetti anticipated that the actual construction of the accelerator ought to take place in the period 1953-58 and that Italy's projected expenditure was a total of 1,500 Million Lire, that is 300 Million Lire per year for the said five years period.

Following pressure from Amaldi, Colonnetti had raised the subject of Italian participation at a meeting of the Italian Commission for UNESCO that was attended by functionnaires of the Ministry for Foreign Affairs as well. Colonnetti urged the Ministry officials to take action in favour of the project, so that Italian government should clarify its position as regards the continuation of Italian participation. Allegedly the Ministry officials declared that their administration was in no position to provide funds. Colonnetti thus promised CNR support for the second phase of the project and willingly accepted to be a delegate of the Italian government at the UNESCO meeting to be held in Paris in December 1951, together with Senatore Casati.

Colonnetti made clear that given the persistent interest for the project on the part of Italian physicists, CNR was willing to provide support for the second phase, that is up to 1952. However, it was clear that expenses to be incurred after 1952 by far exceeded CNR budget possibilities unless CNR were granted special ad hoc funds

Notes: p. 379

for the European project. Colonnetti ended his report by emphasizing the necessity that the government assume its responsibility as soon as possible.

After the Paris UNESCO meeting of 17-21 December 1951, the Italian Delegates Casati and Colonnetti duly informed the Ministry for Foreign Affairs on the conference recommendation of 'Constituer un Conseil de Représentants des pays participants, avec siège à Genève, pour diriger l'exécution du programme défini'.[80] They also informed about the Conference decision 'de créer un Groupe de Travail chargé de préparer un projet d'accord par lequel le Conseil serait créé'. Colonnetti was going to represent Italy at the meeting of the Groupe de Travail to be held in Paris on 3 January 1952, where the problem of the location of the future laboratory was to be discussed. Here Casati and Colonnetti anticipated Italian attitude on the issue of the location announcing the support of the Italian Delegation to the Swiss Delegation proposal. Both Casati and Colonnetti were convinced that Geneva was an excellent choice and a very favourable one for Italy. In case the candidature of Geneva could not succeed, the Italian Delegation was ready to propose Como, although Como's candidature has never been proposed in an official way.

Colonnetti and Casati attended the UNESCO Conference held in Geneva from 12 to 15 February 1952, where the project of convention worked out by the Groupe de Travail nominated in Paris was examined. They duly reported contrasts that arose between the Delegations of Denmark, Britain and Sweden on one side and the other countries' delegations on the other.[81] According to Colonnetti and Casati, the three northern countries showed a tendency to water down the initial project by proposing instead a somewhat 'vaguer' programme of collaboration based on the utilization of already existing research facilities, namely the ones in Copenhagen and Liverpool. They reported that the Italian delegation joined the front of those who stood firmly in favour of the original programme. Casati and Colonnetti ended their letter stressing the need that the Italian parliament ratify Geneva's Agreement for the Constitution of a Council of Representatives of European member states signed on 15 February 1952 by the head of the Italian Delegation, Senatore Casati.

Colonnetti informed also his CNR colleagues on 'Centro Europeo di fisica nucleare' on 17 April 1952.[82] Once again Colonnetti stated that the first phase was over and that the project had entered the second phase, characterized by the actual design of the future laboratory's equipment. CNR had secured Italian participation to the first phase of the project and was ready to continue support for the second phase as well. Following an agreement reached with the Ministry for Foreign Affairs, the Italian share of expenses was 25,000 US Dollars. CNR was ready to anticipate this sum in order to keep Italian connection with the project even in the lack of ratification of Geneva's Agreement by the parliament. In answering a question from Francesco Giordani, President of CNR Chemistry Committee, Colonnetti made clear that no commitment existed yet for Italian participation to the third phase of the project, namely for the actual construction and functioning of the European Center. Giordani was going to be nominated President of the new

Committee for Nuclear Research (CNRN), that was to become the authority responsible for nuclear affairs, which makes his request of more precise information on the part of Colonnetti fully understandable.

Colonnetti and Casati attended the meeting of the Council of Representatives of Member States held in Paris from 5 to 9 May 1952 and duly reported to De Gasperi, who was at that time Minister for Foreign Affairs.[83] In their letter they put due emphasis on the choice of Amaldi as Secretary-General of the Provisional Council. Among other advantages, Amaldi's election meant that part of Italian expenses would be made in Italy, since Amaldi was stationed in Rome. Colonnetti obviously emphasized that such a prestigious success as Amaldi's election was made possible by CNR financial commitment following an agreement with the Ministry for Foreign Affairs. At the end of their letter, Casati and Colonnetti stressed the need of regularizing Italian position also from the juridical point of view by prompt ratification of Geneva's agreement, all the more since such a ratification did not entail any new expenditure on the part of the Treasury.

A few days later, Colonnetti repeated the same message to the members of CNR Scientific Board.[84] He further discussed the financial situation in view of the future expenses to be met in the third phase of the project. Colonnetti said that in approximately 18 months Italy should be ready to pay 300 Million Lire per year for the functioning of the European Center. It was unthinkable that CNR could anticipate expenses for phase three as it did for the previous phases of the project.

Giordani stated that it was urgent to take action in order to avoid the danger that CNR be charged for expenses for phase three as well. He stressed that such a danger was real due to the fact that the Ministry for Foreign Affairs was unable to co-operate with its own funds. Colonnetti replied that he was going to confirm to the government the impossibility for CNR to cover Italian expenses for participation to phase three. He further suggested that the Ministry for Foreign Affairs could propose an ad hoc extra-funding for CNR instead of providing with its own means.

Colonnetti attended the second session of the Council of Representatives of Member States held at Copenhagen on 20-21 June 1952.[85] He reported to Sua Eccellenza Migone of the Ministry for Foreign Affairs, confirming in the first place that CNR was completing the payment for Italian expenses for the second phase of the project. He also complained about the persistent lack of ratification of Geneva's Agreement by the Italian parliament.

At the meeting of CNR Scientific Board of 25 July 1952, Colonnetti notified that CNR had completed the payment for phase two.[86] Immediately Giordani protested that such a payment should have been made by the new Committee for Nuclear Research (CNRN), which was just being established. Giordani announced that CNRN would reimburse CNR for expenses already met in supporting the European project, as soon as its financial situation was better defined. Giordani stressed that the new Committee was to be considered the sole responsible for nuclear research both at the domestic and European level. Apparently that was the first open manifestation of what was later to become known as 'the war of the two beards', since both Giordani and Colonnetti sported long beards. Clearly enough Colonnetti felt particularly strongly about the CERN project, since he was involved in it right from the beginning. Understandably he was not yet prepared to hand over the European project to the new comer, that is CNRN. On the other hand, Giordani was all too eager to have the power and responsibilities of his new 'Comitato' properly defined and better recognized. There is no need to deal here with the 'war of the two beards', except for the fact that it shows the lack of any serious opposition to the CERN project within the Italian scientific establishment. Both the President of the old Consiglio and the head of the new 'Comitato' were most eager to go along with the European project and were even competing between themselves in order to be recognized as the official Italian sponsor of the European venture.

As already stated before, CNR was an already established institution and was therefore better entitled to be considered the Italian interlocutor of the European project. While the juridical and financial status of CNRN was still too much ill defined to allow Giordani to take direct responsibility in the European enterprise.

On 23 October 1952, Colonnetti notified the CNR Scientific Board that Geneva was chosen as the definitive seat of the laboratory at the Amsterdam meeting of the Provisional CERN Council.[87] He also informed that on 30 June 1953, the provisory agreement for the financing of the European Center was due to expire and that a 'definitive' mechanism of financing had to be set up by international agreement. Colonnetti anticipated that such an agreement was already under study and that the Italian share of expenses was estimated around 250 Million Lire per year for a five year period. So huge expenses were to be met only in the following year thus the problem was not an urgent one although Colonnetti foresaw that a special intervention was to be expected from the government.

Colonnetti reported again on 'Consiglio Europeo di Ricerche Nucleari' more than a year later.[88] He repeated once more that CNR was able to ensure participation to the second phase of the project by the paying the Italian share of expenses, thus overcoming the problems raised by the lack of ratification by Italian parliament of the 'convenzione provvisoria', by which he probably meant Geneva's Agreement of 15 February 1952.

Colonnetti declared that he had repeatedly warned the Ministry officials that CNR could act on behalf of the State adminstration only in anticipating relatively small sums, as it had been the case during the first two phases of the project. With the 'definitive' Convention of 1 July 1953, the preparatory stages of the project were over and Italian participation involved an expenditure of around 300 Million Lire per year for a seven year period. That meant that the project could not be budgeted anymore with CNR means. Following contacts with functionaries of the Ministry for Foreign Affairs, an agreement was reached in roughly the following terms: CNR would receive a special extraordinary contribution from the Treasury in order to be able to continue support on behalf of the Italian administration.

Although the information gleaned from CNR 'verbali' is never abundant neither overly explicit, it is nevertheless possible to infer the overall attitude of the Ministry for Foreign Affairs in its negotiations with Colonnetti. Apparently, the Italian functionaries avoided being involved in the project in any official way before ratification by the parliament of the treaties establishing CERN.

As far as ratification was concerned, it is noteworthy that the text of Paris convention of 1 July 1953 was forwarded by Italian Delegate Antonio Pennetta to the Ministry for Foreign Affaires only on 30 September 1953.[89] By December of the same year,[90] Italian government was examining a whole package of treatises, namely Geneva's Agreement of 15 February 1952, Paris Avenant of 30 June and Paris Convention of 1 July 1953.

It took roughly six months before the treatises were presented for ratification to the Italian Senate, on 15 June 1954.[91] After six months they were jointly discussed by the Senate, on 13 December 1954.[92]

The motivations of the Communists' opposition were expressed by the authoritative Senatore Emilio Sereni[93] in a somewhat lengthy and articulate speech. Sereni took issue from the 'relazione' of the Minister for Foreign Affairs, Attilio Piccioni, that apparently was inspired by Amaldi's distinction among the different branches of nuclear research: the purely scientific branch and the applied one, with either military or civilian goals. The Minister had duly stressed that applied research could not be developed in the absence of a consistent effort in pure or 'fundamental' research as well.

Sereni stated that he did not challenge such 'obvious' distinctions, and proceeded to a very general survey of the international political situation, based on American, British and French sources dealing with the problems of nuclear weapons and nuclear energy. He acknowledged that nuclear issues could not be dealt with in a narrow national perspective and that international collaboration for peaceful scientific purposes was a real necessity. However, he elaborated sort of an interpretation of the international division of labour in nuclear research that emphasized Europe's ancillary role in the Western system of alliances.

Drawing upon information gathered in his extended reading, Sereni argued that the US had made an all out effort in military nuclear-technology development. In his opinion, such a concentration of efforts in nuclear-weapons research was accompanied by a relative disregard of nuclear-energy developments, since the energetic basis of US economy was characterized by a large abundance of raw materials and by traditional, that is non nuclear energy sources. In Sereni's opinion, the US had disregarded also pure scientific research in order to concentrate themselves just in military developments.

Sereni went on saying that after the first Soviet nuclear reactor had become operative, the Americans apparently started to fear that the Soviets would be able to supply nuclear-energy technology not only to their allies, but to the Western European countries as well.[94] Hence a renewed interest on the part of the Americans in nuclear-energy technology development.

Notes: p. 379

According to Sereni, the Europeans with the exception of the British, were cut out from nuclear-technology developments addressed to both military and civilian purposes. As a consequence of all this, Sereni argued that from an American point of view, the Europeans' role was in pure or 'fundamental' nuclear research. Thus CERN was envisaged by Rabi in order to pool European brains and direct them toward pure research under American control.

As far as Italy was concerned, Sereni had no objections to international co-operation in pure research. He was aware that Italian scientists would get unique opportunities by participating in the European enterprise. Yet he objected on the ground of the weakness of the Italian national nuclear programme. In his opinion, Italy had not yet developed a sufficiently strong programme for nuclear research and consequently was not in a position to benefit from returns of its involvement in the European laboratory. Furthermore, CERN was likely to favour the brain drain, instead of stopping it, since Italian experts who had received an advanced training at CERN could not utilize their abilities in the relatively backward national laboratories and were thus all the more tempted to stay abroad.

Finally Sereni remarked that in his opinion CERN was but an instrument of the reactionary policy aimed at the division of Europe and the entire world in two opposite blocks. Clearly enough, the sincere Communist resented the fact that, despite Switzerland's well known neutrality and the participation to CERN of several non-Nato countries, the European laboratory in fact provided a further link between Italy and the Western 'system'. Incidentally, this conclusion could hardly be denied by anyone, whether Communist or not.

Towards the end of his speech, Sereni touched upon the problems raised by German rearmament and by German participation to CERN. He declared that he could not avoid the impression that a deep connection existed between the two developments. The German danger was the main theme of a somewhat spirited speech given by Socialist Emilio Lussu,[95] right after Sereni's. One has to remember that the Italian Socialist Party was at that time engaged together with the Communist Party in a strong opposition to the government lead by Christian Democrat Mario Scelba.[96]

Lussu declared that the very presence of the Federal Republic among the member states of CERN was a clear sign of the resurrection of teutonic militarism. Rehearsing a standard argument dear to the far left, Lussu stated bluntly that the 'fresh and new Adenauer's Germany' was the true heir of the old Germany of the Krupp, Keitel and Kesselring. Lussu went on saying that the Federal Republic expected to raise the level of its science and technology by participating to CERN and that was obviously a true statement on the part of Lussu. However he insisted that the ultimate goal of the Federal Republic was German rearmament and the renaissance of German militarism. Christian Democrat Onorevole Caron replied to Lussu that 'one cannot leave out sixty million Germans' because of past wars, thus forgetting about the contributions the German nation brought to the sciences, the arts and civil progress. According to Caron, to ostracize Germany would have been an 'illogical and counterproductive move' against the very interests of peace and of Europe.

As for the national nuclear programme, Caron conceded that Italy had a relatively backward position with respect to other European countries in terms of national laboratories or institutions, although not in terms of qualified manpower. Caron stressed that with the establishment of CNRN the situation was changing rapidly toward a steady improvement. Under-Secretary of State for Foreign Affairs, F.M. Dominedo provided an adequate rebuff of Sereni's argument about American influence on CERN's origin. He said he could not understand why 'external forces' willing to gain control on other forces, should make an effort to 'unify' their subjects, instead of trying to divide them. Thus Dominedo referred indirectly to the old rule: 'divide et impera' that is 'divide and rule'.

The majority of Italian Senate voted in favour of the package of treatises establishing CERN.[97] The following day, 'l'Unità', the official daily of the Italian Communist party carried an article under a flamboyant title: 'Comrade Sereni denounces the warlike goals of the so called European atomic laboratory'.[98] The content of the article was in tune with the title and stressed the section of Sereni's speech in which the laboratory was seen as an integral part of the system of Western alliances.

Apparently Sereni was alarmed by the reporting of his speech by 'l'Unità' and felt it necessary to write a personal letter to Edoardo Amaldi in an effort to clarify his ideas. In a respectful tone, Sereni argued that the main part of his speech at the Senate was devoted to the need of developing nuclear physics in Italy in the first place.[99] Without an adequate planning of development for domestic institutions, Italy was bound to get scarce benefits from CERN. Evidently, Sereni though that Amaldi could not but be very sympathetic with such a point of view. Sereni went on lamenting that the part of his speech in which he dealt with internal problems was not adequately emphasized by 'the press', while his objections to CERN, which were of a more general political character, had been given too much emphasis by the reporters. One might thus have gained the impression that Sereni himself was against international scientific co-operation, which thing was far from being true.

Sereni wrote also an open letter to the director of 'l'Unità' stressing the same point he had touched in his letter to Amaldi.[100] Sereni's reaction to his own party's journal may perhaps be explained in the following terms. No doubt he knew that Amaldi, while not being a Communist at all, was nonetheless a man of the left. Furthermore several physicists of Amaldi's circle were outspoken Socialists and Communists. Sereni's letter was probably meant to ease Amaldi's understandable irritation with the Communist and Socialist physicists of his group, as well as his general feelings of bitter surprise at the attitude of the left parties toward a project that was so dear to him.

Meanwhile right after the ratification of the Conventions by Italian Senate, the

Notes: p. 379

treatises establishing CERN were presented to the Italian parliament on 14 December 1954.[101] The ensuing discussion took place roughly two months later and was not as lively as the one which had taken place at the Senate. The Communists again declared their opposition, while the Socialists took an independent stance. They declared that they would vote in favour of the CERN treatises if the government could assure the parliament that commitment to the European project was not detrimental to a vigorous support of the national nuclear programme.[102] After having declared themselves dissatisfied with the government attitude toward this very issue, the Socialists abstained from voting. After a two days discussion on 21 and 22 February, the final vote came on 23 February.[103]

Christian Democrats and the allied 'center' parties voted in favour both at the Senate and at the Parliament because the CERN project was just a part of their European 'faith'. Thus the Paris Convention of 1 July 1953, together with Geneva's Agreement of 15 February 1952 and Paris Avenant of 30 June 1953, were jointly approved by the Italian parliament on 23 February 1955, and the instruments of ratification of the Convention on the part of Italy were deposited in UNESCO's archive on 24 February 1955.

A few words may be spent on Italian tardiness in approving the package of CERN treatises. The scarcity of documents available does not allow for far-flung interpretations of facts. However, one can possibly surmise that ratification of Geneva's Agreement and Paris Avenant were not deemed strictly necessary by government officials, given Colonnetti's strong commitment to the project. As for the delay in ratifying the Paris Convention, one has to take into account that the text of the Convention was forwarded to the Ministry for Foreign Affairs only on 30 September, that is three months after the signing. Thus it took a relatively short time before the issue was brought to the attention of the Senate in mid-June 1954. After that date, the itinerary followed by the CERN treatises was relatively quick at least by Italian standards, since the final ratification came only eight months after the first presentation of the law at the Senate. As already stated, CERN's European character was no doubt of paramount importance in winning the favour of the ruling parties, led by the arch-European Democrazia Cristiana. So it happened that after having taken so long in promoting nuclear research at the national level, with the establishment of CNRN in 1952, it took a relatively short time to have Italian participation to CERN approved by the Italian parliament.

Notes

1. Segré (1970).
2. Amaldi (1974), 29–32.
3. Amaldi (1979), 379–461.
4. On Bernardo Nestore Cacciapuoti (1913–1979), see Coen (1979) and Segré (1979).

5. CNR was established in November 1923. The great pure and applied mathematician Vito Volterra (1860-1940) can be considered the father of CNR. See Whittaker (1941). For Volterra's political activities see essays by G. Israel, A. Rossi, and F. La Teana in *Atti del III Congresso Nazionale di Storia della Fisica, Palermo, 11-16 Octobre 1982,* F. Bevilacqua and A. Russo eds., CNR, Gruppo Nazionale di Coordinamento perla Storia della Fisica, (1983), Vols.1 and 2. Goodstein (1984).
6. For an assessment of CNR's role between the wars, see Maiocchi (1980).
7. Riordinamento-CNR (1945).
8. Gustavo Colonnetti (Turin, 1886-1968) studied civil engineering and mathematics. In the early twenties, he became dean of Turin's Polytechnic, but was forced to resign for political reasons. A prominent member of Partito Popolare (a pre-war version of Democrazia Cristiana), Colonnetti was a staunch opposer of the Fascist régime. In the fall of 1943 he left Italy and stayed for about a year in Lausanne. Colonnetti's political and organizational activities during his brief exile are discussed in Signori (1983). See also Pozzato (1982).
9. To quote an estimate, one can safely assume that 1 US Dollar was then roughly equivalent to 500-600 Italian Lire.
10. Fis. Nucl. (1945).
11. See Bolla (1952).
12. Nationalization of the electricity industry occurred in 1963 in Italy and was sort of a prerequisite for the Italian Socialist Party (PSI) participation to the so called center-left governments.
13. See Gatti (1982).
14. Bernardini (1974).
15. Amaldi (1949).
16. Amaldi (1947), (1948), (1950a,b), (1951), (1952).
17. Bruno Ferretti, 'Memorandum per Ing. Vittorio De Biasi', 8 July 1950, (MR).
18. For a very personal view of CISE's history and of Italian nuclear programmes see Silvestri (1968).
19. AST-AGC, 83.
20. Pres. CNR (1951).
21. Draft plan for Istituto Nazionale di Fisica Nucleare (AST-AGC, 83).
22. Amaldi (1963).
23. CNRN (1958) and Villi (1976).
24. Verbali del Consiglio di Presidenza del CNR, seduta del 25/7/1952. (CNR)
25. See note 23.
26. Decreto (1952).
27. CNRN (1982).
28. Verbali del Consiglio di Presidenza del CNR, seduta del 20/3/1952 e del 25/7/1952. (CNR).
29. CNRN (1958), Ch.I.
30. Verbali del Consiglio di Presidenza del CNR, seduta del 25/7/1952 (CNR).
31. Verbali del Consiglio di Presidenza del CNR, seduta del 15/6/1952 (CNR).
32. Giulotto (1982).
33. Giulotto (1976). See also Giulotto's report in *Senato della Repubblica, VI Legislatura,* 7 Commissione, Indagine conoscitiva sulla ricerca scientifica in Italia, Resoconto stenografico, 10 seduta, Giovedi 13 Febbraio 1975'. For the present situation, see Rizzuto (1983).
34. See Note 23. For a study of 'l'essor, dans l'après-guerre, des organisations syndicales italiennes regroupant les travailleurs scientifiques des centres de recherches extra-universitaires', like CNR, INFN, CNRN etc. see Cambrosio (1983).
35. Amaldi (1979) and (AST-AGC87-88 and 149).
36. Verbali del Consiglio di Presidenza del CNR, seduta del 13/1/1950 (CNR).
37. An issue raised at the Lausanne Conference was support for the humanities from National Research Councils. Colonnetti discussed the issue in a brief correspondence with Guido Castelnuovo (AST-AGC 87-88). They decided to bring the problem to the attention of the CNR Scientific Board with a

recommendation not to adhere to the Lausanne conference request that national Research Councils extend their activities to the humanities as well. In fact CNR started supporting Geisteswissenschaften as well as experimental and mathematical sciences only from the '60s.

38. See for instance Colonnetti's speech on the occasion of Fermi's visit to Rome in October 1949, reproduced in Colonnetti (1949).
39. CEC (1975).
40. (AST-AGC-153).
41. Amaldi (1977).
42. *Ibid.*, 336.
43. *Ibid.*, 336.
44. See Amaldi (1977), (1979). Interviews of Edoardo Amaldi on the subject of CERN history by M. Gowing (CERN-CHIP-REC 5), and by L. Belloni (tapes deposited in CERN Archives).
45. A copy of Amaldi's 'Quaderno No 1, 27 ottobre 1949 – 31 dicembre 1959' is deposited in CERN Archives.
46. Amaldi's Interview by L. Belloni, 13/12/1982 (CERN).
47. Amaldi's Quaderno No.1 (CERN), p.1.
48. Amaldi, 'Notes for CERN History—from E. Amaldi Diary', (received from Prof. E. Amaldi on 29/1/1981), CERN Archive, p. 1.
49. Amaldi's Interview by L. Belloni.
50. Amaldi's Interview by M. Gowing, (CERN-CHIP-REC 5), p.10.
51. Heilbron, Seidel & Wheaton (1981) and Seidel (1983).
52. Rostagni (1948). For a very short account of the history of Birmingham PS, see Birmingham (1953).
53. UNESCO (1950), 376.
54. E. Amaldi, 'Notes for CERN History—from E. Amaldi Diary' (CERN), p.1.
55. Letter Rabi to Amaldi, 20/5/83 (CERN).
56. Amaldi (1977), 350.
57. This was the International Nuclear Physics Conference held at Harwell, near Oxford, in September 1950 under the auspices of the Atomic Energy Committee of Great Britain.
58. Amaldi (1977), 350.
59. On Mario Alberto Rollier (1909–1980) unpublished speech by Ugo Lucio Businaro, March 1980.
60. Senator Count Alessandro Casati was at that time the Président de la Commission Culturelle et Scientifique de l'Assemblée consultative Européenne in Strasbourg. For a brief biography of A. Casati, see Craveri (1978).
61. Letter Amaldi to Silva, 28/11/1950 (CEC).
62. Letter Rollier to Dautry, 7/11/1950 (MR).
63. Letter Dautry to Silva, 9/11/1950 (CEC).
64. Letter Rollier to Lombardo, 16/11/1950 (MR).
65. Rollier, 'Rapporto preliminare sulla riunione del Groupe d'Etudes des Recherches Scientifiques a Ginevra il 12 Dicembre 1950 al Ministro del Commercio Estero, Milano, 2 dicembre 1950' (MR).
66. Verbali del Consiglio di Presidenza del CNR, seduta del 18/12/1950 (CNR).
67. Rollier, 'Rapporto sulla riunione del Groupe d'Etudes de Recherches Scientifiques del Centre Européen de la Culture, tenutosi a Ginevra il 12 dicembre 1950' (MR).
68. Jean-Paul de Dadelsen, 'Compte-Rendu analytique de la réunion du 12 décembre 1950', Centre Européen de la Culture, Departement des Commissions d'Etudes, Genève, le 18 décembre 1950 (CEC).
69. Giordani was to become the first President of the Italian National Committee for Nuclear Research (CNRN) in 1952.
70. Verbali della Giunta Amministrativa del CNR, seduta del 18/12/1950 (CNR).
71. Most likely, Rollier would have preferred a reactor for the production of radioisotopes for radiochemistry research to an accelerator for the production of high energy particles. Rollier was also very interested in the chemistry of nuclear reactors.

72. Letter Colonnetti to De Gasperi, 18 December 1950. (AST-AGC84).
73. Amaldi (1979).
74. Colonnetti (1965).
75. The US reaction to explosion of the first Soviet atom bomb in 1949 resulted in several such calls by American officials. At a meeting of the American Association for the Advancement of Science, Henry D. Smyth spoke on 28 December 1950 of 'the critical situation' in which the USA were finding themselves and 'then outlined means through which he felt scientific manpower could best be mobilized.' Smyth proposed 'the creation of a Scientific Board responsible to the President, and the establishment of a Student Scientific Corps to provide for the training of new scientists during the 20-year period of international tension and crisis' that he foresaw (from Cleveland-Meeting (1950)). In an open letter addressed to Italian researchers in June 1951 Colonnetti drew attention to the similarities between his and Smyth's proposal. He further remarked that Smyth's proposal had been partially accepted by Mr. Charles Wilson, Head of the Office for Defence Mobilization, who had announced the creation of a committee for scientific researches on defence related problems.
76. Letter Colonnetti to Casati, 18/12/1950 (unsigned copy). (AST-AGC153).
77. Letter Amaldi to Rollier, 11/3/1950; letter Rollier to Amaldi, 13/3/1950 (MR).
78. Istituto Colonnetti (no date).
79. Verbali del Consiglio di Presidenza del CNR, seduta del 24/10/1951 (CNR).
80. Letter Casati and Colonnetti to Ministro per gli Affari Esteri, Paris 21/12/1951 (unsigned copy). (AST-AGC158-159).
81. Letter Casati and Colonnetti to Ministro per gli Affari Esteri, Geneva, 15/2/1952 (unsigned copy). (AST-AGC158-159).
82. Verbali del Consiglio di Presidenza del CNR, seduta del 17/4/1952 (CNR).
83. Letter Casati and Colonnetti to De Gasperi, Paris, 9/5/1952 (unsigned copy) (AST-AGC158-159).
84. Verbali del Consiglio di Presidenza del CNR, seduta del 12/5/1952 (CNR).
85. Letter Colonnetti to Migoni, Copenhagen, 22/6/1952 (unsigned draft manuscript). (AST-AGC158-159).
86. Verbali del Consiglio di presidenza del CNR, seduta del 25/7/1952 (CNR).
87. Verbali del Consiglio di Presidenza del CNR, seduta del 23/10/1952 (CNR).
88. Verbali del Consiglio di Presidenza del CNR, seduta del 17/11/1953 (CNR).
89. *Ibid.*
90. Verbali del Consiglio di Presidenza del CNR, seduta del 13/12/1953 (CNR).
91. Senato della Repubblica (1958).
92. Senato della Repubblica (1954).
93. Mannari (1978). Senatore Sereni was a relative of Bruno Pontecorvo.
94. Here apparently Sereni overlooked the fact that the official announcement of the starting of operations of the first nuclear power plant at Obninsk, near Moscow, came only on 27 June 1954, that is much after the events connected with CERN's origin. See Alekseev (1981).
95. Mattone (1977). On Lussu, see also Fiori (1985).
96. Novacco (1978).
97. Senato della Repubblica (1954).
98. 'Il compagno Sereni denuncia i fini bellicisti del cosiddetto laboratorio atomico europeo', *l'Unità,* 14 dicembre 1954.
99. Letter Sereni to Amaldi, 14/12/1954 (CERN).
100. 'Una lettera del compagno Sereni', *l'Unità,* 16 December 1954.
101. Camera dei Deputati (1953-58).
102. Camera dei Deputati, (1955).
103. *Ibid.,* 16737-16755.

Bibliography for Chapter 10

Alekseev (1981) G.N. Alekseev, 'Energeticheskie epokhi i osnovanye periody razvitiia iadernoi energetiki', Voprosy istorii estestvoznaniia i Tekhniki, **2** (1981), 121-129.

Amaldi (1947) E. Amaldi, 'Centro di studio per la fisica nucleare. Attività svolta durante l'anno 1946', Ricerca scientifica e ricostruzione, **4** (1947), 391-399;

Amaldi (1948) 'Centro di studio per la fisica nucleare e delle particelle elementari. Attività svolta durante l'anno 1947', La ricerca scientifica, **1-2** (1948), 54-60;

Amaldi (1949) E. Amaldi, 'Recenti progressi e prospettive nello sviluppo delle applicazioni dell'energia atomica', Scienza e Tecnica, **10-11-12** (1949), 240-264.

Amaldi (1950a) 'Centro di studio per la fisica nucleare e delle particelle elementari. Attività svolta durante l'anno 1948', La ricerca scientifica, **3** (1950), 269-274.

Amaldi (1950b) 'Centro...Attività... 1949', La ricerca scientifica, **20** (1950), 927-835.

Amaldi (1951) 'Centro...Attività...1950' La ricerca scientifica, **7** (1951), 1149-1160.

Amaldi (1952) 'Centro...Attività...1951', La ricerca scientifica, **6** (1952), 1175-1185.

Amaldi (1963) E. Amaldi, 'L'Istituto Nazionale di Fisica Nucleare', Notiziario del CNEN, I (1963), 9.

Amaldi (1974) E. Amaldi, 'Autobiografia' Scienziati e Tecnologi Contemporanei, Vol I (Milano: Mondadori, 1974). 29-32.

Amaldi (1977) E. Amaldi, 'Personal notes on neutron work in Rome in the 30's and post-war European collaboration in high-energy physics', in C. Weiner (ed.), History of Twentieth Century Physics, International School of Physics Enrico Fermi, LVII Course, (New York: Academic Press, 1977), 294-351.

Amaldi (1979) E. Amaldi, 'The Years of Reconstruction', C.Schaerf (ed.), Perspectives of Fundamental Physics, (Chur: Harwood Academic Publishers, 1979), 379-461.

Bernardini (1974) Gilberto Bernardini, 'Autobiografia', Scienziati e Tecnologi Contemporanei Vol I (Milano: Mondadori, 1974), 117-119.

Birmingham (1953) 'Proton Synchrotron of the University of Birmingham', Nature, Vol. 172 (17 October 1953), 704-705.

Bolla (1952) G. Bolla, 'Il CISE, Centro Informazioni Studi ed Esperienze: Suoi scopi, struttura e risultati', Energia Nucleare, 31 December 1952, 15-20.

Cambrosio (1983) A. Cambrosio, L'emergence du chercheur scientifique: le cas des centres de recherches italiens de l'après-guerre, Thèse présentée à la Faculté des Etudes Supérieures en vue de l'obtention du grade de Philosophiae Doctor (Ph.D.), Institut d'Histoire et de Sociopolitique des Sciences, Faculté des Arts et des Sciences, Université de Montréal, October 1983.

Camera dei Deputati (1953-58) Camera dei Deputati, Legislatura II, 1953-58, Disegni e proposte di legge-Relazioni, Volume XI, N.1329, N.1329-A, N.1330, N.1330-A (Roma: Tipografia della Camera dei Deputati, no date).

Camera dei Deputati (1955) Camera dei Deputati, Legislatura II, Atti Parlamentari Anno 1955, Discussioni, Volume XVII, p. 16571-16610, p. 16673-16687 (Roma: Tipografia della Camera dei Deputati, no date).

CEC (1975) Deux initiatives du CEC, Documents sur les origines du CERN et de la fondation européenne de la culture, (Genève: Centre Européen de la Culture, 1975).

Cleveland-Meeting (1950) 'A Report of the Cleveland Meeting, 26-30 December 1950' Science, Vol. 113, 9 February 1951, p.151.

Bibliography

CNRN (1958)	Comitato Nazionale per le Ricerche Nucleari, Un piano quinquennale per lo sviluppo delle ricerche nucleari in Italia, (Roma: CNRN, 1958).
CNRN (1982)	'Comitato Nazionale per le Ricerche Nucleari', La ricerca scientifica, **10** (1982), 1874.
Coen (1979)	L. Coen, Shalom, 31 May 1979, p. 32.
Colonnetti (1949)	'La visita di Enrico Fermi al Consiglio Nazionale delle Ricerche', La ricerca scientifica, **10** (1949), 1113–1116.
Colonnetti (1965)	Colonnetti, Programma per lo sviluppo della ricerca scientifica in Italia nel decennio 1956-65, Parte I, 'Il Consiglio Nazionale delle Ricerche e la sua attività nell' immediato dopoguerra' (Biella, 1965).
Craveri (1978)	P. Craveri, 'Casati Alessandro', Dizionario Biografico degli Italiani, Vol 21 (Rome: Istituto dell'Enciclopedia Italiana, 1978), 207–211.
Decreto (1952)	'Decreto del Presidente del Consiglio dei Ministri, 26 giugno 1952, Istituzione del Comitato Nazionale per le ricerche nucleari', La ricerca scientifica, **10** (1952), 1882–83.
Fiori (1985)	Giuseppe Fiori, Il cavaliere dei Rossomori (Torino, Einaudi, 1985).
Fis. Nucl. (1945)	'Istituzione di un centro di studio per la fisica nucleare', Ricerca scientifica e ricostruzione, **6** (December 1945), 667–669.
Gatti (1982)	Emilio Gatti, 'Giuseppe Bolla, 1901-1980', Rendiconti dell'Istituto Lombardo di Scienze e Lettere, Parte Generale, Atti Ufficiali, Vol. 11b, 1982, 143–152.
Geymonat Ludovico et al. (1970-76)	Storia del pensiero filosofico e scientifico, (Milano, Garzanti, 1970-76), 7 volumes especially last volume.
Giulotto (1976)	L. Giulotto, 'Sulla struttura degli enti di ricerca nel campo della fisica-Disfunzioni e nodi politici', GNSM-CNR, 1976.
Giulotto (1982)	L. Giulotto, 'L'avvio della ricerca fisica in struttura della materia in Italia: gli anni '40 e '50', Gruppo Nazionale di Struttura della Materia (GNSM), CNR, Rome, 1982.
Goodstein (1984)	J.R. Goodstein, 'The Rise and Fall of Vito Volterra's World', Journal of the History of Ideas, October 1984, 607–617.
Heilbron, Seidel & Wheaton (1981)	J. Heilbron, B. Wheaton and R. Seidel, 'Lawrence and his Laboratory: Nuclear Science at Berkeley', LBL News Magazine, Vol.6, No.3 (Fall 1981).
Istituto Colonnetti (no date)	A ricordo di Gustavo Colonnetti, Consiglio Nazionale delle Ricerche, Istituto di Metrologia G. Colonnetti (Turin, no date).
Maiocchi (1980)	R. Maiocchi, 'Il ruolo delle scienze nello sviluppo industriale italiano', in G. Micheli (ed.), Storia d'Italia, Annali 3, Scienza e Tecnica nella cultura e nella società italiana dal Rinascimento ad oggi, (Turin: Einaudi, 1980), 865–999.
Mannari (1978)	E. Mannari, 'Emilio Sereni', in Franco Andreucci and Tommaso Detti eds., Il movimento operaio italiano, Dizionario Biografico, Roma, Editori Riuniti, (Vol. IV 1978), 608–612.
Mattone (1977)	A. Mattone-G. Melis, 'Lussu Emilio', in Franco Andreucci and Tommaso Detti eds., Il movimento operaio italiano, Dizionario Biografico, Vol. III, (Roma: Editori Riuniti, 1977), 180–191.
Micheli (1980)	Gianni Micheli (ed.), Storia d'Italia, Annali 3, Scienza e tecnica nella cultura e nella società Italiana dal Rinascimento ad oggi (Turin, Einaudi, 1980).
Novacco (1978)	Domenico Novacco, Storia del Parlamento Italiano, Vol. XV, Seconda Legislatura della Repubblica (1953-58), (Palermo: S.F. Flaccovio Editore, 1978).
Pozzato (1982)	E. Pozzato, 'Colonnetti Gustavo', Dizionario Biografico *degli Italiani,* Vol. 27, (Roma: Istituto dell'Enciclopedia Italiana, 1982), 464–466.

Pres. CNR (1951)	'Decreto del Presidente del CNR, 8 agosto 1951, N.599, Istituzione dell'Istituto Nazionale di Fisica Nucleare', La ricerca scientifica, **10** (1951), 1813.
Riordinamento–CNR (1945)	'Riordinamento del Consiglio Nazionale delle Ricerche, Decreto Legislativo Luogotenenziale 1 marzo 1945', published in Gazzetta Ufficiale, 29 March 1945, reproduced in Ricerca scientificia e ricostruzione, 1st July 1945, 14–21.
Rizzuto (1983)	C. Rizzuto, 'Physics research organization, Italian style', Physics Today, **8** (August 1983), 38–43.
Rostagni (1948)	A. Rostagni, 'Laboratori di fisica in Inghilterra', La ricerca scientifica, **5-6** (1948), 532.
Segré (1970)	E. Segré, Enrico Fermi, Physicist, (Chicago: University of Chicago Press, 1970).
Segré (1979)	E. Segré, 'Ricordo di Bernardo Cacciapuoti', Il Giornale, 9 May 1979.
Seidel (1983)	R.W. Seidel, 'Accelerating science: The postwar transformation of the Lawrence Radiation Laboratory', Historical Studies in the Physical Sciences, Vol. 13, Part 2, (1983), 375–400.
Senato della Repubblica (1954)	Senato della Repubblica, II Legislatura Atti Parlamentari, Resoconti delle Discussioni 1953–54, Volume Ottavo, 16 Novembre–28 Dicembre 1954, CCXXV Seduta, 13 dicembre 1954, 8925 (Roma: Tipografia del Senato, 1954).
Senato della Repubblica (1958)	Senato della Repubblica, II Legislatura Atti Interni, Volume VIII, Disegni di legge e relazioni 1953–54, N.584, N.584–A, N.585, N.585–A, (Roma, Tipografia del Senato, 1958).
Signori (1983)	E. Signori, La Svizzera e i fuoriusciti italiani, Aspetti e problemi dell'emigrazione politica 1943–45, (Franco Angeli, Milano, 1983).
Silvestri (1968)	M. Silvestri, Il costo della menzogna. Italia nucleare 1945–1968, (Turin: Einaudi, 1968).
UNESCO (1950)	Actes de la Conférence Générale de l'Organisation des Nations Unies pour l'éducation, la science et la culture, cinquième session, Florence 1950, Comptes Rendus des Débats (Paris, Unesco, Novembre 1950).
Villi (1976)	C. Villi, La fisica nucleare fondamentale in Italia, XXV Anniversario dell'Istituto Nazionale di Fisica Nucleare, 1951–1976 (Roma, INFN, 1976).
Whittaker (1941)	E.T. Whittaker, 'Vito Volterra', Royal Society Obituary Notices, **3** (1941), 690–729.

CHAPTER 11

Germany's part in the setting-up of CERN[1]

Armin HERMANN

Contents

11.1 Up to the UNESCO conference in Paris   384
11.2 Heisenberg's appointment as German delegate   393
11.3 Heisenberg's rôle at the UNESCO conference   399
11.4 The ratification   405
11.5 German positions   413
    11.5.1 The seat of the planned laboratory   413
    11.5.2 The Theoretical Group   414
    11.5.3 German influence and collaboration   416
Notes   421
Bibliography for Chapter 11   429

When the following chapter was written, the Director-General of the Organization was a German, the financial contribution of the Federal Republic was higher than of any other member state, and among staff members the country was well represented. German scientists, engineers and administrators co-operate with scientists, engineers, and administrators of other countries without any difficulties in a real international and European spirit.

When we now go back to the first post-war years, we realize that it was a different period in history. Nearly everyone in Europe had lost dear relatives and friends through a terrible regime and a terrible war, and large parts of the continent were devastated. Despite all bitter feelings the basis for today's European collaboration was established. We may better appreciate the European spirit at CERN, when we realize the difficulties to create it.

We will present our story about the rôle of the Germans in the creation of CERN roughly in a chronological order.

11.1 Up to the UNESCO conference in Paris

From 1933 on, contacts between German physicists and their colleagues abroad had proved increasingly difficult and finally, during the war, all exchange was almost completely interrupted. 'We are naturally very anxious to get in touch with the other European physicists'.[2] This sentence from a letter which Heisenberg wrote to Amaldi in 1946 pertinently describes the general feeling during the post-war years. It not only expresses the desire, which all Germans then shared, to be again accepted as a member of the international community, but also reveals the physicists' wish to learn what new developments in their field had taken place in the world.

Accordingly, the plan to set up a joint European laboratory for nuclear physics was received with strong interest by the German physicists. Political circumstances, however, at first forbade them from taking part in the discussions at which the project was brought up, and until the UNESCO conference in Paris in December 1951 theirs was a passive rôle.

In this first section, we should like to examine which of the Germans had learned of the project, how they did so, and how those concerned reacted. We should likewise be interested in how the pioneers felt about German participation.

After the First World War, the scientists in the Allied countries and those of the Central Powers had conducted a policy of boycott and counterboycott. Although during the Second World War (or rather during Nazi domination), there were unprecedentedly more occurrences which could not but hamper relations, the attitude on both sides was quite different from the start.

The first to hear of the project was Max von Laue. Attending the European Cultural Conference in Lausanne in December 1949, he was enthusiastic at the prospect of establishing European collaboration in physics; something which, since the take-over of the National Socialists, he had only been able to dream about.[3]

A total of twenty personalities had come from the Federal Republic of Germany to Lausanne. In a general sense, they represented the intellectual and cultural life of the country; among them, Christine Teusch, Minister of Culture, Education and Church Affairs of the Land Nordrhein-Westfalen, the editor Peter Suhrkamp, representatives of broadcasting agencies and newspapers, members of the Bundestag and university professors.[4]

The best-known German participant was Carlo Schmid, the first Vice-President of the German Bundestag and Chairman of the Committee for Foreign Affairs. Carlo Schmid had at an early date joined the European Federalist Union and played an important part in the European Movement.[5] His vision of a united Europe and his literary interests linked him with Denis de Rougemont and in particular Jean Paul de Dadelsen; later Carlo Schmid became one of the honorary chairmen of the Centre Européen de la Culture. At the opening of the Cultural Conference, Carlo Schmid made one of the outstanding speeches 'dans un français remarquablement limpide', as the 'Gazette de Lausanne' noted, in which he passionately pleaded for Europe: 'Européens des nations de l'Europe, unissez-vous!'[6]

Nobody among the German participants was probably against the idea of creating a common European institute for nuclear physics, but it is likely that only Max von Laue had a clearly marked interest in such a laboratory.[7] It was also he who, probably shortly after his return, told Werner Heisenberg about the project.[8]

Heisenberg was mainly interested in what was then called 'nuclear physics'[9] and, if only for that reason, the idea of setting up a European laboratory was most likely to attract his keen interest. Moreover, he was at that time President of the Deutsche Forschungsrat, which saw in the representation of German science vis-à-vis other countries one of its principal missions.

We do not know what was said during the conversation between the two Nobel prize winners, but we can infer what the keynote was. When the Deutsche Forschungsrat wrote a memorandum for the Federal Chancellor on the situation of German scientific research, they quoted Max von Laue as principal witness: 'Max von Laue has very pertinently pointed out that, scientific research is today almost the only form in which Germany can still engage in foreign politics. If we stop fighting on this territory, within ten years the world will find us no more interesting than some Bantu-tribe.'[10]

What he meant was that the 'total war', which the National Socialists had so wantonly proclaimed, had ended with a total defeat. The victors had dismembered the country and where, on the remaining territory, mineral resources still existed, as

Notes: pp. 421 ff.

in the Ruhrgebiet, the Allies were in control. The German industry, once in a leading position on the world market, was (at best) capable of producing for the country's own demands, which had become modest. The only field, in which the country had still some value for the world, was in science. The task was therefore to maintain with all available strength this position, so that, in the 'give and take' which make up politics, Germany had something to offer the other countries.

This may come close to what von Laue had meant when speaking of scientific research as a 'form of foreign politics', words which at the time were taken up by the Deutsche Forschungsrat. But before we come to the foreign affairs of the Federal Republic, we have to ask, whether the German scientists were not overestimating their own capacities.

In the first talks between allied and German physicists after the end of the Second World War the question concerning the rank of German science played an important role. 'Complacency was one of the worst enemies the German scientists had', Samuel Goudsmit said at the time: 'Confident of the superiority of German science, they assumed that no nation could equal them, none succeed where they failed.'[11]

This opinion no doubt was still conditioned by the events of war and the National Socialist crimes, and was therefore not balanced. It is however true that before and after the First World War the German scholars were imbued with the high international status which Germany held in the field of science. They found it hard to admit that the expulsion of the Jewish scholars and secondly the disastrous scientific policy of the Third Reich would inevitably leave lasting damage.

It was only through contacts with foreign colleagues, which were again starting to develop after 1945, and visits abroad, that the German scientists became aware of how much ground their country had also lost in scientific research, and this realization often came as a shock. During the celebration of the 50th Anniversary of the Special Theory of Relativity 1955 in Berne, Max von Laue noticed a catastrophic decline in the reputation of German thought.

Let us summarize: After the Cultural Conference in Lausanne, i.e. in the year 1950, Max von Laue and Werner Heisenberg, and with them the majority of their colleagues, had (still) a rather high opinion of the rank held by German science and particularly physics. They not only hoped, but were convinced, that the Federal Republic, even if a have-not in other respects, could still turn this talent to good account.

'Germany needs Europe, but Europe also needs Germany', Konrad Adenauer had been saying since May 1951.[12] The first half of the sentence was obvious: The economically weak and politically powerless Federal Republic, in her situation at the frontier of the Soviet sphere of power, could not stand on its own. As things were, it was important to make clear that the country which depended upon the assistance and friendship of the western democracies had in turn something to offer, and the scholars felt (as said above) that science was an essential asset.

Adenauer had pronounced that the German foreign policy should aim at the Federal Republic becoming as quickly as possible a member of the European community of nations, i.e. a member with equal rights and obligations, and this view was shared by a large majority of the citizens.[13]

We assume that, in the conversations they had together at the end of 1949 the two Nobel prize winners entirely agreed that the project of a joint laboratory would offer the Federal Republic the desired possibility of collaborating as an equally privileged and esteemed partner with other European countries.

In the 'Kommission für Atomphysik' of the Deutsche Forschungsgemeinschaft the opinion of the German physicists was openly expressed: 'Mr Haxel does not consider the physics interest as the main motivation for the German participation. That would rather be on the political side. Mr Heisenberg too feels that to 80% the cost should be considered from the angle of European collaboration.'[14] Although these remarks were made only on 28 February 1953, it is certainly reasonable to suppose, that from the beginning these considerations have determined the attitude of the German physicists.

Hence there had been signs at the Cultural Conference in Lausanne that something was under way in the field of nuclear physics but apparently the news of such events remained within a rather small circle, probably because the project was still considered uncertain and vague.

We know that Wolfgang Gentner only learned about the project in September 1951, during a journey to the United States. Walther Bothe apparently did not hear about it when he attended the International Nuclear Physics Conference in Harwell in September 1950, where Auger (obviously in a smaller circle) reported on it. When Gentner, at the end of 1951, told him about his conversation with Auger in Chicago, the information was apparently new to Bothe.

So it seems that at the beginning from the German physicists only Laue and Heisenberg were informed. The initiative taken by Rabi at the General UNESCO Conference in Florence apparently did not change the situation. There was hardly any mention of it in the newspapers, and probably the three German participants did not attach any importance to the Resolution. There was no scientist among them and one has to suppose that the attention of the Germans was attracted by quite different events and problems.[15]

It may have been Edoardo Amaldi who, although not present at the General Conference of UNESCO, but because of his very close contacts with Rabi and Auger, told Heisenberg. In any case, Amaldi (in a letter to Auger) drew up a short list of physicists from various European countries, who he felt would be interested and at the same time would have some influence within the scientific community and with their governments. The name of Werner Heisenberg was on this list.[16]

As we know, Amaldi's intention to write to all these colleagues was never carried out. Auger left Amaldi's letter with the proposal unanswered and Amaldi did not feel himself authorized to contact them formally on his own initiative.[17]

Notes: p. 422

On 7 November 1950 Auger wrote a letter to Dautry in which he himself prepared a short list of physicists of the various countries.[18] None of the seven physicists mentioned was German or British. The British were missing because Auger had not been able to find a colleague in Great Britain whom he could hope would be sufficiently dedicated to the project. On the other hand, in Germany there was hardly a single physicist who was not interested in the project. But then it was Auger who hesitated. He probably felt that participation by German scientists at that time would add another difficulty to the already existing problems. It seems as if three points had been considered.

Firstly, the Federal Republic was not yet fully accepted as a member of the international family. It was neither a member of the United Nations, nor of UNESCO or IUPAP; although German delegates had worked in the Council of Europe in Strasbourg since the opening of the second meeting of the Advisory Assembly on 7 August 1950, the Federal Republic continued to be excluded from the Council of Ministers. It would certainly be only a question of time for Germany to be admitted as a full member, in the other international and European organizations as well, but undoubtedly nearly everywhere a strong political resistance still existed concerning her admittance. Therefore it must have appeared unwise to put an unnecessary strain on a project as uncertain as the planned European laboratory for nuclear physics, by giving it a pioneer rôle in this respect.

Secondly, nearly all German physicists, with whom one would have to deal, had been involved during the war in the so-called 'Uran-Projekt', concerning the technical utilization of nuclear energy. Even without a clear idea about the work, aims and motives of the Germans, the fact that they had 'collaborated with the Nazis',[19] was sufficient in the eyes of many of their foreign colleagues to disqualify them from international collaboration.

Thirdly, those obstacles of a more psychological nature were supported by solid legal ones: In the Federal Republic, all activity in the field of nuclear physics was forbidden by the laws passed by the Allies. Obviously, one could not simply allow the Germans to do something beyond their border, which they were prohibited from doing in their own country.

When the Centre Européen de la Culture (C.E.C.) in Geneva finally convened the 'Commission de Coopération Scientifique' for the 12th December 1950, Max von Laue was the only German to receive an invitation.[20] As he had participated in the discussions in Lausanne, the C.E.C. regarded him as a member of this Commission.

Max von Laue was known to have always been a declared opponent of the Nazi regime and to have, more than any other scholar in Germany, expressed his opinion very openly; he had never participated in the Uran Projekt. Therefore Auger and Dautry — quite obviously — did not object to his being invited.

However, Max von Laue did not attend the important meeting since he had been invited on the same date to the 50th Anniversary of the Nobel Foundation, and had, as he wrote to Raymond Silva, 'accepted engagements for lectures and speeches there'.[21] It was certainly not an excuse.[22]

Did the meeting of the Commission de Coopération Scientifique on 12 December 1950 actually discuss the participation of the Federal Republic of Germany in the planned European laboratory for nuclear physics? Not in the press release,[23] but in the more detailed minutes,[24] we find a list of West European countries which were considered for collaboration. This list also included 'Allemagne de l'Ouest'. There are three further indications that a participation of the Federal Republic was seriously considered. Thus in the report which Bruno Ferretti made a few days later to the Consiglio Nazionale delle Ricerche in Rome, it was said that for Italy, Germany and other countries a contribution of 12.5% had been envisaged (whereas France was to contribute 30% and Great Britain 3.5%).[25] And again in the corresponding letter of Mario Rollier, the second Italian participant at the meeting in Geneva on 12 December, Germany was mentioned in connection with the financing of the project.[26]

A third indication that a German collaboration had actually been discussed in Geneva was given in a press conference held in a rather different context, namely on the occasion of an international conference on elementary particles in Bombay on 21 December 1950. On the basis of news from the Reuter Agency, the communist paper L'Humanité wrote that Francis Perrin had announced that the participation of German scientists in the planned European Laboratory was being considered. The exact words read: 'Mr Perrin added that the promoters of the project [...] considered associating the Germans in this undertaking and a decision is expected soon'.[27]

It is therefore reasonable to conclude: At the meeting of the 'Commission de Coopération Scientifique' on 12 December 1950 a participation of the Federal Republic was actually contemplated. However, both Auger and his friends probably realized that the decision could not be taken then, but that it would only be a question of time.

This attitude also prevailed during the following months among the persons concerned. It was not before 21 September 1951 that Auger — apparently for the first time — spoke to a German colleague about the project.[28]

One may ask whether the outcome of the Geneva meeting had become known in Germany although there had been no immediate contacts. Three different documents reflected the outcome: the press release, the detailed minutes and in particular the unanimously adopted 'Resolution'.

News agencies and newspapers in many countries had received the press release directly from the Centre Européen de la Culture, but addresses of such agencies in Germany or of correspondents in Switzerland were not known to the C.E.C. For that reason, several copies were sent to Carlo Schmid asking him to pass them on. We have not been able to find out, whether as a consequence German newspapers did in fact report on the matter.

Notes: p. 422

The detailed, eight-page long 'note' of 13 December 1950 was also sent to Carlo Schmid.[29] Again, we have been unable to clarify whether this information got through to the German physicists. There were however no signs of any reactions.

Now, if the direct and private communication had no further effect, what about the official procedure? The resolution of the 'Commission de Coopération Scientifique' of the Centre Européen de la Culture had been addressed simultaneously to UNESCO in Paris and the Committee of Cultural Experts of the Council of Europe in Strasbourg.

As is generally known, Auger's project was advanced very energetically by UNESCO. At that time (first half of 1951) the Federal Republic was not yet a member of the United Nations' cultural organization and hence UNESCO had no reason to send information to Germany.

Since 13 July 1950 the Federal Republic had been an 'associate member' (full member from 2 May 1951) of the Council of Europe. This meant that it could send 36 delegates (elected from among the members of the Bundestag) to the Advisory Assembly (the 'European Parliament'), whereas in the Council of Ministers it had only observer status. In the Committee of Cultural Experts, to which the Resolution of the Centre Européen de la Culture was addressed, the Federal Republic was represented. At the second meeting of this body, on 15 February 1951, the resolution of the C.E.C. was on the agenda; the minutes read as follows: 'In view of the technical and political implications of this item, it was decided to postpone a discussion until the next meeting, when experts would have learnt the opinion of their governments.'[30]

The German members of the Committee of Cultural Experts of the Council of Europe were Rudolf Salat of the cultural department of the Office of the Bundeskanzleramt and Walter Keim, the General Secretary of the Ständige Konferenz der Kultusminister. It can be assumed that both gentlemen reported to their departments. Whether and to what extent this was done and what steps were taken as a result, we have not been able to clarify.

We have always to bear in mind that those who supported the project had not yet finally endorsed the possibility of a German participation. An attack in L'Humanité showed that it was indeed advisable to be careful. The communist paper referred to the Reuter communication from Bombay mentioned earlier, according to which Francis Perrin had given some information about the project. On 22 December 1950 the headline on the front page read: 'Nazi scientists will take part in atomic research in Paris. The dismissal of Joliot-Curie makes sense now.'

The article started with a (fairly) correct report saying that Perrin had mentioned as possible sites of the future international laboratory France, Switzerland, Belgium or the Netherlands. But then the paper continued on the assumption that the planned laboratory would constitute a major part of the Centre for Nuclear Energy in Paris, and combined it with the wildest conjectures:

> The great scientist and great patriot Joliot-Curie was dismissed because his presence made certain measures impossible. Preparations are being made to install Nazi scientists in the Paris laboratories. For what sort of work? The American atomic policy with its tendency towards war makes it easy to predict in what direction the research of the German scientists will go.
>
> From now on, the question will arise about the supremacy of the Nazi scientists over the French scientists. Since, in the American strategy, leadership is a matter of confidence, one could also believe that the German scientists, who have, and for good reasons, the full sympathy of the American masters, will be ahead of their French colleagues. Here you can see just how far a policy of national dishonour can bring a government.[31]

From a communist viewpoint it must have seemed indeed quite obvious that the work of the planned laboratory would concern the utilization of atomic energy, since even the official statement of the C.E.C. contained an indication in this sense due to an error by the minute-writer.[32] The rather bold speculations might well have convinced the reader, particularly if he had the corresponding political orientation.

Since Robert Schuman in May 1950 had given a decisive turn to the French policy towards Germany[33], the French Communist Party protested vehemently against the new orientation, and the communist propaganda made deliberate use of old anti-German resentment.

The scientists had played a very negative role in the relations between France and Germany, contributing to the prejudice existing on both sides. Although during the Age of Enlightenment, Voltaire and with him whole generations of scholars had declared[34] that science was the natural enemy of fanaticism and prejudice, during and after the First World War, the scholars in both countries had been hardly less (and even more) instilled with feelings of national resentment than the average citizen.

From the communist viewpoint, it may therefore have seemed promising to launch a journalistic campaign against the 'plan Schuman nucléaire'.[35] But apparently the propaganda was not very successful. A press attack such as that of L'Humanité at best prompted the supporters of the project to treat the question of the German participation with even greater caution.

In the first half of 1951 the supporters of the project were unanimous that the time was not yet ripe for a German collaboration[36]. At a meeting between officials at the Quai d'Orsay on 18 May 1951, the attitude adopted was a cautious 'wait-and-see':

> It has been recognized in that as regards Germany the question must be left open. Among the factors which play a rôle in the decision we should note
> a) that all work in atomic research and application is forbidden at present and that the power of the Military Government in this field was formally confirmed during the studies concerning the revision of the Occupation Statute.
> b) If Germany should one day regain her freedom of action in this field it would be preferable to see her collaborate in the center of Europe rather than push her to make an independant effort or to collaborate with the Anglo-Saxons or the Soviets.
> c) Western Germany will in all probability join UNESCO in July.[37]

Notes: p. 422

The participants considered it an important, if not essential, condition for the participation of Germany, that the Allies' prohibition on scientific research in Germany should be lifted and that Germany should be admitted to UNESCO. Subsequently the original restrictions on research were indeed to some extent made less stringent and in June 1951 the Federal Republic was able to obtain membership of UNESCO. When on 31 August 1951 the Director-General of UNESCO, Jaimes Torres-Bodet, sent to the Member States the official letter with the invitation to the Conference in Paris, the Federal Republic was among the addressees. The letter asked members to send 'delegates or observers' to the conference in December 1951.

The question of who should act as the promoter of the project in the Federal Republic had already been raised by Amaldi in his letter to Auger of 3 October 1950. It had remained unanswered.

In September 1951, Wolfgang Gentner attended the inauguration ceremony of the Institute for Nuclear Studies in Chicago. On 21 September he accepted an invitation to the house of Enrico Fermi where the physicists celebrated five days in advance the 50th Anniversary of the 'Italian navigator'.[38] Gentner remembered that it was here that he had heard for the first time of the planned European Laboratory.[39] The question Pierre Auger asked him on that occasion was still: 'Which German physicist would have the keenest interest and greatest authority to see the project through in Germany?' Gentner mentioned the name of Otto Hahn but soon after he became doubtful and wrote to Auger on 15 November 1951:

> It is true that Hahn as President of the Max-Planck-Gesellschaft (formerly Kaiser-Wilhelm-Gesellschaft) has considerable influence, but one should not forget that he is basically a chemist and at present mainly concerned with administrative work. I therefore believe it would be preferable to take a nuclear physicist and I would like to propose Professor Bothe of Heidelberg. During a recent visit to Heidelberg I talked with Mr. Bothe about it and got the impression that his attitude in this matter is very positive and I am quite sure that he would use all his influence to convince the German authorities to participate. Bothe is certainly the leading experimental nuclear physicist in Germany. I am personally in very close contact with him so that together we could do everything possible.[40]

But at the time Gentner made these remarks they were already no longer valid. With the official letter of the Director-General of UNESCO to the Federal Chancellor, the project had become a governmental affair.

In a letter of 8 December 1951, Foreign Secretary Walter Hallstein informed the Director-General of UNESCO[41] that the German delegates would be Heisenberg and Hocker, with Heisenberg as head of the delegation. On the same date Hallstein also wrote to Heisenberg and Hocker, informing them about their mission.

There was, however, a problem which was common at that time but seems unimaginable today: Hocker could not obtain an entry visa for France[42] and thus could not participate in the negotiations. Subsequently however, both Heisenberg and Hocker regularly represented the Federal Republic at the meetings of the provisional Council and the Council.

Why did the choice fall on these two men? For an answer we have to examine the self-administration of scientific research in Germany. We shall see that the representation of German science *vis-à-vis* other countries, and in particular the international scientific and cultural organizations, constituted a central point of discussion and controversy for the German scientists.

11.2 Heisenberg's appointment as German delegate

In 1949, the 'Notgemeinschaft der Deutschen Wissenschaft' and the 'Deutsche Forschungsrat' had been established within an interval of two months. Both organizations had set themselves the goal to promote science by soliciting funds—mainly from public sources—which they wanted to utilize for projects considered of first priority.[43]

The Notgemeinschaft continued the successful activity of its precursor organization, founded in 1920. The experiences during the period of the Third Reich with what they called 'politisierende Nichtskönner'[44] had motivated the Notgemeinschaft to stress the idea of self-administration and to reduce, if possible, the influence of the State.

Werner Heisenberg, who had been the driving force behind the creation of the Deutsche Forschungsrat and who was also its President, on the contrary felt that in the middle of the 20th century, the idea of a separation between science and State could only be an illusion. For Heisenberg too, the experiences during the Third Reich had been determining. His personal key experience had been the failure of the German nuclear reactor project,[45] which he quite rightly attributed to the antagonism between State and science. He considered it a vital necessity for the new democratic State, that its highest representatives should have an appreciation for science and his aim was to establish a direct collaboration with the Federal Chancellor. In Konrad Adenauer Heisenberg found someone who actually understood his ideas and a special relationship full of confidence developed between the two men.

Perhaps the concordance between Adenauer and Heisenberg was also based on the fact that the Deutsche Forschungsrat (DFR), unlike the Notgemeinschaft, stressed the aristocratic principle in science. 'The Federal Chancellor saw in the DFR an idealistic counterbalance against the levelling which also menaced in science and the goal to be achieved in his view was a self-administration of research, based on the authority of scientific research'.[46]

But inspite of all the differences, the concepts of the two organizations were still very similar, the differences being mostly only subtleties. It was precisely the similarity of their aims, which caused rivalry and a great deal of undesirable

Notes: pp. 422 ff.

friction. Just as the Forschungsrat on the public side got support from the Federal Government, the Notgemeinschaft turned to the Ministers for Culture of the Länder, so that the opposition between Bund and Länder also came into play here.

More and more scholars found it undesirable that science spoke with two different tongues at a time when it was anxious to eliminate the mental damage caused by the Third Reich and the physical destructions of war and to catch up with the scientific research in other countries. Hence, they first tried to co-ordinate the work of the two organizations, namely clearly to delimit their respective tasks.

On 31 March 1950 the Notgemeinschaft and the Forschungsrat joined up in a 'Arbeitsgemeinschaft', and in the following months this Arbeitsgemeinschaft somehow worked. One of the compromises it reached provided that the representation abroad of German science and the setting-up of priorities should be part of the responsibilities of the Deutsche Forschungsrat.[47]

On 8 February 1950, Ronald G. Fraser, Liaison Officer between UNESCO and ICSU, sent a letter to the executive Vice-President of the 'Notgemeinschaft der Deutschen Wissenschaft' asking him which German institution, in the event of the Federal Republic being admitted into ICSU, would become its official national member; the rules provided that 'Academies and Research Councils' were eligible.[48] Fraser knew the Deutsche Forschungsrat or rather its precursor, the German Scientific Advisory Council, from personal experience; between 1947 and 1949 he had worked with this body as representative of the British Military Government.[49] Maybe that was the reason why he considered the Forschungsrat as best eligible to become the German member in ICSU.

The 'Arbeitsgemeinschaft', in line with the compromise, actually decided the following: The two German representatives in ICSU should be appointed by the Forschungsrat and their names communicated to the Federal Government.[50]

Why were Heisenberg and Hocker chosen as representatives of the Federal Republic in the negotiations concerning the setting-up of CERN? This is the question which interests us here. The key role was played by the institution which at the decisive moment was supposed to be the representative of German science *vis-à-vis* foreign scientific institutions. In 1950 it was the role of the Deutsche Forschungsrat, but at the beginning of 1951 things took a new, and for those involved, dramatic turn.

The 'Arbeitsgemeinschaft' between the Deutsche Forschungsrat und the Notgemeinschaft der Deutschen Wissenschaft had not succeeded in conciliating the opposites between the two institutions. The efforts of the Notgemeinschaft were therefore directed towards what was called a total fusion. The Forschungsrat baulked at this intention, and tried on its part to be recognized by the Federal Government as the official representative of German scientific research especially in dealings with foreign scientific institutions.

In March 1951 the Forschungsrat submitted a request to Adenauer for official acknowledgement, whereupon the Federal Government asked the Ständige

Konferenz der Kultusminister for its opinion and received a reply from its President, Senator Heinrich Landahl. On 11 April 1951 the Senator stated, that the Kultusminister of the Länder could not acknowledge the claim made by the Forschungsrat, 'to be the official representative of German scientific research.'[51] Nevertheless, in a letter of 11 May 1951 to Heisenberg, the Federal Chancellor stated expressly: 'Following our conversation I have the privilege of informing you that for the fulfilment of tasks provided for under article 74 subparagraph 13 (promotion of scientific research), the Federal Government will call on the assistance and the advice of the Deutsche Forschungsrat'.[52]

With this letter Adenauer approved the rules of the DFR which read: 'The functions of the DFR include in particular: [...] The representation of all common interests and claims of German scientific research vis-à-vis governmental authorities and all other public elements of political, cultural and economic life as well as the representation abroad of German scientific research and vis-à-vis the international scientific and cultural organizations and institutions'.[53]

If the letter of the Director-General of UNESCO to the Federal Chancellor, asking him to appoint delegates or observers to the forthcoming UNESCO conference concerning the planned European Laboratory, had been written four months earlier, on 31 May instead of 31 August 1951, it would have needed to be discussed with the President of the Deutsche Forschungsrat, Werner Heisenberg. But the letter was dated 31 August, and at the time it was dealt with in the Foreign Office, in September and October 1951, the Deutsche Forschungsrat had ceased to exist.

Under the continuous pressure from the Notgemeinschaft and the Stifterverband,[54] the self-administration had been fundamentally reorganized, by the fusion between the Notgemeinschaft and the Forschungsrat to form the 'Deutsche Forschungsgemeinschaft' (DFG) on 2 August 1951.

The 'fusion' between the smaller Forschungsrat and the big organization of the Notgemeinschaft basically meant, that the latter continued under a new name, whereas the Deutsche Forschungsrat ceased to exist. Seen more positively (from the point of view of the Forschungsrat), one could also say that the Forschungsrat was 'incorporated' in the Notgemeinschaft, where it formed a new body, the Senate. However the admission of the Forschungsrat was by no means *in corpore*. Out of the 21 members of the Forschungsrat, only 11 became members of the Senate which numbered 27.

Heisenberg had obstinately resisted this development, believing for a long time that Adenauer's support would be a guarantee for the existence of the Forschungsrat. On 2 July 1951 Konrad Adenauer had still held out the prospect of enacting a proper law for the promotion of scientific research, and had recommended to postpone for some time 'the plans concerning a fusion between the Notgemeinschaft der Deutschen Wissenschaft and the Deutsche Forschungsrat'.[55]

Heisenberg however could no longer ignore the strong psychological pressure his colleagues were exerting on him, and felt himself compelled to agree to the fusion. For Heisenberg the end of his Forschungsrat represented a serious personal defeat.[56]

Notes: p. 423

The persons involved acknowledged the sacrifice Heisenberg had made by accepting the fusion. They tried to offer the famous scholar every conceivable possibility of action within the Deutsche Forschungsgemeinschaft. Heisenberg's election as one of the four vice-presidents at the extraordinary General Assembly on 2 August 1951 was part of this endeavour.

More important was the fact that, according to the statutes, the advising of governments and parliaments and the representation of German science in dealings with foreign institutions which used to be a particular concern of the Forschungsrat, had now become the responsibility of the Senate. The setting-up of scientific committees, as already defined and partly initiated by the DFR, was also assigned to the Senate, according to the statutes of the new DFG: 'The Senate takes care of common interests of scientific research, gives his expert advice in scientific matters to governmental authorities and looks after the interests of German scientific research in relation to research abroad .... Within the framework of its competence, the Senate may set up committees whose members do not have to be members of the Senate'.[57]

What were the motivations for the appointment of Heisenberg as head of the German Delegation in the negotiations in Paris? That is still our question. So far we have described Heisenberg's part in scientific policy. Now we come to the processes which led to his appointment. As will be seen, it was also his status, either as delegate or observer, that was concerned. As a simple 'observer', Heisenberg could have been sent to Paris alone; if he were to go as a 'delegate', an expert in organizational and financial affairs would also have been needed. By the time the letter of the Director-General of UNESCO had reached the Federal Foreign Office, it was found there that 'the German representatives' had to be 'selected with the greatest of care'. The Foreign Office asked the Deutsche Forschungsgemeinschaft, the Max-Planck-Gesellschaft and the Federal Ministry of the Interior to send one representative each to a working discussion on 6th November, 1951.[58]

Ludwig Raiser, in charge of the presidency of the DFG,[59] submitted the request, in accordance with the statutes, to the Senate. The newly created body discussed the matter at its meeting of 29 October 1951. All four vice-presidents (Ludwig Raiser, Werner Heisenberg, Walther Gerlach and Eduard Spranger) and 15 out of the 27 members of the Senate were present, as well as Helmut Eickemeyer and Alexander Hocker as officials of the DFG. The minutes of the meeting read:

> Mr. Raiser announces [...], that the Foreign Office has communicated that from 10-12 December 1951 a meeting will be held in Paris on the 'Planning of a European Laboratory for Nuclear Research'. Two German representatives should be present at the discussions, namely one administrative or financial expert and one scientist. [...] The Senate unanimously proposes to appoint Professor Heisenberg as scientific representative from Germany at the meeting in Paris in December. Mr. Heisenberg accepts the proposals with the proviso that Professor v. Weizsäcker from Göttingen or Professor Haxel from Heidelberg may be his substitutes for any committee work subsequent to this first meeting.[60]

Heisenberg was thus proposed by the Senate of the DFG as 'scientific representative' of the Federal Republic of Germany for the negotiations concerning the setting-up of CERN.

Why did the Senate unanimously vote for Heisenberg? Undoubtedly, the uncontested scientific competence of the Nobel prizewinner in the future field of activity of the planned European Laboratory was an important factor, as well as his interest in the project, which he must have expressed. A probably more essential factor may have been a certain gratitude felt for Heisenberg, after the final dissolution of the Deutsche Forschungsrat, a readiness to offer compensation. Even if the word 'gratitude' is not quite right here: One may however certainly say, that his colleagues were prepared to prove Heisenberg that even within the framework of the Deutsche Forschungsgemeinschaft, he would be able to realize his aims with regard to scientific policy.

On 31st October Ludwig Raiser informed the Foreign Office that he would take part himself in the discussion on 6th November and that the Forschungsgemeinschaft was nominating Werner Heisenberg 'as scientific delegate'. There is a handwritten note on this letter: 'Telschow: Heisenberg. He would like to go himself as the administrator'.[61]

This note leads to the conclusion that the Foreign Office had had a telephone conversation with the Secretary-General of the Max-Planck-Gesellschaft, Dr. Ernst Telschow of Göttingen. During this conversation, Telschow had expressly agreed on the appointment of Heisenberg, as may easily be understood. Heisenberg was not only director of one of the most important Max-Planck-Institutes, but also had for many years felt himself closely linked to the Society. Also during this conversation Telschow had clearly put himself forward as the second German representative who, as stated, was to be an expert in administrative and financial matters.

In the discussions at the Foreign Office on 6th November, 1951, 'those present decided to recommend the despatch of a German delegation, regardless of the subsequent decision whether or not the Federal Republic was to take part in the preparations for and the construction of the laboratory'.[62]

The proposed delegate was 'Prof. Heisenberg as scientist' ('he has already signified his agreement via the Forschungsgemeinschaft and the Max-Planck-Gesellschaft') and 'a person to be appointed by the Federal Ministry of the Interior as administrative expert'.[63]

For fear that an official delegation would be empowered to undertake 'considerable financial commitments on behalf of the Federal Republic', the Interior and Foreign Ministries were, however, inclined to be satisfied with a single observer to leave the way open 'for a later active participation of the Federal Republic in the bringing of the project to fruition'. In a note of an official of the Foreign Office it was stated:

> I was still able on Saturday, 24th November, to reach Professor Heisenberg, to whom I put the question 'delegate or observer'. He said that he had no objections to going as an observer, since he would nevertheless be regarded as an equal partner in negotiation in purely scientific matters, regardless of his official status.[64]

Notes: pp. 423 ff.

At the last minute, however, Walter Hallstein decided 'on the basis of fundamental considerations' to send Heisenberg as an official delegate ('although his credentials must contain the proviso that he may make no financial commitments on behalf of the Federal Republic'). Since each delegation was to consist of two members, Dr. Alexander Hocker was also appointed administrative and financial expert. The file record of the Foreign Office shows that Alexander Hocker, the deputy director of the Deutsche Forschungsgemeinschaft, was appointed by the Federal Ministry of the Interior. This corresponds to the recollections of Hocker who has told us that Erich Wende, then Head of the cultural department of the Ministry of the Interior, was highly instrumental in the decision.[65]

We have now explained how the two German delegates to the Paris UNESCO conference were selected. As Heisenberg and Hocker also represented the Federal Republic at the subsequent conferences, we feel that our extensive account is justified. How should the choice of Heisenberg and Hocker be assessed?

The German physicists themselves would probably have chosen someone else for this mission. Gentner's letter to Auger, quoted above, shows that Gentner had Walther Bothe in mind. Bothe, who in 1954, i.e. 24 years after the event, was awarded the Nobel prize for his discovery of nuclear excitation, was the father figure to the German nuclear physicists. In a letter to a colleague of 14 March 1952 Bothe himself made the following comment:

> I don't at all like the idea, that once again (as in those days of the Uranverein) extreme theoreticians shall have the say. I feel that at least a man like Gentner should absolutely sit on this Commission;[66] he should already have participated in the negotiations in Paris and Geneva,[67] not only because of his many friends abroad and his knowledge of languages but mainly because he is the only one in Germany to have actually taken part in creating a particle accelerator (belt-type generator, Heidelberg cyclotron).[68]

We therefore believe that the German nuclear physicists, if the decision had been theirs, would not have elected Heisenberg but someone else—Bothe or Gentner—as German delegate. Nevertheless, the decision to appoint Heisenberg was a lucky one:

1. Heisenberg belonged to the greatest authorities in the field concerned. Even if he was a theoretical physicist and had no experience with the construction of accelerators, he possessed, as hardly anyone else, a general understanding of the whole field of particle physics. It was on account of that particular quality that at the first Council Session on 7-8 October 1954 he was elected Chairman of the Scientific Policy Committee.[69]
2. Heisenberg supported the project very strongly and one of the reasons was certainly that the outcome would be directly relevant for his own work. Already at that time he had a particular goal in mind, which later became his uniform theory of matter. As he wrote in his memoirs:

> I was so extremely interested in those plans because my hope was that experiments at such a big accelerator would provide indications as to whether my assumption was true, that at high-energy collisions between two elementary particles, many such particles are

produced, and whether there are actually many different types of particles, which, likewise the stationary states of the atom, are defined by their symmetry properties, masses and lifetimes.[70]

3. What undoubtedly constituted a handicap for Heisenberg's position in the international community, was the fact that he had participated in the German atomic energy project. But so had — more of less — all the other German physicists who might have been suitable as head of the German delegation.

The choice of Alexander Hocker may also be regarded as a happy one. Alexander Hocker was at the time official at the Deutsche Forschungsgemeinschaft. He was the General-Secretary's deputy and in charge of matters regarding the Senate and the Commissions. Hocker's election emphasizes the strong position of the self-administration of German scientific research at a time when a Ministry for scientific research did not yet exist, but it also clearly shows the importance attached to the project of a European laboratory. Hocker had the reputation of being a 'highly intelligent, strongwilled and ambitious worker, combining great administrative skills with a good knowledge of human nature and a tremendous work potential'.[71]

Hocker's responsibility for the Commissions proved a great advantage. The 'Kommission für Atomphysik' (Atomic Physics Committee) set up by the Deutsche Forschungsgemeinschaft on 29 February 1952, represented the body in which the German physicists defined the attitude to adopt with regard to the planned European laboratory. It was in this Commission that they could directly question not only Heisenberg (and later Gentner) but also Hocker on their impressions gained at the Council Sessions.

Having explained in detail how the two German delegates were chosen and how their selection is to be assessed, we come to our next questions: What was the part of the German delegation at the UNESCO conference in Paris and what were its instructions? We have already mentioned that Hocker was refused the entry visa for France and that Heisenberg was practically alone in representing the Federal Republic in Paris.[72]

What rôle did Heisenberg play at the conference in Paris and what were his instructions?

11.3 Heisenberg's rôle at the UNESCO conference

It was on 9th December that Heisenberg at last received the urgently awaited letter from the Foreign Office setting out the limits of his authority. It had had to be revised owing to the change in his status from that of 'observer' to that of 'delegate'.

Heisenberg had expressly asked the Foreign Office 'to set out precisely in the letter what his attitude to financial and administrative matters should be. In the purely scientific field, of course, he needs no instructions'.[73]

In his letter to Werner Heisenberg, the Secretary of State stressed 'Germany's great scientific and cultural policy interest in the subject of negotiation'. Essentially, the letter said:

> I should like to ask you to give the Foreign Office a detailed report at the end of the conference. In it, would you please not only set out the great scientific significance of the project — which in any event is beyond question — but also take particular account of the following items [...]
> 1. the phraseology of the international agreement(s) or treaty(ies) which are to be signed by the Federal Republic;
> 2. the practical and also, in particular, the economic advantages of the Federal Republic's participation in the project;
> 3. the probable financial commitments;
> 4. the time at which these commitments will fall due.
>
> I should be very grateful for any of your own proposals concerning the Federal Republic's future relations. I should like to stress the fact that our participation in bringing the project about is still open. Provided that it can be done without making any financial commitments, I authorise you to collaborate in the drafting of the working schedule by the conference, but do please abstain from making any political statements.[74]

Heisenberg reported to the Secretary of State on the proceedings of the conference accordingly, on 23rd December, 1951. In his four-page letter Heisenberg spoke about the differences, about the official remarks he himself had made (especially concerning the seat of the laboratory) and about the scientific and economic prospects of the envisaged co-operation.

Heisenberg like others, had noticed two rivalling groups: The younger physicists, mainly from Italy, France, Belgium and Switzerland who had been the initiators of the project and 'who very energetically supported the plan to build a big machine for nuclear physics (an accelerator facility of 3–10 BeV based on the facilities existing in Brookhaven and Berkeley in the United States) and to locate it in Geneva within the framework of a European Institute for nuclear physics to be founded.' The other group, consisting mainly of senior physicists from Great Britain, Scandinavia and the Netherlands, regarded the plan as Utopian and believed that one should 'for the time being refrain from constructing big and very expensive machines.' The elder physicists instead proposed the joint use of the already existing installations, i.e. the accelerators in Great Britain and the Institute for Theoretical Physics in Copenhagen.

Each national delegation made an official declaration. Heisenberg, in his statement, gave his support to any form of European cooperation in the field of nuclear physics, but also declared his particular favour of a new, common research centre. He tried to find a compromise between the two groups and his proposal was to initiate immediately a scientific cooperation to take place at the synchro-cyclotron in Liverpool and the Bohr Institute in Copenhagen, but at the same time 'to pursue energetically the planning of a big machine and the creation of a central institute'.[75]

Heisenberg clearly advocated Geneva as the site for the new European laboratory. In his declaration he said: 'This plan' (i.e. to set up a European centre of nuclear physics in Geneva) 'will probably find much support in our country, on account of the old tradition of neutrality and friendly European cooperation in Switzerland'.[76]

Why did Heisenberg so vehemently advocate Geneva? Since the twenties, Heisenberg, like many theoretical physicists, had been a regular guest in Copenhagen at Niels Bohr's institute. His links with Niels Bohr were those of particularly close friends. Then why did he not, like H.A. Kramers,[77] propose Copenhagen as a site for the future laboratory? The first reason that will come to mind will be a scientific policy approach. The younger physicists, at whose initiative the project was put forward, were in favour of Geneva. The proposal of Copenhagen involved not only a different seat, but also a different and smaller-scale project. Heisenberg, however, was in favour of the 'big solution'.

A much stronger influence in favour of supporting Geneva, we guess, arose from a diplomatic approach by the Swiss Confederation to the Federal government. On 10th December, Dr. Rumpf, diplomatic councillor at the cultural department of the Foreign Office received a visit from August Rebsamen, councillor at the Swiss Embassy. 'On the instructions of his government', the councillor informed him that Switzerland had officially offered itself as the seat of the planned laboratory: 'Switzerland would be very grateful if the German representative [..] would vote for the selection of Geneva. I informed Mr. R. that so far no other government [..] had approached us [..]. In any event, the Federal Republic could not but welcome the choice of a neutral country like Switzerland. I promised Mr. R. that I would instruct the German delegate accordingly'.[78]

In fact, on 12th December, Heisenberg received a letter from the Foreign Office signed by Rudolf Salat: 'It is the current view that only practical reasons should be considered in the choice of the seat of the institute, which is to be for scientific purposes only. A country with a great scientific and technical tradition, like Switzerland, and the international experience and connections of Geneva, are highly favourable factors here. As, moreover, no other approaches have yet been made to the Federal government, it is the Foreign Office's view that, if you consider it justified, following the discussions on the matter at the conference, you should support the Swiss proposal'.[79]

In his report to Walter Hallstein on 23 December 1951, Heisenberg did not discuss the political significance of the project. It had been made clear to him that politics were not his business.

Heisenberg therefore limited his comments to the scientific and economic aspects:

> From the scientific point of view the project is of outstanding importance, as it probably offers the only possibility for Europe to catch up with the big advance America has gained in this branch of nuclear physics. A single European country would hardly be in a position to build the kind of machines that exist in the United States and are planned in the project. However, a direct economic utility for Germany should not be expected from a participation. The undertaking is purely scientific and any practical application is still in a very distant future.[80]

Notes: p. 424

The statement was correct and to the point. Nevertheless, Heisenberg obviously wanted to present an altogether positive view and he mentioned three additional factors pointing to an 'indirect' economic utility in the longer term. Firstly, 'a participation in the project would provide a scientific training for the younger, actively collaborating physicists which could be very useful at a later date with regard to an economic utilization of atomic energy in Germany'. Secondly, 'in the long run this new branch of nuclear physics may of course also open a field of application of similar importance as that of current nuclear physics.' Thirdly, it could be expected 'that the construction of the big machines would at least partly be assigned to German firms'.[81]

The second and third arguments were repeatedly invoked by supporters of the project. However, the interesting and tempting argument, provided it was valid, was the first one. At that time, the utilization of atomic energy was still forbidden to the Federal Republic. In this situation it was precisely Heisenberg who continuously pressed the Federal Government to start with the necessary preparations so that once the permission, expected before long, had been given, work could begin immediately.

In an annex to the 'Deutschlandvertrag' signed on 26 May 1952, the three allied powers authorized the Federal Republic to operate a reactor with a heat capacity of 1500 kW maximum. On 3 October, Adenauer and Heisenberg publicly announced that the Federal Republic would make use of the concession immediately after the entry into force of the Treaty.

At the end of 1951, when Heisenberg wrote his report, he must have reckoned that the question relating to a technical utilization of nuclear energy would soon become relevant to the Federal Republic.[82] One could easily foresee that the lack of specialized scientists and engineers would be among the major problems. Experts could not be produced out of a hat but had to be trained gradually.

Would a collaboration on the setting-up of a research centre for high-energy physics really provide a suitable 'scientific training' for the technical utilization of nuclear energy? Presumably only to a very limited degree. At best, the occupation with problems regarding radiation protection and the design of particle detectors, tasks for which only a smaller portion of physicists would be required, would have allowed to gain experience immediately applicable in nuclear technology.

Bearing in mind however, that at the time research was prohibited by laws of the Allies and that practically no experimental nuclear physics existed in the Federal Republic, one has to conclude that even a limited benefit is a benefit. In any case, that must have been the general feeling among the persons concerned and Heisenberg's argument was certainly not simply a pretext.

In his letter of 23 December 1951 to the Secretary of State, Heisenberg pleaded for a participation of the Federal Republic in the 'second phase' of the work, i.e. the planning of the European laboratory for high-energy nuclear physics.

> I feel that in the name of the DFG I should plead for Germany's participation at least in the second phase. However, in view of the existence of several UNESCO projects, it may be appropriate, before taking a final decision, which, as said earlier, must be taken before 12 February 1952, to have another consultation within the DFG. Should Germany envisage to participate only at a later stage, it would in any case be much more difficult to have a say in the project.

The consultation, if there was any at all, must have taken place within a small circle. On 31 January 1952, Ludwig Raiser wrote to the Secretary of State 'to stress the DFG's interest in the project'.[83]

As early as 30th January, a letter to the Federal Ministry of Finance had been prepared at the Foreign Office as a draft, asking for agreement to the payment of about $ 36,000 for the '2nd phase of the project now starting'. There were two problems to be solved here: should the Federal Republic take part in the work at all, and, if so, which ministry should be responsible.

The above mentioned draft once again summarizes the arguments in favour of the Federal Republic's participation, from the Foreign Office's viewpoint:

> In the opinion of the Foreign Office, the Federal Republic's participation, on an equal footing, in this joint European work on nuclear physics is urgently necessary, not least in view of the restrictions on Federal German atomic research still imposed by the Allies. Moreover, it is essential for a country with the scientific standing of Germany, which has demonstrated its desire for active cooperation in international cultural life by its membership of UNESCO. German participation will benefit both European scientific collaboration and, to at least the same extent, German science and probably also the German economy. Here it must above all be remembered that the Federal Republic's participation will at last be able to provide the so pressingly necessary training for future generations of German scientists. This is all the more necessary as the present restrictions on German atomic research permit no adequate training facilities within the Federal Republic and as the construction of Germany's own machines, even if permitted, would require such vast sums of money as to render it unthinkable.[84]

Thus Heisenberg's arguments were largely adopted. Viewpoints naturally following from the document quoted include the 'important export prospects' for German industry already mentioned by Heisenberg and the possibility of exerting some influence on the choice of the leading staff of the future organization.

Hence, the Foreign Office clearly supported the Federal Republic's participation. Would it also have the responsibility for the planned European laboratory? According to the distribution of responsibilities, science, insofar as it fell under Federal authority, was the province of the Federal Ministry of the Interior and within it, that of the cultural department led by Secretary of State Erich Wende.

On 29th January 1952, Rudolf Salat, head of the cultural department of the Foreign Office, sent Erich Wende a letter couched in cordial terms:

> The idea has been propounded in my department to include, under the special section individual plan IVa, chapter 3, section 44 of the Foreign Office's budget for the costs

> of international UNESCO work, amounting in the current fiscal year to DM 1,250,000, the outlays for special projects abroad planned by UNESCO with German participation, too, in the future. In this way all contributions to UNESCO which have to be sent abroad in the form of foreign exchange, would be uniformly charged under one budget heading of the Foreign Office. However, the UNESCO projects within Germany itself, e.g. the three institutes and the German UNESCO commission, would, of course, remain in the budget of the Federal Ministry of the Interior. This inclusion for accounting purposes of the outlays on UNESCO centres in the budget of the Foreign Office would not, of course, prevent the Ministry from informing the Federal Ministry of the Interior further about projects like the Nuclear Physics Laboratory, the computer centre, etc., and from seeking its collaboration.[85]

Following the basic agreement of the Federal Ministries of the Interior and Finance, the latter asked 'that, owing to the political significance of the matter, the Federal Chancellor should inform at the Tuesday meeting of the Federal Cabinet, the Federal Ministers of the intention [...] to join the agreement on the European atomic research centre. Apparently the budget committee inquires in such cases whether the Federal Cabinet has been given the opportunity to take up a position concerning the project before the binding signature of the agreement'.[86]

At its meeting on 12th February, Secretary of State Hallstein informed the Cabinet 'that Professor Heisenberg will represent the Federal Republic at the [...] conference [...] to be held in Geneva from 12th to 15th February, 1952, [...] and will be authorised to announce that the Federal Republic will join the proposed agreement'.[87] Hallstein 'did not fail' to put the good arguments for German membership before the Cabinet. It agreed.

As a result, on 15 February 1952 Heisenberg could put his signature under the 'Agreement'[88] without the saving clause added by the delegates of most of the other countries 'subject to ratification' or 'sous réserve de ratification'.[89]

But before that, quite a few obstacles had still to be overcome. The working group constituted at the Conference in Paris submitted on 12 February 1952 in Geneva a draft agreement to which Great Britain, Denmark and Sweden made serious fundamental objections, directed, as they already had in Paris, against the creation of a new research centre in Europe.

Once more Heisenberg tried to find a compromise. His report to the Foreign Minister reads:

> The German Delegation, but also the representatives from Yugoslavia, the Netherlands and Greece had an important part to play as mediators. Particularly during the negotiations in the Commission they looked for a compromise to conciliate the opposites. We, i.e. Dr. Hocker and myself, in line with our position already adopted in Paris, declared that we were in favour of creating of a new research center in Geneva but also of constituting the study groups proposed by the experts, in particular a group for theoretical research in Copenhagen. We stressed that the theoretical research group should work in close connection with the other study groups and with the council of national representatives. This proposal was reflected in the final text of the Agreement.[90]

On 29 February 1952 Heisenberg reported before the Senate of the Deutsche Forschungsgemeinschaft 'on the European project of a facility for nuclear physics which had been the subject of the negotiations in Paris (December) and in Geneva (February)'.[91] The meeting was attended by all four Vice-Presidents[92] and 24 out of the 27 members of the Senate. It was clearly a forum of the leading representatives of all branches of German science.

Heisenberg reported to his German colleagues on the two rivalling groups; then, in the indirect speech of the minutes, we read the following astounding statement: 'According to the instructions received from the Foreign Ministry, the German Delegation supported the younger group.'

Then as now it was not a matter of course for a German savant to accept an 'instruction' concerning his own field, even if there were a political dimension to it, and to admit such before colleagues. Was such an 'instruction' indeed issued? And who issued it? There is no mention of it in the Foreign Office's files. It is nevertheless possible that an 'instruction' or recommendation was given orally.

If there was indeed such an 'instruction', it could have been issued only by the Federal Chancellor himself or by Secretary of State Walter Hallstein, for only they had the necessary authority over Heisenberg.

At the time Konrad Adenauer held two offices, that of Federal Chancellor and that of the Federal Foreign Minister. Heisenberg had previously dealt with him in his capacity as President of the Deutsche Forschungsrat; now he talked to him on the matter of the future use of nuclear power.

Walter Hallstein had been ordinary professor for private and company law at the University of Frankfurt when Adenauer called him first into the Chancellor's Office, appointing him as Head of the German Delegation in the negotiations concerning the Schuman Plan.[93] Jean Monnet, the actual initiator of the Schuman Plan, found in him a congenial partner. As Secretary of State in the Foreign Office, Hallstein was in fact the Head of the Ministry, and many regarded him as the future Foreign Minister. He kept the course of the Federal Republic's foreign policy 'persistently and imaginatively towards integration'.[94]

It is therefore possible, although in our view not absolutely probable, that Adenauer or Hallstein did make a certain recommendation. We tend rather to consider that Heisenberg used the term 'instruction'[95] in order to avoid long discussions with his colleagues.

11.4 The ratification

When Heisenberg reported on the 2nd Council Session held in Copenhagen (20–22 June 1952), he expected the Convention to come into force in September 1953 and therefore recommended the provision of 1.8 million Deutschmarks for the fiscal year running from 1 April 1953 to 31 March 1954.[96] The Federal Foreign Office, as the Ministry responsible, was prepared to take the necessary steps for the Federal Republic's accession and to 'include an amount for the participation in the European project in its 1953 draft budget'.[97] However, the idea was dropped for tactical reasons, since it was thought in the Foreign Office that 'a decision concerning the participation of Germany would be taken in May 1953 at the earliest', and 'that the Bundestag should not have occasion to reject this item on the grounds that the Federal Republic did not yet have a binding commitment'.[98]

In his letter of 20 December 1952 to Hallstein, Foreign Secretary of State, Heisenberg pressed for a 'statement in principle by the Federal Government': a corresponding declaration 'was expected from the German Delegation at the next Council Session (Brussels, 12–14 January 1953)'. However, the Secretary of State wrote to Heisenberg on 12 January 1953 that he was not in a position 'to make a final statement on behalf of the Federal Government at this stage':

> Even a decision in principle for the Federal Republic to participate in European atomic research activities—however desirable it seems to me for many, well-known reasons—cannot be taken without the approval of the Federal Ministry of Finance, owing to the significant financial obligations involved [...] I cannot submit the project to the Federal Ministry of Finance until I have received detailed documents relating to the financial commitments [...]. I hope however that this point will be clarified at the forthcoming Council Session [in Brussels...]. I hardly think I need to assure you again of the Federal Government's interest in this collaboration in accordance with the wishes of German scientific circles and also of German industry.[99]

Heisenberg thought that, on the whole, the letter sounded 'relatively positive'.[100] The Session in Brussels did indeed bring the desired 'clarity' concerning the member states' financial commitments. On 14 February 1953 Heisenberg submitted the 'Report on the 4th Session of the European Council for Nuclear Research' (prepared by Alexander Hocker), saying 'that the cost of the project is estimated to a total of 27.3 million dollars, i.e. 3.9 million dollars annually over a period of seven years with subsequent costs of 1.5 million dollars per year':

> According to the proposals made at the last Session, Germany would have to contribute 17.78% of the above sums if the United Kingdom participates in the project. I would be grateful if a statement by the Federal Republic could be made as early as possible.[101]

Hocker, like Heisenberg, put pressure on the Foreign Office. As the Deputy Director of the Deutsche Forschungsgemeinschaft in Bad Godesberg[102] he visited the Foreign Office on several occasions. After a three-and-a-half hour discussion there on 10 March, he wrote, on the following day, a 'Note' that was to be used as a basis for the discussions with the Minister of Finance and the Cabinet, although it 'should actually have been written by the official responsible'.[103]

Previously, at the meeting of the 'Kommission für Atomphysik' held on 28 February 1953, Alexander Hocker had asked for express confirmation of the interest of the German scientific community:

> Mr. Hocker asked the Kommission to state whether German participation in the Geneva Project, which would require an annual contribution of 2.5-3 million Deutschmarks over a period of six to seven years, could be justified from a scientific point of view. Mr. Haxel, stressing the scientific interest of the project, said that in his opinion the political interest of German participation would be of even greater weight. Mr. Heisenberg likewise felt that the expenditure could be justified not only on the grounds of the project's scientific importance but also more particularly from the point of view of European co-operation.[104]

Before we continue with the sequence of events, let us say a word about the 'Kommission für Atomphysik'. It was set up on 29 February 1952 by the Senate of the Deutsche Forschungsgemeinschaft (DFG), with the aim — as the DFG's official report put it — 'of advising the Government and the German Delegation in the negotiations concerning the Federal Republic's participation in the construction of a European Institute for Nuclear Physics'.[105]

The activities of the Kommission, however, went far beyond this task; it was also concerned with the organization of nuclear physics and nuclear technology in the Federal Republic. Moreover, the 'Kommission für Atomphysik' was, as long as it existed,[106] the forum where German physicists, in the confidential atmosphere of a small circle, would exchange and coordinate their opinions and determine the position to be taken in forthcoming negotiations.

It might be of interest to point out that in one case at least, i.e. at the above-mentioned 4th meeting of the 'Kommission für Atomphysik' on 28 February 1953, in which the utilization of atomic energy in the Federal Republic and the German position in the CERN negotiations were discussed, the proceedings marked 'secret' fell into unauthorized hands. An internal British report of 17 April 1953 contained a summary of these proceedings, and the following comment: 'The Germans do not know that we are aware of these proceedings [...] and it would be most undesirable for them to discover our knowledge of the matter.'[107] We do not know whether on the British side this knowledge of the German position actually had any influence or to what extent.

When Alexander Hocker drafted his four-page Note to the Foreign Office, he based himself on the vote taken by the 'Kommission für Atomphysik'. In this Note Hocker explained the scientific, economic and political importance of the project. The scientific arguments were put in the foreground. 'Both for the theoretical and the experimental nuclear physicists', the Note said, 'work at an accelerator that gives an insight into the phenomena of cosmic radiation has now become indispensable.' What that 'insight' meant for the understanding of nature, the Note didn't say, nor did it explain why large accelerators were 'indispensable'.[108]

The decisive argument was: 'The construction of large accelerator facilities is so expensive that no European country, and in particular the Federal Republic, can afford to build such a laboratory on its own. Hence from a financial point of view the project virtually demands European co-operation'.[109]

Notes: pp. 425 ff.

After touching on the project's importance for young scientists, Hocker then examined the economic interests. He only spoke about orders that might be placed with German industry without mentioning (as Heisenberg had done in his letter to Hallstein of 23 December 1951) the possibility of developing important technical applications from high-energy nuclear physics.

Finally Hocker considered the aspect of European co-operation from which, as he wrote, one should not stand aloof, 'as the close contacts between European physicists benefitted this co-operation' which for the first time 'showed tangible and promising signs of taking shape'.

'Please don't be put off by the somewhat exaggerated wording', Hocker said to Heisenberg, the leader of the German Delegation, 'it was chosen in order to produce the desired effect':

> The reason why the scientific interest is mentioned first, before European co-operation (contrary to what had been agreed in the Kommission für Atomphysik) is that 'European co-operation' is no longer so effective with the Federal Ministry of Finance, because for months it has been used as an argument for practically everything.[110]

The Foreign Office immediately adopted the arguments elaborated by Alexander Hocker. The 'Note' of 11 March 1953 became the principal document in the ratification process.

On 17 March 1953 the Foreign Office's budget department drafted a proposal to the Federal Minister of Finance. When it was discussed the officials of the Ministry of Finance expressed reservations 'concerning immediate inclusion in the budget [...] in view of the budgetary regulations (Reichshaushaltsordnung), in particular paragraph 45b (authorization of the expenditure by the legislative bodies); they suggested that the Cabinet be asked for a decision concerning the Federal Republic's accession'.[111]

The Foreign Office immediately drafted a paper for submission to the Cabinet which contained the scientific, economic and political arguments presented by Alexander Hocker using the phrasing of his Note of 11 March. In the subsequent discussions between the Foreign Office, the Federal Ministry of Finance and the Federal Ministry of the Interior Germany's accession was clearly beyond dispute. However views differed significantly as to which ministry should be responsible, and hence whether additional funds should be provided for the European Laboratory or merely taken from the special research funds already granted by the Federal Government.

On 27 March 1953 the Federal Cabinet examined, at its 28th Session, under item 3 of the agenda, the accession of the Federal Republic and on the same day a telegram was sent by the Foreign Office to the German Embassy in Rome: 'Following instructions by Secretary of State please inform Professor Heisenberg by member of the Embassy on Monday 30 March 10.00 a.m. at beginning of meeting of European Council for Nuclear Research [...] that Cabinet has approved the Federal Republic's signature to the Convention for the Establishment of the European Organization for Nuclear Research'.[112]

The ratification

At the Cabinet meeting, the Federal Minister of Justice was asked to examine whether the Convention 'required the approval of the legislative bodies under Article 59, para. 2 of the Basic Law (Grundgesetz)'.[113] The Federal Minister of Justice could give no clear answer; nevertheless 'in agreement with the Foreign Office' he considered it 'necessary to deal with the Convention in accordance with Article 59, para. 2 and initially to sign it subject to ratification'. Heisenberg was given corresponding powers signed by the Federal President and the Federal Chancellor and on 1 July 1953 he signed, in Paris, the 'Convention for the Establishment of a European Organization for Nuclear Research', subject to ratification.

Ratification took some time. First of all new elections were held on 6 September 1953 for the second term of the Bundestag; on 20 October the new Federal Government was set up.[114] At the beginning of December the Foreign Office prepared a bill. The 'Grounds' given in an annex took over almost the exact wording of Hocker's Note of 11 March 1953.

Progress was considerably hampered because of another dispute between the Foreign Office and the Ministry of the Interior concerning the question of competence. Again the Federal Minister of Finance supported the Federal Minister of the Interior. The latter had already been granted 10 million Deutschmarks of special funds for the promotion of research and, according to the Minister of Finance, if the Federal Ministry of the Interior was to become the ministry responsible, the contribution to CERN could be paid out of these funds, in other words without placing a further burden on the Federal budget.

On 7 December 1953 the Cultural Affairs Committee of the Bundesrat under the chairmanship of Nordrhein-Westfalen's Minister of Culture Christine Teusch met in the Parliament Building to discuss the budget for 1954:

> Heading 676 (contributions to cultural organizations other than UNESCO). The only point examined under this item concerned the Federal Republic's contribution to the European Organization for Nuclear Research which gave rise to a detailed discussion. Minister Teusch reported that the Deutsche Forschungsgemeinschaft was worried about the plan of the Federal Minister of Finance to take the necessary 3,000,000 Deutschmarks from funds which had been attributed to the Federal Ministry of the Interior for German fundamental research. The Federal Ministry of the Interior and the Foreign Ministry declared that the question of competence was still not solved. I myself [Rudolf Salat] fully agreed with Minister Teusch who wished the contribution for this Organization to be allocated from the Foreign Ministry's budget. The Committee however disregarded this [...], not being in a position to propose financing.[115]

On 14 December 1953 a meeting took place in the Federal Chancellery at which the questions of competence and financing were to be clarified. The representatives of the Foreign Office were forced on the defensive, as the Ministry of Finance again supported the Ministry of the Interior and the representative of the Ministry of Finance also used as an argument the claim, 'that he had been present during a conversation in which the Secretary of State, Hallstein, had conceded to the Federal Minister of Finance that this amount should be covered by the Federal Ministry of the Interior'.[116]

In the Foreign Office files there is still a 'Note for the Secretary of State' from the budget department. In it, it is proposed 'not to question the provision made for the contribution in the Minister of the Interior's budget planning'.

Did that mean that the Minister of the Interior had won? A note in the files of the Foreign Office of 22 January 1954 indicates that, evidently at the very last minute, Heisenberg, in a private conversation with the Federal Chancellor, succeeded in getting a different settlement accepted. The opportunity for this conversation presented itself on 10 December 1953, when Heisenberg went to the Federal Chancellery in order to receive from Adenauer his letter of appointment as the President of the Alexander von Humboldt Foundation:

> The Federal Chancellor is said to have promised Professor Heisenberg at the inaugural ceremony for the Humboldt Foundation that the contribution would be paid, that the responsibility would remain with the Foreign Office and that the amount in question would be granted as a supplement and not be taken out of the 10 million Deutschmarks which the Federal Ministry of the Interior had been allocated to promote priority programs in German scientific research and which the Forschungsgemeinschaft badly needs.[117]

Two of the three promises mentioned — namely that the amount would actually be paid and that it would not be taken from the funds already granted for research programs — were undoubtedly matters of personal concern for Heisenberg. However, it is not certain whether he also advocated that responsibility should remain with the Foreign Office and if he did, whether it was merely for pragmatic reasons. If the Ministry of the Interior became the department responsible, there was a risk that the special funds shown in its budget for the promotion of scientific research would be entirely or partly used for the payment of the German contribution to CERN.[118]

After the clarification by the Federal Chancellor, the cabinet paper already prepared by the Foreign Office ('Draft Law concerning the Convention of 1 July 1953 for the Establishment of a European Organization for Nuclear Research') could at last be submitted. On 24 February 1954 the Federal Cabinet approved the bill and, in accordance with the Basic Law, transmitted it first to the Bundesrat expressly stating that the Foreign Office was the Ministry responsible.[119]

On 19 March the 'Draft Law concerning the Convention of 1 July 1953 [...]' was called as item 23 on the agenda of the 120th meeting of the Bundesrat and it passed without a report being read or further discussion.[120]

Now the way was open to submit the bill to the Bundestag where the subject was examined on 7 April 1954 in the budget debate when Foreign Office business was discussed. The experts attending as observers were surprised to hear talk about an 'International Committee for Nuclear Research in Berne' instead of the 'European Organization for Nuclear Research in Geneva' as would have been correct:

> The contribution for the International Committee for Nuclear Research in Berne was the subject of a lively debate when the expenditure on scientific research within the Ministry of the Interior was discussed. The Minister of Finance had originally intended to deduct 3 million Deutschmarks from the 10 million DM earmarked for the priority programs in scientific research. The Budget Committee did not follow his proposal but granted a separate amount of 3 million DM for this International Committee leaving the 10 million DM for research programs untouched.[121]

The first reading of the 'Draft Law concerning the Convention of 1 July 1953 for the Establishment of a European Organization for Nuclear Research' took place at the 26th Session of the 2nd German Bundestag on 29 April 1954.[122] The members had received document 394 consisting of the text of the law and detailed justifications.[123] These justifications essentially used almost the exact wording of Hocker's arguments in his Note of 11 March 1953. The bill was transmitted without discussion to the Committee for Foreign Affairs.

On 14 June the Committee adopted the bill unanimously and without debate.[124] On 8 July the plenary Bundestag gave the bill its second and third readings.

The rapporteur, Fürst von Bismarck, stated that 'last Tuesday [on 6 July] the French National Assembly adopted the bill of ratification by a large majority so that, with the German ratification, the legal and financial conditions for the realization of the plans are now fulfilled. The Bundesrat has adopted the bill. May I ask the honourable members of the House to give their assent'.[125]

The bill was unanimously adopted without discussion.

In accordance with the Basic Law, the Bundesrat examined the act once again on 23 July 1954. No objections were raised.[126]

In the meantime Edoardo Amaldi, Secretary-General of the provisional Organization, had become seriously worried. Under its Article XIII the Convention could only enter into force when, in addition to Switzerland as the host country of the planned laboratory, six more countries had deposited their instruments of ratification with the Director-General of UNESCO and if the total of these seven countries' contributions amounted to not less than 75% of the overall budget contributions. This point was exactly the problem.[127]

In France the law concerning the country's accession to the new Organization had come into force on 13 August, and in early September Amaldi received the assurance that the instrument of ratification would be deposited before the end of that month. However Italy, whose contribution amounted to 10.20%, had not yet ratified the Convention and was not expected to do so in the near future, so Germany's accession, with a contribution of 17.7% of the total budget, would be decisive for the Convention to come into force.

On 9 September 1954 Amaldi wrote an urgent letter to Wolfgang Gentner: 'I have no news at all about the progress of the procedure in your country and I am really worried about this question'.[128]

Amaldi sent a copy to Hocker who replied:

> As you know, Parliament approved the Convention before the summer holidays. Unfortunately the necessary formalities take some time; they were delayed because several ministers, the Federal Chancellor and also the Federal President whose signatures are required, were on leave in August. The official in charge in the Foreign Office assured me again today that the instruments would be deposited in Paris before the end of this month. I therefore think that you should not change the date of the Session scheduled for October. Now that I am back again, I can see to the matter myself. I told Preiswerk the same this morning when he called me from Zürich.[129]

The Federal President put his signature to the bill on 17 September 1954; it entered into force on 29 September 1954, i.e. one day after its publication in the Federal Gazette.[130]

For the new Organization, however, only the deposit of the instruments of ratification had legal validity. France and the Federal Republic deposited the instruments on the same day, i.e. on 29 September 1954, thus ensuring the required quota of 75% of the total budget, allowing CERN finally to come into being.

It is very likely that the two countries came to some arrangement. The delaying of the deposit by France where the law had already come into force on 13 August and the unusual speeding up of the process by the Federal Republic which deposited the instruments of ratification on the earliest possible date, namely on the same day the law came into force, speak for themselves. As a matter of fact, a telegram exists which Alexander Hocker sent to Robert Valeur on 24 September in which the deposit of the instruments is mentioned.[131] But we do not know of any documents relating to the detailed arrangements (probably between the Quai d'Orsay and the Foreign Office).[132]

Let us summarize: Although for a time Heisenberg had been worried whether the Federal Government would really decide in favour of accession, following the clear statements of intent made by the scientific community, all political camps had been ready to give their support. The only question was whether additional funds had to be found or whether the expenditure should be met from the resources already allocated for research. Linked to this point was the question which ministry would be responsible. From the documents that still exist one gains the impression that the officials in the ministries were mainly concerned with this problem.

Considering the tenacity with which the officials protected the interests of the Foreign Office, it is surprising how easily responsibility was handed over on 15 October 1955 on the creation of the Ministry for Atomic Affairs. On 24 April 1956 Franz Josef Strauss wrote to the Foreign Office:

> From the fiscal year 1956 onwards, in agreement with the Federal Ministry of Finance, the budget funds for the contribution to CERN shall be shown in budget plan 31 (Federal Ministry for Atomic Affairs)...

The answer simply stated that the Foreign Office accepted the new 'arrangement' as 'CERN is an organization whose aims are purely scientific'.

Even today the German contributions to CERN are shown in the budget of the Federal Ministry for Research and Technology (BMFT) which succeeded the former Federal Ministry for Atomic Affairs. In 1984 the German contribution amounted to 215 million Deutschmarks, i.e. 24.79% of the CERN budget.

11.5 German positions

On the road that led from the first thoughts of European scientific co-operation to the ratification of the Convention on 29 September 1954, when CERN after many birth pangs was launched, there was a constant need for decisions. It was frequently the case (although not the rule) that in some countries special wishes were formulated and again and again the founding fathers showed their skill at finding reasonable compromises.

In the early fifties (as already explained) the Federal Republic of Germany was in a weak position politically, and the Germans generally knew when not to press their views. But even when they had no particular objectives of their own, they had to take a position on those of other countries. To give an example, we shall describe the German position on the siting of the Organization, the related problem of the Theoretical Group and the recruitment policy.

11.5.1 THE SEAT OF THE PLANNED LABORATORY

On 31 May 1952 the Foreign Office asked the Forschungsgemeinschaft whether it considered that the Federal Republic should propose a site in Germany for the planned laboratory. During the negotiations in Paris (17–21 December 1951) and in Geneva (12–15 February 1952) Heisenberg had gained the impression that 'the majority of the government representatives have already decided in favour of Geneva as the seat of the laboratory'. He therefore thought that a German candidature had no chance of being accepted, 'and that it would therefore be preferable to drop it, particularly since we can be quite satisfied if the Laboratory is located in Geneva'.[133]

The Deutsche Forschungsgemeinschaft associated itself with this view.[134]

In the middle of September 1952 the Ambassador of the Netherlands in Bonn contacted the Federal President and the Foreign Office asking them 'to support the proposal of the Dutch Government to establish the International Laboratory for Nuclear Research in Arnhem'.[135] On 24 September the Foreign Office telephoned Heisenberg to inform him and asked him, 'to give favourable consideration to the Memorandum submitted by the Dutch Ambassador'.[136] Heisenberg pointed out 'that the Foreign Office had already sent him a similar proposal from the Swiss Government with the same recommendation'.[137] He asked the Foreign Office to clarify its position which it did on 26 September:

Notes: p. 427

> In our opinion the choice of the location of the Laboratory should primarily be based on practical considerations. Thus the final decision as to which site Germany should vote for must be left to you. Should the preliminary talks reveal aspects of foreign policy, I would ask you to contact the Consulate-General of the Federal Republic in Amsterdam [...], through which if necessary coded telegraphic messages and enquiries could be transmitted to us.[138]

In his 'Report on the 3rd Session of the European Council for Nuclear Research' in Amsterdam, Heisenberg said that Denmark, France, the Netherlands and Switzerland had proposed sites for the laboratory.

> The situation was difficult in as much as the question of the site had been linked with the problem of the future financing of the Theoretical Group in Copenhagen. The risk was that Sweden and Denmark might declare they were no longer interested in co-operation at all, unless the Atomic Centre in Copenhagen, traditionally the most important centre of its kind in Europe, were in some way or another included in the plans. I therefore strongly urged continuing the support given to the theoretical studies in Copenhagen. After a long debate it was decided that, in addition to the construction of the two machines and the laboratory, the theoretical studies would also be continued with the Council's support. Finally France declared that she was prepared to withdraw her candidature for the site of the laboratory if the Council agreed on Geneva. Italy also showed a preference for Geneva. In view of the decision mentioned earlier concerning the continuation of the theoretical studies in Copenhagen, Denmark withdrew her candidature for the laboratory's site [...] I declared on behalf of the Federal Republic, that, in Germany's view, all the proposals were equally acceptable, although the Dutch proposal had for us geographical advantages. France and Denmark having withdrawn their proposals, both the Dutch and the Swiss proposals were very acceptable to the Federal Republic. However, in order to achieve a unanimous decision by the Council on this point, preference should be given to the Swiss proposal.[139]

11.5.2 THE THEORETICAL GROUP

At the first session of the provisional Council held in Paris (5–8 May 1952) the leaders of the four groups were designated. 'Unanimously and by acclamation' the Theoretical Group was entrusted to Niels Bohr; thus it became based in the Institute for Theoretical Physics in Copenhagen.

As quoted, Heisenberg had explained in his 'Report on the 3rd Session of the European Council in Amsterdam', that the question concerning the location of the planned laboratory was linked with the financing of the Theoretical Group in Copenhagen. Heisenberg pleaded in favour of 'continuing the support given to the theoretical studies in Copenhagen'.[140] He probably did so for two reasons: Firstly, he wanted to avoid the risk (already mentioned) that 'Sweden and Denmark might declare they were no longer interested in co-operation at all, unless the Atomic Centre in Copenhagen, traditionally the most important centre of its kind in Europe, were in some way or another included in the plans'.[141] Secondly, Heisenberg was himself a theoretical physicist, and what is more, for many years he had collaborated on friendly terms with Niels Bohr in Copenhagen.[142] So it must have been a matter very dear to his heart to help stimulate the theoretical studies under the direction of Niels Bohr in the context of the planned European co-operation.

At the 4th session of the European Council in Brussels the location of the Theoretical Group was discussed in depth. It was the wish of the Scandinavian countries that Copenhagen should remain the official seat of the Theoretical Group for five years, whereas the majority of the other delegations felt that it was important 'to have a strong Theoretical Group in Geneva'. Probably to deflect confrontation Heisenberg stressed that even more important than the location was 'the enlisting of a senior theoretical physicist for the Theoretical Group as soon as possible, who would also be interested in the large machines and in their potential results. He should work full-time in the group.'

> It therefore seemed appropriate to me to adopt Sweden's proposal and designate Copenhagen as the seat for the Theoretical Group for a period of approximately five years. What should happen at the end of this period could be decided at a later stage. I also pointed out that the work of the Theoretical Group was very inexpensive compared to the expenditure on the other activities. It would take up only a very small percentage of the total budget.[143]

One and a half months later a meeting was held of the 'Kommission für Atomphysik' at which Heisenberg reported to his colleagues on the status of the negotiations:

> Mr. Heisenberg invited the Kommission to examine whether it would not be appropriate to propose to the European Council for Nuclear Research that 10% of the member states' contributions to the European Organization should be used for activities outside Geneva (balloon experiments, work on the machines in Liverpool, Uppsala, etc.). The possibility of co-operation in such fields was provided for in the draft Convention. Although it might be useful for the experimental teams to move to Geneva directly, this did not apply to the same extent for the other scientists taking part in 'other forms of co-operation' [...] After further discussion the Kommission decided to invite the Deutsche Forschungsgemeinschaft to propose to the Foreign Office that the Convention be amended so that 10% of the budget could be used for other forms of co-operation outside Geneva.[144]

This decision was indeed adopted, via the Governing Body of the Deutsche Forschungsgemeinschaft and the Foreign Office, as an official proposal of the Federal Republic to the Council. Heisenberg had expected that his proposal would be supported by the three Scandinavian countries, the United Kingdom and Italy. After clarification of the financial aspects, the motion was in fact approved unanimously by the Council.

The clarification established that 'it is clearly not 10% of the total budget but an amount corresponding to 10% of the *current* expenditure'.

> The budget committee [...] estimated the current expenditure from the eighth year onward at 9 million Swiss francs, and thus the 10% financing for the other forms of co-operation amounted to 0.9 million Swiss francs per year, i.e. at 6.3 million Swiss francs for the first seven years.[145]

This amount was included in the cost estimate.[146] 6.3 million Swiss francs were explicitly earmarked for 'theoretical studies and other forms of co-operation'.

As we have said the first director of the Theoretical Group was Niels Bohr; on 1 September 1954 Christian Møller officially took over this office. From the very first meeting of the Scientific Policy Committee, Felix Bloch, the Director-General, emphasized 'that theory should be gradually moved to Geneva in the course of the coming few years'.[147] The Committee, chaired by Heisenberg, recommended that a theoretical group should be created in Geneva too, 'which would be able to co-operate more closely with experimentalists than the theoretical group in Copenhagen'.[148]

On 1 October 1957 at the end of the five-year period, the work of the Theoretical Group in Copenhagen was brought to an end. On the same day the Scandinavian countries set-up a joint research institute, NORDITA, which was located in Copenhagen.

11.5.3 GERMAN INFLUENCE AND COLLABORATION

In his report on the first negotiations in Paris (17–21 December 1951) Heisenberg had urged that the Federal Republic should co-operate in the planning stage, using the argument that it would otherwise be difficult 'to influence the project'.[149]

Heisenberg himself certainly had great influence.[150] It was not immaterial for the success of CERN that at the first Council Session in Paris (5–9 May 1952) Heisenberg was appointed chairman of a committee whose mandate was to prepare a report 'on the future work of the individual groups, and in particular on the size and the energies of the accelerators proposed'.[151] To a degree this report was the intellectual basis of the whole enterprise.[152] In the report Heisenberg confirmed that 'during the last few years the focus of interest had shifted from nuclear physics to particle physics', and explained the reasons why it was necessary to build accelerators for very high energies.[153]

When the Scientific Policy Committee (SPC) was set up at the first session of the Council after the Convention had come into force, the delegates elected Werner Heisenberg its chairman. He held this office until the end of 1957. The task of the SPC was to assess the research carried out inside and outside the Organization and to make proposals concerning the research activities.[154] The Committee was undoubtedly 'a very important body'.[155]

The second German delegate Alexander Hocker, was as such, a permanent member of the Finance Committee, a body which he chaired in the last year of his mandate.

But what was the situation regarding CERN staff members? In January 1954 Heisenberg noted with satisfaction that 'in spite of the shortage of young scientists in nuclear physics, German participation in the Working Groups' were very active:

> Professor Paul (Bonn) continues to co-operate very actively in the *Synchrocyclotron Group*. Dr. Beyerle (Göttingen), head of the Institut für Instrumentenkunde of the Max-Planck-Gesellschaft has replaced the late Dr. Gund (Erlangen) as technical adviser. Professor Gentner (Freiburg) plays an important part in the planning the *Proton Synchrotron Group*. Full-time staff members of the group include Dr. Schmelzer (Heidelberg), Dr. Lüders (Göttingen), Dr. Citron (Freiburg) and Dr. Geibel (Heidelberg). In the *Theoretical Group* Dr. Haag (München) has taken over from Dr. Lüders (Göttingen)... For the work on the synchrocyclotron in Liverpool Dr. v. Gierke (Heidelberg) has been selected in addition to a Dutchman.[156]

A few months later however Heisenberg complained that the Federal Republic was not adequately represented. His countrymen working with him for the realisation of the laboratory shared this opinion, and it is on their exchange of letters that we based our study.

Reading these letters now (1986) we have to bear in mind that the events could possibly be interpreted differently, and also that there were similar complaints in other countries (for example in France) about an under-representation among the leading positions.

The first German concern was the Laboratory Group. Heisenberg and Gentner believed that Kowarski did not want to take any Germans in his group. In a conversation with Wolfgang Gentner on 9 April 1954, Kowarski used the 'poor excuse' (Gentner), 'that he did not know whom to contact in Germany when there were vacancies in the Laboratory Group'.[157] Thereupon Heisenberg wrote an official letter naming Gentner as the person to contact.[158]

When Frank Goward, who, as Odd Dahl's deputy, had to a large extent been in charge of the PS Group, suddenly died of a brain tumour on 10 March 1954 and Dahl declared that he was not able to invest more than 30% of his time in the work at Geneva, 'the group's management had to be reorganized and a new group leader to be elected'.[159]

Wolfgang Gentner was an official adviser to the PS Group and on 22 and 23 March he discussed the situation with Amaldi, Dahl and Adams:

> We thought that a good temporary solution would be to have Adams run the group with the assistance of Dahl and myself [...] During the discussions Adams suggested by way of a permanent solution that Schmelzer be proposed to the Council as group leader. Particularly from the English side it was underlined that Schmelzer [...] was highly esteemed and very popular. He is also the oldest among the senior physicists working here.[160]

However, in the 9th Session of the provisional Council in Geneva (8-9 April 1954) Robert Valeur objected to the appointment of Schmelzer as Director of the PS

Group. In the view of Odd Dahl 'this was not a personal objection but rather that the Group all the time had insisted on a man with project experience. If the experience of Schmelzer is considered adequate, then the French insisted that they could find just as good or better men in France'.[161]

The de facto arrangement was therefore maintained for the time being with John Adams replacing Goward as Deputy Director. 'Adams was worried', reported Christoph Schmelzer, 'that I might have been too much affected by the whole matter':

> It is a strange feeling to realize that in this European Organization there are first and second class people. As a matter of fact Valeur's arguments against the man without machine, or more correctly without project experience are just a pretext: no-one with the required qualifications was available.[162]

The big machine was considered the heart of the whole enterprise, and many of the participating countries would have liked one of their men to be given the major responsibility for its construction. Since the protagonists of the project in France had already been attacked in their own country—the French having too little and the Germans too much influence[163]—it is possible that Valeur was indeed motivated by political considerations.

When the subject was discussed again several months later at Heisenberg's request, Dahl declared:

> I am getting worried about political appointments in CERN [...] It will be very difficult to build the big machine, and the years are [...] rolling by. One should therefore try very hard to convince everybody that we should only use the best men for the jobs [...] Jobs with political colour, especially in the higher brackets, will quickly have unfavourable reflections in the groups.[164]

Gentner also thought at the beginning of August 1954 that it was too late to change the arrangement: 'Adams has devoted himself with a lot of energy to this task [...] We have decided to propose from our side that at the next Council Session Adams be appointed Group Leader and Schmelzer his deputy'.[165]

The next opportunity to fill one of the senior positions at CERN with a German national came with the creation of the post of 'Chief Administrative and Finance Officer'. On 12 May 1954 Amaldi invited the Council Members to submit proposals.[166] Heisenberg immediately mentioned Alexander Hocker and reported to the Deutsche Forschungsgemeinschaft:

> So far no German has been appointed to one of the senior positions at the Institute in Geneva, and it is therefore very much in the German interest that at least the present vacancy for Director of Administration be filled with a German. However, that will only be possible if we can propose a really qualified man with many years of experience in the field of research administration.[167]

However, the post was given to S.A. Dakin (if only for a limited period initially). In a handwritten letter marked 'personal' and 'confidential' Heisenberg asked for information about what was behind the decision.[168] Amaldi replied that only personal qualifications had played a part in the choice:[169]

> I can assure that there was no objection of any type about Mr. Hocker as person or as German. A priori all the Members of the Selection Committee were in favour to the idea of having a German in such a position.[170]

Several months later, when the same post was again vacant and the German Delegation proposed Dr. Friedrich Rau, Curator of the University of Frankfurt, it became obvious that at least the newly elected Director-General still had some reluctance about a German national. Wolfgang Gentner reported:

> In my conversation with Bloch he told me that it was extremely difficult for him to support a German candidate, unless he could be entirely sure that the person in question had no compromising political past.[171]

At a meeting of the Kommission für Atomphysik on 15 December 1954 Heisenberg brought up the 'political problem' and said that 'as far as the appointment of senior staff was concerned, the Federal Republic had not been given an appropriate share in relation to its contributions':

> As yet, none of the German candidates had been accepted for the post of Division Leaders. For certain applications sent to Geneva the decision were still pending. Mr. Rau (Frankfurt) had a good chance of getting the post of the Director of Administration. Mr. Kowarski had not yet invited Mr. Straub to join his group, but he had appointed a German librarian. Several applications had been received for the post of site engineer. If Mr. Amaldi resigned as Deputy Director-General, as he had said, Mr. Bakker might be promoted to this position, always assuming that he would not then put himself forward for the post of Director-General. Although there was no mistaking the fact that in many countries feelings towards Germany were more hostile than, for example, four years ago — something he had recently noticed himself in the United States — he nevertheless wanted to write to Bloch and suggest that staff appointments to the eight top posts be discussed again. He thought that the Germans might be offered the responsibility for the construction of the small machine.[172]

And so it was. On 1 October 1955 Wolfgang Gentner became leader of the SC Division. The hope of the Kommission für Atomphysik that Gentner would also be appointed Deputy Director-General however did not come to anything. The Committee of Council no longer considered this post necessary.[173] Again, political considerations played a part. Jean Willems, for instance, had declared that he was strictly opposed to the creation of the post of Deputy Director-General but that if it were to be created, France would propose Lew Kowarski.

At first, Gentner, as a German, experienced some difficulty in his division, in particular with his Dutch collaborators, who turned directly to Bakker, the Director-General and their former group leader. However a good working climate was soon established.[174]

As a result of Heisenberg's remarks, repeated again and again since the middle of

Notes: p. 428

1954, to the effect that Germans were not yet represented in the leading positions at CERN, parliamentary questions were asked in the Bundestag on 10 September 1955. The 35 signatories belonged to all parliamentary groups. Three out of the four questions concerned the construction of the first German experimental test reactor near Karlsruhe, the fourth question related to CERN: 'How is the Federal Republic represented at the European Organization for Nuclear Research (CERN) in Geneva?'[175]

But one and a half years passed before the questions could be discussed in the Bundestag. On 22 February 1957 a member of the Bundestag, Mr. Geiger, presented the following arguments:

> The fourth question [...] is already outdated, as Professor Wolfgang Gentner, Director of the Physics Institute of the University of Freiburg im Breisgau, has long since gone to Geneva to take charge of the small machine at the European Organization for Nuclear Research. I would like to point out that at the time when the question was asked [...] those who put it had justified doubts, that Germany might not be permanently and adequately represented at this important international laboratory. Professor Gentner has only signed a contract for two years and we therefore shall soon be faced with the problem of what to do once again.[176]

The Federal Minister for Atomic Affairs, Siegfried Balke, replied to the parliamentary question.[177] He obviously wanted to reassure the members and therefore started by explaining the representation of the Federal Republic in the Council before coming to the real problem:

> The Federal Republic is represented on the CERN Council by Professor Heisenberg and by Dr. Hocker from the Federal Ministry for Atomic Affairs who is also a member of the Finance Committee. Professor Gentner from Freiburg is — as Mr. Geiger has said — leader of the Synchro-Cyclotron Division. Professor Gentner's contract will come to an end in autumn 1957 and he is being asked to agree to an extension for another year. The Federal Government will do its utmost to assure permanent and adequate co-operation and representation of the Federal Republic at CERN. We have good chances of achieving this aim.[178]

The 'permanent and adequate co-operation' referred to by the Minister became a reality. In 1964, at the suggestion of the Director-General Victor Weisskopf, CERN drew up a list of physicists, 'all first-rate scientists' who 'have worked with great success in a leading position here in CERN', and who at the time were going 'to start teaching at German universities'. This list was the following:

> Dr. H. Faissner (Technische Hochschule Aachen)
> Dr. H. Filthuth (University of Heidelberg)
> Dr. J. Heintze (University of Heidelberg)
> Dr. U. Meyer-Berkhout (DESY, Hamburg)
> Dr. G. Weber (DESY, Hamburg)
> Dr. K. Winter (DESY, Hamburg)

A second paper listed the scientific results of the bubble chamber experiments in which teams from the Federal Republic had participated. The conclusion was drawn by Weisskopf himself: 'If one compares this impressive list with the research findings of other countries, one has to conclude that the German results must be placed second and are being surpassed in quantity only by the French results. When making a judgment [...] one has to keep in mind, that this is only the beginning'.[179]

Notes

1. This chapter is based on archival sources to be found mainly in the following archives:
   - Auswärtiges Amt, Politisches Archiv, Bonn (referred to as AA-PA and file number)
   - Bibliothek und Archiv zur Geschichte der Max-Planck-Gesellschaft, Berlin (referred to as MPG), Bothe papers
   - Max-Planck-Institut für Kernphysik, Heidelberg (referred to as MPG/K), Gentner papers
   - Max-Planck-Institut für Physik und Astrophysik, Munich (referred to as MPG/P), Heisenberg papers
   - Deutsches Museum, Sondersammlungen der Bibliothek, Munich (referred to as DM), Max von Laue papers
   - Deutsche Forschungsgemeinschaft, Archiv (referred to as DFG)
   - Council of Europe, Archives (referred to as CE)
   - Centre Européen de la Culture, Archives (referred to as CEC)

   Heisenberg's reports on all sessions of the provisional Council as well as the minutes of the meetings of the 'Kommission für Atomphysik' are to be found in several of these archives, especially in the Bothe-, Gentner-, and Heisenberg-papers.

   The main secondary literature used for the political history of the Federal Republic is Schwarz (1981) and for the history of science administration Stamm (1981) and Zierold (1968).
2. Letter Heisenberg to Amaldi, 9/5/46 (MPG/P).
3. Max von Laue (1879–1960) had received the Nobel prize in 1914; quantum theory and even more so nuclear physics and the developing field of particle physics never appealed to him. What really fascinated him about the project was its political aspect. Unfortunately we were unable to find a report by Max von Laue on the conference in Lausanne. We feel nevertheless that we are allowed to conclude that 'von Laue was enthusiastic at the prospect of establishing European collaboration in physics' from Laue's biography and from what Heisenberg said in his interview with Kowarski and Gowing, namely: 'Von Laue was interested to renew the co-operation between scientists in other countries and scientists in Germany [...] Von Laue was always very much on the side of co-operation between nations.'
4. Mouvement Européen. Conférence Européenne de la Culture, Lausanne. Liste des Membres de la Conférence (CEC).
5. Schmid (1980).
6. Gazette de Lausanne, 9/12/49, p. 6.
7. Von Laue was the only physicist among the German participants. Professor Emil Lehnartz (Münster), a physiological chemist, did not register as member of the 'Commission on European Institutions' (it was in its 'Sub-Committee on Natural Sciences' that the project was discussed), but took part in the 'Commission des Echanges Culturels'. There were no other natural scientists among the German participants.
8. Interview with Heisenberg, Munich, 11/7/73, p. 1f. When the interview took place, Heisenberg did not remember any more that at the time von Laue had talked to him about the Lausanne Conference. He thought, erroneously, that von Laue had been informed by Isidor I. Rabi and concluded that the two Nobel prize winners probably met in 1950 in Lindau at the meeting of Nobel prize winners immediately

8. after Rabi's initiative at the General Conference of UNESCO on 7/6/50. However, the meetings in Lindau only started in 1951.
9. Probably the best idea of what 'nuclear physics' meant in those days, one can get from a textbook written by John M. Blatt and Victor F. Weisskopf, published 1952 under the title 'Theoretical Nuclear Physics'.
10. Eickemeyer (1953), 66. The memorandum was submitted to Chancellor K. Adenauer on 2 April 1951.
11. Goudsmit (1947), p. XI. At the time Goudsmit interrogated many German scientists.
12. Adenauer (1965), 465 and 467.
13. Schwarz (1981), 55.
14. Minutes of the 4th meeting of the Kommission für Atomphysik of the Deutsche Forschungsgemeinschaft on 28 February 1953 (MPG/P). The quotation refers to the draft prepared by Alexander Hocker, which, in this context, is clearer than the final version.
15. Professor Walter Hallstein, Professor Hans Rheinfelder and Dr. Rudolf Salat had the status of 'Advisers' to the Allied High Commission for Germany.
16. Letter Amaldi to Auger, 3/10/50, reproduced in Amaldi (1977), 349–51.
17. Letter Amaldi to Cacciapoti, 16/11/50 (CHS, Amaldi file, CERN). Here Amaldi complains about Auger.
18. Letter Auger to Dautry, 7/11/50 (CHS, Auger file, CERN).
19. 'Most of the German scientists have collaborated with the Nazis, even Heisenberg has [...] worked full blast for these scoundrels—there are a few exceptions, v. Laue and Hahn.' Letter Born to Einstein, 15/7/44, reproduced in Born (1969), 198.
20. Compte-Rendu analytique de la réunion du 12 décembre 1950 (CEC).
21. Letter von Laue to C.E.C., 16/11/50 (CEC).
22. Ivar Waller could not come for the same reason.
23. The 'Communiqué' was written in English and French and published in a shorter version (two pages) and a longer version (three pages).
24. Again two versions exist, both signed by Jean-Paul de Dadelsen: the 'Note sur la résolution du 12 décembre 1950' of 12 December 1950 and the corrected version 'Compte-Rendu analytique' of 18 December 1950 (CEC).
25. Verbali del Consiglio di Presidenza del CNR, seduta del 18/12/50 (CNR).
26. Letter Rollier to Casati and Colonnetti, 20/12/50 (CHS, Amaldi file, CERN).
27. L'Humanité, 22/12/50, p. 2. The statement was apparently correct. At a later meeting at Quai d'Orsay on 18 May 1951 a German participation was again envisaged, but at the same time it was also stated that time was not yet ripe. Francis Perrin did not attend the meeting in Geneva but he probably got all the details from Pierre Auger, who was a friend of his.
28. It was Wolfgang Gentner. See below.
29. Letter de Dadelsen to Schmid, 14/12/50 (CEC).
30. Council of Europe. Committee of Cultural Experts. Second Session. Minutes of the meeting held in Strasbourg on the 15th February 1951. CM/WP III (51) PV3, 26/2/51 (CE-19212).
31. L'Humanité, 22/12/50, p. 1.
32. The press release of the C.E.C. began: 'Pour l'étude et des développements pacifiques de l'énergie atomique, l'Europe occidentale peut organiser un "pool" '.
33. Poidevin (1976).
34. Voltaire (1966), 569.
35. La Tribune des Nations, 18/5/51.
36. This view was apparently shared by Wolfgang Gentner. He said, referring to the future: 'As soon as the question [of participation] becomes acute for Germany.' Letter Gentner to Auger, 15/11/51 (MPG/K).
37. Report on a meeting held 18/5/51. Le Ministre des Affaires Etrangères à Monsieur le Ministre du Budget, no typed date (AEF-76).
38. When Enrico Fermi on 2 December 1941 for the first time set off a fission chain reaction, a telegram was drawn up saying: 'The Italian navigator has landed.'

39. Interview with Gentner, Geneva, 2/10/74, p. 4; Adams (1982), 9.
40. Letter Gentner to Auger, 15/11/51 (MPG/K).—In his interview with Margaret Gowing and Lew Kowarski, Gentner said that when Auger had asked for the name of a German physicist, he had mentioned Bothe and Heisenberg. The letter reveals that Gentner in fact only proposed Bothe. The German experimentalists in nuclear physics considered Walther Bothe as their spokesman. Their relationship with Heisenberg was of a more distant nature. See section 11.2 below.
41. Letter Hallstein to Torres-Bodet, 8/12/51 (UNESCO).
42. From end of June until end of October 1940 Alexander Hocker had been in France as sergeant where he was temporarily detached to the military intelligence branch office in Brest. When Hocker told Heisenberg about his difficulties, Heisenberg assumed it to be the reason for the refusal of the entry visa: 'As soon as the Allies hear the word military intelligence they immediately suspect the worst.' Letter Heisenberg to Hocker, 16/1/52 (MPG/P).
43. Zierold (1968), Eickemeyer (1953).
44. 'Politisierende Nichtskönner' is an expression for people such as Johannes Stark, who were scientifically incompetent, and based their influence and power on their political connections. By direction of the Reichsminister für Wissenschaft, Erziehung und Volksbildung the physicist Johannes Stark, a national socialist, was appointed on 23 June 1934 President of the Notgemeinschaft. Even in his own field the Nobel prize winner was clearly one-sided, with no understanding at all for many other disciplines, in particular the humanities.
45. Heisenberg himself stressed on several occasions that not on any account he had wanted to develop an atomic bomb. As a result of Hitler's lack of understanding for science, no order was actually given. But just as much as Heisenberg was against the idea of having an atomic bomb for Hitler, he was interested in the success of the German reactor development.
46. Eickemeyer (1953), 17.
47. Already on 10 November 1949 an agreement between the Notgemeinschaft and the Forschungsrat had been drafted, which stated that the Forschungsrat was the representative on national level and abroad of German science. However the Notgemeinschaft's Hauptausschuss did not approve the agreement. Zierold (1978), 329.
48. Stamm (1981), 136. Kurt Zierold was at the time Executive Vice-President of the Notgemeinschaft. After the establishment of the Deutsche Forschungsgemeinschaft, he became its General Secretary.
49. Max von Laue (1961), XXXI. In his memoirs Max von Laue says that the assistance of the British Research Branch had been an essential element in the establishment of the Deutsche Physikalische Gesellschaft and in the reconstruction of the Physikalisch-Technische Bundesanstalt in Braunschweig-Völkenrode and that German Physics could not thank strongly enough Dr. Roland Fraser.
50. Stamm (1981), 136.
51. Zierold (1968), 301.
52. Idem—Article 74GG defines fields on which Bund and Länder extend their concurring legislation. Paragraph 13 mentions 'the promotion of scientific research'.
53. Eickemeyer (1953), 86.
54. On 30 September 1949 industrial circles founded the 'Stifterverband für die Deutsche Wissenschaft' which intended to give about two thirds of the donations received to the Notgemeinschaft and the remaining third to other organizations, such as the Studienstiftung des Deutschen Volkes, the Deutsche Akademische Austauschdienst and the Max-Planck-Gesellschaft.
55. Eickemeyer (1953), 86.
56. Hermann (1976), 98; Hermann (1977), 197.
57. Eickemeyer (1953), 112; Zierold (1968), 558f.—The statutes of the former Deutsche Forschungsrat mentioned expressly 'international scientific and cultural organizations' in which the DFR envisaged to represent the Federal Republic.
58. 403-06VI/6204/51 (AA-PA285)

59. The President of the Notgemeinschaft Karl Geiler had already resigned at an earlier date to facilitate the urgently needed fusion with the Forschungsrat. The extraordinary General Assembly of 2 August 1951 elected Otto Flachsbart (TH Hannover) as new President of the Deutsche Forschungsgemeinschaft. He accepted the nomination, but owing to illness his Vice-President Ludwig Raiser had to replace him continuously. In spring 1952 Raiser was elected President.
60. Abridged minutes of the first meeting of the Senate of the Deutsche Forschungsgemeinschaft, 29/10/51. Az. 46/5/51 (DFG).
61. Letter Raiser to Foreign Office (c/o Prof. Dr. Graf Wolf Metternich), 31/10/51. 407-21cVI 7593/51/I (AA-PA285).
62. Note. 407-21cVI/8387/51 (AA-PA285).
63. Letter Foreign Office to Heisenberg (draft), 28/11/51. 407-21cVI/7593/51/II (AA-PA285).
64. Note for Dr. Ringelmann. 407-21cVI/7593/51/III (AA-PA285).
65. Previously Erich Wende had been Secretary of State at the Niedersächsische Kultusministerium in Hannover where Hocker had been one of his collaborators.
66. Reference is made to the Kommission für Atomphysik, set up on 29 February 1952 at a meeting of the Senate of the DFG.
67. Gentner was indeed invited as adviser to all subsequent Council sessions.
68. Letter of Bothe to Regener, 14/3/52 (MPG).
69. It is stated in the Council Minutes, that all the members of the Scientific Policy Committee 'were named simply on grounds of scientific distinctions and with no consideration of nationality'. Minutes of the Council Sessions 1954–55, p. 17.
70. Heisenberg (1967), 279.
71. This is how Kurt Zierold, then Secretary-General of the DFG, described him. Zierold (1968), 289.
72. Heisenberg wrote in his report to Walter Hallstein: 'Mr Noebel, Vice-Consul at the German Consulate in Paris, kindly offered to participate in the final negotiations in Dr. Hocker's place'. Apparently he acted as a purely passive observer.
73. Note 407-21cVI/7593/51/III, 26/11/51 (AA-PA285).
74. Letter Auswärtiges Amt to Heisenberg, 8/12/51 (AA-PA285).
75. Bericht über die Konferenz der UNESCO über die Durchführung von Arbeiten zur Errichtung eines europäischen Laboratoriums für Kernphysik am 17.12.1951, signed Werner Heisenberg (MPG/P).
76. Ibid. Annex 2.
77. Letter Kramers to Auger, 23/8/51 (CHS, Auger file, CERN).
78. 407-21cVI/9587/51, 11/12/51 (AA-PA285).
79. Letter Foreign Office to Heisenberg, 12/12/51, 407-21cVI/9587/51 (AA-PA285).
80. Note 75.
81. Note 75.
82. At that time it was not foreseeable that the lifting of the interdiction would take so long to come, until 5 May 1955. Since about 1951 Heisenberg had continuously pressed for starting the preparations, but Adenauer did not want to risk a conflict with the Western Powers because of this issue, to which he obviously attached less importance. Heisenberg was seriously annoyed and in 1955 refused to represent the Federal Republic at the UN Conference on Peaceful Uses of Atomic Energy held in Geneva in August 1955. His attitude created a considerable public sensation.
83. The letter which Ludwig Raiser signed 'as the President's substitute' has got the reference sign 'Dr. H.' and was probably drafted by Alexander Hocker.
84. Letter of (Foreign Office to Federal Ministry of Finance: 'Our membership [...] of UNESCO would have only a very limited practical value for us, if we were not willing to participate in international cultural cooperation in fields, in which we not only receive, but also give something.' 407-21cVI/00142/52/I (AA-PA285).
85. Letter Foreign Office to Federal Ministry of the Interior, 29/1/52. 407-21cVI/52 (AA-PA286).

86. Notes of the Foreign Office. 407-21cVI/01706/52/III, 7/2/52 (AA-PA286).
87. Note made by Karl Gumbel of 11 February 1952. Documents of the Federal Chancellor's Office. B136/2047.
88. Agreement constituting a Council of Representatives of European States for planning an International Laboratory and organizing other forms of co-operation in Nuclear Research. CERN/Gen/0.
89. Apart from the German Delegation only the Dutch and the Yugoslavian Delegations signed without using the saving clause.
90. Bericht über die Genfer Konferenz für die Gründung eines europäischen Kernphysik-Instituts (12–15 February 1952), signed Werner Heisenberg. The draft was prepared by Alexander Hocker (MPG/P).
91. Minutes of the 3rd meeting of the Senate of the Deutsche Forschungsgemeinschaft, 29/2/52. AZ.46/2/52 (DFG).
92. His illness continued to prevent the President of the DFG from executing his office and he asked Ludwig Raiser, one of the Vice-Presidents, to act as his substitute.
93. Dichgans (1980), 64f.
94. This is the view of the political scientist Hans-Peter Schwarz, Schwarz (1981), 102f.
95. It cannot be excluded that the stronger term 'instruction' was employed by the minute-writer (Erwin Gentz), whereas Heisenberg only mentioned a 'recommendation' of the Foreign Ministry. Nevertheless, it hardly affects our conclusions.
96. Bericht über die 2. Sitzung des Europäischen Rats für kernphysikalische Forschung (Kopenhagen, 20. bis 22 Juni 1952), signed by Werner Heisenberg (MPG/K). The time required for the ratification procedure in the Member States was generally greatly underestimated.
97. Letter Hocker to Heisenberg, 11/8/52 (MPG/P).
98. Letter Hocker to Heisenberg, 4/9/52 (MPG/P).
99. Letter Hallstein to Heisenberg, 12/1/53. 407-21cI/00055/53 (AA-PA287).
100. Letter Heisenberg to Hocker, 26/1/53 (MPG/K).
101. Letter Heisenberg to Hallstein, 14/2/53 (MPG/K).
102. Not far from Bonn. Bad Godesberg is now part of Bonn.
103. Letter Hocker to Heisenberg 12/3/53 (MPG/K).
104. Minutes of the 4th meeting, 28/2/53. The quote is taken from the final version. The draft prepared by Alexander Hocker (before its revision by Heisenberg) is even more explicit. It says: 'Mr Haxel does not consider the physics interest as the main motivation for the German participation. That would rather be on the political side. Mr. Heisenberg too feels that to 80% the cost should be considered from the angle of European collaboration.'
105. Deutsche Forschungsgemeinschaft (1953). — The members of the 'Kommission für Atomphysik' were Fritz Bopp (Munich), Walther Bothe (Heidelberg), Wolfgang Gentner (Freiburg), Otto Haxel (Heidelberg), Werner Heisenberg (Göttingen), Hans Kopfermann (Göttingen), Josef Mattauch (Mainz), Erich Regener (Stuttgart), Wolfgang Riezler (Bonn). Werner Heisenberg acted as chairman. Alexander Hocker attended the meetings as the representative of the Deutsche Forschungsgemeinschaft.
106. When the 'Federal Ministry for Atomic Affairs' (the present Federal Ministry for Research and Technology) was set up on 15 October 1955, the Federal Government appointed the 'Deutsche Atomkommission' (DAK) as its scientific advisory board, which was made up of leading scientists, civil servants and representatives from industry. The DAK established several sub-committees, one of which was for 'Nuclear Physics'. Its composition was identical with that of the 'Kommission für Atomphysik'. The same men under Heisenberg's chairmanship met as the 'Arbeitskreis Kernphysik' set up by the Federal Ministry for Atomic Affairs and as the 'Kommission für Atomphysik' of the Deutsche Forschungsgemeinschaft, until the latter was finally dissolved because the promotion of science in the field of nuclear and high-energy physics was completely taken over by the Ministry.
107. Letter 'KHL' to Male, 17/4/53 (SERC-NP24).
108. Heisenberg had made similar statements the previous year, when at the request of the CERN Council he drew up a report on the 'International Physics Conference' held in Copenhagen from 3 to 17 June 1952.

See Werner Heisenberg: Report on the Scientific conference held in Copenhagen... European Council for Nuclear Research. Second Session... Minutes. Annex III. CERN/Gen/2.
109. Note of 11 March 1953 signed 'Dr. Alexander Hocker, Deutsche Forschungsgemeinschaft' (MPG/P). The pace of the economic boom was unexpected. In fact five years later the Federal Republic was able to start the construction of her own big research laboratory for particle physics. After two and a half years of preparation the Deutsches Elektronen-Synchrotron, known from its abbreviation as DESY, was officially founded.
110. Letter Hocker to Heisenberg, 12/3/53 (MPG/P).
111. Letter from Haushaltsreferat to Kulturabteilung, 18/3/53, 407-21c/VI/6096/53 (AA-PA291).
112. 407-21c/VI, 27/3/53 (AA-PA291).
113. Article 59, paragraph 2 of the Basic Law reads: 'Agreements regulating the political relationships of the Federation or referring to matters of Federal legislation require the approval or the participation of the appropriate legislative bodies in the form of a Federal Law.'
114. CDU/CSU obtained 45.2% of the votes and 243 out of 487 seats. Adenauer formed a new coalition government and also acted as Foreign Minister (until 5 June 1955).
115. 'Aufzeichnung' concerning 'Sitzung des Kulturausschusses des Bundesrats' 400-16/VI/31637/53, 8/12/53 (AA-PA293).
116. 'Aufzeichnung' concerning 'Frage der Zuständigkeit und Beitragszahlung der Bundesrepublik Deutschland zu der Europäischen Organisation für kernphysikalische Forschung gemäss Abkommen vom 1. Juli 1953', 407-21c/VI/32480/53, 16/12/53 (AA-PA293).
117. 'Vermerk' concerning 'Europäische Organisation für kernphysikalische Forschung', 407-21c/VI/23062/53, 22/1/54 (AA-PA293).
118. In Heisenberg's view, the fact that Adenauer was also Foreign Minister was certainly a point in favour of the Foreign Office; Adenauer and Heisenberg were on especially good terms with each other. Against that, Heisenberg and Hocker also thought that the Foreign Office was not particularly effectual. See letter Hocker to Heisenberg, 12/3/53 and letter Heisenberg to Hocker, 16/3/53 (MPG/P).
119. Bundesrats-Drucksache Nr. 65/54 of 26 February 1954. Article 76, paragraph 2 of the Basic Law reads: 'Legislation initiated by the Federal Government shall first be submitted to the Bundesrat. The Bundesrat is entitled to take a position on legislative bills within three weeks.'
120. Verhandlungen des Bundesrates 1954. Stenographische Berichte. Bonn 1954, 69.
121. Letter Hocker to Heisenberg, 28/5/54 (MPG/P), quoting official minutes of 23rd meeting of the second Bundestag held 7/4/54 (Haushaltsdebatte, Haushalt für den Geschäftsbereich des Auswärtigen Amtes) page 816 (B); rapporteur Dr. Vogel (CDU/CSU).
122. Verhandlungen des Deutschen Bundestages. 2. Wahlperiode 1953. Stenographische Berichte. Vol. 19, p. 1141 A.
123. Verhandlungen des Deutschen Bundestages. 2. Wahlperiode 1953. Anlagen zu den stenographischen Berichten. Vol. 28, Drucksache 394.
124. Letter Hocker to Heisenberg, 15/6/54 (MPG/P): 'The Secretary of the Committee indicated, that the subject would probably have to be removed from the agenda. I succeeded in persuading Mr. Grau, the Assistant Secretary, to write and call the attention of the Federal Chancellor to the urgency of the matter. The latter had to attend the Committee's meeting for discussions on the Saar. As a result, ratification was taken as item 1 on the agenda. In order to avoid any further delay, which is particularly likely if the Bundestag does not pass a bill before the summer recess, I suggest that you ask the President of the Bundestag to schedule the second and third reading of the bill in the near future [...] I could then in due course take a similar initiative with the Director of the Bundestag.'
125. Verhandlungen des Deutschen Bundestages. 2. Wahlperiode 1953. Stenographische Berichte. Vol. 20 p.1767 C.—Concerning the ratification in the French National Assembly see Dominique Pestre in chapter 9.
126. The legislative procedure is laid down in Article 77 of the Basic Law. It provides the possibility for the Bundesrat to ask that a bill which has been passed by the Bundestag is re-examined by the so-called 'Vermittlungsausschuss' (mediation committee).

127. A list of the Member States in order of ratification is given in Kowarski (1961), Annex IV, p. 12 and in chapters 7 and 9 of this volume.
128. Letter Amaldi to Gentner, 9/9/54, CERN/2206.
129. Express letter Hocker to Amaldi, 13/9/54 (MPG/P). Hocker had come back from holiday on that day.
130. Bundesgesetzblatt II, p. 1013 and 1132.
131. AEF-77.
132. According to the Foreign Office, a series of documents were 'rashly destroyed' in 1965. Fortunately, in the political archives of the Foreign Office the CERN documents up to the year 1953 inclusive are still available, contrary to earlier fears. Documents for subsequent years, however, seem to have fallen victim to this action.
133. Letter Heisenberg to Salat, 5/6/52, 407-21cVI/08463 (AA-PA287).
134. Letter Deutsche Forschungsgemeinschaft (Ludwig Raiser) to Foreign Office (Rudolf Salat), 10/6/52, 407-21cVI/08700 (AA-PA287).
135. 'Notiz' in the Foreign Office. 407-21cVI/15123/52, 24/9/52 (AA-PA287).
136. Ibid.
137. Ibid.
138. Letter Foreign Office to Heisenberg, 26/9/52. 407-21cVI/15123/52 (AA-PA287).
139. Bericht über die 3. Sitzung des Europäischen Rats für kernphysikalische Forschung (Amsterdam, 4., 6. und 7. Oktober 1952), signed Werner Heisenberg (MPG).
140. Ibid.
141. Ibid.
142. A fellowship from the 'International Education Board', allowed Heisenberg to stay at Niels Bohr's Institute in Copenhagen from September 1924 to March 1925; he worked there again as a lecturer from May 1926 to September 1927, when he was appointed to the University of Leipzig.
143. Bericht über die 4. Sitzung des Europäischen Rats für kernphysikalische Forschung (Brüssel, 12.-14. Januar 1953), signed Werner Heisenberg (MPG).
144. Minutes of the 4th meeting of the Kommission für Atomphysik of the Deutsche Forschungsgemeinschaft, 28/2/53 (MPG).
145. Bericht über die 5. Sitzung des Europäischen Rats für kernphysikalische Forschung (Rom, 30. März bis 2. April 1953), signed Werner Heisenberg (MPG).
146. CERN/Gen/5, p.11.
147. CERN/SPC/1.
148. CERN/128. At this stage the recommendation however only referred to the establishment of a small theoretical group consisting of two or three physicists.
149. Bericht über die Konferenz der UNESCO über die Durchführung von Arbeiten zur Errichtung eines europäischen Laboratoriums für Kernphysik am 17.12.1951, signed Werner Heisenberg (MPG).
150. In section 11.2 we have explained in detail the reason why the appointment of Heisenberg as the German delegate had been a fortunate decision.
151. Bericht über die Ratssitzung zur Gründung eines europäischen Zentrums für Kernphysik (Paris, 5.-9. Mai 1952), signed Werner Heisenberg (MPG).
152. CERN/Gen/2, Annex III.
153. Bericht über die 2. Sitzung des Europäischen Rats für kernphysikalische Forschung (Kopenhagen, 20. bis 22 Juni 1952), signed Werner Heisenberg (MPG).
154. Terms of Reference proposed by the United Kingdom for the Scientific Policy Committee. CERN/82.
155. CERN/106.
156. Bericht über die 8. Sitzung des Europäischen Rats für kernphysikalische Forschung (Genf, 14.-16. Januar 1954), signed Werner Heisenberg (MPG).
157. Letter Gentner to Heisenberg, 10/4/54 (MPG/K).
158. Letter Heisenberg to Kowarski, 13/4/54 (MPG/K).

159. Letter Gentner to Heisenberg, 24/3/54 (MPG/K).
160. Ibid. This completely agrees with Amaldi's description of these talks: 'It was finally agreed that the best solution would be to have Dr. Schmelzer. Everybody in the group has a great consideration for him, not only from a technical point of view, but also for his physical insight in various problems as well as for his human qualities'. Letter Amaldi to Cockcroft, 25/3/54 (CERN/1692).
161. Letter Dahl to Amaldi, 24/7/54 DG/2055.
162. Letter Schmelzer to Gentner, 15/4/54 (MPG/K).
163. See chapter 9. Heisenberg believed, that the French 'were not keen for a German to be in charge of the biggest machine'. Letter Heisenberg to Hocker, 19/7/54 (MPG/P).
164. Letter Dahl to Amaldi, 24/7/54 (CHS, Amaldi file, CERN).
165. Letter Gentner to Heisenberg, 2/8/54 (MPG/K).
166. CERN/1794.
167. Letter Heisenberg to Raiser, 19/5/54 (MPG/K).
168. Amaldi received the undated letter on 23 August 1954.
169. This assertion could be questioned. Alexander Hocker was considered a 'highly intelligent, strong-willed and ambitious worker', whereas in Amaldi's letter he is described 'as rather weak as personality'. Correspondingly, Heisenberg did not believe, 'that the personal impression given by Amaldi was in fact justified'. Letter Heisenberg to Gentner, 27/8/54 (MPG/K).
170. Letter Amaldi to Heisenberg, 25/8/54 (MPG/K).
171. Letter Gentner to Heisenberg, 19/11/54 (MPG/K).
172. Minutes of the Eighth Meeting of the Kommission für Atomphysik, 15/12/54 (MPG).
    By the end of 1954 the following top posts had been filled at CERN:
    Director-General: Felix Bloch (US and CH)
    Director-General adjunct: Edoardo Amaldi (I)
    Director SC and member of Directorate: Cornelis Bakker (NL)
    Director PS: John Adams (UK)
    Director Scientific and Technical Services: Lew Kowarski (F)
    Director Site and Buildings: Peter Preiswerk (CH)
    Director Theoretical Studies: Christian Møller (DK)
    Director of Administration: S.A. Dakin (UK)
173. Minutes of the Ninth Session of the Kommission für Atomphysik, 4/5/55 (MPG).
174. Interview of the author with Wolfgang Gentner on 1 September 1980 in Heidelberg, three days before Gentner's death. John Adams alluded in his obituary to the specific difficulties experienced by Gentner: 'Taking over the SC machine from Bakker in the middle of its construction was no easy job. Bakker had built up a very competent team who were used to working with him since he had directed the work right from the beginning. Gentner found himself sandwiched, so to speak, between the senior members of the SC Division and their old boss who was then the Director-General. It was a situation that required considerable tact and human understanding.' CERN/DOC 82-3.
175. Deutscher Bundestag. 2. Wahlperiode 1953. Drucksache 1657.
176. Verhandlungen des Deutschen Bundestages. 2. Wahlperiode 1953. Stenographische Berichte vol. 35, p. 11050 D.
177. As explained, the responsibility had been transferred to the Federal Ministry for Atomic Affairs which had been newly created on 15 October 1955. At the request of Franz Josef Strauss, its first minister, Alexander Hocker left the Forschungsgemeinschaft for the Federal Ministry for Atomic Affairs.
178. As note 176, p. 11051 D.
179. Memorandum über die Zusammenarbeit Deutschlands mit CERN. CERN archives. G 411.

Bibliography for Chapter 11

Adams (1982)	J. Adams, 'Gentner and CERN', in: *Wolfgang Gentner 1906-1980,* CERN/DOC 82-3.
Adenauer (1965)	K. Adenauer, *Erinnerungen 1945-1953* (Stuttgart: DVA, 1965).
Amaldi (1977)	E. Amaldi, 'Personal Notes on Neutron Work in Rome in the 30's and Post-war European Collaboration in High-energy Physics' in: C. Weiner (ed.), *History of Twentieth Century Physics,* Proceedings of the International School of Physics 'Enrico Fermi', Course LVII, Varenna 1972 (New York: Academic Press, 1977).
Born (1969)	A. Einstein, H. und M. Born, *Briefwechsel 1916-1955* (Munich: Nymphenburger, 1969).
Deutsche Forschungsgemeinschaft (1953)	*Bericht der Deutschen Forschungsgemeinschaft über ihre Tätigkeit vom 1. April 1952 bis zum 31. März 1953* (Bad Godesberg: DFG, 1953).
Dichgans (1980)	H. Dichgans, *Montanunion. Menschen und Institutionen* (Düsseldorf: Econ, 1980).
Eickemeyer (1953)	H. Eickemeyer (ed.), *Abschlussbericht des Deutschen Forschungsrates* (Munich: R. Oldenbourg, 1953).
Goudsmit (1947)	S.A. Goudsmit, *Alsos* (New York: H. Schuman, 1947).
Heisenberg (1967)	W. Heisenberg, *Der Teil und das Ganze. Gespräche im Umkreis der Atomphysik* (Munich: Piper, 1967).
Hermann (1976)	A. Hermann, *Werner Heisenberg. In Selbstzeugnissen und Bilddokumenten* (Reinbek: Rowohlt, 1976).
Hermann (1977)	A. Hermann, *Die Jahrhundertwissenschaft. Werner Heisenberg und die Physik seiner Zeit* (Stuttgart: DVA, 1977).
Hermann (1985)	A. Hermann, 'Max von Laue', in: L. Gall (ed.), *Die grossen Deutschen unserer Epoche* (Berlin: Ullstein, 1985), 45-56.
Kowarski (1961)	L. Kowarski, *An account of the origin and beginnings of CERN,* CERN 61-10.
v. Laue (1961)	M. v. Laue, 'Mein physikalischer Werdegang. Eine Selbstdarstellung', in: M. v. Laue, *Gesammelte Schriften und Vorträge.* Vol. *III.* (Brunswick: Friedr. Vieweg & Sohn, 1961), p. V-XXXIV).
Mensing (1985)	H.P. Mensing (ed.), *Adenauer: Briefe 1949-1951* (Berlin: Siedler 1985).
Müller-Armack (1971)	A. Müller-Armack, *Auf dem Wege nach Europa. Erinnerungen und Ausblicke* (Tübingen: Rainer Wunderlich, 1971).
Poidevin (1976)	R. Poidevin, *Robert Schumans Deutschland- und Europapolitik,* Schriften der Philosophischen Fachbereiche der Universität Augsburg, Nr. 3, 1976.
Schmid (1980)	C. Schmid, *Erinnerungen* (Berne: Scherz, 1980).
Schwarz (1981)	H.P. Schwarz, *Die Ära Adenauer, Gründerjahre der Republik 1949-1957* (Stuttgart: DVA, 1981).
Stamm (1981)	T. Stamm, *Zwischen Staat und Selbstverwaltung. Die deutsche Forschung im Wiederaufbau 1945-1965* (Cologne: Verlag Wissenschaft und Politik, 1981).
Voltaire (1965)	Voltaire, 'Traité sur la tolérance à l'occasion de la mort de Jean Calas', in *Mélanges.* Texte établi et annoté par Jacques Van den Heuvel (Paris: Gallimard, 1965).
Zierold (1968)	K. Zierold, *Forschungsförderung in drei Epochen. Deutsche Forschungsgemeinschaft. Geschichte—Arbeitsweise—Kommentar* (Wiesbaden: F. Steiner, 1968).
Zierold (1978)	K. Zierold, *Lebenserinnerungen.* Unpublished typed script. Written in Bonn around 1978.

CHAPTER 12

Britain and the European laboratory project[1] 1951–mid-1952

John KRIGE

Contents

12.1 The period of detachment   433
    12.1.1 A brief chronology of events   433
    12.1.2 Some reasons for British reticence   435
12.2 The UNESCO May report and the mounting opposition to it on the continent and in Britain   438
    12.2.1 The activities spearheaded by Kramers and Bohr: a quick survey   438
    12.2.2 The activities spearheaded by Chadwick   440
12.3 Forging the alliance: the offer of the Liverpool cyclotron and the Paris conference   445
    12.3.1 The meeting of the British UNESCO Committee   445
    12.3.2 The negotiations over the Liverpool cyclotron   447
    12.3.3 The Paris conference, December 1951   449
12.4 Confronting the new question: Should Britain join the Council of Representatives?   454
    12.4.1 The scientists press for membership   454
    12.4.2 Governmental reactions   458
    12.4.3 The Geneva conference, February 1952   461
12.5 The Cabinet Steering Committee reconsiders the case   462
    12.5.1 A brief chronology of events   463
    12.5.2 The motives of the protagonists   466
    Notes   468

Granted the value of studying various national approaches to the European laboratory project, a piece on the United Kingdom is obviously needed, if only because at the time she was the leading country in nuclear physics in the region, and because of the ongoing vacillation in her relationship with CERN. The reader may however wonder why we have written two chapters on this country, particularly when there is only one on France, Germany, and Italy. There are two reasons for this. The first is the amount and quality of the documentation we have. For example for the fifteen months during 1951–52 when attitudes in the country towards the project changed drastically, we have about fifty letters, memoranda, minutes of meetings, etc. per month. This material is diverse in origin and rich in detail and provides the historian with a fascinating insight into the minds of both scientists and government officials. Archival searches in the other three countries were nowhere near as productive. Given the opportunity we yielded to the temptation, and decided to study Britain at greater length and in depth.

Another consideration which has led us to split the British case is related to the subject matter itself. Around mid-1952 there was a marked change in the terms of the debate, and in the kind and number of physicists participating in it. Until then Britain's scientific policy *vis-à-vis* the project had been more or less exclusively managed by a few members of the scientific elite. This was to change during the second half of 1952, as were the objective conditions under which the physicists refined their views on the scheme. The division into two chapters coincides with this temporal cut.

The overall aim of the study is to explore the gradual transformation in attitude in the United Kingdom towards the European laboratory project, its evolution from relative indifference early in 1951 to qualified commitment late in 1952. This chapter is divided into five main sections. The first two deal predominantly with the scientists' reactions to the scheme, and trace their development from detachment to active opposition. In the next three sections physicists' efforts to win government support for their proposals, and the extent of their success, are described. Speaking generally, if we find Britain drawing closer to 'CERN' membership in this period, fundamental obstacles to her full participation nevertheless remained in June 1952. Chapter 13 explains how these were overcome, and the different choices that that brought in its wake, culminating in the decision that Britain should join CERN. It includes a short discussion on the process leading to the ratification of the Convention and is followed by two appendices. The first describes the constitution and role of the main committees involved in formulating Britain's policy on the European laboratory project. In the second we make a few observations about the decision-making mechanism at the science–government interface.

12.1 The period of detachment

12.1.1 A BRIEF CHRONOLOGY OF EVENTS

1950 was an important year in the early history of CERN. During that year the idea that some kind of scientific co-operation in Europe would be desirable began to be translated into concrete action. A project which had hardly moved beyond the level of vague generalities became progressively refined and articulated so that, by the end of 1950, a fairly precise conception of the scientific core of the European laboratory had been agreed on by some of its major protagonists.

Two events were particularly crucial in this development, the 5th General Conference of UNESCO held in Florence in May–June 1950, and a meeting of a select group of scientists and science administrators under the auspices of the Centre Européen de la Culture (CEC) in Geneva in December 1950. At the first meeting an American-sponsored resolution had called simply for 'regional research centres and laboratories'. Six months later, in Geneva, a 'European Laboratory for Nuclear Physics, centred around the construction of a large instrument for accelerating elementary particles' was envisaged. It was, the delegates hoped, to be the biggest machine in the world, i.e. with an energy of at least 6 GeV.[2]

The main, though by no means the only, driving force behind the refinement of the project was Pierre Auger, himself a cosmic-ray physicist and the Director of UNESCO's Department of Exact and Natural Sciences. Encouraged by the deliberations in Florence, he temporarily adopted the project as his own. He explored it with eminent scientists and science administrators in several European countries, including Britain. He led a discussion on the subject during the Harwell Nuclear Physics Conference in Oxford in September. And having clarified his ideas on the most desirable project, he set out to secure official sanction for it, and financial support for a preliminary study of the scheme.

No British scientists or administrators were actively involved in these early initiatives. However, a formal link between them and the machinery of government was made with the transmission of the resolution passed at the CEC meeting by its secretary, Jean-Paul de Dadelsen, to the Council of Europe. From here it was forwarded to the Overseas Liaison Division (OLD) of the Department of Scientific and Industrial Research (DSIR) on 28 December 1950. Attached was a 'Note on the Resolution of 12th December 1950' written by de Dadelsen, and dated 13 December 1950. It amounted to his draft minutes of the meeting.

These documents,[3] along with a covering note written by the OLD secretariat explaining their background, were discussed at a meeting of the Committee on Overseas Scientific Relations (OSR), chaired by Alexander King, on 10 January 1951. The committee was puzzled by the status of the resolutions and questioned the authority of the CEC to sponsor them. In the event it was felt that all that could be done at this stage was to consider the resolution on its technical merits, and it was

Notes: pp. 468 ff.

agreed that the Nuclear Physics Committee (NPC) 'should be asked to examine the proposals and to express its views on them'.

The matter was passed to J.H. Awbery, the secretary of the NPC. He wrote to Sir James Chadwick on 25 January asking him if he was aware of the scheme to found a European laboratory equipped with a particle accelerator greater than 6 GeV. The chairman's response was unenthusiastic.[4] 'I have heard more than enough about the U.N.E.S.C.O. scheme for a Nuclear Physics Institute', he wrote, adding that he hoped that it would not be necessary to have a meeting to deal with the matter and that 'some modified agreement' among the committee members could be reached by correspondence.

Acting on Chadwick's instructions Awbery circulated a batch of documents to the members of the NPC on 7 February. They included the papers taken by the OSR Committee on 10 January, and an extract from the minutes of that meeting.[5] Awbery also appended parts of a letter written by A.H. Waterfield, H.M. Government's Scientific Attaché in Paris, to the OLD Secretariat, giving details of a further discussion he had had with Auger in mid-January.[6] The members of the NPC were asked to reply to Awbery as soon as possible, stating their opinions on the proposal and on whether Britain should support and join the scheme 'by contributions of money and men'.

On the basis of this documentation, and an exchange of views among themselves,[7] thirteen members of the committee formulated their detailed replies to Awbery. Only one, Cecil Powell, felt that a meeting was desirable to discuss the proposal, and only one, Patrick Blackett, who had been strongly influenced by Sir John Cockcroft, suggested that the proposal be supported with both money and men. Three members of the committee, N. Feather, H.W.B. Skinner, and R. Peierls believed that neither money nor men should be offered, while both Cockcroft and Sir George Thomson, for example, favoured sending British workers to the centre, though were also against supporting it financially.

By the middle of March it was clear that, on balance, the NPC was against Britain taking an active part in the scheme. Chadwick asked Awbery to circulate a document summarizing the views of the individual members of the NPC, and attaching to them the chairman's draft reply to the OSR Committee.[8] No substantial objections were raised to Chadwick's interpretation of the mood of the members, and on 23 April 1951 he replied officially to King.[9] Although the opinion of the members of his committee varied widely, especially in detail, wrote Chadwick, there was a 'very definite balance of opinion that this country should not join directly in the establishment of such a Laboratory and should not promise support either with men or money [...]'. Nevertheless, Chadwick went on, there was a reluctance to 'dissociate ourselves entirely from the scheme' and, if it was adopted, Britain's physicists would 'give informal support by means of advice and help', if asked to do so. Officials in the Ministry of Supply and the DSIR felt likewise: Britain should 'take a friendly interest' in the scheme, giving 'helpful advice' 'through informal contact', and keeping the matter 'fluid so that we could join in later if necessary'.[10]

12.1.2 SOME REASONS FOR BRITISH RETICENCE

Co-operation without commitment: this phrase captures the initial response in Britain to the European laboratory project. No one familiar with the history of Britain's dealings with European schemes from the late forties onwards will be surprised by the reaction. Admittedly the British were not alone in treating the new venture with scepticism, not to say hostility. Amaldi, for example, has pointed out that when he presented the idea at an Executive Meeting of the International Union of Pure and Applied Physics (IUPAP) on 7-8 September 1950, 'most of the people were not prepared, I remember, they were not against it, but it was new to them this idea and then they said "why don't you prepare a report" '. Kowarski noticed similar reactions to Auger's address at the Harwell conference later that month: 'the expressed comments tended to explain why the proposed institute could not possibly work, rather than how it could'.[11] On the other hand, the fact remains that Britain's doubts lasted longer and went deeper than those of her continental allies. It is the reasons for this that we want to explore here.

At the first, and very popular level of analysis, we have a genre of 'psychological' explanations for British reticence *vis-à-vis* collaborative schemes particularly with the continent: it is because she is insular, xenophobic, arrogant ... Thus Sir Ben Lockspeiser, Secretary of the DSIR at the time, has said that 'the reason why we didn't want to join with the Europeans was because we thought that these foreigners were inferior and we had always been top dogs and we were going to be top dogs'.[12] And certainly feelings of this kind were expressed. We have Blackett's incredulity: 'I must say my first reaction is that it [the project to build a European bevatron] is quite crazy', he wrote to Cockcroft on 13 February 1951. 'The idea that, in three years' time, if we voted the money, we could have a better machine than the Americans is too crazy to be taken seriously'. There is Skinner's contempt: the project was 'just one of the high-flown and crazy ideas which emanate from UNESCO'. And an air of condescending resignation in the reaction of the OSR committee on 10 January: the scheme 'appeared to be another example of the general overall confusion in the field of international co-operation'.[13]

At the same level of analysis we have the suspicion that the French were promoting the collaborative venture primarily to further their own sectional interests. Indeed there was much in the information flowing into Britain to suggest this (too narrow) interpretation of French motives, including a comment made by Auger to Waterfield in January that the French were 'proposing to rely on this new centre for the research part of their Atomic Energy Commission programme'.[14] It was remarks like this which led Skinner to identify the European laboratory with a French laboratory. He wrote that 'if the French want to have a nuclear physics research laboratory why don't they go ahead with the cooperation of any other nation interested [...]. Anyhow I am sure that we in this country should have nothing to do with the scheme'. Blackett also picked up the heavy French

Notes: p. 469

involvement: 'If France [...] can afford all that money why don't they finance their present research and build up their Physics again into a decent state', he wrote to Cockcroft. Chadwick's hostility to the scheme was also partly influenced by mistrust of the French, though it took a rather more personal form: 'If there is money to be spent on Nuclear Physics', he wrote, 'we can spend it better than Auger'.[15]

Explanations of a psychological type take us some way but they do not go far enough. More specifically, apart from the inevitable weaknesses in the early organization and articulation of the project itself, there were a number of more 'objective' reasons why scientists and government officials stood off from it.

First and foremost, there was the fact that Britain was already ahead of her European associates in the field of accelerator design, development, and construction. The biggest machine in continental western Europe was a Swedish 200 MeV synchro-cyclotron which was completed towards the end of 1951. By contrast, the Atomic Energy Research Establishment at Harwell was equipped with four major machines, including a travelling-wave linear electron accelerator which was the first of its kind in the world, and a 110 inch, 175 MeV synchro-cyclotron which went into operation in December 1949 and was for a time the most powerful accelerator in Europe. There were even more powerful machines under construction in 1950 in some of the universities—a 300 MeV electron synchrotron at Glasgow, a 156 inch, 400 MeV synchro-cyclotron at Liverpool, and a giant 1 GeV proton synchrotron at Birmingham. Oxford was building a 140 MeV electron synchrotron and, in 1952, Cambridge started to build a 300–400 MeV linac.[16] Granted the number and range of machines with which the country would soon be furnished, proposals that Britain become involved in yet another construction project were bound to be treated with circumspection.

Related to this was the concern that enough time and money had already been spent on constructing major equipment. The universities in particular found it difficult to continue doing experimental research during the many demanding years which it took to build an accelerator. This led Thomson, for example, to wonder whether 'we have already devoted more effort than was wise on these big machines'. 'It is time', he added, 'to stop engineering and do some physics'. In similar vein Chadwick insisted that before tying up more resources in another accelerator project, Britain should 'first put into use the equipment which we have been developing in the past few years'.[17]

Then there was the question of cost. The initial grants to the five universities alone that wanted machines in 1946 had amounted to some £1/2M. By 1950 these had been increased to over £1M and requests for more money were in the pipeline.[18] This large investment in accelerators was tending to skew the whole shape of research not only in physics but in science as a whole, as physicists themselves realized.[19] Chadwick had raised the issue even while his Nuclear Physics Committee was recommending the five-university accelerator programme immediately after the war. By 1950 the dangers were even clearer. Sir Lawrence Bragg, crystallographer, and

Director of the Cavendish Laboratory, wrote that 'we ought now to call a halt, and not start nuclear machines on a large scale at any other centre'. There was, he added, a growing tendency for big projects to dominate physics to its detriment: 'the tail must not wag the dog because it is so large a tail'.[20]

Apart from scientific reasons for British reticence, there were also a number of policy and more general political considerations which engendered a climate that was uncongenial to international ventures. Regarding the former, in the first five years after the war it was stated, and reiterated, that Britain preferred the traditional forms of scientific collaboration—exchange of personnel between existing national laboratories, attendance at conferences, circulation of preprints, etc.—over the 'foundation of new centralized international laboratories in subjects where active research is being carried out on a large scale'.[21] The scheme to build from scratch a European laboratory devoted to the already well-endowed field of nuclear physics ran directly counter to this.

Consider too Britain's domestic and foreign policies in the European arena in the postwar period.[22] Politically, this period was dominated by the Labour Party. Unexpectedly swept to power over Churchill's Conservatives in July 1945, Labour remained in office, until October 1951, though with its majority considerably reduced in a general election held in February 1950.

Labour's immediate objective for Europe was to see solid bonds forged between the continent and the United States. Once it was satisfied that these had been established (e.g. through the Marshall Plan (1947–48) and the setting up of NATO (1947)) the government's interest in European collaboration waned. Labour was hostile to the movements for European unity, it opposed the Schuman plan for a European Coal and Steel Community, and was instrumental in ensuring that the Council of Europe was deprived of any meaningful political power.

Labour's domestic policies were one important factor behind its opposition to these schemes. Having embarked on a wide programme of nationalization, the government was unwilling to sacrifice the newly-won state control over the commanding heights of the economy to a supranational organization. Traditional mistrust of France and Germany, and a feeling that Britain was neither geographically nor historically part of Europe also played a role, and were reinforced in Labour's ranks by a view of their Party as one which was distinctly British, and quite different from continental Socialist parties.

Labour also did not trust the motives of some of the chief advocates of European unity, notably Churchill and the United State's government. The Party was particularly suspicious of American pressure to play a more active role in movements for European integration. Having encouraged the formation of close bonds between the USA and the continent, Labour now feared that America advocated federalism in Europe so that she could reduce her economic and military commitments in the region.

Britain's location on the world political stage also influenced her assessment of

Notes: p. 470

how to deal with her European allies. After the war politicians tended to see Britain as belonging to three overlapping circles, the Commonwealth, the Atlantic community, where she had a 'special relationship' with the United States, and Europe. Britain's interests in these several regions did not always coincide. She feared a loss of influence over the Commonwealth countries, notably those granted independence in the late forties, if she identified herself too closely with the continent. She was even more unwilling to jeopardize the intimacy enshrined in her special relationship with America, a country particularly sensitive to the presence of avowed communists in the French Commissariat à l'Energie Atomique. This led Britain to adopt a policy of 'patronising detachment' towards Europe in the decade or so after the war, and condemned her, in Cockcroft's view, to 'masterful inactivity' in nuclear affairs on the continent until the early fifties.[23]

12.2 The UNESCO May report and the mounting opposition to it on the continent and in Britain

12.2.1 THE ACTIVITIES SPEARHEADED BY KRAMERS AND BOHR: A QUICK SURVEY

The kind of assistance which the members of the Nuclear Physics Committee were prepared to give their continental colleagues was spelt out by Chadwick in his official reply to King on 23 April 1951. Britain, he wrote, could help at the technical level 'by advice and assistance on equipment and its design, possibly by seconding men for *short* periods', and 'by accepting men from the Laboratory for training, experience and research in our own projects'.[24] Consistent with this approach, Cockcroft responded to a request from Auger for a high-frequency expert for 2-6 months by offering him the services of one of his young accelerator builders, Frank Goward, for some ten days.[25] In the event Goward was invited to a meeting of scientific experts convened by Auger and Amaldi in Paris from 23–25 May 1951. The aim of the meeting was to explore the technical, political, and financial ramifications of the decision taken to equip Europe with the biggest accelerator in the world. By way of setting the scene for subsequent developments in Britain, we first quickly describe the results of this meeting, and then go on to summarize the negative reactions to them by influential European physicists. A detailed account of the events described here is given in chapters 4 and 5.

The experts who met in May reaffirmed that the laboratory be centered on a big proton accelerator.[26] They set its energy at between 3 and 5 or 6 GeV—big enough to do important scientific work and to be of interest to all the countries in the region, including Britain, yet not so large as to outstrip the financial and industrial resources of Europe or her scientific and engineering capabilities. It was recommended that the Berkeley Bevatron be copied so as to reduce risks and to enable the project to go ahead that much more rapidly. Costs were estimated at some $20–$25M to be spread over five years, and a four-stage development programme was drawn up covering the technical and organizational aspects of the scheme. The official version of the proceedings was published as UNESCO document 6C/PRG/25, which came to be known as the May report.

An opportunity to discuss this project more widely among scientists, including the cream of Europe's physicists, arose at Copenhagen soon thereafter. During the week of 6–13 July a conference on quantum physics organized by Niels Bohr and Stefan Rozental was followed by the 7th General Assembly of IUPAP. The European laboratory was discussed informally during the conference and, in the light of the 'intricate problems' which it raised, Bohr telegrammed Auger and asked him to attend the IUPAP meeting in person.[27]

There are considerable gaps in our knowledge of what happened during these meetings. However, as reported by Auger some six weeks later, Bohr himself and the eminent Dutch physicist H.A. Kramers, were particularly anxious about the financial implications of the proposed laboratory. Both of them liked the idea of having a European nuclear physics centre. However, they feared that the project would be stillborn if governments were asked to back the May programme, which demanded a large, long-term financial commitment 'just [...] to build a big accelerator of a type now under construction in the U.S.A.', as Kramers was to put it.[28] Would it not be wiser to proceed more gradually, beginning with a less ambitious and expensive project? For instance — and this was Bohr's suggestion — why not first build a small accelerator, its success being used to justify an expansion of effort?

To accommodate these proposals a compromise was reached in Copenhagen. Two accelerators would be built: a 500 MeV electron synchrotron as well as the big bevatron. It was also proposed to set up an institute for advanced studies in nuclear physics to sustain the intellectual life of the laboratory. A new and extended schedule was drawn up reflecting these changes, and assumed the status of an 'official programme' in UNESCO circles.

Auger came away from Copenhagen feeling that 'a satisfactory agreement' had been reached, despite some 'very strong' opposition.[29] In mid-August he began to prepare for a conference of government representatives to be held in Paris in December, at which it was hoped that some $200,000 would be voted for the planning stage of the laboratory. In anticipation of this meeting Auger reconvened his small committee of experts whose task now was to refine the project further.

Around this time, too, Kramers returned to Bohr's Institute of Theoretical Physics in Copenhagen. While there he discussed with Bohr and his staff members the idea that the Institute could serve 'as a most effective nucleus for an international laboratory'.[30] At this stage Kramers apparently accepted that the laboratory would have two machines, as agreed the month before. His main concern was to have it located in Copenhagen. This would provide the basic infrastructure in

Notes: p. 470

personnel and equipment for the international facility which would thus develop 'naturally out of an existing institute (instead of having to build up at a place where there is nothing and nobody [...])'.[31] The prestige associated with Bohr's name was an added attraction.

One or two days after Kramers had started his discussions with Bohr, Sir James Chadwick and his wife happened to arrive in Copenhagen to visit the Bohrs. Kramers took the opportunity to have a talk with Chadwick about his new idea. Sir James later discussed it with Bohr and J.C. Jacobsen, who held a chair in the university and was closely associated with the Institute's affairs. During these talks Chadwick's sympathy for the proposal became apparent, and he returned to Cambridge determined to canvass support for it.[32]

On 23 August 1951 Kramers, still in Copenhagen, wrote to Auger asking for his reactions to the idea of having the new international centre there. He sent a copy of the letter to Chadwick asking him to pass it to Sir Charles Darwin, who had been involved in the discussions about the laboratory during the July meetings. Chadwick's reply was immediate and his support unconditional. On 30 August he wrote:

> My dear Kramers, Many thanks for letting me have a copy of your letter to Auger about your suggestion to create an international centre at Copenhagen. I shall support your proposal strongly.[33]

A detailed exposé of the evolution and refinement of Kramers' proposal has been given in chapter 5. Suffice it to say here that, by the end of October, not only the site of the new laboratory but also its scientific aims were explicitly in dispute. The opponents of the UNESCO project were not willing to take it for granted that the European laboratory should be equipped with accelerators. Indeed they wanted its whole programme of research to be drawn up anew. As Bohr suggested to Auger in a letter of 26 October, an international Board of Directors composed of senior scientific statesmen would formulate such a programme in consultation with a specially selected team of experimental and theoretical physicists. In the light of their recommendations the Board would submit detailed financial estimates to interested governments. The kind of apparatus with which the laboratory was to be equipped would thus be decided with reference to the agreed scientific programme, and not determined in advance as with the UNESCO project. More particularly, the question of whether the centre would have a bevatron was an open one, to be decided by the Board. At best it would be one long-term objective of the European laboratory, and not its main, immediate goal, as envisaged in the May report.

12.2.2 THE ACTIVITIES SPEARHEADED BY CHADWICK

Chadwick's expression of support for Kramers was coupled with an offer to 'take some other steps, in confidence, to prepare people in this country to receive your suggestion favourably'.³⁴ In this section we shall first briefly describe the measures Chadwick actually took to inform his influential associates in Britain about Kramers' proposal. We shall then investigate the initiative in greater depth, organizing the discussion around the following five questions:

  i) How was Kramers' proposal perceived in Britain?
  ii) How far were the British prepared to go in their support of it?
  iii) Why did they support it?
  iv) How important to Bohr and to Kramers was British support for their scheme? and
  v) Did Auger launch a counter-offensive in Britain?

As far as we know Chadwick only contacted two British scientists in connection with Kramers' proposal made to him late in August. These were Sir Charles Darwin, recently retired Director of the National Physical Laboratory, and Sir George Thomson of Imperial College. He wrote to both in mid-September, sending a copy of Kramers' letter to Auger of 23 August, and explaining that he had discussed the project with Bohr in Copenhagen.³⁵ Chadwick made it clear to both his correspondents that he favoured Kramers' idea and suggested that they meet to discuss the matter further. He subsequently had dinner with Darwin and hoped to meet Thomson early in October. He also got in touch with King, chairman of the Committee on Overseas Scientific Relations, told him about Kramers' scheme, and expressed his support for it.³⁶

At the same time Chadwick kept in close contact with Bohr. They informed each other of developments in Britain and on the continent. They understood each other's doubts and hesitations about the organizational and administrative problems raised by using Copenhagen as a pilot centre, and they shared each other's anxieties about the rather hamfisted way in which Ronald Fraser, the Liaison Officer between the International Council of Scientific Unions (ICSU) and UNESCO, was presenting their case in UNESCO circles.³⁷ Fraser was 'completely sold on the Copenhagen scheme', strongly encouraged the British scientists to back it and, in Chadwick's and Bohr's view, was letting his enthusiasm run away with him.³⁸

By the end of September, then, Chadwick had fulfilled his undertaking to Kramers made the month before. And although the group of confidants he had built up around him was small, it contained men whom he could trust and who were influential in both scientific and official circles. Now that we have a picture of this group, we shall turn to the first of our questions, viz.

  i) How was Kramers' proposal perceived in Britain? From the start Chadwick saw the two proposals for launching a European laboratory as radically different from, and incompatible with, one another. On the one hand, there was Auger's scheme, 'so ambitious and divorced from reality as to be quite impracticable', as he wrote to King. To this he was 'strongly opposed'. On the other, there was Kramers' proposal that 'a European laboratory be created in Copenhagen, using Bohr's institute as a basis from which to proceed'. This scheme, he wrote to Thomson,

Notes: pp. 470 ff.

'appeals to me very much. It is certainly practicable, and it is based on facilities, both in men and apparatus, which already exist'.[39]

It is striking that the British immediately polarized the situation. For Chadwick there were two proposals on the table, the one worthy of support, the other not. On the continent, by contrast, Auger and Bohr and Kramers were still in a conciliatory mood. As late as 26 October, when the battlelines had become more clearly drawn, Bohr could write to Auger of 'the great European effort with which, in principle, everyone so deeply sympathises'.[40] This desire to collaborate in a European project was one factor counteracting tendencies towards an open rift between Paris and Copenhagen. The British, lacking these sentiments, were less concerned to find a compromise.

ii) What was the extent of the support for Kramers' proposal in Britain? Firstly, it was not unqualified, for nagging doubts remained about the viability of the whole concept of regional research centres located on the continent. Darwin, for example, was 'not quite convinced that a European laboratory is really desirable'.[41] However, if there was to be one he agreed strongly that it should be in Copenhagen and wrote to Kramers accordingly. And Chadwick himself confided to Thomson that he was 'not strongly attracted to the idea of an international centre', though he was willing to help 'if it is generally desired on the Continent'.[42]

In fact British support for Kramers did not extend beyond backing the fundamental idea of grafting the new laboratory onto Bohr's institute in Copenhagen. Chadwick was kept informed by Bohr of the detailed debate surrounding the implementation of the project, and probably discussed it with his colleagues. But apart from mentioning qualms about the organizational and administrative problems raised by the laboratory, neither Chadwick, nor apparently anyone else in Britain, wished to help the opponents to the UNESCO scheme present a properly worked out proposal to the meeting of government representatives in December, nor to politic widely on its behalf.

Behind the apparently strong signs of support given by Chadwick and others to Bohr and to Kramers, then, there still lurked that reluctance to commit oneself wholly to a European project which we have encountered before. Chadwick and his colleagues did not see themselves as participating fully in a collaborative venture, but rather as the influential benefactors of a scheme intended primarily for continental physicists. On the other hand, it must be said that their interest in this alternative proposal exceeded that which they had shown in Auger's plan. In April the Nuclear Physics Committee could offer only limited technical assistance to the UNESCO project. Now we find Thomson suggesting to Chadwick that 'we should press' the new idea, 'even to the extent of trying to get some financial support from this country', and Sir James asserting that 'we would take some part' in the new European laboratory 'even at some sacrifice'; on one account Chadwick went as far as to tell Bohr that he would be willing to serve on a suitably constituted Board of Governors of the new laboratory.[43]

iii) Apart from a general desire to assist their colleagues across the Channel in a spirit of internationalism, what motives did the group around Chadwick have for offering to support Kramers and Bohr? One important reason was that the scheme that they had put forward was compatible with the traditional forms of international scientific collaboration, and indeed with standing British policy on international laboratories. It involved an extension of an existing facility rather than the establishment of a new laboratory from scratch. It did not make undue demands in personnel or in money, so posing little threat to national research efforts which were short of both. And it granted considerable decision-making powers to delegates from learned scientific societies, and to the elite of physics in general, who did not figure prominently as a group in the UNESCO project. Thus to Chadwick and his associates the Kramers–Bohr scheme was 'reasonable', 'practicable', and 'realistic' by contrast with Auger's alternative.

Other considerations, however, also played a significant role. Chadwick never detached the two projects from their chief protagonists. In supporting Copenhagen as a site he was in fact opposing Auger and expressing his confidence in Bohr and in Kramers. As he wrote to King, 'I have no doubt that if Kramers and Bohr produce a scheme it will be a practicable one', in contrast to Auger's unrealistic and ambitious proposal.[44]

Several factors underpinned Chadwick's solidarity with Bohr and Kramers. They were of the same generation, the 'old guard' of the twenties and early thirties, men who had a feel for the sweep of the discipline of atomic physics, who saw accelerators as part of a larger whole, who were used to operating with small budgets and relatively unsophisticated technology. They baulked at the financial implications of the May report, they were aware of the dangers to physics as such of investing so much money in one branch of it, accelerators, and they were not convinced that the biggest was necessarily the best. Around these shared values mutual bonds of respect and friendship developed. These tied the three closely together and generated the trust revealed in Chadwick's immediate support for Kramers' proposal.

More elusive, but as significant as these personal linkages, was a certain reticence about collaborating in a European venture. This reluctance was shared by several continental countries, notably the Scandinavians. They, like the British, did not feel themselves to be part of Europe, at least not with the fervour of the more central continental states. From a scientific point of view, too, there were shared grounds for their both standing somewhat aloof from the UNESCO-based project. If the British were the leading accelerator constructors in Europe, Bohr had built up the most prestigious theoretical physics institute on the continent, if not in the world. The benefits of Anglo-Danish co-operation in the field of nuclear physics, each complementing the other, were clear, and reinforced the ties between Chadwick and Bohr.

iv) How important was Chadwick's support, and that of the British physicists and authorities in general, to Bohr and to Kramers? By all accounts it was essential, at least to Bohr. Shortly after Kramers first proposed that his institute be used as a pilot centre for the new laboratory, Bohr wrote to Chadwick that 'the support of British physicists for the new idea would certainly more than anything contribute to common confidence in the soundness of the endeavour'.[45] And indeed in subsequent weeks Bohr was loath to take any major initiative without the assurance that the support of European physicists, and of the British in particular, would be forthcoming.[46]

v) Did Auger launch a counter-offensive in Britain? By the beginning of October Bohr and Kramers could count on the support of a small but influential group in Britain centred on Chadwick. Whatever its limitations, its very existence and its determination to do something on behalf of Copenhagen were a threat to the UNESCO project. Did Auger perceive it as such, and did he take concrete measures to counter it?

Apparently not. True, in the first fortnight of October, Auger wrote to Blackett mentioning that two sites were in contention for the new laboratory, and asking him for the names of British scientists and administrators whom he could contact; true, he told King how hostile the French were to Copenhagen as a site; true, he had lunch with Waterfield, giving him a 'very full and frank account of the ins and outs of the affair'.[47] But it is doubtful that these moves signalled a 'counter-offensive' on his part. Auger laid great store by British participation in the European laboratory project. He was simultaneously aware of the apathy in most British quarters to the venture, and of Chadwick's and Thomson's strong support for the Kramers–Bohr proposal. What he probably feared was that the 'debate' in Britain would be monopolized by this one party, and that this would make it increasingly difficult for the country to join a laboratory centred around two accelerators, as agreed in July. Quietly confident at this stage that his preferred scheme would be implemented despite the opposition, our view is that what Auger wanted was to ensure that the relative advantages of both proposals were widely discussed in Britain.

It must be said that even this limited objective was probably too much to hope for. It seems that little, if any, exchange of ideas and information about the project had occurred among British nuclear physicists for some six months. Blackett had seen no paper since those initially circulated to the members of the Nuclear Physics Committee in February, and described himself as 'out of touch' with the attitudes to Auger's scheme among his colleagues.[48] Cockcroft had apparently not corresponded with Chadwick on this matter since March. Furthermore, he appears to have had no knowledge of the steps Chadwick had taken in confidence to support Kramers and Bohr. Thus in mid-October there was a very uneven level of awareness among British nuclear physicists of developments both on the continent and at home. Properly co-ordinated responses to the emerging conflicts abroad were thus out of the question at the time. As a consequence a group like that around Chadwick, determined to advance the cause of Kramers and of Bohr, were left relatively free to shape British policy. This they did very successfully at the meeting of the British UNESCO Committee in November, to which we now turn.

12.3 Forging the alliance: the offer of the Liverpool cyclotron and the Paris conference

To be effective, the opposition to Auger's project in Britain had to be backed by some tangible form of support for the proposals of Kramers and Bohr. To this end, towards the end of October, Sir George Thomson had the idea that Britain could make the Liverpool cyclotron available to the European laboratory in its initial stages. Thomson had originally thought that, since the country could not afford to release personnel, some financial help should be given.[49] A wide range of austerity measures immediately introduced by a newly-elected Conservative government soon killed this idea. To offer the Liverpool machine seemed a suitable alternative 'political' and 'practical gesture'.

The machine at Liverpool, which was still under construction at the time, was a 400 MeV (proton) synchro-cyclotron, and so was suitable for meson physics. Authorization for its construction had been obtained by Chadwick on his return to the university from Washington in 1946. Much of its development, and its completion were entrusted to Prof. H.W.B. Skinner who filled the Lyon's–Jones chair of physics vacated by Chadwick in 1948. The diameter of the envisaged cyclotron grew from an initial 60 inches, first to 120 inches, and then to 156 inches. This inevitably entailed delays in its construction. At the end of 1951 estimates on when it would be ready varied from between about 9 and 18 months. In the event it first operated successfully in 1954, when it was for a while the most powerful accelerator in Europe.[50]

The offer of the Liverpool cyclotron was first officially discussed at a meeting of the British Committee for Co-operation with UNESCO in the Natural Sciences (BUC). Its deliberations, and the resolution it passed, are discussed in the first part of this section. In the second we present the negotiations undertaken by Chadwick with the authorities at Liverpool university, which resulted in their agreement in principle with the scheme. The offer was formally made by Sir George Thomson as British scientific delegate to the conference organized by UNESCO, and held in Paris from 17–21 December 1951. It is described and analyzed in the last part of the section.

12.3.1 THE MEETING OF THE BRITISH UNESCO COMMITTEE

The British UNESCO Committee met in the rooms of the Royal Society on 16 November 1951 with Sir Cyril Hinshelwood in the chair.[51] Chadwick, Thomson, King, and Waterfield were invited to attend. Among its other business, the

Notes: p. 471

committee discussed the 'Proposed establishment of a Regional Nuclear Physics Laboratory'.

Speaking to this item, the two physicists began by distinguishing the 'original proposal' for a 'Centre in Switzerland' from Kramers' proposal that the laboratory be located in Denmark and be 'associated' with Bohr. The former suggestion they deemed 'unrealistic'; the latter 'would command the support of British nuclear physicists'. But what tangible assistance could they give it? British manpower was likely to be occupied for some years on domestic equipment. However, the 'use of the cyclotron at Liverpool' could possibly be offered to the Copenhagen centre instead. The steps necessary to secure such an arrangement were discussed, and it was resolved that Chadwick be nominated British scientific delegate to the Paris meeting, and that

> he convey the goodwill of the United Kingdom towards the proposed establishment of a European Nuclear Physics Laboratory provided that the scheme is realistic and practical and the centre associated with an existing laboratory: that if it is proposed to locate the centre at Copenhagen under the aegis of Professor Niels Bohr the support of British scientists may be promised, and that the delegate be given discretion to refer to the reply, if received, from Liverpool university as a possible form of encouragement.

What significance should be attached to the BUC's requirements that the scheme be 'realistic and practical', and that the centre be 'associated with an existing laboratory'? We shall discuss each of these in turn.

The terms of the first requirement echo the sentiments expressed by Chadwick in his correspondence in September. There they were used to differentiate the proposals of Kramers and Bohr from the project aiming to equip Europe with a big accelerator. Though the minutes suggest that they probably also carried this connotation in the meeting on 16 November we cannot be sure what the committee as a whole felt. Certainly in a report written in February 1952 one of its secretaries, Rackham, said explicitly that the meeting supported the European project provided that it did not involve the building of large, new machines. On the other hand, at the Paris conference Sir George Thomson clearly stated his support for building a big accelerator.

A degree of uncertainty also surrounds the intentions behind the second condition, which merely requires that the centre be associated with an existing laboratory. If that laboratory was Bohr's institute in Copenhagen, British scientific support could be 'promised'. And if not? Presumably, though support could not be promised, the 'goodwill' referred to in the resolution remained, and so too the possibility of help of some kind, the key condition being that the centre be affiliated to an *existing* facility.

In this case the form of words used in the resolution embodied a compromise between a number of different views among the delegates.[52] At one extreme we have Chadwick's opinion as expressed in a letter he wrote to Mountford, Liverpool university's Vice-Chancellor: 'The suggestion about Liverpool is made *on the assumption* that the European centre would be in Copenhagen under the aegis of Niels Bohr' (our emphasis). Rackham thought that this was also Thomson's view, though Waterfield suspected that Sir George 'was less strong about this'. The Scientific Attaché himself felt 'that we should at least give a hearing to the arguments for a centre elsewhere' than in Copenhagen. This was also the view of the DSIR who, in the light of known French opposition to attaching the facility to Bohr's institute, felt that Britain should be prepared to be flexible about location. From these differing points of view a compromise was distilled on 16 November. As expressed by Waterfield in a letter written to Auger, immediately after the meeting it was decided 'that the U.K. would press strongly for a start in Copenhagen, but would be ready to be persuaded by *good* arguments to allow a start elsewhere or in several places'.

To summarize, the discussion in the British UNESCO Committee on the use of the Liverpool cyclotron was exploratory and tentative. While nothing definite was settled, the meeting as a whole looked for ways of making a meaningful gesture of support towards the European centre. A little over six months before the Nuclear Physics Committee had resolved that Britain could offer little more to the project than 'informal support by means of advice and help'. Now she was considering giving the centre some control over the use of a domestic accelerator, under conditions flexible enough to ensure that she did not have to take clear sides in the debate on its location.

The decision on 16 November heralded a change in British policy towards the European laboratory project. It was precipitated initially by Chadwick's determination to support Bohr and Kramers against Auger. However, no sooner did he and Thomson offer the cyclotron in the official forum of the British UNESCO Committee than other, more general, policy considerations were brought into play. When Chadwick set out to negotiate the terms of an agreement with Liverpool he may have thought he was doing so primarily to help Bohr and Kramers. In fact he was also forging a link more profound and enduring than mere technical support and advice. The first cracks had appeared in Britain's stance as detached observer.

12.3.2 THE NEGOTIATIONS OVER THE LIVERPOOL CYCLOTRON

Although the use of the cyclotron at Liverpool was under the control of the university authorities, the machine was actually the legal property of the DSIR. Like other costly scientific installations at British universities the department paid for its construction and maintenance by a special grant earmarked for that purpose. To make the cyclotron available to the European laboratory, therefore, the permission of both the DSIR and the university was required.

The agreement of the former, at least in principle, was readily obtained.[53] At Rackham's request, King discussed the matter with the Secretary of the DSIR, Sir Ben Lockspeiser. He confirmed the department's 'concurrence in the plan so far as the initial discussions are concerned'. If the need for formal contract arrangements arose, the matter would have to be referred to a specialized subcommittee.

The arrangements with Liverpool were taken care of by Chadwick on behalf of the BUC.[54] Correspondence with Skinner and Mountford was followed by a visit to Liverpool on 2 and 3 December, when Chadwick also met with the President of the University Council and the pro-Vice-Chancellor. The result of these discussions, as formally communicated by Mountford to Chadwick on 3 December, was

> that if a central laboratory for nuclear physics is established in Copenhagen, and if it is under the direction of Professor Niels Bohr, the University of Liverpool would be prepared in principle to consider the possibility of collaborating, under agreed conditions, for a trial period of, say, three or four years, with the central nuclear physics laboratory so established.

We can get some idea of the kind of conditions which were acceptable to the university from the correspondence with Chadwick and from the personal notes he made after his two-day visit.[55] They concerned primarily the kind of personnel Liverpool would find acceptable, their relationship to the experimental programme of the cyclotron, and the financial implications of the scheme. We shall treat each of these in turn.

In addition to receiving theoreticians as guests, Liverpool was prepared to accept about three or four research workers of at least Ph.D. standing, with experience in nuclear physics, for a period of a year or more. The name of G. Bernardini, the Italian cosmic-ray physicist who had just left a visiting professorship at Columbia University in New York, was specifically mentioned. The number of places was dictated by considerations of space and the fear that if there were too many senior researchers from abroad it would be difficult not to give them privileged status in the laboratory.

On the other hand the university was not keen to accept less experienced people. Research students were specifically excluded, though it was acknowledged that some technicians might be accepted for training. Skinner, in particular, felt that Liverpool should not be 'picked out' in this regard, and that inexperienced personnel should be distributed between all five of the British universities which had commissioned accelerators.

Then there was the question of communists and political activists. Skinner remarked that 'the laboratory is quite free from political agitation now and I would like it to remain free'. There was a 'good chance' that it would not if Liverpool collaborated in the UNESCO scheme.[56] To allay this and their other fears, the university authorities wanted to have a say in the selection of personnel sent to the laboratory.

As for the experimental programme, Mountford and Skinner emphasized that control over it should be vested in the Head of the Physics Department. There could be no question of some outside body like the European laboratory telling him what to do. It was conceded, however, that the experimental programme could be fruitfully discussed with the group at Copenhagen. And once the machine was running, European workers would have no privileged access to it. They would have to take their turn along with Liverpool researchers, including those junior to themselves.

Finally there were the financial questions. Liverpool assumed that the salaries of the European workers would be paid by the central laboratory. But this was not the only additional cost entailed in their association with the university. The cyclotron would probably be run for two shifts per day instead of just one, resulting in extra expenses in technical staff, power consumption, and possibly auxiliary equipment. The university could not bear these additional costs from its existing budget, and some special arrangement would have to be made with the DSIR.

To summarize, the Liverpool authorities were manifestly unenthusiastic about the collaborative scheme and there is little doubt that Chadwick played a crucial role in winning their reluctant agreement. It seems that the first they heard of the debate in the British UNESCO Committee was a rumour that they may be asked to relinquish control over their cyclotron for five years. This naturally incensed Mountford and Skinner. Though Chadwick managed to reassure them on this score, it is doubtful that he succeeded in persuading Skinner that a formal arrangement with the European centre was needed. He would have preferred a looser scheme of the conventional kind between collaborating laboratories, which would have left him in essential control of arrangements from the British end. In the event it was Chadwick's view, supported by Mountford, that prevailed.

The contrast between Skinner and Chadwick is yet another symptom of the changes afoot in British policy *vis-à-vis* the European laboratory. Skinner's attitude was consistent with his scepticism about the scheme, with the decision taken by the Nuclear Physics Committee in April to support it in the conventional manner, and indeed with standing British policy on international scientific collaboration. It was Chadwick who had changed. On the one hand he was convinced that it was in Liverpool's interest to collaborate formally with a European centre under Bohr. Apart from the prestige associated with such a link, there were the considerable scientific benefits to be gained from a liaison with Europe's leading Institute of Theoretical Physics. In addition, Chadwick was determined to steer the European laboratory project along 'realistic' and 'practicable' lines if he could, and this meant supporting the proposals of Kramers and Bohr, and thwarting the scheme favoured by Auger and his associates. To his mind the offer of the Liverpool cyclotron killed both birds with one stone.

12.3.3 THE PARIS CONFERENCE, DECEMBER 1951

On 31 August 1951 the Director-General of UNESCO invited all member states of the organization to send two delegates, one scientific, the other financial-administrative, to a 'Conference on the Organization of Studies Relating

Notes: p. 471

to the Possible Establishment of a European Nuclear Physics Laboratory'. The conference was to be held in Paris and after two postponements its date was set as 17–21 December.[57]

The official purpose of the conference was to gain the financial and technical support of states for 'stage 2' of the project outlined in the May report and subsequently approved by UNESCO's General Conference. Some $200,000 were needed during 1952 to set up a number of working groups 'to undertake more detailed studies and prepare cost estimates' for a laboratory centred around a giant accelerator. It was stressed that no country participating in this stage of the project would be committed to taking part in setting up a research centre afterwards.

After some interdepartmental correspondence, it was decided that Waterfield should act as Britain's official 'financial and administrative expert' at the meeting. Early in December Chadwick found that college duties precluded his going to Paris as scientific delegate, and he asked Sir George Thomson to replace him. He sent Thomson Mountford's official statement for use in Paris, and some notes on his negotiations with Liverpool which helped give Sir George 'the feel of things there', as he put it.[58]

The Paris conference is discussed in detail in chapter 6. Here we wish merely to present the contextual elements needed to grasp the significance of Britain's offer of the use of the Liverpool cyclotron, formally made by Thomson in his official statement. In what follows we shall first describe the situation in which the British delegates found themselves on arriving at UNESCO House, going on to present Sir George's statement of the United Kingdom's position, and then to analyze in more detail his offer of the use of the cyclotron.

In the six weeks before the conference the attitudes of the groups around Auger and Bohr had hardened. The opposition of the latter to the UNESCO project had become more manifest, and their determination to have the whole scheme reconsidered was matched only by that of Auger and some of the experts he had consulted to press ahead with planning a laboratory centred on one or two big accelerators. This group was united in their opposition to Bohr's view 'that a European Institute of Advanced Nuclear Studies should come first, and any new machines only as an afterthought', as Lew Kowarski put it.[59] Its members wanted at least one machine, most of them wanted it in Geneva, and for a variety of reasons, none of them wanted Bohr as leader of 'their' project.[60]

Cutting across these divisions was the intense interest in what Britain had to offer. Bohr's hopes of meaningful support had been raised by Chadwick who, with tantalizing formality, had informed him in late November that 'It would not be proper for me to tell you in detail what the U.K. statement will be (the draft has indeed not yet been circulated), but I think I may say that it will be in accord with your own views and wishes'.[61] This support was all the more crucial because Kramers did not attend the Paris conference, leaving the Chadwick group as Bohr's only prestigious scientific allies. For their part Auger's associates were keen to have Britain's scientific collaboration at least. They had come to hear of 'a British proposal to organize a European collaboration around the Liverpool synchro-cyclotron', though the details of the offer were obscure to them.[62] However, they were apparently sufficiently impressed by it as to consider withdrawing their proposal to equip the European laboratory with a 500 MeV synchro-cyclotron. The Liverpool machine could be used instead. In the event they decided to go ahead with their original plan at the Paris conference, and to reconsider it early in 1952 in the light of further information on Britain's attitude.

On arriving in Paris Thomson rapidly became aware of the tensions between the delegates.[63] As he reported it, Denmark and Sweden supported Kramers' idea, and were opposed to the construction of a big accelerator, which was deemed essential by France, Italy, and Belgium. Holland and Switzerland were rather divided. Initially Sir George spent a good deal of time with Bohr, an old friend. At Waterfield's insistence, however, he also had lunch with members of the other camp including Cornelis Bakker, Edoardo Amaldi, J. Hendrik Bannier, the Dutch Rapporteur for the conference, and Werner Heisenberg.

It was in this context that the official British statement offering the Liverpool cyclotron was made at the conference.[64] Thomson began by sketching the prevailing situation in Britain regarding accelerators, and its financial consequences. He made it clear that, though his country was keen to collaborate in international scientific ventures, in the present economic climate she could not afford to spend more money on big machines. Since the war one had been completed at Harwell, three more were under construction at universities, and another had been authorized for Cambridge. Now the claims of other expensive branches of physics, like radio-astronomy, and indeed of other branches of science, had to be heard.

All of the British accelerators, said Sir George, would of course 'be available for international co-operation of the old fashioned kind in which private research students from Europe come to work at the universities'. Britain was prepared to go further than this, however. The Liverpool synchro-cyclotron, said Sir George, was nearing completion, and 'should be ready by the late summer or autumn of 1952. It could be used for a period of years to give the Centre a machine of substantial size, over which it would have special rights, both as regards sending men to work with it and as regards the direction of researches to be undertaken'. The details of the offer had to be settled with the Liverpool authorities, although the DSIR, who owned the machine, had agreed in principle. 'There is', Thomson went on 'one necessary condition which is that the Centre should be associated with an existing research organization'. By doing this, the risks associated with the building of a large new machine could be reduced. 'We feel', Thomson added, 'that the laboratory at Copenhagen under Niels Bohr would be an ideal base for such a collaboration'.

In the closing section of his speech Sir George stressed that the envisaged machines would take a long time to build, and that it was important to get started quickly. Britain, he said, wanted to provide Europe with the opportunity to make significant discoveries in high-energy nuclear physics. She also wanted to provide

Notes: pp. 471 ff.

young men with a place 'on this side of the Atlantic' where they could work with high-powered instruments. 'It is because of these young men especially that we feel the problem is an urgent one', said Thomson. He concluded by stressing that the size of its budget was not the only, or the best, measure of the greatness of an institution. 'Men', said Sir George, 'are more important than machines'.

The offer of the use of the Liverpool cyclotron was well received by delegations on both sides of the divide at the conference.[65] Denmark and Sweden liked it of course. From the opposing camp, France welcomed the gesture in her official statement. It was a comparatively cheap way of starting collaboration quickly, and of doing scientific research while the major laboratory was being established. The conference as a whole resolved to accept 'the offer made by the delegation of the United Kingdom to use the Liverpool synchro-cyclotron for 400 MeV protons as an instrument to be operated on a European basis'. In discussion Sir George emphasized that this was a loan, not a gift, and that in the envisaged form of co-operation 'the University of Liverpool would probably be willing to waive their rights' over their accelerator 'for a period of time and within certain limits, which however could not yet be defined [...]'. It was realized that further negotiations with the university authorities and the DSIR were necessary, and Niels Bohr and Paul Scherrer (Switzerland) were asked to handle these on the conference's behalf. The conference also resolved that Bohr's institute in Copenhagen should serve as a nucleus for a new group providing 'theoretical guidance for experimental work' to be carried out with the accelerators, including that in Liverpool.

What did the British delegation achieve at Paris? Reflecting on the conference immediately after it closed, Sir George was well satisfied with their performance.[66] The offer of the Liverpool cyclotron, he wrote to Chadwick, 'was the centre of the meeting [...]. It has prevented us having to take a purely negative attitude, which would have been unpopular, and has thus added to our prestige'. The support promised to Bohr had been given and it looked as though a particularly close and mutually beneficial link between Copenhagen and Liverpool was in the offing.

On the other hand Chadwick's original aim of thwarting the UNESCO project and of 'converting' the French and the Italians had not been achieved. Here Thomson had resigned himself to the realities of the situation as he found it. 'The French and Italian feeling for a big machine is very strong', he wrote to Chadwick, 'and they will have to be allowed to try to raise the money which I do not think that they can do'. In the circumstances Sir George saw no point in opposing them head-on. In his official statement and in discussion he said clearly that he favoured the construction of a giant accelerator, and that the use of the cyclotron was a short-term and intermediate step towards this ultimate objective.[67] At the same time he comforted Chadwick with the thought that the UNESCO project would wither for lack of funds anyway.

Thomson also felt that the offer of the cyclotron had made an 'agreement' or 'compromise' possible between the opposing camps. He probably overestimated its importance. At the end of the Paris conference the two factions remained as determined as ever to pursue their still separate if 'complementary' schemes. For Bohr the conference was a success in that tangible offers of support had been made, including crucially that by Britain, and that no final decision had been taken on how to proceed in 1952. Auger's group would have preferred it to be otherwise. On the other hand time was on their side. The contributions in kind by Britain, and also by Denmark and Sweden, were 'Bohr's'. But some $150,000 had been pledged for the next stage of the project—almost half of it by France, and all of it by five countries who favoured the construction of a bevatron (France, Italy, Belgium, Switzerland, Yugoslavia). It was not difficult to imagine how this cake would be sliced by the conference when it reconvened in Geneva on 12 February 1952.

Indeed on reflection it seems clear that not only had Britain achieved somewhat less than she had hoped for in Paris, but that she had also done so at some cost to her manœuvrability in the further developments around the European centre. Sir George firmly believed that it was time his country made a positive scientific gesture towards the continent. In specifying the terms and conditions of the Liverpool offer he had adopted the flexible and co-operative line advocated by the British UNESCO Committee on 16 November. However, pressured by the enthusiastic delegates, and feeling that here was an opportunity to enhance his country's prestige, he found himself debating it in the language of those already committed to a joint European project. It was construed as a 'contribution in kind', 'to be operated on a European basis'. Theoretical guidance for its experimental program was to be provided by a European 'study group' newly assembled at Bohr's institute in Copenhagen. Whether they liked it or not, the British delegation was being forced to conceptualize its proposals in terms which it might not otherwise have chosen, and which carried with them implications which it did not necessarily agree with.

Whatever its limitations, then, the formal offer of the Liverpool cyclotron profoundly changed Britain's relationship to the European laboratory project. As long as her collaboration was restricted to technical help and advice—as agreed by the Nuclear Physics Committee in April—she could retain her detachment from the scheme, and control the extent of her involvement in it. This was no longer so easy. By making a 'contribution in kind' to the project, her status changed from that of interested observer to that of participant, formally on a par with any state which contributed materially to it. Her scientists were being gradually drawn into the ongoing debate among their continental colleagues on how to secure the future of European nuclear science—something they had avoided until then. And the conclusion they reached no longer concerned them alone. The DSIR was already implicated; soon complex financial and political decisions would have to be taken at the highest levels of government. For better or for worse Britain was now engaged in the European laboratory project. By one of those strange ironies of history, the prize that had eluded Auger and his associates was now almost within their grasp—thanks to the efforts of their opponents.

Notes: p. 472

12.4 Confronting the new question:
Should Britain join the Council of Representatives?

Two major resolutions dealing with the organizational measures required for the next stage of the European laboratory project were passed at the Paris meeting. The first dealt primarily with its scientific and technical aspects, the second with the steps needed to establish an overarching Council of Representatives from participating countries. The conference recommended that a working group be convened to draft an agreement for setting up such a council, and defining its powers and functions, and another for codifying its relationship with UNESCO.

The working group met immediately after the conference and twice more in January.[68] Waterfield was one of the seven people nominated to serve on it. Within a month it had drafted the two agreements asked of it.

On 22 January 1952 the Director-General of UNESCO sent copies of these draft agreements concerning the 'Council of Representatives of States for the Establishment of a European Nuclear Research Centre' to governments officially represented at the Paris conference.[69] The agreement establishing the Council had an annex (to be finalized in Geneva) in which formal offers of money or of contributions in kind were to be recorded. The Director-General pointed out that the Council could well be set up immediately after the reconvened conference, scheduled for 12–15 February. Delegates to Geneva who wished to sign either the agreement or its annex should therefore be vested with the necessary powers by their governments.

In this section we describe the reaction in Britain to these developments. The first part focuses on the scientists who, at a special meeting convened by the Royal Society on 23 January 1952, agreed that Britain should join the Council and drew up a detailed offer of the Liverpool cyclotron, which was to serve as a qualifying contribution in kind. In the second we investigate government attitudes to Council membership as expressed, firstly, in a special meeting called by the DSIR on 6 February, and then in a Cabinet Steering Committee meeting held two days later. The effects of these meetings on the British position at the Geneva conference are described in section 12.4.3.

12.4.1 THE SCIENTISTS PRESS FOR MEMBERSHIP

If Britain was to capitalize on the impact she had made at the Paris meeting she had quickly to finalize the arrangements with Liverpool university, in consultation with Bohr and Scherrer, and to find the money needed for them. A meeting with Mountford, Skinner, and other university authorities was arranged for 9 and 10 January. 'Perfect agreement'—to quote Skinner—was reached between the parties, in what Bohr called 'very pleasant and I think fruitful discussions'.[70]

Bohr had spent a couple of days with Chadwick before going up to Liverpool and, on his advice, did not press the university to make a precise and definite offer. He left satisfied that the authorities were keen to co-operate with him, and with a rough idea of the terms of collaboration which Skinner would find acceptable.

The details of the Liverpool offer were spelt out at a meeting of a newly-constituted Advisory Committee on the European Nuclear Research Centre. This committee was set up at the behest of Sir Cyril Hinshelwood, the Foreign Secretary of the Royal Society, its purpose being to provide expert advice to his British UNESCO Committee on developments around the European project. Apart from Chadwick, its chairman, membership comprised Thomson, Cockcroft, Lockspeiser for the DSIR, and Skinner for Liverpool university. The group, supplemented by Waterfield, met for the first time on 23 January 1952.[71] Among the committee's other business,

> Consideration was given to the terms on which this country could become a member of the Council of representatives by making a contribution in kind, the form of contribution being the provision of facilities on an international basis for work with the synchro-cyclotron at the University of Liverpool.

It went on to recommend that this offer be couched in the following terms:

> That Liverpool University agree to give special consideration to the acceptance of suitable candidates recommended by the Council of representatives. The University could provide places for four experimental physicists for at least a year each, two cyclotron engineers for much shorter periods, and one or two theoretical physicists. The authorities of the University agree to discuss the experimental programme with Professor Niels Bohr's group in Copenhagen and to assist the group, and would undertake as far as possible specific studies agreed upon at such discussions.

The Advisory Committee also resolved that, since Britain was not proposing to make any financial contribution to the establishment of the European centre, a request should be made to the Ministry of Education to arrange 'funds to help with the running costs of the synchro-cyclotron at Liverpool and to maintain either foreign workers at Liverpool or British physicists who work at Copenhagen'.[72]

The most striking feature about these recommendations is that, consistent with the logic of the Paris conference, Chadwick's Advisory Committee now favoured *joining* the Council of Representations, using the Liverpool cyclotron as a contribution in kind. Otherwise little has changed. Indeed there is not a great deal of difference between the terms of the Liverpool offer formulated by the Committee and those agreed on with Chadwick early in December 1951.[73] The recommendation is rather more specific on the kind of personnel acceptable to the university, and on its relationship with Bohr's group. Whereas Skinner was originally reluctant to accept more than three or four people, he now felt able to offer places for about eight. Significantly, too, the committee resolved that money should be found to pay the foreign workers at Liverpool, which had not been considered before. We shall return to this point again.

Though the terms of the offer were much the same as before, the spirit in which it was made by the Liverpool authorities was not. Skinner in particular now felt more

Notes: p. 472

positive about collaboration. As Chadwick had realized, Liverpool could certainly benefit from a formal link with Bohr's institute in Copenhagen. The department was not strong on the theoretical side, and the advantages of formulating a research programme for the new accelerator in consultation with Bohr and his staff were considerable. It was also short of technical personnel and struggling to get the machine going. The two cyclotron engineers, who Skinner was willing to take on immediately, could help commission the accelerator. Hence the resolution of the Advisory Committee that Britain should pay their salaries, and possibly those of other continental workers sent to Liverpool. In short, whereas before the stress had been on 'helping' Bohr, it was now more clearly realized by all concerned that the link between Copenhagen and Liverpool was in their mutual interests.

This shift owes much to the diplomacy of Bohr and Scherrer. They arrived in Liverpool in inauspicious circumstances.[74] Skinner and Mountford first heard of the Paris proceedings from a newspaper report which—following Thomson—claimed that the European Centre would 'supervise' work in 'loaned' laboratories, of which Liverpool was one. Their anger and suspicion were compounded by the terms of the resolution which stated that the cyclotron would be 'operated on a European basis'. Though Chadwick did his best to reassure them, it was in Mountford's words, 'the way in which Bohr, in particular, put the case [which] relieved us of some of the anxieties we had felt'.

This confidence in Bohr permeated the two days of discussion. Indeed whereas in November Skinner's emphasis had been on the dangers of collaboration, now it tended to be on the advantages. But then in November Skinner felt that he was being asked to collaborate with 'UNESCO people'. And, as we have described in section 12.1.2, he was typical of many senior British physicists in disliking the UNESCO project, and in not trusting Auger and his associates, and the French in general. Now he was making arrangements with Bohr and Scherrer, with 'competent people', with men who, like Skinner himself, were both 'obviously doubtful whether the Centre itself is going in the right direction', and who were sceptical about the plans to build a big accelerator. From these shared perceptions of the right way of doing things grew the willingness to co-operate and the conviction that it would be a worthwhile thing to do.[75]

That granted, one should be careful not to exaggerate the importance of the changed atmosphere at Liverpool. After all, at the Paris conference in December Thomson had implied that, in addition to collaborating with the envisaged centre in the 'old-fashioned' way, Britain was hoping to transcend its limitations. Elaborating, he had said that the university 'would probably be willing [partially] to waive their rights' over the cyclotron. On the face of it, the terms of the arrangement agreed to by the Advisory Committee were more restrictive than this, and involved no significant change in policy.

The reaffirmation of past attitudes as expressed through the terms of the offer of the Liverpool cyclotron is to be contrasted with a new determination to participate in the European laboratory project. Indeed one cannot but notice the anxiety in some quarters to ensure that Britain's credentials were in order. For example we have mentioned that Waterfield was invited to attend the Advisory Committee meeting on 23 January. As a member of the working party who had drawn up the draft agreements, he knew the kind of detail required in making the offer of the cyclotron. The form of words used in the recommendation quoted above owes much to him. What is more, two days after the meeting, at Chadwick's request, Rackham telephoned Auger at UNESCO, asking him to confirm 'that if the United Kingdom made contributions in the way of offering facilities at Liverpool University, then this would qualify the United Kingdom as a member of the Council'.[76] In short, whereas in April the year before Britain's physicists had come down firmly against participation in the European laboratory, now they were keen for her to join.

Two of the reasons for this change in attitude have already been mentioned—the scientific advantages of a link-up between Liverpool and Copenhagen, and the determination to support Bohr's rival project. A third reason for the shift is related to this: as a member of the Council, Britain could play a role in shaping the policy of the envisaged facility. This seemed all the more important as its outlines became clearer. Britain's physicists were especially unhappy about the way in which the $150,000 pledged in Paris were to be spent. In Auger's view it would almost all be used for planning the two accelerators. The members of the Advisory Committtee thought that this was a waste of money.[77] The design of the intermediate machine, the 400–500 MeV synchro-cyclotron, was well known, and the details could be obtained from Liverpool or Chicago, where such machines already existed. Chadwick thought that the situation was probably similar in the case of the bevatron—doubtless Ernest Lawrence at Berkeley could supply design data. And anyway the members of the Advisory Committee were 'not convinced on the present evidence that a machine as large as the largest U.S. Bevatron is necessary or desirable'. Why then devote so much money to planning the machines in stage 2? Some of it would be better spent—as Chadwick put it—on 'grants to European physicists to visit Copenhagen and Liverpool, or other suitable European laboratories—that is, for developing cooperation on the basis of existing facilities'.

By accepting the results of the Paris conference, the Advisory Committee found itself trying to achieve its objectives in a context which was in many respects alien to them. The collaboration it wanted between Liverpool and Copenhagen was more or less of the conventional kind, being based on linking existing national facilities. But to achieve this aim the committee found itself seeking membership of a Council of Representatives, one of whose responsibilities was to oversee the planning of two powerful accelerators and to organize a new form of international collaboration appropriate to 'big science'. The committee was not convinced of the need for this. To make matters worse the states who had voted the money in Paris were just those who wanted to see Europe equipped with a big bevatron. Despite the risk of

Notes: p. 472

becoming embroiled in a project they did not like, Britain's physicists, now well and truly ensnared in the framework of the Paris conference, saw no alternative way of serving their interests. Would their government support them?

12.4.2 GOVERNMENTAL REACTIONS

The British government's relationship with the envisaged Council of Representatives was discussed at two meetings held after Chadwick's Advisory Committee had met, and in the week before the Geneva conference. On 6 February a 'Subcommittee' of the DSIR's Committee on Overseas Scientific Relations (OSR) met to consider the implications of making the Liverpool cyclotron available to the researchers recommended by the Council. Their recommendations were forwarded to a meeting of the Cabinet Steering Committee on International Organisations (IOC) which met on 8 February to decide 'whether, as a matter of policy, Her Majesty's Government should support and participate in a European Nuclear Research Centre and become party to the agreement'. What transpired at these meetings and what recommendations were made?[78]

12.4.2.1 The OSR meeting. One of the main purposes of the OSR meeting was to gather together senior physicists and representatives of the government departments involved to formalize the offer of the Liverpool cyclotron, and to consider its financial and security aspects in particular. Scientific opinion was represented by the Assistant Secretary of the Royal Society, and by Cockcroft and Thomson who were specifically invited for that purpose.

Although the Liverpool accelerator was nominally a contribution in kind to the European laboratory, the DSIR had to find money to pay for the additional running costs and, possibly, the salaries of cyclotron engineers. The Treasury made it clear at the outset of the meeting that this would be 'very difficult' in the prevailing financial crisis, particularly since departments were being asked 'to do everything possible to decrease their expenditure on international organizations and their agencies'.

The main argument against this was that the additional cost of the cyclotron was small—estimates varied from £2000 to £7000–8000 per annum—and that the money would be well spent.[79] As Thomson and Cockcroft stressed, there was a long tradition of subsidizing universities engaged in 'international traffic' in science. There were considerable benefits from this and it would be a 'retrograde step' to oppose it in the present case. What is more, the accelerator had cost £1/2M to build and had a life expectancy of ten years. Using it to the full, the DSIR stressed, 'would constitute a more profitable return on the large capital expenditure involved' in its erection. Finally, as the Ministry of Education pointed out, this was 'an excellent way of making a real contribution to the work of UNESCO at negligible cost'. In the light of these arguments the Treasury was willing to consider providing the additional funds for operating the Liverpool cyclotron in conjunction with Bohr's group.[80]

The other major issue discussed by the OSR on 6 February was the question of security. The background to this discussion was the following. One of the articles in the draft agreement drawn up by the working party allowed for the participation in the envisaged Council of Representatives of any European state.[81] In principle this meant that eastern European powers could be admitted, with the prospect that scientists from these countries could work on the Liverpool cyclotron. Indeed Waterfield reported to the DSIR that applications from Czechoslovakia, Poland, and Hungary were 'in the air'.[82] This would have unfortunate effects. It was the policy of the Foreign Office not to make information in the field of nuclear or atomic science, be it classified or unclassified, available to Iron Curtain countries.[83] And there was the danger of a public outcry and of serious political repercussions if it were known that scientists from eastern Europe had been admitted to Liverpool.[84] The effects on American opinion also had to be considered.

Against this, there were grounds for thinking that the security risks were indeed slight. The draft agreement required that states wishing to join the Council had to be willing to co-operate in its work 'on a footing of free reciprocal exchange of persons and scientific information and techniques among its members'. In all probability this would automatically exclude eastern bloc countries. In addition, the work was in nuclear physics, not atomic energy, and so 'fundamental and innocuous'. This, it was said, should be stressed in all press publicity. All the same the Foreign Office wanted a number of amendments to be made to the draft agreement to enable Britain to 'reserve a right of choice in accepting Europeans', 'including undesirables from W. European countries', for work on the Liverpool cyclotron.[85] Apparently then both the financial and security aspects of the link-up with Liverpool were manageable, and the meeting instructed two of its members to draw up a paper requesting the Cabinet Steering Committee 'to authorise the British delegation to sign the Agreement at the meeting in Geneva'.

12.4.2.2 The IOC meeting. The Cabinet Steering Committee on International Organisations (IOC) was a high-level interdepartmental body advising ministers.[86] It was the last stage in the formal bureaucratic decision-making process on the question of whether or not Britain should join the European laboratory project. More than ten government departments were represented at its meeting on 8 February 1952. A.A. Dudley of the Foreign Office was in the chair. All the government officials who were at the OSR meeting two days before, apart from Verry who was ill, were present for part of the agenda, as was Waterfield. No scientists or delegates from learned scientific societies took part in the deliberations.

In what follows, we shall organize our discussion of the meeting around three questions which were of particular concern to the IOC, viz. i) What precisely was Britain committing herself to in signing the draft agreement drawn up by the working party? ii) Could a reversal of Britain's standing policy on international organizations, as entailed by signing the agreement, be justified? and iii) If Britain

Notes: pp. 472 ff.

did not sign what alternative gestures of support could she make? To study these questions we shall use the minutes of the meeting itself, and also reports on it written for Verry and for King by Murray, who was one of those present on behalf of the DSIR. The brief for the United Kingdom delegation to Geneva was prepared by the Ministry of Education taking into account the views expressed by the IOC, and is also used to throw some light on the proceedings.[87]

i) The IOC was puzzled as to whether, in signing the draft agreement, Britain in effect committed herself to participation in a European research centre equipped with accelerators. On the one hand, there was the agreement itself, which implied that she did. It was entitled 'Draft Agreement constituting a Council of Representatives of States *for the Establishment of a European Nuclear Research Centre*' (our emphasis), and in its preamble it expressly stated that the signatories desired, inter alia, 'to establish a European Research Centre to study phenomena involving high energy particles'. On the other hand, at the meeting Waterfield implied that this was an open question, that further studies were necessary, and that 'when these studies had been completed it might be decided that there should be a centre only, or one with its own machines, or even that the project should not be pursued further'.

The committee was not convinced by Waterfield's interpretation. To them it seemed that by signing, Britain would inevitably be drawn into a project demanding far more expenditure than the additional costs of the Liverpool cyclotron, and involving the construction of large new machines which a fortnight before her scientists had said were neither 'necessary' nor 'desirable'.

ii) The chairman made it clear at the start of the meeting that 'there was a standing policy of Her Majesty's Government against joining or promoting the creation of new international organisations unless they were shown to be absolutely necessary'. From as early as 1946 the IOC had expressed a distinct preference for collaboration between existing national laboratories, and the onus was on those who advocated signature of the agreement 'to justify running counter to these accepted principles'.[88]

This they did not do to the committee's satisfaction. The crux of the problem was that, to join the Council, Britain was being asked to sign what amounted to an intergovernmental treaty. This seemed an unnecessarily complex way for her to achieve her modest scientific aims. Even if those behind the project wanted to set up a European research centre—be it to co-ordinate the use of existing national facilities, or to provide a forum for discussion between scientists, or indeed be it equipped with its own costly apparatus—an intergovernmental treaty was unnecessary. There was even less justification for the cumbersome machinery involved if the object of the agreement was simply to set up a Council of Representatives to discuss the *desirability* of such a centre and its form of organization. For these reasons, and in the light of the anxieties described in (i) above, the Cabinet Steering Committee decided that it would not be possible to grant plenipotentiary powers to the British delegation to Geneva.

iii) It was impressed upon the IOC that it 'had to avoid giving the unfortunate impression that the United Kingdom was no longer anxious to co-operate in scientific matters', and that Britain should make some alternative gesture of support at the conference. Accordingly the Steering Committee resolved that, if requested, Britain would agree to collaborate in redrafting the agreement, to which she had some amendments to make. This was not to be seen as committing H.M. Government in any way, even if her amendments were accepted, though the IOC was prepared 'to consider the question again on the basis of the revised proposals which would no doubt emerge from the meeting being held in Geneva the following week'. In addition, the offer of the use of the Liverpool cyclotron remained on the table, subject to final arrangements being made between the relevant authorities, and provided that the offer 'was dissociated from any sort of inter-governmental agreement or new international organization'. The Treasury was willing to consider sympathetically requests for the additional expenditure involved if these conditions were met.[89]

With these discussions Britain's deepening involvement in the European laboratory project was temporarily halted. It took a body like the IOC to do this. Whereas scientists like Thomson and Chadwick and officials in the relevant sections of the DSIR and the Ministry of Education, had been associated with the scheme almost from the start, the Cabinet Steering Committee first assessed it long after it had picked up steam. Not caught up in its ongoing historical development, not sharing its momentum nor trapped by its inner logic,[90] it was freer to question its premises. It did so, and refused to sign.

12.4.3 THE GENEVA CONFERENCE, FEBRUARY 1952

The meeting called by UNESCO in Paris in December reconvened in Geneva under the chairmanship of Prof. Paul Scherrer on 12 February 1952.[91] The British delegation consisted of Waterfield and Thomson, who had represented their country on the previous occasion, and Rackham from the Ministry of Education.[92]

Thomson made it clear at the outset that his delegation had no powers to commit Her Majesty's Government.[93] Speaking personally and as a scientist, however, he stressed that Britain was sincere in its desire to co-operate in European nuclear research. These assurances, coupled with the wish of the conference, and of Auger in particular, to have Britain participate, assuaged doubts about the status of the British delegation, and it was allowed to join in the discussions.

The British team participated in two subcommittees set up to redraft the agreement.[94] Waterfield apparently played a significant role, alongside the Scandinavians, in having the agreement constitute a 'Council of Representatives of European States for Planning an International Laboratory and Organizing Other Forms of Cooperation in Nuclear Research'—a title which juxtaposed the projects championed by the groups around Auger and Bohr, respectively, thereby papering

Notes: p. 473

over the rift that had developed between them. Thomson also formalized the offer of the use of the Liverpool cyclotron.[95]

This item on the agenda was introduced by Bohr and Scherrer who gave a glowing report of their reception at Liverpool, and displayed photographs and drawings of the machines and buildings there. Thomson then made his statement, reading out the terms of the agreement laid down by the Advisory Committee on 23 January, and adding that the offer stood whether or not Britain joined the Council. Sir George went on to criticize the international character of the envisaged body, and to stress that British physicists were not convinced of the 'necessity or even desirability of the ultra large machine'. 'We have no intention of contributing financially to the building or running of such a machine', concluded Thomson.

While Bohr and his allies were doubtless satisfied with the British offer, the accelerator builders were bound to be disappointed. At one stage they had hoped that Britain would cede the Liverpool cyclotron to the European laboratory. Thomson had not only dispelled such optimism once and for all, he had also come out strongly against big accelerators—something he did not do in Paris. But then in Paris it looked very much as though Britain was going to join. Now 'whereas all the other delegates were empowered to sign in one way or another [her's] were casting doubts on the whole structure'.[96]

At noon on 15 February 1952 the Agreement constituting the Council of Representatives was signed by nine delegations, six of them subject to ratification. The annex, itemizing the contributions to the Council, was signed without reserve by seven countries. The United Kingdom delegation was the only one which did not sign either document, or imply that it would soon be in a position to do so.

12.5 The Cabinet Steering Committee reconsiders the case

The British delegates returned from Geneva fully aware that it was merely a matter of time before the Council of Representatives was formally constituted. Embarrassed by their ambiguous status at the meeting, and disappointed by the Steering Committee's refusal to recommend adherence to the Agreement, both Rackham and Thomson took immediate steps to have the decision reversed. The committee had, after all, not closed the door entirely on membership of the Council. On the other hand, it was not going to be easy to get it to change its mind, particularly as long as the physicists, as Thomson had stressed in Geneva, were not prepared to support the construction of an 'ultra-large' accelerator.

In this section we outline the steps taken in government and scientific circles to win support for British membership both in the Steering Committee and among the ministers concerned. It is divided into two main parts. In the first we provide a quick survey of events leading up to the meeting of the Cabinet Steering Committee on

11 June, where the question of whether HMG should join the Council was reconsidered, going on to describe ministerial reactions to its proposals. In the second part we analyze at greater length the arguments for membership in an attempt to grasp the motives of the chief protagonists, to understand why, despite their lack of conviction of the need for a big machine, they continued to press the IOC to reverse its earlier decision.

12.5.1 A BRIEF CHRONOLOGY OF EVENTS

The proceedings in Geneva closed on 15 February. Within a week Rackham had circulated a paper which he hoped to put before the meeting of the IOC on 29 February and Thomson and Hinshelwood had agreed that another official statement of scientific opinion should be made before then.[97] This was done at a meeting of the Executive Committee of the British UNESCO Committee on 28 February, which simply reaffirmed that Britain should join the Council for the now-standard reasons: to facilitate the Copenhagen–Liverpool link and, along with the Scandinavians, to shape the policy of the new organization ensuring that the 'twofold development' of the project allowed for in the Agreement took place.[98]

In the event this rush to have the question reconsidered was of little avail. Rackham's paper, drawn up in haste, met with widespread disapprobation, and was not discussed by the IOC on 29 February.[99] An improved version was circulated early in March, but this time Rackham resolved that he would not put it on the agenda of the Cabinet Committee again before he was reasonably sure of the support of the main departments concerned, notably the DSIR, the Foreign Office, and the Home Office.[100]

Matters were further complicated by the fact that whereas up to now all the activities in which Britain had been involved had occurred under the auspices of UNESCO, the new Council would be more or less independent of the international organization—and hence no longer the direct concern of the Ministry of Education. Accordingly Rackham suggested that the DSIR assume responsibility for the scheme 'particularly since any expenditure arising out of the Liverpool offer would presumably have to fall upon the D.S.I.R. vote'. This matter was discussed within the department in April. The machinery of King's Committee on Overseas Scientific Relations was available and suitable for dealing with such matters. With the approval of Sir Ben Lockspeiser, the Secretary of the DSIR, the department accepted 'to act as agent of Her Majesty's Government with the new Council', *if* the various departments could agree that the United Kingdom join the Council.[101] This did not mean that the DSIR took no interest in the matter until such agreement was reached, and indeed from May 1952 onwards it became increasingly involved in the project.

On Monday 5 May 1952 the Council of Representatives met in Paris for the first time. There had been no change in the government's official position; what is more neither Chadwick, Thomson, nor Cockcroft could attend. Sir James felt it would be 'difficult' for him to be in Paris early on the Monday as he had asked 'the Bishop of Salisbury to dedicate a memorial in [the College] Chapel to [his] predecessor [...] on Sunday afternoon or evening', Sir George 'was laid up in bed with a strained back', and Sir John had 'other engagements'. Thus a somewhat reluctant Waterfield found himself in the 'anomalous' and 'embarrassing' position of being the sole unofficial observer for the leading European country in nuclear physics.[102]

Hopes that this situation would change were soon raised again, only to be dashed. On 9 May Rackham was told by Butler, one of the secretaries of IOC attached to the Foreign Office, that, not withstanding their objections to the Agreement reached in Geneva, there did not appear to be 'fundamental obstacles' against the government's signing it. Rackham duly circulated yet another paper arguing the case for signature, and the matter was placed on the agenda of the Steering Committee meeting of 23 May.[103] Only to be removed once more. On 16 May Cockcroft had a private meeting with Sir Roger Makins in the Foreign Office.[104] Makins had got to know the British physicists in America towards the end of the war, and was to return to Washington as British Ambassador in 1952. According to Sir Roger's record of the conversation, Cockcroft's aim was to point out that 'British scientific opinion was generally in favour of our co-operation' with the Council and to persuade the Foreign Office of the need to join. For his part, Makins simply repeated the objections made at the IOC meeting in February: that the machinery of an intergovernmental agreement was unnecessarily cumbersome for the limited scientific objectives the physicists had in mind, and that by signing Britain would incur 'at least a moral obligation' to contribute to the cost of big machines, about the need for which the scientists were divided. Observer status at Council meetings was therefore sufficient. 'Since Sir John Cockcroft seemed to think that this would be all right', wrote IOC chairman Dudley on 21 May, it seemed unwise for the departments involved to press for signature at the moment.[105]

A period of intense activity followed. On 23 May an informal meeting was held in King's office, and attended by Verry and Murray (DSIR), Rackham (Ministry of Education), Conwy Roberts (Home Office), and Butler (Foreign Office and IOC).[106] Discounting Cockcroft's willingness to accept Makins' arguments as 'not well-considered', the DSIR insisted that 'for various scientific reasons' it was in the government's interest to adhere to the Agreement, and that they would recommend signature. It was agreed that the DSIR should submit a new paper for the IOC. This was circulated a fortnight later on 7 June, for consideration by the Steering Committee the following week.[107]

At the same time the scientists showed a new determination. On 27 May Chadwick's Advisory Committee met.[108] It nominated Chadwick or, failing him, Cockcroft to represent the Royal Society at the second session of the European Council to be held from 20–23 June. It also expressed the hope that Chadwick would be invited to the meeting of the Steering Committee to express the scientist's view—remember that no scientists had been present at the meeting of 8 February.

Immediately afterwards three of the committee members individually contacted Sir Roger Makins: Cockcroft wrote to him on 28 May again stressing that British scientists were very much in favour of adherence to the Agreement; Thomson pleaded the case in person with both him and Lord Cherwell, Prime Minister Churchill's closest confidant on scientific matters, on 30 May; Chadwick wrote to him on 2 June arguing that 'we have more to lose by not signing than by signing'.[109]

The Cabinet Steering Committee met on 11 June. Dudley was again in the chair; this time Chadwick was present. The committee had before it the DSIR paper which recommended signature, as did Sir James himself in his contribution to the meeting.[110] Neither had important new arguments to put before the committee, both reiterated that 'at present United Kingdom scientific opinion was against the building of large cyclotrons' (Chadwick), and both attempted to still Treasury fears about possible long-term commitments by stressing the temporary character of the Agreement, and that Britain's representatives had always made it clear that as things stood 'the United Kingdom could make no financial contribution' to 'the building of machines and the establishment of a laboratory' (Chadwick).

In reply Dudley rehearsed the usual objections of the Foreign Office. But the tone had changed. As Chadwick wrote the next day to Makins, 'I had expected to find sympathetic consideration but not the readiness to meet our views and to help to realise them as was shown by the chairman'.[111] And indeed the Steering Committee approved the DSIR's recommendation that HMG adhere to the Agreement subject to i) the Foreign Office producing a so-called 'document of interpretation' which would include a clear statement of the limited financial commitment assumed by the government (Dudley's suggestion), and to ii) the DSIR obtaining the Lord President's approval and iii) the Treasury obtaining the Chancellor's approval. Clearly then the lobbying had paid off. As Chadwick put it to Makins, he was 'most grateful' to Dudley for adopting such a co-operative attitude, 'and I suspect', Chadwick added, 'I have good reason to be grateful to you'.[112]

How did the ministers concerned react to this recommendation? A week after the Steering Committee had met, and on the eve of the second Council session in Copenhagen, it was clear what their response would be. The Treasury had been informed by the DSIR of what the financial implications of participation would be — 'a trifling sum' during the provisional period, but thereafter the government might find it 'difficult — though not impossible' to contract out of the total scheme, and the Treasury 'ought to be prepared, from some time in 1954 onwards for the next 5 years, to meet an annual expenditure of perhaps £200,000 per annum'.[113] This only hardened the Treasury's resolve. For his part Lockspeiser, asked by King to seek the Lord President's approval, refused to fight the case.[114] 'There would', he felt, 'be little hope of its being successfully contested'.[115] The matter was settled on 23 June when the Chancellor turned down the Cabinet Steering Committee's recommendations; soon thereafter the Lord President let it be known that he too did not like the proposals very much.[116]

Notes: pp. 473 ff.

12.5.2 THE MOTIVES OF THE PROTAGONISTS

With the recommendation of the Cabinet Steering Committee the British government moved significantly closer to membership of CERN, even though the approval of the 'science minister' and the Chancellor was not forthcoming. After all, the complex machinery of government decision-making had been successfully dealt with. There was no longer the need for high-ranking bureaucrats from different departments to secure agreement amongst themselves. The ministers as such were now the main targets, and it was at the level of interministerial consultation that the question would be resolved. The context in which the debate about membership would henceforth be conducted had thus changed markedly.

Of course the most striking thing about this shift was that it occurred *despite* the *weakness* of the scientific case for membership. As Makins pointed out to Cockcroft, British scientists were not convinced of the need for big machines and that 'even if we did not sign the treaty we would do all the things which we would have done if we had signed'.[117] Indeed this was already happening. On 29 April Skinner wrote to Auger stressing how pleased he would be to receive a number of engineers and physicists, including theoreticians, at Liverpool, adding that the necessary arrangements could be made 'without any formality'.[118] At the same time steps were being taken to place British physicists in the CERN accelerator groups: on 14 May Cockcroft proposed Pickavance as a part-time consultant to the SC group (as requested by Bakker a few days before), and Goward for a similar position in Dahl's group.[119] Plainly then it was not necessary to sign the Agreement to achieve the kind of scientific objectives the British physicists had in mind.

The progress made on 11 June also took place *despite* the *vacillation and indecision* of the scientists themselves. For example, Cockcroft told Murray (DSIR) on 6 May that he 'felt very strongly that we ought to take part officially', while ten days later, on 16 May, he was apparently rather easily persuaded to the contrary by Sir Roger Makins.[120] Similarly Chadwick, having argued for membership at the Steering Committee, confided to Makins a month later that he was 'somewhat doubtful about the need for signing the agreement'.[121] Apparently then senior British physicists had no coherent policy, no worked-out position *vis-à-vis* the *official* participation of their government in the European laboratory project. They had a number of ideas more or less developed, a number of sentiments more or less powerful, but no conclusive argument and limited conviction. Why then did they persist in their efforts to persuade their government to sign the Agreement? And whence the determination of the bureaucrats to get the Cabinet Steering Committee to reverse its earlier decision?

At the most general level there was the scientists' need to save face and retain credibility. In January Chadwick's Advisory Committee had recommended that Britain join the provisional Council offering access to the Liverpool cyclotron as a contribution in kind. To reverse that decision now would be difficult. It would be to imply that an error of judgement had been made, it would undermine the scientists' authority, and it would cast doubt on the confidence which the government had in them. As Dudley remarked, when the Foreign Office's reasons for not joining were explained to him, 'Sir George Thomson recognized that a difficult situation had arisen, and', Dudley went on, 'he asked that the Foreign Office should do its best to get the scientists out of it'.[122] For that to happen, though, the physicists' official line had to remain consistent, as indeed it did at the level of committee proceedings, if not in private.

Related to this was the need not to abandon Bohr completely. After all the scientists had originally brought the government into the scheme in order to support him. Personally they were not prepared to be too deeply involved: neither Chadwick nor Cockcroft, for example, attended important meetings, although Bohr frequently implored Chadwick to do so. What they did instead was to try to ensure that Britain's 'official' observers acted 'as far as possible in consultation and in co-operation with the Danish delegation' (Chadwick), while trying to upgrade the status of her representatives by getting the government formally to join the Council.

Refining our analysis a little, we must distinguish between Chadwick's and Cockcroft's reasons for pressing for membership (and probably Thomson's too, though we have not studied his papers). While both doubtless shared the general concerns just mentioned, the need to support Bohr was coupled in Chadwick's mind with the hope of redirecting the emphasis in the European laboratory project away from the construction of big accelerators. It was not in Cockcroft's. On the contrary Sir John had always been sympathetic to the building of an accelerator for the continent, and had released Goward to serve as a technical consultant to the scheme. His association with the venture was given a further impetus by the successful operation of the Cosmotron. Cockcroft visited Brookhaven on 27 March 1952, when he saw the new machine working for the first time.[123] This dispelled any doubts he may have had about its feasibility. It also stimulated his interest in the European project. In a report written after his return he listed costs in detail, adding data specifically needed 'to translate to European costs', and stressing that first results indicated that the cost/GeV of such machines would be considerably less than originally expected.[124] This is not to say that Sir John 'veered suddenly' in favour of his country joining the Council of Representatives, as some have suggested, but simply to indicate that he was even more favourably inclined towards the construction of big machines and even more concerned to retain a stake in what was happening on the continent.[125]

To conclude we wish to say a few words about the determination shown by Rackham (Ministry of Education) and by the DSIR to pursue the case for membership. In Rackham's case he was probably simply doing his job. As secretary of the British UNESCO Committee it was one of his tasks to ensure that its resolutions were translated into practice, and as long as the scientists argued for signature of the Agreement it was up to him to put the case as best he could to the other departments concerned. The same was of course true of a secretary like

Murray in the DSIR. Here though there was an additional factor at work — King's conviction that the scheme was a good one. King spelt out his philosophy in an article published in 1953: 'international collaboration in scientific research and development can be extremely valuable', he wrote, *but* 'should be based on real needs rather than on idealistic or political principles'.[126] Consistent with this attitude he immediately supported the European laboratory project 'in view of the lack of facilities for research on this subject in most European countries', and continued to argue its merits with vigour as a senior official in the DSIR in 1952.[127]

This calls for one final comment. King was attracted to the UNESCO-based project because he favoured regional organizations 'run in partnership by a few countries of approximately similar needs'.[128] Correlatively, the level of interest of British physicists in that project was associated with their perception of the relative lead they had in nuclear physics over their continental associates. In 1951 there was no doubt that they were ahead, and that their needs for the future were apparently markedly different to those of their colleagues across the Channel. Hence of course their initial diffidence to the venture. Matters began to look somewhat different in 1952, however, when it emerged that to remain among the world leaders British physicists had to take positive steps to have access to a big accelerator. It is to the story of how that awareness dawned in the community that we now turn.

Notes

1. The main archival sources used for this and the following chapter on Britain were the files deposited at the Public Record Office, Kew, London, and Sir James Chadwick's personal papers deposited at the Churchill College Archives Centre, Cambridge.

    It is uninteresting to give a complete list here of all the files consulted at the Public Record Office. The following were particularly useful:
    - Series AB6/1074 and 1076 of the United Kingdom Atomic Energy Authority, which is our basic source for Cockcroft's correspondence;
    - Series DSIR 17/559 and 560, covering the Department of Scientific and Industrial Research's dealings with the European laboratory project during 1951 and 1952;
    - Series ED157/302, the Department of Education and Science file on the 'Proposed European Nuclear Physics Laboratory 1951-2';
    - Series FO371/101514-101517, for Foreign Office attitudes in 1952;
    - Series CAB134/943 for the minutes of the meetings of the Cabinet Steering Committee on International Organizations in 1952.

    We have also consulted the Science and Engineering Research Council's files on CERN, and would like to thank them here for permission to do so. In fact we found little in the files for the period up to 1954 that was not already available at Kew. However, we collected a large number of documents for the later period which we shall use for volume II of this study.

    We found several books useful as general background reading. Hendry (1984) for aspects of British physics in the pre-war period; Gowing (1964) and Gowing (1974) for the British atomic energy programme during and after the war; Vig (1968) and Gummett (1980) for more general discussions of the relationship between science and government in post-war Britain; Sked & Cook (1979) for a political history of the country from 1945 to the end of the 70s; Frankel (1975) and Northedge (1974) for analyses of Britain's foreign policy during roughly the same period. Hartland & Gibbons (1972) is an interesting if misleading account of the process whereby Britain joined CERN.

The standard source for comprehensive biographical information on the more eminent (and deceased) scientific personalities is the *Biographical Memoirs of Fellows of the Royal Society*. For a biography of Chadwick which describes his activities as Master of Gonville and Caius in the fifties, see Chadwick (1978).

One potentially interesting source which we could not consult for lack of time was Sir George Thomson's papers. There is an interesting study still to be done of the different perceptions by Chadwick, Cockcroft, and Thomson in 1952 of the European laboratory project.

2. For this paragraph and the next, see chapters 2 and 3.
3. Translations of the CEC documents were provided for the OSR Committee, and were attached to document OSR(51)1, 3/1/51 (PRO-AB6/912). The French originals were obtained from the CEC archives in Geneva. Copies are in the CERN archives. The letter from the Council of Europe to the OLD was obtained from the Council's files. A copy is available in the CERN archives.
4. Letters Awbery to Chadwick, 25/1/51, and reply, 29/1/51 (CC-CHADI, 21/2).
5. Document N.P.C.54, undated (PRO-AB6/912).
6. Document N.P.C. 55, 7/2/51 (PRO-AB6/912). The material in this document was extracted from a letter McMillan to Awbery, 30/1/51 (PRO-DSIR17/559). It is interesting that although the NPC was asked to consider the proposal on its technical merits, Awbery omitted two relevant items mentioned by Auger to Waterfield when drawing up N.P.C. 55. These were that the energy of the accelerator might be as much as 10 GeV, and that the laboratory would do cosmic-ray work 'as a secondary interest'. It is not clear why Awbery chose to omit these items; it is doubtful whether they would have changed the NPC's response in any way.
7. For example, letters Blackett to Cockcroft, 13/2/51, and his reply, 16/2/51, both on (PRO-AB6/912).
8. Letter Chadwick to Awbery, 20/3/51 (CC-CHADI,21/2). Awbery's document summarizing individual views and appending Chadwick's draft letter is document N.P.C.57, undated (PRO-AB6/912).
9. Letter Chadwick to King, 23/4/51 (PRO-DSIR17/559).
10. These remarks were made at the OSR Committee meeting on 10/1/51, see document N.P.C.54, undated (PRO-AB6/912).
11. Amaldi made his remark in an interview with M. Gowing and L. Kowarski on 25/7/73 (CERN), 4. The second quotation is from Kowarski (1967), 5.
12. In an interview with M. Gowing on 5/6/73 (CERN), 26-27. In similar vein King has the following to say about 'the general British psychology with regard to international connections. This was a period of great complacency. There was a national feeling after the experience of the war and with our special relationship with the United States that we were in a superior position to the Europeans, that we had little to learn from them, but that, somewhat patronisingly, we would be glad to assist them' (private communication, 22/10/84).
13. Letter Blackett to Cockcroft, 13/2/51 (PRO-AB6/912). For Skinner see Gowing (1974), vol. II, 227. For the OSR Committee, see note 10. It must be stressed that Cockcroft did not share these attitudes. The scheme, he wrote to Blackett, was 'no doubt framed on the pattern of Brookhaven, which seems to be quite a reasonable project [...]. I do not think one ought to worry too much about certain crazy features of the paper which was circulated'—letter Cockcroft to Blackett, 16/2/51 (PRO-AB6/912). We shall discuss Sir John's attitudes to the European laboratory project at greater length towards the end of the chapter.
14. Letter McMillan to Awbery, 31/1/51 (PRO-DSIR17/559). On the basis of the documents available to the NPC it was not unreasonable to question France's motives for promoting the scheme—as indeed Hartland and Gibbons (1972) have done using domestic sources only. A different picture emerges if one approaches the matter from the point of view of some of the central actors—Auger, Dautry, and de Rougemont for example (see section 3.5 and Krige (1984b), section III). These accounts favour the view recently stated by King that 'French enthusiasm for European cooperation was not, [he] felt at the time, a matter of

chauvinism and a desire to dominate in Europe, as many suspected, but a genuine desire to reconstruct the fabric of European science' (private communication, 22/10/84).

15. Letters Blackett and Skinner, see note 13; letter Chadwick to Cockcroft, 13/3/51 (PRO-AB6/912). On Auger King writes: 'Auger was never really trusted by the British, not only with regard to CERN pre-negotiations, but also on many other UNESCO matters; this was particularly true concerning the Royal Society circles' (private communication, 22/10/84).
16. For British accelerators at the time see Gowing (1974), vol. II, appendix 21; Harwell (1952), chapter 6; this volume section 1.3.3; letter Wallace to Verry, 13/2/53 (PRO-DSIR17/562).
17. Thomson as quoted by Gowing (1974), vol. II, 226; letter Chadwick to Lockspeiser, 2/8/50 (PRO-AB6/774).
18. See *Nuclear Physics Research at Universities*, among the papers prepared for an informal meeting arranged by the DSIR, 23/6/50 (PRO-AB6/774).
19. For example, between 1948 and 1951 special awards by the DSIR to areas of physics, mathematics, and astronomy *other than* nuclear physics and cosmic-ray research amounted to £150,000. In the same three years chemistry, engineering, and biology *together* received some £140,000.
20. *Notes submitted by Sir Lawrence Bragg*, 20/6/50, among the papers for the meeting on 23/6/50—see note 18.
21. The quotation, from a resolution passed by the Royal Society's International Relations Committee in 1948, is to be found in *International Research Laboratories. Draft Brief for U.K. Delegation to ECOSOC*, IOC(50)95, 26/6/50 (PRO-CAB134/405). This paper surveys the situation. For more detail see documents IOC(48)118, 22/6/48 (PRO-CAB134/390); IOC(49) 1st Meeting, 7/1/49 (PRO-CAB134/395).

    British policy in this period, her reluctance to join in a collaborative venture, is akin to that which Salomon attributes to the superpowers: 'Apart from those fields which, by their extranational nature, require joint research efforts (meteorology, oceanography), the two 'big ones' expect from scientific cooperation only what it has always provided—the exchange of ideas and new information. On the other hand, all the other countries are obliged to see in cooperation the indispensable path to the more economic or more rapid achievement of their national objectives'. See Salomon (1971).
22. What follows is based on Frankel (1975), Northedge (1974), Sked & Cook (1980), Laqueur (1982).
23. The first remark is from Northedge (1974), 171. For Cockcroft see Gowing (1974), vol. I, 338.
24. See note 9.
25. Letter Auger to Cockcroft, 2/3/51, and reply, 19/3/51 (PRO-AB6/912). The group at Harwell encouraged the project in other ways too. Cockcroft invited Auger to come to England to discuss the 'general character of the UNESCO project with two or three senior physicists' (letter Cockcroft to Auger, 22/3/51 (PRO-AB6/912)), Goward sent progress reports on the Brookhaven machine to Auger early in June (letter Goward to Auger, 8/6/51 (UNESCO)), etc.
26. The official account of the meeting is *Report by the Director-General on preliminary studies regarding the establishment of a European Regional Laboratory for Nuclear Physics*, 6C/PRG/25, 19/6/51 (UNESCO).
27. Telegram Bohr to Auger, 9/7/51 (UNESCO).
28. Letter Kramers to the Director-General of UNESCO, 24/9/51, reproduced in annex 7 of document UNESCO/NS/NUC/1 (UNESCO).
29. Letter Auger to Rabi, 21/8/51 (UNESCO); letter Rabi to Auger, 15/8/51 (UNESCO).
30. Letter Kramers to Auger, 23/8/51 (UNESCO).
31. *Ibid.*
32. Letters Chadwick to Kramers, 30/8/51 and Chadwick to Darwin, 13/9/51, both on (CC-CHADI,1/13); Chadwick to Thomson, 14/9/51 (CC-CHADI,1/9); Chadwick to King, 17/9/51 (PRO-ED157/302). See also letter Fraser to Auger, 11/9/51 (CC-CHADI,1/3).
33. Letter Chadwick to Kramers, 30/8/51 (CC-CHADI,1/13).
34. *Ibid.*
35. See letters Chadwick to Darwin and to Thomson, note 32.
36. Letter Chadwick to King, 17/9/51 (PRO-ED157/302).

Notes

37. See letters Bohr to Chadwick, 14/9/51; Chadwick to Bohr, around 20/9/51; Bohr to Chadwick 20/9/51 and 25/9/51, all on (CC-CHADI,1/3). For Fraser's activities see letters Fraser to Chadwick, 11/9/51 (CC-CHADI,1/9); Fraser to Auger, 11/9/51 (CC-CHADI,1/3); Chadwick to Fraser, 19/9/51 (CC-CHADI,1/9); Bohr to Fraser, 20/9/51 (CC-CHADI,1/3); Fraser to Chadwick, 24/9/51 (CC-CHADI,1/9); Thomson to Chadwick, 8/10/51 (CC-CHADI,1/10).
38. The words, which are Fraser's own, are in a letter Fraser to Auger, 11/9/51 (CC-CHADI,1/3). Bohr and Chadwick were angry with Fraser for giving the impression that Bohr (not Kramers) was behind the idea that Copenhagen be the site for the new laboratory. This suggested that Bohr was primarily concerned to further his own interests rather than those of European scientific collaboration.
39. Letter Chadwick to Thomson, 14/9/51 (CC-CHAD I,1/9).
40. Letter Bohr to Auger, 26/10/51 (UNESCO).
41. Letter Chadwick to Bohr, about 20/9/51 (CC-CHADI,1/3).
42. Letter Chadwick to Thomson, 14/9/51 (CC-CHADI,1/9).
43. Letter Fraser to Bohr, 11/9/51 (CC-CHADI,1/3); for the quotations from Chadwick and Thomson see letters Chadwick to Thomson, 14/9/51 (CC-CHADI,1/9) and Thomson to Chadwick, 8/10/51 (CC-CHADI,1/10).
44. Letter Chadwick to King, 17/9/51 (PRO-ED157/302).
45. Letter Bohr to Chadwick, 14/9/51 (CC-CHADI,1/3).
46. Fraser claimed that Britain had a crucial role to play at the intergovernmental conference in December: 'Only a strong U.K. delegation, fully briefed to support the Copenhagen scheme, can carry the day so completely that no one can say afterwards that the decision was not genuinely unanimous', letter Fraser to Chadwick, 24/9/51 (CC-CHAD I,1/9). As far as we can see Fraser had his own axe to grind though, and it is difficult to know how reliable his assessment was.
47. Letters Auger to Blackett, 4/10/51 (PRO-AB6/912); letter Waterfield to Verry, 18/10/51 (PRO-DSIR17/559); letter King to Chadwick, 11/10/51 (PRO-ED157/302).
48. Letter Blackett to Auger, 9/10/51 (PRO-AB6/912).
49. Letter Thomson to Chadwick, 8/10/51 (CC-CHADI,1/10).
50. For information in this paragraph see Chadwick (1976), 42-43; Skinner (1960), 263-264; document cited in note 18.
51. The minutes of the ninth meeting of the British Committee for Co-operation with UNESCO in the Natural Sciences, held on 16/11/51, are document C/113(51) or NS(51)20, undated (PRO-ED157/303).
52. The material in this paragraph is drawn from letter, Chadwick to Mountford, 21/11/51 (CC-CHADI,1/8); letter Rackham to Waterfield, 10/12/51 (PRO-ED157/302); letters King to Chadwick, 28/9/51 (CC-CHADI,1/9) and 11/10/51 (PRO-ED157/302); the final quotation is from letter Waterfield to Auger, 16/11/51 (UNESCO).
53. Letter King to Rackham, 28/11/51 (PRO-ED157/302).
54. Letters Mountford to Chadwick, 19/11/51; Skinner to Chadwick, 21/11/51; Chadwick to Mountford, 21/11/51; Mountford to Chadwick, 27/11/51; Chadwick to Skinner, 27/11/51; and Mountford to Chadwick, 3/12/51, all on (CC-CHADI,1/8).
55. See material in previous note, and Chadwick's *Notes on Visit to Liverpool, 2 and 3 December*, and *Arrangements with Liverpool*, undated, both on (CC-CHADI,1/8).
56. Letter Skinner to Chadwick, 21/11/51 (CC-CHADI,1/8).
57. The text of the Director-General's letter is UNESCO document CL/574. The Working Paper for the Conference is document UNESCO/NS/NUC/1. The Draft Final Report is document UNESCO/NS/NUC/9 (Prov.), 21/9/51. The final report was written by J.H. Bannier, and is document UNESCO/NS/99. All are available in (CERN).
58. Letter Thomson to Chadwick, 22/12/51 (CC-CHADI,1/10).
59. Letter Kowarski to Chadwick, 15/12/51 (CC-CHADI,1/11).
60. Their attitude is described by Waterfield in his report *UNESCO Conference on the Establishment of a European Research Laboratory*, 21/12/51 (PRO-ED157/302).

61. Letter Chadwick to Bohr, 26/11/51 (CC-CHADI,1/3).
62. Letter Kowarski to Chadwick, note 59. There is also a reference to the proposal in letter Bakker to Mussard, 11/12/51 (UNESCO). Bakker's letter suggests he got the idea from Cockcroft.
63. See Waterfield's report, note 60. Also Thomson's report, *Notes on a Conference Relating to the Establishment of a European Nuclear Reseach Laboratory, held at U.N.E.S.C.O., Paris, 17th to 20th December*, 20/12/51 (PRO-ED157/302). See also his letter to Chadwick, note 58.
64. The statement is in document UNESCO/NS/99, Annex V, available in (CERN). It was also Annex I to Thomson's official report, see note 63.
65. For details see document UNESCO/NS/99 (CERN).
66. Letter to Chadwick, see note 58.
67. See note 65.
68. To gain an impression of its business see letter, Waterfield to Rackham, 8/1/52, letter Verry to Evans, 11/1/52, and letter Waterfield to Rackham, 31/1/52, all on (PRO-ED157/302).
69. The letter is included in UNESCO's document NS/269.950, which also contains the Draft Agreement and its Annex.
70. For details on this visit see letters Chadwick to Mountford, 1/1/52 (CC-CHADI, 1/8); Hinshelwood to Mountford, 2/1/52 (CC-CHADI,1/5); Chadwick to Bohr, 3/1/52 (CC-CHADI,1/3); Chadwick to Mountford, 8/1/52 (CC-CHADI,1/8); Bohr to Chadwick, 14/1/52 (CC-CHADI,1/3); Skinner to Chadwick, 14/1/52, and Chadwick to Mountford, 19/1/52, both on (CC-CHADI,1/8). See also Chadwick's untitled personal notes, dated '17 January, 1952' (CC-CHADI,1/8).
71. Hinshelwood's initiative is contained in his letter to Chadwick, 1/1/52 (CC-CHADI, 1/5). The minutes of the first meeting of the Royal Society's *Advisory Committee on the European Nuclear Research Centre* on 23/1/52, document NS(52)1, are on (PRO-ED157/302).
72. All of these quotations are from the minutes of the meeting—see note 71.
73. See section 12.3.2, and material in note 55.
74. The press report appeared in the *Manchester Guardian*, 20/12/51; it is available on (CC-CHADI,1/8). There is further relevant material in letters Mountford to Chadwick, 3/1/52 and Chadwick to Mountford, 8/1/52, both on (CC-CHADI,1/8). Mountford's remark at the end of the paragraph is from his letter to Chadwick, 10/1/52 (PRO-ED157/302).
75. For Skinner's old and new attitudes, see his letters to Chadwick of 21/11/51, and 14/1/52, respectively, both on (CC-CHAD I,1/8).
76. For the wording of the offer see letter Verry to Evans, 11/1/52 (PRO-ED157/302). For the quotation see letter Martin to Chadwick, 25/1/52, on the same file.
77. For Auger's view, see letter Martin to Chadwick, note 76. For Chadwick's dissatisfaction see his letter to Bohr, 26/1/52 (CC-CHADI,1/3), and his *Additional Comments on the Projet d'accord*, appended to his letter to Martin, 28/1/52 (CC-CHADI,26/1), from which the concluding quotation in this paragraph is taken. The proceedings of the Advisory Committee are document NS(52)1, see note 71.
78. The minutes of the OSR 'sub-committee' meeting are a document prepared by the Overseas Liaison Division of the DSIR, *European Nuclear Research Centre*, 7/2/52 (PRO-AB6/1074). For the minutes of the IOC Meeting held on 8/2/52, see *Cabinet Steering Committee on International Organizations*, document I.O.C.(52) 2nd Meeting (PRO-CAB134/943). For information on the constitution and composition of these committees, see chapter 13, appendix 1.
79. See the minutes of the OSR meeting—note 78.
80. Letter, Martin to Chadwick, 7/2/52 (CC-CHADI,26/1).
81. A copy of the *Draft Agreement constituting a Council of Representatives of States for the Establishment of a European Nuclear Research Centre*, is found in annex II of UNESCO document NS/269.950 (CERN).
82. Letter, Waterfield to Rackham, 31/8/52 (PRO-ED157/302).
83. See internal memo, Verry to Murray, *Re Security Aspect*, 5/2/52 (PRO-ED157/302). Also document *European Nuclear Research Centre*, drawn up by Saner, 5/2/53 (PRO-FO371/101514).

84. This was the first point discussed in a press report on an interview with Sir George Thomson after the Paris conference—see *Manchester Guardian*, 20/12/51, cutting on (CC-CHADI,1/8). See also internal memo, Verry to Murray, previous note. The remark about American public opinion is explicitly made in the minutes of the IOC meeting—see note 78.
85. The comment about 'undesirables from W. European countries' is from the Foreign Office document cited in note 83.
86. For more detail see chapter 13, appendix 1.
87. Memo, Murray to Verry, *European Nuclear Research Centre*, 9/2/52 (PRO-DSIR17/559); letter Murray to King, 11/2/52 (PRO-DSIR17/551); *Brief for U.K. Representatives* to Geneva conference, undated (PRO-DSIR17/559).
88. The documents spelling out the Foreign Office's standing policy on international scientific research laboratories were summarized for the IOC in document I.O.C.(52)9, Addendum 1, 7/2/52 (PRO-CAB134/944).
89. This is stated by Murray to Verry in his memo, 9/2/52 (PRO-DSIR17/559).
90. For a lucid description of this process see letters Verry to Rossiter, 7/2/52, and Murray to Collier, 8/2/52, both on (PRO-DSIR17/559). See also appended comments on Saner's statement for the Foreign Office—note 83.
91. The draft report on the conference is document UNESCO/NS/NUC/16 (Prov.), (CERN). For a detailed discussion see chapter 6. An official British report on the conference is document NS(EXEC)(52)1, February 1952 (CC-CHADI,26/2).
92. There was considerable difficulty finding a scientific delegate to attend this conference. Thomson initially stated that he could not attend and that he thought Chadwick should go; he, however, was doubtful (see exchange of letters between them 29/1/52, 31/1/52, 1/2/52, all on (CC-CHADI,1/10)). Rackham asked Cockcroft if he was available, but an impending trip to the United States precluded his going (see exchange of letters, 2/2/52 and 5/2/52 (PRO-AB6/1074)). In the end it was a somewhat reluctant Thomson who went to Geneva.
93. See Rackham's 'Outward Savingram. From U.K. Delegation, Geneva, to Foreign Office', 14/2/52 (PRO-ED157/302).
94. Letter Waterfield to Murray, 14/2/52 (PRO-DSIR17/559).
95. The full offer is spelt out in *Text of Sir George Thomson's Statement to the Conference on the Liverpool Cyclotron*, undated (PRO-ED157/302). Other material is available in the official report of the meeting—see note 91.
96. Letter Rackham to Murray, 6/3/52 (PRO-DSIR17/559).
97. See Rackham's draft IOC paper, *European Nuclear Research Council*, provisionally labelled IOC(52)12, 22/2/52 (PRO-ED157/302); letters Martin to Chadwick, 19/2/52 (PRO-ED157/302), and 22/2/52 (CC-CHAD I,26/1).
98. For the minutes of this meeting see document NS(EXEC)52(3), 28/2/52 (CC-CHADI, 26/2).
99. See, for example, letter Verry to Rackham, 25/2/52, and the two enclosures, *Proposals for European Nuclear Research Centre*, 25/2/52, by Verry, and *European Nuclear Research Council*, 23/2/52, by Murray, all on (PRO-ED157/302).
100. See letters Rackham to Butler, 6/3/52 (PRO-DSIR17/559), Rackham to Thomson, 19/3/52 (PRO-ED157/302). We could not find the revised document Rackham circulated at this time, but it was probably only slightly different from document I.O.C.(52)56—see note 103.
101. This and the previous quotation are respectively from letters King to Verry, 25/4/52 (PRO-DSIR17/551), and Rackham to Verry, 7/4/52 (PRO-ED157/302). For other material see, for example, memo Verry to Evans and Brown, 9/4/52, and memo King to Lockspeiser, *European Nuclear Research Centre Proposals*, 23/4/52, both on (PRO-DSIR17/559).
102. For the reasons given by the physicists see letters Chadwick to Martin, 26/4/52 (CC-CHADI,26/1), and Rackham to Waterfield, 1/5/52 (PRO-ED157/302). For arrangements between the DSIR and the Scientific Attaché, see letters Murray to Waterfield, 2/5/52 and 8/5/52 (PRO-DSIR17/559), and

Waterfield to Murray, 9/5/52 (PRO-ED157/302). For Waterfield's report on the first Council session, see his memo *European Nuclear Research Council*, May 1952 (PRO-DSIR17/551).

103. Letter Butler to Rackham, 9/5/52, with enclosed document USE 1241/18, *Comments of Foreign Office Assistant Legal Adviser on the European Nuclear Research Agreement*, undated (PRO-ED157/302). Rackham's revised paper for the Cabinet Steering Committee was document I.O.C.(52)56, *European Nuclear Research Council*, 20/5/52 (PRO-ED157/302).
104. See Makins' *Record of Conversation*, 16/5/52 (PRO-FO371/101515).
105. Letter Dudley to Cowell, 21/5/52 (PRO-ED157/302).
106. For a report on this meeting by Butler, see document E1241/29, 24/5/52 (PRO-FO371/101515).
107. It was document I.O.C.(52)56(Revise), *European Nuclear Research Council, Memorandum by the D.S.I.R.*, 7/6/52 (PRO-DSIR17/551).
108. The minutes of the meeting are document NS(52)1, 18/6/52 (PRO-ED157/302).
109. Letter Cockcroft to Makins, 28/5/52 (PRO-FO371/101515); for Thomson, memo by Dudley, *Nuclear Research Centre*, 30/5/52 (PRO-FO371/101515), and memo to King, *European Nuclear Physics Co-operation*, 6/6/52 (PRO-DSIR17/551); letter Chadwick to Makins, 2/6/52 (CC-CHADI,1/6).
110. For the DSIR paper, see note 107. For the minutes of the meeting on 11/6/52, see document I.O.C.(52) 10th Meeting (PRO-CAB134/943).
111. Letter Chadwick to Makins, 12/6/52 (CC-CHADI,1/6).
112. *Ibid*.
113. Letter Brown to Figgures, 16/6/52 (PRO-DSIR17/551).
114. Letter King to Lockspeiser, 12/6/52, enclosing minutes by Barnard putting the case to the Lord President, 13/6/52 (PRO-DSIR17/551).
115. (Draft) letter Lockspeiser to Chadwick, 20/6/52. See also letter Murray to Verry, 19/6/52, both on (PRO-DSIR17/559).
116. See memo by Verry, *Nuclear Physics Research Council*, 24/6/52 (PRO-DSIR17/559); letter Martin to Chadwick, 27/6/52 (CC-CHADI,26/1). For the Treasury's attitude, see memo by Butler, *European Nuclear Research Council*, 19/6/52 (PRO-FO371/101515).
117. See note 104.
118. Letter Skinner to Auger, 29/4/52 (PRO-ED157/302).
119. Letter Cockcroft to Murray, 14/5/52; letter Bakker to Cockcroft, 8/5/52 (PRO-AB6/1074).
120. Letter Cockcroft to Murray, 6/5/52 (PRO-AB6/1074).
121. Letter Chadwick to Makins, 17/7/52 (CC-CHADI,1/6).
122. Memo by Dudley, *Nuclear Research Centre*, 30/5/52 (PRO-FO371/101515).
123. Letter Cockcroft to Bohr, 5/4/52 (PRO-AB6/1074); memo by Cockcroft, *Notes on the Brookhaven Cosmotron*, 16/4/52 (CC-CHADI,1/2).
124. See Cockcroft's memo, note 123. The basic reason for the decrease in cost/GeV was that preliminary tests on the Cosmotron confirmed that to preserve the beam the magnet gap, and so the magnet weight, could be considerably smaller than the accelerator engineers had originally thought.
125. The claim that Cockcroft 'veered suddenly' in favour of British membership was made by Kowarski during an interview which he and Gowing made with Mussard on 13/9/73 (CERN), 23. Indeed he apparently reported that Cockcroft 'was enthusiastic for Britain to take part in the European scheme for the cosmotron' at the first session of the provisional Council—letter Cockcroft to Waterfield, 14/5/52 (PRO-AB6/1074. In the same letter Cockcroft denied this: 'My attitude has always been', he wrote, 'that we would give whatever technical assistance we could to this project and that I was personally interested in seeing this go forward. I did not however see much possibility of Britain contributing financially to this project in view of our present commitments to our nuclear physics equipment in U.K.'.
126. King (1953), 219–220.
127. See for example, memo King to Lockspeiser, *European Nuclear Research Centre Proposals*, 23/4/52 (PRO-DSIR17/559); for King's immediate support, see letter King to Cockcroft, 14/3/51 (PRO-AB6/912).
128. See his letter to Cockcroft, note 127.

CHAPTER 13

Britain and the European laboratory project[1] mid-1952–December 1953

John KRIGE

Contents

13.1 The emergence of the Harwell group 477
    13.1.1 The Buckland House conference 477
    13.1.2 Refining the case for membership 480
13.2 The polarization of the scientific community 482
    13.2.1 The third meeting of Chadwick's Advisory Committee 482
    13.2.2 The widening debate among the scientists 483
    13.2.3 Group divisions: a brief analysis 486
13.3 Progress at governmental level 487
    13.3.1 The deepening involvement of the DSIR 487
    13.3.2 The reactions of the Lord President and the Chancellor 489
13.4 Refocusing the issue: the 'discovery' of the alternating-gradient principle 490
    13.4.1 A new sense of urgency 491
    13.4.2 Britain's physicists make their choices 492
13.5 Britain joins CERN 495
    13.5.1 The meeting with Amaldi 496
    13.5.2 The Chancellor reconsiders the case 499
    13.5.3 The signature and ratification of the Convention 501
Appendices
    13.1 A short survey of the committees involved in the CERN decision 503
    13.2 Decision-making at the science–government interface: some general comments 506
    Notes 511
    Bibliography for chapters 12 and 13 517

There are several distinctive features about the relationships between Britain's scientists and the European laboratory project in the period up to mid-1952. Firstly, policy was in the hands of very few people: primarily three in fact, Chadwick, Cockcroft, and Thomson. Secondly, all three were highly prestigious and influential elder statesmen of British science. Two had received the Nobel prize and all had been knighted. Thirdly, in putting forward 'British scientific opinion' on the project, they were in fact primarily speaking for themselves. Of course Cockcroft discussed matters with Goward, and probably other members of his staff too, and Skinner was brought into the picture in connection with the Liverpool cyclotron. But the one and only time that a *systematic* attempt was made to canvass the views of the physics community at large was in April 1951, and then they were asked simply to give their opinion in writing (see section 12.1). Indeed, there was very little discussion between Chadwick, Cockcroft, and Thomson themselves let alone with their more junior colleagues.

All of this was to change during the latter half of 1952. In fact even as Sir James Chadwick was claiming to give the Cabinet Steering Committee 'United Kingdom scientific opinion' on 'the building of large cyclotrons', there was a process under way which put in question not only Chadwick's view of what that opinion was, but also his right to speak for it. This process was triggered and accelerated by events occurring outside Britain itself, notably advances of a scientific and technical kind. It first found expression in an informal conference early in June attended by many of Britain's leading physicists and accelerator engineers. It was given further impetus by the discovery of the strong-focusing principle and by related developments within CERN itself. Finally, towards the end of the year, its momentum was such that it brought about a shift in the official policy of the scientists. This in turn changed attitudes in government circles, and just after Christmas 1952 the Chancellor of the Exchequer agreed that Britain join CERN. Her delegates signed the Convention establishing the permanent organization on 1 July 1953, and by the end of the year the instruments of ratification had been deposited at UNESCO House in Paris.

As these introductory comments suggest, in describing these events in this chapter we shall emphasize developments occurring within the scientific community as these were undoubtedly the dominant determinants of change. What is more our treatment of 1953 is somewhat cursory. Little of relevance to our study of how and why Britain joined CERN occurred during this year, and the role that her delegates and scientific advisors played in the life of the provisional Council of Representatives has been dealt with in chapters 7 and 8.

13.1 The emergence of the Harwell group

13.1.1 THE BUCKLAND HOUSE CONFERENCE

On 7 June 1952 over twenty people held an informal conference on 'High energy accelerators for nuclear research' at Buckland House, a country manor near Harwell especially acquired for such purposes. Most of them were experimental physicists and engineers from Harwell and the universities equipped with accelerators. Blackett, it seems, was the only leading cosmic-ray physicist present, Powell, Rochester, and Perkins all being away at Bohr's conference in Copenhagen. On the political side, Lord Cherwell attended the meeting. Cockcroft was there, of course, as was Thomson. Chadwick, although invited, apparently did not attend.[2]

The idea of having such a conference had emerged after discussions between Cockcroft, Skinner, and the Harwell staff. Its aims, as spelt out by Cockcroft in a letter to Chadwick early in May, were to discuss future accelerator policy in Britain and her relationship to the European laboratory project.[3] Two factors in particular lay behind this felt need for reappraisal.

Firstly, as we have already mentioned, there was the success of the Cosmotron.[4] The psychological impact on the British of this advance was at least as important as its purely technical aspects. Remember that no machine bigger than about 500 MeV had yet been built in the world, and that Britain's efforts to construct a 1 GeV machine at Birmingham were beset with difficulties. Remember too that many British physicists had been highly sceptical of the scheme to build a big bevatron on the continent, partly because they feared that it was technically over-ambitious. Now these barriers were down, and the anxieties surrounding the possibility of scaling-up accelerator energies by an order of magnitude were laid to rest.

The second factor calling for a reconsideration of Britain's accelerator policy was quite independent of the first: the maturation of the domestic programme launched immediately after the war. Here two elements converged—on the one hand, the hope that these several machines would soon be working at their design energies; on the other, the need for the government to plan for future expenditure in the five universities equipped with accelerators. Indeed on 7 May Sir John Cockcroft chaired the first meeting of a new committee set up by the DSIR to assess the needs of these universities for the 1952–57 quinquennium.[5] On the next day he wrote his letter to Chadwick proposing to call the conference which was held a month later at Buckland House. In what follows we shall first discuss the contents of the talks given, and the conclusions reached at the meeting, and then say a few words on its significance.

The proceedings of the conference show that considerable care was taken to ensure that the talks given were logically connected with each other and formed a coherent whole. Essentially there were eight of them, framed by an introduction and conclusion by Cockcroft.

In spelling out the aims of the conference in his introductory talk, Cockcroft stressed that it was time to look ahead, and to consider Britain's needs for doing meson work in particular. Two papers of a general kind followed, discussing high-energy research from a theoretical (Peierls) and experimental (Cassels) point of view. For his part, Peierls stressed the importance theoretical physicists attached to having high-energy machines. Their uses were many—the production of V-particles, of $\varkappa$-mesons, and of negative protons, the investigation of nuclear forces using mesons themselves as a tool to explore the nucleus, and the production of events which were not already predicted by existing theory. Considering the devices that Harwell, in particular, should build to do this kind of work, Cassels argued that the choice lay between a proton linear accelerator (linac) of some 500 MeV with a high meson flux or a bevatron in the 10–20 GeV range.

These alternatives were elaborated on in the next four talks. The first pair of papers dealt with the technical aspects of building linear accelerators—a 450 MeV Alvarez-type proton linac, the design of which was being studied at Harwell, and a 300 MeV electron linac envisaged for Cambridge. Skinner and Pickavance then discussed more powerful machines—the former giving some rather general 'Thoughts about a proton synchrotron', while Pickavance discussed the 'Problems involved in constructing a 15 BeV proton synchrotron'.

Skinner began his talk by remarking on how impressed he had been by the 'systematic way' in which the Cosmotron had been designed, 'and the excellent engineering work that had gone into its construction'. He surveyed the cost of its several components, noted that Berkeley was aiming for 6 GeV, and recommended that 'Britain or Europe should have a machine in this energy range', and preferably not less than 10 GeV. Pickavance then explored the machine parameters (injection energy, magnet weight, intensity, cost, manpower ...) for a 10–20 GeV bevatron by extrapolating Brookhaven data. He concluded that both 10 GeV and 20 GeV machines 'were quite feasible', and that the smaller would cost about £3M, the larger somewhat over £5M.

'The Western European Nuclear Research Centre' was the topic of the last pair of papers given by Thomson, who discussed the historical evolution of the project, and by Goward, who described the machines being investigated by Bakker's and Dahl's groups. A wide-ranging discussion followed in which all except perhaps Sir George Thomson thought that it was essential for British physicists to *have access* to a powerful bevatron. She could do this either by building one herself—'Lord Cherwell thought that it was not impossible to get a million pounds a year for a few years' for this purpose—or by making 'a positive contribution to the European effort'. Several possible ways of doing so were considered. It was suggested that Britain 'offer to construct the European machine in this country', which, if accepted, would more or less guarantee British membership. It was proposed that Harwell provide a 10–30 MeV linear injector for the CERN bevatron—'a valuable contribution which only we could readily make'. Some felt that Harwell should go ahead and design a large bevatron anyway, supplying complete information to the European project—which had the further advantage that it put Britain in a strong position if she wished to go it alone.

What were the results of the meeting? Here it must be stressed that its aim was less to reach firm decisions than to survey possibilities. Insofar as one can speak of a majority view at the end of the day it was that Harwell and Cambridge University respectively should go ahead with their proton and electron linacs, and that it was desirable to contribute to the 'European effort'.[6] But this did not mean that anything was settled. On the one hand, it was not clear what form that contribution, if any, should take. On the other, there was some question as to whether the compromises and political and bureaucratic complexities of an inter-European effort were worthwhile. After all, did Britain not have the experience and the personnel to build a 10–20 GeV machine on her own? And had Cherwell not hinted that money could be found for an independent project? Why then embroil the country in a collaborative scheme with the continent who apparently needed her far more than she needed it?

What was the significance of this meeting? That here, for the first time, Britain's nuclear physicists collectively discussed their country's future accelerator policy and the place of the European laboratory in it. And that, for the first time, a group of engineers and physicists based at Harwell had entered the debate, men like Fry, the Head of General Physics, Pickavance, the Head of the cyclotron group, Adams, Cassels, Hereward, and of course Goward. They were in close contact with each other and with Cockcroft—the careful structure of the Buckland House conference is proof enough of that. What is more they were dedicated to the construction of new machines, and determined to remain in the forefront of accelerator physics and technology. Did this mean that Britain would 'inevitably' launch into a new phase of accelerator construction? Certainly not, for of course the Harwell group and their allies had Chadwick (and Thomson) to contend with.

On 11 June, so four days *after* the Buckland House meeting, Chadwick told the Cabinet Steering Committee that Britain's physicists were against the building of large cyclotrons 'because during the next four or five years our available resources should be of sufficient range for adequate research'. The general opinion at Buckland House was just the opposite. Had Chadwick been informed of the results of these discussions? Yes, at least indirectly. For on the day before the Steering Committee met Murray explained to Sir James the line that the DSIR would take, remarking parenthetically that Cockcroft 'was inclined to think that we might derive important advantages' from the construction of a large cosmotron. Beside this, in the margin, Chadwick wrote 'reward for money? perhaps better spent in sending men to US [and/or] Birmingham'. All the old themes then—did British scientists not have enough machines already? was a very big accelerator worth the cost? if it was needed why not devote more resources to the existing domestic programme? or capitalize on the special relationship with America? ... Sir James was not going to change his mind easily.[7]

13.1.2 REFINING THE CASE FOR MEMBERSHIP

The discussion at Buckland House took place among a group most of whom had had no direct dealings with the European laboratory project. The impression they were given of it by those who had was not a particularly encouraging one. In his talk Sir George Thomson naturally stressed the differences of opinion between Bohr and the Scandinavians and those who wanted to build a new research centre equipped with big accelerators. Goward, for his part, remarked that it could be difficult to find money for the scheme, and that people were loath to commit themselves to it full-time until it was properly launched. Not surprisingly then, the discussion about Britain's relationship to the venture was somewhat unfocused. Clearly there was little need for those who wanted access to a big bevatron to refine their thinking on membership of the European laboratory as long as its future was felt to be so uncertain.

This situation was soon to change however, at least for two of the Harwell physicists who had been at Buckland House, Jim Cassels and Donald Fry. Along with Goward, Cassels attended the last week of the international conference arranged by Bohr at his Copenhagen institute. Here he (and Goward) presented reports on the machines described at the meeting in England ten days before; he also took part in detailed discussions on the programme of the laboratory and the accelerator projects to be undertaken by Bakker's and by Dahl's groups. Fry, for his part, attended the second session of the Council on 20 and 21 June 1952 as an official observer representing the Royal Society. What they found made a considerable impact on them.[8]

On the 23 June Cockcroft wrote to Chadwick to say that he had 'just seen D.W. Fry on his return from Copenhagen' and that he thought it would be a 'good plan' to call 'an early meeting' of Chadwick's Advisory Committee. On 24 June Cassels drafted a paper over five pages long entitled 'Great Britain and the European Laboratory' which Cockcroft circulated widely to nuclear physicists (and to Lord Cherwell) a week later. On 25 June Fry's report on the proceedings of the Copenhagen conference was ready to be sent to the members of Chadwick's Advisory Committee.[9] We shall devote the remainder of this section to a discussion of these two documents.

The papers had very different aims. Fry's was primarily intended as an official report on the Copenhagen session, to which description he appended four of his own brief 'observations regarding Great Britain's position with respect to the Council'. Cassels' paper, by contrast, was a structured argument for full British membership of the organization. It was divided into five sections describing successively the programme of the envisaged laboratory, the scientific, technical, and practical considerations underlying the Copenhagen decision to equip it with a proton synchrotron of 10–15 GeV and a synchro-cyclotron of 600 MeV, the scheme's chances of success, the policy of Great Britain towards it and, finally, the financial cost of her joining.

The most important result of their participation was that both Cassels and Fry came away from Copenhagen without any doubt that CERN, as it was now officially called, stood a good chance of success whether or not Britain participated. It was noted that the scientists had accepted the machine programme without dissent, that all ten member states of the provisional Council had produced money for the design stage, and that France had urged delegates to press their governments to put aside money for the main scheme in 1953. Scandinavian opposition, in particular, had thus effectively melted away. 'From the point of view of the European Laboratory', wrote Cassels, 'the most serious effect of Great Britain's abstention [would] probably be that the synchrotron would take perhaps 2 years longer to complete and might be for 10 instead of 15 GeV'.[10] This had a number of consequences.

To begin with, there was the problem of Britain's remaining competitive. It had been argued at Buckland House, and confirmed at Copenhagen, that 'very high energy physics in the 10–20 GeV region [was] certain to be taken over by accelerator laboratories'. America was known to be planning machines to enter this energy range. Europe too was now likely to have one. If Britain's physicists had no access to a big bevatron, by 1960 they would probably 'have to be content to take no part in a field of research accessible not only to the U.S.A. but also to [...]', and Cassels listed the ten member states of the provisional CERN. It was one thing to be second to America. It was another to be eleventh after Yugoslavia.[11]

This competitive argument assumed an added poignancy in Britain's case. True, she was in danger of losing her position among the world leaders in a particular branch of nuclear physics. But there was more to it than that. For she also risked failing to capitalize on the breakthroughs which her cosmic-ray physicists were making and which could give her a competitive edge over her rivals. In the late 1940s Powell's group at Bristol had made important discoveries in meson physics—only to have the π-meson field appropriated by physicists working with American accelerators. Now the groups at Manchester and Bristol had again played a major role in the discovery of a host of new particles—particles whose properties would undoubtedly be investigated in the bevatrons being planned. Without access to such a machine Britain would simply repeat the mistake she had made in the late forties when, for lack of an accelerator powerful enough to do meson physics, she had had to stand by and see the initiative pass beyond her shores.

A third consequence of the realization that the European laboratory would go ahead with or without Britain's participation was that she was forced to reappraise the bargaining power of the contributions in kind she was making—the provision of expert consultants to the study groups and the offer of access to the Liverpool cyclotron. As Fry put it—and the point was reiterated by Cassels—the Liverpool proposal was 'not of dominating importance when compared with the very ambitious plans which the Council [was] launching'. And while the technical assistance was appreciated, her physicists could hardly expect to be offered a 'full

Notes: p. 512

share' in the facilities of the big accelerator if Britain had not made any financial contribution to it.[12]

By the force of such arguments Fry and Cassels were led to conclude that Britain either had to build a big bevatron herself, or participate fully in the European programme. The former left it at that. Cassels, on the other hand, went further, arguing that Britain should not try to go it alone. Such a policy would jeopardize the domestic programme—if implemented it would probably be at the expense of the proton linac envisaged at Harwell. It would also be wasteful—because Britain would probably not have enough first class researchers to exploit fully all machines available (including her own bevatron) at the end of the fifties. And it would be of little competitive value—as Cassels put it 'Great Britain and the European laboratory, acting independently, [were] not likely to compete very effectively with the U.S.A.'—the implication being that if they pooled their resources they could reach the higher energies under discussion across the Atlantic.[13] Cassels ended his paper by quickly mentioning the cost of participating—£285,000 per annum for eight years—and stressing that much of that could probably be spent in Britain.

Before discussing the impact of these documents we want to say a few words on the Liverpool cyclotron. In its heyday the offer of access to it was seen as a means of fostering scientific co-operation with the continent, as a cheap way of giving Britain a say in the affairs of the European laboratory project, and as a political move in support of Bohr. Now, a mere six months later, it was stripped of all but its scientific interest, and that was limited. In the interim it had been the subject of seemingly endless negotiations, both with Skinner and the Liverpool authorities, who were initially rather suspicious of the whole idea, and between the DSIR, the Treasury, and the Home Office, who had to settle the financial and security aspects of the arrangements. Here one sees bureaucracy at its worst—the additional sum of money involved in running the machine 20% longer to accommodate non-British physicists was no more than £7,000–8,000 per year, the security risks were grossly exaggerated. And to what avail? Early in July Skinner privately and without any formality accepted two engineers from Bakker's group (F. Bonaudi and F. Krienen).[14] And the cyclotron itself only worked in March 1954!

13.2 The polarization of the scientific community

13.2.1 THE THIRD MEETING OF CHADWICK'S ADVISORY COMMITTEE

In response to Cockcroft's call for an early meeting, Chadwick's Advisory Committee gathered for the third time on 16 July 1952. Its four scientific members—Chadwick, Cockcroft, Skinner, and Thomson—were present. King stood in for Lockspeiser and Fry attended by invitation. All of the scientific members had received Cassels' paper and Fry's report on the Copenhagen conference. One of them—Skinner—had already let both Cockcroft and Chadwick know that he was impressed by Cassels' arguments and that he felt that there was 'no practical alternative to the policy of joining in with the European laboratory', despite his personal preference that Britain 'go ahead on its own'.[15]

After the usual preliminaries, Chadwick reported on the meeting of the Cabinet Steering Committee held on 11 June and Fry on the Copenhagen session of CERN. Those present then dealt with two basic questions: should there be a bevatron in Europe? and should Britain sign the agreement establishing the provisional CERN?

According to the minutes, on the first question the committee 'agreed that it was too early to make any recommendation'. This anodyne formulation concealed the very real differences in the meeting, however. As Cockcroft wrote to Peierls, the permanent scientific membership of the committee was 'evenly divided on the question of the large machine. Chadwick and Thomson [were] strongly opposed to this project, but [were] favourable to a cooperation with the Bohr group'. In adopting their stance, as Cockcroft went on to point out, his colleagues chose to ignore the opinion of 'the majority of the nuclear physicists in this country [who] agreed that it was desirable to have a high energy machine such as a bevatron in Europe'.[16]

On the second question the committee also reaffirmed its standing policy, again passing a resolution urging signature subject to the usual qualifications regarding limited financial liability. As Cockcroft put it to Peierls, 'we'—so presumably he and Skinner too—were not prepared to make any commitments beyond the planning stage. Here again there were differences of opinion though. This time it was Chadwick who found himself in a 'minority of one', suggesting that it may not be necessary to sign the agreement *at all,* arranging instead 'for the cooperation we have in mind to be sponsored by the Royal Society'.[17]

Determined opposition to the big machines, a resolution calling on the government to sign the agreement while drawing back from full engagement—what is striking about this meeting is that nothing has changed *despite* the important transformations occurring within the physics community itself, and the case made by Cassels and Fry. Plainly, Chadwick and Thomson were not prepared to yield on the question of the bevatron. But why did Cockcroft and Skinner let them have their way? Perhaps because *in the context of this committee* the issue of whether or not a bevatron was needed was tied up with the question of Britain's relationships with the European laboratory. Cockcroft and Skinner were convinced of the need for a powerful accelerator; they were still not sure whether it was preferable for Britain to build her own or to join in the collaborative venture. Better to leave matters as they stood until the community knew what it wanted scientifically *and* organizationally.

13.2.2 THE WIDENING DEBATE AMONG THE SCIENTISTS

It comes as no surprise that, in the wake of the decision taken by Chadwick's committee, those who wanted a change in British policy were provoked into action.

In particular Chadwick and Cockcroft—and perhaps Thomson and Skinner too—received a number of letters written soon afterwards. On 31 July Pickavance sent Cockcroft the draft of a very extensive letter he proposed to send to Chadwick; Sir James received a slightly amended version dated 12 August. On 1 August Sir Harrie Massey (University College, London) wrote similar letters to both Cockcroft and Chadwick. On 14 August Rudolf Peierls (Birmingham University) wrote to Cockcroft. All recommended that Britain participate fully in the European laboratory project.[18]

What new arguments were deployed here? Two in particular, both touching on questions of direct concern to Chadwick: could the cost of a bevatron be justified? and how best could the future of younger people engaged in nuclear physics research be stimulated and sustained?

At the Cabinet Steering Committee Chadwick had stated that he was not sure it was worth spending money now on very large machines: anyway, Britain's available resources would satisfy her needs for the next four or five years. Both Pickavance and Massey took issue with this. It was no use, Pickavance wrote, insisting that Britain had 'already spent enough money on nuclear physics'. On the contrary the post-war domestic accelerator programme had been 'badly planned' and Britain had not had value for money as a result. A list of deficiencies followed. The Harwell (170 MeV) cyclotron was an engineering success, but not powerful enough to do meson physics. Liverpool was—but would enter the field 'at least 5 years after Berkeley, at a time when much of the cream will have been skimmed'. As for the rest of the machine programme—it had been 'even less satisfactory', and problem-ridden. Massey agreed, attributing the deficiencies to 'the dispersion of effort and the consequent slow rate of construction', and a failure to recognize that there were 'a few things which a good engineer [could] do better than a good physicist'. This did not mean that further expenditure in the field was unjustified. On the contrary, it was time, concluded Pickavance, 'to recognize our former errors' and to decide anew on what machines we needed 'from considerations of engineering and nuclear physics'.

Pickavance deployed two other arguments dealing with this question of cost. He considered the objection, explicitly attributed to Powell, that money available for 'future British efforts in the physics of very high energies' should be 'concentrated on cosmic rays' rather than accelerators in view of the 'remarkable success' of the field 'in recent years'. Pickavance did not dispute that cosmic-ray physics deserved support: he simply reiterated the now-familiar scientific reasons why the 'complementary' technique of high-energy accelerators should be funded in parallel. In addition, Pickavance tried to show that the actual amount involved if Britain joined CERN (expected to be £285,000 annually for 7 or 8 years on the DSIR vote) was little compared with the annual 'direct expenditure from Government funds on civil scientific research' (of which it was about 1.8%) and with the research and development expenditure of the Ministry of Supply (0.3%). These comparisons suited Pickavance's case of course; others would have been less convincing, and as far as we know this *kind* of argument was never used again, probably because of its perils.[19]

What of the educational argument? Here both Pickavance and Peierls stressed the attraction in terms of intellectual challenge and prestige exerted on 'many young physicists' and 'most students' by the now-blossoming domain of high-energy physics using accelerators. It was the 'most exciting field of work', and they would be 'stimulated by the prospect of being in the front rank of the development of the fundamental side of the subject'. If Britain failed to exploit this opportunity she was not only likely to lose permanently her better experimentalists to the European laboratory. There would also be 'a sterilising influence on the whole of nuclear physics' in the country, and 'through this, on many fields of technology which have no direct relation to high energy physics'. Indeed for Pickavance these wider effects of participation in accelerator physics provided a key reason, a 'more specific justification', for the heavy expenditure involved, 'especially in view of the importance in the public mind of all work which is in any way associated with atomic energy'.

How did Chadwick react to these arguments? By ignoring the *scientific* case for building a bevatron which Pickavance, like Cassels, had spelt out at some length.[20] Apparently this argument was more than outweighed in his mind by two other considerations: that the universities at least were satisfactorily equipped already—and here he dismissed Pickavance's criticism in a few lines—and that the country was 'close to bankruptcy'—if there was money to spare, wrote Chadwick, the Treasury might feel that there were 'more urgent matters than this', a view with which he 'should be inclined to agree'.

Two last remarks before concluding. This exchange of letters serves as a warning against anachronistically assuming that the demise of cosmic-ray physics followed rapidly on the production of mesons with the Berkeley accelerator early in 1948. By mid-1952 it was still by no means 'obvious' to men like Chadwick and Powell that Britain *had* to have access to a big accelerator if she wanted to stay 'in the front-line of high-energy nuclear physics'—cosmic rays would do.[21] In fact it must not be forgotten that the claim used by the proponents of powerful accelerators that their device 'complemented' cosmic rays was made from a *defensive* position: they were trying to secure a foothold in a terrain that had been successfully occupied by a rival for at least three decades and which was going through a particularly ebullient phase in the early fifties.

This brings us to our second point. Massey's letters at least were symptomatic of a change in thinking in some university circles of their future role in accelerator physics. When in 1946 the Nuclear Physics Committee was laying the foundations of the country's post-war accelerator programme it 'did not ask what apparatus the nation must possess and where it should be located'—it simply funded machines at the five universities that wanted them.[22] Result: to quote Massey, a 'dispersion of effort and the consequent slow rate of construction'. The lesson to be learnt from these mistakes—and Massey stressed this—was that accelerators had to be built 'on a communal basis'. The Brookhaven model was beginning to take hold in Britain.

13.2.3 GROUP DIVISIONS: A BRIEF ANALYSIS

In the summer of 1952 the British physics community was polarized into two main groups around the question of whether or not it was necessary to have a large bevatron (10–20 GeV) in Europe. At one extreme we have Chadwick and Thomson, at the other we have an active group at Harwell whose main spokesmen were Cassels and Pickavance. Three features typify this group. They were a generation younger than Chadwick and Thomson: the oldest was Fry (b. 1910), the youngest was Cassels (b. 1924), the average age was about 35. Many of them (Fry, Goward, Adams, Hine, ...) had worked on the development of radar during the war where they acquired skills readily transferable to the accelerator domain. And they constituted the most experienced nucleus of accelerator builders in Europe—Fry built the first-ever travelling wave linear accelerator, Pickavance and Adams designed and constructed the 170 MeV Harwell synchro-cyclotron, Goward was active in the development of electron synchrotrons, etc. For these men doing experimental physics meant working in teams around heavy items of equipment. They responded to the technical challenge posed by a major engineering project, they could spend hundreds of thousands of pounds without batting an eyelid. They were the children of the new age of big science.

Sharing this group's conviction of the need for a big bevatron we have Cockcroft and Skinner. In terms of age (both in their fifties) and of status (members of the establishment, influence in government circles) they, and particularly Cockcroft of course, were close to the Chadwick–Thomson pole. In terms of professional formation however they had much in common with the young group at Harwell. Cockcroft—manager of a major scientific research establishment, convinced of the importance of big accelerators, insistent that to build them you needed a skilled team of specialists and the infrastructural support of a big institution. Skinner—worked on accelerators and radar during the war, was Cockcroft's chief physicist until 1949, when he took over from Chadwick at Liverpool university, completed the construction of the 450 MeV cyclotron. Then there was Blackett, cosmic-ray physicist, the same age as Cockcroft and, like him, having strong personal links with the younger members of the continental physics community. Initially opposed, then more or less indifferent, to the European laboratory project, Blackett was now coming round to the view that accelerators had an important role to play in the 10–20 GeV region. Finally, there were the nuclear physicists in the universities, including the theoreticians like Peierls, who felt that his views were 'fairly representative' of those held in this sector—and who was in favour of having access to a big machine through British membership of CERN.[23]

It is important to realize that the divisions among British physicists around the question of whether or not a bevatron was needed were not correlated with views on the question of whether or not she should sign the temporary CERN agreement. The latter had, by virtue of the historical process we have described, become *the* issue between some influential members of the community and their government; here Cockcroft and Skinner were aligned with Chadwick and Thomson. Between physicists themselves, however, it was of secondary importance, at least after Buckland House, and the community was polarized around the *more or less independent* question of the bevatron. Here, unlike Chadwick and Thomson, the bulk of the physics community was convinced that Britain should participate in the construction of a new generation of accelerators. But there was as yet no clear consensus among them on how best to do that nor, more specifically, on whether anything more than technical support need be given CERN.

Crucially, the main reason for this was that Britain had the resources in expertise, personnel, and money to consider building a big accelerator *on her own*. This put her in a unique position in western Europe. One of the key factors accounting for the birth of CERN was the recognition in countries like France and Italy that the only way in which their scientists could have access to big and expensive scientific equipment was through a European-wide collaboration. Britain's physicists shared neither their needs nor their enthusiasm: if they agreed to join CERN it would not be because they felt they had no other choice.

13.3 Progress at governmental level

13.3.1 THE DEEPENING INVOLVEMENT OF THE DSIR

Officials in the DSIR were disappointed, though not defeated, by the Chancellor's decision on 23 June to refuse the recommendation made by the Cabinet Steering Committee that HMG sign the agreement establishing the 'provisional CERN'. Initiatives to reverse the decision, however, came to an abrupt halt after a visit by Fry to the department on 24 June. His report on developments at the second Council session in Copenhagen made it clear that any chance Britain may have had of actually impeding or even defeating the CERN accelerator programme was now lost. If she persisted in this she would find herself politically isolated in the Council with a bargaining chip—the Liverpool cyclotron—of little value. With that, of course, the main underlying reason for signing the temporary agreement evaporated. 'The issue now to be faced', concluded Verry after his meeting with Fry, 'was whether or not the U.K. would be prepared to join the Council and play a full part, including, presumably, making a substantial contribution towards the £8,000,000 programme'.[24]

With these ideas in mind King attended the meeting of Chadwick's Advisory Committee on 16 July, where he became aware of the deep division between the leading scientists over the question of the big accelerator. On the following day he chaired a meeting of his Committee on Overseas Scientific Relations. King found himself in a dilemma. He was aware that the majority of British physicists wanted access to a big accelerator, and that 'it would obviously be cheaper to contribute to the cost of one constructed' by CERN 'than to build one independently in the U.K.'. Yet no matter how logical it seemed to King that Britain should join fully in the project, the fact remained that Chadwick's committee had not mandated him to pursue this course of action. In the event King convinced himself that a case for signature could be made, subject to the usual qualifications about the uncertain need for big accelerators, and he duly took up the matter with Lockspeiser. On 8 August Sir Ben, who had refused to fight for membership six week earlier, wrote a long memo to the Lord President asking him to reconsider his initial reluctance to join CERN.[25]

Why had Lockspeiser shifted his position? Doubtless King had impressed on his Secretary that the majority of Britain's physicists were determined to have access to a big accelerator, and that the European option was the cheaper way forward. If the DSIR acted now it could more readily steer the scientific community towards this alternative when the physicists collectively agreed on the need for a bevatron—as they inevitably would. Translating these concerns into concrete policy, Lockspeiser suggested to the Lord President that Britain could try to get the Council to adopt the OEEC system whereby a state could contribute to the running expenses of an organization and attend its meetings without being obliged to participate in all its technical schemes—or to contribute financially to them. This principle—that one only paid for what one wanted oneself—became a standard thread in Britain's dealings with CERN throughout 1952 and was embodied, albeit in a different shape, in the convention establishing the permanent organization (see section 8.2.1).

Another factor which probably swayed Sir Ben was the attitude of Cherwell. At Buckland House Cherwell had expressed a clear interest in building a big bevatron, though of course his preference was for Britain to have her own project. Nevertheless, he was prepared to support the DSIR and his senior physicists in pressing the case for membership both with the Lord President and the Chancellor. Lockspeiser could hardly continue to refuse to argue the merits of membership with his minister knowing that Cherwell was doing so.[26]

A third consideration underlying Sir Ben's move whose weight is difficult to assess is that of ambition—ambition for his department and, by extension, for himself. As we mentioned above, the DSIR had recently set up an advisory committee chaired by Cockcroft to plan expenditure for the next quinquennium in those universities having accelerators. Late in July Lockspeiser and Hinshelwood agreed that the functions of Chadwick's Advisory Committee on CERN (joint Royal Society/Ministry of Education) should also be duly taken over by Cockcroft's—Sir Ben, wrote Verry, being 'quite determined that his own Nuclear Physics Committee [should] be the responsible advisory body', and not one nominated by the Royal

Society.[27] These moves occurred at a time when the DSIR found itself again slipping into second place behind the scientific services of the defence departments which, in response to the 'cold war', had received a new injection of funds notably in the fields of supersonic aircraft, guided missiles, and the use of atomic energy.[28] Here was a way to redress the balance, to ensure that the DSIR had a stake in the most glamorous and prestigious area of civil science—and that its senior officers had an importance and authority commensurate with the significance of the field and of CERN.

13.3.2 THE REACTIONS OF THE LORD PRESIDENT AND THE CHANCELLOR

When he received Lockspeiser's minute the Lord President (Woolton) discussed its contents with Cherwell and with Blackett. Both of them impressed upon him that they, as well as Cockcroft and his Harwell staff, were convinced that a bevatron was needed. The minister was won over. On the 20 August he dictated a minute in King's presence inviting the Chancellor of the Exchequer (R.A. Butler) to alter his earlier decision. A week later he had the reply: no.[29]

What were Woolton's arguments? And Butler's reactions to them?

After giving the usual list of influential supporters of membership of the Council—the Royal Society, Cherwell, the people at Harwell, 'my departmental advisers',—Woolton advanced two specific arguments for membership. Firstly, (and this, wrote Woolton, was the argument that had convinced him), there was a 'very practical application of scientific knowledge on this subject which may well come to have a profound effect on our economic life', and if Britain was to retain her world lead in the field it was essential that her physicists be 'informed and stimulated by discussion [...] with scientists of other Nations on this Council'. Secondly, the British scientists felt, wrote Woolton, 'that if they were official members of this Committee, they might be able to secure that it should be built in this country and paid for by joint contribution'. These arguments call for several comments.

To begin with, it is striking (but not surprising) that at this, the highest level of government decision-making, there is no reference to the scientific interest of building a 10–20 GeV accelerator: V-particles and mesons have no place in Woolton's minute. It is rather the perceived economic and technical fallout, and the prospect of having the laboratory on British soil—two arguments which were also used to considerable effect in France (see section 9.4), for example—which were the main attractions of membership.

But—and this is our second comment—what are we to make of Woolton's proposal to build CERN in Britain? His minute, it will be remembered, was written on 20 August 1952, thus three weeks after the official deadline for offering a site for the laboratory. As far as we know Woolton realized that the deadline had passed, and was not alone in thinking that Britain could still offer to host CERN—Pickavance had made the same proposal in his letter to Chadwick of 12 August.[30] Considering how seriously governments on the continent took the deadline, how was it possible for British scientists and officials to do otherwise?

Their attitude, we suspect, was symptomatic of the view that Britain was indispensable to the European laboratory. This view was voiced by some scientists (although 'the Europeans' would not like the idea of having CERN in Britain 'they might be willing to accept it as the only means of bringing Britain into the scheme', wrote Pickavance on 12 August) and in government circles ('The other countries were so anxious to secure our co-operation that we could probably join on our terms' said King on 17 July).[31] This myth of indispensability, in turn, can be traced back to British 'superiority'—both in the psychological sense (see section 12.1.2) and in the more objective sense that, in western Europe, she was *the* leading nation in the field of nuclear physics and in terms of global political influence.

Finally, what was the Chancellor's reaction to Woolton's arguments? Needless to say he was not persuaded by them—in fact he ignored them. But the context in which he phrased his now-familiar objections is worth quoting, all the more so in view of what we have just said. 'I agree', wrote Butler, 'to the sending of observers to the Council. I would expect that such observers would, by their prestige and by the fact that the European countries will eventually want our help and our money, be able to achieve our main aims [...]'. For the time being, Butler went on, 'we ought to keep ourselves as free as we possibly can [...]'.[32]

13.4 Refocusing the issue: the 'discovery' of the alternating-gradient principle

'During a visit to America on behalf of CERN in August 1952, it has been suggested by the Brookhaven team that we should consider building a European Proton Synchrotron following a new principle that they are working on'. Thus did Frank Goward begin a technical paper dated 22 August 1952—so written immediately after his return—in which he explained the theory behind the alternating-gradient or strong-focusing principle. 'Blackboard descriptions' at Brookhaven, wrote Goward, indicated that, by suitably arranging magnet sections, the cost/GeV of an accelerator could be dramatically reduced: 'a 30 BeV. machine could be constructed with little more complication than the present 3 BeV. machine, i.e., within a $10 million budget, approximately'.[33]

Dahl's group, of which Goward was deputy-head, decided without further ado to abandon their scaled-up cosmotron project and to study the possibility of building a 30 GeV strong-focusing proton synchrotron. By the end of September Jakobsen, at Bohr's institute in Copenhagen, had built a model which confirmed that in principle the idea was sound.[34] Our aim in this section is to describe the reactions in Britain to the new concept and to explore the role it played in reshaping her policy *vis-à-vis* CERN.

13.4.1 A NEW SENSE OF URGENCY

Perhaps the first point to be clear about is that, in contrast to Dahl's group and the CERN Council, the discovery of the alternating-gradient principle did not precipitate a dramatic shift of policy in Britain. For some, like Pickavance, it *reduced* British options: if she wished 'to act honourably' she could not immediately build her own strong-focusing machine since Goward had visited Brookhaven on CERN's behalf. For others, like Verry, it actually *broadened* them: Britain, he wrote, might 'now have to consider whether or not it should go for a machine much larger than 10 Gev., e.g. a 30 Gev. for the same money, or continue to be content with the proposed 10 Gev. machine for about one-third of the present estimated cost'. From this of course Cherwell drew the further conclusion that the discovery of strong focusing actually *weakened* the case for British membership of CERN.[35] Her scientists had argued at Buckland House that she must have access to an accelerator of energy not less than 10 GeV. Why not invest the money she would otherwise spend on CERN to build her own machine of say 15 GeV?

If anything, then, the immediate effect of the discovery of strong focusing in Britain was to confirm existing attitudes on CERN membership at the 'extremes' (Pickavance, Cherwell), and to increase the uncertainty on what to do next in the 'centre'. Consistent with this the brief prepared for the two British observers to the third Council session (Blackett and Fry) contained no surprises. For the present the government was content to keep a watchful eye on developments. To ensure that the status of the (Royal Society) observers remained intact, the two delegates were told that they could discreetly mention that Britain was considering making a modest financial contribution to the Council's work—intended of course to compensate for the loss in bargaining power of the Liverpool cyclotron.[36]

Blackett and Fry duly attended the Council session in Amsterdam from 4–7 October and, as usual, wrote a lengthy report for the Royal Society on the proceedings.[37] In it they stressed that Britain's main interest in CERN lay in the plan (now official) to build a 10–30 GeV proton synchrotron. Their concluding recommendation was unambiguous, not to say ominous: if the government '[did] not intend to join CERN fully', they wrote, it was 'highly desirable that the decision not to do so [was] made quickly and definitely [...]'.

There were three related reasons for this renewed sense of urgency. At the Amsterdam session Blackett and Fry were 'strongly urged, particularly by Professor Bohr, that the *early* adherence of the UK was exceedingly desirable' if the Council were to consider British scientists for key posts in the organization. Amaldi confirmed that if Britain wanted to shape CERN's policy in any significant way 'it will be necessary for us to have made up our minds to join before Christmas'. As the organization picked up momentum Britain's risk of missing the boat altogether was increasing.[38]

Then there were the implications of Britain's *not* joining on her domestic programme. If she decided to build her own bevatron a British design group would

Notes: p. 513

have to be formed 'immediately'. Its nucleus would obviously have to be provided by Harwell. But Harwell already had its pet project: a 600 MeV high-intensity proton linac. Could the Harwell team take on both? Or would they have to forego their linac? And were they willing to build a bevatron instead?

Most fundamentally, there was the question of competition, the 'eleventh-after-Yugoslavia' syndrome. Blackett and Fry pointed out that Dahl's group was already ahead of the British and, they warned, 'this lead [was] likely to increase rapidly during the next year', particularly since the Americans were keen to collaborate. To procrastinate now would be fatal. 'If nothing is done', they wrote, 'we stand the chance of neither having our own machine nor full access to the European one'. Such a setback 'would be a bad blow to British physics, and deeply resented by many British physicists [...]'.

It is important to realize that the competitive situation caused considerable alarm among British physicists because of the rather special features of the discovery of strong focusing. The idea was *very rapidly* disseminated throughout the major (western) groups interested in building accelerators. Within about two weeks the principle was known in Brookhaven and Berkeley and in CERN and Harwell.[39] Then there was the *radical nature* of the innovation, the fact that it did not involve scaling-up existing devices but designing a new kind of machine. This meant that past experience and expertise on accelerator construction were relatively less important than the ability to launch and sustain a major research and development project. In other words, as Pickavance put it, the innovation produced 'the interesting situation that the Europeans now start level with the Americans', a sentiment echoed by Blackett and Fry: the discovery, they remarked, 'unexpectedly allows the U.K. and/or Europe the possibility of competing on not too unequal terms with the U.S.A. [...]'.[40] The discovery of strong focusing, then, was a great leveller: at one blow it drastically cut back Britain's lead in accelerator construction over the rest of Europe. With it went the luxury of procrastination which her advanced position until then had allowed.

13.4.2 BRITAIN'S PHYSICISTS MAKE THEIR CHOICES

In the light of Blackett and Fry's report, steps were immediately taken to have the scientists reconsider the situation. Two committees were at hand for this purpose: Chadwick's Advisory Committee which had still not been wound up and Cockcroft's DSIR Committee. After some discussion it was decided to hold a meeting of the former, enlarged to include most of the members of Cockcroft's committee so as 'to get together all those who could reasonably claim to have a say on the scientific aspects of the matter'—and to make sure that Chadwick was consulted.[41]

Several initiatives were taken to prepare the ground in anticipation of the meeting. On 21 October Skinner wrote to Cockcroft comparing the costs of building a bevatron in Britain and on the continent. On 22 October there was a policy meeting about the bevatron at Harwell. On or before 31 October Fry circulated a memo entitled 'British Accelerator Policy and the European Nuclear Physics Laboratory', specifically to aid discussion of the issues involved. After giving a few scientific and technical notes on big accelerators, Fry discussed construction requirements in terms of manpower and costs, and the implications of Britain building her own big machine and of participating in the European laboratory.[42]

On 1 November 1952 the fourth and last meeting of Chadwick's Advisory Committee was held at the Cavendish Laboratory, Cambridge.[43] Sir James was in the chair; the other three scientific members, Cockcroft, Skinner, and Thompson were all present. Eleven more physicists attended by invitation, including Blackett, and Cassels, and Fry, Goward, and Pickavance from Harwell. After considerable discussion, the minutes tell us, the committee resolved:

> i) that it was the unanimous view of all present that it was necessary for the progress of British nuclear physics that British nuclear physicists should have access to a proton synchrotron of the size and character projected by the European Council for Nuclear Research;
>
> ii) that it was the unanimous view of all present that (i) could best be achieved by the United Kingdom cooperating in the European scheme and signing the agreement with the European Council for Nuclear Research; and
>
> iii) that it be a recommendation to Council that the Royal Society propose to the Treasury that, in view of the contributions in kind to be made by this country to the European scheme, an annual commitment of £250,000 for eight years be undertaken as the British contribution to the European Council.

Two main resolutions then, and the first *official* affirmation by Britain's physicists that they needed access to a big accelerator, and that this aim was best achieved by membership of CERN.[44] What considerations led them to advocate this course of action?[45]

(i) Why did Britain's physicists need 'access to a proton synchrotron of the *size and character*' projected by CERN? Primarily because

- they wanted to work on the most powerful accelerator which was *technically feasible*. This is particularly clear from Fry's memo, for he suggested that the choice lay between building an 'old-style' 15 GeV machine—roughly the limit to which one could scale up the Cosmotron—and a higher energy strong-focusing machine. However, the only reason why the 15 GeV conventional device was injected into the argument was to accommodate doubts about the viability of the new principle—it was there in the event that the method was 'ultimately found to be impractical for any reason' to quote Fry.

- they wanted to remain *competitive*. As Fry put it, 'Another kind of minimum energy which any new machine should have is set by the proton synchrotrons at present nearing completion'. This meant that the energy of a conventional machine should be substantially above that of the 6.4 GeV Berkeley bevatron, in fact 'the design energy should not be less than 15 GeV'. It also meant that any

Notes: p. 514

purely British strong-focusing machine should be at least 30 GeV. This was the energy of Dahl's device. There was talk of the Americans going to 50 GeV. 'If these plans mature', wrote Fry, 'then any other machine which is started now should also aim to be in this same energy region if the objective with which it is built is to keep in the forefront of research'.

Does this mean that the specific scientific work one could do played no role in setting the parameters of a desirable machine? Of course not: only that such considerations were not of major importance, and that the higher one pitched the energy, the further one probed into the unknown, the vaguer and more imprecise they (obviously) became. Cosmic-ray results, as Fry (like Cassels and Pickavance before him) said, indicated that a *minimum* machine energy of 10 GeV was needed. It was plausible to argue that this should be increased to 15 GeV to improve meson yields. With further increases in the energy all one could say was that it was 'reasonable to hope that several lines of research which [were] qualitatively new [might] be opened up [...]'. More detailed than that one could not be, at least at this stage of development of the field.

(ii) Why was it thought that these aims could best be achieved by membership of CERN? Here the main issues at stake were:
- the question of *cost*. Skinner calculated that the European bevatron would cost some £4M. He and Fry estimated that the same device, including its building (£0.5M) would cost some £3–3.5M to construct at home, the cheaper price reflecting the lower costs for equipment and materials, particularly steel, in the United Kingdom. This was to be compared with the £2M foreseen for Britain's contribution to CERN. Skinner found the situation tempting: 'So for very roughly 50% more outlay we could have the bevatron for ourselves', he wrote. This was the most optimistic estimate, however. The consensus seems to have been that it would cost Britain roughly twice as much to build her own machine.[46]
- the need to *preserve the domestic programme*. Using Brookhaven figures as a basis, Fry calculated that at least 25 scientific staff and anything from 50–100 other staff (technicians, draughtsmen, and 'skilled industrials') would be needed to build a big bevatron. It was agreed that if Britain was to launch a project of this scale Harwell would have to play a leading role. However, the AERE was finding it difficult to recruit staff, particularly in some of the lower grades. It could only build the bevatron quickly enough to remain competitive by abandoning other projects, notably the proton linac. Many at Harwell were loath to do this, including Cassels and Pickavance—the latter had remarked in June that Harwell 'ought to build the linear accelerator no matter what happens'.[47] By contrast, if Britain joined in the construction of the CERN bevatron she only had to provide some fifteen people which, said Fry, was easily done. By pursuing this option her physicists had the best of both worlds: they could have access to a bevatron without disrupting their domestic programme.

– the need for the physicists to *hedge their bets*. Britain's physicists were very conscious of the fact that the alternating-gradient principle was a new and radical development. It was, in Fry's view, 'still only about 70% certain' that it could be 'successfully applied', and he took the possibility of failure seriously enough to include a discussion of a 15 GeV 'old-style' accelerator in his paper prepared for the Cavendish meeting. This placed the scientists in a dilemma. On the one hand, they felt the need to proceed with caution, not to overcommit themselves in terms of money and personnel. On the other hand, they wanted to compete, to stay in the forefront of research, and this meant that they could not afford to wait and see whether or not the principle would work in practice: if it *did*, it would be too late to catch up. By joining CERN the dilemma was resolved: Britain's physicists could participate in the research, development, and (hopefully) construction of the most powerful accelerator then on the drawing-board, without being overcommitted in the short run. If the venture failed, the domestic programme remained intact, and resources could be diverted into a 'conventional' 15 GeV scaled-up cosmotron.

To conclude we want to pick up a point we made earlier: that British physicists who wanted access to a big bevatron had to make a *choice* —to collaborate with their fellow Europeans *or* to build the machine at home. To all intents and purposes this latter option was denied their continental colleagues. Yet, interestingly enough, the considerations which *ultimately* persuaded Britain's physicists to participate fully in CERN (cost, manpower, protection of the domestic programme, competition ...) were not markedly different from those invoked by their colleagues in the member states. This has important methodological implications.[48] If we had tackled the question of why Britain joined CERN by looking at the *outcome* and trying to list the reasons why the decision was finally taken, at the expense of studying the *process* which led up to it, we risked failing to see two important things: the *divisions* within the physics community which were (of course) 'resolved' when the unanimous decision was taken on 1 November, and the *differences* between the situation in Britain and that on the continent. In short the specificities of our object of study would have eluded us.

13.5 Britain joins CERN

The recommendations made by Chadwick's committee on 1 November removed one of the biggest single obstacles in government circles to British membership of CERN—the lack of consensus in the physics community, notably among its most senior members, on the need for having a big accelerator. This did not settle matters, however; there was still a good deal of work to be done. Cherwell, though invited to the meeting at the Cavendish did not attend: he would be disappointed that his physicists had resolved not to have a solely British project. A new Lord

President, the Marquess of Salisbury, took over from Lord Woolton early in December and he would have to be put in the picture. And even with their support there was no guarantee that the Chancellor would reverse his stance—if, wrote Chadwick, 'it [was] a question of money, [Butler's] decision [was] hardly open to argument'.[49] Either way a quick decision was now imperative: Amaldi's first report to member states informing governments of the financial and organizational implications of CERN membership, and a first draft of the Convention establishing the permanent body were in the post, and were due to be discussed at the fourth Council session in mid-January 1953.

The first two sections below discuss the developments occurring at the highest levels of British government which culminated in the Chancellor's decision on CERN membership towards the end of year. Section 13.5.1 is primarily chronological; section 13.5.2 is more analytical. The third section is devoted to the Convention and its ratification. Many of the issues we discuss in this section have been dealt with at some length in chapters 7 and 8; our treatment of them will be correspondingly curtailed here.

13.5.1 THE MEETING WITH AMALDI

On 6 November the Council of the Royal Society met and discussed the recommendations of Chadwick's committee. A week later the Secretary of the Royal Society, Sir David Brunt, transmitted them to the Chancellor, spelling out briefly the arguments for membership, and asking him to 'consider them as a matter of urgency'.[50] This letter found its way to Cherwell, who convened a meeting of high-ranking departmental officials on 20 November—Lockspeiser attended for the DSIR, and the Treasury, the Foreign Office, and the Lord President's Office were all represented. 'There was', wrote Verry, 'apparently a long general discusssion on what should be the next step, and finally it was agreed that, in principle, the United Kingdom would be well advised to participate in the new Centre [...]'.[51] Lockspeiser was left to pursue matters further. He proposed that Amaldi, Secretary-General of the provisional CERN, be invited to Britain, that the government's attitude be fully explained to him and his opinion 'be sought about how far we can hope to get the benefits of the Centre with the least possible liability'.[52] If agreement could be reached Lockspeiser and Cherwell would report to the Treasury and ask the Chancellor to approve official British participation in CERN. On 21 November Lockspeiser sent his official letter of invitation to Amaldi, who replied on 1 December: he would be in London on 11 and 12 December stopping off on his way to conferences in Brookhaven and Rochester.[53]

On the morning of 11 December Amaldi met with Lockspeiser, King, Blount (for the Royal Society), Verry, and Cockcroft (for part of the time).[54] After lunch with Cockcroft and Adams, Cockcroft and Lockspeiser took him to see Cherwell. According to Amaldi, there was a sharp exchange of views at this meeting, which lasted no more than about ten minutes. Amaldi was then driven to Harwell by Adams, where he had dinner with members of the accelerator group, including Adams, Fry, Goward, and Lawson. The next morning was again spent at the DSIR, on this occasion with King, Verry, and Waterfield. 'The U.K. provisional first re-draft' of the CERN Convention 'was gone through in detail', and 'the reasons for the various U.K. amendments' were explained. After dealing with certain administrative matters at the DSIR and visiting Imperial College, Amaldi left for New York on 13 December. In what follows we shall first discuss the meetings held at the DSIR, and then comment on the significance of Amaldi's encounter with Cherwell.

Two relatively minor procedural points and two major points of policy were discussed with Amaldi by the DSIR officials. Concerning the former, Britain wanted the system whereby her share of the budget (assessed at 35% on some variants) was calculated to be discussed.[55] She also wanted to be free to send official observers to Council meetings and to sign the Convention establishing the permanent organization *without* being party to the Agreement constituting the provisional CERN. Neither of these, it seemed, raised any serious difficulties.

The first major issue of policy which the British raised with Amaldi concerned limiting the scope of the laboratory to research on high-energy particles, in particular that which could not be done in *existing* European laboratories or institutes. This was at the specific request of Cockcroft and Chadwick, who feared—and Amaldi confirmed this at the meeting on the 11th—that some Council members favoured extending the laboratory's activities well beyond this.[56] Speaking personally Amaldi agreed that this might result in a waste of effort; when discussing the British version of the Convention on the 12th he 'fully understood and accepted' the corresponding stipulations in its Article II.

The other major point was less easily dealt with. It concerned the OEEC principle so dear to King which we mentioned earlier: the idea that Britain could contract out from those projects which were of no interest to her, 'by which, in fact, we meant that we did not want to contribute towards the cost of the 600 MeV machine', wrote Verry. Amaldi thought that, in this form, this idea would encounter strong opposition in the Council. What is more, recent news from America indicated that this energy was likely to be important for meson research, so that, as Cockcroft said, 'it would be of benefit to the U.K. [scientists] to have access to such a machine'.[57] With that the hope that Britain could contract out of CERN's 600 MeV accelerator was dropped. Instead it was agreed that Britain insert in the proposed redraft of the Convention a provision enabling individual member states to 'contract out of *subsequent* projects or activities which might be approved by the Council beyond the initial programme' (our emphasis).[58]

The general impression one gets from reading Verry's memoranda on Amaldi's meetings with the DSIR is that they were conducted in an atmosphere of goodwill and of readinesss to compromise on both sides. For his part Amaldi realized the

Notes: pp. 514ff.

tremendous importance the other member states attached to having Britain in the scheme, and did not want to raise unnecessary obstacles to her membership. The negotiators from the DSIR, for their part, got virtually all they wanted—and notably *some* kind of restriction on financial liability—their success, it should be added, being primarily due to the fact that Amaldi at least welcomed many of their suggestions.

What of Amaldi's encounter with Cherwell? Amaldi has described this on several occasions, of which the following is a typical example.[59] It is taken from a lecture given some 20 years after the event:

> Lord Cherwell appeared to be very clearly against the participation of the U.K. in the new organization. As soon as I was introduced in his office he said that the European laboratory was to be one more of the many international bodies consuming money and producing a lot of papers of no practical use. I was annoyed and answered rather sharply that it was a great pity that the U.K. was not ready to join such a venture which, without any doubt, was destined to full success, and I went on by explaining the reasons for my convictions. Lord Cherwell concluded the meeting by saying that the problem had to be reconsidered by His Majesty's Government [...].
>
> A few weeks later Sir Ben Lockspeiser wrote an official letter asking that the status of observer be given to the U.K. in the provisional organization, and the D.S.I.R. started to regularly pay "gifts", as they were called, corresponding exactly to the U.K. share calculated according to the scale adopted by the other eleven countries.

This account, which has assumed the status of a 'standard version' of the events, is open to a number of criticisms. For one thing, after Amaldi's meetings in London there were no drastic changes in British policy on CERN nor, in particular, of the kind that Amaldi identifies. The request that Britain have the status of observer was made the day Amaldi *arrived* in Britain—and not 'a few weeks later'. (It was agreed to in a letter from Amaldi to Lockspeiser on 20 December.[60]) Similarly, the first 'gift' of £5,000 from the DSIR was authorized on the eve of the Amsterdam session of the Council early in October. (It was officially notified to Amaldi by Lockspeiser on 17 November and was paid by the DSIR on 25 November.[61]) Thus the decisions to be observers at the provisional CERN and to make financial contributions to the organization were made before, even long before, Amaldi met Cherwell in London. Finally the value of the two 'gifts' made (totalling £10,000) amounted to less than 12% of the budget of the 'planning' stage: they accordingly had no connection with the UK share of the budget to the interim period and beyond.[62]

We also have reason to doubt Amaldi's version of his encounter with Cherwell—as we saw earlier, Cherwell had *already* agreed, however reluctantly, that Britain participate in CERN at an interdepartmental meeting on 20 November.[63] What then are we to make of Amaldi's story? Perhaps it is accurate—in which case we must conclude that Cherwell deliberately misled Amaldi at the start of the meeting into thinking that he was against British membership. More plausibly, though, the stress that Cherwell apparently placed on financial matters probably reflected his strong feeling that the Convention be so phrased as to allow its signatories to contract out of projects. Was this perhaps not the point he was driving at in his interview with Amaldi? And did the meeting not end quickly because he was assured that a provision of this kind could be made without undue difficulty?[64]

Our aim in criticizing Amaldi's story is not to score points off him: his accounts of the origins of CERN have provided useful points of reference for us. Nor is it to expose, yet again, the fallibility of human memory. Rather it is to underline that one of our main tasks in this project has been to construct, laboriously, a history of CERN *against* a host of misleading anecdotes embedded in the collective memory of those who built it. It is a theme we return to again in our conclusion to this volume.

13.5.2 THE CHANCELLOR RECONSIDERS THE CASE

The officials of the DSIR were enormously reassured by the visit of Amaldi. A meeting of all interested departments was arranged for 19 December, and a new draft of the Convention was drawn up. Three days later, on 22 December, Lockspeiser submitted it as an annex to a minute to his new minister, in which he argued that Britain should join CERN.[65] It contains few surprises: it details the results of the meeting with Amaldi, the equipment the laboratory would have and the importance the physicists attached to it, the costs of membership (calculated on the assumption that Britain would pay 25% of the announced total bill of 120 MSF). To conclude, Lockspeiser insisted that there were no major security risks involved—the laboratory would do 'non-secret, publishable, fundamental research, and no military research or applications would be undertaken'. And somewhat unusually for arguments used at this time, he appealed to the political repercussions if Britain continued to stand aside—this would 'leave the most unfortunate impression about the interest of the U.K. in Western European scientific life and result in those countries becoming increasingly dependent on the U.S.A. for guidance in nuclear research'.

On the same day the Marquess of Salisbury drafted a minute to Butler, to which he attached Lockspeiser's minute.[66] It basically recapped the Secretary of the DSIR's arguments, came down in favour of Britain joining CERN, and asked the Chancellor to agree. On 29 December 1952 Butler replied as follows:

> Provided that the necessary safeguards are accepted and duly embodied in a convention, I am willing for the U.K. to join the European Nuclear Research Organization, incurring the potential financial commitment estimated at £348,000 a year for seven years and thereafter operating expenses of the order of £134,000 a year.[67]

The Chancellor, then, had finally accepted that Britain join CERN as a fully participating member—'in view', as he put it in a letter on 25 November, 'of the clearly expressed view of British scientists that our physicists should have access to the more powerful machines now being planned [...]'.[68] Yet while it is plain that *without* this consensus Butler had been most unwilling to engage his government formally with the European laboratory, it would be too simple to attribute his decision on 29 December 1952 to this factor *alone:* a number of other important considerations also played a role.

Firstly, one must not forget the main results of the meeting with Amaldi: the conviction that Britain's interests could be protected in the Convention without her demands provoking undue opposition from the existing member states of the Council.

Then there was the positive attitude of the scientists themselves, and of the project's erstwhile opponents in particular. It was not simply a question of Chadwick and Thomson accepting the decision taken on 1 November: what is striking about Chadwick, for example, is how constructive he immediately became. For example, he wrote to Lockspeiser in mid-December suggesting that he or Thomson invite Amaldi to Cambridge adding that 'we are ready to help if we can be of use in any way'. And when, early in December, John Lawson at Harwell disturbed everyone by showing some of the practical limitations of the alternating-gradient principle, Chadwick responded by saying that while he hoped that this would induce a 'more sensible attitude' to the idea in the CERN group, he also hoped that they would not go to the other extreme—'these things', he wrote to Cockcroft, 'generally work out better than the many difficulties and obstructions lead one to expect'.[69]

Nor should one overlook the role of Cherwell, difficult to pinpoint from our documents, but undoubtedly crucial. As Paymaster-General in Churchill's government, Lord Cherwell was supposed to co-ordinate scientific research and development. More importantly, though, he was a kind of super-minister, part of Churchill's 'overlords' experiment whereby, between 1951 and 1953, he grouped a number of ministers together under a single head.[70] This only increased Cherwell's already considerable influence, and no major step could be taken in November and December without consulting him, and having him on one's side.

Refocusing a little our analysis of the factors shaping Butler's decision, let us look more closely at the amount of money involved. How did it compare with current expenditure in this area of science? And what of the fears that the Treasury could not support an investment on this scale given the prevailing financial priorities and the state of the economy?

The first point to stress is that the amount authorized by the Treasury (on the DSIR vote) of £348,000 a year was considerably more than the annual DSIR expenditure on nuclear physics in all five of the universities having accelerators. For the five years ending 31 March 1953 this amounted to about £240,000 per year; for the following quinquennium it was anticipated to be some £225,000 annually.[71] Or, to give another point of reference, the annual contribution to CERN was some 7% of the total DSIR vote for the year 1950/51 (£4.9M).[72] What is more, it was accepted by Lockspeiser that, as in all such projects, the provisional estimates were likely to be exceeded by as much as 50%; at his meeting with Cherwell on

20 November a total CERN contribution of £3M was mentioned (£429,000 for 7 years).[73] Comparatively speaking then, the contribution to CERN involved a major escalation in the expenditure on high-energy physics by the government. Several factors helped to soften the blow.

For one thing, it was becoming accepted in government circles that scientific research in general, and accelerator physics in particular, were expensive, and expansive. In 1950/51 the annual DSIR budget for civil research and development was double that in 1945/46. Five years later it was 1.5 times greater again.[74] The escalation of expenditure on accelerators as such was even more dramatic in some cases. The Liverpool cyclotron was originally budgetted for at £60,000. By October 1952 its estimated cost of construction had rocketted to £520,000—almost a factor of nine.[75] Any shock which government departments may have had over the high cost of such machines, and the voracity of their appetite for funds had long since worn off. Britain's science administrators, in other words, were becoming used to the financial demands of big science.

A second factor easing the way to this expenditure was the new priority given to rearmament and weapons development: indeed Vig has called the period from 1951 to 1957 Britain's 'defence science' era.[76] High-energy physics benefitted from the corresponding increase in investment in civil and military nuclear development, with which it was associated. This is not to say that applications were expected in either sector in the short term. On the other hand it was surely no coincidence that the strongest team of accelerator builders in Europe was at the Atomic Energy Research Establishment. Who could say what possible rewards would not be reaped from this esoteric domain? A look at recent history was enough to raise expectations.

Finally, there was the overall improvement in the British economy: indeed it entered something of a boom period between mid-1952 and mid-1955.[77] A deficit on the balance of payments of nearly £700M inherited by Churchill's government in October 1951 had become a surplus of £300M by the end of 1952. Equity share prices more than doubled in the three-year period ending in July 1955. This upward turn in the economic indicators doubtless encouraged Butler, who had something of a reputation for extravagance and for being willing to toy with new ideas, to authorize the expenditure for CERN.[78]

13.5.3 THE SIGNATURE AND RATIFICATION OF THE CONVENTION

Her Majesty's Government was officially represented by observers for the first time at the fourth session of the provisional CERN Council which was held in Brussels from 12 to 14 January 1953. The honours fell to Sir John Cockcroft and Sir Ben Lockspeiser, who were warmly received by the other delegations. From this time onwards Britain was regularly represented at the meetings of the provisional Council even though she never officially joined it, and her administrators and scientists began to play an increasingly important role in the organization. Sir Ben

Notes: pp. 515 ff.

Lockspeiser was appointed chairman of the Interim Finance Committee, while early in 1954 John Adams, the Harwell engineer, was nominated to head the all-important scientific group charged with building the proton synchrotron.

The immediate concern of the United Kingdom delegation during the first six months of 1953 was to establish a working group to make financial and administrative arrangements for the transition to the permanent CERN, and to ensure that the government's wishes were formally enshrined in the Convention.[79] Concerning the latter, as Amaldi expected, the British had little difficulty in persuading other delegations to accept that the scientific objectives of the laboratory be closely defined, and that provision be made for a member state to contract out of projects other than those specified in the initial programme of activities. They did, however, encounter stiff resistance to a demand from the Foreign Office that no *explicit* limitation of CERN membership to European states be made in the Convention. Briefly—for this matter is discussed at considerable length in section 8.2—the Foreign Office wanted to avoid giving the impression that CERN had an 'anti-American bias', and to leave the door open to Commonwealth membership. The move was unpopular in many quarters because it appeared to vitiate the entire European character of the organization, and was specifically opposed by France and Switzerland for more specific reasons of state. However, 'after a certain amount of adjustment and horse dealing' the Convention establishing CERN was signed on 1 July 1953 by Sir Ben Lockspeiser on behalf of Her Majesty's Government.[80]

For the next fifteen months the main task of the Council was to keep the organization going until the requisite number of states had ratified the Convention. This involved acquiring parliamentary approval, followed by the formal step of depositing the instruments of ratification at UNESCO House in Paris. Britain was the first of the signatories to complete this process. The United Kingdom's instruments were deposited in Paris on 30 December 1953, some six weeks before the next state, Switzerland did so, and one day before Norway had even *signed* the treaty. The Convention itself came into force on 29 September 1954 with its formal ratification by France and by Germany.

The sequence of the ratification process prompts a question: how was it that, of all the member states, Britain, who had been the least willing to commit herself to the European laboratory, should be the first to ratify the Convention establishing it?

Paradoxically, one reason for this is connected to the very fact that Britain had taken so long to make up her mind. As a result when the Chancellor agreed to join CERN towards the end of 1952, it was as a member of the permanent organization. And by the time the Convention was signed on 1 July 1953, all the relevant government departments and ministers, notably the Treasury and the Foreign Office, had *already* agreed that, at least as far as the executive was concerned, Britain could participate in the scheme. By contrast, for the other member states, formal government commitment did not stretch beyond membership of the provisional CERN. For them signature of the Convention was a declaration of intent, and complex bureaucratic hurdles, particularly concerning questions of financing, still had to be crossed before ratification was possible.[81]

Another factor facilitating ratification was the simplicity of the procedure itself. A White Paper containing the Convention was, as the Marquess of Salisbury explained, 'laid before Parliament for the customary twenty-one days. If Members of either House wish[ed] to object, they [could], of course, do so at any time during that period. The Convention would normally be ratified at the expiration of that period'.[82] In practice what this meant was that a Convention—in fact an international treaty—of this type could be ratified without the Members of Parliament being called on collectively to give it even their nodding assent, let alone to discuss it. This in fact happened: the Convention was laid before both Houses shortly after Parliament assembled on 4 November 1953, no objections were raised to it, it was 'automatically' ratified, and the instruments of ratification were deposited at UNESCO House by the end of the year.

This leads to a third remark: that not simply were there no objections raised in Parliament to Britain joining CERN—the issue was barely discussed at all. Before the Convention was signed the general attitude in Whitehall was that it was 'desirable to say as little as possible at this stage'—and as little as possible was said.[83] On 4 November 1953, Earl Jowitt did question the Lord President briefly about CERN, his main concern being, he said, that Parliament could still discuss the matter, his main fear being that Britain was 'already committed to this because UNESCO has pronounced on it'.[84] Reassured that parliamentary sovereignty had not been infringed, neither Jowitt nor any other Member felt the need to question the proposals of the executive. In this their behaviour was consistent with the role played by Parliament in matters of science policy at least until the early sixties, if not beyond, a role described by one commentator as involving the 'vicious circle of a growth in executive power, a tradition of withholding information from Parliament, the lack of research services for Members of Parliament and apathy of the latter in the face of the resultant parliamentary situation [...]'.[85]

Appendix 13.1

A short survey of the committees involved in the CERN decision

The aim of this appendix is to assist the reader who may be confused by the several committees involved in the making of British policy *vis-à-vis* the European laboratory project. The information we give is not comprehensive: *it is restricted to the years 1951–52,* and is based on what we have gleaned from the primary sources we have collected.

1. *British Committee for Co-operation with UNESCO in the Natural Sciences* (we also call it the British UNESCO Committee, or the BUC)

Role: This was a committee established in 1947 by the Royal Society 'to co-operate with UNESCO in the field of the natural sciences on behalf of the United Kingdom, and to make recommendations to the Council of the Royal Society and to the Minister of Education'.[86]

Chairman: Sir Cyril Hinshelwood, Foreign Secretary of the Royal Society

Secretariat: H.C. Rackham (Ministry of Education)
D.C. Martin (Royal Society)

Membership: Representatives of various government departments, including the DSIR, of the Royal Society, and of several national scientific committees.

Attached to the BUC was an *Executive Committee* which met very infrequently. During our period it was used on 28 February 1952. Its chairman and secretariat were those of the main committee.

2. *Cabinet Steering Committee on International Relations* (abbrev. IOC; we also call it the Cabinet Steering Committee)

Role: This was the highest-level committee involved in the decision on whether or not Britain should join CERN. Its recommendations were passed directly to the ministers concerned.

Chairman: A.A. Dudley (Foreign Office)

Secretariat: K.S. Butler (Foreign Office)
D.J.M. Brenton

Membership: Representatives of some dozen government departments, notably several Treasury officials.

3. *'Chadwick's Advisory Committee'* (full name: British Committee for Co-operation with UNESCO in the Natural Sciences, Advisory Committee on the European Nuclear Research Centre)

Role: The committee was set up by the BUC in January 1952

> i) to advise on the proposals for the establishment of a European Nuclear Research Laboratory and in particular to advise the British Committee on the draft agreement setting up a Board of representatives to govern the Laboratory, and
> ii) to report to the British Committee for Cooperation with UNESCO in the Natural Sciences.[87]

It met for the fourth and last time on 1 November 1952.

Chairman: Sir James Chadwick
Secretariat: D.C. Martin (Royal Society)
Members: J.D. Cockcroft, B. Lockspeiser (DSIR), H.W.B. Skinner, G. Thomson.

4. *'Cockcroft's Advisory Committee'* (full name: Department of Scientific and Industrial Research, Advisory Council, Scientific Grants Committee, Nuclear Physics Sub-Committee)

Role: The committee was set up in May 1952 to deal with applications for special grants from the DSIR for nuclear physics at the universities equipped with accelerators (Birmingham, Cambridge, Glasgow, Liverpool, and Oxford), for the 1952–57 quinquennium.[88] Later in the year its role was expanded to include advising on CERN.

Chairman: Sir John Cockcroft
Secretariat: I.G. Evans (DSIR)
Members: P.M.S. Blackett, L. Bragg, D. Brunt (Royal Society), P.I. Dee, N.F. Mott, F.E. Simon, G. Thomson, A. Trueman.

5. *Nuclear Physics Committee* (abbrev. NPC)

Role: This committee was set up in or around 1946 by the Ministry of Supply 'to make recommendations regarding the programme of nuclear physics to be pursued in this country as a whole', in particular insofar as it concerned the necessary machines and capital expenditure in the universities. It was disbanded in July 1951, some of its functions being taken over by Cockcroft's Advisory Committee (q.v.).[89]

Chairman: Sir James Chadwick
Secretariat: J.H. Awbery (Ministry of Supply)
Members: (At least) P.M.S. Blackett, E. Bretscher, J.D. Cockcroft, Collie, C.G. Darwin, N. Feather, O. Frisch, R. Peierls, C.F. Powell, J. Rotblat, E.T.S. Shire, H.W.B. Skinner, G. Thomson.

6. *Committee on Overseas Scientific Relations* (abbrev. OSR)

Role: This committee was set up in March 1948, its aim being 'to consider and advise on questions of United Kingdom Government policy in matters of overseas scientific relations'. It was to maintain close liaison with the Cabinet Steering Committee (q.v.) and was empowered to make recommendations to the Lord President (the minister responsible for civil scientific affairs) or to individual government departments about ways of improving scientific relations in general. The committee was situated in the Overseas Liaison Division of the DSIR.[90]

Notes: p. 516

Chairman: Dr. Alexander King (DSIR)
Secretariat: H.L. Verry (DSIR)
   A.R.M. Murray (DSIR)
Members: In the main, representatives of several government departments, the Research Councils, and the Royal Society.

Appendix 13.2

Decision-making at the science–government interface: some general comments

The events described in this and the preceding chapters, and the documentation on which they are based, can be used to throw some light on the mechanisms whereby 'science policy' *vis-à-vis* the European laboratory project and CERN was made in Britain. As our analysis indicates, this evolved through two quite distinct phases: the 'bureaucratic' phase, when the project was subject to the logic of the government's committee structure, and a 'ministerial' phase when the responsibility for the decision passed to the highest levels of the executive. The meeting of the IOC on 11 June 1952 with which we ended chapter 12 was the watershed between these two phases.

In what follows we want to investigate the 'bureaucratic' phase of the decision-making process a little more carefully. To this end we shall first present a somewhat idealized and oversimplified model of the process and shall then critically transform it to bring it more in line with our evidence.[91] To conclude we shall comment on the role of the scientists in policy-making, a role conventionally, but, we shall argue, misleadingly thought of as being advisory.

1. THE RATIONALISTIC MODEL

The first preconception to shed is that of thinking of policy as being made by 'the government', as if that were a homogeneous, monolithic entity. 'The government' is a complex interlocking network of relatively autonomous departments or, more precisely, of committees which can be located primarily in a department or can straddle the divide between two or more departments.[92]

Within each department there are, of course, hierarchies of authority, and it is encumbent on officials at each level to judge which decisions they can take and which have to be referred to higher levels. More important though is that the department itself, and particular committees within it, have specific areas of responsibility. These areas are jealously guarded, and since their boundaries are not clearly defined they are subject to negotiation and dispute between different committees and departments.

What of the scientists? In our case their official positions were expressed either through a committee located within a department—like the Nuclear Physics Committee of the Ministry of Supply—or in a 'joint' committee, like the British UNESCO Committee, where the secretariat was shared by the Royal Society and the Ministry of Education. Indeed, in terms of the framework just presented a body like the Royal Society may itself be regarded as akin to a 'government department', alongside and continually interacting with official nonscientific departments.

That granted, we immediately see that it is misleading to think of 'scientists' and 'the government' as two more or less independent groups. Many scientists in our study are *part of* government in the sense that they are integrated into the decision-making mechanisms of the state, and are subject to the same logic as other elements in it. In particular they serve on committees with their own restricted areas of responsibility and authority—this being the scientific aspects of an issue in their case. Both of these points require further elaboration.

It should be remembered that, when speaking of 'the scientists', and notably those who are 'part of' government we are effectively speaking of no more than half-a-dozen senior British physicists—Chadwick, Thomson, Cockcroft, Skinner, and, towards the end of our study, Blackett and perhaps Fry.[93] Of course other physicists were aware of developments at various times, and indeed a large body of them made their views known towards the end of 1952. But this does not alter the fact that at the *interface* between 'science' and 'government' the community was 'spoken for' by a small and not necessarily representative elite.

The authority which this elite had derives from several sources. There is their knowledge, their scientific expertise—theirs is a *cognitive* authority.[94] Reinforcing this, there is the status which an individual has in the scientific community, as indicated by awards and prizes, membership of the Royal Society, professorship in a university, and so on. The influence of men like Chadwick and Cockcroft was further enhanced by their success in managing scientific projects both during and after the war—and by the contacts and friendships they had thereby formed at the highest levels of government.

Turning now to the formal decision-making process itself, two cardinal principles underlie it. Any issue 'input' to the mechanism has to be channelled to that department and committee into whose area of responsibility it falls. If the issue is a complex one of the kind generated by the emergence of 'big science', however, it transcends the limits of authority of any one department. In this case it is broken down into its components, and assessed by each department from that aspect which concerns it directly. Secondly, existing policy serves as a basic point of reference for the line adopted by the department. This does not mean that standing policy is the only, or the determining factor in the decision taken. Other arguments for or against a particular course of action can be adduced. Yet the general concern persists to remain consistent with policies hallowed by time or, in the case of the Treasury, formulated in the light of the prevailing politico-economic situation, and only to reverse them if strong arguments for doing so are brought forward.

Notes: p. 516

In terms of this 'model' a decision taken by a committee is first to be understood as being in line with, or involving a change of, standing policy on the issue or aspect of the issue under discussion. Changes in policy need to be justified by arguments. If the committee is interdepartmental each department will argue its case from the standpoint of its policy on that aspect of the issue which concerns it most. The ultimate decision of the committee is the 'sum' of the several 'policy-vectors' mobilized during the course of the debate by various participants.

This description captures some of the more prominent features of the mechanism whereby decisions were taken by the OSR Committee on 6 February 1952 and the IOC Committee two days later. At the first the issue was whether additional money could be found for the Liverpool cyclotron, at the second whether Britain should sign the intergovernmental agreement. Prevailing policy of both the Treasury and the Foreign Office was to discourage increasing British involvement in international organizations. Against this the Ministry of Education argued that 'H.M. Government belonged to, and *had hitherto supported,* the United Nations and its various agencies', and in the light of this policy the department was *'naturally* concerned to support strongly any constructive proposals towards the implementation of UNESCO's programme ...'.[95] The DSIR noted that it had *'strongly backed'* regional research schemes in UNESCO and in the OEEC. As for the use of the Liverpool cyclotron 'on a European basis', it remarked that *'No new principle* in University administration [was] involved, since it [was] *regular policy* of British Universities to accept foreign research workers and D.S.I.R. to admit such workers to institutions for which it [was] financially responsible'.[96] Even the scientists articulated their case in these terms. The minutes of the OSR meeting record that Thomson and Cockcroft 'pointed out that during *the past fifty years* it had become *common practice* for Universities to offer facilities for research to scientists from other countries ...'[97] (our emphasis throughout this paragraph). The 'vector-sum' of these policy considerations (reinforced by other arguments of course) was that the Treasury, but not the Foreign Office, was prepared to reconsider its prevailing policy in this case.

Our analysis reveals how complex is the decision-making mechanism in British government. The disaggregation of an issue into its component parts, their distribution among several departments each with its own limited area of responsibility and regular timetable of committee meetings, produces delays, misunderstandings, and missed opportunities. On the other hand there is an inbuilt flexibility in the system, in the sense that a negative decision on an issue by one department is not necessarily final and can be reversed. In particular those aspects of the issue which are consistent with standing policy in another department remain unscathed, and provide a basis for relaunching an offensive when circumstances are more propitious. Thus proponents of British membership of the European centre came away disappointed but not dispirited from the IOC meeting. They would put their case again, using new and hopefully more powerful arguments than the time before.

Indeed the tenacity with which bureaucrats in the Ministry of Education and the DSIR pursued the case for membership with the IOC after the first meeting on 6 February 1952 is quite remarkable. As we mentioned in section 12.5.1, the paper that was laid before it in June had been written and rewritten in the light of ongoing interdepartmental discussion and criticism, the arguments becoming increasingly more coherent and insistent. And when, at the end of August, the Chancellor turned down the IOC recommendations that HMG sign the provisional agreement, Verry *immediately* drafted a long memo analyzing its arguments and suggesting that the reply had 'in fact presented us with several useful loopholes [...] such as to justify a further approach to the Chancellor' which, Verry hoped, would 'dispose' of his 'latest objections'.[98]

Though our simple model throws some light on the decision-making process it is limited, notably in that it is over-rationalistic: it implies that policy decisions are simply the outcome of discussion and argument in which considerations for and against a particular course of action are balanced against each other to produce a 'vector-sum'. Factors of other kinds also influence the course of events and shape their final outcome.

Firstly, there are conflicts between departments precipitated by the 'border wars' we mentioned above. Different departments resent intrusions by others into their areas of responsibility. When these occur it is less the 'intruders' arguments for a particular course of action that are assessed than its right to have acted as it did, all be those actions in accordance with the intruding department's interpretation of its own standing policy. Typically, in the background papers prepared by the Foreign Office for the IOC meeting in February, the offer of the use of the cyclotron, made at Paris with the support of the Ministry of Education and the DSIR, was criticized. It seems to me, wrote K. Butler, 'that a contribution, whether in money or in kind, which is regarded as a "qualifying" contribution' presupposes that Britain wishes to participate in the centre. Butler went on to challenge the authority of the British UNESCO Committee to advocate the change of policy implied by this.[99]

Secondly, there are the ambitions of the departments or units within departments themselves, those 'separate empires, each with its own internal loyalties, and each concerned to further—or at least to protect—its own sphere of influence', as Zuckerman has put it.[100] We have suggested that the willingness of the DSIR to put the case for CERN membership may have been due to considerations of this kind: indeed it may also explain Verry's tenacity in the face of the Chancellor's refusals of which we spoke a few moments ago.

Finally, and crucially, a too-rationalistic model overlooks the effects of lobbying, the use of the informal, personal contacts at the highest levels of authority to influence or even bypass the committee structure. This was extensively used by the scientists and, as Pestre has noted in section 9.3.5, it is typically resorted to at two stages of the decision-making process:

(i) The stage *before* a specific proposal is subject to scrutiny at the formal level.

Notes: pp. 516ff.

Typically, in our case, before the proposal to offer the use of the Liverpool cyclotron was put on the agenda of the BUC, Chadwick 'informally' solicited support among his scientific colleagues and indeed in some government circles for the Kramers–Bohr scheme. He agreed with Thomson that the offer of the use of the accelerator would be a good way of expressing that support. It was only then, when the scientists had a concrete proposal to put forward, that the formal decision-making mechanism was 'triggered'.

(ii) The period *after* a negative decision is reached. In this connection we remember the sustained lobbying to which Sir Roger Makins was subjected by Chadwick, Cockcroft, and Thomson during the weeks and months after the IOC turned down the request that Britain join the provisional CERN. The outcome of that pressure was the reversal of policy at the IOC meeting on 11 June 1952, after which the question of CERN membership passed beyond the purview of government committees into the hands of departmental secretaries and ministers.

Granted the limitations of the model, and of the importance of lobbying in shaping policy, it may well be asked whether a rationalistic approach is of any interest at all: would a study of pressure-group activities not be enough? We think not. Indeed our model is useful as an antidote to an 'over-sociological' interpretation of events, as demanding that the role of argument and debate in policy-making be taken seriously. Nor can it be claimed that what debate there was, was rendered irrelevant by pressure-group tactics. On the contrary it was immensely important, if only because it threw the key issues into relief, weighed their importance, and exposed just how vulnerable the scientific case for membership was. By virtue of that vulnerability the amount of progress which could be made by argument at the formal level of committees was heavily circumcribed. Correlatively, the limits of a rationalistic model are not only intrinsic to the model itself: they also reflect the poverty of the arguments themselves.

2. THE ROLE OF THE SCIENTISTS

Related to the two stages of the lobbying process we have two aspects of the scientists' behaviour: as *triggering* the decision-making mechanism, and as *'product-champions'* within it.

By describing scientists as triggering official machinery we mean primarily that unless, and until, scientists regard a particular course of action to be in a country's scientific interest, officialdom is loath to take any novel initiatives in the field of science policy. In this sense its role is reactive, and dependent first and foremost on what scientists judge to be worthy of support. Thus, to go back a little, when in April 1951 the Nuclear Physics Committee decided that there was little to be gained by Britain participating in the European laboratory project, no further action on the matter was taken officially, although a man like King, for example, would have favoured a more positive approach in the light of his views on regional research centres. No pressure was put on the scientists to change their minds, however. Correlatively, it was only after Thomson and Chadwick had suggested the offer of the use of the Liverpool cyclotron that the complex machinery described above was brought into play, and the further articulation and even reversal of prevailing science policy was on the agenda.

By describing the scientists as product-champions we mean that they took the cause of CERN membership to heart and fought for it in the face of considerable opposition in the government. They persisted in this although the original grounds for getting embroiled in the scheme—to try to have the laboratory based in Copenhagen, and to delay the construction of big machines—were quickly eroded, and they persisted although they had no convincing reply to the objection that Britain's scientific interests, as they defined them, were just as well protected without signature of the provisional Agreement. In short, having once embarked on a course of action, and notwithstanding periodic misgivings, Cockcroft, Chadwick, and Thomson continued to defend it unanimously in public and to lobby for it in private.

It is a commonplace that, particularly since the war, scientists have become a 'part of' government, and that they play an important role in the formulation of government policy. More often than not, though, that role is regarded as *advisory,* as the giving of expert opinion on the scientific aspects of an issue, and of being at most indirectly responsible for the decisions ultimately taken. This clearly fails to capture the essentials of the scientists' role in shaping policy in this case. The relatively passive detached image of an advisor bears no relation to the reality we have described: a reality in which the scientists decided what they wanted, were determined to fight for it, and solicited the support of officials at many levels of the state apparatus to achieve their ends.

Notes

1. The sources used for this chapter are those listed in note 1 of chapter 12.
2. The proceedings of the Buckland House meeting were summarized by M. Snowden in an informal *Report on the conference on high energy accelerators for nuclear research, Buckland House, 7th June 1952* (CC-CHADI, 26/2). We do not have a list of those who attended this conference; the names we have are of those who gave papers and participated in the discussion. The invitation to Chadwick is in letter Cockcroft to Chadwick, 8/5/52 (CC-CHADI, 1/2)
3. See Cockcroft's letter, note 2.
4. For details see chapter 12, note 124.
5. The unconfirmed minutes of the first meeting of the DSIR Advisory Committee, Scientific Grants Committee, Nuclear Physics Sub-committee, held on 7/5/52 are attached to those of its second meeting held on 4/6/52 (PRO-DSIR17/451). The foundations of this committee were laid at two informal meetings arranged by the DSIR on 23/6/50 and 20/11/50 to deal with nuclear physics research at universities (PRO-AB6/774). For further information on the committee see appendix 1 of this chapter.

6. For Pickavance's view of what was achieved at the conference see his memo *Accelerator Programme*, 10/6/52 (PRO-AB6/1014).
7. The minutes of the Cabinet Steering Committee are document I.O.C. (52) 10th Meeting, 11/6/52 (PRO-CAB134/943). For Murray's letter to Chadwick with Sir James' annotations see the version in (CC-CHADI, 1/7).
8. The talks given by the British delegates to the conference are included in its 'proceedings', Conference - Copenhagen (1952). In their capacity as observers both Cassels and Fry wrote reports after the meeting. Cassels' informal report entitled *The European Laboratory, Report on Discussions at Copenhagen, 16th –19th June, 1952*, undated, is on (PRO-AB6/1014). For Fry we have his *Report to the Royal Society upon the second session of the Council for European Nuclear Research held in Copenhagen 20th and 21st June, 1952*, 25/6/52 (CC-CHADI,1/2).
9. Letter Cockcroft to Chadwick, 23/6/52 (CC-CHADI,1/2); discussion paper by J.M. Cassels, *Great Britain and the European Laboratory*, 24/6/52 (PRO-AB 6/1074); report by Fry, see note 8. For the list of those to whom Cassels' paper was circulated see the names written on a memo Fry to Cockcroft, *Great Britain and the European Laboratory*, 27/6/52 (PRO-AB 6/1074).
10. Cassels' paper, note 9, 4.
11. *Ibid.*, p.5.
12. Fry's report, note 8, 4.
13. Cassels' paper, note 9, 5.
14. In a letter Skinner to Auger, 29/4/52 (PRO-ED157/302), Skinner informed Auger that he would be 'very pleased to have the assistance of one or two physicists or engineers' to help at Liverpool, repeated the offer to accommodate four scientific researchers, and added that he would also be 'glad at any time to receive one or two theoretical physicists [...]'. The arrangement to have them in the laboratory could, he felt sure, be made 'without any formality'. He confirmed that Bonaudi and Krienen were coming at CERN's expense in a letter Skinner to Vernon, 2/7/52 (PRO-DSIR17/551).
15. The minutes of the meeting of Chadwick's committee on 16/7/52 are document NS(52)3, undated (PRO-ED157/302). Skinner's letters are Skinner to Cockcroft, 3/7/52 (PRO-AB6/1074) and Skinner to Chadwick, 7/7/52 (CC-CHADI,1/8). The quotations are from the first and second of these letters, respectively.
16. Letter Cockcroft to Peierls, 18/8/52 (PRO-AB 6/1074).
17. Letter Chadwick to Makins, 17/7/52 (CC-CHADI, 1/6).
18. (Draft) letter Pickavance to Chadwick, 1/8/52, with a covering note to Cockcroft, 31/7/52 (PRO-AB6/1074); letter Pickavance to Chadwick, 12/8/52 (CC-CHADI,1/14); letters Massey to Chadwick and to Cockcroft, 1/8/52 (CC-CHADI,1/2); letter Peierls to Cockcroft, 14/8/52 (PRO-AB6/1074).
19. For example, things looked less rosy if one compared the annual cost for CERN membership with the average expenditure by the DSIR on nuclear physics for the quinquennium ending 31/3/53, viz. £240,000 per year. See DSIR Advisory Council, Nuclear Physics Sub-committee, Report A.C.(51.2)49, 16/7/52 (PRO-DSIR 17/449).
20. Letter Chadwick to Pickavance, 18/8/52 (CC-CHADI,1/14). We do not have a reply to Massey.
21. We do not know Powell's views in detail. Whatever doubts he may have had in July about the wisdom of investing in a big machine were apparently dispelled by September, however—see letter Fry to Cockcroft, 1/10/52 (PRO-AB 6/1074).
22. Gowing (1974), vol. II, especially 212-236. The quotation is from p. 225.
23. Peierls' claim that his views were 'fairly representative' is made in his letter to Cockcroft, see note 18.
24. This paragraph is based on a memo by Verry, *Nuclear Physics Research Council*, 24/6/52 (PRO-DSIR17/559), from which the quotation is also taken.
25. The minutes of the OSR Committee meeting held on 17/6/52 are document O.S.R. (52) 2nd Meeting (PRO-DSIR17/486). On 6/8/52 King sent Verry a draft of the sort of minute Lockspeiser wanted to send

to the Lord President, asking for his comments (PRO-DSIR17/551). The version the Lord President received was *United Kingdom Participation in Council for Nuclear Research,* 8/8/52 (PRO-DSIR17/551).

26. That Cherwell was fully briefed on the scientists' official position is clear from letter Chadwick to Makins, 17/7/52 (CC-CHADI,1/6). Both the scientists and officials in the DSIR repeatedly advised the Lord President to consult with Lord Cherwell—see, for example, minutes of the 2nd meeting of the OSR, and the draft of the minute from Lockspeiser to Cherwell dated 6/8/52, note 25. Cherwell's support for membership is explicitly referred to in the Lord President's minute to the Chancellor of 20/8/52 (see note 29).

27. For the correspondence between Hinshelwood and Lockspeiser on this matter, see letters Hinshelwood to Lockspeiser, 22/7/52 and 2/8/52, and Lockspeiser to Hinshelwood, 30/7/52, all in (PRO-DSIR17/551). The quotation is from a memo by Verry, *Nuclear Research Centre,* 31/7/52 (PRO-DSIR17/559), in which he comments on this exchange, and remarks that Sir Ben was 'taking very strong exception' to Hinshelwood's proposal that the DSIR invite 'the Royal Society to constitute a new Advisory Committee' (his letter of 22/7/52) to replace Chadwick's.

28. Melville (1962), pp. 44-45.

29. The consultations leading up to the drafting of the Lord President's minute are described in a memo from King to Lockspeiser, *European Nuclear Physics Council,* 21/8/52 (PRO-DSIR17/551). For Woolton's minute to Butler, *Chancellor of the Exchequer,* 20/8/52, see (PRO-FO371/101516). For Butler's reply, *Lord President of the Council,* 27/8/52, copied 28/8/52, see (PRO -DSIR17/551).

30. We surmise that Woolton learnt that the deadline had passed from King, who had been at the meeting of Chadwick's Advisory Committee on 16/7/52, where the deadline was explicitly mentioned (see minutes of the meeting, note 15), and who was with the Lord President when he drafted his minute to the Chancellor. For Pickavance's letter, see note 18.

31. He made this remark at the OSR meeting—for minutes see note 25.

32. The quotation is from Butler's minute referred to in note 29.

33. All quotations are from Goward's *New Brookhaven Cosmotron Scheme Using Alternating + and − n Values,* 22/8/52 (CC-CHADI,26/2). The paper is some six pages long, and has six sections: introduction and summary, principle of scheme, betatron oscillations, synchrotron oscillations, factors being studied, and tentative machine design. There is a short appendix and four pages of technical diagrams. Continental and British physicists were thus given a very comprehensive introduction to the alternating-gradient principle almost immediately after it was 'discovered' in Brookhaven.

34. See appendix 2, *The Design of the Proton Synchrotron,* written by Fry, of the *Report to the Royal Society on the 3rd Session of the European Council for Nuclear Research (C.E.R.N.) held at Amsterdam, 4 and 6 October 1952,* by Blackett and Fry, undated, but received by Cockcroft on 16/10/52 (PRO -AB6/1074). Fry reports on Jakobsen's work in this appendix. It is also alluded to in a report from the British Scientific Attaché in Stockholm, who visited Bohr's Institute with Alexander King on 29/9/53 (PRO-DSIR17/561).

35. For Pickavance, see his letter to Chadwick, 15/9/52 (CC-CHADI,26/2). The remark by Verry is from a minute sheet, ref. 969-4-2, 26/9/52 (PRO-DSIR17/560). Cherwell's view is described in another memo written by Verry, ref. 969-4-2, 1/10/52 (PRO-DSIR 17/560).

36. The brief was prepared at a meeting on 1 October between Lockspeiser, Cherwell, Makins, and a Treasury representative, see memo by Verry, ref. 969-4 -2, 1/10/52 (PRO-DSIR17/560). The (undated) brief itself is entitled *European Nuclear Research Council. Note for Professor Blackett's use at Amsterdam* (PRO-DSIR17/551).

37. See note 34. Unless otherwise stated, the quotations below are from Blackett and Fry's report.

38. Bohr's assertion that Britain should join quickly was also made to King when the latter visited him late in September—see letter King to Verry, 29/9/52 (PRO -DSIR17/560)—and was immediately passed on to Sir Ben in an 'urgent and personal' memo from Verry to Lockspeiser, 30/9/52 (PRO-DSIR17/551). Amaldi's view is quoted from letter Fry to Cockcroft, 1/10/52 (PRO -AB6/1074).

39. The news was spread in Europe by Dahl and Goward (see note 33). For the USA we know that the CERN visitors were in Brookhaven in the week beginning August 4 and then went on to Berkeley the following week where, according to Dahl, 'Goward reported on the new development in Brookhaven. Berkeley had had but vague rumours prior to our arrival, and his presentation caused definite interest', document CERN-PS/S4, *Report to the Secretary, covering visit to USA by Goward, Wideröe and Dahl in August 1952,* 25/8/52 (CERN-DG20551).
40. Letter Pickavance to Chadwick, 15/9/52 (CC-CHADI,26/2); Blackett and Fry report, note 34, p. 8.
41. For some of the correspondence on this see letter Brunt to Lockspeiser, 17/10/52 (PRO-DSIR17/551); two letters from Chadwick to Cockcroft, 23/10/52 (CC-CHADI,1/2); memo to Evans, 24/10/52 (PRO-DSIR17/453); letter Chadwick to Martin, 25/10/52 (CC-CHADI, 26/1); letter Lockspeiser to Playfair, 31/10/52 (PRO-DSIR17/551) from which the quotation is taken. This correspondence reveals the importance which Lockspeiser, though not Chadwick himself, attached to having Sir James present at the meeting—for obvious reasons.
42. Letter Skinner to Cockcroft, 21/10/52 (PRO-AB6/1074), where reference is also made to the Harwell meeting to be held the next day. Fry's *British Accelerator Policy and the European Nuclear Physics Laboratory,* was received in Cockcroft's office on 31/10/52 (PRO-AB6/1074).
43. The minutes of the fourth meeting of Chadwick's 'Advisory Committee on the European Nuclear Research Centre', held on 1/11/52, are document C/105 (52), (PRO-ED157/302). The eleven physicists over and above the four regular members who attended were Blackett, Bragg, Cassels, Dee, Frisch, Fry, Goward, Moon, Pickavance, Shire, and Simon. Cherwell was invited but did not attend.
44. It will be noted that the total amount of money called for in the third resolution is £2M. This was in fact 1/6 less than the estimate made of Britain's share of the CERN budget. As the resolution implies, the reduction was to be *in lieu* of her contributions in kind which, it was hoped, could be offset against the £2.4M that was her due (see letter Brunt to Chancellor of the Exchequer, 13/11/52 (PRO-DSIR17/551)). Amaldi strongly resisted this proposal when he visited London on 11 and 12 December (see section 13.5.1 below), and the final sum authorized by the Chancellor was in fact some £2.4M (see section 13.5.2 below).
45. All references to views of Fry and Skinner in what follows are from the documents quoted in note 42.
46. For the consensus see Brunt's report of the meeting in his letter to the Chancellor, note 44.
47. For the Pickavance quotation see his memo dated 10/6/52, note 6. Skinner remarked that Cassels was 'concerned about dropping the Linear Accelerator' in his letter to Cockcroft, 21/10/52 (PRO-AB6/1074).
48. Rudwick (1985), chapter 1 draws attention to the 'methodological necessity' of narrative accounts which reveal the *process* whereby conclusions are arrived at. See also Allison & Morris (1975).
49. Letter Chadwick to Bohr, 12/11/52 (CC-CHADI,1/3).
50. Letter Brunt to Chancellor of the Exchequer, 13/11/52 (PRO -DSIR17/551).
51. See memo by Verry, *European Nuclear Research Council,* 21/11/52 (PRO-DSIR17/560).
52. *Ibid.*
53. Letter Lockspeiser to Amaldi, 21/11/52 (PRO-DSIR17/551); letter Amaldi to Lockspeiser, 1/12/52 (CERN-DG20924).
54. The basic sources used for the meetings are memos by Verry, *European Nuclear Research Organization—C.E.R.N.,* 11/12/52, and *European Nuclear Research Organization (C.E.R.N.),* 12/12/52, to which was appended (?) an unsigned document, marked 'Not for circulation', dated 12/12/52, and entitled *European Council for Nuclear Research*—all in (PRO-DSIR17/551). These give very detailed information on what transpired on 11 and 12 December. Amaldi has described his movements during his visit in his diary, of which there is a copy in the CERN archives, in his interview with Gowing and Kowaski made in 1974 (see note 1, chapter 12), and, for example, in Amaldi (1977), 339. The quotations in this paragraph are taken from Verry's memo of 12/12/52.
55. Letter Collier to Verry, 2/12/52 (PRO-DSIR17/560) derives this figure on the assumption that Britain would be assessed on the UNESCO basis, which, he insisted, would be 'unfair'.

56. The alarm was first raised by Cassels in a letter to Cockcroft from Copenhagen, 11/11/52, which was copied to Chadwick. He reported that 'there [was] some pressure to use the proposed laboratory at Geneva not only for high energy physics but also for conventional nuclear and even atomic physics'. Both Cockcroft and Chadwick felt that the Convention should be so worded that the laboratory was restricted to doing high-energy physics research—see letter Cockcroft to Martin, 20/11/52 (PRO-DSIR17/560); letters Chadwick to Lockspeiser, 29/11/52 and 8/12/52 (PRO-DSIR17/551).
57. This quotation is from Verry's memo on the meeting of the 11th—see note 54. In the back of Cockcroft's mind was probably a remark by Fry that 'recent information from America' had indicated that 600 MeV was an important energy, and that Fermi had advised the Europeans to aim for it' (letter Fry to Cockcroft, 1/10/52 (PRO-AB6/1074)). The reason for Fermi's interest was doubtless associated with his recent detection of the first resonance in particle physics.
58. The quotation is from Verry's memo on the meeting of the 11th—see note 54. The implications of this arrangement on the drafting of the Convention are discussed at greater length in section 8.2.1, which also studies British motives at greater length than is possible here.
59. Amaldi (1977), 339.
60. Letter Amaldi to Lockspeiser, 20/12/52 (CERN-DG20924).
61. See memo Verry, ref. 969-4-2, 1/10/52 (PRO-DSIR17/560) and letter Smith to Collier (in the Treasury), 9/12/52 (PRO-DSIR17/551); letter Lockspeiser to Amaldi, 17/11/52 (PRO-DSIR17/560); letter Lockspeiser to Brunt, 30/12/52 (CC-CHADI,1/1) in which the date of payment is specified as 25 November.
62. The total budget of the 'planning' stage was some 1 MSF. Britain's two gifts amounted to some 122,000 SF—see minutes of the sixth Council session, CERN/GEN/10. Her contribution to the permanent organization was 23.84%. Although Britain's official status remained that of an observer during the interim period, she still paid 23.84% of the total funds raised by the interim Council. This, and not the donations paid during the planning stage, was doubtless what Amaldi had in mind in his account.
63. As reported by Verry in his memo of 21/11/52—see note 51.
64. The 'great importance' Cherwell attached to Britain having the power to contract out of projects is stressed by the Marquess of Salisbury in a minute to the Chancellor, 22/12/52 (PRO-FO371/101517). The only other information we have on the encounter of Amaldi with Cherwell is a letter from Cockcroft to Chadwick of 16/12/52 which (characteristically) gives no hint that the meeting was a particularly fiery one (CC-CHADI,1/2).
65. For those invited to the meeting see memo Murray to Verry, 15/12/52 (PRO-DSIR17/560); for the *United Kingdom Provisional Re-Draft* of the Convention see copy in (PRO-DSIR17/551). The rest of this paragraph is from minute Lockspeiser to Lord President, 22/12/52 (PRO-DSIR 17/560).
66. Minute Salisbury to Chancellor of the Exchequer, 22/12/52 (PRO-FO371/101517).
67. Letter Butler to the Most Hon. The Marquess of Salisbury, 29/12/52 (copy on PRO-DSIR17/551).
68. Letter Butler to Brunt, 25/11/52 (PRO-DSIR17/551).
69. Letter Chadwick to Lockspeiser, 13/12/52 (CC-CHADI,1/1); letter Chadwick to Cockcroft, 10/12/52 (CC-CHADI,1/2); for Lawson's *The Effect of Magnet Inhomogeneities on the Performance of the Strong Focussing Synchrotron*, 2/12/52, and (revised) 10/12/52, see (CC-CHADI,1/2).
70. For a few remarks on Churchill's 'overlords' experiment, see Sked & Cook (1979) 114. The 'official' biography of Cherwell is Cherwell (1961).
71. See the unconfirmed minutes of the first meeting of the DSIR's Nuclear Physics Sub-committee held on 7/5/52—see note 5.
72. Science Policy (1966), appendix II.
73. See memo by Verry, note 51.
74. See note 72, and Gummett (1980), Table 2.1 for the DSIR figures.
75. See (unconfirmed) minutes of the fourth meeting of the DSIR Advisory Council's Nuclear Physics Sub-committee, 28/10/52, and letter Brunt to Evans, 9/11/52 (PRO-DSIR17/453).

76. Vig (1975), 65-66.
77. Sked & Cook (1979), 116-120.
78. *Ibid.* for these remarks on Butler. They quote Macmillan as saying that 'Rab is one of those men who cannot cook without meat! I can cook with bread and water'.
79. For a fuller discussion of the activities of the British during the first half of 1953, and the debates around the drafting of the Convention in particular, see sections 7.5 and 8.2.1, respectively.
80. Letter Lockspeiser to Cherwell, 14/7/53 (PRO-DSIR17/554).
81. For further details see section 9.4.1 and chapter 10.
82. Parliamentary Debates (Hansard) (House of Lords), **184**, (5/11/53), column 177.
83. The quotation is from a *Note for Ministers' Guidance,* 30/1/53 (PRO-DSIR17/561). Typical of the discussion in Parliament on the question is the following exchange between a Member and Churchill in October 1952 (Parliamentary Debates (Hansard) (House of Commons), **505**, No. 151 (23/10/52), column 1273):

    'Mr. Peart: Is the right hon. Gentleman aware that the decision not to participate [in CERN] seems strange—it may be the right policy—in view of the flamboyant speeches the Prime Minister himself made on European unity?

    The Prime Minister: Well, there are a lot of strange things in the world'.
84. Earl Jowitt raised his questions in his reply to the Queen's Speech on 4/11/53—Parliamentary Debates (Hansard) (House of Lords), **184** (4/11/53), columns 42-43. For the Lord President's answer, see note 82, columns 175-177.
85. Both Vig (1975), 98-100, and Albu (1963) have described the difficulties Parliament faces in debating questions of science policy and some measures taken to improve the situation. The quotation is from Albu (1963), 19.
86. Letter Boatwright to Thompson, 14/9/51 (PRO-ED157/302).
87. Document NS (52) 1, 30/1/52 (PRO-ED157/302).
88. See (DSIR) *Advisory Council Nuclear Physics Committee 2nd meeting, 4th June 1952* (PRO-DSIR17/451).
89. See Gowing (1974) vol. II, 220, 227. The quotation is on page 220.
90. The terms of reference of the OSR Committee are spelt out in document O.S.R. (48) 1, 15/3/48. Its formal relationship to the Overseas Liaison Division of the DSIR is specified in document O.S.R. (48) 3, 18/3/48. Both are in (PRO-DSIR17/485).
91. This model is a slightly revised version of the one we have described elsewhere, Krige (1984c) section IV. Philip Gummett and Roy Edgley (private communications) and Dominique Pestre (section 9.3.5) have made helpful criticisms of it. For a more general criticism of rationalistic models of decision-making see Allison (1971), Allison & Morris (1975).
92. A clear account of government structure as required for a discussion of science policy matters is presented by Gummett (1980), chapter 1. For a detailed description see Chester & Willson (1957). This point may seem banal, but as has been stressed, 'That the "government" is in fact a loose collection of organizations and people is readily apparent. [...] But explanations and predictions in terms of this simplification nevertheless tend to collect the activities of these people and organizations into one box called "the government" [...]', Allison & Morris (1975), 104.
93. Snow (1961) has forcefully drawn attention to the small number of high-ranking scientists and officials involved in important decisions.
94. This concept is now common currency in the sociology of scientific knowledge: see, for example, the session on 'Perspectives on cognitive authority in the sciences' in Callebaut et al. (1984), 251-261. See also Bourdieu (1975).
95. The quotation is from the minutes of the OSR 'sub-committee' meeting held on 6/2/52, *European Nuclear Research Centre,* 7/2/52 (PRO-AB6/1074). See also document I.O.C. (52) 9, 7/2/52 (PRO-ED157/302).

96. Memo Verry to Rackham, *Notes for Agenda,* 1/2/52 (PRO-ED157/302), and document I.O.C. (52) 9—see note 95.
97. See note 95.
98. Memo Verry to Secretary, 29/8/52 (PRO-DSIR17/551).
99. Comments by Butler appended to (?) a memo from Saner to Makins, *European Nuclear Research Centre,* 5/2/52 (PRO-FO 371/101514).
100. Zuckerman (1980), 7.

Bibliography for chapters 12 and 13

Albu (1963)	A. Albu, 'The Member of Parliament, the Executive and Scientific Policy', *Minerva,* 2 (1963), 1-20.
Allison (1971)	G.T. Allison, *Essence of Decision* (Boston: Little, Brown and Company, 1971).
Allison & Morris (1975)	G.T. Allison and F.A. Morris, 'Armaments and Arms Control: Exploring the Determinants of Military Weapons', *Daedelus,* **104** (1975), 99-130.
Amaldi (1977)	E. Amaldi, 'Personal Notes on Neutron Work in Rome in the 30's and Post-war European Collaboration in High-Energy Physics', in C. Weiner (ed.), *History of Twentieth Century Physics,* Proceedings of the International School of Physics 'Enrico Fermi', Course LVII, Varenna 1972 (New York: Academic Press, 1977).
Bourdieu (1975)	P. Bourdieu, 'The Specificity of the Scientific Field and the Social Conditions of the Progress of Reason', *Social Science Information,* **14** (1975), 19-47.
Callebaut et al. (1984)	W. Callebaut, S.E. Cozzens, B.-P. Lecuyer, A. Rip, and J.P. van Bendegem, *George Sarton Centennial,* University of Ghent, Belgium, 14-17 November 1984 (Ghent: Communication and Cognition, 1984).
Chadwick (1976)	H. Massey and N. Feather, 'James Chadwick, 20 October 1891-24 July 1974', *Biographical Memoirs of Fellows of the Royal Society,* **22** (1976), 11-70.
Chadwick (1978)	M.J. Prichard and J.B. Skemp (eds.), 'Mastership of Sir James Chadwick', *Biographical History of Gonville and Caius College Vol. VII,* Appendix I (Cambridge: Gonville and Caius College, 1978).
Cherwell (1961)	The Earl of Birkenhead, *The Prof. in Two Worlds. The Official Life of Professor F.A. Lindemann, Viscount Cherwell* (London: Collins, 1961).
Chester & Willson (1957)	D.N. Chester and F.M.G. Willson, *The Organization of British Central Government 1914-1956* (London: George Allen and Unwin, 1957).
Conference-Copenhagen (1952)	O. Kofoed-Hansen, P. Kristensen, M. Scharff, and A. Winter (eds.), *Report of the International Physics Conference,* sponsored by the Council of Representatives of European States, Institute for Theoretical Physics, Copenhagen, 3-17 June, 1952.
Frankel (1975)	J. Frankel, *British Foreign Policy 1945 -1973* (London: Oxford University Press, 1975).
Gentner (1961)	W. Gentner, 'The CERN 600 MeV Synchrocyclotron at Geneva', *Philips Technical Review,* **22** (6 March 1961), 142-149.
Gilpin (1962)	R. Gilpin, *American Scientists and Nuclear Weapons Policy* (Princeton, N.J.: Princeton University Press, 1962).

Gowing (1964)	M. Gowing, *Britain and Atomic Energy 1939-1945* (London: Macmillan, 1964).
Gowing (1974)	M. Gowing, *Independence and Deterrence. Britain and Atomic Energy, 1945-1952. Vol. I. Policy Making. Vol. II. Policy Execution,* (London: Macmillan, 1974).
Gummett (1980)	P. Gummett, *Scientists in Whitehall* (Manchester: Manchester University Press, 1980).
Hartland & Gibbons (1972)	J. Hartland and M. Gibbons, 'Britain Joins CERN: An Analysis of the Decision Process', unpublished paper, Department of Liberal Studies in Science, Faculty of Science, University of Manchester, November 1972, 61 pp.
Harwell (1952)	*Harwell, the British Atomic Energy Research Establishment, 1946-1951* (London: Her Majesty's Stationery Office, 1952).
Hendry (1984)	J. Hendry, *Cambridge Physics in the Thirties* (Bristol: Adam Hilger, 1984).
Jay (1955)	K.E.B. Jay, *Atomic Energy Research at Harwell* (London: Butterworths, 1955).
King (1953)	A. King, 'International Scientific Co-operation: its Possibilities and Limitations', *Impact of Science on Society,* **4** (1953), 189-218.
Kowarski (1961)	L. Kowarski, *An Account of the Origin and Beginnings of CERN* (Geneva: CERN 61-10, 1961).
Kowarski (1967)	L. Kowarski, *The Making of CERN: Memories and Conclusions,* unpublished, 17 pp., March 1967 (Kowarski Interview file, CERN).
Krige (1983)	J. Krige, *The Influence of Developments in American Nuclear Science on the Pioneers of CERN* (Geneva: CERN, CHS-1, 1983).
Krige (1984a)	J. Krige, *Launching the European Laboratory Project: Britain's Importance to it and the Obstacles to her Participation in 1950* (Geneva: CERN, CHS-4, 1984).
Krige (1984b)	J. Krige, *Britain's Physicists Respond to the European Laboratory Project: January-June 1951* (Geneva: CERN, CHS-6, 1984).
Krige (1984c)	J. Krige, *The Change in Policy of British Physicists towards the European Laboratory Project and their Government's Reaction to it: July 1951-February 1952* (Geneva: CERN, CHS-11, 1984).
Laqueur (1982)	W. Laqueur, *Europe since Hitler. The Rebirth of Europe* (Harmondsworth: Pelican Books, 1982).
Melville (1962)	H. Melville, *The Department of Scientific and Industrial Research* (London: George Allen and Unwin, 1962).
Northedge (1974)	F.S. Northedge, *Descent from Power. British Foreign Policy 1945-73.* (London: George Allen and Unwin, 1974).
Ronayne (1984)	J. Ronayne, *Science in Government* (Victoria, Australia: Edward Arnold, 1984).
Rudwick (1985)	M.J.S. Rudwick, *The Great Devonian Controversy* (Chicago: University of Chicago Press, 1985).
Salomon (1971)	J.-J. Salomon, 'The *Internationale* of Science', *Science Studies,* **1** (1971), 23-42.
Science Policy (1966)	Council for Scientific Policy, *Report on Science Policy,* Cmnd. 3007 (London: Her Majesty's Stationery Office, 1966).
Sked & Cook (1979)	A. Sked and C. Cook, *Post-War Britain. A Political History* (Harmondsworth: Penguin Books, 1975).
Skinner (1960)	H. Jones, 'Herbert Wakefield Banks Skinner, 1900-1960', *Biographical Memoirs of Fellows of the Royal Society,* **6** (1960), 259-267.

Snow (1961)	C.P. Snow, *Science and Government* (London: Oxford University Press, 1961).
Vig (1968)	N.J. Vig, *Science and Technology in British Politics* (London: Pergamon Press, 1968).
Vig (1975)	N.J. Vig, 'Policies for Science and Technology in Great Britain: Postwar Development and Reassessment', in T.D. Long and C. Wright (eds.), *Science Policies of Industrial Nations* (New York: Praeger, 1975).
Wilkinson (1979)	D. Wilkinson, 'Events surrounding the construction of Nimrod', in J. Litt (ed.) *Nimrod. The 7 GeV Proton Synchrotron* (Didcot: Science Research Council, 1979), 7–20.
Zuckerman (1980)	Lord Zuckerman, 'Science Advisers and Scientific Advisers', *Proceedings of the American Philosophical Society,* **124** (1980), 1-15.

PART V

Concluding remarks

CHAPTER 14

The how and the why of the birth of CERN[1]

John KRIGE and Dominique PESTRE

Contents

14.1 A brief narrative account   524
14.2 A classical interpretation of CERN's origins   525
14.3 The first group of actors: the physicists   529
14.4 The second group of actors: high-level science administrators and some diplomats   530
14.5 The activities of the 'CERN lobby'   532
14.6 The reactions of the member states   535
14.7 The motivations in governmental circles   536
14.8 CERN, an American puppet?   537
14.9 CERN, an organization of military importance?   539
   Notes   543

Now that we have come to the end of our detailed description of the emergence of CERN it is opportune to draw together the threads of our tale, and to extract what we have learnt from our study. To this end we first quickly remind the reader of the main events leading to the laboratory's establishment at the end of 1954. We then go on to explore the factors which made this possible, taking the opportunity to raise some more general questions of a socio-historical kind, questions which are as difficult to answer as they are easy to ask.

14.1 A brief narrative account[2]

Towards the very end of 1949, in the aftermath of President Truman's announcement of the explosion of the first Soviet atomic bomb, several personalities associated with nuclear matters in Europe began to think seriously about the possibilities for multinational co-operation in this area. The most important of the first initiatives was that of Raoul Dautry, Administrator-General of the French Commissariat à l'Energie Atomique. At the European Cultural Conference in Lausanne in December 1949 he had a resolution passed recommending that studies be undertaken for the creation of a European institute for nuclear science 'directed towards applications in everyday life'. Six months later Isidor I. Rabi, American Nobel prizewinner and co-founder of the Brookhaven National Laboratory, put a resolution to the annual conference of UNESCO. It invited the states who so wished to create one or more regional European laboratories, including one in nuclear science. This was adopted by the General Assembly on 7 June 1950 (as we described in chapter 2).

Two small groups took up these proposals in the following months. One comprised a handful of specialists in classical nuclear physics (people like Kowarski in France or Preiswerk in Switzerland), or in cosmic rays (most notably, of course, Amaldi in Italy and Auger in France, the latter also being Director of UNESCO's Department of Exact and Natural Sciences). The other group was composed of three important administrators of science—Raoul Dautry, Gustavo Colonnetti (who was President of the Italian Consiglio Nazionale delle Ricerche), and Jean Willems (Director of the Belgian Fonds National de la Recherche Scientifique). In December 1950 a first gathering of scientists and administrators organized by Auger and Dautry at the seat of the Centre Européen de la Culture in Geneva proposed that the biggest accelerator in the world (i.e. about 6 GeV) be constructed. A reactor was ruled out for political reasons, notably the problems posed by industrial applications and military interests (chapter 3).

In May, October, and November of the following year (1951) Auger, along with a number of scientific consultants, further refined the project advocated in Geneva (chapter 4). In December 1951 their recommendations were submitted to a European intergovernmental conference officially called by UNESCO, but in fact orchestrated by Auger himself. After lengthy discussions which reflected serious differences of opinion among scientists (chapter 5), the conference proposed that a provisional organization be established. It was endowed with $200,000 and given 18 months to present potential member states with worked out technical, organizational, and financial plans. The formal Agreement embodying these proposals was signed on 15 February 1952 by all those present, with the exception of the United Kingdom. Early in May, with the $200,000 guaranteed, and five signatures ratified, the Agreement entered into force (chapter 6).

On 5 May 1952 the provisional Council met for the first time. The technical groups to design the accelerators and plan the laboratory were set up. In October Geneva was adopted as the site for the laboratory, and it was decided to construct a 25–30 GeV proton synchrotron embodying the new alternating-gradient principle. This meant that a research and development effort—with its associated risks—was needed, and that the machine would take some five or six years to build. In January 1953 the British government was represented officially in the Council for the first time, and the discussion of the text of the Convention establishing the permanent organization began in earnest. On 1 July 1953 this Convention was signed by eight of the eleven member states of the provisional CERN and by the United Kingdom.

A period of waiting ensued. The last two ratifications needed for the Convention to enter into force were deposited some 15 months later on 29 September 1954. During the hiatus much was done: around October 1953 the group designing the big accelerator was installed in Geneva, and an embryonic administrative structure was set up; from January to March 1954 protracted and complex negotiations around the top post of Director-General led to the nomination of Felix Bloch in April; in May excavation work began on the site at Meyrin. In summary, during the 15-month interim period, the organization asked for, and got, 9.2 MSF from the member states, the scientific groups expanded steadily (numbering some 120 people gathered from all over Europe in September 1954), and the work proceeded to such an extent that, when the Council of the permanent organization first met on 7 October 1954, it was asked to approve the award of major building and technical contracts (chapters 7 and 8).

14.2 A classical interpretation of CERN's origins

With the main chronological steps in the emergence of CERN fresh in our minds, let us now turn to an analysis of how the organization came into being. We are not the first to do this of course.[3] Indeed in this volume we have already come across some of the available accounts which, at various levels of specificity, dealt with events which are supposed to have made CERN's history—from Rabi's story of what led him to put his resolution in Florence, at one extreme of detail, up to more global analyses of the origins and beginnings of CERN at the other. Yet while this legacy is undoubtedly useful in orientating one in the documents and in bringing questions to the fore, it can also be something of a trap: a trap because by virtue of the standing of those who have passed it down (Auger, Amaldi, Kowarski, etc.) and by virtue of the repetition of the same interpretations in the same terms, this legacy has become progressively transformed, it has gradually solidified into a quasi-mythical account of origins, a kind of 'standard version' of events, a taken-for-granted interpretation of how the organization was launched.

No one writing the early history of CERN can then avoid or evade these already-available histories. From time to time we have challenged some of them dealing with rather particular events, like Amaldi's meeting with Cherwell. Here, in the conclusion, we want to do something rather different, and to shift the focus of our analysis to the level of the more general interpretations. Among these, one has caught our attention, namely that first spelt out by Kowarski in a little report published by CERN in 1961, and entitled 'An account of the origin and beginnings of CERN', and further developed by him a decade later in a lecture given at the International School of Physics 'Enrico Fermi' at Varenna Sul Lago di Como.[4] If we choose Kowarski it is because he was undoubtedly one of the most insightful commentators on the emergence of CERN, and because in these texts he more or less explicitly articulated a kind of model to account for the birth of CERN, a model which is widely accepted and frequently drawn on. It is this model that we want to describe and then criticize.

The first point to note, according to Kowarski, is the large number of so-called big science organizations set up in the advanced countries after 1945 'in response to various forces, among which [...] national prestige, in its widest implications, [was] the most determining factor'.[5] Initially such institutes were established 'on strictly national lines', notably in the most glamorous of fields, nuclear physics, where the constraints of secrecy limited the scope for wider forms of collaboration. Two factors, however, worked in favour of a more open policy, in favour of the creation of large, supranational institutes. On the one hand, there were the pro-European tendencies, the growing push, initiated by Schuman, towards some form of European association—and Kowarski reminds us that the idea of having a Coal and Steel Community, 'the first sizable manifestation of European technological unity', was first put forward in 1950. On the other hand, there was the recognition 'that in post-war Europe to set up a truly worth-while big institute was a hard task for any single nation', notably in a field like nuclear physics where demands far exceeded the capacity of most western European states. In other words, for Kowarski, CERN was a kind of resultant of two forces, that embodied in the pro-Europeans searching for a prestigious mascot or symbol—and what better than the nuclear—and that represented by 'those European physicists who had some first experience, either wartime work or post-war work mainly in America, and knew something about these new forms of organization in science', and 'who were eager to create an institution capable of giving them the necessary equipment'.

'Next point:', Kowarski goes on, 'Why high-energy physics'. Simply because it shared in the glamour of nuclear physics—most of the powerful accelerators envisaged or under construction in 1950 were located in nuclear physics institutes like the Brookhaven National Laboratory or the British Atomic Energy Research Establishment—, but was free of the nuclear 'problem'—since no applications were in view, particularly of a military kind. There was then satisfaction all round, among the physicists because this domain of the nuclear field was the most promising as far as fundamental research was concerned, among the politicians and science administrators, because they could demonstrate European unity in a form at once tangible, prestigious, and without risk.

This way of presenting matters is not completely false—on the contrary, it is useful and pertinent at a particular level of analysis. On the other hand, it is plainly too rough-hewn, too simple, 'too good to be true'. We could illustrate this with many examples drawn from this book. Four will suffice here.
  i) The scientists were neither united in their aims, nor clearly aware of wherein their 'best interests' lay. In fact until the summer of 1952, some of the most influential physicists in Europe—Bohr, Kramers, Chadwick, Thomson,...—felt that the whole question of European scientific co-operation should be approached from a different angle, and were not at all convinced that the best thing to do was to try to build the biggest accelerator in the world;[6]
  ii) The pro-European spirit among the politicians was neither as determined nor as widespread as Kowarski would lead us to believe. Of course Dautry, Colonnetti, and Willems were authentic Europeans, and this must not be forgotten. But nor should we forget that the ardour for European initiatives in the Quai d'Orsay in 1952 was not shared by the French parliament in 1954, and that the European argument was seldom if ever used in Britain and the Scandinavian countries—on the contrary;[7]
  iii) Until about 1953–54, with the possible exception of the United Kingdom, the idea of having a policy for science was only in its infancy in Europe. As a result people like Auger, Colonnetti, Randers, or even Lockspeiser, were relatively free to follow the lines of action they chose, and a man like de Rose could become involved in the project not because he had to, but because he wanted to, because it was the kind of thing he was interested in. Related to this is the complexity in the attitudes of the states which, while allowing scope for independent initiatives on science matters, showed little enthusiasm for financing the 'European nuclear mascot' in 1951;[8]
  iv) The choice of high-energy physics was not self-evident. Of course we should

Notes: p. 543

not underestimate the myth of the nuclear, but nor should we overlook the fact that other domains of science like astrophysics and computing were mentioned as possible candidates for international co-operation, and that as late as December 1952 there were those who would have liked CERN to do low- as well as high-energy physics. In fact, in explaining why CERN was dedicated to high-energy physics, we must not forget the role played by personal and national scientific interests—the fact that Auger and Amaldi were cosmic-ray physicists, or that the British were reluctant to pay for what they did not need, for example.[9]

At a purely empirical level, then, Kowarski's image of an historical 'encounter between two drives [...]: the scientists' search for new ways of acquiring large-scale equipment, and the statesmen's search for domains [...] in which [...] to produce tangible manifestations of European unity' cannot be sustained beyond a first approximation.[10] There were actually *several* groups of actors, several forces at work, each having its own interests, stated or not, for joining or opposing the project—or for remaining indifferent to it. This is not to say that Kowarski's approach is of no use—it does allow for a 'quick, imaginative sorting out of a problem of explanation or analysis [...] requiring a minimum of information'.[11] On the other hand, it is methodologically flawed for it tends to see CERN as *the rational solution to a well-defined problem,* as the answer to the pre-given question of how to equip Europe with a prestigious collaborative institute in big science, as an *optimum choice* made by actors who were fully conscious of all the issues at stake. Within this cluster of assumptions it seems that CERN *could not but* have been established, its birth is imbued with necessity, its history is suffused with reason.

As a working hypothesis the assumption of a degree of inevitability may be defensible if one were to write the history of the atomic energy establishments created in the developed countries in 1945–46, bodies like France's CEA or Britain's Harwell. As Gilpin, Salomon, and others have emphasized, with the explosion of the bomb science moved from the periphery to the centre of the political process.[12] The governments of major powers had little choice but to develop their own atomic-energy programmes if they wished to retain their influence. In the case of CERN, however, there was no such compulsion, and the situation was far more fluid, indeterminate. Here it is probably preferable to start from the assumption that an element of chance, of conjunctural coincidence, was involved in the birth of the organization. Here particularly we must avoid what Marrou calls the 'idols of reason', we must not suppose 'the real to be fully intelligible', we must appreciate 'the wisdom hidden in Hamlet's famous words to Horatio: *There are more things in heaven and earth...*'.[13]

These sentiments have inspired our account of the emergence of CERN in this volume. If we have immersed ourselves in the historical detail, it is because we have learnt that a fruitful way of countering our own (and others') *a priori* conceptions was to try to come to grips with the 'messiness of the actual process' of creation before analyzing the reasons for it, to study the how before the why.[14] In applying this lesson here we again *begin* by focusing our attention on a *description,* recapitulating the activities of the main protagonists, their supposed or recognized motives, and their relationships with wider groups (networks of friends, government apparatuses, and so on). Then, expanding our scope, we look more closely at the role of the states in the decision-making process, and at motivations in political circles. Finally, arriving at the highest level of generality, we will be in a position to respond to two of the more global explanations of why CERN came to be without falling prey to over-facile interpretations.

14.3 The first group of actors: the physicists

At the core of those who pushed the European laboratory project ahead for more than two years, until December 1951, we have encountered some dozen physicists, most of them experimentalists. Grouped around Auger and Amaldi, we have remarked that they were rather young (about 40 years old), specialists in cosmic-ray physics or divisional heads in the new nuclear science institutes created after the war, and people not without influence in their own countries. Their *main concern* was to do pure science at the research frontier, and at a level comparable to that of the United States. This meant that they had to have equipment which was at once sophisticated and costly. Hence, after a year in which various alternatives were considered, the idea of building the biggest accelerator possible took hold. However, and this is a second important element, this group remained *relatively small and isolated,* and was notably lacking in support from the European scientific establishment until around mid-1952—a point all too often glossed over in the 'standard view'.[15]

Why was this so? The major bone of contention was the question of whether a new generation of powerful accelerators was really necessary or desirable. Here one must take care not to be anachronistic: the feeling of being in a race, the relentless competition between groups and rival laboratories, the need to go to higher and higher energies and to build ever more powerful machines, was not self-evident. In Bohr's view one should take such steps only if they had been recommended 'after most thorough considerations' by a group of experts. For Kramers there was no point in turning to governments for money 'just [...] to build a big accelerator of a type now under construction in the U.S.A.', a concern shared in some political circles. The costs of the envisaged project, noted the Dutch Minister of Education, Arts, and Sciences in October 1951, would be 'twice the annual contribution of the participating countries to the UNESCO budget', so bringing 'almost insurmountable difficulties' for many of the interested European countries. Hence the resistance to the Amaldi–Auger scheme from the summer of 1950 to that of 1952.[16]

Notes: p. 543

Around mid-1952 the doubts began to melt away for two main reasons. The Brookhaven National Laboratory's 3 GeV 'Cosmotron' came into operation and showed that accelerator construction costs could be drastically reduced in future (notably by reducing the magnet gap which had been overdesigned). In addition there was a growing appreciation of the scientific importance of the kind of physics which one could do with very powerful accelerators, particularly among the younger generation of researchers and the students. As a result a very large part of the European physics community aligned themselves with the big-machine project, and the initial small group became the nucleus of a much larger conglomeration.[17]

In Britain the situation differed markedly from that on the continent. During the second half of 1952 the struggle to have the need for big machines recognized was waged against influential members of the indigenous scientific establishment, notably Chadwick who effectively controlled the domestic decision-making process for nuclear-physics research. In most continental countries the opposition was far sparser, generally far less entrenched in the machinery of government, and so far less influential. Taking the continent as a whole the pockets of resistance were much more patchy and uncoordinated than in the United Kingdom: one could navigate one's way through them rather than having to confront them head-on. Granted this latitude Auger and Amaldi were quickly able to solicit support—and money—for their project in some countries, notably Belgium, France, and Italy, remaining relatively unperturbed by the opposition it raised in others.[18]

One thing remains to be said. Notwithstanding the various kinds of centrifugal tendencies *within* the scientific community, when putting their project to governments its members sought to present a united front. There is no better evidence for this than the arrangements agreed on at the UNESCO-sponsored conference in December 1951: the device adopted on that occasion was to pass a resolution *juxtaposing* the two projects—planning a new laboratory on the one hand, coordinating existing European activities on the other. In this way the conference could be 'saved', and the nuclear scientists could show governments their collective determination to have the Treasuries finance the schemes they wanted.[19]

14.4 The second group of actors: high-level science administrators and some diplomats

As we know, another group of people associated with the European project from the start worked in close collaboration with the physicists. It comprised top state officials, people like Colonnetti, Dautry, and Willems. Their importance during the years of gestation (1949–early 1952) derived from their position at the head of influential scientific institutions and from their 'social weight': Gustavo Colonnetti, President of the CNR from 1944, was Christian Democratic representative in the Italian National Assembly in 1946; Raoul Dautry, one of the two leading figures (with Frédéric Joliot) in the creation of the CEA, was head of the French railways in the 30s, Minister of Armament before the war, of Reconstruction in 1945; Jean Willems, Director of his country's Fonds National de la Recherche Scientifique from its foundation in 1928, became first President of the Belgian Institut Interuniversitaire de Physique Nucléaire after the war.[20]

If these men allied themselves with the project it was for a *number* of reasons less specific than those of the scientists. At a first level of analysis we can, with Kowarski, identify two: a strong pro-European sentiment—indeed Dautry was President of the French Executive Committee of the European Movement, and all three, it will be remembered, were active members of de Rougemont's Centre Européen de la Culture—and an awareness of the needs and post-war importance of big science—to which they had become sensitized by virtue of their functions. It would be too simple to leave matters there, however. One must not forget that, as men in the higher echelons of government, they all—including some others like de Rose—were used to taking initiatives which they felt to be in the *national* interest, and were well aware of the importance of atomic affairs to it. So we have found Dautry, in anticipation of the CEC meeting in December 1950, getting in touch with Italian circles who favoured setting up an atomic-energy commission in their country. After that meeting we have seen Colonnetti urging Italian participation in the planning stage of the European laboratory on Prime Minister de Gasperi with the argument that science and scientists had to be mobilized for national defence. And we have also seen de Rose, who had participated in negotiations at the United Nations about the international control of atomic energy, actively involved in Kowarski's and Perrin's earlier attempts to develop contacts with Kramers and others in Europe. For people like this the new venture was one of a number of interconnected moves with long-term strategic significances for their respective countries—a point we shall return to later.[21]

With the establishment of the provisional CERN this small group of people (excepting Dautry, who died in August 1951), was joined by colleagues having leading positions in other European scientific organizations. Some were men rather younger, rather lower in the hierarchy, but influential all the same: Alexander Hocker, for example, who was Deputy-Director of the Deutsche Forschungsgemeinschaft, J. Hendrik Bannier, President of the Dutch organization for fundamental research, or Robert Valeur, senior civil servant in the Ministry of Foreign Affairs in Paris, who took over de Rose's role in CERN when the latter was nominated French Ambassador to Madrid in February 1952. Others had the political and administrative status of a Colonnetti or a Dautry, a man like Sir Ben Lockspeiser, Secretary of the Department of Scientific and Industrial Research (UK), for example.

Looking at the main scientific and political protagonists, one cannot but be struck by its *remarkable stability,* by the *remarkable continuity* in the composition of the official delegations which held the destiny of CERN in their hands. Consider the two UNESCO-sponsored conferences (December 1951/February 1952) and the nine sessions of the provisional CERN (May 1952–April 1954). One finds that at least one delegate from almost every member state officially represented his country on at least ten of these eleven occasions—Verhaege and Willems for Belgium, Nielsen for Denmark, Perrin for France, Heisenberg and Hocker for Germany, Colonnetti for Italy, Bannier for the Netherlands, Waller for Sweden, Picot and Scherrer for Switzerland, Dedijer for Yugoslavia. From 1953 onwards they were joined by Valeur for France, Pennetta for Italy, Lockspeiser for the United Kingdom ... [22] Similarly, the key scientific posts in the provisional organization were occupied by physicists who had been closely associated with the scheme from early in 1951, or even before. Secretary-General Amaldi had been in contact with Auger, Dautry, and Rabi since mid-1950. Kowarski, Dahl, and Bakker, three of the four group leaders, had all served on Auger's team of consultants during 1951, as had two of their deputies, Preiswerk and Goward. Only Adams, who took over from Goward when he died in March 1954, joined the squad much later. In fact, we find more and more people coming in, and virtually no one of importance leaving. Layer after layer is added to the tiny nucleus which originally formed around Auger and Amaldi, until by 1954 it had grown into a solid and rather cohesive unit.

How do we explain the formation of such a closely-knit group of heterogeneous people? How do we understand such stability—and such dedication? The prestige of nuclear physics in the early 1950s, the novelty and importance of what was at stake, the potential that the new venture had, all of these could not but have made the venture attractive and challenging. Here was an opportunity to be part of a collaborative scheme in *nuclear* physics encompassing no less than a *dozen* European states—only six had joined the European Coal and Steel Community. Here was a chance to put European science back on the map, to compete with the American monolith by building the biggest accelerator in the world—10 to 15 GeV (and later 30 GeV) compared to 6 GeV for Berkeley and 1 GeV for Birmingham. This was no ordinary job, either for the scientists who were to design, construct, and use the machines, or for the administrators who were responsible for preparing the ground for the permanent organization. From whence the feeling of adventure, of doing something really significant for science, for Europe, ... and for one's own future. As J. Hendrik Bannier recently mentioned to us, his role in these early days gave 'an important stimulus to [his] career, both nationally and in Europe'.

14.5 The activities of the 'CERN lobby'

The dominant image which we have of European bodies like the European Economic Community or Euratom is one of organisms prey to the dictates of national interests, where delegates are strictly mandated by their governments to follow a particular line of action.[23] What is striking in the case of CERN—though this will come as no surprise now—is that the image is inverted. In a word here we have a Council of Representatives whose members behave more like the delegates of CERN to their governments, than like the delegates of their governments to CERN. Reading the internal documents and the correspondence one finds not only the same names, not only a rather permanent group of people, but also a strong *esprit de corps,* a determination to succeed at any cost: from as early as 1950 there was CERN on the one hand, the world at large on the other.

Of the many instances illustrating this behaviour we shall mention only a few.

(i) As late as October 1952 it was still being suggested that it was possible for governments to veto the planned European laboratory, and that a special meeting would be called 'to take a final decision regarding the construction' of the facility. In fact this option was never exercised. The provisional CERN Council delegates thrashed out the text of a convention for the permanent body in consultation with their respective domestic officials, and signed it themselves on behalf of their governments without holding a plenary conference. The 'phase' of drawing up the plans for the laboratory was continuous with, and indistinguishable from, the 'phase' of their implementation.[24]

(ii) In October 1953 the Council was informed by CERN's scientific directorate that the costs of the envisaged proton synchrotron—the single most expensive item of equipment—were likely to be much higher than originally expected. The first reaction of the delegates was not to try to understand why the costs had gone up, nor to ask the scientists if cuts were possible elsewhere. In essence it was to try to find the best arguments with which to 'sell' the increase to their governments. One delegate thought that it was unnecessary to be specific now and that if, after two years or so, the programme appeared to be financially unrealistic, 'Governments would certainly understand the necessity to raise their contributions'. Others felt that the member states 'should be frankly informed of the probable cost of the planned machine' as soon as was feasible. Collectively though the government representatives—something of a misnomer in this case—agreed that the scientists should 'decide the energy of the synchrotron from a purely scientific point of view' and that they would arrange the financial side of things with their domestic bureaucracies once the tenders for the machine were in.[25]

(iii) At the eighth meeting of the Interim Finance Committee in June 1954 the senior scientists and the architect impressed on the delegates that if valuable time was not to be lost, it was necessary to proceed with work on the site. During discussion strong differences of opinion arose about how far it was 'legitimate or wise to incur expenditure which much exceeded interim planning, particularly while the date of entry into force [of the Convention] was still problematical'. Nevertheless the committee declared that it 'fully realized the desirability' of this, if one was 'to avoid costly dislocation of the machine programmes and a bad moral effect on the staff'. Accordingly, it soon overcame its scruples, and the work on the site proceeded at an even greater pace during the summer of 1954.[26]

Notes: p. 543

The measures adopted by the CERN lobby in their dealings with the governments are thus easily imagined. In the initial period they followed a 'minimalist' approach, i.e. they adopted postures which 'seemed to close off the least number of future alternatives, [...] that left the most issues still undecided'. As Schilling appropriately puts it, they found 'how to decide without really choosing'.[27] In this way they sought to postpone controversial and potentially divisive issues, all the while drawing governments more deeply into the project. For example, we have the protagonists—and notably Auger, from the end of 1950—insisting on the provisional, reversible, nature of the 'commitments' called for from the member states. When, in May 1951, de Rose and Dautry managed to have France provide 2 MFF for Auger's study office, they explained that this involved no undertaking for 'phase 2' of the scheme, the creation of the provisional CERN. Similarly, both the French and the British pointed out to their governments from early 1952 onwards that becoming members of the provisional CERN did not, so they said, entail any obligation to participate in the construction of the accelerators.[28] The other side of the coin is that, even while claiming that the future was still open, the CERN lobby took a number of steps which considerably narrowed any government's room for manoeuvre in the future—for example, by agreeing to start work on the site six months before the Convention was ratified.[29]

Another 'tactical' aspect of the Council's behaviour towards governments was that it always tried to 'de-ideologize', to 'depoliticize', the nature of the decisions taken: one was not building the world nor reconstructing Europe, one was simply constructing a scientific laboratory. Hence we find a number of provisions inserted in the Convention early in 1953 which (retrospectively) appear rather unusual—no national quotas in the recruitment of personnel (in practice the national rivalry for posts in the directorate was considerable, but for them only), no attempt to correlate the value of industrial contracts awarded with a country's level of contribution to the CERN budget (a policy which was applied until the seventies), no right of veto in the Council including the vote on the budget (decisions were usually to be taken by simple majority, and by a majority of two-thirds on some particularly thorny issues).[30]

It only remains to be said that it was precisely *by* 'depoliticizing' and 'de-ideologizing' their project that the CERN lobby could exercise the degree of autonomy which it had. If the Council had both the support *and* the freedom which it wanted it was because it simultaneously managed to capitalize on the nuclear myth, the anxieties of the cold war, the desire to reconstruct Europe, *and* to detach the project from such concerns: after all CERN would only do 'pure' science. Thus it could both win the backing of governments by exploiting the nuclear mascot or prevailing diplomatic attitudes, and keep the project out of their hands, by stressing that it was a purely technical scheme, with no real political stakes involved. It thereby created the space in which to push ahead unhindered with the establishment of the laboratory.

14.6 The reactions of the member states

With these considerations in mind, let us now look a little more closely at the attitude of the European governments. The first point to appreciate is that although they intervened regularly in the creation of CERN, they did so in a very particular way, playing a role quite different to—and of course less important than—that which they were then adopting in the formation of the European Coal and Steel Community or, a few years later, in the launching of Euratom. The main motives for collaboration in the latter pair were economic and political, and the authorities called all the shots. Euratom, for example, was supposed to fuse national research and development efforts in the application of atomic energy, and to do so within the framework of the Treaty of Rome. The scheme was conceived at the level of government heads or of their immediate subordinates, and was subsequently implemented and controlled by accredited European executives. In the creation of CERN, on the contrary, the events were subject to a quite different logic. The actors whom we have described were its genuine promoters—the necessary complement of this fact being that governments were placed in an essentially *reactive, which is not to say passive, role*.

Considering what we now know, to illustrate the *reactive* aspect we need only consider Council nominations. Certainly politicians Colonnetti, Willems, and de Rose, and scientists Perrin (the High Commissioner of the CEA who replaced Joliot in 1951), Bohr (the 'father' of European atomic physics), and Cockcroft (the Director of Harwell) were recognized by their governments as official representatives in the Council. But not because they had been 'specially' chosen for the role. Their dedication to the project, and the relative lack of sustained involvement in scientific and technical affairs in most state apparatuses rendered their 'nomination' a mere formality—they more or less 'selected' themselves.

Most of the reasons which led governments to find themselves in this reactive position have already been given—the myth of the nuclear, the 'pure' nature of the research, the prestige of the physicists, the tactical attitudes adopted by the CERN lobby, and so on. However, one element particularly worth highlighting is the looseness, the virtually complete absence, of governmental machinery on the continent to deal specifically with scientific affairs in the early fifties.[31] On the one hand this meant that the project could be steered forward by its dedicated advocates through the pores of the prevailing structure. It also meant that there were few if any in the government with the experience or expertise to assess for themselves what was being presented to them ready packaged. In this respect CERN was fortunate in being the *first* post-war European collaborative scientific venture. A decade later when scientists connected with CERN tried to pull it off again, tried to set up a comparable body, the European Space Research Organization (ESRO), the member states were on their guard, and laid down a number of restrictions in advance—a fixed total budget for the first eight years, a 'just return' on contracts, etc. Governments also learn.[32]

Notes: p. 543

Though reactive, the state apparatuses were not *passive*. When the need arose (inadequate scientific justification, matters of prestige or of foreign policy), they clearly made their presence felt. Remember the initial refusal of the UK Cabinet Steering Committee to agree that Britain join CERN because Chadwick and Thomson were not convinced of the need for the bevatron. We have also seen examples in the negotiations over the clause in the Convention admitting new member states and over the location of the laboratory. In such cases the project was subjected to the logic of the foreign policies being pursued by the major protagonists, to Switzerland's desire to preserve her neutrality, to Britain's wish not to do anything which might jeopardize her special relationship with the United States, or to France's determination to have the laboratory geographically within the orbit of her influence.[33]

14.7 The motivations in governmental circles

Now that we have described the how of CERN's emergence, we are in a position to step back a little, to reflect on some of the reasons why it came to be accepted and finally financed. In this section let us consider the *arguments actually used* during the decision-making process in each country, as revealed in internal memoranda, minutes of committees, parliamentary debates, and so on.

The first thing to be said is that, at least as far as the earliest countries which supported the project were concerned and at the level of 'official' arguments, Kowarski's 'convergence of two forces' model fits rather well. In France in particular, but also probably in at least Italy and Belgium, two arguments were repeatedly used in support of the scheme: the national backwardness in accelerator technology (which was clear for all to see), and the desire to contribute to European development along the lines of the Schuman and Pleven plans. Kowarski's interpretation also captures something of the spirit in Germany. Here the dominant opinion was that, while the scheme was of undoubted scientific importance for a Germany prohibited from doing nuclear research by her victors, it was even more important for her to participate from a political point of view. This would accelerate her re-entry into normal interstate relationships. As Heisenberg put it, '80% of the cost should be considered from the angle of European collaboration'.[34]

It is when we turn to countries like Britain that the model is less convincing. Here the scientific argument was overwhelming—and that only late in 1952 when the domestic physics community saw in membership the most economical way of gaining access to a 25-30 GeV accelerator. A European connection was not only muted: it was positively disavowed—as far as the Foreign Office was concerned the United Kingdom 'should be able to show to the world that the Organization had in fact no political significance as a European body'.[35] Kowarski's model is thus limited in that it too quickly generalizes from particular situations, understandably from that in his own France. What is more, and unfortunate for us, Kowarski's two forces always worked in the same direction, so that it is impossible to weight clearly the relative importance of either—if only one state had been both technologically advanced in accelerator construction *and* enthusiastic for European unity![36]

Finally, as in the case of individuals, other motives were given for supporting the European laboratory project. Some were common to nearly all countries—the competitive argument, i.e. the chance to catch up with the United States, the financial savings, or the stimulus to domestic industry—others were more specific. Switzerland, for example, was pleased that the laboratory would be in Geneva, a consideration which also mattered to France since, being on her border, the organization would be exposed to French language and culture. We could give other examples, though our selection would be biassed by the fact that we have studied only four countries in depth. However, when all is said and done—and this is our first strong claim concerning the question of why the CERN project was supported and financed—the scientific interest of the laboratory was of cardinal importance. It was always stressed, and it apparently always carried the greatest weight. This is certainly true in the case of Britain and the Scandinavian countries, but it also applies to France, where parliament agreed that the country join CERN even while it was rejecting the Pleven Plan for a European Defence Community.[37]

So far we have remained at the level of the reasons given by the main actors. Obviously, this is not enough and it would be naïve to think that such arguments tell the whole story, that they were always taken at face value either by those who advanced them or by those at whom they were directed. In all politico-diplomatic relations what is left unsaid, what forms the shared framework of unspoken assumptions, often counts heavily. In addition, there are always broader forces at work in human behaviour, forces which elude an actor's clear light of consciousness. In what follows we would like to discuss two such considerations, both of which have enjoyed a measure of popularity at one time or another. According to the one CERN was above all a child of American foreign policy, a response to American aims (and needs) in the context of the cold war. According to the other it was the military interest in the medium term which, all denials notwithstanding, 'really' inspired CERN's paymasters to back the venture.

14.8 CERN, an American puppet?

The essence of the argument of those who support the first hypothesis is that there is one true father of CERN, to wit Isidor I. Rabi, one of the key figures in the American scientifico-political establishment. They point out that in 1949 Rabi was nominated by the State Department to serve on a commission set up to make recommendations about the role of science in American foreign policy. Its findings were submitted in April 1950, so shortly before Rabi's resolution at the UNESCO meeting was voted (7 June 1950). The Berkner report, as it was called, stressed the importance to the United States of *basic research* done beyond its shores, 'American pre-eminence thus far [having been] in the application of scientific discovery'. Yet although 'foreign scientific progress' was to the country's 'practical advantage', little had been done to foster it. 'We have come to the aid of our allies in support of their economies. We are providing military aid. What has been done in support of science and technology which form the foundation of security and of future welfare?'. Here, so the argument runs, lies the rationale for Rabi's intervention in Florence.[38]

The proof of American involvement in the origins of CERN is thus established. And if further evidence were needed, CERN's geo-political co-ordinates would provide it: it was an Atlantic organization with Sweden and Switzerland added, a body comprising a number of powers none of which, apart from these two, could 'be considered as neutral *vis-à-vis* the Soviet Union', as a British official in Berne put it.[39] But what of Yugoslavia's participation, one may ask? If anything this confirms that CERN was part of the West's offensive in the technological cold war. Did Tito not officially request the western powers for military equipment in April 1951, did Yugoslavia not receive $50 M of American economic aid in August, and did the country not sign a military treaty with the United States on 14 November? It is hardly surprising then that the country was represented at the intergovernmental conferences in December 1951 and February 1952, and later at CERN itself![40] Anyway, those who launched CERN recognized its American paternity: on 15 February 1952 they collectively telegrammed Rabi with the good news: 'WE HAVE JUST SIGNED THE AGREEMENT WHICH CONSTITUTES THE OFFICIAL BIRTH OF THE PROJECT YOU FATHERED AT FLORENCE. MOTHER AND CHILD ARE DOING WELL, AND THE DOCTORS SEND YOU THEIR GREETINGS'.[41]

This line of argument, developed notably by the European Communist Parties in the 1950s,[42] is not without substance, and is certainly no more simplistic than the orthodox version which pretends that the organization was absolutely neutral politically speaking. And indeed no one can deny the dependence, both economic and political, of western Europe on the United States at the start of the 50s. The cold war was at its height. NATO was taking shape—the nomination of General Eisenhower as Supreme Allied Commander Europe occurred simultaneously with the first definition of what was to be the CERN project between September and December 1950—as did the idea of having a European Defence Community incorporating West Germany, proposed by René Pleven on 23 October 1950. CERN could not be an exception, could not but be in one of the two camps—and no one thought things could be otherwise at the time.[43]

At a certain level of analysis, then, it is not unreasonable to argue that CERN was a component of the scientific system set up by the USA in its competition with the East. On the other hand one must take care to avoid two things:

(i) giving this an *instrumental* meaning, speaking in terms of 'American manipulation': there was not *one* consciously articulated project in 1950, which subsequently developed 'naturally' as the chicken from the egg. CERN is rather to be understood as the hybrid resultant of diverse interests, scientific as well as political, individual as well as national, national as well as European, and so on;

(ii) *reducing* CERN to it, thinking of CERN *only* as a component of the system set up by the West. For many, for an Amaldi or a Dautry, for example, CERN was *also* a card played in the competition between Europe and America, and it was *just as* important in this aspect as in the other. And paradoxically it is this that explains the interest aroused in Europe by Rabi's resolution: it was immediately interpreted as a sign that the United States was *loosening* the security constraints that it had imposed, that Europe was being encouraged to invest, in whatever way she liked, in this domain of research. Hence the importance of the vote on 7 June 1950 and the retroactive attribution of paternity to Rabi.[44]

14.9 CERN, an organization of military importance?

In the explanatory scheme articulated by Kowarski which we earlier gave in outline, the author stressed the difference, at the start of the 50s, between the high-energy physicists, who were interested in studying 'the nature of the particles themselves, a kind of knowledge for which no possible applications were in sight [...]', and the politicians who, in the light of their recent experience, hoped for 'desirable applications'. The ensuing 'wholehearted and unquestioning' support for CERN, Kowarski goes on, sprung from 'a remarkable case of double connivance'. The physicists 'chose not to disillusion these vague expectations too bluntly', while the politicans accepted that 'if the physicists want to keep their purity and say that they are far from all applications, all right, we understand'. This 'slight dose of hoax enabled the politicians and the physicists to have it both ways so decisively [...]', concludes Kowarski.[45]

This way of accounting for events raises a number of questions: Is it true that the physicists *knew* that particle physics had no (military) applications in view? Was the hope that there *would be* (such) applications the main, or most fundamental reason why governments gave CERN their 'wholehearted' support? And are we to assume that Article II.1 of the Convention, 'The Organization shall have no concern with work for military requirements and the results of its experimental and theoretical work shall be published or otherwise made generally available', is purely cosmetic? In short, around 1950, what were the assumptions and hopes of a military kind associated with launching a laboratory equipped with a giant accelerator?

We cannot raise these questions without some trepidation. Our experience has

Notes: p. 544

been that to suggest to European high-energy physicists that their work may be of military relevance is to provoke hostility, even anger. In CERN the subject is more or less taboo, the matter being further complicated here by the political implications for the member states. If we grasp the nettle, it is in the hope that, at least as far as the launching of CERN is concerned, informed debate will replace defensive reaction.[46]

To begin with, it is crucial to bear in mind that there are (at least) three possible ways in which a major research facility can be of interest for military (or industrial) purposes: in terms of the scientific *results* produced, in terms of the (new kinds of) scientific *equipment* and *techniques* it develops, and in terms of the *pool of expertise* it generates and sustains. Far too many debates on the question of the 'applications' of basic science are restricted to the first of these aspects only; in 'pure' big science where thousands of people spend millions of Swiss francs or whatever working at the leading edge of science and technology, a far richer concept of 'application' is called for, and will be deployed in what follows.

As for 'useful' scientific results, it seems likely that nobody expected these in the short-term, even if British scientists (according to Major-General G.J. Keegan Jr.) had invented the concept of using particle beams as weapons as early as 1944–45.[47] On the other hand, five or six years after Hiroshima, a 'who-knows?' attitude certainly prevailed, not only in political circles, but also among the scientists. For example, it was a belief of this kind, a belief in the 'ultimate value' of 'untargeted' research which led the Department of the Navy's Office of Naval Research to launch a huge scheme primarily for financing research in American universities in the immediate post-war period. 'Our contracts emphasize the fundamental nature of the research', wrote the Director of the Office in 1946, 'and are carefully designed to preserve the freedom of inquiry and action so essential to the spirit and methods of research work'.[48] More prosaically, we are reminded of Colonnetti's letters to de Gasperi and Casati, or Sir Ben Lockspeiser's remark in December 1952 to his 'science minister'—that it could not be denied that 'the research in the European laboratory could be applied to military uses'.[49]

For their part the scientists were not exactly ignorant of how matters stood. In a somewhat cynical aside in 1952 Erwin Schrödinger remarked that although one could not 'kill anybody' with one nuclear particle, 'their study promises, indirectly, a hastened realization of the plan for the annihilation of mankind which is so close to all our hearts'. More soberly, at the end of 1951 Werner Heisenberg used arguments strictly analogous to those of Colonnetti or Sir Ben: 'in the long run', he said, 'this new branch of nuclear physics may of course also open a field of application of similar importance as that of current nuclear physics'.[50] Was he just telling his officials what he thought they wanted to hear, as Kowarski would have us believe? Or was he not also speaking from conviction with one eye on the possibilities? Of course, Heisenberg himself may not have been remotely interested in such activities—like Amaldi and Auger, whose motivations lay elsewhere, as we have stressed. But this is beside the point. What we want to bring out is the *possibility* of an expectant attitude even among the scientists at the time.

What of our second mode of application? Here it is certain that, at least in some quarters in the United States, a close association between accelerator technology and military needs was built up during and after the war. Ernest Lawrence, the Director of the Radiation Laboratory at Berkeley, was actively involved in the scheme which produced most of the U-235 for the Hiroshima bomb. For this he used a bank of electromagnetic separators or 'calutrons'. With the announcement of the explosion of the Soviet bomb in 1949 Lawrence and his laboratory once more entered weapons development. In March 1950 he was authorized to build a high-intensity linear accelerator designed to produce polonium for use in the weapons programme and in radiological warfare ('Mark I'). This was to serve as the prototype for a gigantic 'production accelerator' ('Mark II'). Mark II was approved a few months later, after the outbreak of the Korean war. And although it was never built, the whole programme laid the foundations for the Livermore Weapons Laboratory which became a branch of Lawrence's Radiation Laboratory at Berkeley.[51]

The degree of militarization of European accelerators, and accelerator builders, bears no comparison with that of some of their American counterparts, of course. On the other hand, granted the developments across the Atlantic, these devices, no matter where they were built, inevitably took on a military/industrial dimension, whether one liked it or not. In some institutional settings this was a *conscious imperative*, as in Harwell, for example. There, for instance, every accelerator had to have 'a defensible project orientation', and the construction of the 600 MeV proton linac—the project, we will remember, that the staff were so reluctant to jettison—was justified by 'a report from Berkeley that if you bombarded heavy elements with protons of 600 MeV something like 30 neutrons came out'—neutrons which could be used to produce fissile elements.[52] In other places, by contrast, and at CERN in particular, the situation was radically different, and no military/industrial imperative was at work. This did not prevent the builders of the proton synchrotron being *unwittingly* entangled in the military net. To their dismay, for the first six months of 1954 they had no official news of the parallel development of the Brookhaven alternating-gradient synchrotron, an unexpected situation as the two groups had collaborated very closely until then. Recent historical research has revealed that the American Department of Defence was behind the hold-up: believing that the new focusing principle could be useful for the development of a beam weapon, they imposed security restrictions on the further dissemination of information about it.[53]

This brings us to the third kind of interest which a laboratory like CERN could have had for its paymasters: as source of a highly-trained cadre of scientists and engineers. Here the evidence is clear: from the start the physicists called upon to justify Auger's project insisted that the laboratory would produce scientific and technical experts endowed with skills which were transferable beyond the limits of particle physics. Of course, the only aim of the laboratory was to do pure science, and of course it would help to stop the brain drain in this domain. But being a laboratory at the leading edge of nuclear-physics research, it would also be 'in its scope a universal laboratory', 'a training centre' for other disciplines (chemistry, medicine ...), a laboratory 'creating a type of research worker adaptable to industrial research in the home countries'. Heisenberg was even more precise: 'participation in the project', he told his government, 'would provide a scientific training for the younger, actively collaborating physicists which could be very useful at a later date with regard to an economic utilization of atomic energy in Germany'. In short, high-energy physics, being one of the most complex and sophisticated of scientific practices, would generate a pool of expertise which could be drawn on by various technologically advanced sectors of the economy, including those of interest to the so-called 'military-industrial complex'.[54]

We see then that the innocence which Kowarski attributes to the scientific community was certainly not shared by all its members—if it ever existed at all. Does this mean that European politicians and scientists all behaved like Lawrence? No, this would be ridiculous, as should be evident from what we have said in this volume and in this conclusion. Reasons differed between scientists and politicans, between members of the CERN Council and parliaments involved in the decision: there was no *one* motive for launching CERN, and among the many, our feeling is that military-oriented ones were of minor importance, even at governmental level. Perhaps some people saw, or at least argued for, the laboratory from a primarily military perspective, their perception being heightened by the context of the cold war. Some remarks made here and there, often in passing, could be interpreted in this way. So too could Cherwell's objection to having the laboratory in Copenhagen: it would be too close and vulnerable to the Soviet Union.[55] All the same, one cannot escape one simple fact, a fact situated at the core of our analysis: whatever may have been the secret hopes of some regarding military 'benefits', the very nature of CERN itself *demanded* that the laboratory was not involved as such in business of this kind—being intergovernmental, it had to remain neutral, 'de-ideologized', as we have said. Perhaps its results could be used for military purposes, perhaps it could develop new accelerating techniques which would be snapped-up by the generals, perhaps it could be a training ground, a centre of excellence, whose products could be used by militarized industries all over the world, as with more or less all scientific activities. But in itself, CERN had to renounce such linkages; its very existence, its very success as a European laboratory, largely depended on its doing so. Hence the insistence on Article II.1 of the Convention; hence the public character imposed on the work done at Geneva; hence the fact that later, though not that much later, CERN came to be like a university laboratory, opening its doors to researchers from western and eastern Europe, from the United States and from the Soviet Union, as well as to busloads of sightseers who come onto the site with a minimum of formalities.

Notes

1. The bibliography for this section is at the end of part III of this volume.
2. No detailed references are given in this section; we prefer to indicate the sections in the main body of the text in which a particular event is dealt with.
3. Global interpretations of events shaping CERN's history which have already been published are Adams (1955), Amaldi (1955a,b, 1977), Auger (1975), Belloni (1986), Brouland (1970), CEC (1975), CERN (1979a, b, c, d, 1984), Herzog (1977), Jungk (1968), Kowarski (1955, 1961, 1973, 1977a,b), Krige (1983), Mussard (1974, 1979), Nyberg & Zetterberg (1977), Pestre (1984b,c), Petrucci (1959), Salomon (1964b, 1968), and Wentzel (1953). Accounts of more specific events, like Adams (1968) on Odd Dahl's role, are also listed in the bibliography.
4. Kowarski (1961, 1977a).
5. Unless otherwise stated, all quotations are from Kowarski (1977a).
6. See all of chapter 5, and most of chapter 12.
7. See sections 2.2 and 2.4, and sections 9.5, 12.5, and 13.5.
8. See sections 9.2, 9.5, and 13.3. Also section 5.1.
9. See chapter 1 and sections 2.1, 2.3, 3.2, 13.3, and 13.5.
10. Kowarski (1961), 1.
11. Allison (1971), 254. For more detail on these methodological issues see Allison (1971), Allison & Morris (1975), Armacost (1969), Coulam (1977), and Schilling (1961).
12. Gilpin (1962), Salomon (1970a), and Greenberg (1971), for example.
13. Marrou (1961), 1529.
14. Allison & Morris (1975), 121.
15. For more detail see section 5.4.
16. The quotations and a more careful analysis are to be found in section 5.1.
17. See sections 13.1 and 13.2.
18. See section 13.2, where Chadwick's ability to overrule the opinion of the majority of British physicists emerges clearly. For the latitude which Auger had, see section 6.4 for example.
19. On 'saving' the conference, see sections 6.1 and 6.3.
20. The biographies at the end of this book give further personal information on these people.
21. Sections 2.2 and 2.4, 3.2, and 3.4 deal with these points.
22. For the list of delegates see appendix 7.4.
23. See Cadres Juridiques (1970), and Polach (1964).
24. For the quotation see CERN/GEN/3, 25/10/52, 4. See also section 7.7.
25. See section 8.5. The quotations are from the minutes of the seventh Council session, CERN/GEN/12, 20/3/54.
26. See the minutes of the eighth meeting of the Interim Finance Committee, CERN/IFC/46 (CERN-CHIP 10025).
27. Schilling (1961), 24.
28. See sections 5.3, 6.4, 9.3, and 12.4.
29. See sections 7.9 and 8.3.
30. The Convention is in CERN (1955). See also section 7.12.
31. For general analyses see Long & Wright (1975) and Ronayne (1984). On France see sections 9.1 and 9.2.
32. On ESRO and related institutions see Schwarz (1979) and Williams (1973).
33. See sections 8.2 and 9.4.
34. For France see section 9.3 and 9.6. For Germany see chapter 11, from which the quotation is also taken.
35. For Britain see sections 12.5 and 13.5. The quotation is from a confidential memo written by Verry, 16/4/53 (SERC-NP24).
36. This point is elaborated in section 5.4.

37. For Britain, see note 35. For France see sections 9.5 and 9.6.
38. Rabi's role at Florence is critically discussed in section 2.5. The quotations are from Berkner Report (1951).
39. Letter Macdermot to Somers Cocks, 18/11/52 (PRO–DSIR17/560).
40. For some background see, for example, Laqueur (1982), and the detailed chronology at the end of this volume.
41. Telegram, Alfvén, Amaldi, Auger, Bakker, Bohr, Casati, Dahl, Dedijer, Dupouy, Gustafson, Heisenberg, Jacobsson, Kowarski, Mercier, Nielsen, Perrin, Preiswerk, de Rose, Savić, Scherrer, Waller, and one other (illegible), to Rabi, 15/2/52 (UNESCO).
42. On Communist opposition to the laboratory, see sections 8.1 and 9.5. See also Vitale (1974).
43. See Grosser (1972), NATO (1971), OTAN (1953), and the chronology at the end of this volume.
44. See section 2.5. For Amaldi's disappointment over the removal of the word 'European' from the Convention, see section 8.2.
45. Kowarski (1977a).
46. Concerning the military potential of CERN, see Grinevald et al. (1984). Concerning the taboo, among many examples, see 'Job for CERN', *Financial Times,* 11/4/85.
47. Quoted by Wade (1977).
48. Glantz (1978), 111. See also Godement (1978–79).
49. For Colonnetti, see Belloni (1986), 78; for Lockspeiser, section 13.5.
50. See E. Schrödinger, 'What is Matter?', in Kaufmann (1980), 11–16; for Heisenberg see chapter 11.
51. Heilbron, Seidel & Wheaton (1981), Greenberg (1971), 226, Vogt (1985), 3836. According to Pickering (1985) and Schweber (1985), military connections even influenced the way theoretical physics developed in the 1950s in the United States.
52. See D. Wilkinson, 'Events Surrounding the Construction of Nimrod', in Nimrod (1979), 7–20. On the Harwell staff's attitudes, see section 13.4.
53. See section 8.5 and Seidel (1986).
54. For Heisenberg see section 11.2. For the first set of quotations see annex 1 of UNESCO/NS/NUC/1, 19/12/51, and section 4.3. For general information see Glantz (1978), Mercer (1978), and Pickering (1985).
55. Letter Chadwick to Bohr, 3/10/52 (NBI).

APPENDIX 1

Who's who in the foundation of CERN

Armin HERMANN

In the following pages the reader will find 65 biographies of people who played a role in the foundation of the European laboratory for elementary particle physics in Geneva. Our aim is to give short and precise information on the educational background of these personalities, their career and their functions in the foundation process. We include data about their positions in the permanent organization: we do not, however, list honorary degrees, distinctions (except for the Nobel prize), or memberships in academies (except for the presidency).

The length of each article depends solely on the information we though necessary and is in no way meant to indicate 'relative importance'. Additional biographies of CERN pioneers who worked for the organization in the fifties and sixties, will appear in a similar appendix to Volume II.

Adams, Sir John (1920–1984), worked during the war in the Radar Laboratories of the Ministry of Aircraft Production. Thereafter, until 1953, he worked at the Atomic Energy Research Establishment at Harwell on the design and construction of a 180 MeV synchro-cyclotron. He came to CERN in September 1953 and was appointed director of the PS division in 1954. He served as interim Director-General after the death of Cornelis Bakker (23 April 1960) until Victor F. Weisskopf became new Director-General on 1 August 1961.

From 1961–66 Adams worked as director of the Culham Fusion Laboratory. From 1966–71 he was member of the Board of the United Kingdom Atomic Energy Authority. In 1971 he came back to CERN and served until 1975 as Director-General of Laboratory II, responsible for the design and construction of the SPS accelerator. From 1976–80 he was executive Director General.

Alfvén, Hannes (b. 1908) was awarded the Nobel prize in 1970 (together with Louis Néel) 'for fundamental work in magnetohydrodynamics with fruitful applications in different parts of plasma physics'. He has been a professor at the Royal Institute of Technology in Stockholm since 1940; his field was first the theory of electricity, then, from 1945–63 electronics, and since 1963 plasma physics. He was a member of the UNESCO 'board of consultants' which helped Auger develop the plan for the European laboratory, and he took part in the two sessions of the UNESCO conference in December 1951 and February 1952, and in the 1st, 2nd and 4th sessions of the interim Council. In all of these meetings he served as official Swedish delegate, except for the 2nd session of the provisional Council, which he attended as a substitute.

Amaldi, Edoardo (b. 1908), worked on spectroscopy, cosmic rays, nuclear and elementary particle physics, and became a professor of experimental physics at Rome University in 1937. He retired in 1978. From 1948-54 Amaldi was vice-president and from 1957-60 president of IUPAP. He was chairman of the Comité Scientifique et Technique of Euratom from 1958-59, of the Scientific Committee for Physics of the Solvay Foundation from 1968-78, chairman of the Scientific Programme Committee of the European Space Agency (ESA) from 1980-81, and of the Space Science Advisory Committee of ESA from 1982-83.

He worked from its early beginnings for the realization of CERN, served as Secretary-General of the provisional organization (May 1952-September 1954), Deputy Director-General from 1954-55, chairman of the Scientific Policy Committee from 1958-60, Italian delegate to the CERN Council from 1959-71, and its president from 1970-71. In 1963 he was president of the European Committee for Future Accelerators (ECFA) and from 1968-69 chairman of the 300 GeV Steering Group.

Auger, Pierre (b. 1899), worked on atomic physics ('Auger effect'), X-rays, neutrons, and cosmic rays ('Auger showers'). He became professor of physics at the Sorbonne in 1937 and worked from 1942-44 as head of the physics division of the Anglo-Canadian project on atomic energy. From 1945-48 he was director of Higher Education in France and Commissaire Scientifique of the Commissariat à l'Energie Atomique. From 1948-59 he served as director of UNESCO's Department of Exact and Natural Sciences, was responsible for the first feasibility studies for a European research centre and initiated the UNESCO conference in Paris (December 1951) and Geneva (February 1952), which led to the 'Agreement constituting a Council [...] for planning an International Laboratory [...]'. After other positions in space research he served as Director-General of the European Space Research Organization (ESRO) from 1962-67.

Bakker, Cornelis Jan (1904-1960), was nominated professor and director of the Zeeman Laboratory of Amsterdam University and also director at the Institute of Nuclear Physics of the Dutch nuclear research center in 1946.

He was a member of the UNESCO 'board of consultants' which helped Auger develop the plans for the European laboratory and served (together with Bannier) as Dutch delegate at the UNESCO conference in Paris and Geneva (December 1951/February 1952). At the first session of the provisional Council he was nominated head of the SC group. When the permanent organization came into existence, Bakker was made a member of the (tripartite) directorate and director of the SC division. On 1 September 1955 he succeeded Felix Bloch as Director-General of the organization.

Cornelis Bakker was killed in an aeroplane accident on 23 April 1960.

Bannier, Jan Hendrik (b. 1909), studied physics at Utrecht and, when the

Department of Education was reorganized after the liberation of the Netherlands, became deputy director of the Division of Higher Education and Research. He took part in the creation of the Nederlandse Organisatie voor Zuiver-Wetenschappelijk Onderzoek (Z.W.O.), the Netherlands Organization for the Advancement of Pure Research, and served as secretary of its Preliminary Board and its preliminary Director. When he was nominated director of Z.W.O. in 1950, Bannier gave up his position in the Department of Education. He remained director of Z.W.O. for two decades and served from 1970–78 as vice-president of its Council.

He worked as rapporteur for the two sessions of the UNESCO conference in Paris (December 1951) and Geneva (February 1952), and was a member of the working group which formulated the Agreement. He served without interruption until 1977 as a delegate of his country in the provisional and permanent Council, was chairman of the provisional Council from its 4th to its 6th session (all in 1953), chairman of the CERN Finance Committee from 1958–60, and president of the Council from 1965–67.

Blackett, Patrick M.S. (1897–1974), was awarded the Nobel prize in 1948 for the invention of the counter controlled cloud chamber and his discoveries with it. He became professor at Birkbeck College, London in 1933, at Manchester University in 1937, and at Imperial College, London in 1953. Blackett was chairman of the Council of the Department of Scientific and Industrial Research (DSIR) from 1955–60 and president of the Royal Society from 1965–70.

He served for CERN as a member of the Scientific Policy Committee from 1954–59.

Blewett, John P. (b. 1910) worked during the war on radar and radar counter measures in the Research Laboratory of the General Electric Company (U.S.A.). After the war he joined the newly established Brookhaven National Laboratory where, from 1947–52, he helped to direct the design, construction, and operation of the 3 GeV Cosmotron. From 1952–61 he was co-designer and builder of the 30 GeV AGS.

During 1953–54 he accepted an invitation to assist in the establishment of CERN and was given leave of absence by the Brookhaven Laboratory. In July 1953 he went with his former wife, M. Hildred Blewett, to Bergen, Norway, where he worked with Odd Dahl and Kjell Johnsen on the first designs of the PS, paying particular attention to the PS magnet system. In September 1953 he and his wife moved to Geneva where they put Brookhaven's experience in accelerator design at the disposal of the newly-assembled CERN PS group.

From 1954–78 he was a Senior Physicist at Brookhaven. He retired in 1978.

Blewett, M. Hildred (b. 1911) joined the newly established Brookhaven National Laboratory in 1947 where she worked on the design, construction and early operation of the Cosmotron. In 1953–54, she spent some months first in Bergen and then in Geneva, with the PS group for whom she prepared the first preliminary cost estimate. Back at Brookhaven, she worked on the design, construction and early operation of the AGS, and assisted for three months in 1959 with the running-in of the CERN PS. Mrs Blewett joined the Argonne National Laboratory in 1964 working on the early operation and the planning of experiments at the ZGS. During 1966–67, she was with the group at CERN who were making preliminary plans for the 300 GeV accelerator, giving most of her attention to planning experiments. In 1969 she returned to CERN and collaborated on the running-in and operation of the Interacting Storage Rings (ISR) and in its experimental programme. From 1969 to 1976 she was Secretary of the Intersecting Storage Rings Committee (ISRC). Before retiring from CERN in 1977, she was one of the team who were making the initial plans for LEP.

Bloch, Felix (1905–1983), was awarded the Nobel prize in 1952 jointly with Edward Purcell 'for the development of new methods for nuclear magnetic precision measurements'. From 1934 until his retirement he was at Stanford University except for the war years, in which he worked for the Manhattan Project. Bloch was the first Director-General of CERN serving roughly for one year from the official foundation until 31 August 1955.

Bohr, Niels (1885–1962), won the Nobel prize in 1922 'for his studies on the structure of atoms and the radiation emanating from them'. From 1917 onwards he was professor at the University of Copenhagen and director of the Institute for Theoretical Physics from its foundation in 1920. He worked as consultant for the Anglo-American atomic energy project in 1943–45 and was, from its foundation in 1954, chairman of the Danish atomic energy commission.

He was nominated director of the theoretical group at the first meeting of the provisional Council, a position he retained until he was replaced by Christian Møller on 1 September 1954. He took part in all meetings of the provisional Council (first as adviser to the Danish Delegation, then as group director), and was Danish delegate to the Council in 1954–55.

Capron, Paul (1905–1978), became a professor at the Université Catholique de Louvin in 1943, was responsible for the Nuclear and Radio-chemistry Laboratory and taught general physics, nuclear chemistry, and nuclear physics. From 1962–69 he was 'Conseiller Scientifique' to the Rector of his university, a position in which he had to supervise all academic matters. He retired in 1974.

Capron was a member of the Conseil d'Administration du Centre d'Energie Nucleaire (CEN) at Mol from 1959–65, and a member of the Comité des Experts du Conseil National de la Politique Scientifique from 1960–65, where he was particularly active in the working groups developing nuclear research programs.

He was also a member of the Commission Scientifique of the Institut Interuniversitaire de Physique Nucleare (IIPN) and its successor, the Institut Interuniversitaire des Sciences Nucleares (IISN) from its foundation in 1947. It was in this capacity that he took part in the meeting of the Centre Européen de la Culture on 12 December 1950, and in 1951 he was a member of the UNESCO board of consultants which helped Auger to develop the plans for the European laboratory.

Casati, Alessandro (1881–1955) was nominated Senator on 1 March 1923 and served as Minister for Public Education from 1 July 1924 to 3 January 1925, when he resigned due to conflicts with the Fascists. He re-entered public life only after the liberation of Rome. In June 1944 he became Minister of War in Bonomi's cabinet, and served as such also in Bonomi's second cabinet until 1945. Subsequently he became a member of Consulta Nazionale, President of Consiglio Supremo di Difesa and, in 1948, President of Consiglio Superiore della Pubblica Istruzione. Casati was a member of the Italian delegation to the Assembly of the Council of Europe until 1953, and chairman of its Commission of Cultural Relations. He also served as a member of the Italian delegation to UNESCO and took part in the UNESCO General Conference held in Florence from 22 May to 17 June 1950.

Having established close relations with Denis de Rougemont at the European Cultural Conference in Lausanne (8–12 December 1949), he took part in the meeting of the Commission of Scientific Cooperation of the Centre Européen de la Culture on 12 December 1950 in Geneva. He was official delegate of his country (together with Gustavo Colonnetti) at both sessions of the UNESCO conference in Paris (December 1951) and Geneva (February 1952) and at the first session of the provisional Council.

Chadwick, James (1891–1974), was awarded the Nobel prize for the discovery of the neutron. From 1919 he was Rutherford's closest collaborator at Cavendish Laboratory in Cambridge. He became professor at the University of Liverpool in 1935, and worked for the Anglo-American atomic energy project as head of the British atomic energy mission. After the war he initiated the construction of the 400 MeV Liverpool synchro-cyclotron for protons, which was finally inaugurated in 1954. From 1948–58 he was Master of Gonville and Caius College, Cambridge.

When the Royal Society set up an Advisory Committee on CERN matters early in 1952, Chadwick was made chairman. In this capacity he played a decisive role in British participation.

Cockcroft, John D. (1897–1967), won the Nobel prize in 1951 jointly with Ernest T.S. Walton for the study of nuclear reactions induced by protons and α-particles. He was director of the Anglo-Canadian project on atomic energy from 1944–46, and of the Atomic Energy Research Establishment at Harwell from 1946–59. He allocated three of his Harwell staff, John Adams, Frank Goward and Mervyn Hine, to co-operate in the construction of the CERN proton synchrotron, and sent them to Geneva in September 1953 to join the PS group.

Cockcroft was a member of the U.K. delegation (which had observer status) at the 4th, 5th, 7th and 8th sessions of the provisional Council, and member a of the CERN Council and of the Science Policy Committee until 1961.

Colonnetti, Gustavo (1886–1968), graduated in civil engineering and mathematics and worked mainly on elastic solids. He was professor of engineering in Genoa, Pisa and from 1919 onwards at the Turin Polytechnical Highschool. He became a member of the Consulta Nazionale (1945–46) and of the Assemblea Costituente (1946–48), which prepared the new republican constitution. From 1944–56 he served as president of the Consiglio Nazionale delle Ricerche (CNR).

Dahl, Odd (b. 1898), took part as early as 1932 in the construction of Van de Graaff generators at the Carnegie Institution. After ten years in Washington he returned to Norway. He took a post at the Christian Michelsens Institute in Bergen, where he worked until he retired. He joined the UNESCO board of consultants in April 1951 to help Auger develop the plans for a European laboratory, and served as adviser of the Norwegian delegation at the UNESCO conference in Paris and Geneva (December 1951/February 1952). At the first session of the provisional Council Dahl was nominated director of the PS group. He could only work for CERN for 30% of his time, and did not move to Geneva. When the permanent CERN started, Dahl resigned and was replaced by Adams.

Dakin, Samuel Arthur French (1909–1978) worked as staff officer, principal and assistant secretary at the Board of Trade in London, and served, on loan from the Board of Trade, from October 1954 to September 1955 as director of administration for the Organization. In November 1958 he returned to CERN as director of administration after Jean Richmond. When on 1 January 1961 the laboratory's new internal organization was established, Dakin also became a member of the directorate. In September 1963 he returned to England to work again in the Board of Trade and stayed there until 1969, when he took up an appointment as consultant with the National Research Development Corporation. He retired in 1972.

Dautry, Raoul (1880–1951) was director-general of the French national railways (later the S.N.C.F.) from 1928–37, ministre de l'armement from 1939–40, and ministre de la reconstruction et de l'urbanisme from 1944–46. He directed the Commissariat à l'Energie Atomique (CEA) as administrateur général from 1946, and had numerous honorary positions, such as president of the Comité exécutif français of the European movement from 1948, president of the Fondation Nationale de la Cité Universitaire (from 1948), vice-president du Conseil de

Direction of the Centre Européen de la Culture in Geneva (from 1950). Dautry was a founder-member of the College of Europe at Bruges in 1951.

He died unexpectedly on 21 August 1951 at Lourmarin (Vaucluse), where he had been elected major in 1945, and where he always spent his vacations.

Ferretti, Bruno (b. 1913), became associate professor for theoretical physics in Milan in 1948, and a professor in Rome in 1951. He took part in the meeting of the Commission of Scientific Cooperation of the Centre Européen de la Culture on 12 December 1950, for which he had prepared, together with Amaldi, a concise plan as a basis for discussion. He declined, however, an offer to work in Paris for Auger's 'bureau d'études'.

On 1 January 1957 he became head of the theoretical group at CERN and on 1 October that year director of the theoretical study division. After two years in Geneva he went to Bologna, where he held the chair of theoretical physics until his retirement. Among his numerous contributions to nuclear and elementary particle physics, quantum field theory and crystal physics are ideas for determining the parity of charged pions and the non-conservation of baryon numbers.

Fry, Donald William (b. 1910), joined the Royal Aircraft Establishment in Farnborough in 1936 and worked on a two-way VHF communication system for fighter aircraft. In 1940 he moved to radar work at Swanage (later Malvern).

After the war Fry became a member of the staff of the Atomic Energy Research Establishment, leading a group interested in the acceleration of particles to high energies. In 1946 Goward (a member of his team) demonstrated with Barnes the synchrotron principle, and in 1947 Fry with Harvie, Mullett, and Walkinshaw demonstrated the travelling-wave electron linear accelerator. He became Head of the General Physics Division at the A.E.R.E. at Harwell in 1949. Its work included further studies of particle acceleration methods. In 1950 he was appointed Chief Physicist at the A.E.R.E., served from 1954-58 as Deputy Director, and from 1959 to his retirement in 1973, as Director of the Atomic Energy Authority's Establishment engaged in nuclear-reactor development at Winfrith in Dorset.

During the formative years of CERN Fry discussed with Cockcroft the way the U.K. could contribute. He took part in the 2nd, 3rd, 6th and 9th sessions of the provisional Council and was his country's delegate to the Council in 1956 (together with Sir Ben Lockspeiser).

Funke, Gösta Werner (b. 1906), was lecturer and professor of physics until 1945, when he became secretary-general of the Atomkommitten Stockholm (the Swedish Atomic Energy Committee), a position which he held until 1959. From 1945-72 he also served as secretary-general of the National Science Research Council. He was a member of the Swedish National Committee for UNESCO from 1957-60 and of numerous other committees concerning Nordic or international collaboration in science. He served as the Swedish delegate in the permanent Council from its foundation in 1954 until 1971, was chairman of the Finance Committee from 1962–64 and president of the Council from 1967–69.

Gentner, Wolfgang (1906–1980) worked 1936-46 with Walther Bothe at the Kaiser-Wilhelm-Institut für Medizinische Forschung in Heidelberg and became professor of physics in Freiburg in 1946. He played an active part in the foundation of CERN and served as adviser of the German delegation in the provisional and permanent Council. He became director of the SC division in 1955, and director of research in 1958. He left CERN in 1960 when he was appointed director of the Max-Planck-Institut für Kernphysik in Heidelberg.

Gentner served as chairman of the Science Policy Committee from November 1968 to the end of 1971 and was president of the Council from 1972–74.

Goldschmidt-Clermont, Yves (b. 1922), worked on elementary particles, cosmic rays and nuclear instrumentation and completed his 'Docteur ès Sciences Appliquées' at the Université Libre in Brussels in 1950. He became a member of Kowarski's Laboratory group in 1953 (later the Scientific and Technical Services Division) and served as Deputy Division Leader in the Data Handling Division from 1961–64.

From 1979–82 he was Scientific Assistant to the CERN Director-General and Secretary of the Research Board, and in 1983/84 Deputy Division Leader of the Experimental Physics Division.

Goward, Frank Kenneth (1919–1954) joined the Telecommunications Research Establishment in Malvern in 1942 and worked in the field of antenna design for ground-based radar. After the war he became a staff member of the Atomic Energy Research Establishment initially at Malvern and later (from 1951) at Harwell. In 1946, together with Barnes, he was the first to demonstrate the synchrotron acceleration principle by converting a small betatron. He was responsible for the design of a 30 MeV synchrotron at Malvern, and contributed notably to early synchrotron development.

Goward joined the UNESCO board of consultants in April 1951 to help in the design of the European accelerators. When the provisional Organization was set up he worked, first as a consultant for the PS group, and then (from early 1953) as its deputy director. In October 1953 he was granted leave of absence from Harwell, and moved with his family to Geneva. Goward became seriously ill in February 1954 and had to return to England. He died on 10 March 1954.

Gustafson, Torsten (b. 1904) was professor of theoretical physics in Lund from 1939 until his retirement in 1970. He also served as a member of the Atomkommitten Stockholm, the Swedish Atomic Energy Committee, from its foundation in 1945 until 1964.

He was a member of the Swedish delegation at the UNESCO Conference in Paris and Geneva (December 1951/February 1952), and was present at all sessions of the provisional Council.

Heisenberg, Werner (1901–1976), won the Nobel prize in 1932 'for the establishment of quantum mechanics'. He worked as professor of theoretical physics from 1927 in Leipzig and from 1942 in Berlin. After the war he directed the 'Max-Planck-Institut für Physik and Astrophysik' in Göttingen and in Munich (after 1958), until he retired in December 1970.

Heisenberg had many important honorary positions. He was president of the Deutsche Forschungsrat (DFR) as long as it existed (1949/52), chairman of the Kommission für Atomphysik of the Deutsche Forschungsgemeinschaft (DFG) from its foundation in 1952, and president of the Alexander von Humboldt-Stiftung, again from its foundation in 1953.

de Hemptinne, Marc, Comte (1902–1986), worked mainly on molecular spectroscopy and was, from 1931 until his retirement in 1971, professor at the Université Catholique de Louvain and director of its Centre de Physique Nucléaire.

He served as a member of the 'commission de physicochimie et électrochimie' of the Fonds National de la Recherche Scientifique (FNRS) from 1948–69 and as its president 1957–69, and as a member of the commission scientifique of the Institut Interuniversitaire de Physique Nucléaire (IIPN) from June 1947 to November 1948. He held the same position from November 1951 to October 1968 in IIPN's successor, the Institut Interuniversitaire des Sciences Nucléaires (IISN). From 1951–73 he was a member, and from 1971–73 president, of the Conseil Scientifique of the Institut Royal Météorologique, administrateur of the Centre d'étude de l'énergie nucléaire at Mol (C.E.N.) from 1954–63, and président of its conseil from 1957–63.

He took part in the first session of the UNESCO Conference in Paris (17–20 December 1951) as a delegate of his country and served in the same capacity in the CERN Council 1954–57.

Hine, Mervyn G. (b. 1920). After two years at Cambridge (1939–41) he worked on microwave radar at the RAF Radar Establishment until 1945. He then gained a PhD in nuclear physics at Cambridge, and joined the Atomic Energy Research Establishment at Harwell, where he collaborated closely with John Adams. He came with him to CERN in September 1953 to work on the construction of the PS.

When the new internal organization was established on 1 January 1961 Hine became a member of the directorate with responsibility for applied physics. He was involved extensively in long-term planning and budgeting, as well as in computing and accelerator-policy formation. When CERN's structure was changed again on 1 July 1966, he continued this work, serving as director of the Applied Physics

Department, and held this position until the end of 1971. Thereafter he was concerned with many initiatives on data handling and computer networks at CERN, and internationally in collaboration with the Commission of the European Communities (C.E.C.).

Hocker, Alexander (b. 1913), finished his legal studies with the two examinations and his thesis at Leipzig. He was deputy director of the Deutsche Forschungsgemeinschaft from its foundation in 1949, and from 1956-61 Ministerialrat and Ministerialdirigent at the Bundesministerium für Atomfragen. From 1961-69 he served as a member of the directorate for the Kernforschungsanlage Jülich and was Director-General of the European Space Research Organization from 1971-74. He was the German delegate in the provisional Council and the Council until 1961, and in his last year chairman of the Finance Committee.

Holtsmark, Johan Peter (1894-1975), was nominated professor at the Norwegian Institute of Technology in Trondheim at the age of 29 and came to Oslo University in 1942. He worked mainly on spectroscopy, nuclear physics, and acoustics.

He was his country's delegate to the provisional Council (and took part in all sessions except the 4th and the 9th) and to the Council from 1957-60.

Hylleraas, Egil A. (1898-1965) was appointed professor of theoretical physics at the University of Oslo in 1937, where he remained all his life. His early work concentrated on the helium atom; later on he published on positronium, magnetic properties of atoms and molecules, and collision problems. He edited the Norwegian periodical 'Fra Fysikkens Verden' from its foundation in 1939 until 1956.

Hylleraas took part in the two sessions of the UNESCO conference on the 'establishment of a European nuclear research laboratory' in December 1951 and February 1952 as a member of the Norwegian delegation (together with Odd Dahl) and in the 2nd, 4th, 5th and 8th sessions of the provisional Council. He was a member of the Council from the foundation of the organization until 1957.

Jacobsson, Malte Ferdinand (1885-1966), worked in theoretical philosophy and psychology and published a fundamental study on 'Psykisk Kausalitet' in 1913. As professor for philosophy at Gothenburg University from 1920-34 he was engaged in local politics to stimulate economic activity and served as Governor of the Country of Gothenburg and Bohus from 1934-50. He held a considerable number of important honorary positions in science administration and government, such as chairman of the board of Chalmers University of Technology (1934-39) and of the Business Highschool, both in Gothenburg. Jacobsson was, from its foundation in 1945 until 1958, chairman of the Atomkommitten Stockholm, the Swedish Atomic Energy Committee, and from 1947-56 chairman of the board of the state-owned

Atomic Energy Company. He took part, as a delegate of his country, at the UNESCO conference in Paris (December 1951) and Geneva (February 1952), as well as at the 2nd and 5th sessions of the provisional Council.

King, Alexander (b. 1909), did his DSc at the Royal College of Science in London. From 1932-40 he was a demonstrator and senior lecturer in physical chemistry at Imperial College, London. In 1942 he was appointed deputy science adviser to the Ministry of Production, and he served from 1943-47 as Head of the U.K. Scientific Mission in Washington and Scientific Attaché at the British Embassy. From 1946-47 he was scientific adviser to the British delegation on atomic energy to the United Nations and from 1947-50 Head of the Lord President's Scientific Secretariat and Secretary of the Advisory Council on Scientific Policy (ACSP), and from 1950-56 Chief Scientific Officer in the Department of Scientific and Industrial Research (DSIR), where his duties included being chairman of the Committee on Overseas Scientific Relations.

King left Britain in 1956 to work for the OECD, and was its Director-General of Scientific Affairs from 1968-74. From 1974-84 he was chairman of the International Federation of Institutes for Advanced Study (IFIAS). In 1984 King became President of the Club of Rome of which he was a founder member.

Kowarski, Lew (1907-1979), undertook jointly, with Halban and Joliot at the Collège de France, the famous experiments on uranium fission. On 17 June 1940 he brought, together with Halban, the French stock of heavy water to Britain. Kowarski worked until 1944 in Cambridge, then in Montreal, and constructed the ZEEP (Zero Energy Experimental Pile). In 1946 he became director of the Département de Physique et Technologie at the Commissariat à l'Energie Atomique (CEA).

He took part as adviser of the French delegation in both sessions of the UNESCO conference (Paris/Geneva, 1951/52), and at the first session of the provisional Council. At this session he was nominated director of the laboratory group. In the permanent organization he was director of the Scientific and Technical Services Division (1954-60) and of the Data Handling Division (1961-63). In 1963 he took leave of absence from CERN and was professor of nuclear engineering at Purdue University in Lafayette, Indiana. He came back in 1965 and served later as a consultant.

Kramers, Hendrik Anthony (1894-1952), collaborated with Niels Bohr in Copenhagen from 1916-26, the last two years in the post of lecturer. He then became professor of theoretical physics in Utrecht, and in Leiden in 1934, where he stayed until his death. When the UN Atomic Energy Commission was founded in 1946, Kramers was elected chairman of the Scientific and Technological Committee. From 1946-50 he was president of the International Union of Pure and Applied Physics (IUPAP).

Kramers took part in the meeting of the Commission of Scientific Cooperation of the Centre Européen de la Culture in Geneva on 12 December 1950, as well as in the discussions about the planned European laboratory the following year. He was unable to attend the UNESCO conference in Paris (December 1951) and Geneva (February 1952) and died on 24 April 1952.

Leprince-Ringuet, Louis (b. 1901), was professor of physics at the Ecole Polytechnique from 1936–69, and professor of nuclear physics at the Collège de France from 1959–72, commissaire at the Commissariat à l'Energie Atomique from 1951–71, member of the Académie française (from 1966) and the Académie des Sciences (from 1949). From its foundation in 1954 he was vice-chairman of the Science Policy Committee for ten years and its chairman from 1964 until 1966.

Lockspeiser, Sir Ben (b. 1891), made his career in the scientific civil service, which he entered in 1916. He took a prominent part in planning Britain's air defence before and during World War II, and was knighted in 1946 and appointed Chief Scientist to the Ministry of Supply with responsibily for co-ordinating research on military and aeronautical programmes. In May 1949 he succeeded Edward Appleton as Secretary to the Department of Scientific and Industrial Research (DSIR).

From its 4th session in Brussels (12–14 January 1953) Sir Ben took part in all meetings of the provisional Council, where his country had the status of an observer, and served as chairman of the Interim Finance Committee. He was elected the first president of the Council in October 1954 and remained in this office until 1957.

Mercier, André (b. 1913). Born and educated in Geneva, he became associate professor (1939) and then professor (1947) of theoretical physics at the University of Berne, where he worked until he retired. His major field was gravitation theory. To commemorate the 50th anniversary of Einstein's special theory of relativity in 1955 he organized a 'Congress on General Relativity and Gravitation', which became a regular institution called GRG-Congresses for short.

Mercier was president of the Swiss Physical Society from 1951–53, and worked in collaboration with Paul Scherrer, Albert Picot and others for the realization of CERN and its implementation in Geneva. He took part in both sessions of the UNESCO conference in Paris (December 1951) and Geneva (February 1952) as a member of the Swiss delegation and attended the first session of the provisional Council in Paris (May 1952) as a substitute.

Møller, Christian (1904–1980) worked on diffraction problems, special and general relativity, and conservation laws of energy and momentum. He became professor of mathematical physics at Copenhagen university in 1943.

He took over as director of the Copenhagen-based theoretical study group of CERN from Niels Bohr on 1 September 1954, and was director of the Nordic

Institute for Theoretical Atomic Physics (NORDITA) from its foundation on 1 October 1957 until his retirement in 1971. He served as a member of the Scientific Policy Committee from 1959–72.

Mussard, Jean A. (b. 1912). Born in Brussels, the son of an old Geneva family, he did his degree of Diplomingenieur at the Eidgenössische Technische Hochschule, Zürich, and joined the UNESCO staff in 1948. When Auger created the 'bureau d'études' (study office) in April 1951 to plan the European Laboratory, he became a sort of secretary. He worked closely with Amaldi and Kowarski, and served as secretary of the Drafting Committee for the Convention (signed 1 July 1953). Mussard was then charged with other UNESCO projects, and worked notably as head of the secretariat of the Commission préparatoire européenne de recherches spatiales (COPERS). In 1964 he was nominated director of the secretariat of the European Space Research Organization (ESRO), and worked from 1970–72 in the same capacity for the UN Conference on the Human Environment held in Stockholm in 1972. He retired in 1972 and now lives in Geneva.

Nielsen, Jakob (1890–1959), was professor of mechanics at the Technical High School in Copenhagen from 1925–51, and professor and director of the mathematical institute of the University of Copenhagen from 1951–56. He served as secretary-general of the Royal Danish Academy of Sciences from 1946–59.

He was a Danish delegate at both sessions of the UNESCO conference in Paris (December 1951) and Geneva (February 1952), in the provisional Council, and the Council until the end of 1955. At the first session of the CERN Council, on 7 October 1954, he was elected one of the two vice-presidents, and held this position until the fourth session (19–20 December 1955), the last he attended.

Penney, Richard W. (b. 1918), worked from 1937 for the War Office and joined the Royal Army Ordnance Corps in 1940. After demobilization he was Higher and Senior Executive Officer at the Ministry of Supply.

From October 1953 to October 1954 he worked for the provisional Organization and received, when CERN was officially founded, a full time contract as personnel officer, in which capacity he developed the first CERN staff rules and procedures. He left CERN on 31 December 1964 to become Deputy Secretary to the United Kingdom Air Transport Licensing Board and in 1968 Head of Personnel of the European Space Operations Centre in Darmstadt. In 1969 he joined the International Civil Aviation Organization, Montreal, from which he retired as Chief of Staff Administration in December 1982. He has also been a consultant to several United Nations Agencies in Geneva, New York, Rome and West Africa.

Perrin, Francis (b. 1901), worked on statistical quantum mechanics and nuclear chain reactions. He was professor at the Collège de France from 1946–72 and, after the dismissal of Joliot in 1951, became High Commissioner of the Commissariat à l'Energie Atomique (until 1970). He was Membre de l'Institut from 1953. Perrin served as French delegate in the provisional Council and the Council until 1972, and member of the Science Policy Committee from 1960-75.

Picot, Albert (1882-1966), son of a 'juge fédéral' and an old Geneva family, was a lawyer in Geneva until 1931, when he was elected Conseiller d'Etat. He led several government departments and served as president of the Council in 1938, 1944 and 1947.

From 1946-54 Picot was in charge of the Département de l'instruction publique (the educational and cultural department) and was also, from 1949-55, deputy for Geneva at the Conseil des Etats in Berne. From the start he was strongly in favor of the European laboratory and its location in Geneva. He took part as a member of the Swiss delegation at the two sessions of the UNESCO conference in Paris and Geneva and in all sessions (except for the 2nd) of the provisional Council. He was also (together with Paul Scherrer) Swiss delegate to the Council from the foundation of the organization in 1954 until October 1958, when he was succeeded by Aymon de Senarclens.

Powell, Cecil Frank (1903-1969), was awarded the Nobel prize in 1950 'for his development of the photographic method in the study of nuclear processes and for his discoveries concerning mesons'. He worked from 1928 until his death (immediately after his retirement as head of the physics department) at Bristol University, and was nominated professor in 1948. He was chairman of the Science Policy Committee of the Science Research Council from 1965-68, president of the World Federation of Scientific Workers from 1956 until his death, and a founder member of the Pugwash Movement.

In CERN he served as a member (1959-60) and as chairman (1961-63) of the Science Policy Committee.

Preiswerk, Peter (1907-1972), studied physics in Berlin and joined the laboratory of Frédéric and Irène Joliot-Curie in 1934, where he worked on slow neutrons. In 1936 he took part in the construction of the cyclotron at the Eidgenössische Technische Hochschule (ETH) at Zurich and later published many papers on nuclear spectroscopy. He was nominated professor at the ETH in 1950.

On 12 December 1950 he was present at the meeting of the Commission of Scientific Cooperation of the Centre Européen de la Culture in Geneva. 'This event proved a turning point in his career', Nature wrote in the obituary, 'and he devoted the rest of his life to the realization of this European scientific venture'. He belonged to the small board of consultants which, from May 1951, developed the first plans with Auger, and took part in both sessions of the UNESCO conference (December 1951/February 1952) as a member of the Swiss delegation. He worked as deputy director of the laboratory group for the provisional CERN and as director of the Site and Buildings Division for the permanent organization. He joined the SC Division in 1958 and was leader of the Nuclear Physics Division from 1 January 1961 to 1 February 1970.

Rabi, Isidor Isaac (b. 1898), won the Nobel prize in 1944 for his measurements of magnetic moments of elementary particles and nuclei. From 1930 until his retirement in 1967 he was professor at Columbia University in New York except for the years 1940–45, when he was a member of staff and associate director at M.I.T. Radiation Laboratory.

Rabi played an active part in the foundation of the Associated Universities, Incorporated (which created and governed Brookhaven National Laboratory), served as trustee from the beginning in 1946, as president (1961–62), and as chairman of the board of trustees (1962–63). He was also a member of the General Advisory Committee of the Atomic Energy Commission (AEC) from its foundation in 1946.

He had many eminent positions in science administration and science policy. He was a member of the U.S. National Committee for UNESCO 1950–53 and 1958–63, as well as a member of the U.S. delegation at the 5th General Conference in Florence (May/June 1950), at which he proposed resolution No. 2.21. He was also a member of a small UN committee (chaired by Dag Hammerskjold) which prepared the UN Conferences on the Peaceful Uses of Atomic Energy in 1955, 1958 and 1964. This developed into the Advisory Committee for the International Atomic Energy Agency. He served as a member of the Science Advisory Committee of the Ballistic Research Laboratory, the Naval Research Advisory Committee, the Science Committee of NATO, and the President's Science Advisory Committee (PSAC) of which he was chairman from 1956–57.

Randers, Gunnar (b. 1914). After his studies in astrophysics at Oslo University he went to the United States, and was research fellow at Mt. Wilson Observatory from 1939–40, and instructor at the University of Chicago from 1940–42. He took part in the Allied war effort as science officer at the British Ministry of Supply and operational analysis officer of the US Air Force.

From 1948–71 he held the position of managing director of the Norwegian Institute for Atomic Energy, and was director of the Joint Establishment for Nuclear Energy Research of Norway and the Netherlands at Kjeller, near Oslo. On his initiative the first international conference on atomic energy, especially on heavy water reactors, was held at Kjeller from 11–13 August 1953.

He took part in the meeting of the Commission of Scientific Cooperation of the Centre Européen de la Culture in Geneva on 12 December 1950, and served as an adviser of the Norwegian delegation to the 3rd session of the provisional Council in Amsterdam (4–7 October 1952).

Regenstreif, Edouard (1915-1979). Born in Rumania, he settled in France in 1934 and became a citizen in 1947. He wrote his thesis at the Ecole Normale Supérieure on electron optics in 1951, and was asked by Auger to help in the preparatory work for the planned European laboratory. He visited several research centres and wrote detailed reports.

When CERN was established he worked in the PS division both in research and in administration until the end of 1963. From then on he was professor at the University of Rennes.

Rollier, Mario Alberto (1909-1980), worked on X-ray analysis of chemical structure, radio-chemistry, and chemistry of nuclear reactions. He became assistant (1934) and associate professor at the Polytechnical High School in Milan. He played an active role in the resistance, joined the social-democratic party (PSDI) in 1947 and became a consultant in nuclear matters for the Ministry for Foreign Trade in 1949. Later he was professor of chemistry in Cagliari (1956-60) and Pavia (1960-79).

As a founder member of the European Federalist Movement in Italy he had close relations to the Centre Européen de la Culture, and took part in the meeting of the Commission of Scientific Cooperation on 12 December 1950 in Geneva.

de Rose, François Comte de Tricornot (b. 1910), started his diplomatic career in 1937 and held several posts abroad and at the Quai d'Orsay. From 1946-50 he was a member of the French delegation to the U.N. in New York, and from 1956-61 directed the services des pactes, des affaires atomiques et des affaires spatiales at the Foreign Office in Paris. From 1970-74 he served as permanent representative of France in NATO. He received the honorary title of 'Ambassadeur de France' in 1974. In 1976 he became President Director-General of the société Pathé-Cinéma et de Scama (France).

He was elected chairman of the first intergovernmental conference in Paris (December 1951), which led finally to the creation of CERN, and was president of the Council from 1958-60.

Rougemont, Denis de (1906-1985), a renowned Swiss writer in French, he won many literary prizes. He played an active role in the European movement and was, from its foundation in 1950, director of the Centre Européen de la Culture in Geneva, and its honorary president from 1976. He was president of the executive committee of the Congress for Freedom of Culture from 1951-66. In 1963 he became director and professor at the Institut Universitaire d'Etudes Européennes in Geneva.

Scherrer, Paul (1890-1969), developed with Peter Debye the famous Debye-Scherrer method (X-ray analysis of crystal powder), and was appointed director of the physics institute of the Eidgenössische Technische Hochschule (ETH) at Zürich in 1927, where he worked until his retirement in 1960.

When according to his proposal the 'Schweizerische Studienkommission für Atomenergie' was founded in 1946, he was nominated president, and he also served as president after its transformation into the 'Kommission für Atomwissenschaften' of the Nationalfonds until 1960.

He was his country's delegate to the first UNESCO meeting in Paris and Geneva (1951–52), of all sessions of the provisional Council (1952–54), and of the Council until 1961. He was also chairman of the first three sessions of the provisional Council (1952–53), and member of the Scientific Policy Committee from its foundation in 1954 until 1963.

Schmelzer, Christoph (b. 1908) did his PhD with Max Wien at Jena University on the absorption of electromagnetic waves in liquids and solutions, and continued this research from 1936–39 in the United States. He worked during the war on electromagnetic waves in the decimeter range, and came into contact with accelerators when he started working with Walther Bothe in Heidelberg in 1949. At its third session in Amsterdam on 4–7 October 1952 the provisional Council agreed to his appointment as senior staff member for the PS group. He made notable contributions with his ideas on beam self-control. Schmelzer left CERN in 1960 to take over a chair of applied physics at Heidelberg University, and served as chairman of the Scientific Council of DESY from 1959–65. From 1970–78 he was managing director of the Gesellschaft für Schwerionenforschung (GSI) in Darmstadt. He had worked out the concept for the Universal Theory Ion Linear Accelerator (UNILAC) between 1958 and 1960, while still a member of the CERN staff.

Skinner, Herbert W.B. (1900–1960) worked on radar and electromagnetic isotope separation during the war, and joined the Atomic Energy Research Establishment at Harwell in 1946, where he held the position of Chief Physicist. In 1949 he became professor at Liverpool University, where a 400 MeV synchro-cyclotron was completed in 1951. He was nominated adviser to the CERN synchro-cyclotron group in 1952.

Steiger, Rudolf Bernhard (1900–1982), acquired his degree of a 'Diplomarchitekt' at the Federal Institute of Technology in Zurich. He had his own office from 1924 (in Zurich from 1925), and specialized in hospitals, business- and congress-centres as well as in regional- and city-planning. In 1951 he received an honorary medical doctor's degree at Zurich University 'for his great merits in planning and construction of the new Canton's Hospital'. Steiger occupied high positions in the board (Zentralvorstand) of the Federation of Swiss Architects (Bund Schweizer Architekten) and the board (Vorstand) of the association of engineers and architects of Zurich (Zürcher Ingenieur- und Architektenverein). From 1964–74 he was chairman of the Planners Association of Switzerland, and founding member (1958) and president (1963-69) of the Regional Planning Association of Zurich and Surroundings.

He was appointed chief architect at the ninth session of the provisional Council (8-9 April 1954), and worked for CERN together with his son Peter (b. 1928) until 1960.

Thomson, George Paget (1892-1975) was awarded the Nobel prize in 1937 jointly with Clinton J. Davisson for the discovery of electron diffraction. He worked at the University of Aberdeen as Professor of Natural Philosophy from 1922, and was appointed Professor of Physics at Imperial College London in 1930.

In Spring 1939, Thomson managed, with the help of Sir Henry Tizard to secure a ton of uranium oxide for Britain on the world market. When he realized, in April 1940, the possibility of a chain reaction with fast neutrons, he formed and steered the famous M.A.U.D. Committee. In August 1941 he went to Ottawa as head of the British Scientific Office in Canada. He returned to England a year later, becoming Deputy Chairman of the Radio Board and Scientific Adviser to the Air Ministry.

In 1952, Thomson went back to his birthplace Cambridge, as Master of Corpus Christi, the college which he had attended as a young fellow. In the post-war years he worked in a great number of honorary positions both for his country and for the international scientific community.

Trumpy, Bjørn (1900-1975), worked on molecular, atomic and nuclear physics, cosmic radiation and terrestrial magnetism, and became professor at the geophysics institute of Bergen's museum in 1935. From 1936-40 he served, additionally, as secretary-general of the Society for the Advancement of Science.

From 1943-48 he was director of Bergen's Museum, and worked for the creation of a university. When Bergen University was founded he became its first rector and served for two periods from 1948 to the end of 1953. In 1952 he was nominated full professor, and when the Nuclear Research Laboratory was founded, he became its head.

Trumpy had many honorary positions: From 1955 he was a member of the board of the Christian Michelsen Institute and its chairman from 1968-70. He was also, from its foundation in 1951, a member of the board of the Joint Establishment for Nuclear Energy Research of Norway and the Netherlands at Kjeller. He worked as a member for the Royal Norwegian Council for Scientific and Industrial Research and for the Norwegian Atomic Energy Council. From 1954 -57 he was president of the Norwegian Physical Society.

Trumpy served as Norwegian delegate at the 4th, 5th, 7th and 9th sessions of the provisional Council, and in the same capacity at most sessions of the CERN Council from 1960-67.

Valeur, Robert (1903-1973), studied economics and law in Lyon, went to England in

1926, and to the United States in 1930, where he stayed until 1945. He was 'Gaulliste de la première heure' and one of the founders of the movement 'France for ever' in the U.S.

After the liberation he entered the French diplomatic service, was consul general at Sao Paulo until 1951, and became head of the Department of Cultural Relations of the Quai d'Orsay at the beginning of 1952. In this capacity, he represented his country as permanent delegate to UNESCO. He also presided over the Committee of Cultural Experts of the Council of Europe from 1953-54. In 1954 he was appointed director of the Information and Cultural Services of the French Embassy in Washington. Later he was ambassador in Ecuador and Columbia.

He took part in the second session of the UNESCO conference in Geneva in February 1952, and in all sessions of the interim Council, in the first session as a substitute, in all others as a delegate of his country. He was chairman of the Drafting Committee for the convention and served also as chairman of the Interim Council from the seventh session in Geneva (October 1953) until the foundation of the permanent Organization.

Verhaege, Julien Léon (1905-1972), completed his thesis in 1927 at Gent University and stayed there all his life, becoming a professor in 1947. He served as a member of the Scientific Commission of the Institut Interuniversitaire de Physique Nucléaire (IIPN) in 1948, and was its president from 1950-51. After the reorganization of the IIPN into the Institut Interuniversitaire des Sciences Nucléaires in 1952, he was again a member of its Scientific Commission, and later (when there were three scientific commissions) chairman of the commissions on 'low energy' and 'high energy', and vice-president of the Council. Verhaege took part in the meeting of the Commission of Scientific Cooperation of the Centre Européen de la Culture in Geneva on 12 December 1950, represented his country at the UNESCO conference (at the first session in Paris with de Hemptinne, at the second with Willems) and attended every session of the provisional Council (with Willems).

Waller, Ivar (b. 1898), worked on lattice dynamics, X-ray diffraction, atomic and nuclear physics, and was nominated professor of theoretical physics at Uppsala University in 1934, and stayed in this position until his retirement in 1964. From 1947-65 he was a member of the Atomkommitten Stockholm, the Swedish Atomic Energy Committee. Waller took part in the two sessions of the UNESCO conference in Paris and Geneva (December 1951/February 1952) as a member of the Swedish delegation, and in the same capacity in all sessions of the interim Council. He was also a Swedish delegate in the Council until 1965, when he was replaced by Gösta Ekspong.

Wergeland, Harald Nicolai (b. 1912), was professor of theoretical physics at the Norwegian Institute of Technology, from 1946 until his retirement in 1979 and chairman of the Trondheim seminar of theoretical physics from 1952-63. He was nominated president of the Royal Norwegian Society of Science and Humanities in 1959 and remained in this position until the end of 1965.

Wergeland served for many years as a member of the Royal Norwegian Council for Scientific and Industrial Research. He was one of the proponents of Nordic collaboration in theoretical nuclear physics, and worked as the Norwegian representative in the board of the Nordic Institute for Theoretical Atomic Physics (NORDITA) from its foundation on 1 October 1957 until the end of 1974. He was its chairman from 1973-74. He took part as a member of the Norwegian delegation in the second session of the UNESCO conference on the 'establishment of a European nuclear research laboratory' in Geneva (12-15 February 1952) and at the 3rd and 9th sessions of the provisional Council; he served as an adviser at the 6th session. He was a Norwegian delegate to the CERN Council in 1956.

Wideröe, Rolf (b. 1902). Born in Oslo he studied electrical engineering in Germany. As a student in Karlsruhe he conceived independently of J. Slepian the principal ideas of the Betatron, and for his Hamburg Dr. Ing. degree he built the first linac in 1928, using a paper of G. Ising as his only basis. In 1943/44 he developed a small 15 MeV betatron, the first in Europe, at the C.H.F. Müller X-ray factory in Hamburg.

In August 1946 he started the construction of betatrons for medical and industrial applications with Brown-Boveri (BBC) in Baden (Switzerland). He was head of the BBC radiation laboratory until 1961. Wideröe worked for CERN as a part-time consultant for the PS group from its foundation, with special responsibility for 'accelerator design and power'. He accompanied O. Dahl and F. Goward on their famous trip to Brookhaven National Laboratory on 4-9 August 1952 when the idea of AG focusing was presented for the first time.

Willems, Jean (1895-1970), was secrétaire of the Free University of Brussels from 1919-28. In 1928 he became director of the Fonds National de la Recherche Scientifique en Belgique and of the Fondation Universitaire. When in 1947 the Institut Interuniversitaire de Physique Nucléaire (IIPN) was founded he was nominated president, and remained in this position when the IIPN was reorganized in 1952 into the Institut Interuniversitaire des Sciences Nucléaires. Willems held many other leading positions in science administration.

He was always Belgium's delegate in the provisional Council and the Council until 1969, served as chairman of the Finance Committee from 1954-57, as vice-president of the Council from 1958-60, and as its president from 1961-63.

APPENDIX 2

Chronology of events

John KRIGE and Dominique PESTRE

General*   **Proposals for international laboratories**

1945

March 1/September 5
Principle of phase stability for accelerators (Veksler, McMillan)

June
26 - Signature of United Nations (UNO) charter

July
5 - Labour Party wins the election in Britain

August
6 - Atomic bomb dropped on Hiroshima

October
18 - Statute creating the French Commissariat à l'Energie Atomique (CEA)
25 - UNO comes into being
29 - British decision to establish an Atomic Energy Research Establishment (AERE) announced

November
1 - First session of the conference organized by UNO dealing with international intellectual cooperation, which led to the creation of UNESCO

1946

January
20 - Resignation of General de Gaulle in France
24 - The UN Atomic Energy Commission (UN AEC) set up by the General Assembly of UNO

---

* In column **General,** the scientific events are italicized. When scientific results are given, the date is generally that of receipt of the paper by the journal which published it. Ulrike Mersits collaborated in drawing up this part of the chronology.

Chronology of events 569

General

Early 1946
Foundation of Brookhaven National Laboratory (BNL)

June
6 and 19 - American and Soviet proposals on the control of atomic affairs (UN AEC)

July
22/27 - *First post-war international conference on fundamental particles held in Europe (Cambridge)*

September
19 - Churchill's speech declaring that he is in favour of a 'United States of Europe' (Zurich)

October

November
First operation of the 184-inch synchro-cyclotron (380 MeV, α) at Berkeley

December
31 - *The mesotron is not the Yukawa meson (Conversi, Pancini, Piccioni)*

1947

May
24 - *First photographs of (π, μ) decay by Lattes, Muirhead, Occhialini, and Powell*

June
2/4 - *First 'Shelter Island' conference*
5 - Marshall proposes his European Recovery Program in Harvard
18 - *Measurement of Lamb shift by Lamb and Retherford*
27 - *First calculation of Lamb shift in non-relativistic QED*

Proposals for international laboratories

2/3 - UN Economic and Social Council (ECOSOC): French resolution on the establishment of UN research laboratories

General

August
Gleep, first UK experimental reactor, goes critical at Harwell

September
22 - Sixteen western European countries accept the Marshall plan

October
5 - Creation of the Cominform in Belgrade

December
20 - *First two photographs of V^0 particles by Rochester and Butler*

1948
January

March
8 - *Energies of the Bevatron (6 GeV) and Cosmotron (3 GeV) decided on*
9 - *Detection of artificially-produced π-mesons at Berkeley (Gardner, Lattes)*

March 30/April 2
Second 'Shelter Island' conference

April
16 - Marshall Plan: The Organization for European Economic Co-operation (OEEC) set up by 16 countries

May
7/10 - Congress for Europe, The Hague, which gathered several pro-European movements

June
23 - Start of the Berlin blockade

Proposals for international laboratories

23 - UN ECOSOC: report of the Secretary General on establishing UN research laboratories

General	Proposals for international laboratories
July *Bepo, second UK experimental reactor, goes critical at Harwell* August November 2 - Truman elected President of the United States December 11 - Preliminary talks in Washington about NATO 13 - *Nobel Prize: P.M.S. Blackett 'for his development of the Wilson cloud chamber method'* 15 - *Zoe, first French experimental reactor, goes critical at Fort de Châtillon (near Paris)* **1949** January 15 - *A tau-meson (later known as K_{π_3}) is identified in cosmic radiation (Brown et al.)* April 4 - Signature of NATO treaty in Washington 11/14 - *Third 'Shelter Island' conference* May 5 - Signature of the Statute creating the Council of Europe 23 - The fundamental law (*Grundgesetz*) of the Federal Republic of Germany enters into force July 28 - New policy of co-operation (Canada–UK–USA) in atomic affairs	 10 - UN ECOSOC: Discussion of the report presented on 23 January

General	Proposals for international laboratories

August

	4 - UN ECOSOC: Table of proposals regarding the establishment of UN research laboratories
24 - NATO treaty enters into force	16/24 - UN ECOSOC: Meeting of the committee of experts on international laboratories

September

15 - Konrad Adenauer elected first Federal Chancellor

23 - Truman announces the first Russian atomic test (26 August)

October

1 - Mao Tse-tung proclaims the People's Republic of China	Preparations for the European Cultural Conference (Lausanne, December). Dautry in charge of the scientific section

7 - Constitution of GDR enters into force; Otto Grotewohl Prime Minister

November

	Kowarski's *Note on European atomic co-operation*
11 - *Inauguration of the Dutch synchro-cyclotron in Amsterdam*	11 - Amsterdam meeting (B, DK, NL, S, CH) on European atomic co-operation

December

UK 110-inch synchro-cyclotron operates (175 MeV, p)

	8/12 - European Cultural Conference organized by the European Movement in Lausanne: proposal for an Institute of Nuclear Science in its application to daily life
12 - *Nobel Prize: H. Yukawa 'for his prediction of the existence of mesons'*	
22 - Truman declares that the USA would not be indifferent to aggression against Yugoslavia	

General

1950

January
31 - American decision that an effort be made to determine the technical feasibility of an H-bomb.

February
9 - McCarthy's first speech in Wheeling (West Virginia)
23 - Labour Party wins the election in Britain

March
6 - V^0 particles confirmed by 34 new events *(Seriff et al.)*
19 - Congress of the Peace Movement: Stockholm declaration

April
28 - Joliot dismissed from his position in the French CEA
28 - *Evidence for the artifical production of neutral π-meson with the electron-synchrotron at Berkeley (Steinberger, Panofsky, Steller)*

May
Publication of Berkner report in USA on *Science and Foreign Relations*
9 - Schuman proposes his plan based on a pooling of German and French resources in coal and steel. The proposal is open to other European countries

June

25 - Outbreak of the Korean War

European laboratories

Winter - Idea of European laboratories further discussed among Italian physicists

Spring - New versions of Kowarski's *Note on the creation of a co-operative organism for atomic research in western Europe*
Visit of de Rose and Perrin to Kramers to discuss this proposal. They return discouraged

19 - UN ECOSOC: report of the committee of experts on international laboratories

7 - UNESCO, 5th General Assembly, Rabi's resolution on regional (in fact western European) research centres. Auger entrusted with its implementation

General	**European laboratories**

July

7 - Petsche, French minister of finance, presents a note inspired by Kowarski's to the OEEC Council

13 - West Germany enters the Council of Europe as associate member

September

7/8 - IUPAP Executive Committee: Rabi's resolution referred to by Amaldi

7/13 - Oxford–Harwell Nuclear Physics Conference: Auger leads a debate on a possible European laboratory for nuclear physics

October

3 - Suggestion by Amaldi of a IUPAP-UNESCO collaboration

7/8 - Inauguration of the Centre Européen de la Culture (CEC)

14/19 - Decision by Auger (UNESCO) and Dautry (CEC) to hold a joint meeting on 12 December

26 - Pleven plan for a European Army (EDC)
27 - Superior Headquarters of Allied Powers in Europe (SHAPE) established
Windscale's No. 1 pile goes critical in Britain

November

13 onwards - Letters of invitation for the meeting of 12 December sent by the CEC

December
11 - *Nobel Prize: C.F. Powell 'for his development of the photographic method of studying nuclear processes'*

12 - Meeting at the CEC headquarters in Geneva: proposal to build a European laboratory centred around a giant particle accelerator; UK and Germany—for different reasons—not represented

16 - *First Rochester conference*

General

European laboratories

18 - Consiglio Nazionale delle Ricerche (CNR): Italy offers 2 million lire (14,000 SF) for the study office to be set up at UNESCO

18 - Italy: Colonnetti writes to de Gasperi about the European laboratory

19 - France: de Rose contacted by Auger for money

19/29 - Confusion in the press about the CEC meeting

1951
January

Early January - Close contacts between Auger, the French (de Rose, Dautry) and the Italians (Colonnetti, Amaldi)

8 - Letter de Rougemont (CEC) to Director-General (DG) of UNESCO to establish a common study office

20/27 - Refusal of UNESCO to establish the study office with the CEC

February

10 - Belgium: Verhaeghe informs Auger that the IIPN (Institut Interuniversitaire de Physique Nucléaire) supports UNESCO project

13 - Sweden: Waller shows great interest in the project

15 - The conference on the European Defence Community opens in Paris

21 - Circular letter by Auger to leading physicists in Europe asking for help and advice

23 - Belgium: Willems (IIPN) offers 50,000 BF (4,400 SF) for the study office

March

6 - A ministry of Foreign Affairs created in Bonn; Adenauer first Minister of Foreign Affairs

General	UNESCO-sponsored project
16 - West Germany member of European Council of Ministers	
	End of March/beginning of April: Leading European physicists agree to help Auger by sending experts for short periods
31 - *Two types of V^0 particles (Λ^0 and K^0) recognized*	
April	
8 - Josip Tito asks western countries for war material	
18 - Schuman plan on ECSC signed by B, F, FRG, I, L, NL	
	Mid-April: the British NPC (Nuclear Physics Committee) resolves that Britain could help but should not participate directly in the UNESCO project
	25 April/3 May: Amaldi in Paris to organize the work with Auger. It is impossible to set up the study office
May	
2 - West Germany full member of the Council of Europe	2 - Circular letter by Auger convening a meeting of European consultants to draw up a precise project
12 - Experiment on H-bomb by Americans on Eniwetok atoll	
	18 - French official decision to support Auger's action. Two million FF (25,000 SF) are offered
	23/25 - First consultants meeting: proposal to build a 3-6 GeV proton synchrotron (PS); estimated cost of the lab: $20-25M, (85-107MSF); strong commitment of Kowarski (F), Preiswerk (CH), Verhaeghe (B)
June 18/July 11 UNESCO 6th General Assembly: Japan and West Germany become member states	UNESCO General Assembly: Decision to continue the studies for a European laboratory for nuclear research; decision to set up an International Computation Centre (UNESCO - UN ECOSOC project)

Chronology of events 577

| **General** | **UNESCO-sponsored project** |

28/30 - Beginning of Regenstreif's survey of high-energy facilities in Europe: trip to Switzerland. Scherrer described as rather cold

July

4/7 - Regenstreif's trip to Belgium: general approval of UNESCO's project; British hostility debated at length

5 - Auger asks Weisskopf to convince Scherrer (and Bohr) of the soundness of the project

11/13 - 7th General Assembly of IUPAP in Copenhagen: first opposition (Kramers, Bohr); consensus of a kind apparently reached: a synchro-cyclotron will also be built and construction of the two machines will be staggered in time

16/18(?) - Regenstreif's trip to Norway

Dutch–Norwegian pile goes critical at Kjeller (N)

August

19 - Circular letter by Auger: he proposes to set up a technical committee (Bakker or Heyn, Dahl, Perrin, Preiswerk) to prepare the reports needed for an intergovernmental conference

23 - First letter from Kramers: he suggests Bohr's Institute to serve as 'pilot institute'

28 - F, UK, and USA grant $50M in economic aid to Yugoslavia

31 - UNESCO: official letters sent to the member states convening an intergovernmental conference in December 1951 on the organization of a European laboratory for nuclear research

31 - Second letter from Kramers: he mentions serious misgivings in Dutch official circles

September

First week - Regenstreif's trip to the Netherlands

Second week - Regenstreif's trip to Britain

General	UNESCO-sponsored project
14 - F, UK, and USA ministers of Foreign Affairs decide to admit West Germany to the European army	14 - Bohr asks Chadwick what the British physicists' attitudes might be about Kramers' proposal
15 - NATO decides to admit Greece and Turkey	
17/20 - *Chicago colloquium in honour of Fermi's 50th birthday*	Chicago: first contact of Auger with a German physicist (Gentner) about the project
	24 - Third letter from Kramers: project too expensive; working document to be redrafted; Bohr's Institute to be the core of the European laboratory
October	
	4/5 - Contacts of Auger with several people in Britain (Blackett, King, Waterfield)
	11 - Letter Auger to Kramers suggesting a 6 to 10 GeV PS
	12 - Letter Malte Jacobsson (S) to UNESCO DG: S sustains Kramers' proposal
	24 - Letter Dutch Minister of Education to UNESCO DG: NL sustains Kramers' proposal
25 - Conservatives win the election in Britain	
	26 - Letter Bohr to Auger: he proposes that the whole project be rethought
	26/27 - Second meeting of Auger's consultants: Bohr and Kramers' suggestions ignored; a 'bevatron' is reaffirmed as first priority; Dahl (N) and Bakker (NL) definitely in favour of UNESCO's scheme
31 - NL ratifies the Schuman plan	
November	
14 - Signature of a military treaty between the USA and Yugoslavia	
	15 - Gentner proposes that Auger contact Walter Bothe in Germany
	17/18 - Third meeting of consultants: the decisions of October are confirmed
	20 - Yugoslavia informs the UNESCO DG that she will take part in the intergovernmental conference

General

December

11 - *Nobel Prize: Cockcroft and Walton 'for their pioneering work on the transmutation of atomic nuclei by artificially accelerated atomic particles'*
13 - F: the National Assembly ratifies the Schuman plan

1952
January

11 - West Germany: the Bundestag ratifies the Schuman plan
11/12 - *Second Rochester conference; Fermi postulates the existence of a resonance level $I = 3/2$ in the 100–200 MeV region—the first resonance in physics*

UNESCO-sponsored project

28/29 - Norway: inauguration of the Dutch–Norwegian reactor: tension peaks between the UNESCO consultants group and Bohr and Kramers

4/6 - Mussard calls for a last-minute meeting of consultants on the eve of the intergovernmental conference
5/8 - Regenstreif's trip to Sweden

14 - Fourth meeting of consultants
17/21 - Intergovernmental conference called by UNESCO: partially successful thanks to the mediation of the Dutch delegation; the conference will resume in February; the British offer the use of the Liverpool synchro-cyclotron

3/4 - First meeting of the working group on the Draft Agreement establishing a Council of Representatives of States
5/12 - Trip Bohr/Scherrer to UK to secure the use of the Liverpool machine for the European laboratory

14/16 - Second meeting of the working group

General	UNESCO-sponsored project
	18/19 - Meeting of consultants (Amaldi, Bakker, Dahl, Goward, Kowarski, Preiswerk, Verhaeghe, Auger) 24 - UNESCO: the documents prepared by the various groups are officially sent by the DG to the member states
February	
	5 - Danish amendments: for coordination and promotion of existing activities rather than for the immediate creation of a new centre 11(?) - Swedish amendments in line with those of the Danes 11 - Italian amendments to counter the Danish and Swedish ones 11 - Official notification that Britain will not take part in the Council to be set up 12/15 - 2nd session of the Conference: a Council is set up; nine states (including Denmark, Greece, and Sweden) sign the Agreement. The Federal Republic of Germany, the Netherlands and Yugoslavia sign without reservation
March 11 - *First evidence for a charged hyperon* Ξ^- 27 - *Cockcroft at Brookhaven 'when the protons were going round the first 200 times' in the Cosmotron*	
April	
	2 - Belgium signs the Agreement 11 - France ratifies the Agreement
22 - First tactical atomic bomb exploded in Nevada	
May	
	2 - Sweden ratifies the Agreement 5 - Norway signs the Agreement

General

26 - Adenauer, Schuman, Eden, and Acheson in Bonn: full sovereignty given to West Germany
27 - Treaty creating the European Defence Community signed in Paris (B, F, FRG, I, L, NL)

June
Cosmotron starts working at 2 GeV
3/17 - International physics conference at Bohr's Institute in Copenhagen, aiming 'to provide useful information for the planning work of the Council'
12 - Publication of bubble-chamber principle (Glaser)

July
Brookhaven scientists conceive what will be known as the alternating-gradient or strong-focusing principle

August

October
3 - First British atomic weapon test at Monte Bello

CERN

5/8 - First session of (provisional) Council: heads of the four study groups nominated: Dahl, Bakker, Kowarski, Bohr; Amaldi, Secretary-General

20/21 - Second session of Council in Copenhagen: recommended to study a 600 MeV SC and a 10–20 GeV PS which 'could be constructed along the lines of the Brookhaven cosmotron'; Council agrees to 'sponsor co-operation in the field of cosmic rays'; name 'CERN' officially agreed

4/10 - Courant, Livingston and others discuss the strong-focusing principle with Dahl, Goward, and Wideröe who are visiting Brookhaven

General

November
1 - American H-bomb test at Eniwetok
4 - Eisenhower elected President of the United States
19 - Spain admitted to UNESCO

December
2 - *Deleterious effect of magnet inhomogeneities on the performance of a strong-focusing synchrotron demonstrated (Lawson)*
11 - *Nobel Prize: F. Bloch and E. Purcell 'for their development of new methods for nuclear magnetic precision measurements'*
18/20 - *Third Rochester conference*

1953
January

28 - *First photographs of 'superproton' (Σ^+) in cosmic radiation (Bonetti et al.)*

February

25 - De Gaulle rejects EDC

March
5 - Death of Stalin

CERN

4/7 - Third Council session: Geneva decided as site for the laboratory; first discussion of CERN draft convention; agreed to design a strong-focusing PS 'of about 30 GeV'

25 - First British donation of £5,000 paid to provisional CERN

12/14 - Fourth Council session: British government officially represented at CERN; Swedish initiative to establish independent theory group at Copenhagen strongly opposed

17 - Scandinavian delegates meet with representatives of Iceland and Finland to discuss setting up an independent theoretical institute (which became NORDITA)

March 30/April 2 - Fifth Council session: further discussion of the convention

Chronology of events 583

General **CERN**

April
29 - *Production of V^0 particles (K-mesons) by the Cosmotron (Fowler et al.)*

April 30/May 1 - First meeting of the Administrative and Financial Working Group to prepare for the transition to the permanent CERN

May

5 - Publication of the Second Report to Member States providing 'all essential information' on the organizational and financial implications of CERN membership
27/28 - Referendum in Geneva: two-thirds of those voting accept to have CERN in the Canton

June

29/30 - Sixth Council session: committee to nominate the principal officers of CERN set up; Supplementary Agreement prolonging the Agreement constituting the Council signed by all members bar F and N, who hope to do so shortly

July

1 - Signature, subject to ratification, of the Convention for the establishment of a European Organization for Nuclear Research by nine states (B, F, FRG, G, I, NL, S, UK, Y)
1/2 - First meeting of Interim Finance Committee

6/12 - *International conference on cosmic rays in Bagnères-de-Bigorre (F)*
16 - *Protons accelerated to 1 GeV in Birmingham synchrotron*

17 - Switzerland signs the Convention

27 - Armistice ends Korean war

31 - Contract signed with the architect Steiger

General	CERN

General

August
12 - Explosion of first Soviet H-bomb
August 21/November 16
First steps towards the concept of strangeness (Gell-Mann, Nakano, and Nishijima)

September
8 - USSR recognizes Austrian republic

October

26/28 - *Conference on the alternating-gradient proton synchrotron at the University of Geneva*

November
10 - *Associated production of strange particles shown experimentally; first photographs of what will be called Σ^-, both with the Cosmotron (Fowler et al.)*

December

1954
January
10 - *US AEC announces that a 25 GeV alternating-gradient PS will be built at Brookhaven*

CERN

5 - First members of PS group arrive at Geneva and move into premises at the university's Institute of Physics

29–31 - Seventh Council session: design energy of the PS reduced to 25 GeV; appointment of the first three members of an administrative nucleus approved; decision on the nomination of the Director-General postponed to the next session

23 - Denmark signs the Convention
30 - UK deposits the instruments of ratification of the Convention
31 - Norway signs the Convention

Chronology of events

General	**CERN**
	14 - Eighth Council session: further 884,300 SF appropriated for February, March, and April; Nominations Committee reports that it was finding it 'difficult to propose a name' for the first Director-General
25/27 - *Fourth Rochester conference*	
February	
	12 - Switzerland deposits the instruments of ratification of the Convention
April	
1 - *Protons accelerated to 6.1 GeV in Berkeley Bevatron*	
	5 - Denmark deposits the instruments of ratification of the Convention
	8/9 - Ninth Council session: Bloch officially proposed as first Director-General
21 - USSR announces that she will join UNESCO	
May	
7 - Fall of Diên Biên Phu	
	17 - Work started on the site at Meyrin
June	
15 - Formal establishment of a European Atomic Energy Society 'to promote co-operation in nuclear energy research and engineering'; representatives from B, F, I, NL, N, S, CH, UK	15 - The Netherlands deposits the instruments of ratification of the Convention
	23/24 - Eighth meeting of the Interim Finance Committee: invitations for tender for the SC magnet frame and coils ready; procedure for a major increase in expenditure after July agreed on
July	
	7 - Greece deposits the instruments of ratification of the Convention
	15 - Sweden deposits the instruments of ratification of the Convention
	19 - Belgium deposits the instruments of ratification of the Convention

| **General** | **CERN** |

August
9 - Greece, Turkey, and Yugoslavia sign a mutual defence agreement

September

29 - France and Germany deposit the instruments of ratification of the Convention

October
3 - The group of nine recommend that Germany be admitted to NATO; FRG agrees not to manufacture atomic, chemical, or biological weapons

4 - Norway deposits the instruments of ratification of the Convention

7/8 - First session of the Council of the European Organization for Nuclear Research

Name index *

Amaldi, Edoardo   8, 70, 155, 168-170, 172, 174, 192, 194, 201, 384, 529
    and Auger, 1950   85, 99, 101, 104, 362, 387, 574-576
    and Bloch   225, 266-272, 291
    and CERN assets   226
    and cosmic-ray research   218, 219
    and his difficulties at the intergovernmental conference   181
    and his meeting with Cherwell, December 1952   496, 498, 499
    and his visit to Paris, April 1951   125, 126, 128, 129
    and Italy   354-371, 375
    and IUPAP, 1950   85, 98, 104, 105, 362
    and IUPAP, 1951   138, 139, 148
    and Rabi   83, 86, 87, 98, 362
    and the CEC meeting in Geneva   107, 109, 111, 117, 362, 363
    and the consultants meetings   130, 135-137, 158-160, 163, 188, 580
    as candidate for D-G   262, 263, 266-268, 291
    as Secretary-General   161, 166, 211, 263, 271, 371, 411, 581

Auger, Pierre   308, 316, 338, 343, 529, 577
    and Dautry and the CEC   101-109, 126, 127, 574, 575
    and his candidacy for D-G   262, 291
    and the acronym CERN   212
    and the Bohr-Kramers scheme   149, 150, 153-157, 162, 163, 166, 167, 172-174
    and the British   106, 107, 156, 157, 438, 444, 450, 451, 470, 578
    and the CEC meeting in Geneva   109-112, 113-117, 364-367
    and the consultation with European physicists, 1950   85, 86, 99-101, 387, 388, 392, 398, 433
    and the European Movement   77
    and the French project   69-79, 320, 321
    and the Germans   106, 158, 159, 387, 578
    and the group of consultants   127-130, 135-137, 158-160, 188, 575, 576, 578
    and the idea of a giant bevatron   107, 108, 112
    and the intergovernmental conference   180, 181, 184, 186, 190, 192, 193, 199, 577
    and the IUPAP conference in Copenhagen   138-140, 148
    and the Rabi resolution   83, 85, 87, 89, 90
    and the UNESCO bureaucracy   202
    and the UNESCO study office   111, 112, 125-127, 129, 130, 158
    and UNESCO   65-67, 99, 114, 126, 127, 134, 135, 140, 141

---

* Only names in the text are included, not those in the notes, bibliographies, and appendices. The running list was drawn up by Armin Hermann.

Bakker, Cornelis J.   22, 105, 168, 169, 181
   and the consultants meetings   127–130, 158–160, 188, 577, 578, 580
   and the development of the SC   216, 217
   and the Dutch physicists   139, 140, 159, 169, 172, 173
   and the Nobel prizewinners   273
   as candidate for D-G   226, 262–264, 268–270, 272
   as head of the SC group   161, 165, 211, 212, 226, 264, 276, 581

Bohr, Niels   9, 105, 106, 214, 401
   against a reactor, 1950   88, 364, 367
   and Auger's critiques   154–156, 166, 167
   and Bloch   264, 265, 273, 291
   and Chadwick and the British   148–150, 244, 440–444, 454–456, 578
   and Fraser and ICSU   149, 150
   and his alternative programme   152, 153, 156, 157, 169–171, 578
   and his anxieties at the IUPAP conference in Copenhagen   137, 139, 148
   and his ongoing opposition, 1952   243–245
   and Kramers' idea   150–152, 577
   and Nordita   246, 286
   and the Committee for other Forms of Co-operation   218, 235
   and the consultants meetings   158, 159, 162, 164, 579
   and the intergovernmental conference   181, 184, 185, 189, 190, 192, 193, 196, 198, 199
   and the Oslo meeting   163, 167–169, 187, 579
   and the visit of Kramers and Chadwick, summer 1951   148, 149
   and the visit to Liverpool university   187, 454–456, 462, 579
   as head of the Theory Group   165, 211, 217, 218, 414–416, 581

Chadwick, James   188, 476, 504, 505, 530
   accepts the need for a big accelerator   493, 500
   and the Buckland House meeting   477, 479
   and the Cabinet Steering Committee   465, 484
   and the CEC resolution, 1950   107, 434
   and the Liverpool SC   445–449, 454–457
   and the lobbying for membership   465, 510
   and the recommendation to join, January 1952   455, 466, 467
   and the support for Bohr and Kramers   148–150, 152, 156, 169–171, 192, 193, 440–449, 454–457, 467, 483
   and the visit to Bohr, summer 1951   148, 149, 440

Dahl, Odd   155, 168
   and Goward   213, 279
   and his manner of working   279
   and the AG principle   213, 216, 217, 274, 581
   and the arrival in Geneva   223
   and the consultants meetings   127, 128, 130, 135, 158–160, 188, 577, 578, 580
   and the difficulties with the Council   418
   and the Norwegian politicians   172, 173, 181
   as head of the PS Group   161, 165, 211, 212, 223, 269, 279, 417, 418, 581

Kowarski, Lew   102, 167, 194, 308, 310, 317, 338, 419
   and British suspicion   283, 284, 293

and German suspicion 417
and his interpretation of CERN history 526–528, 536, 537, 539
and the Bohr–Kramers opposition 156, 168–171
and the consultants meetings 130, 158–160, 188, 323, 580
and the criticisms of the French Communists 336, 338
and the French project 66–68, 69–72, 74, 79–81, 89, 115, 318, 320, 572, 573
and the Rabi resolution 86, 87, 99, 100
and the report on the site 239, 329, 330
as head of the Laboratory Group 161, 165, 211, 417, 581

Waterfield, Anthony H., *reports on:*
Auger's feelings in October 1951 115, 171, 172
the CEC meeting in Geneva 113
the February 1952 session of the conference 195, 196
the Working Group for the February 1952 session of the conference 192
the fourth Council session, January 1953 218, 235

Abragam, Anatole 270
Adams, Sir John 216, 223, 227, 269–271, 278–281, 417, 418, 497
Adenauer, Konrad 374, 386, 387, 394, 395, 402, 405, 410
Alfvén, Hannes 130, 132, 155, 156, 158, 159, 169
Alvarez, Luis Walter 17
Amaldi, Ginestra 360
Anderson, Carl David 29
Anderson, Herbert 8, 39
Armenteros, Rafael 29
Awbery, James Henry 434, 505
Balke, Siegfried 420
Bannier, Jan Hendrik 76, 77, 173, 181–183, 186, 194, 199, 222, 241, 269, 451, 531, 532
Bergeron, Paul Jean 321
Bernal, John Dew Desmond 361
Bernardini, Gilberto 6, 68, 98, 356, 362, 443
Bethe, Hans Albrecht 14, 33, 35
Beyerle, Konrad 417
Biasi, Vittorio de 357
Bishop, George Robert 137
Bismarck, Philipp von 411
Blackett, Patrick M.S. 6, 29, 107, 114, 155, 239, 240, 262, 434, 435, 444, 477, 486, 489, 491–493, 505, 507
Blewett, John P. 217, 223, 279, 280, 281
Blewett, M. Hildred 217, 223, 279, 281
Bloch, Felix 225, 226, 264–271, 273, 416, 419, 524
Blount, Bertie Kennedy 496
Bohm, David 33

Bohr, Aage 36
Bolla, Giuseppe 355
Bonaudi, France 482
Bothe, Walter 211, 387, 392, 398
Born, Max 9, 10
Bradner, Hugh 281
Bragg, Sir Lawrence 436, 505
Breit, Gregory 33
Brenton, D.J.M. 504
Bretscher, Egon 505
Brobeck, William Merrison 16, 17, 137
Broda, E. 8
Broglie, Lous-Victor Prince de 72–75, 337, 338, 360
Broglie, Maurice Duc de 317
Brueckner, Keith Allan 39
Brunt, Sir David 496, 505
Buckingham, Richard Arthur 10
Bujard, Eugène 164
Butler, Clifford Charles 28, 43
Butler, K.S. 464, 504, 509
Butler, Richard A. 489, 490, 496, 499–501
Cacciapuoti, Nestore Bernardo 99, 125, 354, 366
Capron, Paul 109, 113, 126, 130, 160, 366
Cardan, Jérome 115
Caron, Onorevole 374
Casati, Alessandro 104, 181, 233, 362, 363, 366, 369–371, 540
Casimir, Hendrik 162, 217, 245, 262, 265, 269
Cassels, Jim 245, 248, 278, 478, 480–483, 485, 486, 493, 494

Castelnuovo, Guido 363, 366
Cherwell, Lord (=Frederick Lindemann) 215, 241, 465, 477, 488, 489, 495–498, 500, 526, 542
Christophilos, Nicolas 23
Churchill, Sir Winston 215, 437, 500, 501
Citron, Joachim Jörk Anselm 417
Clementi, Alberto de 181
Cockcroft, Sir John D. 19, 76, 77, 99, 105, 114, 127, 128, 130, 139, 158, 160, 161, 167, 170, 172, 219, 225, 234, 245, 248, 249, 261, 262, 267, 268, 272, 273, 278–280, 284, 434, 435, 438, 444, 455, 458, 463–467, 476–480, 482, 484, 486, 488, 489, 492, 493, 496, 497, 500, 501, 505, 507, 508, 510, 511, 535
Cogniot, Georges 337, 340, 341
Collie, Carl Howard 505
Collins, Malcelm Frank 136, 137
Colonnetti, Gustavo 68, 72, 75, 77, 78–83, 101, 103, 105, 109, 111, 113, 125, 135, 158, 159, 172–174, 181, 186, 199, 200, 222, 225, 233, 261, 355, 357, 360, 362–373, 376, 524, 527, 530–532, 535, 540
Conversi, Marcello 25, 34, 37, 43
Cosyns, Max 105
Courant, Ernest David 23, 217
Crivon, R. 114
Dadelsen, Jean Paul de 126, 363, 364, 385, 433
Dakin, Samuel Arthur French 228, 272
Darlington, Cyril 72, 104
Darrow, Karl Kelchner 33
Darwin, Sir Charles Galton 101, 149, 150, 170, 440–442, 505
Dautry, Raoul 68, 69, 73, 75–79, 86, 97, 100–105, 107–109, 111, 113, 115–118, 125, 172–174, 199, 200, 304, 307–309, 317–321, 326, 327, 330, 342, 343, 360, 363, 388, 524, 527, 530, 531, 534
De Gasperi, Alcide 103, 200, 367, 368, 371, 531, 540
De Witt, Cécile (née Morette) 113
Dedijer, Stefan 532
Dee, Philip Ivor 172, 505
D'Espagnat, Bernard Georges 270
Dirac, Paul Adrien Maurice 9, 35–37
Donzelot, Pierre 321
Dudley, Sir Alan Alves 251, 459, 464, 465, 467, 504
Dunn, Leslie Clarence 13
Dupouy, Gaston 100, 321, 326, 327, 343
Dyson, Freeman John 38

Eickemeier, Helmut 396
Einstein, Albert 69
Eisenhower, Dwight D. 538
Evans, Ivor G. 505
Extermann, Richard C. 223
Faissner, Helmut 420
Fano, Ugo 98
Feather, Norman 8, 434, 505
Fermi, Enrico 8, 10, 33, 34, 37, 39, 158, 354, 356, 392, 526
Ferretti, Bruno 6, 86, 88, 98, 107–111, 113, 117, 125, 129, 161, 356, 357, 362–367, 389
Feschbach, Herman 33
Feynman, Richard Philips 33, 36–38
Fierz, Markus 161
Filthut, Heinz 420
Fraser, Ronald 149, 150, 152, 155, 394, 441
Frisch, Otto Robert 505
Fry, Donald William 128, 161, 234, 239, 480–483, 486, 487, 491–495, 497, 507
Funke, Gösta Werner 164, 243
Gaillard, Felix 323, 327, 329–331, 343
Gardner, Eugene 26
Gaulle, Charles de 307
Geibel, Kurt 417
Geiger, Karl Martin 420
Géheniau, Jules 161
Gentner, Wolfgang 159, 387, 392, 398, 399, 411, 417–420
George, André 72, 76, 78, 104
Gerlach, Walther 396
Gierke, Gerhart Otto Julius von 417
Gilbert, Etienne 115
Gilpin, William Cecil 528
Giordani, Francesco 366, 370–372
Girardet, André 233
Goldschmidt, Bertrand L. 100, 308, 318, 320
Goldschmidt-Clermont, Yves 283, 284
Goudsmit, Samuel Abraham 98, 386
Goward, Frank Kenneth 130, 158–162, 188, 213, 223, 224, 268, 270, 274, 278–280, 417, 418, 438, 466, 467, 476, 478, 480, 486, 491, 493, 497
Grivet, Pierre 129
Groeneveld Meijer, N.E. 224
Groot, Sybren Ruurds de 161, 233
Groves, Leslie R. 11
Guéron, Jules 69, 100, 308, 318, 320
Guggenheim, Edward Armand 334, 342
Gund, Albrecht 417

Gustafson, Torsten   163, 164, 233
Haag, Fred G.   417
Hahn, Otto   11, 159, 392
Halford, Aubrey Seymour   114
Hallstein, Walter   392, 398, 401, 404–406, 408, 410
Hansen, Hans Marius   181, 232
Haxel, Otto   387, 396, 407
Heintze, Joachim   420
Heisenberg, Werner   10, 105, 106, 181–183, 194, 212, 219, 222, 225, 233, 261, 262, 264, 273, 384–387, 392 –410, 412–416, 418–420, 451, 532, 536, 540, 542
Heitler, Walter   6, 10, 36
Hemptinne, Marc Comte de   104, 105, 109, 186
Hereward, Hugh   479
Heyn, Frans Adriaan   22, 130–134, 158–161
Hine, Mervyn G.   223, 278, 279, 486
Hinshelwood, Sir Cyril   455, 463, 488, 504
Hocker, Alexander   181, 232, 233, 259, 392–394, 396, 398, 399, 404, 406–409, 411, 412, 416, 418–420, 531
Holtsmark, Johan Peter   233
Hondros, Demetrios   183
Huber, Otto   128, 136
Huxley, Julian   65, 83
Hylleraas, Egil A.   181, 233
Jacobsen, Jacob Christian   163, 164, 181, 217, 232, 440, 490
Jacobsson, Malte Ferdinand   150, 164, 182, 185, 190, 193, 194, 243
Janossy, Lajos   6
Johnsen, Kjell   223
Joliot-Curie, Frédéric   102, 105, 106, 115, 126, 307–309, 316, 317, 327, 328, 337–339, 361, 535
Joliot-Curie, Irène   308, 316, 336–339, 390, 391
Joxe, Louis   321, 327, 343
Jowitt, Earl William Allen   503
Keegan, G. John   540
Keim, Walter   389
King, Alexander   76, 77, 154, 155, 244, 248, 433, 434, 438, 441, 443, 444, 447, 460, 464, 465, 468, 482, 487 –490, 492, 496, 497, 506, 510
Klein, Oskar   37
Koch, Agathon John Hjalmar   161
Kouyoumzelis, Theodore George   233
Kramers, Hendrik Anthony   33, 34, 68, 76, 77, 81, 82, 101, 105, 109, 111, 117, 127–128, 130, 133, 138, 139, 148–159, 162–172, 174, 180, 181, 183, 192, 193, 238, 319, 401, 439–447, 449–451, 510, 527, 531

Krienen, F.   482
Krige, John   137, 188, 325, 326
Kruyt, M.   77, 83
Lamb, Willis Eugene   33, 35
Landahl, Heinrich   394
Lattes, Cesare Mansueto Giulio   26
Laue, Max von   72, 104, 109, 385–388
Laugier, Henri   65, 173
Lawrence, Ernest Orlando   16, 26, 106, 457, 541
Lawson, John David   497, 500
Leprince-Ringuet, Louis   6, 28, 32, 117, 125, 316, 317, 327, 338
Lescop, René   320
Lhéritier, Philippe   28
Levy, Paul   77
Livesey, D.L.   8
Livingston, M. Stanley   23
Lockspeiser, Sir Ben   76, 77, 215, 221, 222, 226, 228, 234, 245, 248, 256–258, 268, 270, 272, 282, 284, 435, 447, 455, 463, 465, 482, 488, 489, 496, 498, 500–502, 505, 527, 531, 532, 540
Lombardo, Ivan Matteo   83, 104, 363
Longchambon, Henri   341
Lucet, Charles Ernest   329
Lüders, Gerhart Claus Friedrich   417
Lussu, Emilio   374
MacInnes, Duncan Arthur   33
Maheu, René   201
Makins, Sir Roger   464–466, 510
Manneback, Charles   72, 73, 104
Marrou, Henri Irénée   528
Marshak, Robert Eugene   33, 35, 38, 40, 217
Marshall, Leona   8
Martin, Sir David Christie   504, 505
Massey, Sir Harrie Stewart Wilson   10, 484
Maud, John   83
McMillan, Edwin M.   14, 17
Mendès-France, Pierre   315, 341
Meyer-Berkhout, Ulrich   420
Migone, Bartolomeo   371
Moch, Jules   341
Møller, Christian   10, 161, 217, 228, 416
Monnet, Jean   405
Montel, Paul   72
Morette, Cécile   117
Mott, Sir Nevill Francis   149, 505
Mountford, Sir James Frederick   446, 448, 449, 454, 456
Muralt, Alexander Ludwig von   83, 89
Murray, Raymond L.   460, 464, 466, 468, 479

Mussard, Jean A. 131, 134–136, 140, 155, 159, 160, 163, 168, 169, 180, 187, 201, 221, 226, 248, 271
Naegelen, Maurice 341
Nakicenovic, Slobodan 234
Needham, Joseph 65
Neumann, John von 33
Nielsen, Jakob 149, 150, 181–183, 186, 188, 194, 232, 243, 532
Nordsieck, Arnold Theodore 33
Occhialini, Guiseppe 8
O'Ceallaigh, Cormac 30
Oliphant, Sir Mark 20, 21, 172, 362
Oppenheimer, J. Robert 11, 33, 34, 36–38, 67
Pais, Abraham 33, 39, 40
Pancini, Ettore 25, 34, 37, 43
Panofsky, Wolfgang 27
Papayannis, J. 233
Parodi, Alexandre 81, 329–331, 343
Paul, Wolfgang 417
Pauli, Wolfgang 9, 33, 39, 217
Pauling, Linus Carl 33
Peierls, Sir Rudolf 434, 478, 483, 485, 486, 505
Pennetta, Antonio 233, 268, 373, 532
Penney, Richard W. 224, 230
Perez, Joseph Jean Camille 76, 77
Perkins, Donald Hill 477
Perrin, Francis 67, 79, 81, 100, 102, 105, 106, 125, 126, 130, 138, 158, 160, 161, 170, 201, 222, 225, 232, 261, 262, 264, 265, 268, 304, 308, 317–319, 321, 323–326, 329–331, 335, 337, 339, 343, 389, 531, 532, 535
Persico, Enrico 356
Pestre, Dominique 240, 505
Peterlin, Anton 234
Piccioni, Attilio 373
Piccioni, Oreste 25, 34, 35, 37, 43
Pickavance, Thomas Gerald 161, 278, 280, 478, 484–486, 490–494
Picot, Albert 186, 194, 233, 532
Placzek, George 98
Pleven, René 306, 536, 538
Pontecorvo, Bruno 37, 98
Powell, Cecil Frank 6, 8, 25, 27, 35, 36, 43, 217, 219, 434, 477, 484, 505
Preiswerk, Peter 109, 113, 127, 128, 130, 135, 136, 137, 139, 140, 155, 158–161, 168, 170, 172, 194, 227, 283, 524, 532
Proca, Alexander 10, 161
Puppi, Giampietro 37

Rabi, Isidor Isaac 15, 16, 33, 68, 82–89, 98, 99, 103, 104, 137, 139, 140, 156, 169, 210, 242, 319, 322, 323, 336, 362, 374, 387, 524, 526, 532, 537–539
Rackham, Harold C. 192, 446, 447, 457, 461–464, 467, 504
Raiser, Ludwig 396, 397, 403
Ramsey, Norman Foster 15
Randers, Gunnar 104, 105, 113, 127, 130, 157, 158, 169, 170, 172–174, 527
Rau, Friedrich 419
Rebsamen, August 401
Regenstreif, Edouard 129, 130, 135–139, 153, 154, 159, 160, 223
Retherford, Robert C. 35
Rieux, Jean-Pierre 342
Roberts, Conwy 464
Rochester, George Dixon 28, 43, 477
Rollier, Mario A. 83, 104, 109, 362, 363, 366–368, 389
Rose, François Comte de Tricornot de 67, 79, 81, 82, 86, 113, 125, 183, 186, 199, 232, 304, 318–321, 323–327, 329–331, 339, 342, 343, 527, 531, 534, 535
Rose, M.E. 14
Rosenblum, Salomon 316
Rosenfeld, Léon 9, 10
Rossi, Bruno 31, 33, 98, 354
Rotblat, Joseph 505
Rougemont, Denis de 69, 73, 74, 76–78, 101, 103, 109, 126, 363, 366, 385, 531
Rozental, Stefan 138, 217, 439
Rumpf, Paul Gunther 401
Salat, Rudolf 389, 401, 403, 409
Salomon, Jean-Jacques 528
Sandys, Duncan 72
Savic, Pavle 234
Scelba, Mario 374
Schein, Marcel 32
Scheinmann, Lawrence 309
Scherrer, Paul 76, 77, 104, 105, 109, 127, 128, 130, 136, 137, 139, 158, 186–188, 194, 216, 225, 233, 244, 261, 262, 273, 452, 454, 456, 461, 462, 532
Schilling, Warner R. 534
Schmelzer, Christoph 417, 418
Schmid, Carlo 385, 389, 390
Schrödinger, Erwin 540
Schuman, Robert 173, 202, 319, 343, 391, 405, 437, 526, 536

Schwinger, Julian Seymour   33, 35, 36, 38
Segré, Emilio   354
Serber, Robert   26, 33
Sereni, Emilio   373, 374
Seriff, A.J.   29
Severi, Francesco Buonaccorso   83
Shire, E.T.S.   505
Siegbahn, Karl Manne   104, 105, 109, 161
Silva, Raymond   101, 362, 363, 389
Silvestri, Mario   355
Simon, Sir Francis Eugen   76, 505
Skinner, Herbert W.B.   114, 172, 448, 449, 454–456, 466, 476–478, 482, 483, 486, 487, 492–494, 505, 507
Snyder, Hartland S.   23
Soustelle, Jacques   341
Spaak, Paul-Henri   72
Spranger, Eduard   396
Staub, Hans   136, 161
Steiger, Rudolf Bernhard   224, 228, 258
Steinberger, Jack   27
Steller, J.   27
Strassmann, Fritz   11
Straub, Ferenc Bruno   419
Strauss, Franz Josef   411
Stueckelberg de Breidenbach, Baron Ernest C.G.   10
Suhrkamp, Peter   385
Supek, Ivan   234
Surdin, Maurice   161
Svedberg, Theodor   127
Teller, Edward   33, 34
Telschow, Ernst   397
Teusch, Christine   385, 409
Thirring, Hans   158
Thomson, Sir George Paget   83, 85, 149, 150, 169–171, 181, 182, 185, 187, 188, 191, 434, 436, 441, 442, 445 –447, 450–453, 455, 456, 458, 461–463, 465, 467, 476–478, 480, 482–484, 486, 487, 493, 500, 505, 507, 508, 510, 511, 527, 536
Tito, Josip   538

Tizard, Sir Henry   99
Tomonaga, Sin-Itiro   35, 38
Torrès-Bodet, James   149, 150, 392
Trueman, A.   505
Truman, Harry S.   71, 82, 524
Trumpy, Bjørn   233
Uhlenbeck, George Eugene   33
Valeur, Robert   224, 226, 232, 259, 263–271, 273, 329, 331, 342, 343, 412, 417, 418, 531, 532
Vassiliou, N. Hadji   233
Veksler, Vladimir   14
Verhaeghe, Julien Léon   109, 113, 126, 136, 139, 155, 158–160, 194, 232, 532
Verry, H.L.   226, 251, 265, 460, 496, 497, 506, 509
Vig, Norman J.   501
Vleck, John Hasbrouck van   33
Voltaire (François Marie Arouet)   391
Walen, R.   234
Waller, Ivar   104, 105, 109, 113, 127, 128, 130, 158, 222, 233, 532
Weisskopf, Victor F.   33, 34, 98, 137, 139, 245, 316, 420, 421
Weber, Gustav   420
Weizsäcker, Carl Friedrich Freiherr von   396
Wende, Erich   397, 403
Wentzel, Gregor   10, 36
Wergeland, Harald Nicolai   233
Wernholm, O.   158, 159, 161
Wheeler, John Archibald   33
Widerøe, Rolf   139, 158, 161, 162, 213, 274
Wigner, Eugene   36
Wilkinson, Denys Haigh   8
Willems, Jean   68, 72, 75, 76, 100, 101, 104, 105, 109, 111, 126, 130, 135, 172, 173, 199, 200, 222, 232, 268, 282, 419, 524, 527, 530–532, 535
Wilson, John Graham   6, 7
Winter, Klaus   420
Wöffler   136
Woolton, Earl Frederick James   489, 490, 496
Yukawa, Hideki   6, 7, 25, 26, 37
Zinn, Walter   8

Thematic subject index*

Conferences and meetings
Fundamental particles, Cambridge, 1946   5–10, 33, 569
UN ECOSOC, 1946/50   64–66
UN AEC 1946/48   66–68
Quantum mechanics of the electron, Shelter Island, 1947   33–35, 569
Theoretical physics conference, Pocono Manor, 1948   26, 36, 37, 570
Cosmic radiation, Bristol, 1948   6, 27
Particules Elémentaires, Conseil Solvay, 1948   26, 50
European conference, The Hague, 1948   68, 571
Fundamental physics, Oldstone-on-the-Hudson, 1949   37, 38
Inauguration of the cyclotron, Amsterdam, 1949   70–72
European cultural conference, Lausanne, 1949   69, 72–75, 112, 318, 360, 385, 387, 524, 572
First Rochester conference, 1950   38, 574
Nuclear physics conference, Harwell, 1950   85, 99, 362–365, 387, 433, 574
UNESCO, 5th General Assembly, 1950   82–88, 362, 524, 573
IUPAP Executive Committee, 1950   85, 98, 101, 362, 435, 574
Centre Européen de la Culture, Geneva, 1950   109–116, 238, 362–366, 388, 389, 433, 524, 574
Problems in quantum physics, Copenhagen, 1951   138
UNESCO, 6th General Assembly, 1951   131, 135, 140, 141, 576
IUPAP, 7th General Assembly, 1951   138–140, 387, 577
Intergovernmental conference, Paris, 1951   135, 180–197, 214, 232–234, 392, 393, 396, 398–400, 449–453, 525, 579
Meson physics, Rochester, 1952   39, 40, 579
High energy nuclear physics, Rochester, 1952   41
Intergovernmental conference, Geneva, 1952   194–197, 214, 232–234, 370, 404, 461, 462, 525, 580
Buckland House, 1952   477–480
International physics conference, Copenhagen, 1952   212, 480, 581
Cosmic rays, Bagnères-de-Bigorre, 1953   24, 31, 32, 583
On the alternating-gradient proton synchrotron, October 1953,   275

Institutions
Académie des Sciences, Paris   337, 338
AERE (Harwell)   19, 20, 215, 223, 276, 279, 280, 436, 477–479, 486, 492, 494, 528, 541, 568
Atomkommitten (S)   113, 243, 256
Brookhaven National Laboratory   15–17, 67, 84, 89, 111, 136, 137, 169, 213, 223, 274, 281, 361, 486, 490, 569, 584
Berkeley (Radiation Laboratory)   16, 17, 106, 131, 137, 213, 281, 361, 514, 569, 573

* This index is intended only to highlight the more important items in the text. Unlike the name index, it is not meant to be exhaustive.

CEA   22, 67, 68, 79, 115, 118, 304, 307-309, 310-313, 317-321, 338, 528, 568, 573
CEC   72, 77, 78, 101-109, 110-112, 114-116, 126, 127, 173, 174, 200, 388-391, 524, 531, 574, 575
CISE   355-357
Collège de France, Paris   316, 317
CNR   68, 125, 126, 321, 354-373, 389, 530, 575
CNRN   256, 358, 371, 372, 376
CNRS   76, 310-313, 315, 320, 321, 338
Council of Europe   68, 77, 111, 114, 388, 390, 433, 437, 571
Deutsche Forschungsgemeinschaft (DFG)   387, 395-397, 399, 403, 405-407, 415, 419
Deutscher Forschungsrat (DFR)   393, 395, 396
DSIR   20, 76, 114, 215, 216, 244, 433, 447, 457, 458, 463, 468, 487-489, 497-501, 531
Ecole Polytechnique, Paris   314
ECSC   68, 173, 202, 306, 333, 339, 532, 535, 573, 576
EDC   68, 173, 202, 306, 333, 336, 340-342, 537, 538, 574, 575, 581, 582
ELDO   236
ESRO   236
Euratom   89, 200, 532, 533, 535
European Movement   68, 69, 72-75, 76-78, 101, 173-174, 200, 318, 385, 531
Faculté des Sciences, Paris   310, 311, 314, 315
FOM   131
Fonds National de la Recherche Scientifique   531
Geneva University, Institut de Physique   223
ICSU   149, 153, 394
IIPN   126, 531
IISN   68
INFN   358, 359
Institute for Theoretical Physics, Copenhagen   212, 217, 218, 238, 240, 400, 401, 449
IUPAP   76, 102-104, 137, 170, 362, 365
JENER   163, 167
Liverpool University   194, 235, 276, 447-449, 452, 454-457
MPG   392, 397
NATO   305, 336, 437, 538, 571, 572, 578, 586
Nordita   246
Notgemeinschaft der Deutschen Wissenschaft   393-395
OEEC   68, 81, 248, 305, 319, 488, 497, 574
Royal Society   234, 239, 445, 455, 458, 464, 480, 483, 488, 491, 496, 502-506
UN AEC   66-68, 318, 319, 568, 569
UN ECOSOC   64-66, 112, 569-573
UNESCO   65, 66, 99, 100, 102-104, 110-112, 114, 126, 219, 221, 320, 321, 359, 369, 390-392, 394-396, 403, 404, 409, 411, 435, 568, 585
    study office, 1951   125-130, 132, 158-160, 363-366
    consultants meetings, 1951   130-134, 157-162, 163-166, 186, 188, 190, 201, 438, 524, 532, 576, 578-580
    see also Auger
UNO   64-68, 568
Uppsala University   235, 276
ZWO   76, 149, 181, 256

National dimensions
Austria   110, 129, 136, 158

Belgium   71, 84, 100, 104, 105, 110, 154, 197, 221, 231, 232, 235, 258, 536, 585
    and the CEC   68, 69, 72, 75-77, 99, 113
        financial support for Auger, 1951   126, 136, 158, 172-174, 180-182, 186, 195, 575, 580
        situation in mid-1951   22, 135
    see also Willems
Denmark   71, 85, 106, 110, 129, 169-174, 180, 181, 185, 188, 189, 192, 193, 195, 197, 219, 221, 231, 232, 246, 259, 578, 580, 582, 585
    Copenhagen as a site   153-155, 186, 214, 239, 240, 414, 440-452
    Theory Group   243, 244, 414-416
    see also Bohr
France   22, 64-66, 68, 70-72, 77, 79-82, 84-86, 93, 100, 103-105, 110, 113, 117, 118, 125, 126, 128, 129 172-174, 180-182, 214, 220-222, 226, 227, 231, 232, 235, 239, 240, 255, 268, 390-392, 536, 537, 579
    see also chapter 9
Federal Republic of Germany   23, 84, 103, 106, 110, 129, 158, 159, 173, 174, 181, 193, 195, 197, 221, 226, 227, 231-233, 243, 259, 306, 322, 324, 349, 536, 571, 574-576, 578, 579, 581, 586
    see also chapter 11
Greece   110, 129, 180, 183, 221, 222, 231-233, 258, 259, 585
Italy   23, 68, 69, 72, 75-77, 84-86, 98, 99, 103-105, 107, 110, 112, 113, 117, 118, 125, 126, 128, 158, 172-174, 180-182, 189-191, 195, 197, 221, 222, 227, 231-233, 235, 258, 536, 582
    see also chapter 10
The Netherlands   22, 77, 81, 82, 84, 85, 104, 105, 110, 111, 127, 128, 131, 135, 150, 152, 158, 172-174, 197, 221, 226, 231-233, 235, 259, 269, 577-579, 585
    Amsterdam meeting, 1949   70-72, 572
    Arnhem as a site   214, 239-241, 285, 413, 414
    attitudes at the intergovernmental conference   180, 181, 183-185, 193, 195
    Kramers' opposition   81, 111, 138-140, 148-152, 156, 157, 169-171, 573, 577, 578
    see also Bakker and Bannier
Norway   22, 104, 105, 110, 127, 128, 135, 158, 172-174, 195, 197, 221, 231-233, 243, 246, 577, 579, 582, 586
    difference between scientists and government   172-174, 180, 181, 193
    see also Dahl and Randers
Sweden   21, 22, 71, 104, 105, 110, 127, 128, 132, 135, 169-174, 192, 197, 221, 231-233, 235, 246, 582, 585
    financing the interim CERN   256, 257
    scale of contributions   221, 231
    support for Auger, 1951   113, 119, 128, 158, 575
    support for Bohr, 1951/2   150, 172-174, 180-182, 185, 189, 190, 195, 218, 243, 578, 580
Switzerland   22, 71, 77, 84, 85, 104, 105, 110, 127, 128, 154, 158, 197, 220, 221, 228, 230, 231-233, 235, 250-252, 259, 401, 537, 585
    domestic conflict over Geneva as a site   214, 241-243, 583
    Geneva as a site   186, 214, 239-241, 330, 370, 400, 401, 413, 414
    neutrality   241-243, 250-252, 333, 401
    situation in mid-1951   135-137, 577
    support for Auger, 1951   113, 135, 137, 172-174, 180-182, 195
United Kingdom   19-21, 70, 71, 77, 79, 82, 84, 85, 99, 107, 110, 114, 115, 118, 127, 128, 135, 139, 150, 161, 169-174, 180-183, 185, 187, 191, 192, 195, 215, 216, 220-222, 227, 231, 232, 234, 235, 240, 248-252, 254, 259, 332, 333, 525, 535, 580, 581, 584
    see also chapters 12 and 13
Yugoslavia   110, 119, 129, 181, 182, 197, 221, 231, 232, 234, 243, 248, 538, 572, 576-578, 586

Themes-general

Accelerators
    Bevatron   17, 18, 23, 43, 44, 84, 112, 132, 136, 137, 160, 198, 438, 457, 570, 585
    CERN PS   24, 112, 131, 155, 156, 160, 161, 164, 165, 182, 185, 212, 213, 216, 217, 273-282, 576, 581
    CERN SC   21, 160, 161, 185, 212, 216, 217, 273, 274, 276-278, 457
    Cosmotron   16-18, 23, 43, 44, 84, 98, 110, 112, 137, 160, 198, 212, 215, 237, 467, 477, 478, 530, 570, 581
    Liverpool SC   21, 182, 184, 187, 190, 211, 218, 400, 415, 436, 445-459, 462, 481, 482, 501
    other European accelerators   21-23, 135, 172, 315, 317, 322, 436, 583
    Uppsala SC   21, 182, 185, 211, 218, 415, 436
    see also technology of accelerators, experiments with accelerators

CERN/European laboratory project, 132
    and administrators and diplomats   256-258, 260, 261, 307, 342, 343, 401, 530-532, 534
    and attitudes of the member states   261, 307, 319, 320, 324, 331-334, 535, 536, 540
    and cosmic rays   218, 219, 228, 247, 266, 270, 284
    and possible military interests   228, 230, 242, 322, 368, 369, 499, 501, 527, 531, 539-542
    asymmetrical relations with   230, 542
    as centre of expertise   135, 485, 540-542
    as furthering narrow French interests   108, 117, 118, 322-324, 330, 331, 435, 436, 469
    as furthering national interests   124, 125, 322, 323, 331, 368, 374, 416-421, 526, 531
    as furthering one's career   108, 532
    as having a scientific spirit   219, 228, 263, 265, 266, 270
    as multipurpose laboratory   135, 248, 364-367, 460, 497
    as rebuilding Europe   117, 131, 171, 318, 407, 532
    as rebuilding European science   111, 130, 138, 156, 171, 187, 252, 318, 341, 401, 527, 532, 534
    as revealing national susceptibilities and interests   159, 166, 248, 329-334, 337, 348, 494, 495, 497
    as reversing the brain drain   111, 131, 171, 337, 374, 452, 542
    as stimulating European industry   116, 162, 230, 318, 368, 401, 489, 537, 541, 542
    contract policy   229, 236
    cost estimates   132, 133, 161, 162, 220, 282-284
    cost of the PS   131, 160, 161, 165, 213, 217, 274-276, 281, 282, 494, 533, 576
    early accommodation   223, 224, 227
    excavating the site   224, 225, 227, 236, 257-259, 585
    like a university   228, 229, 542
    national staff quotas   229, 417, 420
    salary scales   229, 230, 266
    time scheduling   132, 133, 139, 155, 156, 222, 282

CERN, Committees and Groups
    Administrative and Financial Working Group   235, 502
    Director/Secretary General   161, 165, 192, 196, 197, 211, 225, 226, 228, 229, 261-273, 371, 525, 585
    Executive Group   211, 214, 223, 226, 245, 253, 254, 257, 261-263, 284, 291
    Interim Finance Committee   215, 224, 225, 235, 253-259, 502, 583, 585
    Laboratory Group   161, 165, 183-185, 196, 211, 224, 227, 235, 239, 257, 260, 283, 284, 417
    Nominations Committee   225-227, 261-273
    PS Group   161, 165, 183-185, 190, 196, 211, 213, 216, 217, 223, 224, 227, 255, 260, 274-276, 278-281, 417, 418
    SC Group   161, 165, 183-185, 190, 196, 211, 216, 217, 224, 227, 260, 276-281, 417, 419, 482
    Theory Group, Copenhagen   161, 165, 184, 190, 196, 211, 217, 228, 243-246, 414-416

Theory Group, Geneva   218, 228, 243–246, 270, 416
Working Group to draft provisional agreement   188, 370, 461

CERN, Council of Representatives of States   184–186, 188, 192, 194–197, 210, 212, 215, 216, 226, 501, 502, 532–534
   first session, 1952   210, 211, 215, 232–234, 371, 414, 416, 581
   second session, 1952   211, 212, 215, 219, 371, 405, 581
   third session, 1952   213–215, 218, 219, 239–241, 414, 582
   fourth session, 1953   218, 243, 245, 250, 406, 415, 582
   fifth session, 1953   216, 217, 408, 582
   sixth session, 1953   218, 220, 225, 245, 252, 583
   seventh session, 1953   217, 219, 275, 276, 282, 584
   eighth session, 1954   254, 255, 585
   ninth session, 1954   225, 254, 255, 257, 258, 268, 269, 417, 585
   permanent Council, first session, 1954   226, 271, 398, 586

Communist attitudes   82, 106, 241, 242, 266, 309, 335–339, 361, 372–376, 389–391, 538

Convention
   admission of new states   220, 221, 250–252, 332, 334
   basic and supplementary programmes   219, 220, 246–249
   financial protocol to   220, 221
   ratification of   210, 221, 222, 226, 227, 232, 236, 253–255, 258, 339–342, 373–376, 407–411, 525, 584–586
   signature of   210, 215, 221, 222, 227, 409, 583, 584
   Supplementary Agreement   222, 234, 253–257

Cosmic ray and accelerator physics, relationship between   14, 26, 27, 32, 38, 41–44, 219, 484, 485

East–West tension   84, 103, 192, 197, 220, 250, 306, 322, 332, 333, 336, 459, 534, 537, 538, 541, 573, 574, 583

Europe
   economic situation   222, 223, 305, 315, 570
   European ideal and construction   68, 69, 72, 117, 182, 200, 249, 250–252, 322, 326, 333, 339, 369, 376, 526, 527, 536, 539
   regional European alliances/differences   70, 71, 74, 172–174, 214, 240, 241, 267, 268, 322, 332, 333, 370, 375
   scientific situation   19–23, 247–249, 304, 309–317, 322

Experiments
   with accelerators   26–28, 32, 34, 43, 44, 356, 478, 494, 581
   with cosmic rays   5–8, 16, 25, 28–32, 34, 42–44, 316, 356, 481

Financial contributions,   110, 183, 328, 329, 367–372, 484, 485, 499–501, 514
   contributions in kind   191, 197, 453, 455
   for the interim CERN   252–261, 369, 370, 403–405, 525
   for the UNESCO study office   103, 112, 113, 125, 126, 321, 322, 365–367, 575, 576
   offered at the intergovernmental conferences   182, 186, 193, 197
   scale of contributions   221, 231, 389, 406, 413
   United Kingdom 'gifts'   215, 491, 498, 582

Forces
    electromagnetic (QED)   9, 34–38, 43, 569
    strong (nuclear)   7–10, 34, 35
    weak (universal Fermi interaction)   10, 37

Governmental machinery   227, 261, 305, 307, 315, 319–321, 323–328, 334, 339, 340, 346, 408–412, 458–461, 502–511

Methodology   210, 249, 325–328, 338, 495

Particles
    heavy mesons   6, 30, 31, 39
    hyperons   30, 31, 580
    mu-mesons   6, 7, 24–28, 34, 35, 45
    pi-mesons (mesotron, Yukawa meson)   6, 7, 24–28, 35, 39, 49, 569, 570, 572, 573
    the first resonance   39, 515
    strange particles   28, 39, 584
    V-particles   28–31, 39–41, 570, 573, 576, 583

'Rabi' resolution, 1950,   82–84, 89, 90, 98, 103, 210, 319, 524, 573
    interpretations of   84–87, 242, 336, 360, 372, 373, 537–539
    reactions to   87, 88

Reactors/Nuclear Energy   15, 19, 48, 49, 67, 80, 81, 84, 89, 98, 107, 115, 116, 131, 308, 318, 319, 355–357, 364, 365, 388, 402, 524, 528, 531, 570–572, 574, 579

Scientists and governments,   71, 73, 80–82, 84, 90, 100, 103, 116, 118, 129, 191, 192, 199, 200, 304, 318, 327, 328, 401, 406, 507, 510, 511, 526, 527, 530, 539
    Auger's philosophy on   103, 108, 109, 117, 118, 129, 141, 142, 184, 200–202, 319, 320, 323, 325, 534
    Bohr and Kramers' philosophy on   149–153, 184
    in the Council   532, 533
    national attitudes   128, 129, 134, 159, 197, 324, 325, 327, 418
    scientific consultants' philosophy on   133–135, 138, 164, 320
    the CERN 'lobby'   201, 202, 261, 282, 326, 327, 331, 531–534

Scientists, divisions between,   100, 101, 111, 196, 212, 278, 279, 283, 284, 338, 339
    Auger's tactical attitude   159, 162, 166, 167, 193, 530
    communist opposition, in France   115, 324, 327, 328, 336–339, 391
    in 1951, over Auger's project   138–140, 148–157, 159, 162–164, 166–174, 187–193, 195, 196, 201, 434–437, 440–444, 453, 456, 527, 529
    in 1952, between the Executive Group and Bohr   214, 240, 245
    in 1952, in Britain   479, 483–487, 530
    over the choice of a Director-General   226, 263, 269, 273

Scientists, presenting a united front,   321, 322, 325
    at the intergovernmental conference   167, 188, 400, 404, 405, 530
    in Britain   464, 465, 493, 499, 500, 511
    in dealing with governments   202, 534

Site problem   111, 112, 121, 131, 132, 151–155, 162, 164–166, 186, 213, 214, 238–246, 324, 329–331, 370, 400, 413, 414, 446, 447

Technology of accelerators
    alternating-gradient focusing   23, 24, 213, 215–217, 222, 245, 273–282, 490–492, 581, 582, 584
    principle of phase stability   13–15, 19, 568

United States of America,   82, 83, 88, 89, 110, 137, 161, 305, 322, 537–539, 571, 573, 576, 580, 582
    competition with   70, 71, 79, 82, 111, 112, 118, 157, 171, 318, 361, 362, 401, 482, 492–494, 529, 537
    nuclear physics programme   15–19, 26, 43, 541, 571
    support for Auger   88, 136, 145, 162, 205, 206
    support for PS Group   213, 217, 274, 279, 281, 541
    war research and its consequences   8, 9, 11–13, 42, 46